2002

INLAND FLOOD HAZARDS

This edited volume presents a comprehensive review of issues related to inland flood hazards. The volume addresses physical controls on flooding, flood processes and effects, and responses to flooding from the perspective of human, aquatic, and riparian communities. It is a thoroughly interdisciplinary treatment, drawing on the expertise of atmospheric scientists, biologists, botanists, civil engineers, geographers, geologists, and hydrologists.

Individual chapter authors are recognized experts in their fields, who draw on examples and case studies of inland flood hazards from around the world. As flood-related damages to human communities and river ecosystems continue to increase, the type of review discussions presented in this volume become of increasing importance to our efforts to mitigate flood hazards. This volume is unique among treatments of flood hazards in that it addresses how the nonoccurrence of floods, in association with flow regulation and other human manipulation of river systems, may create hazards for aquatic and riparian communities.

This volume will be a valuable resource for everyone associated with inland flood hazards: professionals in government and industry, and researchers and graduate students in civil engineering, geography, geology, hydrology, hydraulics, and ecology.

Ellen E. Wohl is an Associate Professor of Geology in the Department of Earth Resources, Colorado State University.

INLAND FLOOD HAZARDS

HUMAN, RIPARIAN, AND AQUATIC COMMUNITIES

Edited by

Ellen E. Wohl

Colorado State University

PUBLISHED BY THE PRESS SYNDICATE OF THE UNIVERSITY OF CAMBRIDGE
The Pitt Building, Trumpington Street, Cambridge, United Kingdom

CAMBRIDGE UNIVERSITY PRESS
The Edinburgh Building, Cambridge CB2 2RU, UK http://www.cup.cam.ac.uk
40 West 20th Street, New York, NY 10011-4211, USA http://www.cup.org
10 Stamford Road, Oakleigh, Melbourne 3166, Australia
Ruiz de Alarcón 13, 28014 Madrid, Spain

First published 2000

Printed in the United States of America

Typefaces Melior 10/12 pt. and Optima *System* LaTeX 2_ε [TB]

A catalog record for this book is available from the British Library.

Library of Congress Cataloging in Publication Data
Inland flood hazards: human, riparian and aquatic communities / edited by Ellen E. Wohl.
p. cm.
ISBN 0-521-62419-3 (hb)
1. Floods. 2. Flood control. I. Wohl, Ellen E., 1962–
GB1399 .I54 2000
363.34′93 – dc21 99-052944
CIP

ISBN 0 521 62419 3 hardback

Contents

FLOOD HAZARD MITIGATION STRATEGIES

Preface

This volume on inland flood hazards is designed to provide both a general reference on flood hazards inland from coastal regions and, within each chapter, a comprehensive review of existing knowledge through specific case studies. The volume addresses floods and how the presence or absence of floods may constitute a hazard to human as well as nonhuman communities along the river corridor. The authors of the various chapters are geologists, civil engineers, geographers, meteorologists, and biologists. In asking each of them to write about some aspect of inland flood hazards, my intent has been to encourage an interdisciplinary dialogue of the hazards associated with changing flood regimes in river basins around the world. Three drainage regions – the Colorado River basin of the United States and Mexico, the Tone River basin of Japan, and the lower drainages of the Ganges and Brahmaputra Rivers in Bangladesh – are used throughout this volume to exemplify the various aspects of inland flood hazards. As increasing human population density results in more manipulation of natural processes, and more pressure on river systems, it is imperative that we understand how hazards arise from natural and regulated flows along rivers. The steadily growing economic damage and loss of human life associated with floods, on the one hand, and the reduction in biological diversity and increase in number of endangered species, on the other hand, are the best justifications for studying inland flood hazards.

I would like to thank each of the individuals who helped to make this volume possible. The chapter authors were a pleasure to work with, and they provided helpful reviews of other chapters in this volume. Chapter reviews were also provided by Michael Abbott, Robert Behnke, Robert Brakenridge, Stephen Burges, Michael Douglas, Benjamin Everitt, Gerald Galloway, Janet Heiny, Cliff Hupp, Robert Jarrett, W. Carter Johnson, Donald Knight, Frank Magilligan, Claudio Meier, Burrell Montz, John Moore, Jim O'Connor, John Pitlick, LeRoy Poff, Sara Rathburn, Dan Rosbjerg, Andrew Simon, and Christopher Waythomas. Catherine Flack and Matt Lloyd of Cambrige University Press have worked on this volume since its inception, and I thank them for their assistance in what turned out to be a long process.

Ellen E. Wohl

Contributors

Gregor T. Auble
Midcontinent Ecological Science Center
United States Geological Survey

Victor R. Baker
Department of Hydrology and Water Resources
The University of Arizona

David S. Biedenharn
Engineering Research and Development Center
U.S. Army Corps of Engineers

Daniel A. Cenderelli
Department of Geological Sciences
University of Alabama

L.L. Ely
Department of Geology
Central Washington University

Jim B. Finley
Shepherd Miller, Inc.

Jonathan M. Friedman
Midcontinent Ecological Science Center
United States Geological Survey

Eve Gruntfest
Department of Geography and Environmental Studies
University of Colorado

Douglas Hamilton
Hydrologic Consultant
Irvine, California

K.K. Hirschboeck
Laboratory of Tree-Ring Research
University of Arizona

Alejandro Joaquin
Hydrologic Consultants
Irvine, California

Melinda J. Laituri
Department of Earth Resources
Colorado State University

R.A. Maddox
Cooperative Institute for Mesoscale Meteorological Studies
University of Oklahoma

Dorothy Merritts
Geosciences Department
Franklin and Marshall College

Leal A.K. Mertes
Department of Geography
University of California at Santa Barbara

Jorge A. Ramírez
Civil Engineering Department
Colorado State University

Clair B. Stalnaker
Midcontinent Ecological Science Center
U.S. Geological Survey

Jery R. Stedinger
School of Civil and Environmental Engineering
Cornell University

Chester C. Watson
Department of Civil Engineering
Colorado State University

Edmund J. Wick
TETRA TECH, Inc.

Ellen E. Wohl
Department of Earth Resources
Colorado State University

Richard S. Wydoski
U.S. Fish and Wildlife Service (retired)

INTRODUCTION

CHAPTER 1

Inland Flood Hazards

Ellen E. Wohl
Department of Earth Resources
Colorado State University

Floods occur on all rivers and are vital in shaping the physical characteristics and biotic communities of rivers. Floods may also create hazards for humans who live along rivers, and human flood-hazard mitigation measures in turn may create hazards for nonhuman aquatic and riparian communities. This volume is designed to explore how various types of flood hazards develop and how they may be mitigated. This introductory chapter provides a context for topics that are discussed in greater detail in subsequent chapters. Floods are discussed in the context of the riparian corridor, which includes landforms and biotic communities within the 100-year floodplain. Numerous physical, biological, and chemical processes connect the channel and the riparian corridor. These processes are spatially differentiated within a drainage basin, as are flood hazards. Humans have been attempting to mitigate floods with dams, levees, and other structures for more than 2000 years. These measures have been only partially successful, as indicated by the trends of increasing flood damage in many countries during the 20th century. These trends are partly the result of land-use changes and increasing human populations. Clearance of upland forests, grazing, crops, and urbanization have all contributed to increased flood discharges. The most effective response to these changes in flood hazards will come from interdisciplinary approaches that recognize the interconnections among floods, river channels, and riparian corridors. The Colorado River basin in the western United States, the Tone River basin of Japan, and the lower portion of the Ganges and Brahmaputra Rivers in Bangladesh are used throughout this volume to illustrate various aspects of inland flood hazards.

Introduction

Floods are an integral part of the dynamics of any river channel. Because humans historically have settled in river valleys, floods have created hazards for human communities for millennia. Human attempts to reduce these hazards by altering flood hydrology and hydraulics in turn have created hazards for the nonhuman aquatic and riparian communities that have adapted to natural flood regimes. As a result, the presence as well as the absence of floods may create hazards to some type of community along a river. The purpose of this volume is to explore how these hazards develop and how they may be mitigated. This introductory chapter is designed to provide a context for the topics covered in the volume.

The chapter begins with a brief discussion of floods and flood hazards, followed by an explanation of the riparian corridor that surrounds a river and how this corridor is affected by floods. Spatial differentiations of geomorphic processes, biota, and human land use within a drainage basin are discussed in relation to the different forms of flood hazards throughout the basin. An historical review of human responses to flood hazards and of how human land-use practices have altered river systems leads to a discussion of contemporary approaches to flood hazard mitigation. Finally, the three representative drainage regions that are discussed throughout this volume are introduced.

Floods and Flood Hazards

A flood is defined most simply as the flow of a larger-than-average volume of water along a river channel. Floods are most commonly described in terms of discharge relative to channel morphology (e.g., overbank flood) or of estimated recurrence interval (e.g., 100-year flood). The perception of what constitutes a flood may vary for people living along different rivers or for people along the same river over decades or centuries. Human perception of river processes is a vital component of flood hazards, for people's response is based on their conception of a process (Slovic et al., 1974; White, 1974; Laituri, Chapter 17, this volume). Schumm (1994) has noted that inappropriate responses to fluvial processes may arise from a perception of stability (which leads to the conclusion that any change is not natural), from a perception of instability (which implies that change will not cease), or from a perception of excessive response (which suggests that changes will always be major). Flood-control structures such as levees or dams may also create a misleading sense of security when people assume that floods will no longer occur (Ericksen, 1974). In any of these cases, if the human population of a drainage basin does not understand the interacting physical, chemical, and biological processes that govern river behavior and has no knowledge of historical change along the river, that population may respond ineffectively to hazards associated with flooding.

A key component of human perception and response to hazards is attitude toward extreme events, such as floods. Floods may be regarded as aberrations – catastrophes – that either dominate the system or have little effect on normal system functioning. Alternatively, floods may be regarded as infrequent but regular events that lie on a continuum of flow magnitude and frequency. Various conceptual models of river processes developed by geomorphologists (Wolman and Miller, 1960; Wolman and Gerson, 1978) and biologists (Junk et al., 1989; Bayley, 1991) have been based on each of these differing perceptions, as have individual (Slovic et al., 1974) and governmental (Visvader and Burton, 1974) responses to flood hazards.

All natural rivers are characterized by floods. Flood hydroclimatic context may vary dramatically among river basins. Some basins, such as those in the seasonal tropics, may have regular "flood seasons," whereas other basins may have mixed flood populations resulting from several different types of meteorological circulation systems (Hirschboeck, 1988; Hirschboeck et al., Chapter 2, this volume). Similarly, some rivers have a much greater hydrologic variability. One measure of hydrologic variability is the coefficient of river regime, which is the ratio of maximum to minimum discharge. Rivers such as the Thames or the Rhine have coefficients well under 100, whereas the coefficient for the Tone River of Japan exceeds 900 (Japanese River Bureau, 1985). The variability in flood magnitude and frequency along a particular river will exert an important influence on how humans in that river basin perceive and respond to floods and on how aquatic and riparian communities are shaped by floods.

Flood hazards vary as a function of hydroclimatic regime, position within the drainage basin, and human alteration of the drainage basin (Figure 1.1). Hydroclimatic regime will determine the temporal and spatial distribution of precipitation falling on a drainage basin (Hirschboeck et al., Chapter 2, this volume). How this precipitation concentrates to create a flood in turn will be governed by hillslope, drainage basin, and channel characteristics (Wohl, Chapter 6, this volume). Slope characteristics will control infiltration and the rate at which water reaches stream channels. The drainage basin relief, drainage area and contributing area, drainage density, stream order, and channel geometry will then control how rapidly water moves along the channels of a stream network (Bull, 1988; Kochel, 1988b; Patton, 1988a,b). The magnitude, frequency, and duration of water movement along channels in turn influences the movement of sediment, nutrients, and contaminants: Channels dominated by frequent flows have different patterns of floodplain sedimentation than those dominated by infrequent flows (Brakenridge, 1988). Human land-use activities may affect all aspects of drainage basin and channel characteristics that control the rate of water movement and thus flood conditions (Dunne and Leopold, 1978; Kresan, 1988; Wohl, Chapter 4, this volume).

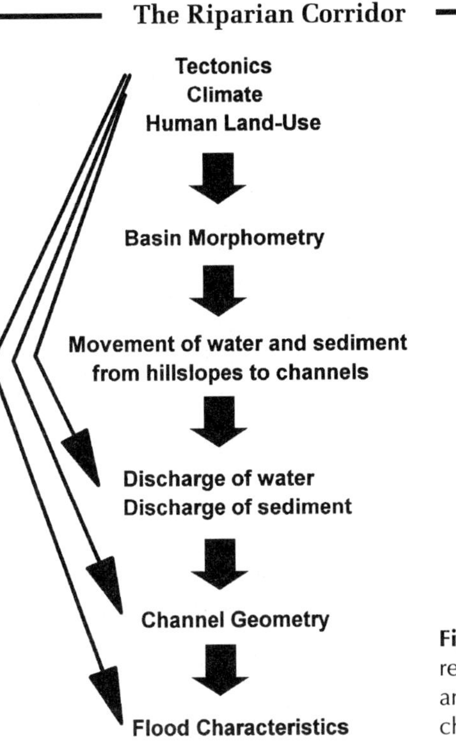

Figure 1.1. Schematic representation of interrelations among controls on flood characteristics.

The Riparian Corridor

Floods constitute a hazard when they alter the existing riparian corridor in a manner that is perceived by humans as being dangerous or deleterious. The riparian corridor includes the bottomland, floodplain, and streambank communities within the 100-year floodplain along inland waterways (Swift, 1984). The riparian corridor has numerous vital connections to the low-flow channel (Sedell and Froggatt, 1984; Gregory and Gurnell, 1988; Gregory et al., 1991; Large and Petts, 1992; Friedman et al., 1996; Hupp and Osterkamp, 1996). Forest vegetation growing along this corridor attenuates incoming sunlight, thus influencing ground surface and water temperatures (Barton et al., 1978; Petts, 1990a); contributes organic material of different sizes to the channel, thus providing nutrients, habitat, and roughness elements that may alter flow hydraulics and sediment movement (Fisher and Likens, 1973; Keller and Swanson, 1979; Keller and Tally, 1979; Heede, 1981; Triska et al., 1982; Maser and Sedell, 1994; Gurnell and Gregory, 1995; Shields and Gippel, 1995; Thompson, 1995); provides low-velocity, shallow-water nursery habitat for young fish (Petts, 1990a; Wydoski and Wick, Chapter 9, this volume); increases out-of-bank roughness, causing attenuated flood peaks and accumulation of sediment and nutrients (Zimmerman et al., 1967; Lewin, 1994; McKenney et al., 1995); and increases bank stability via the root network (Zimmerman et al.,

1967; Welcomme, 1979; Vannote et al., 1980; Naiman and Decamps, 1990; Fetherston et al., 1995; Nakamura, 1995). Lateral cutting by channels, as well as overbank scour and deposition, destroy existing riparian communities and set the stage for development of new communities by invading plants (Swanson et al., 1982; Johnson, 1992, 1994; Friedman and Auble, Chapter 8, this volume).

Reaches of wider valley floor may be particularly important as "buffer zones" that reduce variance in water and sediment discharge through time as well as associated variance in water chemistry and nutrient flux (Petts et al., 1989b; Large and Petts, 1994; Webb and Walling, 1994; Mertes, Chapter 5, this volume). Subsurface water may be stored in the more extensive hyporheic zone of wider reaches; in rivers with coarse gravels, the lateral hyporheic zone may extend out to 3 km from the wetted channel (Stanford and Ward, 1988). Variation in the local exchange of flows between the channel and the hyporheic zone may produce temporally shifting concentration gradients of dissolved oxygen, nitrate, and ammonium in the subsurface waters that form important controls on aquatic invertebrates and riparian vegetation (Triska et al., 1990). Surface waters may spread out and move more slowly along wide reaches during floods (Gustard, 1994; Lewin, 1994). Sediment may be stored during periods of high sediment discharge or removed during periods of low sediment discharge (Nakamura, 1989; Lewin, 1994). Dunne et al. (1998) found that sediment transport along the Amazon River valley involves exchanges between the channel and floodplain that in each direction exceed the annual flux of sediment out of the river. In terms of flood hazards, riparian vegetation has an interdependent relationship with channel shape and thus with efficiency of conveyance of floodwaters (Friedman et al., 1996; Hupp and Osterkamp, 1996): Vegetation influences channel form by increasing the roughness and shear strength of the bed and banks, reducing the velocity of floodwaters and increasing sedimentation. Vegetation also responds to changes in channel form caused by changes in the discharge of water and sediment – for example, with increasing vegetation density or a change in vegetation type as a result of increased sedimentation.

The riparian biological community changes progressively in a downstream direction. The idea of a biological river continuum (Vannote et al., 1980) emphasizes the importance of longitudinal linkages within a drainage basin (Minshall et al., 1985). The river continuum is based on the concept that the physical variables within a river system present a continuous gradient of physical conditions to which the constituent biological populations respond. These responses result in a continuum of biotic adjustments that produces consistent patterns of loading, transport, utilization, and storage of organic matter along the length of a river (Vannote et al., 1980) (Figure 1.2). In contrast, the flood pulse concept of river-forest interaction (Junk et al., 1989) focuses on lateral linkages between the floodplain and the channel and emphasizes the strong and positive influence of seasonal floods on the productivity of the fluvial system (Swanson and Sparks, 1990). Both theories of riverine ecosystems imply that the biota occupying the riparian corridor have adapted to the unique physical and chemical processes of that corridor, including repetitive flooding (Kozlowski, 1984b; Poff and Ward, 1989, 1990), and to the specific flood regime of that reach of the channel (Hupp, 1988; Davies et al., 1994;

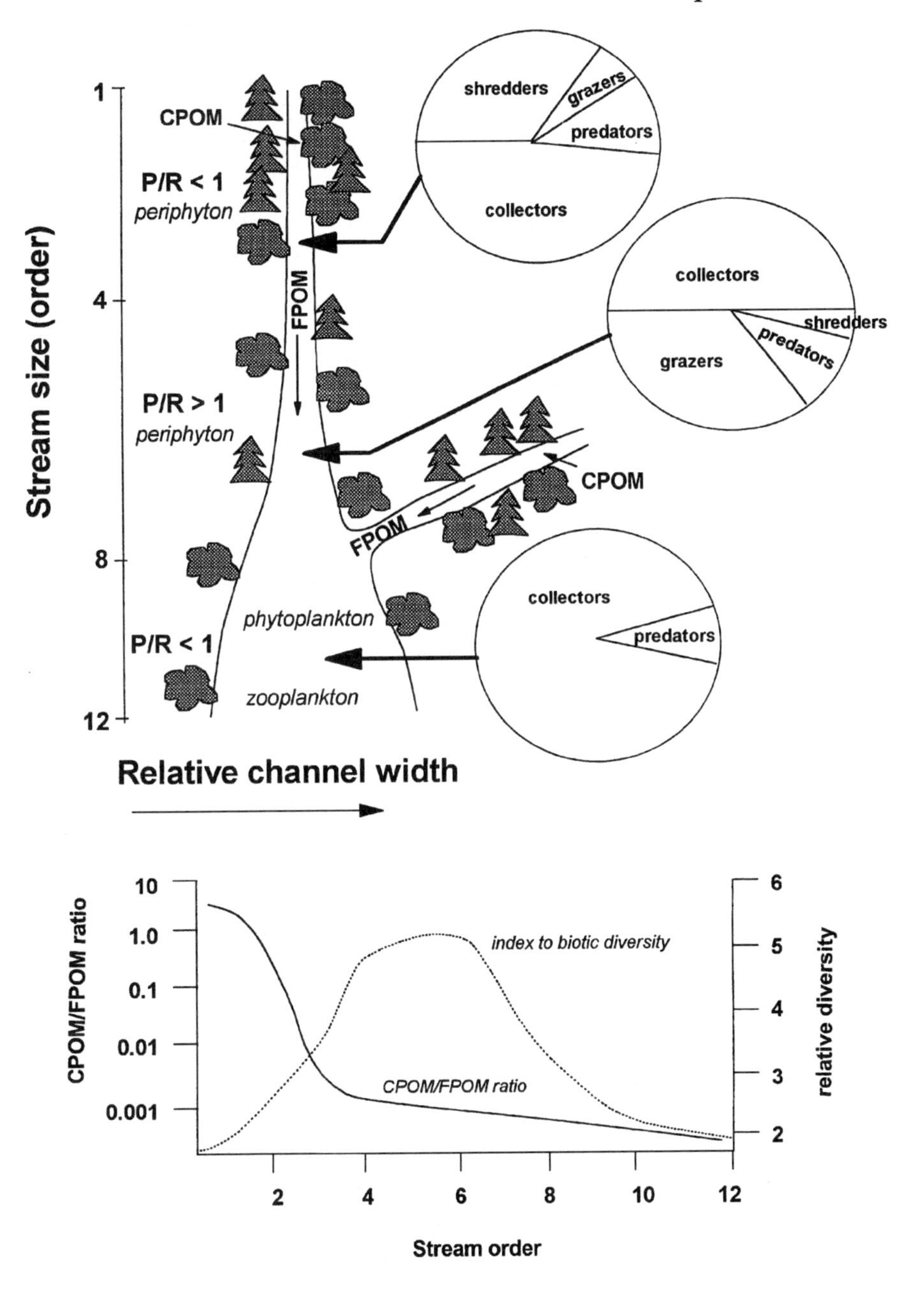

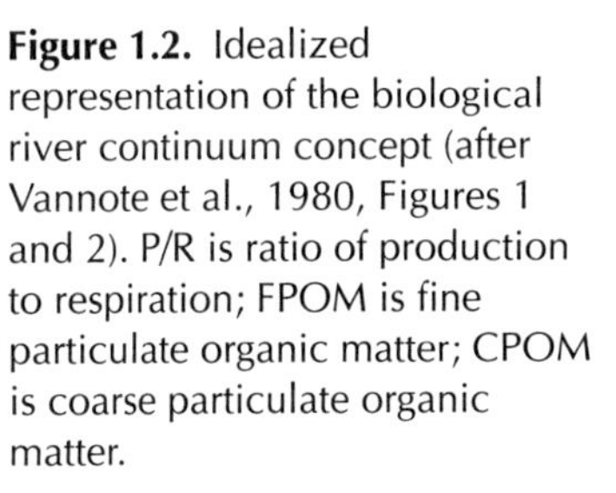

Figure 1.2. Idealized representation of the biological river continuum concept (after Vannote et al., 1980, Figures 1 and 2). P/R is ratio of production to respiration; FPOM is fine particulate organic matter; CPOM is coarse particulate organic matter.

Power et al., 1995b). Humans who occupy the riparian corridor generally attempt to reduce flood hazards by migrating out of the corridor during flood season, by restricting land use within the corridor, or by restricting flooding of the corridor. Each of these human responses, particularly that of altering flood regime, affects the other species that occupy the riparian corridor (Brooker, 1981; Bravard et al., 1986; Petts et al., 1989a; Dister et al., 1990; Malanson, 1993; Walker, 1994). Both field data and model simulations indicate that the most complex and diverse ecosystems are maintained only in riparian environments that fluctuate because of flooding (Power et al., 1995a). Thus, it becomes important to consider

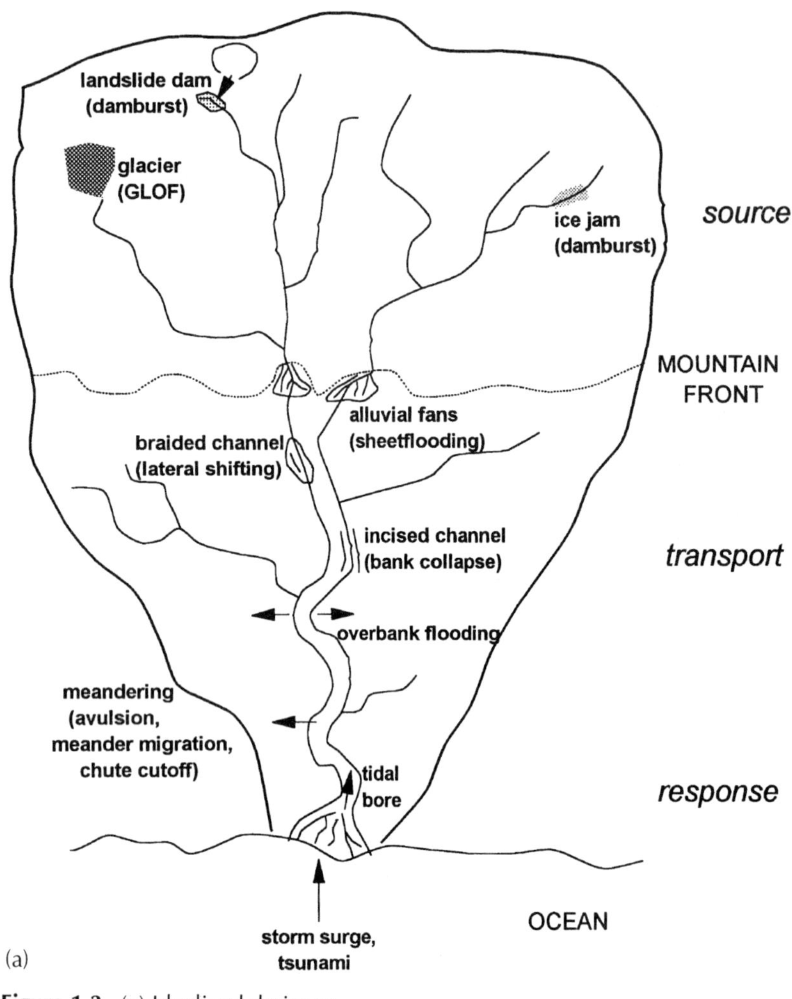

Figure 1.3. (a) Idealized drainage basin, illustrating spatial distribution of flood hazards (after Schumm, 1977). (b) Idealized longitudinal basin profile (after Montgomery and Buffington, 1993). The source portion of the basin is dominated by hillslope processes, with large amounts of colluvial sediment introduced directly into the channel by hillslope mass movements (debris flows, landslides) and by slope wash. Channels within the transport portion of the basin serve primarily as conduits for the movement of water and sediment downstream. Channels within the response portion of the basin respond most readily to changes in water and sediment discharge from upstream.

flood processes, and the hazards they create, within the context of the entire riparian ecosystem (e.g., Parker et al., 1996).

Spatial Differentiation within a Drainage Basin

The spatial distribution of hazards within a basin may be discussed by using the idealized drainage basin of Figure 1.3. The types of hazards present along a given length of channel will be partly a function of drainage basin relief, channel slope, and sediment supply and transport (Ohmori and Shimazu, 1994). Hazards in the upper portion of a drainage basin are often associated with hillslope instability. Mass movements such as debris flows and landslides may alter slope characteristics enough to change rainfall-runoff relations and thus peak flood discharge (Agata, 1994). Mass movements also rapidly introduce large volumes of sediment to the channel, which may cause channel aggradation and out-of-bank flooding, or may temporarily dam the channel and facilitate a damburst flood. Misidentification of debris-flow deposits as fluvial sediments may lead to overestimation of flood discharge (Costa, 1988b).

The uplands of many drainage basins in the middle and high latitudes are dominated by a snowmelt flood regime, with a reasonably predictable, low-magnitude, long-duration flood peak during the snowmelt season (Burt, 1994). Uplands in the tropical latitudes, or in warm deserts, may have floods with larger discharge peaks of shorter duration associated with various types of rainfall.

The aquatic biota of upland or headwaters channels are adapted to abundant terrestrial inputs of coarse particulate organic matter from riparian vegetation as well as cooler water temperatures produced by shading from this vegetation (Vannote et al., 1980). Aquatic invertebrates are primarily shredders and collectors, and fish tend to be cool-water, invertivorous species. Species diversity is relatively low (Ward, 1994).

Human land-use activities that may affect flood characteristics in the upper portion of drainage basins commonly include deforestation, mining, road construction, reservoir construction, and urbanization. Each of these activities may alter slope stability and thus the volume and distribution of water and sediment entering stream channels.

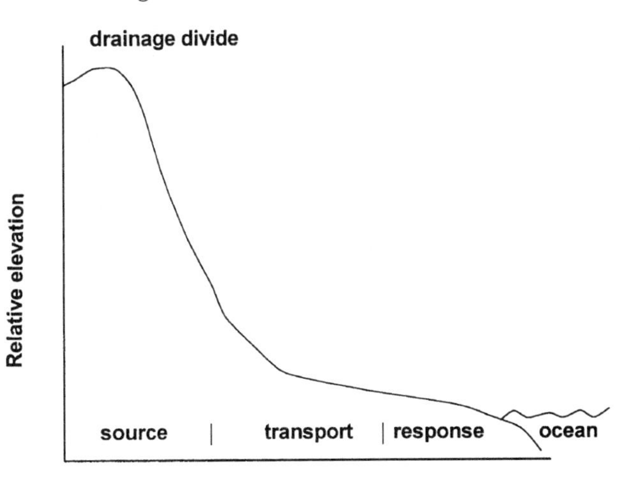

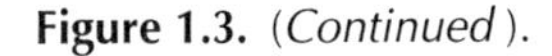
Figure 1.3. *(Continued)*.

Flood hazards along the channels located in the central portion of a drainage basin are dominated by overbank flooding and bank instability. These hazards may affect only a small portion of the basin area if the channels are confined by uplands or by valley walls. In the case of a broad, low-relief surface such as the Amazon Basin, overbank flooding may affect thousands of square kilometers (Kozlowski, 1984a), with the floodwaters from one channel effectively mingling with the floodwaters of adjacent channels. Large floods along the central region of a basin are generally associated with widespread rainfall produced by large tropical storms, monsoonal depressions, mesoscale convective clusters, and low-pressure fronts (Gupta, 1988; Hayden, 1988).

Aquatic biota that live in the central portion of a basin are more dependent on autochthonous primary production by algae and rooted vascular plants and on organic transport from upstream than are communities in the headwaters. Fine particulate organic matter is abundant, and aquatic invertebrates are mainly collectors and grazers. Fish species are warm-water piscivores and invertivores, and both species diversity and fluctuations in water temperature are maximized (Vannote et al., 1980).

Human land-use activities common to the central portion of drainage basins include agriculture (crops, grazing), deforestation, road construction, urbanization, reservoir construction, channelization and levee construction, flow diversion, and destruction of wetlands.

Flood hazards in the lower portion of a basin may be caused by conditions downstream from the basin terminus as well as by upstream precipitation or dam bursts. Where a smaller basin is tributary to a larger basin, floodwaters from the main channel may backflood the lower portion of the tributary. Where a drainage basin enters a large reservoir or the ocean, storm surges, tidal bores, and tsunamis may cause backflooding of the distal end of the drainage basin. The distributary zone of a drainage basin that forms a delta or an alluvial fan may be particularly prone to widespread sheetflooding, where the interfluves between channels are effectively drowned by floodwaters.

Aquatic invertebrates present along channels at the lower end of drainage basins tend to be mostly collectors that filter nutrients from transport or gather them from sediments. Fish are warm-water piscivores, invertivores, and planktivores; species diversity is lower than in channels of the middle reaches of the basin (Vannote et al., 1980).

Land-use activities within the lower portion of a drainage basin are similar to those of the central basin. In extreme cases, agriculture and urban development may be particularly intense in the lower basin, as in Bangladesh, which has the highest population density in the world – 125 million people inhabit 144,000 km^2, of which 70,000 km^2 are the deltas of the Ganges and Brahmaputra Rivers. Sixty-six percent of Bangladesh is floodplain.

The geomorphic concept of a drainage basin (as opposed to an isolated channel segment) and the biologic concepts of a river continuum and flood pulses emphasize spatial and temporal exchanges: hillslopes with channels, upstream and downstream portions of channels, surface and subsurface water, floodplains, and channels. Knowledge of such exchanges is critical to understanding how channel processes operate and how they may respond to changes in controlling variables. Effective flood hazard mitigation must also be integrated throughout an entire drainage basin (Marchi et al., 1996). Leuchtenburg (1953) summarized the problems resulting from uncoordinated attempts at flood control in the Connecticut River valley. The Connecticut River basin includes portions of the states of Vermont, New Hampshire, Massachusetts, and Connecticut. Responsibilities for water resources within the basin were assigned to several federal agencies, so that (i) flood hazards increased as dikes designed in association with flood-control dams were emplaced years before the dams were actually built, (ii) flood-control dams designed without regard to pollution control exacerbated existing water-quality problems, and (iii) numerous disagreements arose over resource use and power development among state and federal agencies as well as among the states.

One of the goals of this volume is to present flood hazards not as isolated problems but as a manifestation of channel processes that occur within, and must be addressed from, a drainage basin representing interacting physical, chemical, and biological systems.

History of Flood Hazards and Flood Mitigation

Flood hazards were initially perceived by humans as those processes that occur during floods that directly threatened humans or their property. Archeological and historical records provide a lengthy chronicle of such flood hazards: The abandonment of the city of Mohenjo-Daro along the Indus River (Pakistan) circa 1700 B.C., after 800 years of occupation, appears to have been driven partly by devastating floods (Nautiyal, 1989). During their campaign against the Babylonians in 689 B.C., the Assyrians dammed the Euphrates River to build up a reservoir of water and then breached the dam to send a devastating flood downstream to destroy the city of Babylon (Brooker, 1981).

Historical flood records may take the form of physical high-water marks, such as the stone inscriptions relating to 5000 years of Nile River flood levels in Egypt (Bell, 1970). Flood marks date to A.D. 223 on the Huang He (Yellow River) and to A.D. 1153 on the Chang Jiang (Yangtze River) of China (Baker and Kochel, 1988). Historical records may also take the form of maps or written accounts (Hooke and Kain, 1982), such as those describing the floods that occurred circa 747 B.C. along the Nile River (Hoyt and Langbein, 1955). Using historical chronicles, Yao (1942, 1943) compiled a record of more than 3000 floods in China between 206 B.C. and A.D. 1911. In Europe, historical reports dating back to A.D. 1306 record a series of highly damaging floods along the upper Rhine River (Kern, 1992).

These human records may be augmented by botanical and geological indicators of past floods (Stedinger, Chapter 12, this volume; Baker, Chapter 13, this volume). In central India, for example, historical flood records extend to A.D. 1291, but sedimentary deposits produced by floods indirectly record flood magnitude and frequency over the past 5000 years (Kale et al., 1994).

Human responses to perceived flood hazards also have a long tradition. The oldest constructed dam discovered thus far dates to circa 2900 B.C. in Wadi El-Garawi, south of Cairo, Egypt (Smith, 1971; Costa, 1988a). Li Ping initiated an extensive system of irrigation canals and flood-control structures more than 2100 years ago in the Szechuan region of China (Yao, 1943). A much later increase in the frequency of recorded floods and droughts during the Yuan dynasty (A.D. 1234–1367) has been attributed in part to the destruction of irrigation and drainage systems along the Yellow River by Mongol invaders (Chu, 1926).

Levee construction along the Yodo River in Japan began in the 4th century A.D. (Japanese River Bureau, 1985), and, in an effort to control channel erosion (Japanese Ministry of Construction, 1993), a Japanese imperial decree of A.D. 806 prohibited cutting trees along riverbanks. The Edo era in Japan (17th to mid-19th centuries) was characterized by rapid population increase and extensive land development and flood control projects. The entire channel of the Tone River, the largest river in Japan, was rerouted more than 100 km to the east so that the channel bypassed the Kanto Plain and the site of the city of Edo (Tokyo) (Japanese River Bureau, 1985).

In Europe and Britain, systematic flood-control works were generally not undertaken until the 19th century (e.g., Froehlich et al., 1977; Petts, 1990b, 1994; Burrin and Jones, 1991), although localized efforts and nonstructural hazard mitigation programs were undertaken much earlier. Since the 12th century, French legislation has recommended channel maintenance and flood reduction by clearing large woody debris from channels (Piégay and Gurnell, 1997). Flood embankments were built along the Po River in the 14th century (Braga and Gervasoni, 1989). Channels draining into the Adriatic Sea near Venice were rerouted at the start of the 16th century (Cosgrove, 1990). As early as 1780, the Duke of Savoy promoted the embankment of the left bank of the Rhone River to protect rural communities from overbank flooding, causing up to 4 m of channel degradation (Bravard et al., 1994). People who lived along the Rhone soon realized the dangers of piecemeal channel alteration. A French law enacted in 1858 forbade any new flood embankments upstream of Lyon, which was damaged by a large flood in 1856 (Bravard et al., 1994). Systematic channel stabilization, aimed at improving navigation and reducing the damages from winter floods, began during the late 17th century along the River Trent of central England (Petts et al., 1992). Flood-control measures including channel deepening and widening, field drainage, and pump drainage began as early as 1630 and 1727 in various eastern counties of England (Wood, 1981; Butlin, 1990).

In the United States, local attempts at flood control generally began soon after regions were settled. For example, New Orleans was settled by the French about 1717, and levee construction began in 1718 (Watson and Biedenharn, Chapter 14, this volume). Supreme Court decisions and congressional actions relating to navigation and flooding date to the first decades of the 1800s, but a federal flood-control program was not initiated until about 1850, when flood protection became a key component of the navigation program on the lower Mississippi River. Up to 1936, federal expenditures for flood control were minor, totaling only about $400 million. Major channels such as the Mississippi River had already been dramatically altered by flood-control works, however. By the 1930s, the river between St. Louis and the Gulf of Mexico had been shortened by 370 km through elimination of cutoffs and levee construction (Rasmussen, 1994). After a series of severe droughts and floods during the 1930s, the Flood Control Act of 1936 authorized 211 flood-control projects that affected nearly every state in the union (Barrows, 1948; Hoyt and Langbein, 1955). The Act initiated a massive effort, ranging from construction of dams, reservoirs, and diversion channels through management of watershed lands (contour strip cropping, terracing, reforestation, erosion-control structures, and public-land acquisition). This effort included construction of the first massive dams (e.g., the 221-m-high Hoover Dam on the Colorado River) and integrated development of entire river basins (e.g., the Tennessee Valley Project) (Petts, 1994).

The U.S. Flood Control Act of 1936 reflected an understanding that (i) uncoordinated local efforts could not satisfactorily control flood hazards; (ii) direct federal appropriations, with partial payment of costs by beneficiaries, was the most feasible approach; and (iii) it was desirable to promote multipurpose development throughout the nation (Hoyt and Langbein, 1955). The Bureau of Reclamation, the U.S. Army Corps of Engineers, the Soil Conservation Service, the Tennessee Valley Authority, and other agencies received allocations for flood control, and by 1950 the annual federal investment in flood control had reached $600 million. Such expenditures were justified by the fact that 7% of the national population lived on the 3% of the total U.S. land area that was subject to occasional inundation and flood hazards, with 80% of these people in urban areas (Hoyt and Langbein, 1955). The shift toward engineering-intensive flood control on large spatial scales that occurred in the United States during the late 1930s through the 1950s had affected most nations by the 1970s and is still occurring today.

In general, flood-control measures focus on (i) regulating flow along a channel, primarily by using dams, reservoirs, and flow diversions; (ii) restricting overbank flooding by increasing the channel's capacity to contain floodwaters, through the use of levees, training structures, channelization, channel clearing, and dredging (Vanoni, 1975); (iii) restricting bank collapse and lateral channel movement by stabilizing the channel with revetments (Vanoni, 1975); and (iv) reducing runoff concentration to channels through forest-fire control, revegetation, weather modification, and modification of cropping practices (Beyer, 1974). During the past two to three decades, increasing use has been made of "soft engineering" approaches, such as bank stabilization with living vegetation or construction of pool and riffle sequences (Brookes et al., 1983; Gregory, 1987; Hey, 1994). Socioeconomic and political responses to flood hazards have included flood insurance, flood warning systems, floodproofing of structures, disaster relief, floodplain mapping and land-use zoning, and tax writeoffs (Beyer, 1974; Kunreuther, 1974; Edelen, 1976; FEMA, 1983, 1986; Gruntfest, Chapter 15, this volume; Hamilton and Joaquin, Chapter 18, this volume).

Despite this extensive history of flood-related damages and mitigation responses, floods continue to constitute a major natural hazard throughout the world. In the United States, for example, an average of 146 people die each year from floods (NOAA, 1992). NOAA has estimated that average annual economic losses from floods in the United States have been increasing at about 4% per year in real terms (National Science Foundation,

1980), and countries as diverse as Indonesia (Hadimoeljono, 1992) and Japan (Japanese River Bureau, 1985) have had similar trends of increasing flood damage during the 20th century. As of 1980, floods accounted for 25% to 33% of average annual monetary loss from geologic hazards in the United States and for 80% of the annual loss of life from such hazards (Costa and Baker, 1981). These data were calculated before the 1993 Mississippi River flood, which caused damages in excess of $12 billion (NOAA, 1994; Galloway, 1995). Hoyt and Langbein began their 1955 book on floods with a prologue that listed disastrous floods, ending as follows:

> Somewhere in the United States, Year 2000 plus or minus. Nature takes its inexorable toll. Thousand-year flood causes untold damage and staggering loss of life. Engineers and meteorologists believe that present storm and flood resulted from a combination of meteorologic and hydrologic conditions such as may occur only once in a millennium. Reservoirs, levees, and other control works which have proved effective for a century, and are still effective up to their design capacity, are unable to cope with enormous volumes of water involved. This catastrophe brings home the lesson that protection from floods is only a relative matter, and that eventually nature demands its toll from those who occupy flood plains. (Hoyt and Langbein, 1955, p. 4)

Except for the "staggering loss of life," this account is a very good description of the 1993 Mississippi River flood. Table 1.1 provides a partial list of devastating floods. Such a list is necessarily incomplete because of the thousands of damaging floods that have occurred throughout the world during the centuries in which humans have kept records. Reading the more complete listing for the period 1900–1952 contained in the work of Hoyt and Langbein (1955), the cumulative losses to floods in the United States alone are staggering (Figure 1.4a). A running total of flood damage worldwide may now be accessed from the Dartmouth Flood Observatory at the Internet address: http://www.dartmouth.edu/artsci/geog/floods/.

The frequency of damaging floods increases toward the present time, as reflected in Table 1.1 and Figure 1.4, as a result of three factors: (i) more thorough and complete records of floods because of better preservation of more recent historical documents and introduction of systematic river discharge measurements, (ii) steady encroachment of people into riparian corridors as world population increases inexorably, and (iii) accelerating rates of landscape change, as a result of human activities, that promote flooding.

Channel Change and Land-Use Patterns

The archeological record provides early examples of changes in channel characteristics and flood regime resulting from human land-use practices. Extensive archeological and sedimentological studies in Greece have revealed a period of widespread and substantial channel aggradation from late-Classical times until almost the present. Climatic changes (Vita-Finzi, 1969) and forest clearance and the rise of animal husbandry (Butzer, 1974, 1980) contributed to this aggradational episode (Davidson, 1980). Analogous episodes of widespread alluviation associated with increasing land use have also been reported in Spain (Van Zuidam, 1975; Harvey, 1978),

Table 1.1. *A partial listing of devastating floods. (Sources: Hoyt and Langbein, 1955; Beyer, 1974; NOAA, 1983; Costa, 1988a; Dartmouth Flood Observatory, 1997)*

Location	Date	Cause	Damages
Nile River, Egypt	ca. 747 B.C.	rainfall	unspecified
Mississippi River	March 1543	rainfall	unspecified
China	1642	rainfall	300,000 dead
James River, USA	May 1771	rainfall	city of Richmond, Virginia destroyed; 150 drowned
Coastal California	Dec. 1861–Jan. 1862	rainfall	unspecified
Central Colorado	May–June 1864	rainfall	19 drowned
Connecticut River, USA	May 1874	reservoir failure	$1 million damages; 143 dead
Yellow River, Honan, China	1887	rainfall	900,000+ dead
Johnstown, Penn., USA	May 1889	reservoir failure	2200 dead
Upper Mississippi River basin, USA	May–June 1903	rainfall	$40 million damages; >100 dead
Yangtze River, China	1911	rainfall	100,000 dead
Lower Mississippi River basin, USA	March 1912	rainfall	$70 million damages
Mississippi River basin, USA	March–April 1913	rainfall	$154 million damages; >470 dead
Texas, USA	Dec. 1913	rainfall	$9 million damages; 177 dead
Central Colorado, USA	June 1921	rainfall	$25 million damages; 120 dead
Texas, USA	Sep. 1921	rainfall	$19 million damages; 224 dead
Florida, USA	Sep. 1926	storm surge	350 dead
Lower Mississippi River basin, USA	March 1927	rainfall	$300 million damages; 313 dead
Southern California, USA	March 1928	reservoir failure	$15 million damages; >350 dead
Yellow River, China	1933	dike failure	18,000 dead
Northeastern USA	March 1936	rainfall	$270 million damages; 107 dead
Mississippi River basin, USA	Feb.–March 1937	rainfall	$420 million damages; 137 dead
Northeastern USA	Sep. 1938	storm surge and rainfall	$37 million damages; 500 dead
Yellow River, China	1938	dike breach	890,000 dead
Kansas River, USA	July 1951	rainfall	$1 billion damages; 50 dead
Manchuria, China	Aug. 1951	?	5000+ dead
Kazvin District, Iran	Aug. 1954	?	2000+ dead
Belluno, Italy	Oct. 1963	dam overtopped	2000+ dead
Gujarat, India	Aug. 1968	?	1000 dead
West Virginia, USA	Feb. 1972	reservoir failure	$10 million damages; 125 dead
South Dakota, USA	June 1972	rainfall	$164 million damages; 237 dead
Colorado, USA	July 1976	rainfall	$56 million damages; 141 dead
Gujarat, India	1979	dam failure	up to 10,000 dead
Southern Thailand	Nov. 1988	rainfall	$11 million damages
Philippines	Nov. 1991	rainfall	8000 dead
Mississippi River basin, USA	June–July 1993	rainfall	$20 billion damages; 48 dead
Central China	1996	?	$113 million damages; 1509 dead
Vietnam and Thailand	Nov. 1997	rainfall	$472 million damages; 313 dead

from the upper Danube Basin of central Europe (Butzer, 1974, 1980), and from the Vistula Valley of Poland (Froehlich et al., 1977; Mycielska-Dowgiallo, 1977). In Britain, much of the landscape before the past 5000 years was extensively forested, and forest removal caused substantial changes along river channels, including phases of aggradation (Crampton, 1969; Gregory, 1983; Robinson and Lambrick, 1984; Gregory and Gurnell, 1988; Busch et al., 1989). These changes mirror river metamorphosis and altered flood regimes resulting from natural climatic changes (Leopold, 1951; Dury, 1964, 1965; Schumm, 1968; Graf, 1983; Knox, 1988; Hall, 1990; Nanson et al., 1995) or from changes intrinsic to channel processes (Patton and Schumm, 1975, 1981). The changes induced by land-use often occur more rapidly than natural changes, sometimes with unfortunate consequences for human, riparian, and aquatic communities.

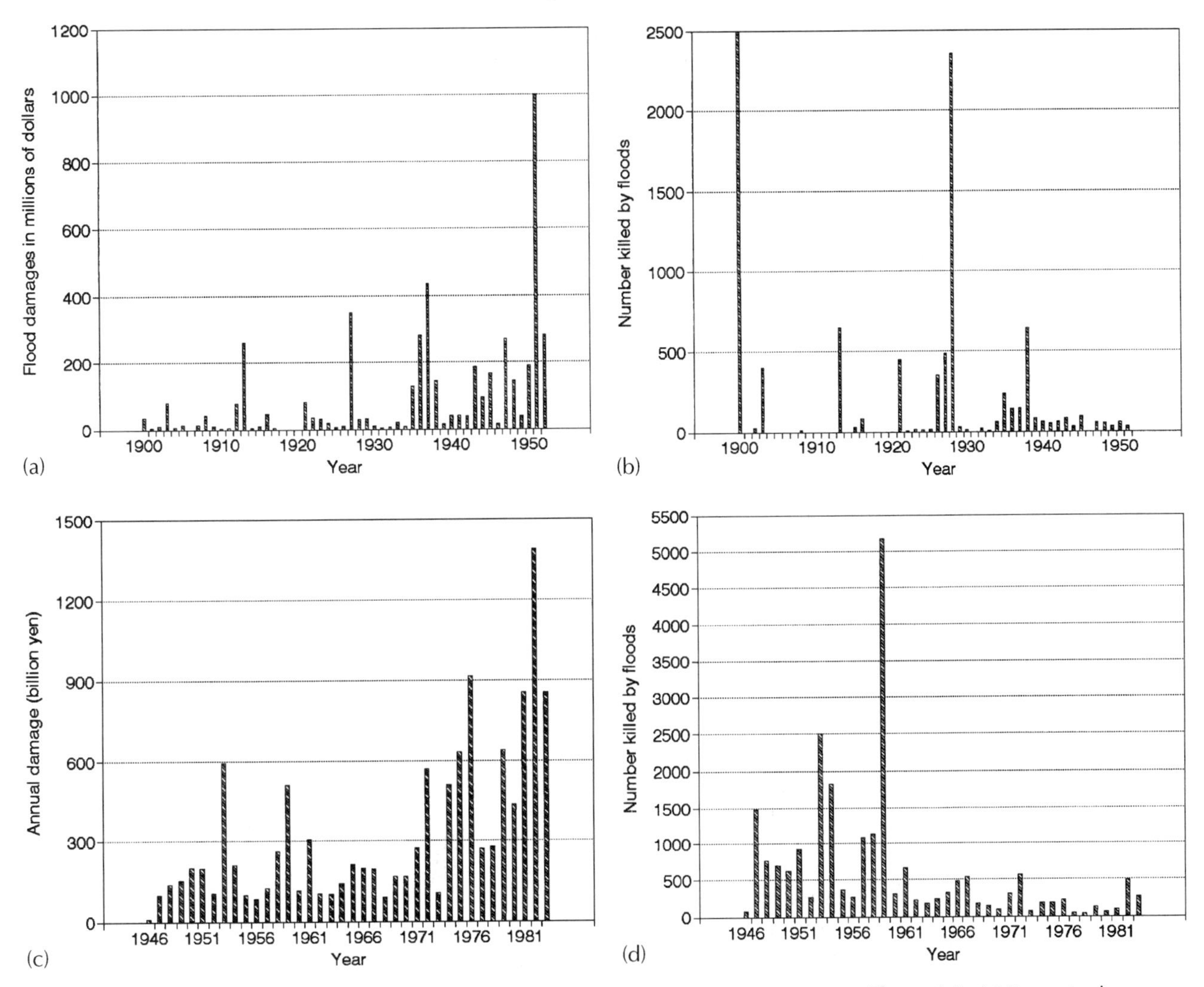

Figure 1.4. (a) Property damage from floods in the continental United States, 1900–1952 (source Hoyt and Langbein, 1955). Monetary damages represent minimum figures, as they include only the major floods in any given year and have not been adjusted for inflation. (b) Loss of life from floods in the continental United States, 1900–1952 (source Hoyt and Langbein, 1955). For scaling reasons, the figure does not include the total of 6000 deaths resulting from floods in 1900. (c) Property damage caused by flooding in Japan, 1946–1983 (source Japanese River Bureau, 1985). (d) Loss of life from floods in Japan, 1946–1983 (source Japanese River Bureau, 1985).

Numerous examples of historical changes in channels and flow regime resulting from human land use can be cited; Table 1.2 summarizes a few case studies. Of particular importance to understanding such interactions is the realization that human-induced landscape change has been steadily accelerating since the Industrial Revolution, as human population density and technology have exploded across the planet's surface. Despite increased levels of awareness among the general public during the past few decades about the importance of naturally functioning ecosystems (egs. Hey et al., 1990; Mochizuki and Jikan, 1993; Seki and Takazawa, 1993; Hey, 1994), human-induced landscape change will continue to accelerate in many parts of the world for at least the next few decades. Therefore, it is important to briefly review the effects of various types of land use on flood regimes, a topic covered in more detail in Chapter 4 of this volume.

Land-use activities may be classified as indirect or direct with respect to stream channels (Park, 1977). Indirect activities are those that alter the discharge of water and/or sediment entering the stream channels, generally by altering the infiltration characteristics of slopes within a drainage basin. Examples of indirect land uses and their effects on flood regimes

Table 1.2. *Representative examples of changes in channel characteristics and flow regime resulting from human land use*

Location	Land use	Response of flow regime/channels	Reference
Western Great Plains, USA	irrigated agriculture 1880–1960	peak flood magnitude decreased, baseflow increased; channels grew narrower and more sinuous, with establishment of dense riparian vegetation	Williams, 1978; Nadler and Schumm, 1981; Eschner et al., 1983
Vistula Basin, Poland	deforestation, followed by various types and extents of crops; 1800–1970	change in channel pattern from meandering to braided with increased sediment yield; incision of channels following flood-control works (straightening, shortening, & embankments); general increase in average flood discharge	Froehlich et al., 1977; Klimek, 1987; Starkel, 1988
Piedmont of Maryland, USA	agriculture followed by reforestation, and then urbanization 1830–1965	sediment yield increased with agriculture, decreased under forest cover, peaked sharply and then decreased with urbanization; channel alternately aggraded and scoured, with attendant flooding	Wolman, 1967; Wolman and Schick, 1967
Colorado River basin, USA	dams 1960-present	annual peak discharge decreased, water temperature decreased, sediment load decreased, bankfull channel width decreased	Graf, 1978; Williams and Wolman, 1985; Andrews, 1986
Sierra Nevada region, California, USA	placer mining 1853–1884	aggradation, channel avulsion, and flooding, followed by channel degradation	Gilbert, 1917; Knighton, 1989; James, 1989
Yellow River basin, China	deforestation, agriculture	already large suspended sediment load increased, exacerbating channel aggradation, overbank flooding, and channel avulsion; necessitating levees and channel stabilization	Clark, 1982

are listed in Table 1.3. Timber harvest, for example, causes increased water yield to channels from a fixed precipitation input, as a result of reduced evapotranspiration, faster snowmelt, and surface compaction due to road construction (Bosch and Hewlett, 1982). Timber harvest also may increase sediment discharge as a result of vegetation removal; slope compaction and more erosive surface runoff, especially along unpaved roads; and increased occurrence of mass movements (Johnson and Beschta, 1980; Reid and Dunne, 1984; Sidle et al., 1985; Trimble, 1988). Increased sediment discharge, which occurs within days to months after timber harvest, may cause channel aggradation and loss of conveyance, leading to more frequent out-of-channel floods. Over a period of months to years, flood magnitude and frequency may increase as precipitation that formerly moved slowly from the basin slopes into the stream channel network moves more rapidly off the slopes (Gentry and Lopez-Parodi, 1980).

Direct land-use activities are those that occur within the stream channel and directly alter the characteristics of water and sediment movement along the channel. Among the examples listed in Table 1.3 is placer mining. By disrupting the sediment forming the channel bed and banks, placer mining may enhance sediment mobility, causing downstream aggradation that reduces channel capacity and increases the frequency of out-of-channel floods (Gilbert, 1917; James, 1989; Leigh, 1994). The mining also may decrease the stability of the channel banks, promoting lateral movement and hazards caused by bank collapse during floods (Hilmes and Wohl, 1995), or channel incision (Graf, 1979).

Channel response to land-use activities may occur as adjustments of channel cross section (size, width/depth ratio, composition of bed and

Table 1.3. *Examples of land-use activities and their effects on flood regime*

Land-use activity	Effect on hillslope or channel processes	Effect on flood regime	Sample reference
Indirect			
Timber harvest/ deforestation	increases water and sediment yield from slopes by reducing infiltration capacity	increased flood magnitude and frequency	Bosch and Hewlett, 1982; Reid and Dunne, 1984; Sidle et al., 1985; Trimble, 1988; Yao et al., 1994
Agriculture			
crops	increased water and sediment yield due to lower vegetation density, more frequently disturbed ground surface, ditch networks, subsurface tile drain systems, ground compaction by farm machinery	increased flood magnitude and frequency	Trimble, 1988; Boardman et al., 1994; Prestegaard et al., 1994
Grazing	increased water and sediment yield due to surface compaction and vegetation removal; may cause channel incision	increased flood magnitude and frequency	Gregory and Walling, 1973; Whitlow, 1985; Trimble, 1988; Trimble and Mendel, 1995
Destruction of wetlands	by reducing complexity of floodplain, reduces lag time and increases rapidity of flood wave movement downstream; may cause channel incision	increased flood magnitude and frequency	Walling, 1987
Urbanization	construction phrase increases sediment yield from slopes; once construction is complete, sediment yield decreases and water yield greatly increases due to reduced infiltration capacity	increased flood magnitude and frequency	Wolman, 1967; Leopold, 1968; Park, 1977; Dunne and Leopold, 1978; Gupta, 1984; Boardman et al., 1994
Introduced exotic riparian vegetation	generally causes channel narrowing via aggradation along the banks	increased flood magnitude	Graf, 1978
Direct			
Placer mining	destabilized channel, greater sediment mobility promotes channel aggradation; greater channel mobility promotes avulsion	increased flood magnitude and frequency, greater lateral channel movement	Gilbert, 1917; Graf, 1979; James, 1989; Leigh, 1994; Hilmes and Wohl, 1995
Sand and gravel mining	as above, or sediment depletion may cause channel and coastal erosion	as above	Klimek and Starkel, 1974; Uda et al., 1995
Channelization for flood control or navigation	straighter, more uniform channel conveys flood water more rapidly downstream; headward erosion may cause aggradation downstream	more peaked flood hydrograph	Keller, 1976; Winkley, 1982; Brookes, 1994
Construction of levees	containment of floodwater within channel may promote scour and bank erosion, more rapid movement of floodwave	increased flood magnitude	Marchi et al., 1996
Construction of reservoirs	depending on operation, magnitude, frequency, and sediment concentration of flows are altered – may cause changes in channel morphology and associated changes in riparian and aquatic communities	variable	Petts, 1977; Williams, 1978; Brooker, 1981; Eschner et al., 1983; Gore, 1994; Harper, 1994; Walker, 1994
Flow diversion	depending on timing and magnitude of flow removed or added, may cause channel narrowing or widening	variable	Richards and Wood, 1977; Williams, 1978; Eschner et al., 1983
Removal of woody debris	by reducing channel complexity and boundary roughness, increases conveyance of flood waters downstream	increased flood magnitude	Bisson et al., 1987; Sedell and Froggatt, 1984; Shields and Smith, 1992
Grazing	grazing along channel destroys bank vegetation and compacts banks, creating wide, shallow channel	increased out-of-bank flooding	

bank sediments), of channel pattern, or of drainage network (extent, density, shape) (Gregory, 1976, 1983). Such changes may be quantified by using a channel change ratio (CCR) of final channel dimension to initial channel dimension, but channel response may vary widely. For example, the average CCR value for 180 channels downstream from urban runoff was 3.38, but CCR values ranged from 0.13 to 15.0 (Gregory, 1987). Such variability makes it difficult to predict the consequences of land-use activities for flood hazards.

When assessing the impacts of both indirect and direct land-use activities, it is important to remember that a drainage basin functions as an interconnected system rather than as a series of independent channel links. If changes in the ratio of water and sediment discharge in the upper channels of a basin are of sufficient magnitude and duration, the lower channels in the drainage network will also respond. In some cases, a human response to a hazard in one portion of the basin will create hazardous conditions elsewhere in the basin. Channel incision and beach erosion resulting from sediment control provide a prime example. Numerous sediment-retention (sabo) dams have been built along the upper reaches of channels in Japan, with the intention of reducing hazards associated with debris flows and floods. As sediment accumulates behind these dams, channels downstream become laterally and vertically unstable as a result of bed and bank scour by sediment-impoverished water flows. Humans often respond to this channel instability by stabilizing the channel with riprap, soil cement, or other bed and bank protection, which in turn creates greater sediment depletion downstream. Beach erosion ultimately results from this decrease in alluvial sediment supply (Uda and Yamamoto, 1994), which in turn may create more hazardous conditions along the coast during storms or floods.

In many drainage basins, hazards associated with human-induced channel change are greatest during floods, because floods are episodes of relatively rapid change within channels and floodplains (Hack and Goodlett, 1960; Walling et al., 1992; Osterkamp et al., 1995; Uda et al., 1995; Wohl, Chapter 6, this volume). The effects of land-use changes that occur over a century may accumulate to produce dramatic channel change during a flood lasting only a few days (Kresan, 1988). Similarly, chemical contaminants that have gradually accumulated in sediments along the riparian corridor or adjacent uplands may be suddenly mobilized and transported downstream during a flood, as discussed by Finley (Chapter 7, this volume), Lewin et al. (1977), and Webb and Walling (1994).

At the most comprehensive level, the general effect of indirect human land-use activities during the past 200 years has been to greatly increase the volume and rate of movement of water from drainage basin slopes to stream channels by reducing slope permeability and infiltration through deforestation, destruction of wetlands, and urbanization. Similarly, the general effect of direct land-use activities has been to alter naturally complex, irregular, or sinuous channels toward more straight, uniform channels that convey water and sediment more rapidly downstream. Because the net effect of these changes has been to move water more rapidly from slopes to channels as well as downstream through the channels, it is likely that floods in a given drainage basin will have shorter rising limbs and higher peak discharges, irrespective of any changes in climate. When these trends are combined with increasingly higher densities of human population along riparian corridors, the potential for even greater flood hazards in the future becomes clear. In addition, both indirect and direct land uses affect aquatic and riparian biota by altering the chemical and physical processes that shape the habitat to which these organisms are adapted. In the United States, for example, channel modification, water impoundment, floodplain clearing for agriculture, and urbanization together reduced an original 30–40 million hectares of riparian habitat by approximately 75%, to 10 million hectares, by the early 1980s (Brinson et al., 1981; Swift, 1984). In a world where truly natural environments are practically nonexistent, the rapidity and extent of change along riparian corridors is driving many species to extinction.

Recognition of these problems has led to a contemporary emphasis on flood-hazard mitigation in many societies. Such mitigation may be considered to have at least three facets: (i) urban planning, using structural alterations of slopes and channels, floodplain mapping, land-use regulation, and flood insurance zoning (Edelen, 1976; FEMA, 1983), floodproofing of structures (FEMA, 1986), and flood warning systems (NOAA, 1994) to reduce human exposure to hazardous flood processes; (ii) flood forecasting, with the aim of estimating both the magnitude and frequency of potentially hazardous floods that may occur as a result of present drainage basin and meteorological conditions (NOAA, 1994; Dilley and Heyman, 1995; Killingtveit, 1996), and the characteristics of floods occurring after predicted future land-use and climatic changes in a basin; and (iii) understanding the effect of changing flow regimes on riparian and aquatic ecosystems (Brooker, 1981; Milner et al., 1981; Johnson et al., 1995; Marston et al., 1995; Scott et al., 1996) to be able to ensure at least minimum flow requirements for these ecosystems (Bullock and Gustard, 1992; Stalnaker, 1992; Karim et al., 1995). Each of these subjects is treated in greater detail in the remainder of this volume: aspects of urban planning are discussed in Chapters 10, 14, 15, 17, and 18; aspects of flood forecasting are discussed in Chapters 2, 3, 5, and 11–13; and biotic responses to changing flow regime are covered in chapters 8, 9, and 16.

Flood Hazard Research

The topics discussed in this volume derive from a long and continuing problem with flood hazards. Systematic, scientific approaches to identifying and mitigating flood hazards are a relatively recent phenomena, however. Natural hazard research was initiated by Gilbert White and his students in the 1960s (Gares et al., 1994). They established a research paradigm based on five components: assessing the risk from a natural event, identifying adjustments to cope with the hazard, determining people's perception of the event, defining the process by which people choose adjustments, and estimating the effects of public policy on the process of choice. Numerous variations to this approach have subsequently been developed, primarily by social scientists (Gares et al., 1994).

Social scientists often emphasize the complexity of analyzing human responses to hazards, with the responses being a function of both individual and cultural traits (Saarinen, 1974; Slovic et al., 1974), government policies (Brown, 1972; Visvader and Burton, 1974), and historic occurrence of hazards (Beyer, 1974). The physical, chemical, and biological processes during a flood may be equally complex and difficult to predict. Modeling of rainfall-runoff relations is problematic because these are spatially heterogeneous and dependent on antecedent conditions (Ramirez, Chapter 11, this volume). Similarly, flood and overbank flow hydraulics are complex and not amenable to simulation by the one-dimensional hydraulic models that are readily available. Curves of stage, sediment concentration, and wave speed versus discharge all have a break at bankfull discharge (e.g., Knight, 1996), and these hydraulic complexities translate into uncertainties when modeling channel change. During a flood the main channel may have subcritical flow while flow over the levees is supercritical. These spatial variations in hydraulic force produce spatial variations in erosion and deposition. Thus, natural scientists and engineers must also play a central role in hazard mitigation (Dunne, 1988).

Just as a drainage basin represents a complexly interdependent system, so must effective flood hazard mitigation be based on understanding derived from exchanges of knowledge among different academic disciplines and approaches. To cite two examples: (i) for the past few decades flood predictions have been based on either extrapolation from small, common floods or on models that represent theoretical generalizations. Both of these approaches focus on mathematical methodology and have been dominated by statisticians and civil engineers (Baker, 1994). But research during the past 20 years has demonstrated that flood data derived from a geological approach, which focuses on the history and processes of actual floods, may prove extremely valuable in constraining flood predictions (Stedinger and Cohn, 1986; Stedinger and Baker, 1987; Kochel, 1988a; Webb et al., 1988; Wohl et al., 1994; Baker, Chapter 13, this volume), particularly in situations in which climatic variability has produced nonstationarity in the hydrologic time series (Hirschboeck, 1988; Ely et al., 1993; Hirschboeck et al., Chapter 2, this volume). (ii) The hazards to riparian and aquatic ecosystems associated with changes in flood regime following river regulation were not widely recognized during the 1945–1971 period of extensive dam construction in the United States (Brooker, 1981). Flow regime downstream from the dams was regulated to enhance flood control, navigation, crop irrigation, or hydroelectric power generation. However, numerous studies since 1971 have demonstrated that altered flow regimes also change channel patterns of sedimentation, as well as water chemistry and temperature, and thus affect aquatic and riparian vegetation, macroinvertebrates, fish, and all the other species that depend on them (Petts, 1984). Redesign of regulated flow regimes to minimize the hazards caused by altered flood magnitude and frequency requires interdisciplinary work by hydrologists, engineers, aquatic ecologists, limnologists, and geomorphologists. One of the challenges of interdisciplinary work is becoming familiar with the vocabulary and conceptual framework of other disciplines. Ecologists and botanists, for example, regard riparian vegetation as a biological community composed of individuals and species with complex responses to controlling physical, chemical, and biological variables. Civil engineers often regard riparian vegetation as effectively inert hydraulic roughness elements that may be manipulated to control flow distribution. Effective mitigation of flood control hazards to both human and nonhuman communities obviously will require more effective communication between ecologists and engineers.

The primary goal of this volume is to encourage an interdisciplinary dialogue of the hazards associated with changing flood regimes in river basins around the world. Between 1988 and 1992, floods accounted for the third largest number of human casualties worldwide caused by natural disasters (hurricanes and earthquakes were first and second, respectively) (Gares et al., 1994). At the same time, the number of dams on the world's rivers exceeded 10,000, with major reservoirs storing more than 5000 km^3, or 12.5% of annual global land runoff to the oceans (Walling, 1987). Currently, dams greater than 15 m high are being completed throughout the world at a rate of approximately 500 per year (Brookes, 1994). By the year 2000, it is estimated that more than 60% of total streamflow in the world will be regulated (Petts, 1989). Massive projects are entirely transforming large river basins. Runoff from the entire Caroni River basin (>100,000 km^2) of Venezuela, for example, will be impounded by nine dams with a combined surface area of 9300 km^2, or 1% of Venezuela's land area (Petts, 1990b). The Narmada Project of India involves 30 large dams and over 3000 small structures within the 98,800-km^2 Narmada basin (Petts, 1990b). Engineering projects on such a scale also create massive ecological and social disruptions. Creation of the Tucurui Dam in Brazil destroyed 2160 km^2 of tropical rainforest (Monosowski, 1984). The Volta River scheme of Ghana covered 8500 km^2 (5% of the country's total area) and forced 80,000 people to resettle (Goldsmith and Hildyard, 1984; Graham, 1984). The Pa Mong project in Vietnam displaced 450,000 people, and the Three Gorges Dam in China displaced 1.4 million people (Goldsmith and Hildyard, 1984).

To summarize the problem, human societies are engaged in a balancing act. On the one hand, we have alterations that are made to drainage basins in an effort to accommodate increasing numbers of people and increasing material affluence and levels of technology for those people. On the other hand, we have the recognition that humans must protect natural functions of drainage basins, as expressed through slope and channel processes, both to reduce the cost and hazards to human communities when those processes no longer operate and to preserve the nonhuman species that depend on those slope and channel processes.

Representative Drainage Regions

Three drainage regions, which are used throughout this volume as an example of the topics discussed, serve to illustrate the issues touched on in this introductory chapter. The first is the Colorado River basin in the western United States, the second is the Tone River basin in central Japan, and the third is the lower portion of the Ganges and Brahmaputra Rivers included in the country of Bangladesh. These regions have been chosen to illustrate three divergent cases in which increasing human populations have dramatically altered river regimes, enhancing flood hazards to human and riparian communities.

Colorado River Basin

The Colorado River drains approximately 640,000 km^2 of arid and semiarid lands in the southwestern United States and northwestern Mexico. This region was the last area of the contiguous United States to be explored by European Americans, and its rivers and mountains were the last to be

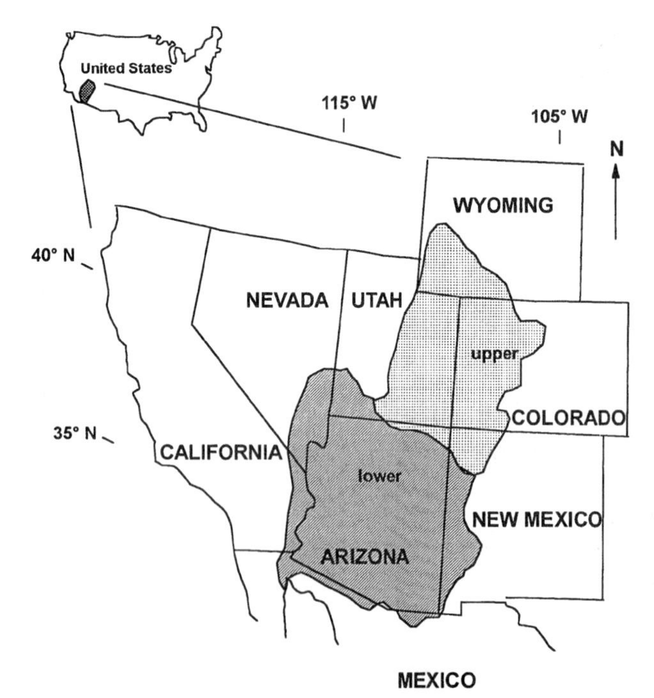

Figure 1.5. Location of the Colorado River drainage basin (shaded) (after Diaz and Anderson, 1995, Figure 4).

named (some as late as 1872) (Fradkin, 1981). The Colorado River drainage basin is usually divided into the upper and lower basins (Figure 1.5). The upper basin receives 34 cm of mean annual precipitation, and the lower basin receives 30 cm. In addition, evaporative losses from the upper basin are 60% of those in the lower basin (Waggoner and Schefter, 1990). Much of the Colorado River's runoff is derived from snowpack in the upper basin, which has a mean elevation of 2195 m, compared with 1100 m for the lower basin. Peak recorded discharge at the Grand Canyon was 8500 $m^3 s^{-1}$ in 1974 (Unesco Press, 1976).

Both the upper and lower basins are regions of rapid population growth (Figure 1.6), and the Colorado River provides much of the water for this growth. More water is diverted from the Colorado River basin than from any other river basin in the United States (Fradkin, 1981). Twelve major reservoirs within the basin contain 74 billion m^3 (60 million acre-feet) of usable storage, approximately a 4-year supply. Water usage is apportioned among civil use (water withdrawn for public and private uses); irrigation; consumptive use and irrigation conveyance losses (water lost to the system through processes such as evapotranspiration and hence not available for further use); and hydroelectric (power generation). The increase in civil use (Table 1.4) reflects increased population in the region, whereas other uses have leveled off or decreased slightly since 1980 (Diaz and Anderson, 1995). The major dams along the Colorado River act to: (i) decrease peak flow magnitude and frequency (mean annual flood magnitude has been reduced 60% below Hoover Dam, for example; Dolan et al., 1974) and (ii) reduce downstream sediment load (9.87 million m^3 of bottom sediment were scoured from the channel downstream of Glen Canyon Dam during dam construction and initial operation, 1956–1965; Pemberton, 1976). These changes in turn affect fluvial and riparian systems downstream of the dams. The Colorado River basin contains 13 national parks and monuments and numerous endemic ecosystems. Currently, four of the six dominant native fish species in the Colorado River are listed as threatened or endangered (Harvey et al., 1993). At the same time,

Figure 1.6. Population growth in the six most populous states of the Colorado River drainage basin, 1935–1995 (after Diaz and Anderson, 1995).

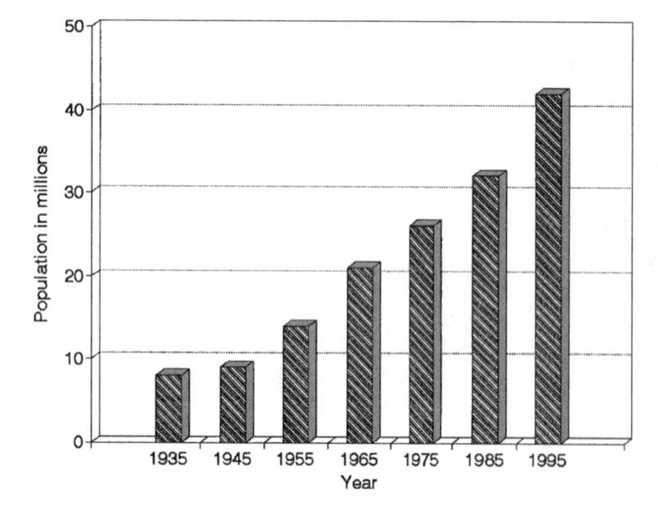

> more than half the population of the West, in all but the four most northwestern of the eleven western states, is directly dependent to some extent on Colorado River water. Additionally, the Northwest is tied into the electrical grid, part of whose power is generated by Colorado River water. (Fradkin, 1981, p. 31)

As of 1992, approximately 6.6 million people, or 10 people per km^2, lived within the basin. Although the Colorado River basin is sparsely populated by

Table 1.4. *Water-use statistics for the Colorado River basin (after Diaz and Anderson, 1995, Table 1)*

Year	Arizona	California	Colorado	Nevada	New Mexico	Utah	Entire basin
Ratio of the amount of water used for civil supply and consumptive losses to total for irrigation and conveyance losses							
1960	0.05	0.10	0.05	0.05	0.06	0.09	0.08
1990	0.16	0.17	0.07	0.14	0.12	0.17	0.14
Percentage of state water usage supplied by Colorado River							
1995	22%	7%	16%	5%	12%	21%	
Change in per capita consumption for civil use, 1980–1990							
	+10%	−15%	−42%	+7%	+22%	−49%	−20%

world standards, the extensive modification of the basin's flow regime has created numerous flood-related hazards for aquatic and riparian communities (Wydoski and Wick, Chapter 9, this volume).

Tone River Basin

The Tone River drains 16,840 km^2 of central Japan. Flowing southeastward from elevations of 1900 m in the Japan Alps, the channel formerly entered the Pacific Ocean at Tokyo Bay. After establishment of the Tokugawa Shogunate in 1590, the main channel of the Tone River was diverted more than 100 km to the east to empty into the Pacific at Choshi (Figure 1.7). This massive diversion project was undertaken partly to prevent flooding in the city of Edo (Tokyo), one of the world's largest cities at that time, and to open new lands for rice cultivation on the Kanto Plain. The new diversion channel began at a width of 12.6 m when first excavated and was widened to 18 m in 1654, to 50 m between 1688 and 1703, and to 70 m in 1809 (Uzuka and Tomita, 1993).

The Tone River has an exceptionally steep longitudinal gradient for such a large river, with an average channel slope of 0.003 (Japanese River Bureau, 1985). The river also has a high unit discharge (3.3 m^3 s^{-1} km^{-2} compared with 0.06 m^3 s^{-1} km^{-2} for the Rhine River, for example), a large ratio of maximum-to-minimum discharge; a peaked flood hydrograph (short, steep rising, and falling limbs), and high sediment loads (Japanese River Bureau, 1985). Mean annual precipitation for the Tone River basin is 1350 mm (Hukunari and Hiroki, 1993), and more than 90% of the historic floods along the river have been associated with typhoons (Uzuka and Tomita, 1993). Table 1.5 lists damages associated with major floods along the Tone River since 1935.

Flood damage in this river basin runs high in part because of human settlement patterns. Most of the Japanese people live in high densities on narrow plains along the river valleys and coasts (Nakano et al., 1974). As of 1980, population density in the basin was 970 people per km^2 (Japanese River Bureau, 1985), making it one of the most densely populated regions in the world. Before the early 1900s, the water of the Tone River was used primarily for agriculture, navigation, and fisheries. Subsequently, hydroelectric power generation and urban and industrial water supplies

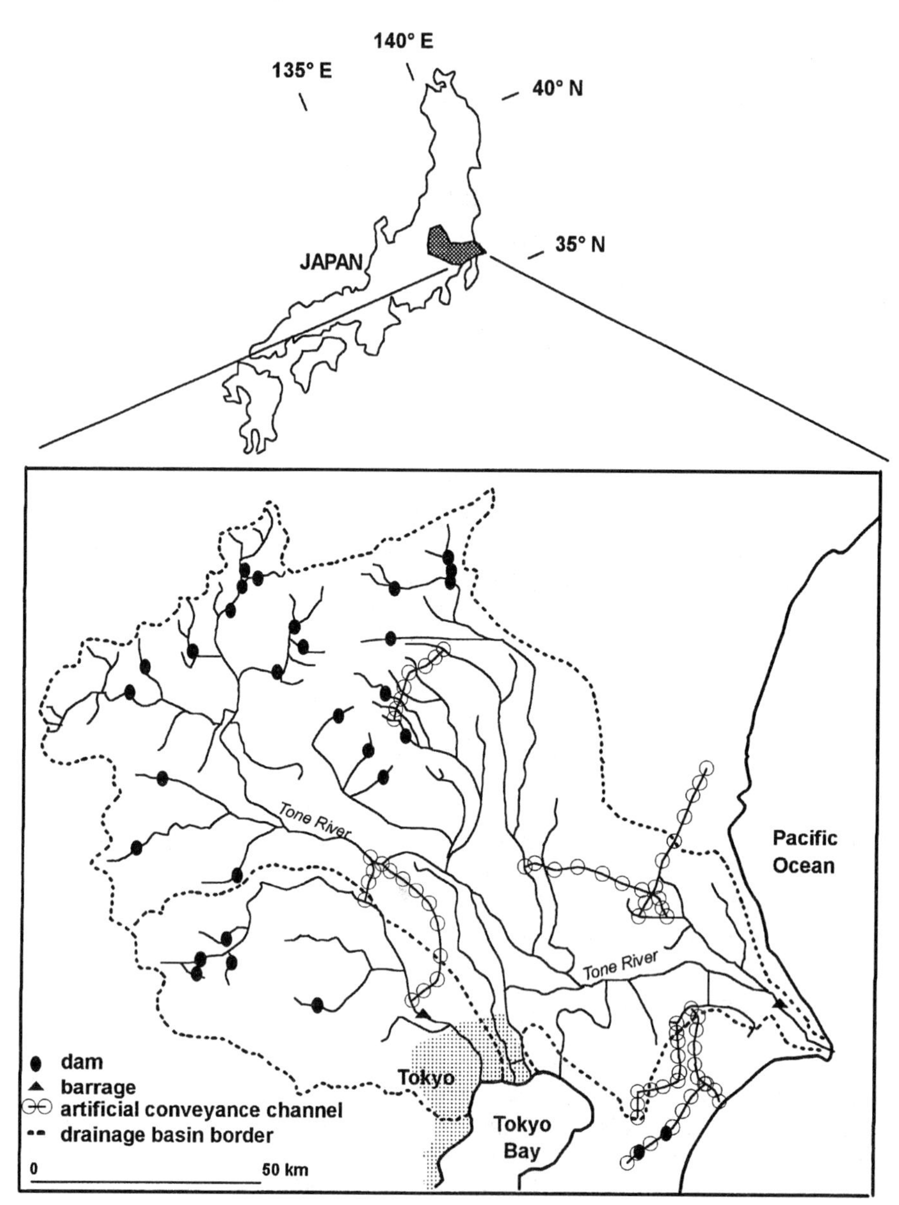

Figure 1.7. Location map of the Tone River drainage basin, Japan, with water resources structures (after Japanese River Bureau, 1985).

have taken increasingly greater amounts of the Tone's flow, reflecting the population growth of the six prefectures drawing water from the basin's channel. Population in these prefectures increased from 17.7 million in 1955 to 30.6 million in 1990 (Hukunari and Hiroki, 1993). The daily supply for the cities of Tokyo, Chiba, and Saitama alone is 7.65 million m^3, or 0.06% of the basin's 13 to 14 billion m^3 of annual runoff. Ninety-one hydroelectric power plants across the basin generate 3.8 million kW of power, and 3250 intake gates supply 250,000 hectares of irrigated farm land (Hukunari and Hiroki, 1993), so that the Tone River drainage basin may be reasonably compared with a vast plumbing system (Figure 1.7). Nevertheless, the Tokyo region, where the nation's administrative and economic functions as well as one-third of the population are concentrated, has been subject to water shortages nearly every 3 years.

Table 1.5. *Damages caused by major floods along the Tone River, Japan (after Uzuka and Tomita, 1993, Tables 2.1.1 and 2.1.2). Damage costs for the 1995 Kobe earthquake are included for the sake of comparison*

Date	Discharge[1] (m^3/s)	Monetary damage[2]
Sep. 1935	9030	59,463
June and Sep. 1938	2850	73,430
Sep. 1941	8990	40,355
Sep. 1947	17,000	277,613
Sep. 1948	—	77,890
Aug. 1949	9680	55,984
Sep. 1950	6320	68,023
Sep. 1958	9730	48,644
Aug. 1959	9070	30,425
June 1966	5880	13,348
Aug. 1971	2460	15,420
Sep. 1972	4380	9020
1995 Kobe region earthquake: 9,926,800,000,000 yen in damages; 6279 deaths; more than 34,000 people injured		

[1]Discharge as measured at Yattajima.
[2]Monetary damage in millions of yen at 1976 prices.

Flood hazards to riparian and aquatic communities along the Tone River, as along the Colorado River, are the result of alterations to flow regime and channel morphology associated with river engineering. The cumulative effect of sabo (erosion control) dams, channelization, and water diversion on rivers throughout Japan has been to reduce fish habitat to the point where such formerly abundant and widespread species as the Japanese charr and masu salmon are now endangered (Takahashi, 1989, 1990; Takahashi and Higashi, 1984). The species composition of many aquatic communities has shifted from anadromous to fluvial fish species and to species adapted to the warm, shallow water and fixed bed of concrete-lined channels. Riparian communities have been endangered by the reduction in overbank flooding and lateral channel movement.

Located in a wet climate with a highly industrialized, urbanized, and rapidly growing population, the Tone River basin exemplifies an aspect of flood hazards different from those that affect the Colorado River basin.

Bangladesh

Bangladesh is a country of rivers. The region drains almost 75% of the total Himalayan runoff in a period of approximately 3 months (Bandyopadhyay and Gyawali, 1994). Sixty-six percent of the country's 144,000 km^2 is occupied by the floodplains and deltas of the Ganges, Brahmaputra, Meghna, and other rivers (Figure 1.8). As a result, the country's geomorphology is dominated by fluvial processes and deposits. The eastern and southeastern portion of Bangladesh, the hill country, forms about 10% of the country. Here, steeply dissected hills of sedimentary rocks rise to a maximum elevation of 1003 m (Johnson, 1975). The remainder of the country is composed of benchlands (alluvial terraces) that rise up to 30 m

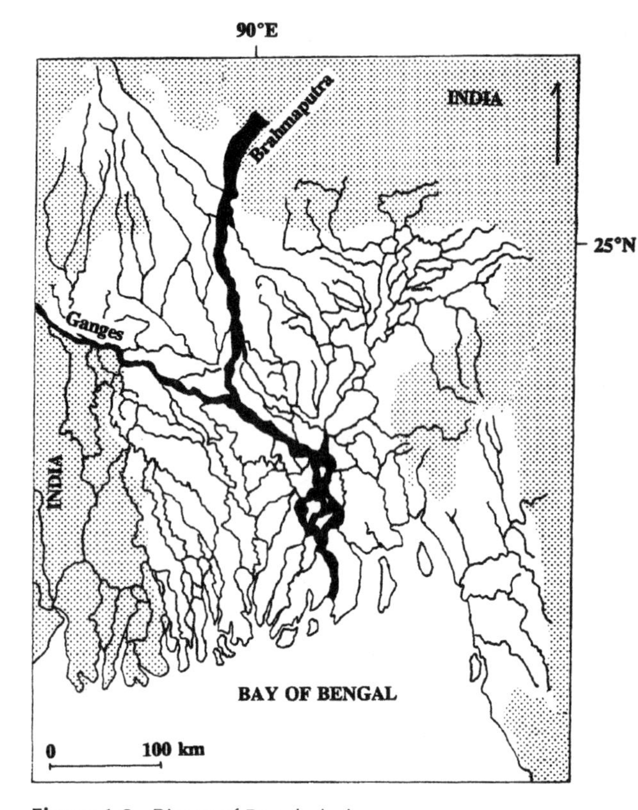

Figure 1.8. Rivers of Bangladesh.

above the surrounding plains and of the alluvial plains at elevations of sea level to 20 m (Mahmood, 1969).

Mean annual rainfall varies from 1500 mm in the west, to over 3000 mm in the southeast, with 70–85% of the rain falling during the June–October monsoonal wet season (Johnson, 1975). Cyclones also cross into Bangladesh from the Bay of Bengal during April–May and October–November. The cyclones bring high winds and storm surges that in turn cause widespread flooding; such a storm and flood killed 138,000 people in April 1991 (Katsura et al., 1992). Flooding also occurs as a regular part of the wet season. Daily precipitation in July averages 13 mm, and the drainage of local surface water is impeded when minor tributaries are ponded by backwaters from the major rivers (Johnson, 1975). When the major rivers peak in August or September, one-third of the country floods, and approximately a third of this area has water depths greater than 1 m (Johnson, 1975).

Floods bring benefits as well as damage to Bangladesh. With over 900 people per km^2, Bangladesh is the most densely populated country on Earth. Between 80% and 90% of these people grow rice or jute for a living (Oliver, 1978), and the seasonal monsoonal inundation covers an area more than 200 km wide in the central portion of the country, damaging crops and structures, and interrupting communications. Flood-control measures include embankments and dikes, but these have had limited success. Although hydroelectric dams have been built on smaller rivers, the three major rivers remain unregulated. Of these, the Ganges River has a gradient of 0.06 and a recorded peak discharge of 73,200 m^3 s^{-1} in 1961; the Brahmaputra River flows at a gradient of 0.03 and peaked at 76,600 m^3 s^{-1} in 1970; and the Meghna River peaked at 14,800 m^3 s^{-1} in 1966 (Unesco Press, 1976).

As a densely populated country in a developing region of the world, Bangladesh exemplifies yet another aspect of flood hazards relative to the Colorado and Tone River basins.

Summary

Every point on Earth not covered by seawater falls within a drainage basin. Therefore, human activities on any part of the Earth may ultimately affect the slope and channel processes that control flow regime along the many channels that drain the land surface. The riparian corridors along these channels form a diverse environment that sustains a disproportionately large number of species (Risser, 1990; Mertes et al., 1995) as well as a wide array of human land-use activities. As human activities have increasingly affected slope and channel processes, we have put both ourselves and other species at risk from hazards associated with increases or decreases in flood magnitude and frequency. Existing hazards will continue to increase as human populations increase, and they may be augmented by such regional- or global-scale changes in hydroclimatology as those associated with greenhouse warming of the planet's atmosphere (Wohl, Chapter 4, this volume). It is easy to argue that an understanding of flood hazards is vital to preserving both biotic and human communities through effective, rationally planned hazard mitigation programs. The difficulty lies in grappling with the diverse, extensive hazards associated with floods. The remainder of the chapters in this volume summarize various aspects of flood hazards, beginning with the physical controls on flooding, continuing through flood processes and effects, and to responses to flooding. We must have at least a preliminary understanding of the causes and effects of floods before we can hope to perform the balancing act between alteration and protection of drainage basins.

Acknowledgments

Sara Rathburn provided a thorough and helpful review of this chapter.

References

Agata, Y. (1994). Change in runoff characteristics of a mountain river caused by a gigantic failure and debris flow. *Proceedings of the International Symposium on Forest Hydrology*, Oct. 1994, pp. 359–366. Tokyo: International Forest Society.

Andrews, E.D. (1986). Downstream effects of Flaming Gorge Reservoir on the Green River, Colorado and Utah. *Geological Society of America Bulletin*, **97**, 1012–1023.

Baker, V.R. (1994). Geomorphological understanding of floods. *Geomorphology*, **10**, 139–156.

Baker, V.R. and Kochel, R.C. (1988). Flood sedimentation in bedrock fluvial systems. In *Flood geomorphology*, eds. V.R. Baker, R.C. Kochel, and P.C. Patton, pp. 123–137. New York: Wiley and Sons.

Bandyopadhyay, J. and Gyawali, D. (1994). Himalayan water resources: ecological and political aspects of management. *Mountain Research and Development*, **14**, 1–24.

Barrows, H.K. (1948). *Floods: their hydrology and control.* New York: McGraw-Hill Book Company, 432 pp.

Barton, D.R., Taylor, W.D., and Biette, R.M. (1978). Dimensions of riparian buffer strips required to maintain trout habitat in southern Ontario streams. *North American Journal of Fisheries Management*, **5**, 364–378.

Bayley, P.B. (1991). The flood-pulse advantage and the restoration of river-floodplain systems. *Regulated Rivers*, **6**, 75–86.

Bell, B. (1970). The oldest records of the Nile floods. *Geographical Journal*, **136**, 569–573.

Beyer, J.L. (1974). Global summary of human response to natural hazards: floods. In *Natural hazards: local, national, global*, ed. G.F. White, pp. 265–274. New York: Oxford University Press.

Bisson, P.A., Bilby, R.E., Bryant, M.D., Andrew, C.D., Grette, G.B., House, R.A., Murphy, M.L., Koski, K.V., and Sedell, J.R. (1987). Large woody debris in forested streams in the Pacific Northwest: past, present, and future. *Proceedings of the Symposium on Streamside Management: Forestry and Fishery Interactions*, eds. E.O. Salo and T.W. Cundy, pp. 143–231. Seattle: University of Washington Press.

Boardman, J., Ligneau, L., de Roo, A., and Vandaele, K. (1994). Flooding of property by runoff from agricultural land in northwestern Europe. *Geomorphology*, **10**, 183–196.

Bosch, J.M. and Hewlett, J.D. (1982). A review of catchment experiments to determine the effect of vegetation changes on water yield and evapotranspiration. *Journal of Hydrology*, **55**, 3–23.

Braga, G. and Gervasoni, S. (1989). Evolution of the Po River: an example of the application of historic maps. In *Historical change of large alluvial rivers: western Europe*, eds. G.E. Petts, H. Möller, and A.L. Roux, pp. 113–126. Chichester: John Wiley and Sons.

Brakenridge, G.R. (1988). River flood regime and floodplain stratigraphy. In *Flood geomorphology*, eds. V.R. Baker, R.C. Kochel, and P.C. Patton, pp. 139–156. New York: John Wiley and Sons.

Bravard, J.P., Amoros, C., and Pautou, G. (1986). Impact of civil engineering works on the successions of communities in a fluvial system: a methodological and predictive approach applied to a section of the Upper Rhone River. *Oikos*, **27**, 92–111.

Bravard, J.P., Roux, A.L., Amoros, C., and Reygrobellet, J.L. (1994). The Rhone River: a large alluvial temperate river. In *The Rivers Handbook: Hydrological and Ecological Principles*, eds. P. Calow and G.E. Petts, **1**, 426–447. Oxford: Blackwell Scientific Publications.

Brinson, M.M., Swift, B.L., Plantico, R.C., and Barclay, J.S. (1981). *Riparian ecosystems: their ecology and status.* U.S Fish and Wildlife Service, FWS/OBS-81/17, 154 pp. Washington, DC.

Brooker, M.P. (1981). The impact of impoundment on the downstream fisheries and general ecology of rivers. In *Advances in applied biology*, ed. T.H. Coaker, **6**, 91–152. London: Academic Press.

Brookes, A. (1994). River channel change. In *The Rivers Handbook: Hydrological and Ecological Principles*, eds. P. Calow and G.E. Petts, **2**, 55–75. Oxford: Blackwell Scientific Publications.

Brookes, A., Gregory, K.J., and Dawson, F.H. (1983). An assessment of river channelization in England and Wales. *The Science of the Total Environment*, 27, 97–112.

Brown, J.P. (1972). *The economic effects of floods: investigations of a stochastic model of rational investment behavior in the face of floods.* Lecture Notes in Economics and Mathematical Systems, vol. 70. Berlin: Springer-Verlag, 87 pp.

Bull, W.B. (1988). Floods; degradation and aggradation. In *Flood geomorphology*, eds. V.R. Baker, R.C. Kochel, and P.C. Patton, pp. 157–168. New York: John Wiley and Sons.

Bullock, A. and Gustard, A. (1992). Application of the instream flow incremental methodology to assess ecological flow requirements in a British lowland river. In *Lowland floodplain rivers: geomorphological perspectives*, eds. P.A. Carling and G.E. Petts, pp. 251–277. Chichester: John Wiley and Sons.

Burrin, P.J. and Jones, D.K.C. (1991). Environmental processes and fluvial responses in a small temperate zone catchment: A case study of the Sussex Ouse Valley, southeastern England. In *Temperate palaeohydrology*, eds., L. Starkel, K.J. Gregory, and J.B. Thornes, pp. 217–252. Chichester: John Wiley and Sons.

Burt, T.P. (1994). The hydrology of headwater catchments. In *The Rivers Handbook: Hydrological and Ecological Principles*, eds. P. Calow and G.E. Petts, **1**, 3–28. Oxford: Blackwell Scientific Publications.

Busch, D., Schirmer, M., Schuhardt, B., and Ullrich, P. (1989). Historical changes of the River Weser. In *Historical change of large alluvial rivers: western Europe*, eds. G.E. Petts, H. Möller, and A.L. Roux, pp. 297–321. Chichester: John Wiley and Sons.

Butlin, R. (1990). Drainage and land use in the Fenlands and Fen-edge of northeast Cambridgeshire in the seventeenth and eighteenth centuries. In *Water, engineering and landscape*, eds. D. Cosgrove and G. Petts, pp. 54–76. London: Belhaven Press.

Butzer, K.W. (1974). Accelerated soil erosion: a problem of man-land relationships. In *Perspectives on environment*, eds. I.R. Manners and M.W. Mikesell, pp. 57–77. Washington, DC: Association of American Geographers.

Butzer, K.W. (1980). Holocene alluvial sequences: Problems of dating and correlation. In *Timescales in geomorphology*, eds. R.A. Cullingford, D.A. Davidson, and J. Lewin, pp. 131–142. Chichester: John Wiley and Sons.

Chu, K. (1926). Climate pulsations during historical times in China. *Geographical Review*, **16**, 274–282.

Clark, C. (1982). *Flood.* Alexandria, Virginia: Time-Life Books, 176 pp.

Cosgrove, D. (1990). Platonism and practicality: hydrology, engineering and landscape in sixteenth-century Venice. In *Water, engineering and landscape*, eds. D. Cosgrove and G. Petts, pp. 35–53. London: Belhaven Press.

Costa, J.E. (1988a). Floods from dam failures. In *Flood geomorphology*, eds. V.R. Baker, R.C. Kochel, and P.C. Patton, pp. 439–463. New York: John Wiley and Sons.

Costa, J.E. (1988b). Rheologic, geomorphic, and sedimentologic differentiation of water floods, hyperconcentrated flows, and debris flows. In *Flood geomorphology*, eds. V.R. Baker, R.C. Kochel, and P.C. Patton, pp. 113–122. New York: John Wiley and Sons.

Costa, J.E. and Baker, V.R. (1981). *Surficial geology: Building with the Earth.* New York: John Wiley and Sons, 498 pp.

Crampton, C.B. (1969). The chronology of certain terraced river deposits in the southeast Wales area. *Zeitschrift fur Geomorphologie*, **13**, 245–259.

Davidson, D.A. (1980). Erosion in Greece during the first and second millennia B.C. In *Timescales in geomorphology*, eds. R. A. Cullingford, D.A. Davidson, and J. Lewin, pp. 143–158. Chichester: John Wiley and Sons.

Davies, B.R., Thoms, M.C., Walker, K.F., O'Keeffe, J.H., and Gore, J.A. (1994). Dryland rivers: their ecology, conservation and management. In *The Rivers Handbook: Hydrological and Ecological Principles*, eds. P. Calow and G.E. Petts, **2**, 484–511. Oxford: Blackwell Scientific Publications.

Diaz, H.F. and Anderson, C.A. (1995). Precipitation trends and water consumption related to population in the southwestern United States: A reassessment. *Water Resources Research*, **31**, 713–720.

Dilley, M. and Heyman, B.N. (1995). ENSO and disaster: droughts, floods and El Niño/Southern Oscillation warm events. *Disasters*, **19**, 181–193.

Dister, E., Gomer, D., Obrdlik, P., Petermann, P., and Schneider, E. (1990). Water management and ecological perspectives of the Upper Rhine's floodplains. *Regulated Rivers: Research and Management*, **5**, 1–15.

Dolan, R., Howard, A., and Gallenson, A. (1974). Man's impact on the Colorado River in the Grand Canyon. *American Scientist*, **62**, 392–401.

Dunne, T. (1988). Geomorphologic contributions to flood control planning. In *Flood geomorphology*, eds. V.R. Baker, R.C. Kochel, and P.C. Patton, pp. 421–438. New York: John Wiley and Sons.

Dunne, T. and Leopold, L.B. (1978). *Water in environmental planning.* W.H. Freeman and Company. San Francisco, CA.

Dunne, T., Mertes, L.A.K., Meade, R.H., Richey, J.E., and Forsberg, B.R. (1998). Exchanges of sediment between the flood plain and channel of the Amazon River in Brazil. *Geological Society of America Bulletin*, **110**, 450–467.

Dury, G.H. (1964). Principles of underfit streams. *U.S. Geological Survey Professional Paper 452-A.* Washington, DC: U.S. Government Printing Office.

Dury, G.H. (1965). Theoretical implications of underfit streams. *U.S. Geological Survey Professional Paper 452-C.* Washington, DC: U.S. Government Printing Office.

Edelen, G.W., Jr. (1976). *National program for managing flood losses.* U.S. Geological Survey Open-File Report, 30 pp. Washington, DC: U.S. Government Printing Office.

Ely, L.L., Enzel, Y., Baker, V.R., and Cayan, D.R. (1993). A 5000-year record of extreme floods and climate change in the southwestern United States. *Science*, **262**, 410–412.

Ericksen, N.J. (1974). Flood information, expectation, and protection on the Opotiki floodplain, New Zealand. In *Natural hazards: local, national, global*, ed. G. F. White, pp. 60–70. New York: Oxford University Press.

Eschner, T.R., Hadley, R.F., and Crowley, K.D. (1983). Hydrologic and morphologic changes in channels of the Platte River Basin in Colorado, Wyoming, and Nebraska: A historical perspective. *U.S. Geological Survey Professional Paper 1277A.* Washington, DC: U.S. Government Printing Office.

FEMA (Federal Emergency Management Agency). (1983). *Questions and answers on the National Flood Insurance Program.* Washington, DC, 26 pp.

FEMA. (1986). *Floodproofing non-residential structures.* Washington, DC, 199 pp.

Fetherston, K.L., Naiman, R.J., and Bilby, R.E. (1995). Large woody debris, physical process, and riparian forest development in montane river networks of the Pacific Northwest. *Geomorphology*, **13**, 133–144.

Fisher, S.G. and Likens, G.E. (1973). Energy flow in Bear Brook, New Hampshire: an integrative approach to stream ecosystem metabolism. *Ecological Monographs*, **43**, 421–439.

Fradkin, P.L. (1981). *A river no more: the Colorado River and the West.* New York: Alfred A. Knopf, 360 pp.

Friedman, J.M., Osterkamp, W.R., and Lewis, W.M., Jr. (1996). The role of vegetation and bed-level fluctuations in the process of channel narrowing. *Geomorphology*, **14**, 341–351.

Froehlich, W., Kaszowski, L., and Starkel, L. (1977). Studies of present-day and past river activity in the Polish Carpathians. In *River channel changes*, ed. K.J. Gregory, pp. 411–428. Chichester: John Wiley and Sons.

Galloway, G.E., Jr. (1995). New directions in floodplain management. *Water Resources Bulletin*, **31**, 351–357.

Gares, P.A., Sherman, D.J., and Nordstrom, K.F. (1994). Geomorphology and natural hazards. *Geomorphology*, **10**, 1–18.

Gentry, A.H. and Lopez-Parodi, J. (1980). Deforestation and increased flooding of the Upper Amazon. *Science*, **210**, 1354–1356.

Gilbert, G.K. (1917). Hydraulic mining debris in the Sierra Nevada. *U.S. Geological Survey Professional Paper 105.* Washington, DC: U.S. Government Printing Office.

Goldsmith, E. and Hildyard, N. (1984). *The social and environmental effects of large dams.* San Francisco, California: Sierra Club Books, 404 pp.

Gore, J.A. (1994). Hydrological change. In *The rivers handbook: hydrological and ecological principles*, eds. P. Calow and G.E. Petts, **2**, 33–54. Oxford: Blackwell Scientific Publications.

Graf, W.L. (1978). Fluvial adjustments to the spread of tamarisk in the Colorado Plateau region. *Geological Society of America Bulletin*, **89**, 1491–1501.

Graf, W.L. (1979). Mining and channel response. *Annals of the Association of American Geographers*, **69**, 262–275.

Graf, W.L. (1983). The arroyo problem – palaeohydrology and palaeohydraulics in the short term. In *Background to palaeohydrology*, ed. K.J. Gregory, pp. 279–302. New York: John Wiley and Sons.

Graham, R. (1984). Ghana's Volta Resettlement Scheme. In *The social and environmental effects of large dams*, eds. E. Goldsmith and N. Hildyard, **2**, 131–139. Camelford: Wadebridge Ecological Center.

Gregory, K.J. (1976). Changing drainage basins. *Geographical Journal*, **142**, 237–247.

Gregory, K.J. (1983). Human activity and palaeohydrology: A review. In *Palaeohydrology of the temperate zone*, ed. S. Kozarski, pp. 73–80. Quaternary Studies in Poland, 4. Poznan, Poland: Polish Academy of Sciences.

Gregory, K.J. (1987). River channels. In *Human activity and environmental processes*, eds. K.J. Gregory and D.E. Walling, pp. 207–235. Chichester: John Wiley and Sons.

Gregory, K.J. and Gurnell, A.M. (1988). Vegetation and river channel form and process. In *Biogeomorphology*, ed. H.A. Viles, pp. 11–42. Oxford: Basil Blackwell.

Gregory, K.J. and Walling, D.E. (1973). *Drainage basin form and process: a geomorphological approach.* London: Edward Arnold, 458 pp.

Gregory, K.J., Swanson, F.J., McKee, W.A. and Cummins, K.W. (1991). An ecosystem perspective of riparian zones. *BioScience*, **41**, 540–551.

Gupta, A. (1984). Urban hydrology and sedimentation in the humid tropics. In *Developments and applications of geomorphology*, eds. J.E. Costa and P.J. Fleisher, pp. 240–267. Berlin: Springer-Verlag.

Gupta, A. (1988). Large floods as geomorphic events in the humid tropics. In *Flood geomorphology*, eds. V.R. Baker, R.C. Kochel, and P.C. Patton, pp. 301–315. New York: John Wiley and Sons.

Gurnell, A.M. and Gregory, K.J. (1995). Interactions between semi-natural vegetation and hydrogeomorphological processes. *Geomorphology*, **13**, 49–69.

Gustard, A. (1994). Analysis of river regimes. In *The rivers handbook: hydrological and ecological principles*, eds. P. Calow and G.E. Petts, **1**, 29–47. Oxford: Blackwell Scientific Publications.

Hack, J.T. and Goodlett, J.C. (1960). Geomorphology and forest ecology of a mountain region in the central Appalachians. *U.S. Geological Survey Professional Paper 347*, 66 pp. Washington, DC: U.S. Government Printing Office.

Hadimoeljono, M.B. (1992). *Flood control management in river basins: strategies for integration.* Ft. Collins, Colorado: Unpublished PhD dissertation, Colorado State University, 244 pp. U.S. Governemnt Printing Office, Washington, DC.

Hall, S.A. (1990). Channel trenching and climatic change in the southern U.S. Great Plains. *Geology*, **18**, 342–345.

Harper, P.P. (1994). La Grande Riviere: A subarctic river and a hydroelectric megaproject. In *The rivers handbook: hydrological and ecological principles*, eds. P. Calow and G.E. Petts, **1**, 411–425. Oxford: Blackwell Scientific Publications.

Harvey, A.M. (1978). Dissected alluvial fans in southeast Spain. *Catena*, **5**, 177–211.

Harvey, M.D., Mussetter, R.A., and Wick, E.J. (1993). A physical process-biological response model for spawning habitat formation for the endangered Colorado River squawfish. *Rivers*, **4**, 114–131.

Hayden, B.P. (1988). Flood climates. In *Flood geomorphology*, eds. V.R. Baker, R.C. Kochel, and P.C. Patton, pp. 13–26. New York: John Wiley and Sons.

Heede, B.M. (1981). Dynamics of selected mountain streams in the western United States of America. *Zeitschrift fur Geomorphologie*, **25**, 17–32.

Hey, R.D. (1994). Environmentally sensitive river engineering. In *The rivers handbook: hydrological and ecological principles*, eds. P. Calow and G.E. Petts, **2**, 337–362. Oxford: Blackwell Scientific Publications.

Hey, R.D., Heritage, G.L. and Patteson, M. (1990). *Design of flood alleviation schemes: engineering and the environment.* London: Ministry of Agriculture, Fisheries and Food.

Hilmes, M.M. and Wohl, E.E. (1995). Changes in channel morphology associated with placer mining. *Physical Geography*, **16**, 223–242.

Hirschboeck, K.K. (1988). Flood hydroclimatology. In *Flood geomorphology*, eds. V.R. Baker, R.C. Kochel, and P.C. Patton, pp. 27–49. New York: John Wiley and Sons.

Hooke, J.M. and Kain, R.J.P. (1982). *Historical change in the physical environment: a guide to sources and techniques.* London: Butterworth Scientific, 236 pp.

Hoyt, W.G. and Langbein, W.B. (1955). *Floods.* Princeton, New Jersey: Princeton University Press, 469 pp.

Hukunari, K. and Hiroki, K. (1993). Water resources development planning – the Tone River. *Journal of Hydroscience and Hydraulic Engineering* (River Engineering, Special Issue), **VI**, 103–119.

Hupp, C.R. (1988). Plant ecological aspects of flood geomorphology and paleoflood history. In *Flood geomorphology*, eds. V.R. Baker, R.C. Kochel, and P.C. Patton, pp. 335–356. New York: John Wiley and Sons.

Hupp, C.R. and Osterkamp, W.R. (1996). Riparian vegetation and fluvial geomorphic processes. *Geomorphology*, **14**, 277–295.

James, L.A. (1989). Sustained storage and transport of hydraulic gold mining sediment in the Bear River, California. *Annals of the Association of American Geographers*, **79**, 570–592.

Japanese Ministry of Construction. (1993). *Sabo.* Kobe, Japan: Ministry of Construction, 36 pp.

Japanese River Bureau. (1985). *Rivers in Japan.* Tokyo,

Japan: River Bureau, Ministry of Construction, Japan, 92 pp.

Johnson, B.L.C. (1975). *Bangladesh*. New York: Harper and Row, 104 pp.

Johnson, M.G. and Beschta, R.L. (1980). Logging, infiltration and surface erodibility in western Oregon. *Journal of Forestry*, **78**, 334–337.

Johnson, W.C. (1992). Dams and riparian forests: case study from the upper Missouri River. *Rivers*, **3**, 229–242.

Johnson, W.C. (1994). Woodland expansion in the Platte River, Nebraska: patterns and causes. *Ecological Monographs,* **64**, 45–84.

Johnson, W.C., Dixon, M.D., Simons, R., Jenson, S., and Larson, K. (1995). Mapping the response of riparian vegetation to possible flow reductions in the Snake River, Idaho. *Geomorphology*, **13**, 159–173.

Junk, W.J., Bayley, P.B., and Sparks, R.E. (1989). The flood-pulse concept in river-floodplain systems. In *Proceedings of the International Large River Symposium*, ed. D.P. Dodge, pp. 110–127. Canadian Special Publication of Fisheries and Aquatic Sciences 106. Ottawa, Canada: National Research Council of Canada.

Kale, V.S., Ely, L.L., Enzel, Y., and Baker, V.R. (1994). Geomorphic and hydrologic aspects of monsoon floods on the Narmada and Tapi Rivers in central India. *Geomorphology*, **10**, 157–168.

Karim, K., Gubbels, M.E., and Goulter, I.C. (1995). Review of determination of instream flow requirements with special application to Australia. *Water Resources Bulletin*, **31**, 1063–1077.

Katsura, J., Hayashi, T., Nishimura, H., Isobe, M., Yamashita, T., Kawata, Y., Yasuda, T., and Nakagawa, H. (1992). Storm surge and severe wind disasters caused by the 1991 cyclone in Bangladesh. *Japanese Ministry of Science, Education, and Culture, Research Report No. B-2*, 101 pp. Tokyo: Japanese Ministry of Science etc.

Keller, E.A. (1976). Channelization: environmental, geomorphic and engineering aspects. In *Geomorphology and engineering*, ed. D.R. Coates, pp. 115–140. State University of New York at Binghamton.

Keller, E.D. and Swanson, F.G. (1979). Effects of large organic debris on channel form and fluvial processes. *Earth Surface Processes*, **4**, 361–380.

Keller, E.D. and Tally, T. (1979). Effects of large organic debris on channel form and fluvial processes in the coastal redwood environment. In *Adjustments of the fluvial system*, eds. D.D. Rhodes and G.P. Williams, pp. 169–197. Dubuque, Iowa: Kendall/Hunt.

Kern, K. (1992). Restoration of lowland rivers: the German experience. In *Lowland floodplain rivers: geomorphological perspectives*, eds. P.A. Carling and G.E. Petts, pp. 279–297. Chichester: John Wiley and Sons.

Killingtveit, A. (1996). Flood regimes and flood prevention in Norway: Lessons learnt from the 1995 flood. In *Recent trends of floods and their preventive measures*, pp. 57–61. Preconference Proceedings, Foundation of Hokkaido River Disaster Prevention Research Center, Sapporo, Japan.

Klimek, K. (1987). Man's impact on fluvial processes in the Polish Western Carpathians. *Geografiska Annaler*, **69A**, 221–224.

Klimek, K. and Starkel, L. (1974). History and actual tendency of floodplain development at the border of the Polish Carpathians. *Abhandlungen die Akademie der Wissenchaften in Gottingen, Mathematic-Physic Klasse*, **III**, 29, 185–196.

Knight, D.W. (1996). Conveyance and sediment transport capacity of flood channels. In *Recent trends of floods and their preventive measures*, pp. 71–76. Preconference Proceedings, Foundation of Hokkaido River Disaster Prevention Research Center, Sapporo, Japan.

Knighton, A.D. (1989). River adjustment to changes in sediment load: the effects of tin mining on the Ringarooma River, Tasmania, 1875–1984. *Earth Surface Processes and Landforms*, **14**, 333–359.

Knox, J.C. (1988). Climatic influence on upper Mississippi Valley floods. In *Flood geomorphology*, eds. V.R. Baker, R.C. Kochel, and P.C. Patton, pp. 279–300. New York: John Wiley and Sons.

Kochel, R.C. (1988a). Extending stream records with slackwater paleoflood hydrology: examples from west Texas. In *Flood geomorphology*, eds. V.R. Baker, R.C. Kochel, and P.C. Patton, pp. 377–391. New York: John Wiley and Sons.

Kochel, R.C. (1988b). Geomorphic impact of large floods: review and new perspectives on magnitude and frequency. In *Flood geomorphology*, eds. V.R. Baker, R.C. Kochel, and P.C. Patton, pp. 169–187. New York: John Wiley and Sons.

Kozlowski, T.T. (1984a). Extent, causes, and impacts of flooding. In *Flooding and plant growth*, ed. T.T. Kozlowski, pp. 1–7. Orlando, Florida: Academic Press.

Kozlowski, T.T., ed. (1984b). *Flooding and plant growth.* Orlando, Florida: Academic Press, 356 pp.

Kresan, P.L. (1988). The Tucson, Arizona, flood of October 1983. In *Flood geomorphology*, eds. V.R. Baker, R.C. Kochel, and P.C. Patton, pp. 465–489. New York: John Wiley and Sons.

Kunreuther, H. (1974). Economic analysis of natural hazards: an ordered choice approach. In Natural hazards: local, national, global, ed. G.F. White, pp. 206–214. New York: Oxford University Press.

Large, A.R.G. and Petts, G.E. (1992). *Buffer zones for con-*

servation of rivers and bankside habitat. R&I Project Record 340/5/Y, Loughborough, England: National Rivers Authority.

Large, A.R.G. and Petts, G.E. (1994). Rehabilitation of river margins. In *The rivers handbook: hydrological and ecological principles*, eds. P. Calow and G.E. Petts, eds., **2**, 401–418. Oxford: Blackwell Scientific Publications.

Leigh, D.S. (1994). Gold mining's impact on north Georgia floodplains: Mercury contamination and floodplain sedimentation. In *Effects of human-induced changes on hydrologic systems.* Proceedings, American Water Resources Association Symposium. Bethesda, Maryland: American Water Resources Association.

Leopold, L.B. (1951). Rainfall frequency: an aspect of climatic variation. *Transactions, American Geophysical Union*, **32**, 347–357.

Leopold, L.B. (1968). Hydrology for urban land planning – a guidebook on the hydrological effects of urban land use. *U.S. Geological Survey Circular 554.* Washington, DC: U.S. Government Printing Office.

Leuchtenburg, W.E. (1953). *Flood control politics: The Connecticut River valley problem, 1927–1950.* Cambridge: Harvard University Press, 339 pp.

Lewin, J. (1994). Floodplain construction and erosion. In *The rivers handbook: hydrological and ecological principles*, eds. P. Calow and G.E. Petts, **1**, 144–161. Oxford: Blackwell Scientific Publications.

Lewin, J., Davies, B.E., and Wolfenden, P.J. (1977). Interactions between channel change and historic mining sediments. In *River channel changes*, ed. K.J. Gregory, pp. 353–367. Chichester: John Wiley and Sons.

Mahmood, A.Z.S. (1969). *Introducing East Pakistan.* Dacca, East Pakistan: Barnali Printers and Publishers, 118 pp.

Malanson, G.P. (1993). *Riparian landscapes.* Cambridge, England: Cambridge University Press, 296 pp.

Marchi, E., Roth, G., and Siccardi, F. (1996). Flood control and forecasting in the Po River basin. In *Recent trends of floods and their preventive measures*, pp. 68–70. Preconference Proceedings. Sapporo, Japan: Foundation of Hokkaido River Disaster Prevention Research Center.

Marston, R.A., Girel, J., Pautou, G., Piegay, H., Bravard, J.P., and Arneson, C. (1995). Channel metamorphosis, floodplain disturbance, and vegetation development: Ain River, France. *Geomorphology*, **13**, 121–131.

Maser, C. and Sedell, J.R. (1994). *From the forest to the sea: the ecology of wood in streams, rivers, estuaries, and oceans.* Delray Beach, Florida: St. Lucie Press, 200 pp.

McKenney, R., Jacobson, R.B., and Wertheimer, R.C. (1995). Woody vegetation and channel morphogenesis in low-gradient, gravel-bed streams in the Ozark Plateaus, Missouri and Arkansas. *Geomorphology*, **13**, 175–198.

Mertes, L.A.K., Daniel, D.L., Melack, J.M., Nelson, J.B., Martinelli, L.A., and Forsberg, B.R. (1995). Spatial patterns of hydrology, geomorphology, and vegetation on the floodplain of the Amazon River in Brazil from a remote sensing perspective. *Geomorphology*, **13**, 215–232.

Milner, N.J., Scullion, J., Carling, P.A., and Crisp, D.T. (1981). The effects of discharge on sediment dynamics and consequent effects on invertebrates and salmonids in upland rivers. In *Advances in applied biology*, ed. T.H. Coaker, **6**, 153–220. London: Academic Press.

Minshall, G.W., Cummins, K.W., Petersen, R.C., Cushing, C.E., Bruns, D.A., Sedell, J.R., and Vannote, R.L. (1985). Developments in stream ecosystem theory. *Canadian Journal of Fisheries and Aquatic Sciences*, **42**, 1045–1055.

Mochizuki, T. and Jikan, S. (1993). Environment-friendly water resources development facilities. *Journal of Hydroscience and Hydraulic Engineering* (Special Issue), **VI**, 153–174.

Monosowski, E. (1984). Brazil's Tucurui Dam: development at environmental cost. In *The social and environmental effects of large dams*, eds. E. Goldsmith and N. Hildyard, **2**, 191–200. Camelford: Wadebridge Ecological Center.

Montgomery, D.R. and Buffington, J.M. (1993). *Channel classification, prediction of channel response and assessment of channel conditions.* Report TFW-SH-10-93-002, SHAMW Committee of the Washington State Timber/Fish/Wildlife Agreement, 84 pp.

Mycielska-Dowgiallo, E. (1977). Channel pattern changes during the Last Glaciation and Holocene, in the northern part of the Sandomierz Basin and the middle part of the Vistula Valley, Poland. In *River channel changes*, ed. K.J. Gregory, pp. 75–87. Chichester: John Wiley and Sons.

Nadler, C.T. and Schumm, S.A. (1981). Metamorphosis of South Platte and Arkansas Rivers, eastern Colorado. *Physical Geography*, **2**, 95–115.

Naiman, R.J. and Decamps, H., eds. (1990). *The ecology and management of aquatic-terrestrial ecotones.* Carnforth: Parthenon Publishing.

Nakamura, F. (1989). Buffer effect of a wide riverbed section on sediment discharge. *Journal of the Faculty of Agriculture, Hokkaido University*, **64**, 56–69.

Nakamura, F. (1995). Structure and function of riparian zone and implications for Japanese river management. *Transaction, Japanese Geomorphological Union*, **16**, 237–256.

Nakano, T., Kadomura, H., Mizutani, T., Okuda, M., and Sekiguchi, T. (1974). Natural hazards: report from Japan. In *Natural hazards: local, national, global*, ed. G.F. White, pp. 231–243. New York: Oxford University Press.

Nanson, G.C., Barbetti, M., and Taylor, G. (1995). River stabilisation due to changing climate and vegetation during the late Quaternary in western Tasmania, Australia. *Geomorphology*, **13**, 145–158.

National Science Foundation. (1980). *A report on flood hazard mitigation*. Washington, DC, 253 pp.

Nautiyal, K.P. (1989). *Protohistoric India*. Delhi, India: Agam Kala Prakashan, 209 pp.

NOAA (National Oceanic and Atmospheric Administration). (1983). *Flash flood awareness packet*. Washington, DC, 43 pp.

NOAA. (1992). *Flash floods and floods*. Washington, DC, 12 pp.

NOAA. (1994). *The great flood of 1993*. Washington, DC, 270 pp.

Ohmori, H. and Shimazu, H. (1994). Distribution of hazard types in a drainage basin and its relation to geomorphological setting. *Geomorphology*, **10**, 95–106.

Oliver, T.W. (1978). *The United Nations in Bangladesh*. Princeton, New Jersey: Princeton University Press, 231 pp.

Osterkamp, W.R., Hupp, C.R., and Schening, M.R. (1995). Little River revisited – thirty-five years after Hack and Goodlett. *Geomorphology*, **13**, 1–20.

Park, C.C. (1977). Man-induced changes in stream channel capacity. In *River channel changes*, ed. K.J. Gregory, pp. 121–144. Chichester: John Wiley and Sons.

Parker, G., Cui, Y., Imran, J., and Dietrich, W.E. (1996). Flooding in the Lower Ok Tedi, Papua New Guinea due to the dispersal of mine tailings and its amelioration. In *Recent trends of floods and their preventive measure*, pp. 62–64. Preconference Proceedings. Sapporo, Japan: Foundation of Hokkaido River Disaster Prevention Research Center.

Patton, P.C. (1988a). Drainage basin morphometry and floods. In *Flood geomorphology*, eds. V.R. Baker, R.C. Kochel, and P.C. Patton, pp. 51–64. New York: John Wiley and Sons.

Patton, P.C. (1988b). Geomorphic response of streams to floods in the glaciated terrain of southern New England. In *Flood geomorphology*, eds. V.R. Baker, R.C. Kochel, and P.C. Patton, pp. 261–277. New York: John Wiley and Sons.

Patton, P.C. and Schumm, S.A. (1975). Gully erosion, northwestern Colorado: A threshold phenomenon. *Geology*, **3**, 88–90.

Patton, P.C. and Schumm, S.A. (1981). Ephemeral-stream processes, implications for studies of Quaternary valley fills. *Quaternary Research*, **15**, 24–43.

Pemberton, E.L. (1976). Channel changes in the Colorado River below Glen Canyon Dam. *Proceedings, Federal Third Interagency Sedimentation Conference*, pp. 5-61–5-73.

Petts, G.E. (1977). Channel response to flow regulation: The case of the River Derwent, Derbyshire. In *River channel changes*, ed. K.J. Gregory, pp. 145–164. Chichester: John Wiley and Sons.

Petts, G.E. (1984). *Impounded rivers: perspectives for ecological management*. Chichester: John Wiley and Sons, 326 pp.

Petts, G.E. (1989). Historical analysis of fluvial hydrosystems. In *Historical change of large alluvial rivers: western Europe*, eds. G.E. Petts, H. Moller, and A.L. Roux, pp. 1–18. Chichester: John Wiley and Sons.

Petts, G. (1990a). Forested river corridors: a lost resource. In *Water, engineering and landscape*, eds. D. Cosgrove and G. Petts, pp. 12–34. London: Belhaven Press.

Petts, G. (1990b). Water, engineering and landscape: development, protection and restoration. In *Water, engineering and landscape*, eds. D. Cosgrove and G. Petts, pp. 188–208. London: Belhaven Press.

Petts, G.E. (1994). Rivers: dynamic components of catchment ecosystems. In *The rivers handbook: hydrological and ecological principles*, eds. P. Calow and G.E. Petts, **2**, 3–22. Oxford: Blackwell Scientific Publications.

Petts, G.E., Large, A.R.G., Greenwood, M.T., and Bickerton, M.A. (1992). Floodplain assessment for restoration and conservation: linking hydrogeomorphology and ecology. In *Lowland floodplain rivers: geomorphological perspectives*, eds. P.A. Carling and G.E. Petts, pp. 217–234. Chichester: John Wiley and Sons.

Petts, G.E., Moller, H., and Roux, A.L., eds. (1989a). *Historical changes of large alluvial rivers: western Europe*. Chichester: John Wiley and Sons.

Petts, G.E., Thoms, M.T., Brittan, K., and Atkin, B. (1989b). A freeze-coring technique applied to pollution by fine sediments in gravel-bed rivers. *The Science of the Total Environment*, **84**, 259–272.

Piégay, H. and Gurnell, A.M. (1997). Large woody debris and river geomorphological pattern: examples from S.E. France and S. England. *Geomorphology*, **19**, 99–116.

Poff, N.L. and Ward, J.V. (1989). Implications of streamflow variability and predictability for lotic community structure: a regional analysis of streamflow patterns. *Canadian Journal of Fishery and Aquatic Sciences*, **46**, 1805–1818.

Poff, N.L. and Ward, J.V. (1990). Physical habitat template of lotic systems: recovery in the context of historical pattern of spatiotemporal heterogeneity. *Environmental Management*, **14**, 629–645.

Power, M.E., Parker, G., Dietrich, W.E., and Sun, A. (1995a). How does floodplain width affect floodplain river ecology? A preliminary exploration using simulations. *Geomorphology*, **13**, 301–317.

Power, M.E., Sun, A., Parker, G., Dietrich, W.E., and Wootton, J.T. (1995b). Hydraulic food-chain models. *BioScience*, **45**, 159–167.

Prestegaard, K.L., Matherne, A.J., Shane, B., Houghton, K., O'Connell, M., and Katyl, N. (1994). Spatial variations in the magnitude of the 1993 floods, Raccoon River basin, Iowa. *Geomorphology*, **10**, 169–182.

Rasmussen, J.L. (1994). Management of the Upper Mississippi: A case history. In *The rivers handbook: hydrological and ecological principles*, eds. P. Calow and G.E. Petts, pp. 441–463. Oxford: Blackwell Scientific Publications.

Reid, L.M. and Dunne, T. (1984). Sediment production from forest road surfaces. *Water Resources Research*, **20**, 1753–1761.

Richards, K.S. and Wood, R. (1977). Urbanization, water redistribution, and their effect on channel processes. In *River channel changes*, ed. K.J. Gregory, pp. 369–388. Chichester: John Wiley and Sons.

Risser, P.G. (1990). The importance of land/land-water ecotones. In *The ecology and management of aquatic-terrestrial ecotones*, eds. R.J. Naiman and H. Decamps, pp. 7–22. Carnforth: Parthenon Publishing Company.

Robinson, M.A. and Lambrick, G.H. (1984). Holocene alluviation and hydrology in the upper Thames basin. *Nature*, **306**, 809–814.

Saarinen, T.F. (1974). Problems in the use of a standardized questionnaire for cross-cultural research on perception of natural hazards. In *Natural hazards: local, national, global*, ed. G.F. White, pp. 180–184. New York: Oxford University Press.

Schumm, S.A. (1968). River adjustment to altered hydrologic regime – Murrumbidgee River and paleochannels, Australia. *U.S. Geological Survey Professional Paper 598*. Washington, DC: U.S. Government Printing Office.

Schumm, S.A. (1977). *The fluvial system*. Chichester: Wiley.

Schumm, S.A. (1994). Erroneous perceptions of fluvial hazards. *Geomorphology*, **10**, 129–138.

Scott, M.L., Friedman, J.M., and Auble, G.T. (1996). Fluvial process and the establishment of bottomland trees. *Geomorphology*, **14**, 327–339.

Sedell, J.R. and Froggatt, J.L. (1984). Importance of *streamside* forests to large rivers: the isolation of the Willamette River, Oregon, USA, from its floodplain by snagging and streamside forest removal. *Verhandlungen International Vereinigung Limnology*, **22**, 1828–1834.

Seki, K. and Takazawa, K. (1993). Projects for creation of rivers rich in Nature – toward a richer natural environment in towns and on watersides. *Journal of Hydroscience and Hydraulic Engineering* (Special Issue), **6**, 87–101.

Shields, F.D., Jr. and Gippel, C.J. (1995). Prediction of effects of woody debris removal on flow resistance. *Journal of Hydraulic Engineering*, **121**, 341–354.

Shields, F.D., Jr., and Smith, R.H. (1992). Effects of large woody debris removal on physical characteristics of a sand-bed river. *Aquatic Conservation: Marine and Freshwater Ecosystems*, **2**, 145–163.

Sidle, R.C., Pearce, A.J., and O'Loughlin, C.L. (1985). *Hillslope Stability and Land Use*. Washington, DC: American Geophysical Union.

Slovic, P., Kunreuther, H., and White, G.F. (1974). Decision processes, rationality and adjustment to natural hazards. In *Natural hazards: local, national, global*, ed. G.F. White, pp. 187–205. New York: Oxford University Press.

Smith, N. (1971). *A history of dams*. London: Peter Davies.

Stalnaker, C.B. (1992). Evolution of instream flow habitat modelling. In *The rivers handbook: hydrological and ecological principles*, eds. P. Calow and G.E. Petts, **2**, 276–286. Oxford: Blackwell Scientific Publications.

Stanford, J.A. and Ward, J.V. (1988). The hyporheic habitat of river ecosystems. *Nature*, **335**, 64–66.

Starkel, L. (1988). Tectonic, anthropogenic and climatic factors in the history of the Vistula River valley downstream of Cracow. In *Lake, mire, and river environments during the last 15,000 years*, eds. G. Lang and C. Schlüchter, pp. 161–170. Rotterdam: A.A. Balkema.

Stedinger, J.R. and Baker, V.R. (1987). Surface water hydrology: historical and paleoflood information. *Reviews of Geophysics*, **25**, 119–124.

Stedinger, J.R. and Cohn, T.A. (1986). Flood frequency analysis with historical and paleoflood information. *Water Resources Research*, **22**, 785–793.

Swanson, F.J. and Sparks, R.E. (1990). Long-term ecological research and the invisible place. *BioScience*, **40**, 502–508.

Swanson, F.J., Gregory, S.V., Sedell, J.R., and Campbell, A.G. (1982). Land-water interactions: the riparian zone. In *Analysis of coniferous forest ecosystems in the western United States*, ed. R.L. Edmonds, pp. 267–291. Stroudsburg, Pennsylvania: Hutchinson Ross Publishing Company.

Swift, B.L. (1984). Status of riparian ecosystems in the United States. *Water Resources Bulletin*, **20**, 223–228.

Takahashi, G. (1989). Status of charr and masu salmon

management in Japan; a call for conservation guidelines. *Physiology and Ecology Japan*, **1**, 683–690.

Takahashi, G. (1990). On the consistency of conservation of freshwater fish and 'sabo' works in streams. *Bulletin Institute of Zoology, Academic Sinica*, **29**, 105–113.

Takahashi, G. and Higashi, S. (1984). Effect of channel alteration on fish habitat. *Japanese Journal of Limnology*, **45**, 178–186.

Thompson, D.M. (1995). The effects of large organic debris on sediment processes and stream morphology in Vermont. *Geomorphology*, **11**, 235–244.

Trimble, S.W. (1988). The impact of organisms on overall erosion rates within catchments in temperate regions. In *Biogeomorphology*, ed. H.A. Viles, pp. 83–142. Oxford: Basil Blackwell.

Trimble, S.W. and Mendel, A.C. (1995). The cow as a geomorphic agent – A critical review. *Geomorphology*, **13**, 233–253.

Triska, F.J., Sedell, J.R., and Gregory, S.V. (1982). Coniferous forest streams. In *Analysis of coniferous forest ecosystems in the western United States*, ed. R.L. Edmonds, pp. 292–332. Stroudsburg, Pennsylvania: Hutchinson Ross Publishing Company.

Triska, F.J., Duff, J.H., and Avanzino, R.J. (1990). Influence of exchange flow between the channel and hyporheic zone on nitrate production in a small mountain stream. *Canadian Journal of Fisheries and Aquatic Sciences*, **47**, 2099–2111.

Uda, T. and Yamamoto, K. (1994). Beach changes caused by obstruction of longshore sand transport: an example of the Hidaka Coast in Hokkaido. *Coastal Engineering in Japan*, **37**, 87–106.

Uda, T., Takahashi, A., and Fujii, M. (1995). Bar topography changes associated with a dredged hole off the Niyodo River mouth. *Coastal Engineering in Japan*, **38**, 63–88.

Unesco Press. (1976). *World catalogue of very large floods.* Paris, France, 424 pp.

Uzuka, K. and Tomita, K. (1993). Flood control planning – case study of the Tone River. *Journal of Hydroscience and Hydraulic Engineering* (Special Issue), **6**, 5–20.

Vannote, R.L., Minshall, G.W., Cummins, K.W., Sedell, J.R., and Cushing, C.E. (1980). The river continuum concept. *Canadian Journal of Fisheries and Aquatic Sciences*, **37**, 130–137.

Vanoni, V.A. (ed.) (1975). *Sedimentation engineering.* ASCE Task Committee, Sedimentation Committee of Hydraulics Division, 531–546. New York, NY.

Van Zuidam, R.A. (1975). Geomorphology and archeology: evidences of interrelation at historical sites in the Zaragoza region, Spain. *Zeitschrift fur Geomorphologie*, **19**, 319–328.

Visvader, H. and Burton, I. (1974). Natural hazards and hazard policy in Canada and the United States. In *Natural hazards: local, national, global*, ed. G.F. White, pp. 219–231. New York: Oxford University Press.

Vita-Finzi, C. (1969). *The Mediterranean valleys.* Cambridge: Cambridge University Press.

Waggoner, P.E. and Schefter, J. (1990). Future water use in the present climate. In *Climate change and U.S. water resources*, ed. P.E. Waggoner, pp. 19–39. New York: John Wiley and Sons.

Walker, K.F. (1994). The River Murray, Australia: a semi-arid lowland river. In *The rivers handbook: hydrological and ecological principles*, eds. P. Calow and G.E. Petts, **1**, 472–492. Oxford: Blackwell Scientific Publications.

Walling, D.E. (1987). Hydrological processes. In *Human activity and environmental processes*, eds. K.J. Gregory and D.E. Walling, pp. 53–85. Chichester: John Wiley and Sons.

Walling, D.E., Quine, T.A., and He, Q. (1992). Investigating contemporary rates of floodplain sedimentation. In *Lowland floodplain rivers: geomorphological perspectives*, eds. P.A. Carling and G.E. Petts, pp. 165–184. Chichester: John Wiley and Sons.

Ward, J.V. (1994). A mountain river. In *The rivers handbook: hydrological and ecological principles*, eds. P. Calow and G.E. Petts, **1**, 493–510. Oxford: Blackwell Scientific Publications.

Webb, B.W. and Walling, D.E. (1994). Water quality II. Chemical characteristics. In *The rivers handbook: hydrological and ecological principles*, eds. P. Calow and G.E. Petts, **1**, 73–100. Oxford: Blackwell Scientific Publications.

Webb, R.H., O'Connor, J.E. and Baker, V.R. (1988). Paleohydrologic reconstruction of flood frequency on the Escalante River, south-central Utah. In *Flood geomorphology*, eds. V.R. Baker, R.C. Kochel, and P.C. Patton, pp. 403–418. New York: John Wiley and Sons.

Welcomme, R.L. (1979). Fisheries ecology of floodplain rivers. New York: Longmans.

White, G.F. (1974). Natural hazards research: concepts, methods, and policy implications. In *Natural hazards: local, national, global*, ed. G.F. White, pp. 3–16. New York: Oxford University Press.

Whitlow, J.R. (1985). Dambos in Zimbabwe: A review. *Zeitschrift fur Geomorphologie*, Supplement Band **52**, 115–146.

Williams, G.P. (1978). The case of the shrinking channels – the North Platte and Platte Rivers in Nebraska. *U.S. Geological Survey Circular 781*. Washington, DC: U.S. Government Printing Office.

Williams, G.P. and Wolman, M.G. (1985). Effects of dams and reservoirs on surface-water hydrology –

changes in rivers downstream from dams. *National Water Summary 1985, U.S. Geological Survey*, pp. 83–88. Washington, DC: U.S. Government Printing Office.

Winkley, B.R. (1982). Response of the Lower Mississippi to river training and realignment. In *Gravel-bed rivers*, eds. R.D. Hey, C.R. Thorne, and J.C. Bathurst, pp. 659–681. Chichester: John Wiley and Sons.

Wohl, E.E., Webb, R.H., Baker, V.R., and Pickup, G. (1994). Sedimentary flood records in the bedrock canyons of rivers in the monsoonal region of Australia. *Colorado State University Water Resources Paper 107*, 102 pp. Ft. Collins, CO.

Wolman, M.G. (1967). A cycle of sedimentation and erosion in urban river channels. *Geografiska Annaler*, **49A**, 385–395.

Wolman, M.G. and Gerson, R. (1978). Relative scales of time and effectiveness of climate in watershed geomorphology. *Earth Surface Processes*, **3**, 189–208.

Wolman, M.G. and Miller, J.P. (1960). Magnitude and frequency of forces in geomorphic processes. *Journal of Geology*, **68**, 54–74.

Wolman, M.G. and Schick, A.P. (1967). Effects of construction on fluvial sediment, urban and suburban areas of Maryland. *Water Resources Research*, **3**, 451–464.

Wood, T.R. (1981). River management. In *British rivers*, ed. J. Lewin, pp. 170–195. London: George Allen and Unwin.

Yao, H., Hashino, M., and Yoshida, H. (1994). Analyzing effects of deforestation and afforestation on stream flow by using a physically-based conceptual model. *Journal Japan Society of Hydrology and Water Resources*, **7**, 196–203.

Yao, S. (1942). The chronological and seasonal distribution of floods and droughts in Chinese history 206 B.C.–1911 A.D. *Harvard Journal for Asiatic Studies*, **6**, 273–312.

Yao, S. (1943). The geographical distribution of floods and droughts in Chinese history 206 B.C.–1911 A.D. *Far Eastern Quarterly*, **2**, 357–378.

Zimmerman, R.C., Goodlett, J.C., and Comer, G.H. (1967). The influence of vegetation on channel form of small streams. *International Association of Scientific Hydrology: Symposium on River Morphology, Publ. 75*, pp. 255–275. Wallingford, UK: IAHS Pubs.

PHYSICAL CONTROLS ON FLOODING

CHAPTER 2

Hydroclimatology of Meteorologic Floods

K.K. Hirschboeck
Laboratory of Tree-Ring Research
University of Arizona

L.L. Ely
Department of Geology
Central Washington University

R.A. Maddox
Cooperative Institute for Mesoscale Meteorological Studies
University of Oklahoma

Using the framework of flood hydroclimatology, we present an overview of precipitation systems that generate floods. We also discuss the role of antecedent climatic and hydrologic conditions and highlight factors that amplify the hazards associated with flood occurrence. Finally, we address flood variations over long-term time scales and examine the implications of hydroclimatic flood variations for the analysis and management of flood hazards.

Scale is a key component of flood causality because the way precipitation is delivered in space and time affects the type of flood and its accompanying hazards. Precipitation systems that occur at one scale are strongly interconnected with systems at other scales, and larger scale processes set the stage for activity at smaller scales. Across all scales, the *persistence* of a precipitation system is a key element for generating exceptionally large floods. Most major floods are characterized by a synergistic combination of atmospheric, hydrologic, and drainage-basin factors that intensify the event.

Over long-term time scales, regional flooding variability identified from historical records and paleoflood information may be linked to persistent synoptic to macroscale patterns of atmospheric circulation. Certain regional responses to teleconnections [e.g., El Niño/Southern Oscillation (ENSO)] that are detectable in precipitation records often are not as strong in flood records because of the complexities of flood causality. Emerging temporal and spatial relationships among flood hydroclimatic regions globally can augment information from specific regions to evaluate potential flood hazards under changing climatic conditions.

Introduction

Meteorologic floods arise when weather, climate, and hydrology work together in ways that produce greater-than-average amounts of runoff, exceeding the capacity of stream channels. The timing and spatial distribution of precipitation with respect to a river basin are key determinants of whether a flood occurs. Hence, to understand the origins of meteorologic floods, the temporal and spatial scales of the atmospheric processes that lead to floods must be addressed in conjunction with the temporal and spatial scales of the hydrologic processes that operate in specific drainage basins. This chapter focuses primarily on the atmospheric processes that generate floods, but we also address the issue of synergy between atmospheric and hydrologic processes in flood development.

The meteorologic and climatologic causes of different types of flooding have been discussed by many authors (Ward, 1978; Maddox et al., 1979, 1980; Hayden, 1988; Hirschboeck, 1987a, 1988, 1991, 1996; Doswell et al., 1996). We will not replicate either the comprehensive scope or specific concentration of these earlier works. The goal of this chapter is to present an overview of the meteorologic and climatologic processes that directly and indirectly cause floods, underscoring factors that amplify the hazards associated with flood occurrence. In addition, we highlight flood variations over long-term time scales (decadal-to-millennial) and examine the implications of long-term flood information for perception, analysis, and management of flood hazards.

Flood Hydroclimatology: A Context for Understanding Flood Causation

The atmospheric causes of flooding can best be understood within the framework of *flood hydroclimatology*. Flood hydroclimatology is an approach to analyzing floods from the perspective of the temporal context of their history of development and variation and the spatial context of the local, regional, and global atmospheric processes and circulation patterns from which the floods develop (Hirschboeck, 1988). This approach is based on identifying the meteorologic causes of floods, but it also seeks to address this meteorologic-scale activity within a broader spatial and temporal, climatic perspective. In this way flood variability can also be examined in terms of antecedent conditions, regional relationships, large-scale anomaly patterns, global-scale controls, and long-term trends (Hirschboeck, 1988).

Meteorologic Processes That Directly Cause Flooding

Simply stated, meteorologic floods occur when excessive precipitation over a watershed cannot be fully accommodated by the basin's internal storage reservoirs and drainage network. The way this above-average precipitation is delivered – in space and time – affects both the type of flood that occurs and its accompanying hazards. The intensity of flood-causing precipitation and the temporal and spatial distribution of the precipitation with respect to a drainage basin depend on the kinds of storm systems that deliver the precipitation. Storms and storm systems can be classified in a number of ways: for example, Houze (1981) provides a comprehensive overview of the structures of atmospheric precipitation systems. We focus on storms and weather systems that deliver exceptionally large amounts of precipitation and describe different types of storm systems from the perspective of both their causative mechanisms and their spatial and temporal scales of influence, as represented by storm size and duration.

The scales of meteorological phenomena associated with heavy rainfall relate directly to the area affected by the precipitation and therefore to the type and extent of possible resultant flooding (Figure 2.1). Generally speaking, larger scale precipitation processes (i.e., those of **macroscale** and **synoptic scale**) tend to produce moderate to heavy rainfalls over fairly large regions (e.g., precipitation from one or more extratropical cyclones that affects a major river system, such as the Ohio River or Upper Mississippi River basin). The floods associated with macroscale and synoptic scale atmospheric processes tend to develop over tens of hours to days and affect large geographic regions. These large-scale floods have the potential to produce extreme property and crop damage and to cause tremendous economic losses. In regions of the world with efficient communication networks and public streamflow prediction and warning systems, floods on the larger scales pose only a small threat to human life. This is not true for less-developed countries where loss of lives may be significant.

In contrast to larger scale atmospheric processes, **mesoscale** and **storm scale** (Figure 2.1) processes are much shorter lived and can produce extreme amounts of rainfall over very localized areas (e.g., a portion of a state, a city, a small drainage basin) within a few hours or less. Floods caused by mesoscale and storm-scale rainfall tend to produce rapid runoff and

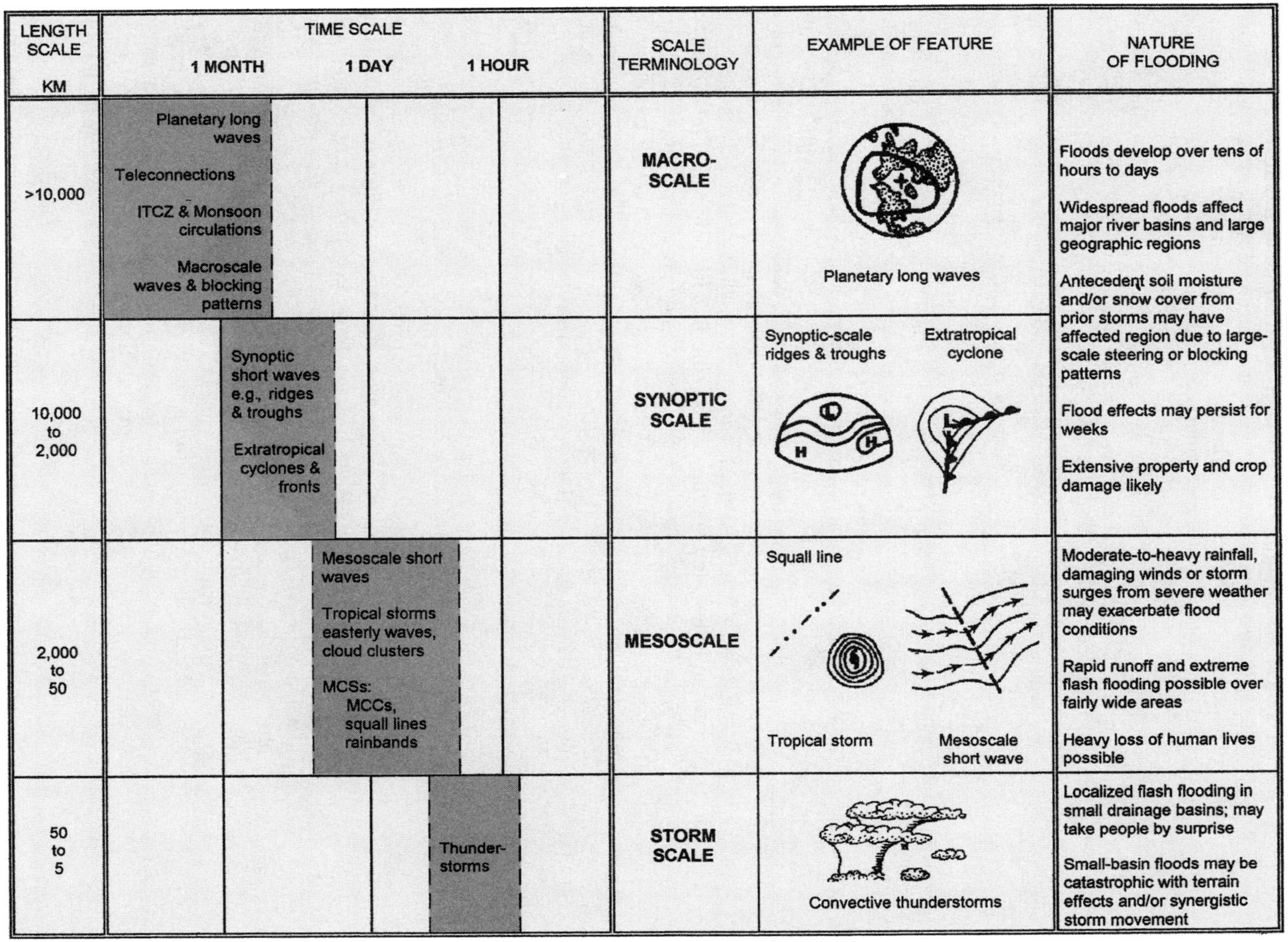

LENGTH SCALE KM	TIME SCALE (1 MONTH – 1 DAY – 1 HOUR)	SCALE TERMINOLOGY	EXAMPLE OF FEATURE	NATURE OF FLOODING
>10,000	Planetary long waves Teleconnections ITCZ & Monsoon circulations Macroscale waves & blocking patterns	MACRO-SCALE	Planetary long waves	Floods develop over tens of hours to days Widespread floods affect major river basins and large geographic regions Antecedent soil moisture and/or snow cover from prior storms may have affected region due to large-scale steering or blocking patterns
10,000 to 2,000	Synoptic short waves e.g., ridges & troughs Extratropical cyclones & fronts	SYNOPTIC SCALE	Synoptic-scale ridges & troughs Extratropical cyclone	Flood effects may persist for weeks Extensive property and crop damage likely
2,000 to 50	Mesoscale short waves Tropical storms easterly waves, cloud clusters MCSs: MCCs, squall lines rainbands	MESOSCALE	Squall line Tropical storm Mesoscale short wave	Moderate-to-heavy rainfall, damaging winds or storm surges from severe weather may exacerbate flood conditions Rapid runoff and extreme flash flooding possible over fairly wide areas Heavy loss of human lives possible
50 to 5	Thunder-storms	STORM SCALE	Convective thunderstorms	Localized flash flooding in small drainage basins; may take people by surprise Small-basin floods may be catastrophic with terrain effects and/or synergistic storm movement

Figure 2.1. A simple subdivision of atmospheric phenomena that influence the hydroclimatology of heavy rains and floods (modified from Orlanski, 1975; Hirschboeck, 1987a). The time scale refers to the typical duration or lifetime of specific phenomena. The length scale refers to the typical horizontal scale of specific phenomena.

extreme flash flooding. These events often cause major damage in conjunction with loss of human lives. Because of their short temporal and small spatial scales, prediction of these events is extremely difficult. Countries in which weather service agencies have been established may use radar and other observing technologies to detect the heavy rainfall associated with mesoscale and storm-scale events and to provide warnings of impending flash floods, but even this may not be enough to prevent deaths. Countries without such technologies are even more prone to disastrous flash-flooding scenarios and major loss of lives from these smaller-scale weather systems.

It should be noted that whenever precipitation systems of any scale develop over, or move into, regions of complex terrain, the character of the precipitation can be both increased and localized. The reasons for this are complicated but relate to two processes: (1) terrain causes increased upward motion whenever low-level winds flow upslope, and (2) mountains can influence the movement of precipitation systems, often causing them to slow or become nearly stationary. Orographic enhancement of precipitation is important in regions of complex terrain worldwide – from the intermontane western United States to the Himalayan foothills – and orographically enhanced precipitation events have resulted in many extreme floods (e.g., Matthai, 1969; Maddox et al., 1978; Smith et al., 1996).

In the following overview of specific flood-causing precipitation systems, we begin with smaller-scale systems and progress to larger-scale systems, arranging our discussion around atmospheric scaling terminology adapted from a framework advanced by Orlanski (1975) (see Figure 2.1). However, it is important to realize that precipitation systems that occur at one scale are strongly tied to systems and processes at other scales and that in most cases the larger-scale processes set the stage for atmospheric activity at smaller scales.

Smaller-Scale Precipitation Systems: Storm and Mesoscale

Storm-scale and mesoscale features that produce heavy rains are almost always convective in nature. Convective storms form within an atmosphere that is conditionally unstable. If an air parcel becomes saturated, the heat released when water vapor condenses causes the parcel to be warmer and less dense than its surrounding environment. The parcel then rises rapidly, much like a cork released under water. Sometimes deep convective clouds can reach heights well over 15 km above the ground. When lightning occurs within or from convective clouds, thunderstorms or thundershowers are said to be in progress. Intense convective storms can produce prodigious rainfall with amounts of tens to hundreds of millimeters occurring over small areas (e.g., tens of square kilometers) during periods of minutes to several hours. These heavy convective rains are usually, but not necessarily, accompanied by lightning.

Storm-Scale Systems. Isolated thunderstorms can deliver localized rainfall rates and amounts sufficient to produce local flash flooding. The most intense rain rates originate within isolated storms that occur in larger-scale synoptic environments with very high moisture contents in concert with very light winds aloft. These storms can produce rainfall rates of more than 100 mm/h and persist for up to an hour, although the most intense rain is likely only over areas of a few tens of square kilometers. Some isolated thunderstorms, called heavy precipitation supercells, can produce rain rates of this intensity for several hours (Weisman and Klemp, 1986). Flash flooding is likely if the intense rainfall produced by isolated thunderstorms is concentrated within a small drainage basin or if the storm moves slowly across the basin. When the rain from such storms falls into basins with a very impervious surface character, as is often the situation in sparsely vegetated, semiarid, desert, and mountainous regions of the world, the flood hazard increases dramatically. The intensity of the rain and rapid runoff can cause severe erosion and carry huge amounts of debris, clogging channels, culverts, and bridge openings and exacerbating the damage. Because these types of floods often occur in small or ephemeral watersheds in rural or remote areas, inhabitants, motorists, campers, and hikers may be taken by surprise. In canyons and mountainous regions, heavy thunderstorm rains in upstream areas may affect unsuspecting residents, hikers, and campers in downstream regions where rain has not occurred. This scenario took place in the tragic Antelope Canyon flash flood in northern Arizona during the summer of 1997 when 11 hikers died (Burstein, 1997).

Isolated thunderstorms can also be triggered by the interaction of winds near the ground and mountainous terrain. If winds aloft are weak, a series of very intense storm cells can form in the same location and then move slowly across the same region. This type of situation, although relatively rare, can produce prodigious rainfall totals and severe flash flooding. The floods at Rapid City, South Dakota, in 1972 and in the Big Thompson Canyon in Colorado during summer 1976 were of this type, and rain amounts approached 400 mm in less than 6 h (Maddox et al., 1978). A more recent example is the Rapidan flood in the central Appalachians, which produced 600 mm of rain in a 6-h period from a series of heavy convective showers (Smith et al., 1996). In some island locations (e.g., Hawaii, Taiwan), where winds from the ocean encounter abrupt mountain slopes, unusually moist conditions can cause nearly stationary convective showers that rain so heavily they produce significant flash flooding (Schroeder, 1977; Chen and Yu, 1988). Often these types of orographically forced convective storms occur at night, which further increases the flood hazard. Finally, even in small basins affected by isolated convective storms, under certain synergistic combinations of atmospheric and hydrologic processes, extremely large flash floods can occur. An example is the Eldorado Canyon, Nevada, flash flood of 1974, which was exceptionally large because the convective storm moved progressively downstream through the canyon, causing its intense rainfall to be superimposed on flood waters generated earlier and arriving from upstream (Glancy and Harmsen, 1975).

Mesoscale Systems. Convective clouds, particularly thunderstorms, do not typically evolve as isolated entities. Rather, they tend to organize into narrow lines or bands or into clusters or complexes of individual storms (Hobbs, 1978). These structures tend to be determined by the character of the atmospheric wind fields within which the convective clouds are occurring, by the degree of conditional instability present, and by the size and strength of the synoptic-scale upward motion field within which they are embedded. It is this organized aspect of convective storms that allows mesoscale precipitation systems to be characterized by lifetimes of a number of hours and to generate heavy rainfall over large areas. The heavy rain potential of these systems increases dramatically when they move slowly or become nearly stationary. In a very broad sense, mesoscale convective systems in the midlatitudes can be classified as *precipitation bands, squall lines*, and *mesoscale convective complexes* (MCCs). In tropical regions, mesoscale convective weather systems frequently occur as *convective cloud clusters* associated with easterly waves or tropical disturbances, and *tropical storms* (including hurricanes and tropical cyclones – see Gray, 1968; Pielke, 1990). In some subtropical and tropical regions squall lines and MCCs are also noted (Laing and Fritsch, 1997).

Precipitation Bands. The precipitation areas of synoptic-scale extratropical cyclones are often characterized by mesoscale rainbands that can assume a variety of orientations (Hobbs, 1978). These rain (or snow) bands contribute intensified amounts of precipitation to the overall weather event. The bands result when the circulations within the parent cyclone produce an unstable thermal stratification that will support convection. Mesoscale precipitation bands occur most frequently during the cool season over the oceans and adjacent coastal regions. The convection typically does not grow deep enough to produce lightning, unless the underlying ocean is much warmer than the air (e.g., over the warm Gulf Stream or Kuroshio currents). Occasionally, intense bands of snowfall develop over and downwind of large continental lakes, such as the North American Great Lakes. These bands can produce extreme snow accumulations – amounts greater than 500 mm in a matter of hours – and thus exacerbate snowmelt flooding later in the season. In general, however, mesoscale precipitation bands do not pose immediate local flooding threats; rather, they contribute to the overall synoptic-scale precipitation pattern.

Squall Lines. Long, narrow lines of intense thunderstorm cells are termed squall lines (see Hane, 1986). Squall lines are also usually associated with larger synoptic-scale cyclones and their attendant fronts (the boundary zones separating air masses having distinctly different temperature, wind, and moisture characteristics). The immediate local impact of squall lines may be great because of tornadoes, hail, or damaging wind gusts at the surface. They are most frequent over the continents and usually occur in the warm, moist air ahead of cold fronts. Thus, in the United States, they occur most often over portions of the country where there is a direct source of warm, moist, unstable air (e.g., the Gulf of Mexico) to the south of extratropical cyclone tracks. Squall lines sometimes form independently of synoptic cyclones (e.g., along the slopes of mountain ranges, along mesoscale air mass boundaries, and in tropical and subtropical latitudes).

The main weather threat of squall lines is the locally severe and damaging phenomena produced by intense thunderstorms. Squall lines tend to move rapidly with the winds aloft and thus affect any given location only briefly. Rain rates can be very intense (50–150 mm/h), but the duration of the rains usually are not long enough to cause serious local flooding. An exception occurs when squall lines stall and become nearly stationary (Chappell, 1986). This can happen if the line becomes oriented nearly parallel to the winds aloft. This situation allows many intense thunderstorm cells within the line to move across the same region, leading to very large rainfall amounts (sometimes greater than 300 mm) that accumulate over a few hours. Stationary squall lines therefore pose a great danger of local flash flooding and serious threats to life and property. The hazardous nature of these events is increased because squall lines, if they become stationary, tend to do so at night (Anderson and Arritt, 1998) when flood warnings may not be received.

Mesoscale Convective Complexes. MCCs are large, organized convective cloud systems that lack the distinct linear structure of squall lines (see Maddox et al., 1986). The precipitation structures embedded within MCCs can be very complicated, consisting of short squall lines, rainbands, clusters of thunderstorms, and individual intense thunderstorms as well as widespread areas of steady, light-to-moderate rainfall. Several of these structures are usually present simultaneously in the mature MCC. These systems tend to be less directly related to extratropical cyclones than the other small-scale systems discussed above. They usually occur during the warm season when extratropical cyclones and fronts are weak or have shifted poleward. MCCs occur over many areas of the globe, in both middle and low latitudes (Figure 2.2), where key ingredients are all present. These regions tend to be distant from synoptic-scale storm tracks downwind (with respect to prevailing middle-level winds) from significant mountain ranges and are usually affected frequently by low-level jets, which are strong inflows of near-surface winds and high humidity from lower latitudes that provide abundant water vapor to fuel the convection. MCCs frequently deliver rain rates and amounts similar to those produced by stationary squall lines. These features are often slow moving; hence, they pose a significant threat of heavy rains and local flash flooding (e.g., the damaging Johnstown flash flood of 1978) (Hoxit et al., 1982). They can also produce large hail, high winds, and extremely frequent lightning, which increases their hazardous character. If, over a period of days to weeks, a succession of MCCs develops in, and tracks across, roughly the same area, the regional flood hazard increases significantly (Fritsch et al., 1986). It was

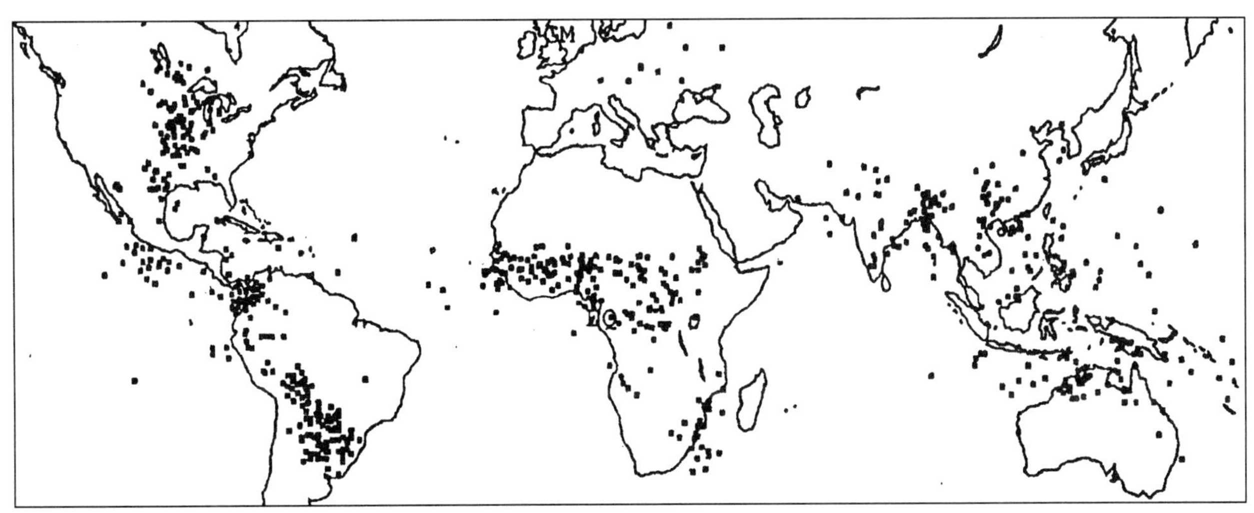

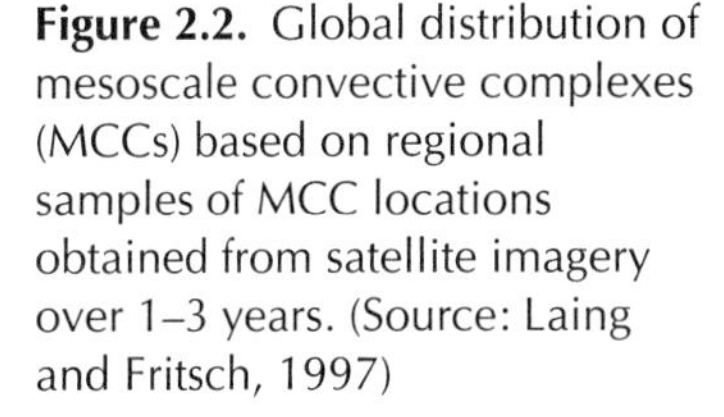

Figure 2.2. Global distribution of mesoscale convective complexes (MCCs) based on regional samples of MCC locations obtained from satellite imagery over 1–3 years. (Source: Laing and Fritsch, 1997)

through this scenario that MCCs contributed to the Great Flood of 1993 in the Upper Mississippi River basin (Junker et al., 1995).

Tropical Systems. In the tropics, short-wave troughs in the easterly wind flow (i.e., easterly waves) are typically manifested as convective cloud clusters. Clusters pose little flooding threat unless they become nearly stationary or occur with unusual frequency over the same area. Some convective clusters interact with the environment in ways that lead to an increase in strength of the system. This local intensification can lead to tropical disturbances and more fully developed tropical storms and tropical cyclones (hurricanes, typhoons, etc.) that can amplify to great intensity and persist for days. The greatest hazards from tropical cyclones arise from the damaging winds and disastrous storm-surge flooding they produce in low-lying coastal areas. They can deliver huge amounts of rainfall (i.e., greater than 500 mm) if they stall or move slowly, causing severe flooding threats, particularly over the small drainages of mountainous ocean islands. Further, when storms move inland they sometimes become very slow moving and interact with local terrain features (e.g., Carcena and Fritsch, 1983). The principal threat becomes widespread heavy rains along the storm's path. Such events, for example hurricane Agnes in the eastern United States during summer 1972 (see Bailey et al., 1975), can produce widespread flooding including localized flash floods and large-basin floods. These events pose a severe risk to property and human life. Land-falling tropical storms also often produce numerous tornadoes, adding a severe weather threat to that of the strong winds and heavy rains. Finally, under certain circumstances, tropical storms can play a role in generating floods in areas distant from the storm itself or in areas affected by the dissipating stages of the storm (Figure 2.3c). This occurs when large amounts of water vapor associated with the decaying storm support development of other precipitation systems, such as one of the mesoscale convective systems described above.

Larger Scale Precipitation Systems: Synoptic and Macroscale

The larger-scale processes that produce heavy rains are often associated with anomalous and persistent middle-level atmospheric wave patterns, slow moving or stagnant features such as blocking anticyclones and cutoff lows, or nearly stationary synoptic frontal zones. When such features also import air, usually from lower latitudes, having very high absolute moisture contents, expansive regions of heavy precipitation can occur (Means, 1954). Rain amounts of 50 to 150 mm can be delivered over a period of 2 or 3 days over regions spanning thousands of square kilometers. Macroscale features of the atmosphere, such as large-scale wave patterns aloft, tend to help set the stage for the occurrence of heavy rain events on synoptic and smaller scales. In addition, large-scale interactions between the atmosphere and the earth's surface also occur at the macroscale and can affect extensive regions of the globe on seasonal or longer time scales by means of monsoonal circulations and ocean-atmosphere teleconnections, such as the El Niño/Southern Oscillation (ENSO).

Synoptic-Scale Systems. At the synoptic scale, ridge and trough short-wave patterns aloft coupled with slow-moving pressure centers (both highs

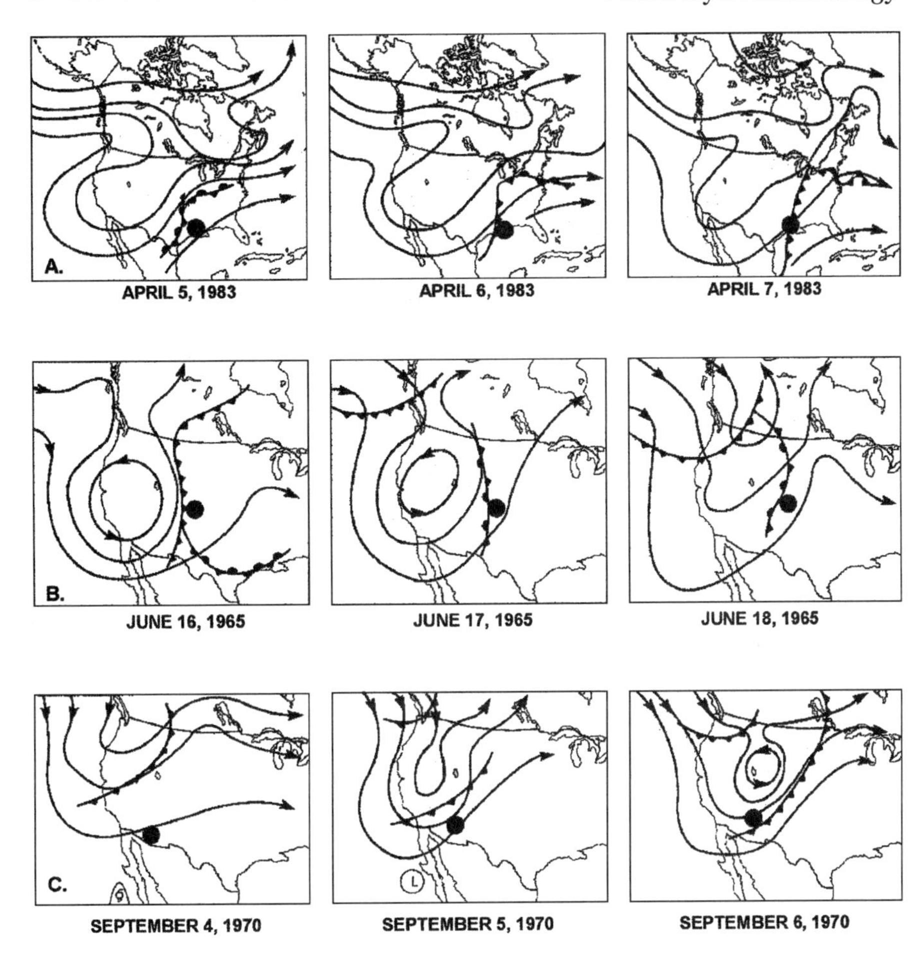

Figure 2.3. Three-day sequences of persistent or slowly changing large-scale circulation patterns associated with some severe floods in the United States. The direction of middle-level flow is depicted with arrows; surface frontal positions are also shown. Black circles indicate where flooding occurred. (A) Quasi-stationary trough-and-ridge pattern and surface frontal activity led to significant flooding in southern Louisiana. (B) Cutoff low within quasi-stationary large-scale trough over western United States steered warm unstable air northward into eastern Colorado, where it interacted with terrain. This, coupled with a shortwave perturbation on June 17, led to an episode of extreme flooding in east-central Colorado. (C) Interaction between a slow-moving large-scale trough, a surface front, and moisture from dissipating tropical storm Norma, led to severe flash flooding in central Arizona. (Modified from Hirschboeck, 1991)

and lows) and associated fronts can cause very heavy rains (see, for example, Matsumoto et al., 1971). This often occurs with continued inflow of very moist air from low-level jets, causing persistent rains along frontal zones. Synoptic-scale heavy rains can occur along and ahead of cold fronts or along and to the cool side of warm and stationary fronts. This type of situation can also indirectly contribute to heavy rains by setting up an environment very conducive to the development of smaller-scale precipitation systems. The key aspect of synoptic heavy rain events is that the synoptic pattern is changing slowly enough to allow winds from low latitudes to import large amounts of water vapor into the weather system (see Figure 2.3). The floods associated with synoptic weather features are usually widespread, so flooding occurs over large river systems during an extended period. However, this does not imply that very localized regions of enhanced rain, which generate local flash floods, cannot be embedded within synoptic events. It is also important to remember that intense synoptic systems in the winter can set the stage for future flooding by depositing very heavy snowfalls over large areas. The macroscale and synoptic scales often interact in the winter to establish a persistent, weeks-to-months storm track that allows many events to build large snow packs.

Perhaps the most important synoptic-scale features associated with heavy rains are *extratropical cyclones* (i.e., lows, low pressure systems, or depressions) and their associated waves (i.e., troughs) in the temperature and wind patterns aloft. These flow perturbations, generally called *short-wave troughs*, can occur on both the synoptic and mesoscale, depending on wavelength. They are distinctly different from macroscale waves in that they tend to move with the wind flow aloft and to evolve through a life cycle of formation, intensification, and decay during periods of several days to a week or so. It is these short-wave troughs that most often produce upward motions strong enough to generate regions of precipitation or to trigger formation of smaller-scale, more intense, weather features. Fronts associated with extratropical cyclones are also capable of producing heavy rains. It is important to note that atmospheric fronts tend to be both synoptic and mesoscale in nature. This is because fronts can extend horizontally for several thousand kilometers, whereas the temperature, humidity, and wind contrasts that define the front are confined most clearly in zones across the front that are only several hundred kilometers in extent. Synoptic cyclones and their attendant fronts can produce broad areas of general upward motion in the atmosphere or trigger the development of embedded smaller-scale weather features and storms. Often, the heaviest rains occur in extratropical cyclones in regions where both extensive precipitation, due to the cyclone's general lifting of the air, and embedded small-scale, intense rains occur (Maddox et al., 1979). On rare occasions, a synoptic scale short-wave trough can move over a region of surface high pressure without causing development of a surface cyclone. If such a feature imports moist, humid air aloft, widespread heavy rains and snows still occur, even though there is no distinct surface cyclone, or low pressure center, present. During the cold season extratropical cyclones often produce wide bands of very heavy snowfall.

Macroscale Systems. Most major regional floods are associated with circulation anomalies that can be observed in the large-scale wave patterns of middle-level geopotential height fields (e.g., 700 or 500 mb level). These wave patterns aloft may range in scale from the synoptic to the macroscale depending on wavelength. Macroscale waves can support the development, persistence, or sequential recurrence of smaller-scale precipitation systems that produce floods. Moreover, the most extreme and widespread regional floods usually evolve from macroscale circulation anomalies that are characterized by exceptional persistence in their embedded synoptic-scale wave patterns (Figure 2.3). In particular, circulation anomalies involving quasi-stationary patterns such as blocking ridges and cutoff lows in the middle-level flow are especially likely to be associated with extremely large meteorologic floods (Hirschboeck, 1987a, 1991). Periods when the patterns of macroscale troughs and ridges in the middle-level atmospheric circulation remain blocked in approximately the same configuration for several days can lead to nearly stationary frontal patterns, allowing extended episodes of storm activity and precipitation (Figure 2.3a). Persistent wave patterns can also result in long periods of flow impinging on mountain ranges. If the winds are carrying high amounts of moisture, then heavy rainfall can occur along upwind slopes (Figure 2.3b). In general, however, heavy rains during periods with persistent circulation patterns occur in direct association with synoptic- and smaller-scale weather features embedded within the flow regime (Figure 2.3b), or interacting with it (Figure 2.3c).

In addition to macroscale wave patterns aloft, there are other macroscale systems or processes that can influence flood activity over extensive or distant regions of the globe. One such system is located in the tropical latitudes (~15°N to ~15°S), where convective rainfall is common throughout the year, within the *intertropical convergence zone.* This macroscale feature reflects the zone where the trade winds from the northern and southern hemispheres converge. It is characterized by extensive cloudiness and embedded easterly wave disturbances with clusters of deep convective clouds and rains (Gray, 1968). Tropical storms sometimes develop to the poleward side of this zone as well. Thus, the intertropical convergence zone essentially provides a globe-encircling moist and favorable environment for the occasional development of flood-producing heavy rains, which are usually delivered by smaller-scale features.

Other macroscale processes that affect flooding at continental or global scales derive from interactions between the atmosphere and the earth's surface (land or ocean) and occur as *monsoonal circulations.* Several regions of the world are affected by monsoon flow regimes. Simply stated, these are circulations characterized by cold continental air flowing off the continent during the winter and warm humid air from a tropical or subtropical ocean flowing onto the continent during the summer. Unusually strong summer monsoons can lead to very widespread and extreme amounts of rainfall. This is particularly true when a strong summer monsoon flow regime develops in regions with steep gradients of elevation, such as the Indian subcontinent, where the northeastward onshore flow encounters first the western Ghats and then the rapidly rising mountain massifs of the Himalayas. Monsoonal precipitation regimes are strongly seasonal. For example, in the Narmada River basin in central India (Figure 2.4), 90% of the 1250-mm mean annual rainfall falls during 4 months of the year (June through September) (Dhar et al., 1985). There also can be a great deal of interannual variability in the monsoon precipitation, as seen in the all-India monsoon rainfall record (Figure 2.5).

Although monsoon circulations can be viewed as macroscale features that are driven by large-scale atmosphere–ocean interactions, the monsoon rainfall itself is not delivered by a single large precipitation system but is composed of a series of discrete rainfall events that operate at smaller scales (Webster, 1987) Synoptic and mesoscale features embedded in the macroscale monsoon circulation play an important role in determining variations in interannual rainfall totals and in delivering heavy rainfall sufficient to cause floods. At times, MCCs develop within the general monsoon flow and affect monsoonal regions of Africa, India, Asia, Australia, and all the Americas (Laing and Fritsch, 1997) (see Figure 2.2), enhancing the likelihood of flooding.

Disturbances in the tropical easterly flow also can contribute to heavy rainfall during the monsoon season. An example of this type of feature can be seen in the low-pressure systems and monsoon depressions that form in the Bay of Bengal and move west-northwestward across India from June through September. These can lead to significant flooding in the interior of the subcontinent when their tracks move far inland along the axis of an east–west-oriented drainage basin, such as the Ganges or Narmada. For example, in the Narmada River basin, large floods can occur when several low-pressure systems track successively over the upper part of the basin or when a single system moves downstream with a trajectory

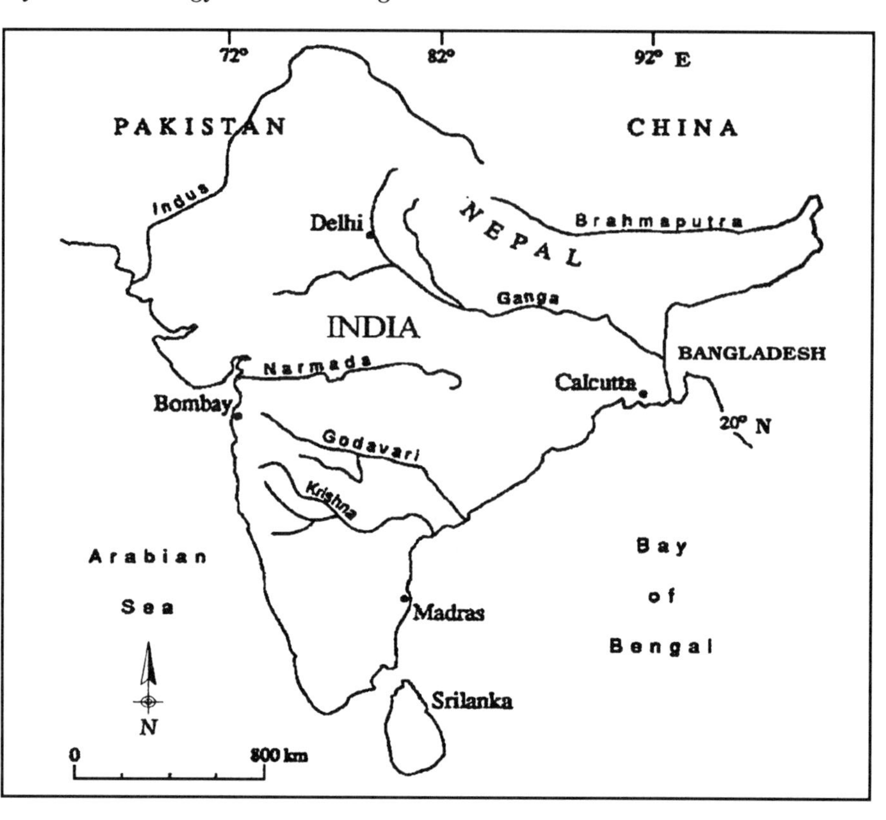

Figure 2.4. Map showing major rivers in India.

that is oriented along the axis of the basin, superimposing intense rainfall on the downstream-moving flood wave (Hirschboeck et al., 1996). In the postmonsoon period, intense tropical cyclones may also deliver heavy rains to the Indian subcontinent, further contributing to the flood hazard.

Monsoon rainfall also varies spatially – both intra- and interannually – due to shifts in larger scale features, such as the middle-level monsoon trough that forms over India. The position of this feature varies throughout the monsoon season, resulting in shifts in the location of heavy rainfall associated with monsoon disturbances embedded in the trough (Das, 1987; Bhalme and Mooley, 1980; Singh et al., 1988). Despite interannual variations in the strength of the Asian monsoon, some areas under its influence experience flooding on nearly an annual basis. In Bangladesh, for example, the discharges of the Brahmaputra and Ganges River basins (Figure 2.4) lead to a periodic inundation of the heavily cultivated and populated low-lying, delta regions of the combined basins. However, within the upstream portions of these basins, a strong spatial variation in the receipt of monsoon rainfall has been observed by some researchers. When the Ganges River basin to the west receives lower monsoon rainfall, the Brahmaputra River (and its lower tributary, the Meghna River) tend to receive above average rains and vice versa (Bandyopadhyay et al., 1997). Most of the severe flood years in Bangladesh (e.g., 1974, 1987, 1988) appear to be influenced by heavy rains and runoff from the Brahmaputra and Meghna basins, not the Ganges (Bandyopadhyay et al., 1997). Moreover, rainfall patterns in the upstream Himalayan parts of these basins appear to have less impact on severe Bangladesh flooding than does a synergism

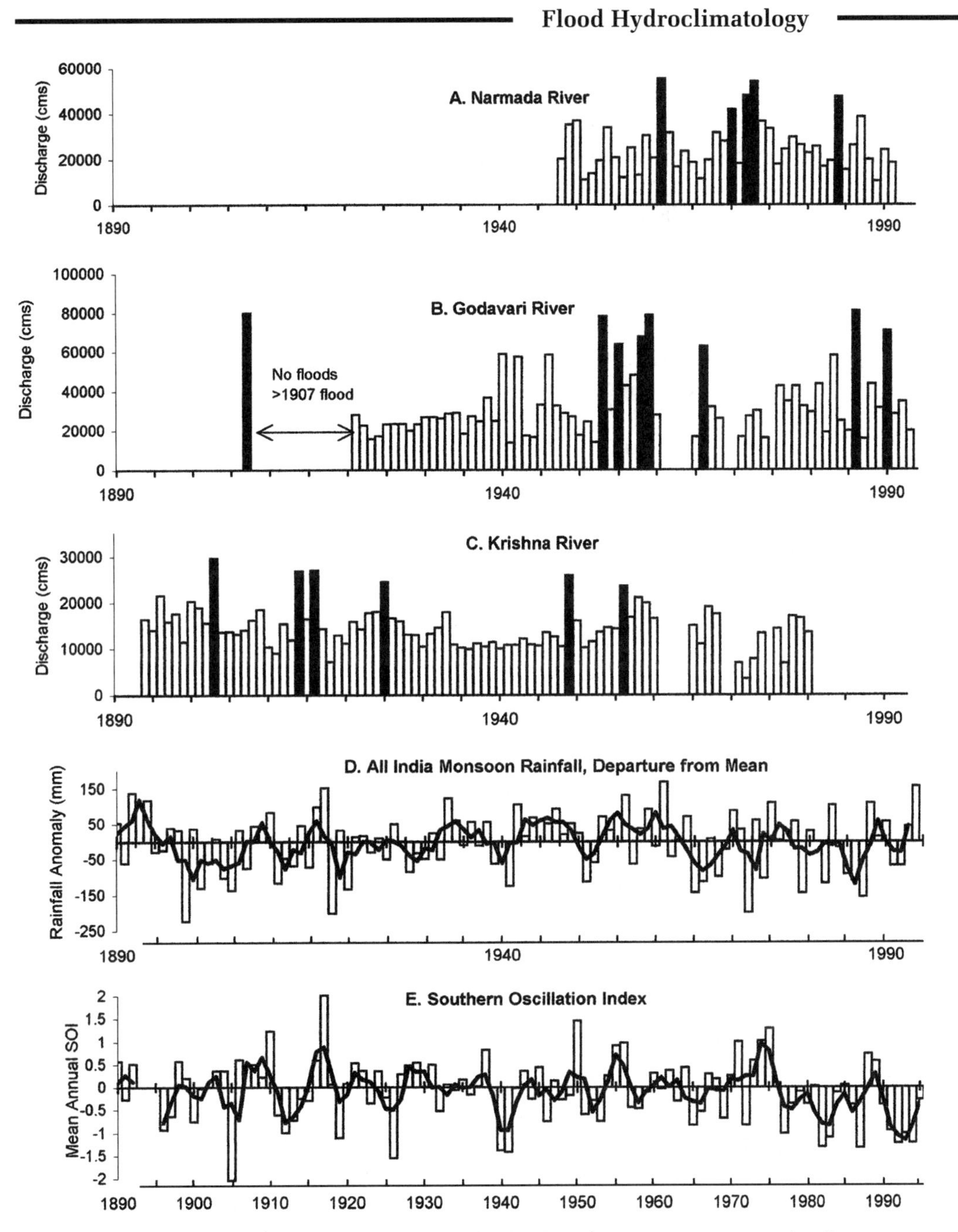

Figure 2.5. Peak annual gaged discharge records for the three largest rivers in central India compared with standardized all-India monsoon rainfall and the Southern Oscillation Index (SOI) over the last century. The period of above-average monsoon rainfall and moderate SOI in the middle of this century was one of few large floods on these rivers, as indicated by these gage records and paleoflood records (see text for more detail). (A–C) Dark bars highlight the largest five to seven floods on each river. Discharge data for the Narmada, Godavari, and Krishna Rivers compiled from UNESCO (1969–79, 1976), Rodier and Roche (1984), and Indian National Institute of Hydrology personal communication (1991, 1994). (D) All-India monsoon rainfall shows annual departure from 1874 to 1994 mean [data from Parthasarathy et al. (1993) and B. Parthasarathy, Indian Institute of Tropical Meteorology); dark line is 3-year running mean. (E) Annual variations in SOI, based on standardized data from NOAA Climate Prediction Center; dark line is 3-year running mean.

between more locally heavy monsoon rainfall (e.g., note the high concentration of MCCs in this area in Figure 2.2), combined with climatologic and hydrologic factors already in place in the lower reaches of the river systems. These flood-enhancing factors include high groundwater tables, low relief, infrastructure-related drainage impediments, and those occasions when a simultaneous inflow of above-normal discharge occurs from both the Ganges and Brahmaputra Rivers (Bandyopadhyay et al., 1997).

Because monsoons arise from large-scale seasonal land–ocean–atmosphere interactions, there is some evidence that they also may be linked globally through *teleconnections*. Teleconnections are statistically defined atmospheric and oceanic interactions between widely separated regions. For example, the Southern Oscillation Index (SOI) defines a teleconnection based on an inverse relationship in sea level pressure between two locations in the tropical Pacific Ocean (Darwin minus Tahiti pressure). Variations in this index are associated with the occurrence of anomalously warm (El Niño) or cold (La Niña) sea-surface temperatures in the Pacific. The combined atmosphere and ocean components of the phenomenon (ENSO) can be viewed spatially as macroscale phenomena, especially because various other responses throughout the globe have been statistically linked to ENSO variability. Furthermore, the influence of ENSO can be translated from tropical to extratropical latitudes and affect macroscale planetary long-wave patterns, which may in turn drive other teleconnections, such as the Pacific North American teleconnection pattern (Horel and Wallace, 1981). Temporally, teleconnections operate over a range of time periods. The ocean–atmosphere processes and interactions involved generally develop over a period of weeks to months and may persist for a year or longer. Hence, even though the influence on flooding of a teleconnection such as ENSO may be manifested at short time scales through meteorological processes that produce heavy rains in given areas of the world, teleconnections themselves operate on much longer climatic time scales and are discussed in more detail later.

Antecedent Climatic and Hydrologic Factors That Indirectly Cause Flooding

Floods are relatively rare and each one tends to arise from a unique set of ingredients. For example, the key atmospheric ingredients that are needed for flash-flood-producing storms are sustained high rainfall rates generated by forced ascent of air containing substantial water vapor (Doswell et al., 1996). However, even when sustained heavy rainfall occurs from some of the meteorologic systems described earlier, it may not always be enough to cause flooding without a synergistic combination of factors related to climatic and hydrologic conditions already in place in a drainage basin. To predict floods effectively, meteorologic information must be coupled with a knowledge of the rainfall-runoff processes, surface and subsurface hydrologic processes (Ramirez, Chapter 11, this volume), and land-use properties of the basin of interest. In addition, antecedent conditions must be considered to determine the likelihood of flooding from a given rainfall amount. Antecedent climatic and hydrologic factors that can influence runoff and subsequent flood occurrence are the degree of soil saturation and the level of shallow groundwater tables, which are affected by earlier precipitation events; the amount of snow accumulation from previous storms; the rate of snow melt, which depends on the length and degree of warm intervals; and the depth of frozen soil, which is related to the timing and severity of prior cold spells. In large basins with dams and reservoirs, the amount of available reservoir storage is another important factor that can influence future flooding.

Soil Moisture and Subsurface Water. Soil moisture, soil saturation, and groundwater levels have a variable influence as antecedent determinants of flooding when coupled with a heavy rainfall event. Saturated soils are not a required precursor to severe flooding. For example, flash floods commonly occur when runoff is conveyed to stream channels rapidly as *Hortonian overland flow*, regardless of soil moisture content (Dunne, 1983). Hortonian flow occurs when rainfall intensity exceeds the infiltration capacity and is associated with arid regions; urbanized areas; shallow, impermeable or disturbed soils; bedrock exposures; and sparsely vegetated areas. Meteorologic factors also play a role in whether Hortonian flow takes place. Heavy, convectively driven rainfall from thunderstorms can be so intense that it falls faster than it infiltrates, thereby generating overland flow, even where soils are unsaturated or the overall rainfall totals are not very high. This is common in arid and semiarid regions where investigators have found that thunderstorm-generated runoff is not very sensitive to antecedent soil moisture (Goodrich et al., 1994; Michaud and Sorooshian, 1994). It also has been noted that significant flash floods in arid regions of the western United States can be generated with much less rainfall than is required to generate a similar flash flood in wetter parts of the country (Maddox et al., 1979). In more humid regions, Hortonian flow can occur when soils are wet from previous storms (but not yet saturated) because the wet soils can decrease the infiltration capacity. In addition to Hortonian overland flow, runoff may rapidly enter stream channels as *subsurface storm flow*, which occurs during storm events when water infiltrates and very quickly moves laterally through the soil to enter the stream channel (Dunne, 1983). Because of the rapidity of both the Hortonian and subsurface storm flow processes during intense rains from convectively driven meteorologic processes, floods are possible when soils are not saturated or even wet. Hence, in some environments, antecedent soil moisture conditions may or may not significantly affect flood occurrence.

In certain hydroclimatic settings, saturated soils and shallow groundwater tables do play a critical role in causing and exacerbating floods. When rain falls onto saturated areas adjacent to streams, runoff can be generated rapidly at the surface as *saturation overland flow* (Dunne, 1983). During sustained periods of rainfall, the areas contributing saturation overland flow within a drainage basin grow larger as shallow water tables rise to the surface. Saturation overland flow is most likely to occur in watersheds that contain thin soils, concave hillslopes adjacent to wide valley bottoms, and perennial flow. It is also likely to be limited to humid areas and occur only after a lengthy episode of antecedent precipitation or when soil moisture is at a maximum at the end of the snowmelt season. The role of soil moisture, saturated soils, and shallow groundwater tables is most evident when storms have repeatedly rained on a watershed so that nearly all natural and artificial surface and subsurface storage reservoirs are full. Under such conditions, even a small amount of rainfall can result in a major flood hazard.

Snow and Snowmelt. In cold climates and at high elevations, where snow accounts for a significant portion of the total annual precipitation, the magnitude of the seasonal snow accumulation and its rate of melting are important antecedent factors that influence flooding. The stored snowpack can be released as runoff gradually or episodically throughout the winter and spring seasons, but when it is released rapidly during an abrupt thaw the flood hazard is increased. When portions of rivers have been frozen, ice breakup and ice jams (Cenderelli, Chapter 3, this volume) also present a severe backwater flood hazard, as occurred during the 1997 floods on the Red River of North Dakota. It is during and right after the spring snowmelt season that soil moisture reservoirs are usually filled to capacity, compounding the flooding problem from saturation overland flow. At such times, and during intermittent thaws throughout the winter, the presence of frozen ground, which prevents infiltration, is another flood-enhancing factor. Rapid, flood-producing snowmelt will take place over a span of several exceptionally warm and sunny days, especially when a persistent high-pressure anomaly, often associated with a blocking wave pattern, has established itself over an area previously visited by a sequence of cold low-pressure systems that precipitated large volumes of snow.

Some of the largest snowmelt-related floods occur when rain falls on an antecedent snow cover. The effect of rain on snow depends on the characteristics of the snowpack, including its temperature and the temperature of the rain itself (Harr, 1981). Early in a snow season before the snow compacts, rain falling on snow can be absorbed by a snowpack, adding to its total water content but not producing snowmelt. Later in the season when the snow is more dense, rainfall is less able to be absorbed. If the rain falling on the snow is warm, it is believed to accelerate the melt rate; however, the heat transferred to the snowpack during condensation of water vapor on the snow's surface may be another source of heat for the snowmelt process (Harr, 1981). In either case, it is during a prolonged, warm rainfall event that snowmelt can be a major contributor to flood runoff. During such an event, a large amount of water from both the current rainfall and the accumulated water content from the prior season's snowfall events can be rapidly introduced into a drainage system. The threat of severe flooding is compounded when the basin also contains areas of both frozen and saturated soils.

Flood Hydroclimatology and the Long-Term Perspective

It is possible to forecast or monitor a developing flood situation over meteorologic and short climatic time scales, but other approaches are needed to address flood hazards over longer time scales. Hydrologists have long used time series of gaged flood-peak data to evaluate the probability of occurrence of a flood of a given magnitude. Flood frequency analysis (i.e., the statistical analysis of observed flood events to estimate flood magnitudes at different recurrence intervals) constitutes a basic component of flood hazard zone determination in the United States and many other countries

(Stedinger, Chapter 12, this volume). A basic underlying assumption in this statistical approach to evaluating flood variability is that the data values in a flood time series represent an array of flood information that "is a reliable and representative time sample of random homogeneous events" (U.S. Water Resources Council, 1981, p. 6). Implicit in this assumption is the notion that flood-flow values are stationary and not affected by climatic trends over time and that the record represents a sample of time-homogeneous events, even though the floods may have originated from different types of hydrometeorologic causes. From a physical-process perspective it has been acknowledged that climatic variability may indeed affect flood-flow values in a nonrandom way and that flood series often are not homogeneous but composed of mixed distributions of events originating from different causal mechanisms (U.S. Water Resources Council, 1981; Hirschboeck, 1985, 1987b, 1988). However, despite the widespread recognition of these possible violations of the random homogeneous events assumption, the practical implications of these concerns for flood hazard perception, assessment, prevention, and readiness are only beginning to be explored. From a statistical test standpoint, violations in the randomness and time-homogeneity assumptions often are not evident; yet from the perspective of flood hydroclimatology, important and useful information for flood hazard management can be obtained from both a hydrometeorologic and hydroclimatic appraisal of a gaged flood record. In addition, flood evidence prior to the observed record (Baker, Chapter 13, this volume) can be used to extend our hydroclimatic understanding even further back in time. In the following sections we address how the flood hydroclimatology framework can be used to explore some of the underlying meteorologic and climatic causes of long-term flooding variability.

Climatic Information in Gaged Records: Multiple Mechanisms for Generating Floods

In many areas of the world, especially in the middle latitudes, floods in a single drainage basin can be generated by different types of precipitation events. The type of event that occurs depends in part on the season but also may depend on factors related to the large-scale circulation environment. In the middle latitudes of North America, for example, individual convective storms and mesoscale convective systems are predominant summer flood producers, whereas extratropical cyclones and their associated fronts are more likely to generate winter, spring, or fall floods. Tropical storm flooding is generally limited to summer and fall. In lower latitudes and warmer climates with sufficient moisture, convective storm systems may prevail year-round. Because the intensity and duration of heavy rainfall varies in different types of precipitation systems, the flood peaks in a given stream may reflect this, resulting in hydroclimatically defined mixed distributions in the overall probability density function of flood peaks (Hirschboeck, 1985, 1987b). Figure 2.6 illustrates this for two rivers in the Arizona portion of the Lower Colorado River basin, both of which experience flooding from a variety of precipitation systems (Hirschboeck, 1985, 1987b), which have been grouped into three general categories: storm-scale or mesoscale convective precipitation systems (e.g., mostly summer thunderstorms), tropical storm-related precipitation systems,

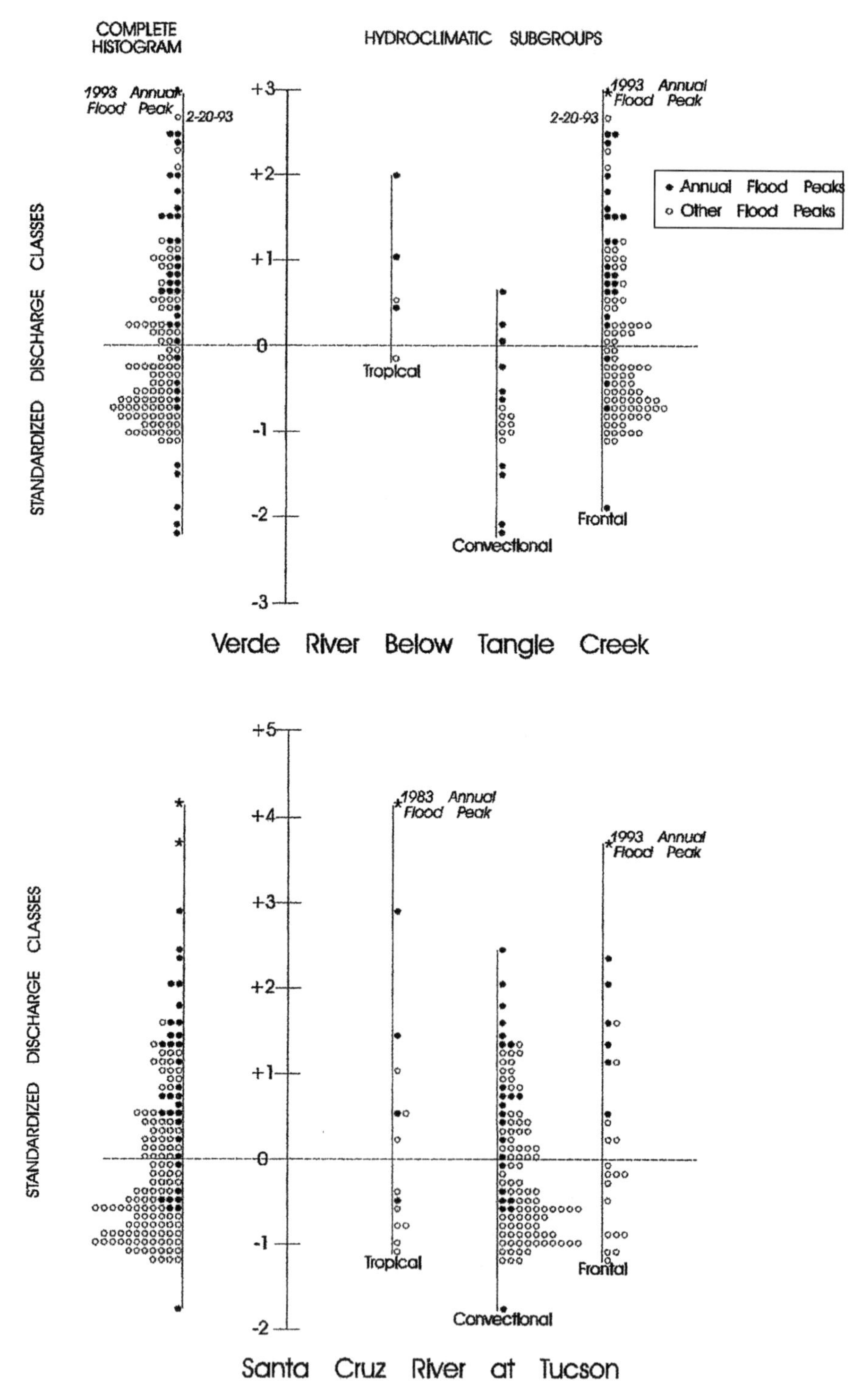

Figure 2.6. Decomposition of two flood records from the lower Colorado River basin according to the type of precipitation system that produced each flood. Peak discharges are displayed in the form of standardized dimensionless *z* scores for comparison. Annual flood peaks are shown as solid circles and all other peaks above base are shown as open circles. The period of record is 1950–1985; the extreme flood peaks of winter 1993 are also plotted. Verde River data are for the Verde below Tangle Creek (USGS gage no. 09508500). Santa Cruz River data are for the Santa Cruz at Tucson (USGS gage no. 09482500). (Source: House and Hirschboeck, 1997)

and synoptic-scale extratropical cyclone–frontal precipitation systems (House and Hirschboeck, 1997).

The distribution of annual and peak-above-base floods for the Verde River in central Arizona shows the dominance of synoptic–frontal floods for shaping the river's overall flood probability distribution. The Verde basin is situated at a latitude that is regularly influenced by winter storms and frontal passages and these cold-season storms yield seasonal precipitation totals that are larger than those received by the basin in summer from convective events. Hence, in the Verde, floods produced by synoptic–frontal events are more frequent than those produced by either of the other types of systems. These winter synoptic–frontal floods are also responsible for the largest peaks of record. Furthermore, when floods in the Verde are produced by rare tropical storm-related precipitation systems, the size of these events tends to be larger than the smaller-scale convective storm events, larger than the mean flood of the series, and comparable to some of the higher-magnitude frontal floods. Compared with the Verde, the Santa Cruz River is located in southern Arizona – south of the typical winter storm track – and the most frequent flood-producing precipitation systems are small-scale summer convective systems, which are especially active in southeastern Arizona. Note, however, that the largest floods in the Santa Cruz record were produced by tropical storm-related systems or winter synoptic–frontal systems, even though these types of events occurred less frequently.

The separation of these two relatively short flood records (~35 years) into their hydroclimatically defined components adds process-based meteorological and geographical information, which can supplement a purely statistical analysis of the flood peaks in these rivers by revealing the kinds of weather systems that produce different magnitudes of floods in each. Such an approach can also be used to evaluate the causes of flood variations over time – by examining changes in the decade-to-decade frequencies of floods produced by different types of storms in several streams throughout a region (Hirschboeck, 1985, 1987b) or by evaluating a long flood record of a single stream in detail (Webb and Betancourt, 1992).

On a larger scale, regional floods are often repeatedly associated with characteristic synoptic and macroscale atmospheric circulation patterns (Hirschboeck, 1985, 1991). For example, composite maps of daily 700-mb pressure heights over the eastern North Pacific Ocean and western North America show that severe winter floods in six different subregions of southern California, Arizona, Utah, and Nevada are all associated with a low 700-mb height anomaly off the California coast and a high 700-mb height anomaly near Alaska. Relatively minor shifts in the position of these anomalies appear to control which subregions experience floods (Ely et al., 1994). Hence, an analysis of variations in the position, intensity, or frequency of these large-scale atmospheric circulation patterns can improve our understanding of the underlying hydroclimatic causes for the spatial distribution of flooding.

Climatic Information in Gaged Records: Teleconnection Signals

The effect of ENSO on interannual variations in regional precipitation and streamflow has been a focus of a variety of research efforts across the globe (e.g., Ropelewski and Halpert, 1986, 1987; Cayan and Peterson, 1989; Gregory, 1989; Redmond and Koch, 1991; Cayan and Webb, 1992; Lins, 1997). However, the potential connection between these ENSO-related interannual variations in climatic conditions and the frequency or severity of extreme flood events is more difficult to document. In at least one area of the world the ENSO–precipitation–flood relationship is straightforward: along arid northern coastal Peru all but one recent strong El Niño event – that of 1976 – corresponded to both high rainfall and flood peaks (Waylen and Caviedes, 1986; Wells, 1990). But in other regions of the world that exhibit an ENSO link with precipitation, a corresponding relationship to floods may not be evident. High seasonal precipitation alone is not enough to cause a severe flood because of the synergism needed for many specific hydrologic and meteorologic factors to occur in just the right combination and timing to create the flood peak. In one extremely long gaged record from 1824 to 1973, low annual peak discharges on the Nile River resulting from below-average Indian–African monsoon precipitation showed a correlation with strong El Niño years (Parthasarathy and Pant, 1984; Quinn, 1992). However, the relationship between high peak annual floods and interannual climatic variations is rarely as straightforward as that between years of low discharge and drought. As noted above, there are only a few areas of the world where floods and precipitation are integrally linked to a single climatic variable or causal mechanism to the extent that a strong, consistent correlation appears on an interannual scale. Most regions are affected by multiple hydroclimatic mechanisms for generating floods and this complicates the relationship of floods with the interannual variations of a single regional climate component.

In the Arizona portion of the lower Colorado River basin a positive correlation exists between El Niño conditions and precipitation from winter extratropical cyclones and fronts and early fall tropical storms (Andrade and Sellers, 1988; Redmond and Koch, 1991). Because of the demonstrated effect of El Niño on the types of precipitation systems that produce the largest floods in Arizona (see Figure 2.6), one would logically expect a correlation between large floods and El Niño years. The single largest floods on some of the rivers in the lower Colorado River basin have indeed occurred during strong El Niño events (Cayan and Webb, 1992), although the record-breaking floods in Arizona in winter 1993 were associated with only a moderate El Niño (House and Hirschboeck, 1997). In general, there appears to be some connection between floods and El Niño in the lower Colorado River basin, even though the sample size may prohibit a statistically significant correlation (Webb and Betancourt, 1992; Ely, 1997). Still, not all El Niño years exhibit identical regional responses in precipitation and flooding (Schonher and Nicholson, 1989). Large floods have occurred during non-El Niño years, and some strong El Niño years (e.g., winter 1997–98) have not been characterized by significant flooding in the lower Colorado River region.

In another ENSO-teleconnected region, central India, summer-monsoon precipitation shows a positive correlation with the positive SOI (non-El Niño) phase of ENSO (Parthasarathy and Pant, 1984). However, the largest historic floods on the major rivers in central India (Narmada, Tapi, Godavari, Krishna) do not occur exclusively during the years of high monsoon precipitation or strong positive SOI (Ramaswamy, 1987; Dhar and

Nandargi, 1993). In fact, several of the largest floods in the region were actually associated with El Niño conditions and years of rainfall deficiency (Ely et al., 1996; Kale et al., 1997). As noted earlier, major inland floods in this region are caused by low-pressure systems and monsoon depressions that move west-northwestward from the Bay of Bengal and over the Indian subcontinent (Mooley and Shukla, 1987). Although these systems occur during the monsoon season, there does not appear to be a strong link between their flood-producing capabilities and ENSO.

These examples show that a relationship between ENSO and flooding is geographically variable, and, in those areas where such a relationship can be detected, it may not be consistent or statistically robust. Nevertheless, because there is an established connection between ENSO variability and seasonal precipitation in several areas of the world, there is considerable interest in using this information to issue climatic and hydrologic forecasts up to a season in advance. As in the case of the Nile River discharge, such forecasts may be far more reliable for low-discharge years than for extreme flooding years (Quinn, 1992); yet, some applications for flood forecasting are possible. For example, in river basins of the western United States, where annual peak flooding depends on the previous winter's accumulated snowpack at high elevations and where a strong ENSO sensitivity to winter precipitation has been identified, long-range hydrologic forecasts for the upcoming flood season can be issued. However, in any given year, the circulation patterns that regulate the rate of snowmelt will ultimately determine the severity of flooding that occurs in such basins. If we are to improve our understanding of the linkages between regional flooding and teleconnections such as ENSO, we need to understand the full range of variability that can exist within regional ENSO–precipitation–flood relationships. One means of doing this is to examine these relationships over much longer time periods than gaged flood records provide.

Bridging from Gaged Records to Longer Periods

Projected anthropogenic climatic changes as well as natural shifts in climatic conditions are potentially capable of affecting hydrologic systems to a degree that exceeds the range of conditions experienced within the relatively short observational records of most rivers (Houghton et al., 1996). Flood hazards are no exception to this prediction. Past or future changes in climatic conditions can be described in terms of changes in long-term means of precipitation and temperature, but the heterogeneity of the weather processes that compose climate dictates that there will be significant variance from these means, including changes in the frequency or intensity of extreme flood-generating storms or droughts. Floods have more severe effects when superimposed on already high streamflows, and thus a change in average seasonal climate and streamflow conditions could be associated with increased flood hazards. Paleohydrologic research has indicated that large changes in flood frequency and magnitude on river systems in several different hydroclimatic regions have been contemporaneous with small changes in mean temperature and precipitation over the past few thousand years (Chatters and Hoover, 1986; Knox, 1993; Ely et al., 1993; Ely, 1997). Traditional statistical approaches to estimating flood magnitudes and frequencies may break down over long time periods when the circulation patterns and processes that drive flooding variability change or shift over time, rendering assumptions of stationarity and random homogeneous time series invalid. In their place, a hydroclimatic understanding of the sources of flood variability may provide alternative or supplementary ways to estimate the likelihood of extreme floods. For example, in the lower Colorado River basin, where research has been conducted on links between circulation patterns and floods in the gaged record, additional research on the long-term spatial and temporal patterns in the occurrence of large floods in this region shows that a consistent relationship between floods and climatic variations appears to hold true for at least the past 5000 years (Ely, 1997).

Decadal-Scale Variations over the Twentieth Century. The period of historical climatic records, generally the last century at most, has been characterized by regional variations in temperature and precipitation persisting for periods up to several decades (Diaz and Quayle, 1980; Barry et al., 1981; Balling and Lawson, 1982; Webb and Betancourt, 1992). Many of these decadal-scale climatic fluctuations have coincided with shifts between two general states of macroscale wave patterns – one dominated by more frequent occurrences of zonal flow and the other dominated by more frequent meridional flow patterns, manifested as either recurring or persistent ridges and troughs. From the 1930s through the 1950s in the northern hemisphere, zonal flow occurred somewhat more often, whereas the periods before and after experienced a tendency toward more frequent periods of meridional flow (Dzerdzeevski, 1969; Kalnicky, 1974; Hirschboeck, 1988; Webb and Betancourt, 1992). Changes in climatic and hydrologic systems have been observed at similar times in several parts of the globe, including a shift from warmer to cooler North Atlantic and North Pacific sea-surface temperatures around 1960 (Folland et al., 1986; Namias et al., 1988; Gordon et al., 1992; Slowey and Crowley, 1995); generally low variability in SOI and the frequency and intensity of El Niño events from the 1920s to the mid-1950s (Elliott and Angell, 1988; Webb and Betancourt, 1992; Whetton et al., 1990); and increased summer monsoon rainfall in India from 1930 to 1964 (Parthasarathy et al., 1987). Other decadal-scale climate variations have also been observed. In the early part of the century, southern California and the lower Colorado River basin in the southwestern United States were wetter than average, and the northwestern part of the country was drier than average during the same time period (McGuirk, 1982; Fritts, 1991). In contrast, the middle of this century was distinguished by low winter precipitation in the lower Colorado River basin (Balling and Lawson, 1982). This well-documented tendency for opposite patterns of precipitation and streamflow in the southwestern and northwestern United States correlates with variations in ENSO and has been linked to persistent large-scale atmospheric circulation states (Bradley et al., 1987a, 1987b; Meko and Stockton, 1984; Cayan and Peterson, 1989; Redmond and Koch, 1991; Lins, 1997).

These prominent decadal-scale climatic fluctuations over the twentieth century affect seasonal precipitation and therefore could have an impact on flooding as well. In the southwestern United States, if floods varied directly with patterns in decadal precipitation and streamflow, it is possible that differences in flood frequency in individual rivers might be detected when comparing climatically different subperiods of the twentieth century. Webb and Betancourt (1992) divided the annual flood series of the Santa Cruz River in southern Arizona into three periods: 1915–29, 1930–59, and 1960–86. Although the mean discharges for the three periods were not significantly different, the variance for 1960–86 was significantly greater than for the preceding periods and the variance for 1930–59 was the lowest of the three. Webb and Betancourt attributed this pattern to the increased magnitude and frequency of floods from winter frontal storms and tropical cyclones and the decreased frequency of summer floods after 1960. In the same region, floods that exceeded the 10-year recurrence interval discharge on 20 rivers in the lower Colorado River basin were tabulated to determine whether any regional-scale variations existed in the temporal distribution of floods (Ely, 1992). No significant change was found in the frequency of winter floods from 1900 to 1988, although the peak discharges of winter floods were generally higher in the periods 1905–41 and 1965–88 than in the intervening decades. Over the period 1920–88, the frequency of floods from tropical storms also remained fairly constant, except for a prominent absence of floods from this storm type from 1940 to 1950. In contrast, the number of summer floods in this region was slightly greater during the middle part of the century, from 1919 to 1955, than after 1955. Although the patterns in the occurrence of 10-year floods are not particularly strong for all three types of precipitation systems, they do indicate that the findings from the Santa Cruz River (Webb and Betancourt, 1992) are valid across a larger hydroclimatic region and they support the expected response to the decadal variations in climate and atmospheric circulation over this century.

Many regional and global oceanic and atmospheric phenomena that influence the potential for floods on an interannual scale, such as ENSO or the North Atlantic oscillation, also hold the potential to influence longer-term temporal patterns in the occurrence of floods over decades to centuries through links with the atmospheric circulation patterns discussed earlier. Examining the likelihood of clusters of floods within multiple-year periods dominated by a particular set of climatic conditions conducive to flooding is more successful than attempting to assess the occurrence of floods on an annual scale. For example, although the effect of individual El Niño events on flood magnitude may vary among El Niño years, winter floods and tropical storms in the lower Colorado River basin show a much more consistent relationship with multiple-year periods dominated by negative SOI values (Ely et al., 1993).

Long-term flooding variations have also been observed in some rivers in India during the twentieth century. Few to no large floods occurred on the three largest rivers in central India during the middle third of the century from about 1926 to 1949 (Figures 2.4 and 2.5). Interestingly, this decrease in extreme floods overlaps with an extended period of above-average monsoon precipitation from the 1930s through the 1950s (Figure 2.5). The number of large floods increased after 1950 and continued into a period of greater variability in both monsoon precipitation and ENSO activity that began around 1960 and that has lasted to the present (Elliott and Angell, 1988; Whetton et al., 1990; Ely et al., 1996). This heightened variability may have enhanced the conditions conducive to the occurrence of extreme floods through changes in atmospheric circulation or other unidentified factors. With the scarcity of long discharge records for other rivers in

India, it is impossible to determine with certainty whether the decadal-scale patterns observed on these three rivers are related to regional climatic factors or are random. In addition, construction of two dams on the lower Krishna River has probably diminished the flood peaks in the latter part of the record. However, evidence from paleoflood records on these rivers suggests that the recent cluster of large floods is unusual in the long-term experience of the rivers. Floods equivalent in magnitude to those in recent decades have not occurred in at least the past 300–400 years on the Narmada River and in the past 650 years on the Godavari River. The frequency of extreme floods since 1960 is unprecedented in the 1700-year paleoflood record on the Narmada River (Ely et al., 1996). The relationship between the recent floods and paleofloods on the Krishna River is unclear because of the presence of dams on the river.

As mentioned earlier, neither abundant seasonal monsoon precipitation nor a positive SOI phase is correlated with the pattern of severe floods in central India. The floods are directly caused by synoptic-scale low-pressure systems and monsoon depressions from the Bay of Bengal and late-season tropical cyclones. A decrease in the frequency of tropical cyclones in the Indian Ocean from 1951 to 1986 (Raper, 1993) suggests that cyclone frequency alone is not sufficient to bring about this type of change in the magnitude and frequency of extreme floods. Much debate has arisen about the possible effects of future global warming on the frequency, intensity, and geographic range of tropical cyclones (Henderson-Sellers et al., 1998). However, evidence from areas such as India and the United States, where many extreme floods are associated with tropical cyclones, indicates that factors other than tropical storm frequency may be more important for predicting the impact of future climate variations on regional flood hazards. A synergism between the tropical cyclones and their pathways with respect to drainage basins is a key factor that leads to severe floods in tropical-storm regions and this relationship can evolve independently of any change in cyclone frequency. However, in some areas, more frequent storm occurrence may increase the likelihood of tropical storms taking pathways that produce major floods. A good example is the increased occurrence of floods from tropical cyclones during El Niño years in the southwestern United States. Typically, the greatest number of eastern North Pacific tropical storms are generated in midsummer and the number tapers off toward fall. However, it has been observed that many of the floods from tropical storms in the Southwest occur during El Niño years in the late summer and fall seasons. One explanation is that the higher overall frequency of tropical storms during El Niño years increases the probability of a few being caught in late-season meridional flow and steered northward.

Lessons from the Paleorecord: Flood Variations on Centennial-to-Millennial Time Scales

The rare, catastrophic floods that characterize the extremes of the hydrological system are the least understood and the most difficult to study by traditional methods of direct observation (Baker, 1987). Paleohydrologic data provide the best means to capture the long-term range of variability exhibited by these rare events over time scales of centuries or greater. A variety of paleohydrologic techniques exist for examining different aspects of the hydrologic system (Costa, 1987; Jarrett, 1991; Wohl and Enzel, 1995; Baker, Chapter 13, this volume). By combining the paleoflood data with independent paleoclimatic evidence of mean precipitation and temperature conditions over the same period, one can begin to examine the relationship between floods and climate change over long time scales (Knox, 1993; Ely, 1997; Enzel and Wells, 1997). This link is best established in regions where only one or two major storm types are associated with the largest floods in the modern record and where these flood-generating storms are consistently associated with a specific large-scale atmospheric circulation pattern or global phenomenon such as ENSO.

A growing body of paleoflood evidence from diverse hydrologic and climatic systems around the world indicates that many rivers have experienced variations in the magnitude or frequency of large floods at centennial to millennial time scales. Comparisons of the temporal and spatial patterns in paleoflood and paleoclimatic data have revealed that the magnitude and frequency of large floods in several regions around the world are associated with Holocene climatic variations (Hassan, 1981; Knox, 1983, 1993; Chatters and Hoover, 1986; Enzel et al., 1989; Smith, 1992; Ely et al., 1993; Rumsby and Macklin, 1994; Benito et al., 1996). It must be stressed that long-term changes in mean temperature or mean precipitation do not directly cause changes in flood frequency; rather, a change in mean conditions signals a change in large-scale circulation patterns, air mass boundaries, or storm trajectories that may have influenced the occurrence of flood-generating conditions (Knox and Kundzewicz, 1997).

A regional chronology of paleoflood records covering the past 5800 years on 19 rivers in the lower Colorado River basin of Arizona and southern Utah shows that clusters of large floods in this region coincide with intervals of cool wet climate, global neoglacial advances, high regional lake levels, and an increased frequency of strong El Niño events (Ely, 1992, 1997; Ely et al., 1993). The frequency of extreme floods has varied in response to these shifting climatic conditions and is clearly nonrandom in time. Large floods were relatively frequent in this region from 3800 to 2200 B.C. and after 400 B.C., with particularly prominent peaks in magnitude and frequency from 3600 to 3200 B.C., from A.D. 900 to 1100, and after A.D. 1400. In sharp contrast, the periods from 2200 to 400 B.C. and A.D. 1200 to 1400 were marked by significant decreases in the number of large floods on virtually all the rivers in the study. The intervals of extreme floods coincided with cool, wet, climatic conditions in the region, whereas the sharp decrease in large floods from A.D. 1200 to 1400 directly coincided with the warm, dry conditions during the Medieval Warm Period. Over at least the past 1000 years a positive relationship has existed between paleoflood frequency and variations in the frequency of strong El Niño events (Anderson, 1992). The same relationships between floods and climatic conditions that are seen at interannual or decadal scales in the Colorado River basin are apparent over these longer time scales. The studies listed above show that extreme floods are not random in time but that the climatic factors that influence the occurrence of extreme floods in the short term also exert control over the response of floods to longer-term climatic variations.

Comparison of paleoflood chronologies from different regions can highlight consistent, predictable, long-term similarities or differences in the

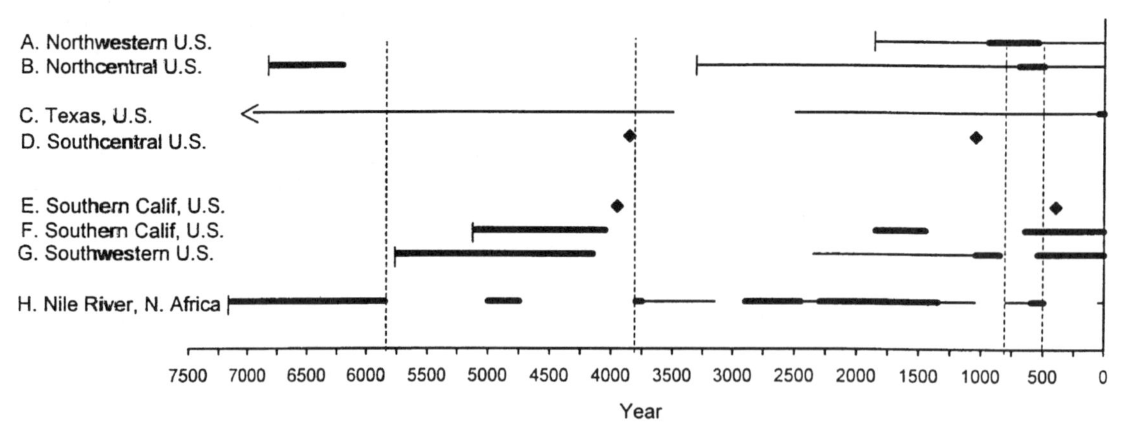

Figure 2.7. Comparison of historical and paleoflood chronologies from Nile River and several regions of North America. The flood record on the Nile River (H) bears a negative correlation with strong El Niño years (Quinn, 1992). Periods with an increased (decreased) frequency of large floods on the Nile River correspond to periods of decreased (increased) floods in the southwestern United States (E–G) and, to a lesser extent, increased (decreased) floods in the northwestern and northcentral United States (A and B). The timing of floods in the southcentral United States (C and D) appears similar to that of the southwestern United States, but the results are inconclusive because of the sample size and resolution of the data set. Solid heavy lines, periods of frequent large floods; solid thin lines, periods of moderate-sized floods; blank spaces, periods of few to no large floods; diamonds, single paleoflood; vertical dashed lines, times of major changes in Nile River record. Years are calibrated to calendar years before 1950 (Stuiver and Reimer, 1993). Data sources: A (Chatters and Hoover, 1986); B (Knox, 1993); C (Kochel et al., 1982; Patton and Dibble, 1982); D (McQueen et al., 1993); E (Enzel and Wells, 1997); F (McGill and Rockwell, 1998); G (Ely, 1997); H (Hassan, 1981; Said, 1993).

occurrence of floods among distinct hydroclimatic regions. As discussed earlier, individual floods can result from multiple causes that may or may not be related to the same set of persistent, large-scale atmospheric or oceanic conditions. Therefore, in attempting to identify climate-related patterns in paleoflood records from different areas, analysis must begin with regions that show very straightforward connections to present climatic phenomena, such as ENSO, or that exhibit strong positive or negative correlations with seasonal streamflow or precipitation in another region. The paleoflood record from the lower Colorado River basin provides examples of several different interregional hydroclimatic links (Figure 2.7). Discharge records in the lower Colorado River basin and the Nile River demonstrate an opposite response to El Niño years and this is reflected in centennial-scale variations in flood characteristics over the past 1000 years. Periods of exceptionally low Nile River floods from 3900 to 1840 B.C. [5850 to 3790 before present (BP)], A.D. 930 to 1090 (1020 to 860 BP), and in the past 300 years (Hassan, 1981; Said, 1993) correspond with prominent peaks in flood magnitude and frequency in the lower Colorado River basin (Ely et al., 1993), whereas the period with very few years of low floods from A.D. 1090 to 1470 (860 to 480 BP) overlaps with the sharp decline in large floods in the lower Colorado River basin during the Medieval Warm Period.

In another example, one of the dominant spatial patterns of hydroclimatic variation in North America is the out-of-phase relationship in precipitation and streamflow between the northwestern and southwestern parts of the United States (Meko and Stockton, 1984; Redmond and Koch, 1991). This pattern is partially attributed to the strong opposite response of these two regions to variations in the intensity of large-scale atmospheric and oceanic phenomena in the Pacific, particularly ENSO and the Pacific North American pattern (Cayan and Peterson, 1989). The highest frequency of large floods in a 1900-year record from the Columbia River in the northwestern United States was from A.D. 1020 to 1390 (Chatters and Hoover, 1986), which is the interval of least frequent floods in the lower Colorado River basin over the same period (Figure 2.6). A similar asynchronous pattern occurs between these southwestern United States paleofloods and a 7000-year flood record from the upper Mississippi River basin in the north-central United States (Knox, 1993; Knox and Kundzewicz, 1997). In southwestern Texas, on the other hand, frequent, moderate-sized floods during a humid climate from ~3400 to 2600 calendar years BP coincided with the virtual lack of large floods on the southwestern rivers, whereas the drier periods before and after this time experienced sporadic, but larger floods (Patton and Dibble, 1982). Thus, the clusters of the largest floods coincide in time, although the climatic conditions associated with the floods differ. The relationships between all these areas suggest that persistent large-scale atmospheric circulation patterns that increase the potential flood hazards in the southwestern and possibly south-central United States have decreased the potential for floods in the northwestern and north-central parts of the country over the same time periods (Ely, 1992; Knox and Kundzewicz, 1997).

In several studies, clusters of particularly frequent severe floods have been attributed to enhanced variability during periods of climatic transition, such as the change from warmer to cooler climate in many parts of North America and Europe sometime between A.D. 1250 and 1500 (Ely, 1992; Knox, 1993; Benito et al., 1996). The frequency of floods in the upper Mississippi River valley in the north-central United States increased between A.D. 1250 and 1400 (Knox, 1993), whereas floods in the southwestern United States and Spain increased after A.D. 1400 to 1500 (Ely, 1992; Benito et al., 1996). Although the periods of increased flood frequency are out of phase between the north-central United States and the other regions, they still may be responding to different aspects of this transitional period, either the end of the warm period or the beginning of the cool period. This sort of intriguing pattern can be examined and verified as the database of long-term records grows across the globe.

Contradictions in the expected patterns of floods and climate change between regions introduce an interesting point. Obvious hydroclimatic links among separate paleoflood records may follow modern examples during certain periods in the past, whereas at other times floods in the same or different regions appear to respond to climatic conditions in different ways. For example, several periods of high and low floods occurred on the Nile River (Said, 1993) during the long interval with no depositional record of large floods in the southwestern United States from about 4100 to 2300 years ago (Figure 2.6). Very long paleoflood records from relict flood levees at tropical latitudes in Australia suggest that the magnitude and frequency of extreme floods could increase under two dramatically different climate conditions: the much cooler, wetter conditions prior to the last glacial maximum and the warmer, wetter conditions of the mid-Holocene (Nott et al., 1996). The challenge therefore is to develop enough data sets in a variety of climatic regions around the globe that these periods of coincidence and divergence in the hydrologic responses to climate begin to emerge. In this way, the flood records can actually begin to point toward specific shifts in large-scale circulation patterns, airmass boundaries, and storm tracks that might not be apparent from changes in mean climatic conditions, providing insight into how the global climate system has operated at different periods in the past and how hydrologic systems in different regions could respond to future climate changes.

Concluding Remarks

Understanding the hydroclimatic controls on variations in the magnitude and frequency of extreme floods is critical for flood-frequency forecasting.

As shown in the examples in this chapter, the spatial and temporal distribution of large floods in a variety of regions around the world are influenced to a large extent by persistent, anomalous patterns in hemispheric to global-scale atmospheric and oceanic circulation that have an influence on flood-generating precipitation systems. However, it is these smaller-scale precipitation systems, operating at relatively short time scales, that directly cause floods. Moreover, in many cases, floods tend to develop from a unique set of ingredients that involve a synergism between meteorologic, climatic, hydrologic, and drainage basin factors. To understand the causes of floods fully, we attest that both short-term meteorologic processes and longer-term climatic processes must be addressed with a flood hydroclimatology perspective.

Extreme floods are by nature rare, and the gaged record from a single river provides insufficient data to characterize the frequency of these events accurately or to evaluate the maximum flood potential in the context of the long-term history of the river. Geologic evidence of paleofloods from the past several centuries to millennia reveals patterns in the magnitude and frequency of the largest floods that are not apparent from the relatively short stream-gage records in the western United States. Examining the linkages between climate and floods across various temporal and spatial scales will improve our ability to anticipate the local and regional flood hazards that can be expected under a range of climatic conditions, an important component of long-term planning in the face of uncertain future climate scenarios. The understanding gained through establishing these connections provides practical input for floodplain development and management; evaluation of the safety, design, and economic feasibility of dams and other structures; and the understanding and management of riparian ecosystems.

One main conclusion that can be drawn from our overview of flood-causing meteorologic and antecedent climatologic processes is that the manner in which above-average precipitation is delivered in space and time affects both the type of flood that occurs and its accompanying hazards. Hence, the scale of a precipitation system is an important component of flood causality. However, precipitation systems that occur at one scale are strongly interconnected with systems at other scales, and larger-scale processes tend to set the stage for activity at smaller scales. Across all scales of activity, persistence – or the slow-moving nature of a precipitation system – appears to be a key element for producing exceptionally large floods.

Over the long term, differences in paleoflood patterns within and between regions have enormous implications for flood-frequency forecasting and river management. Long-term variations in climate create changes in the overall flooding regime and cannot be used to predict the occurrence of individual floods in any given year or even in a span of multiple years. Understanding the extent of the impacts of past climatic variations on flooding can aid in assessing the potential economic or societal implications of future change. For example, past changes in climate and flood regimes may have contributed to the demise of ancient societies such as the Anasazi and Hohokam of the southwestern United States through an increased frequency of severe floods that caused deep incision of alluvial channels and/or destruction of floodplain irrigation systems (Huckleberry, 1993). Could this type of change have a substantial impact on modern societies? Our river management is designed around expectations of a predictable range of flood hazards centered largely on statistical extrapolations based on the historical record of a river over a period of decades. A change in the long-term flood regime may render aspects of this infrastructure no longer applicable, leading to economic impacts at the least, and possibly extensive loss of life and property. Stationarity in a flood series means that the probability distribution of any random variable, such as the mean discharge or variability, is constant through time. Nonstationarity, or nonrandom changes in the flood distribution, create significant problems for traditional statistical methods of flood frequency analysis. Examination of the past relationships between floods and climate places the current flood hydroclimatic conditions into perspective and reveals any evidence of nonstationarity in the record. For example, is a region currently undergoing a period of frequent or infrequent flooding compared with the long-term paleorecords? How would the addition of the paleoflood information affect statistical flood-frequency analyses and risk assessments based solely on the modern flood record? On some rivers, paleoflood information has revealed pronounced nonstationarity in the flood series over long time periods (Baker, 1987). Since the inception of the gaged record in 1948, the Narmada River in India has experienced several high-magnitude floods (Figure 2.5). Both the magnitude and frequency of these floods stand out as an anomaly compared with the 1700-year paleoflood record (Ely et al., 1996). The design of the Narmada Sagar Dam currently under construction was based on the gaged record, which by chance in this case incorporates the largest floods in the long-term history of the river and is probably an adequate representation of the maximum floods that are likely to occur in the future. On rivers where past floods were larger or more frequent than present, the gaged records could grossly underestimate the floods that could be reasonably expected with a moderate shift in future climatic conditions. Knowledge of the paleoflood history could avert the potential disaster of underdesigning major structures on the river.

The flood–climate relationship on a river is also an effective means of evaluating the feasibility of the theoretical models of worst-case scenario floods commonly used for design purposes. On the Verde and Salt Rivers in central Arizona the largest paleofloods in at least the past 1000 years were slightly less than twice the magnitude of the largest floods in the gaged records before 1993 (Ely and Baker, 1985; Partridge and Baker, 1987). The theoretical probable maximum flood, calculated by the Bureau of Reclamation for dam design on these rivers, was four to five times greater than the maximum paleoflood on each river (Baker et al., 1987). Armed with this knowledge, one can better evaluate the utility of the enormous public expenditure in designing for a theoretical flood that is significantly greater than the worst floods that have occurred under the range of climatic conditions over the last millennium. In other cases, such as the Colorado River itself, the probable maximum flood was only slightly larger than the largest paleoflood (O'Connor et al., 1994).

A possible solution to these problems of choosing the appropriate design flood for structures or floodplain development is to identify past climatic periods characterized by the highest flood frequencies or magnitudes and use them as a basis for future flood hazard assessment in a region. Incorporating the range of past natural hydroclimatic variability into a long-range flood-frequency forecast model is critical for an accurate assessment of the range of effects future climatic change might have on flood hazards for that specific region. The emerging patterns of similarities and differences among flood hydroclimatic regions globally can aid tremendously in augmenting evidence from any single region in evaluating the potential flood hazards under scenarios of changing climate, from annual to millennial scales.

Acknowledgments

We wish to thank our reviewers, E.E. Wohl, J.M. Friedman, and M.W. Douglas for their insightful comments and suggestions. This work was funded by NASA Earth Science Enterprise Grant NAGW-3498 and NOAA Paleoclimatology Grant GC-94-504 (Hirschboeck), and NSF Grants EAR-9202498 and EAR-9104489 and the Faculty Research Fund at Central Washington University (Ely).

References

Anderson, R.Y. (1992). Long-term changes in the frequency of occurrence of El Niño events. In *El Niño: Historical and Paleoclimatic Aspects of the Southern Oscillation*, ed. H.F. Diaz and V. Markgraf, pp. 193–200. Cambridge: Cambridge University Press.

Anderson, C.J. and Arritt, R.W. (1998). Mesoscale convective complexes and persistent elongated convective systems over the United States during 1992 and 1993. *Monthly Weather Review*, **126**, 578–599.

Andrade, E. and Sellers, W. (1988). El Niño and its effect of precipitation in Arizona. *Journal of Climatology*, **8**, 403–410.

Bailey, J.F., Patterson, J.L., and Paulhus, J.L.H. (1975). Hurricane Agnes rainfall and floods, June–July 1972. U.S.G.S. Professional Paper 924. Washington, DC: U.S. Government Printing Office.

Baker, V.R. (1987). Paleoflood hydrology and extraordinary flood events. *Journal of Hydrology*, **96**, 79–99.

Baker, V.R., Ely, L.L., O'Connor, J.E., and Partridge, J.B. (1987). Paleoflood hydrology and design applications. In *Regional Flood Frequency Analysis*, ed. V. Singh, pp. 325–338. Boston: D. Reidel.

Balling, R. and Lawson, M. (1982). Twentieth century changes in winter climatic regions. *Climatic Change*, **4**, 57–69.

Bandyopadhyay, J., Rodda, J.C., Kattelmann, R., Kundzewicz, and Kraemer, D. (1997). Highland waters – a resource of global significance. In *Mountains of the World – a Global Priority*, eds. B. Messerli and J.D. Ives, pp. 131–155. London: The Parthenon Publishing Group.

Barry, R.G., Kiladis, G., and Bradley, R.S. (1981). Synoptic climatology of the western United States in relation to climatic fluctuations during the twentieth century. *Journal of Climatology*, **1**, 97–113.

Benito, G., Machado, M.J., and Pérez-González, A. (1996). Climate change and flood sensitivity in Spain. In *Global Continental Changes: The Context of Palaeohydrology*, ed. J. Branson, A.G. Brown, and K.J. Gregory, pp. 85–98. Geological Society of London Special Publication No. 115.

Bhalme, H.N. and Mooley, D.A. (1980). Large-scale droughts/floods and monsoon circulation. *Monthly Weather Review*, **108**, 1197–1211.

Bradley, R., Diaz, H., Eischeid, J., Jones, P., Kelly, P., and Goodess, C. (1987a). Precipitation fluctuations over northern hemisphere land areas since the mid-19th century. *Science*, **237**, 171–75.

Bradley, R., Diaz, H., Kiladis, G., and Eischeid, J. (1987b). ENSO Signal in continental temperature and precipitation records. *Nature*, **327**, 497–501.

Burstein, J. (1997). "1 dead, 10 missing in canyon flood: Guide found at bottom is only survivor," Article in *The Arizona Daily Star*, Tucson, Arizona, Thur. 14 Aug. 1997.

Carcena, F. and Fritsch, J.M. (1983). Forcing mechanisms in the Texas Hill Country flash floods of 1978. *Monthly Weather Review*, **111**, 2319–2332.

Cayan, D. and Peterson, D. (1989). The influence of North Pacific atmospheric circulation on streamflow in the West. In *Aspects of Climate Variability in the Pacific and the Western Americas*, ed. D.H. Peterson, pp. 375–397. Washington, DC: American Geophysical Union Geophysical Monograph 55. Washington, DC.

Cayan, D. and Webb, R.H. (1992). El Niño/Southern Oscillation and streamflow in the western United States.

In *El Niño: Historical and Paleoclimatic Aspects of the Southern Oscillation*, ed. H. Diaz and V. Markgraf, pp. 29–68. Cambridge: Cambridge University Press.

Chappell, C.F. (1986). Quasi-stationary convective events. In *Mesoscale Meteorology and Forecasting*, ed. P. Ray, pp. 289–310. Boston: American Meteorological Society.

Chatters, J.C. and Hoover, K.A. (1986). Changing Late Holocene flooding frequencies on the Columbia River, Washington. *Quaternary Research*, **26**, 309–320.

Chen, G.T.-J. and Yu, C.-C. (1988). Study of low-level jet and extremely heavy rainfall over northern Taiwan in the Mei-Yu season. *Monthly Weather Review*, **116**, 884–891.

Costa, J.E. (1987). A history of paleoflood hydrology in the United States, 1800–1970. In *History of Hydrology*, eds. E.R. Landa and S. Ince, pp. 49–67. History of Geophysics, vol. 3. Washington, DC: American Geophysical Union.

Das, P.K. (1987). Short- and long-range monsoon prediction in India. In *Monsoons*, eds. J.S. Fein and P.L. Stephens, pp. 549–578. New York: John Wiley and Sons.

Dhar, O.N., Mandal, B.N., and Mulye, S.S. (1985). Some aspects of rainfall distribution over Narmada basin up to Sardar Sarovar Damsite in Gujarat. Proceedings from National Seminar-cum-Workshop on Atmospheric Sciences and Engineering, Jadavpur University, 186–193.

Dhar, O.N. and Nandargi, S. (1993). The zones of severe rainstorm activity over India. *International Journal of Climatology*, **13**, 301–311.

Diaz, H. and Quayle, R. (1980). The climate of the United States since 1895, spatial and temporal changes. *Monthly Weather Review*, **108**, 249–266.

Doswell, C.A. III, Brooks, H.E., and Maddox, R.A. (1996). Flash flood forecasting: an ingredients-based methodology. *Weather and Forecasting*, **11**, 560–581.

Dunne, T. (1983). Relation of field studies and modeling in the prediction of storm runoff. *Journal of Hydrology*, **65**, 25–48.

Dzerdzeevski, B. (1969). Climate epochs in the Twentieth Century and some comments on the analysis of past climates. In *Quaternary Geology and Climate, Pub. No. 1701*, ed. H. Wright, pp. 49–60. Washington, DC: National Academy of Sciences.

Elliott, W. and Angell, J. (1988). Evidence for changes in Southern Oscillation relationships during the last 100 years. *Journal of Climate*, **1**, 729–737.

Ely, L.L. (1992). *Large Floods in the Southwestern United States in Relation to Late- Holocene Climatic Variations*. Ph.D. dissertation, University of Arizona, Tucson. 326 pp.

Ely, L.L. (1997). Response of extreme floods in the southwestern United States to climatic variations in the Late Holocene. *Geomorphology*, **19**, 175–201.

Ely, L.L. and Baker, V.R. (1985). Reconstructing paleoflood hydrology with slackwater deposits: Verde River, Arizona. *Physical Geography*, **6**, 103–126.

Ely, L.L., Enzel, Y., Baker, V.R., and Cayan, D.R. (1993). A 5000-year record of extreme floods and climate change in the southwestern United States. *Science*, **262**, 410–412.

Ely, L.L., Enzel, Y., Baker, V.R., Kale, V.S., and Mishra, S. (1996). Changes in the magnitude and frequency of Late Holocene monsoon floods on the Narmada River, Central India. *Geological Society of America Bulletin*, **108**, 1134–1148.

Ely, L.L., Enzel, Y., and Cayan, D.R. (1994). Anomalous North Pacific atmospheric circulation and large winter floods in the southwestern United States. *Journal of Climate*, **7**, 977–987.

Enzel, Y., Cayan, D.R., Anderson, R.Y., and Wells, S.G. (1989). Atmospheric circulation during Holocene lake stands in the Mojave desert: evidence of regional climate change. *Nature*, **341**, 44–48.

Enzel, Y. and Wells, S.G. (1997). Extracting Holocene paleohydrology and paleoclimatology information from modern extreme flood events: an example from Southern California. *Geomorphology*, **19**, 203–226.

Folland, C.K., Palmer, T.N., and Parker, D. (1986). Sahel rainfall and worldwide sea temperatures, 1901–1985. *Nature*, **320**, 602–607.

Fritsch, J.M., Kane, R.J., and Chelius, C.H. (1986). The contribution of mesoscale convective weather systems to the warm season precipitation in the United States. *Journal of Climate and Applied Meteorology*, **25**, 1333–1345.

Fritts, H.C. (1991). *Reconstructing Large-Scale Climatic Patterns from Tree-Ring Data*. Tucson: University of Arizona Press.

Glancy, P. and Harmsen, L. (1975). A hydrologic assessment of the September 14, 1974 flood in Eldorado Canyon, Nevada. *USGS Professional Paper 930*. Washington, DC: U.S. Government Printing Office.

Goodrich, D.C., Schmugge, T.J., Jackson, T.J., Unkrich, C.L., Keefer, T.O., Parry, R., Bach, L.B., and Amer, S.A. (1994). Runoff simulation sensitivity to remotely-sensed initial soil water content. *Water Resources Research* **30**, 1393–1405.

Gordon, H., Whetton, P., Pittock, A., Fowler, A., and Haylock, M. (1992). Simulated changes in daily rainfall intensity due to the enhanced greenhouse effect: implications for extreme rainfall events. *Climate Dynamics*, **8**, 83–102.

Gray, W.M. (1968). Global view of the origin of tropical disturbances and storms. *Monthly Weather Review*, **96**, 669–700.

Gregory, S. (1989). Macro-regional definition and characteristics of Indian summer monsoon rainfall, 1871–1985. *International Journal of Climatology*, **9**, 465–483.

Hane, C.E. (1986). Extratropical squall lines and rainbands. In *Mesoscale Meteorology and Forecasting*, ed. P. Ray, pp. 359–89. Boston: American Meteorological Society.

Harr, R.D. (1981). Some characteristics and consequences of snowmelt during rainfall in western Oregon. *Journal of Hydrology*, **53**, 277–304.

Hassan, F.A. (1981). Historical Nile floods and their implications for climatic change. *Science*, **212**, 1142–1145.

Hayden, B.P. (1988). Flood climates, In *Flood Geomorphology*, eds. V.R. Baker, R.C. Kochel, and P.C. Patton, pp. 13–26. New York: John Wiley and Sons.

Henderson-Sellers, A., Zhang, H., Berz, G., Emanuel, K., Gray, W., Landsea, C., Holland, G., Lighthill, J., Shieh, S., Webster, P., and McGuffie, K. (1998). Tropical cyclones and global climate change: a post-IPCC assessment. *Bulletin of the American Meteorological Society*, **79**, 19–38.

Hirschboeck, K.K. (1985). *Hydroclimatology of Flow Events in the Gila River Basin, Central and Southern Arizona*. Ph.D. dissertation, University of Arizona, Tucson. 335 pp.

Hirschboeck, K.K. (1987a). Catastrophic flooding and atmospheric circulation anomalies. In *Catastrophic Flooding*, eds. L. Mayer and D.B. Nash, pp. 23–56. Boston: Allen and Unwin.

Hirschboeck, K.K. (1987b). Hydroclimatically-defined mixed distributions in partial duration flood series. In *Hydrologic Frequency Modeling*, ed. V.P. Singh, pp. 199–212. Dordrechet: D. Reidel Publishing Company.

Hirschboeck, K.K. (1988). Flood hydroclimatology. In *Flood Geomorphology*, eds. V.R. Baker, R.C. Kochel, and P.C. Patton, pp. 27–49. New York: John Wiley and Sons.

Hirschboeck, K.K. (1991). Climate and floods. In *National Water Summary 1988–89 – Floods and Droughts: Hydrologic Perspectives on Water Issues*, pp. 67–88. U.S. Geological Survey Water-Supply Paper 2375.

Hirschboeck, K.K. (1996). Floods. In *Encyclopedia of Climate and Weather*, ed. S.H. Schneider, pp. 308–311. Oxford, UK: Oxford University Press.

Hobbs, P.V. (1978). Organization and structure of clouds and precipitation on the mesoscale and microscale in cyclonic storms. *Reviews of Geophysics and Space Physics*, **16**, 741–755.

Horel, J. and Wallace, J. (1981). Planetary-scale atmospheric phenomena associated with the Southern Oscillation. *Monthly Weather Review*, **109**, 813–829.

Houghton, J.T., Meira Filho, L.G., Callander, B.A., Harris, N., Kattenberg, A., and Maskell, K. (eds.) (1996). *Climate Change 1995: The Science of Climate Change IPCC Report*. New York: Cambridge University Press, 572 pp.

House, P.K. and Hirschboeck, K.K. (1997). Hydroclimatological and paleohydrological context of extreme winter flooding in Arizona, 1993. In *Storm-Induced Geological Hazards: Case Histories from the 1992–1993 Winter Storm in Southern California and Arizona*, eds. R.A. Larson and J.E. Slosson, pp. 1–24. Boulder, Colorado: Geological Society of America Reviews in Engineering Geology, v. XI.

Houze, R.A. Jr. (1981). Structures of atmospheric precipitation systems: a global survey. *Radio Science*, **16**, 671–689.

Hoxit, L.R., Maddox, R.A., Chappell, C.F., and Brua, S.A. (1982). Johnstown – Western Pennsylvania storm and floods of July 19–29, 1977. U.S. Geological Survey Professional Paper 1211. Washington, DC: U.S. Government Printing Office.

Huckleberry, G.A. (1993). *Late-Holocene Stream Dynamics on the Middle Gila River, Pinal County, Arizona*. Ph.D. dissertation, University of Arizona, Tucson. 135 pp.

Jarrett, R.D. (1991). Paleohydrology and its value in analyzing floods and droughts. In *National Water Summary 1988–89 – Floods and Droughts: Hydrologic Perspectives on Water Issues*, pp. 105–16. U.S. Geological Survey Water-Supply Paper 2375. Washington, DC.

Junker, N.W., Schneider, R.S., and Scofield, R.A. (1995). The meteorological conditions associated with the Great Midwest Flood of 1993. In *Preprints, 14th Conf. On Weather Analysis and Forecasting*, pp. (J4) 13–17. Washington DC: American Meteorological Society.

Kale, V.S., Mishra, S., and Baker, V.R. (1997). A 2000-year palaeoflood record from Sakarghat on Narmada, Central India. *Journal Geological Society of India*, **50**, 283–288.

Kalnicky, R. (1974). Climatic change since 1950. *Annals of the Association of American Geographers*, **64**, 100–112.

Knox, J.C. (1983). Responses of river systems to Holocene climates. In *Late Quaternary Environments of the United States*. Vol. 2, *The Holocene*, ed. H. Wright, pp. 26–41. Minneapolis: University of Minnesota Press.

Knox, J.C. (1993). Large increases in flood magnitude in response to modest changes in climate. *Nature*, **361**, 430–432.

Knox, J.C. and Kundzewicz, Z.W. (1997). Extreme hydrological events, palaeo-information and climate change. *Hydrological Sciences Journal*, **42**, 765–779.

Laing, A.G. and Fritsch, M.J. (1997). The global population of mesoscale convective complexes. *Quarterly Journal of the Royal Meteorological Society*, **123**, 389–405.

Lins, H.F. (1997). Regional streamflow regimes and hydroclimatology of the United States. *Water Resources Research*, **33**, 1655–1667.

Maddox, R.A., Canova, F., and Hoxit, L.R. (1980). Meteorological characteristics of flash flood events over the western United States. *Monthly Weather Review*, **108**, 1866–1877.

Maddox, R.A., Chappell, C.F., and Hoxit, L.R. (1979). Synoptic and meso-α scale aspects of flash flood events. *Bulletin of the American Meteorological Society*, **60**, 115–123.

Maddox, R.A., Hoxit, L.R., Chappell, C.F., and Caracena, F. (1978). Comparison of meteorological aspects of the Big Thompson and Rapid City flash floods. *Monthly Weather Review*, **106**, 375–389.

Maddox, R.A., Howard, K.A., Bartels, D.L., and Rodgers, D.M. (1986). Mesoscale convective complexes in the middle latitudes. In *Mesoscale Meteorology and Forecasting*, ed. P. Ray, pp. 390–413. Boston: American Meteorological Society.

Matsumoto, S., Ninomiya, K., and Yoshizumi, S. (1971). Characteristic features of the Baiu front associated with heavy rainfall. *Journal of the Meteorological Society of Japan*, **49**, 267–281.

Matthai, H.F. (1969). *Floods of June 1965 in South Platte River Basin, Colorado*. U.S. Geological Survey Water Supply Paper 1850-B. Washington, DC: U.S. Government Printing Office.

McGill, S. and Rockwell, T. (1998). Ages of Late Holocene earthquakes on the Central Garlock Fault near El Paso Peaks, California. *Journal of Geophysical Research*, **103(B4)**, 7265–7279.

McGuirk, J. (1982). A century of precipitation variability along the Pacific Coast of North America and its impact. *Climatic Change*, **4**, 41–56.

McQueen, K., Vitek, J., and Carter, B. (1993). Paleoflood analysis of an alluvial channel in the south-central Great Plains: Black Bear Creek, Oklahoma. *Geomorphology*, **8**, 131–146.

Means, L.L. (1954). A study of the mean southerly wind-maximum in low levels associated with a period of summer precipitation in the middle west. *Bulletin of the American Meteorological Society*, **35**, 166–170.

Meko, D.M. and Stockton, C.W. (1984). Secular variations in streamflow in the Western United States. *Journal of Climate and Applied Meteorology*, **23**, 889–897.

Michaud, J. and Sorooshian, S. (1994). Effect of rainfall sampling errors on simulations of desert flash floods, *Water Resources Research*, **30**, 2765–2775.

Mooley, D. and Shukla, J. (1987). *Characteristics of the Westward-Moving Summer Monsoon Low Pressure Systems Over the Indian Region and Their Relationship with the Monsoon Rainfall*. College Park, Maryland: Center for Ocean-Land-Atmosphere Interactions, University of Maryland.

Namias, J., Yuan, X.C., and Daniel, R. (1988). Persistence of North Pacific sea surface temperature and atmospheric flow patterns. *Journal of Climate*, **1**, 682–703.

Nott, J.F., Price, D.M., and Bryant, E.A. (1996). A 30,000 year record of extreme floods in tropical Australia from relict plunge-pool deposits: implications for future climate. *Geophysical Research Letters*, **23**, 379–382.

O'Connor, J.E., Ely, L.L., Wohl, E.E., Stevens, L.E., Melis, T.S., Kale, V.S., and Baker, V.R. (1994). 4000-year record of large floods on the Colorado River in the Grand Canyon. *Journal of Geology*, **102**, 1–9.

Orlanski, I. (1975). A rational subdivision of scales for atmospheric processes. *Bulletin of the American Meteorological Society*, **56**, 527–530.

Parthasarathy, B., Kumar, K.R., and Munot, A.A. (1993). Homogeneous Indian monsoon rainfall: variability and prediction. *Proceedings of the Indian Academy of Sciences*, **102** (March), 121–155.

Parthasarathy, B. and Pant, G.B. (1984). The spatial and temporal relationships between the Indian summer monsoon rainfall and the Southern Oscillation. *Tellus*, **36A**, 269–277.

Parthasarathy, B., Sontakke, N.A., Monot, A.A., and Kothawale, D.R. (1987). Droughts/floods in the summer monsoon season over different meteorological subdivisions of India for the period 1871–1984. *Journal of Climatology*, **7**, 57–70.

Partridge, J.B. and Baker, V.R. (1987). Paleoflood hydrology of the Salt River, Arizona. *Earth Surface Processes and Landforms*, **12**, 109–125.

Patton, P.C. and Dibble, D. (1982). Archeologic and geomorphic evidence for the paleohydrologic record of the Pecos River in west Texas. *American Journal of Science*, **282**, 97–121.

Pielke, R.A. (1990). *The Hurricane*. New York: Routledge.

Quinn, W.H. (1992). A study of SO-related climatic activity for A.D. 622-1990 incorporating Nile River flood data. In *El Niño: Historical and Paleoclimatic Aspects of the Southern Oscillation*, eds. H.F. Diaz and V. Markgraf, pp. 119–149. Cambridge: Cambridge University Press.

Ramaswamy, C. (1987). *Meteorological Aspects of Severe Floods in India, 1923–1979*. Meteorological Monograph Hydrology, No. 10. New Delhi: Indian Meteorological Department.

Raper, S.C. (1993). Observational data on the relationships between climatic change and the frequency and magnitude of severe tropical storms. In *Climate and Sea-Level Change: Observations, Projections, and Implications*, eds. R.A. Warrick, E.M. Barrow, and T.M.L. Wigley, pp. 192–212. New York: Cambridge University Press.

Redmond, K. and Koch, R. (1991). ENSO vs. surface climate variability in the western United States. *Water Resources Research*, **27**, 2381–2399.

Rodier, J. and Roche, M. (1984). *World Catalog of Maximum Observed Floods*. Wallingford, UK: International Association of Hydrological Sciences Publication 143. 354 pp.

Ropelewski, C. and Halpert, M. (1986). North American precipitation and temperature patterns associated with El Niño-Southern Oscillation (ENSO). *Monthly Weather Review*, **114**, 2352–2362.

Ropelewski, C. and Halpert, M. (1987). Global and regional scale precipitation patterns associated with the El Niño/Southern Oscillation. *Monthly Weather Review*, **115**, 1606–1626.

Rumsby, B. and Macklin, M.G. (1994). Channel and floodplain response to recent abrupt climate change: the Tyne Basin, Northern England. *Earth Surface Processes and Landforms*, **19**, 499–515.

Said, R. (1993). *The River Nile: Geology, Hydrology and Utilization*. Oxford: Pergamon Press.

Schonher, T. and Nicholson, S. (1989). The relationship between California rainfall and ENSO events. *Journal of Climate*, **2**, 1258–1269.

Schroeder, T.A. (1977). Meteorological analysis of an Oahu flood. *Monthly Weather Review*, **105**, 458–468.

Singh, N., Soman, M.K., and Krishna Kumar, K. (1988). Hydroclimatic fluctuations of the upper Narmada catchment and its association with break-monsoon days over India. *Proceedings of the Indian Academy of Sciences (Earth and Planetary Sciences)*, **97**, 87–105.

Slowey, N.C. and Crowley, T.J. (1995). Interdecadal variability of Northern Hemisphere circulation recorded by Gulf of Mexico corals. *Geophysical Research Letters*, **22**, 2345–2348.

Smith, A. (1992). Holocene palaeoclimatic trends from palaeoflood analysis. *Palaeogeography, Palaeoclimatology, Palaeoecology*, **97**, 235–240.

Smith, J.A., Baeck, M.L., and Steiner, M. (1996). Catastrophic rainfall from an upslope thunderstorm in the central Appalachians: the Rapidan storm. *Water Resources Research*, **32**, 3099–3113.

Stuiver, M. and Reimer, P. (1993). Extended ^{14}C data base and revised CALIB 3.0 age calibration program. *Radiocarbon*, **35**, 215–230.

UNESCO (1969–79). *Discharge of Selected Rivers in the World: A Contribution to the International Hydrological Decade*, v. 1–3. Paris: Unesco Press.

UNESCO (1976). *World Catalogue of Very Large Floods*. Paris: Unesco Press.

U.S. Water Resources Council (1981). *Guidelines for Determining Flood Flow Frequency*. Bulletin 17B, Washington, DC: U.S. Water Resources Council.

Ward, R. (1978). Floods, *A Geographical Perspective*. New York: John Wiley. and Sons

Waylen, P.R. and Caviedes, C.N. (1986). El Niño and annual floods on the North Peruvian littoral. *Journal of Hydrology*, **89**, 141–156.

Webb, R.H. and Betancourt, J.L. (1992). *Climatic Variability and Flood Frequency of the Santa Cruz River, Pima County, Arizona*. U.S. Geological Survey Water-Supply Paper 2379. Washington, DC: U.S. Government Printing Office.

Webster, P.J. (1987). The variable and interactive monsoon. In *Monsoons*, eds. J.S. Fein and P.L. Stephens, pp. 269–330. New York: John Wiley and Sons.

Weisman, M.L. and Klemp, J.B. (1986). Characteristics of isolated convective storms. In *Mesoscale Meteorology and Forecasting*, ed. P. Ray, pp. 331–358. Boston: American Meteorological Society.

Wells, L.E. (1990). Holocene history of the El Niño phenomenon as recorded in flood sediments of northern coastal Peru. *Geology*, **18**, 1134–1137.

Whetton, P., Adamson, D., and Williams, M. (1990). Rainfall and river flow variability in Africa, Australia and East Asia linked to El Niño – Southern Oscillation events. *Geological Society of America Symposium Proceedings*, 1, 71-82*ENSO-Related Floods and Droughts*, vol. 1.

Wohl, E.E. and Enzel, Y. (1995). Data for Palaeohydrology. In *Global Continental Palaeohydrology*, eds. K.J. Gregory and L. Starkel, pp. 23–59. New York: John Wiley and Sons.

CHAPTER 3

Floods from Natural and Artificial Dam Failures

Daniel A. Cenderelli
Department of Geological Sciences
University of Alabama

Natural and artificial dam failures have the potential to cause considerable damage and loss of life because the sudden release of stored water generates floods that are substantially larger than rainfall or snowmelt floods. During the past two centuries, artificial and natural dam-failure floods have caused the deaths of at least 30,000 people, millions of dollars in property damage, and severe alteration of channels. Additionally, the failure of mine tailings dams that store toxic waste can have a severe environmental impact on flooded areas. The potential risks and hazards associated with flooding from natural and artificial dam failures are steadily increasing because of increased human settlement and development in close proximity to dams and the construction of artificial dams at less than ideal sites. To minimize the potential risks and hazards associated with flooding from natural and artificial dam failures, predictive equations and models have been developed to better predict the discharges of dam-failure floods. This chapter reviews floods from natural and artificial dam failures to illustrate the downstream effects on stream channels and adjacent communities. Additionally, empirical and physically based equations used to predict the potential peak discharge at the breach for various types of dam failures are reviewed.

Introduction

Throughout the world there are thousands of lakes of various sizes behind natural dams that were impounded by a variety of geological and geomorphic processes, including lava flows, landslides, glaciers, glacial moraines, river ice, and woody debris (Hutchinson, 1957). In the past 5000 years, humans have built artificial dams to create reservoirs for irrigation, water supply, hydroelectric power, flood control, recreation, and containment of mining waste (Smith, 1971; International Commission on Large Dams, 1973; Jansen, 1980; United States Committee on Large Dams, 1994). Because dams commonly store large volumes of water, dam failures can have dire consequences for communities downstream from dams as well as dramatically modifying downstream channels and floodplains. Throughout the world, natural and artificial dam failures have produced floods that have claimed the lives of thousands of people and caused millions of dollars in property damage. Additionally, dam-failure floods have significantly altered downstream channels by erosion and deposition (Gruner, 1963; Jansen, 1980; Costa, 1988; Costa and Schuster, 1988). Floods from dam failures are commonly much larger than snowmelt or rainfall-runoff floods (Costa, 1988) and can travel tens to hundreds of kilometers from the source areas, sometimes inundating areas that were perceived to be safe from flooding. Because of the large volumes of water potentially impounded by natural dams, their failures have produced some of the largest and most destructive floods on Earth (Costa, 1988). This chapter briefly reviews and summarizes the formation and failure mechanisms of natural and artificial dams and the downstream behavior and hazards associated with dam-failure floods. Some notable examples of dam-failure floods are discussed to illustrate the downstream effects on stream channels and adjacent communities. Finally, the most recent updated empirical and physically based equations used to predict the potential peak discharge at the breach for various types of dam failures are discussed and used to evaluate the discharge hydrographs of dam-failure floods.

Floods from Natural Dam Failures

Closed Basins

Throughout the world, tectonic processes at a variety of scales have formed topographically closed basins (Hutchinson, 1957; Smith and Street-Perrott, 1983). Wetter climatic conditions in the western United States during the Pleistocene caused approximately 100 closed basins to be partially or completely filled with water (Smith and Street-Perrott, 1983; Benson and Thompson, 1987), resulting in several lakes overtopping drainage divides and flowing into adjacent basins [e.g., Pleistocene Lake Bonneville (Gilbert, 1878; Malde, 1968; O'Connor, 1993) and Pleistocene Lake Manix (Meek, 1989)]. When Lake Bonneville breached its preflood divide composed of gravel and cobble alluvium, approximately 5000 km^3 of water were released into the Snake River drainage as the outlet incised 100 m into the divide (O'Connor, 1993). The principal failure mechanism was probably groundwater seepage and piping through the alluvial dam associated with a prolonged high stand of Lake Bonneville (O'Connor, 1993). The peak discharge at the divide was estimated to be 1 million m^3/s (O'Connor, 1993); 500 km downstream from the divide, the peak flow attenuated to about 0.9 million m^3/s (Jarrett and Malde, 1987; O'Connor, 1993). The magnitude and duration of this flood is remarkable, especially compared with the largest known historical flood of 1340 m^3/s (Jarrett and Malde, 1987). Erosional and depositional features along the Snake River are evidence of this extreme flood and its effects on the landscape (Malde, 1968; O'Connor, 1993).

Volcanic Dams

Lava flows or pyroclastic flows that extrude into and extend across narrow valleys commonly dam drainages and form lakes. Lava dams have been documented in many volcanic regions of the world, including the western United States (Fuller, 1931; Stearns et al., 1938; Hamblin, 1994), western Canada (Brown, 1969), New Zealand (Cotton, 1944), Japan (Hutchinson, 1957; Costa, 1988), and Mexico (Costa, 1988). Hamblin (1994) gives a detailed description of the formation and destruction of 13 major lava dams between 0.44 and 1.8 million years ago along the Colorado River in the Grand Canyon, USA. Using modern discharge data and geometric estimates of the lava dams, Hamblin (1994) calculated that the lakes behind the lava dams were 60 to 700 m deep and took 2 days to 23 years to fill before overtopping.

Historical accounts of floods produced by the breaching of lava-flow dams are few. Near Asamo Volcano, Japan, a lava-dam failure flood killed 1200 people approximately 200 years ago (as reported by Costa, 1988). During the 28–29 March 1982 eruption of El Chichon Volcano in Mexico, a lava flow dammed the Rio Magdalena, and about 2 months later the lava dam failed and flooded the Rio Magdalena Valley, causing extensive property damage (as reported by Costa, 1988).

Craters or calderas at the summit of volcanoes may contain water. Hutchinson (1957) provides numerous examples of lake-filled craters in different volcanic regions of the world. Waythomas et al. (1996) noted that these lakes can be hazardous if the crater or caldera rim fails and releases the impounded water down steep volcanic slopes. Such an event occurred about 3400 years ago in southwestern Alaska, when a large lake situated in the Aniakchak Caldera breached the caldera rim and rapidly released an estimated 3.7×10^9 m^3 of water with a peak discharge of about 1 million m^3/s (Waythomas et al., 1996). The resulting flood created a deep gorge in the caldera rim, deposited boulder bars and produced scabland topography along the flood route, and formed a large alluvial fan where the valley widens from 800 to 5000 m approximately 21 km from the caldera (Waythomas et al., 1996). A more recent example of the failure of a lake-filled crater occurred at Mount Ruapehu, New Zealand, on 24 December 1954 (Houghton et al., 1987). The collapse of an ash and ice segment of the crater rim caused the rapid release of a large volume of water from the lake, which quickly transformed into a debris flow that caused considerable erosion and deposition in the Whangaehu Valley as well as destroying a railroad bridge, derailing a train, and killing 151 people.

Landslide Dams

Landslides occur in a wide range of geomorphic settings throughout the world; however, the frequency and magnitude of landslides are higher in mountainous regions where earthquakes and extreme precipitation may initiate slope failures. The term landslide has been used to describe a wide range of slope processes such as debris slide, debris flow, and debris avalanche that move large volumes of sediment from valley slopes to the valley bottom. If the sediment transported by a landslide extends across the valley bottom and is of sufficient height, the drainage system can be blocked, causing a lake to form upstream.

Based on the classification of 184 landslide dams, Costa and Schuster (1988) showed that 85% of the landslide dams result from mass movements from valley slopes or tributaries that block the main valley drainage. Not all landslide dams fail, but when they do fail it is usually soon after formation. Costa and Schuster (1988) examined 73 cases in which landslide dams failed and noted that 22% failed within 1 day of formation, 50% failed within 10 days of formation, 56% failed within 1 month of formation, 80% failed within 6 months of formation, and 85% failed within 1 year of formation. They concluded that the duration of a landslide dam before failure depends on (1) the height, length, and geotechnical characteristics of the dam; (2) the rate of seepage through the landslide dam; and (3) the rate of inflow into the landslide-dam lake. Although landslide dams are typically composed of poorly sorted, uncompacted sediment with high rates of seepage, most landslide dams fail when the dam is overtopped by the water it impounds (Schuster and Costa, 1986; Costa and Schuster, 1988). When the rate of seepage through the landslide dam is less than the rate of inflow into the landslide-dam reservoir, the lake level behind the landslide dam rises and eventually overtops the dam. Once water begins to flow over the surface of the landslide dam, the dam is quickly breached because erosion weakens the structure (Schuster and Costa, 1986; Costa and Schuster, 1988).

The floods resulting from landslide dam failures can be geomorphically destructive as well as lethal to communities (see Schuster and Costa, 1986; Costa, 1988; and Costa and Schuster, 1988 for references). One of the most catastrophic landslide-dam failure floods occurred on the Indus River in the Karakoram Himalayas, northern Pakistan, in early June 1841 (Mason, 1929; Code and Sirhindi, 1986). Mason (1929), who summarized earlier descriptions and accounts, reported that this flood resulted from breaching of an approximately 300-m-high earthquake-triggered landslide dam. The resulting flood traveled hundreds of kilometers downstream, inundating hundreds of villages in the Indus River valley and killing thousands of people. The Karakoram Himalaya region in northern Pakistan is highly prone to landslides and landslide damming; Hewitt (1998) documented at least 73 rock slides or rock avalanches that formerly dammed the Indus River and its tributaries.

Evans (1986) described 18 landslides that dammed rivers in narrow valley segments of the Coast Mountains in western Canada. Ten of these landslide dams eventually failed and released large volumes of water downstream. The resulting floods destroyed property, eroded and deposited large quantities of sediment, and severely affected salmon migration routes. Two of these floods (6 July 1891, North Pacific Cannery; 28 October 1921, Britannia Beach) inundated nearby villages, and 50 to 77 people died.

In the mountainous terrain of south-central China, Tianche et al. (1986) described the formation of eight landslides this century. Six of the eight landslide dams failed, and the other two would have failed if the dams had not been stabilized to prevent breachings. The most devastating of these landslide-dam failures occurred in 1933 when three landslide dams, triggered by an earthquake on 25 August 1933 and within 5 km from each other, systematically failed by overtopping and subsequent breaching. When the furthest downstream landslide dam was breached, 400 million m^3 of water were released into the Min River Valley, causing extensive flooding that resulted in the deaths of 2423 people.

Glacier Dams

Lakes associated with glaciers can form in a variety of ways (see Hutchinson, 1957; Marcus, 1960; Costa and Schuster, 1988; and Walder and Costa, 1996 for examples). Small lakes can form on the surface of glaciers in closed depressions. These supraglacial lakes are typically small and not particularly hazardous (Costa and Schuster, 1988). Large, subaerial glacier-dammed lakes form when major valley glaciers block tributary drainages and at sites where tributary-valley glaciers extend into glacier-free valleys and obstruct the drainage. Large bodies of water can also form beneath and within glaciers (Young, 1980; Haeberli, 1983; Driedger and Fountain, 1989; Björnsson, 1992).

During the last glacial maximum, continental ice sheets blocked drainages, resulting in ice-dammed lakes. Many of these ice dams failed and released large volumes of water, producing continental scale erosional and depositional features in areas such as the Channeled Scabland region of the northwestern United States (Bretz, 1969; Baker, 1973), the Båldakatj area of Swedish Lapland (Elfström, 1987), and the Atlay Mountains of south-central Siberia (Baker et al., 1993). Flood features in the Channeled Scabland region were created when an ice dam that impounded Pleistocene Lake Missoula repeatedly failed (Baker, 1973; Waitt, 1985). Approximately 2200 km^3 of water were released in each flood, with the peak discharge of largest flow exceeding 17 million m^3/s near the dam failure (O'Connor and Baker, 1992). The failure of an ice-dammed lake in the Atlay Mountains region resulted in a flood with an estimated peak discharge of 18 million m^3/s (Baker et al., 1993). These floods were likely the largest floods to have occurred on Earth (Baker et al., 1993).

Floods from glacier-dammed lakes can occur in most mountainous glaciated regions, albeit with smaller discharges than those during the Pleistocene when glaciation was more extensive. Such floods have been reported in Iceland (Thorarinsson, 1939, 1953; Arnborg, 1955; Björnsson, 1992), Norway (Aitkenhead, 1960; Tvede, 1989), Switzerland (Haeberli, 1983); Pakistan (Mason, 1929; Hewitt, 1982), China (Yongjian and Jingshi, 1992; Xiangsong, 1992), Peru (Lliboutry et al., 1977), Argentina (Nichols and Miller, 1952), the United States (Richardson, 1968; Post and Mayo, 1971; Sturm and Benson, 1985; Mayo, 1989; Driedger and Fountain, 1989), and Canada (Marcus, 1960; Jackson, 1979; Young, 1980; Clarke, 1982; Blown and Church, 1985; Desloges et al., 1989; Desloges and Church, 1992; Evans and Clague, 1994; Clague and Evans, 1997). Refer to Walder and Costa (1996) for an extensive summary list of glacial-lake outburst floods.

The earliest records of outburst floods from glaciers are from Iceland in the 13th century (Thorarinsson, 1939). Consequently, the Icelandic term *jökulhlaup* is commonly used to describe the sudden and rapid release of water impounded by, within or under, or on the surface of a glacier. Jökulhlaups are typically several times larger than annual snowmelt or rainfall floods. Although most jökulhlaups occur in remote regions, human occupation and development in glaciated regions have increased the hazards from jökulhlaups in some areas. For example, jökulhlaups have caused loss of life as well as eroded and deposited large quantities of sediment that have damaged farms, buildings, bridges, and roads in Iceland (Björnsson, 1992). In Switzerland, there have been approximately 100 jökulhlaups since the 1500s, killing almost 200 people and damaging many alpine villages (Haeberli, 1983).

The failure of glacier dams that form subaerial, subglacial, and englacial lakes can generate dangerous floods. Glacier-dammed lakes can drain either through subglacial tunnels, by overtopping the dam at the valley margin and glacier contact, or by complete rupture of the ice dam (Thorarinsson, 1939, 1953; Post and Mayo, 1971; Nye, 1976; Young, 1980; Clarke, 1982; Haeberli, 1983; Sturm and Benson, 1985; Sturm et al., 1987; Costa, 1988; Costa and Schuster, 1988; Driedger and Fountain, 1989; Walder and Costa, 1996). Most subaerial glacier-dammed lakes drain through one or more tunnels beneath the glacier. Walder and Costa (1996) reviewed the mode of drainage of 117 subaerial glacier-dammed lakes and found that 100 of the jökulhlaups drained through subglacial tunnels. In general, as a glacier-dammed lake fills to near capacity, water begins to drain beneath the glacier when the hydrostatic pressure equals the weight of glacier dam at a point(s) of weakness between the glacier and valley floor or within the glacier (Marcus, 1960; Nye, 1976; Clarke, 1982). Once water begins to flow beneath the glacier, energy dissipated by the flowing water melts the ice and enlarges the subglacial passageways, which in turn causes the magnitude of the jökulhlaup to steadily increase (Marcus, 1960; Nye, 1976; Clarke, 1982). This positive feedback mechanism continues until the lake nearly or completely drains. When drainage ceases, most subglacial tunnels are blocked by roof collapse or sealed by the plastic flow of the glacier (Marcus, 1960; Nye, 1976; Clarke, 1982; Björnsson, 1992), resulting in refilling of the lake. This process of lake drainage and refilling can occur on an annual or irregular basis and frequency is strongly controlled by glacier activity, closure of the subglacial tunnels, and rate of lake refilling.

Englacial and subglacial lakes are particularly hazardous because their formation, growth, and releases of water cannot be observed (Haeberli, 1983; Driedger and Fountain, 1989; Björnsson, 1992). In Switzerland, 30% to 40% of the jökulhlaups were from releases of water from englacial or subglacial lakes (Haeberli, 1983). Haeberli (1983) suggested that the jökulhlaups from subglacial and englacial lakes in Switzerland occurred in steep glaciers with slopes ranging from 7.5° to 30°. Similarly, all the outburst floods from glaciers on Mt. Rainer, USA, originate from englacial or subglacial lakes beneath steeply sloped glaciers (Richardson, 1968; Driedger and Fountain, 1989). Driedger and Fountain (1989) hypothesized that stepped bedrock terrain beneath the steep glaciers forms subglacial cavities that store and periodically release water. In Iceland, subglacial lakes form beneath glaciers that overlie geothermal areas and active volcanoes (Björnsson, 1992). Continuous melting of ice by geothermal heat at the base of the glacier causes subglacial lakes to form in cavities beneath the ice, which may be periodically released through subglacial tunnels, causing jökulhlaups (Björnsson, 1992).

Many researchers (e.g., Young, 1980; Haeberli, 1983; Sturm and Benson, 1985; Blown and Church, 1985; Evans and Clague, 1994; Clague and Evans, 1997) have reported that the frequency and magnitude of jökulhlaups have decreased during the past 150 to 200 years because of overall thinning and retreat of alpine glaciers since the culmination of the Little Ice Age. Nevertheless, jökulhlaups are still common and typically have discharges several times larger than snowmelt or rainfall floods. Consequently, jökulhlaups can cause substantial channel changes because of the large quantities of sediment that are eroded, transported, and deposited during a given flood (e.g., Richardson, 1968; Post and Mayo, 1971; Hewitt, 1982; Haeberli, 1983; Desloges et al., 1989; Mayo, 1989; Björnsson, 1992; Desloges and Church, 1992; Clague and Evans, 1997). For example, Desloges and Church (1992) showed that the 1984 jökulhlaup from Ape Lake in British Columbia, Canada, had an estimated peak discharge of 1500 m^3/s and exceeded the mean annual climatic flood in the region by a factor of 2.5. The average width of the main channel widened by a factor of 2 because of lateral erosion, and floodplains aggraded up to 0.75 m from sediment deposited by the jökulhlaup.

In some instances, jökulhlaups transform from a water flood to a hyperconcentrated flow or debris flow where readily available sediment is eroded by and incorporated into the flood flow, thereby increasing the sediment concentration of the flow (Richardson, 1968; Jackson, 1979; Desloges and Church, 1992). Jackson (1979) documented that three of the four debris flows that buried railways and highways in British Columbia, Canada, were initiated by drainage of a lake impounded by Cathedral Glacier. The 1978 debris flow was the most damaging because it deposited approximately 200,000 m^3 of sediment, which blocked railway and highway routes and derailed a freight train.

Moraine Dams

Moraine-dammed lakes form where terminal or recessional moraines block drainages; generally the blocked drainage is meltwater from the now smaller glacier that formed the moraine (Hutchinson, 1957). During the past 200 to 250 years, numerous moraine-dammed lakes formed in glaciated mountainous regions because of the retreat of glaciers since the Little Ice Age. The Little Ice Age was a period of cooler climate between 1500 and 1850 when many alpine glaciers advanced to their furthest positions in the Holocene (Grove, 1988). Since the mid-19th century culmination of the Little Ice Age, the global annual mean temperature has increased at least 1°C (Jones et al., 1986) and most alpine glaciers have thinned and retreated from their advanced Little Ice Age positions in various mountainous regions of the world (Costa and Schuster, 1988; O'Connor and Costa, 1993; Evans and Clague, 1994). Many moraines deposited by alpine glacial advances during the Little Ice Age have impounded meltwater from the present-day glaciers.

Moraines that form alpine lakes are prone to failure because the moraines (1) have steep, unvegetated slopes that are unstable; (2) consist of poorly sorted sediment that may be noncompacted and noncohesive; and (3) may have a melting ice core with only a thin cover of sediment (Costa and Schuster, 1988). Lake levels behind moraines are controlled by inflow rates from glacier meltwater, snowmelt, and precipitation and the outflow rates of seepage through the moraine or by channels flowing across the top of the moraine (Costa and Schuster, 1988). The most common failure mechanism of moraine dams is by overtopping and subsequent breaching (Costa and Schuster, 1988). Overtopping and breaching can occur when excessive runoff from glacier meltwater, snowmelt, and/or intense precipitation causes the rate of inflow into the lake to exceed the rate of outflow from the lake by seepage through the moraine dam or by overflow in channels on the surface of the moraine, which may lead to surface and/or subsurface erosion of the moraine, causing it to fail (Costa and Schuster, 1988). Moraine-dammed lakes are susceptible to avalanches or falls adjacent to glacier fronts and steep rock faces, which can produce waves that overtop and breach the moraine dam (Blown and Church, 1985; Costa and Schuster, 1988). Less common failure mechanisms of moraine dams are melting of frozen moraine sediment (Yesenov and Degovets, 1979) and melting of buried ice in a moraine (Fushimi et al., 1985). Melting of frozen sediment and ice causes the moraines to collapse and then be overtopped and breached by water from the lake (Yesenov and Degovets, 1979; Fushimi et al., 1985).

Moraine-dam failures and resulting floods have been reported in most glaciated mountainous regions (see Costa and Schuster, 1988, for

Figure 3.1. Upstream view of the Imja Khola valley (foreground) and the Nare Khola valley (background) showing the erosional and depositional effects of the moraine dam failure flood that occurred in the Mount Everest region, Nepal, in 1977. The river valley bottom in the foreground is between 50 to 70 m wide. Photo was taken in October 1995.

additional references). Moraine-dam failure floods have discharges substantially larger than those from snowmelt or rainfall and can travel tens to hundreds of kilometers downstream as water floods, hyperconcentrated flows, or debris flows. These floods have locally modified channel systems and, in some cases, have killed thousands of people and destroyed property. In the Cordillera Blanca of Peru, 4 of the 10 large moraine-dam lakes that formed around the turn of the century drained between 1938 and 1950 when the moraine dams failed (Lliboutry et al., 1977). The most devastating of these floods occurred on 13 December 1941 when excessive precipitation caused Lake Palcacocha to overtop and breach the moraine dam (Lliboutry et al., 1977). The resulting flood destroyed a portion of the city of Huaraz and killed more than 6000 people (Lliboutry et al., 1977). Another lethal moraine-dam failure flood that occurred in Peru on 17 January 1945 destroyed the town of Chavin de Huantar and killed a large, but unknown, number of people (Lliboutry et al., 1977).

There have been at least eight moraine-dam failures in the Canadian Rockies since 1945 (see Evans and Clague, 1994, for references). Of these eight moraine-dam failures, four produced water floods and four produced debris flows (Evans and Clague, 1994). Because the outburst floods in the Canadian Rockies occurred in remote settings, there were no human fatalities (Evans and Clague, 1994). However, each of the outburst floods modified channels for tens of kilometers downstream from the breached moraine. For example, on 19 July 1983 an ice fall from the Cumberland Glacier plunged into Nostetuko Lake, generating waves that overtopped and breached the moraine dam. Approximately 6.5 million m^3 of water were released with an estimated peak discharge of 9100–11,000 m^3/s (Blown and Church, 1985). Along the outburst-flood route, the flood eroded, transported, and deposited large quantities of sediment and woody debris (Blown and Church, 1985). As noted by Evans and

Figure 3.2. Upstream view of the Bhoti Kosi Valley showing the erosional and depositional effects of the moraine-dam failure flood that occurred in the Mount Everest region, Nepal in 1985. The valley in the foreground is 200 to 250 m wide and is located about 16 km downstream from the breached moraine. A nearly completed hydroelectric dam located near the fan on the right valley margin was destroyed by the flood. Photo was taken in October 1995.

Clague (1994), the channel was severely altered by this outburst flood and may take decades to return to its preflood condition.

In the Cascade Range of Oregon and Washington, 11 debris flows have been triggered by the breaching of Little Ice Age moraines (O'Connor et al., 1993, 1994, in press). The abundance of unconsolidated sediment and steep slopes caused the water released from the lakes to transform into debris flows. Several of the larger debris flows had discharges in excess of 300 m^3/s and traveled as far as 10 km from the breached moraine before transforming into sediment-laden floods. Existing moraine-dammed lakes in the region are susceptible to breaching and the subsequent flood(s) are a potential hazard to downstream communities.

Eleven outburst floods from the failure of moraine dams have been reported in the Himalaya Mountains of Nepal and Tibet, China, since 1935 (Watanabe et al., 1994; Mool, 1995). These floods have caused loss of life and have destroyed bridges, trails, pastures, farms, and hydroelectric plants for tens of kilometers along the flood routes (Mool, 1995). Extensive erosion of channels and valley slopes along outburst-flood routes removed much of the vegetation, making these areas susceptible to accelerated erosion during the monsoon season (Figures 3.1 and 3.2) (Ives, 1986; Vuichard and Zimmermann, 1987). On 4 June 1985, a moraine-dammed lake failed when an ice avalanche into the lake generated waves that overtopped and breached the moraine dam, causing 5 million m^3 of water to drain from the lake (Vuichard and Zimmermann, 1986, 1987; Ives, 1986). The resulting flood had an estimated peak discharge of 2350 m^3/s 4 km from the breach and was 60 times larger than annual snowmelt and monsoon floods (Cenderelli and Wohl, 1997; Cenderelli, 1998). Twenty-eight kilometers downstream from the breach, the peak discharge of the flood attenuated to 1375 m^3/s but was still 7 times larger than annual snowmelt and monsoon floods (Cenderelli and Wohl, 1997; Cenderelli, 1998). Using photographs taken before and after the flood, Vuichard and Zimmermann

(1987) estimated that 3 million m^3 of sediment were eroded from the channel and valley slopes along the upper 40 km of the flood route.

River Ice Jams

Cold winter environments in countries such as Canada, Russia, the northern United States including Alaska, Norway, Sweden, and Poland have rivers that can freeze and accumulate ice along segments of the river, forming ice dams (see Hoyt and Langbein, 1955; Bolsenga, 1968; Smith, 1980; Church, 1988; and Gerard and Davar, 1995, for references). Ice jams can cause severe flooding that can damage and destroy roads, bridges, buildings, houses, and farmland. For example, Carlson et al. (1989) reported that in 1988 the U.S. Army Corps of Engineers Cold Regions Research and Engineering Laboratory estimated that ice jam floods cause approximately $100 million of damage on an annual basis in the United States.

Ice jams can be classified as freeze-up jams or breakup jams. Freeze-up jams are typically small accumulations of ice that occur immediately downstream from steep, high-energy reaches during the early winter months (Smith, 1980; Church, 1988). Freeze-up jams usually disintegrate slowly and are not particularly hazardous (Smith, 1980; Church, 1988; Beltaos, 1995). In contrast, breakup jams can be quite massive and are potentially hazardous when they release suddenly during late winter or early spring thaws (Smith, 1980; Church, 1988; Beltaos, 1995). Breakup jams typically form along channels at sharp bends, shallow reaches, reductions in gradient, and natural or artificial constrictions where flushing ice forms a jam that thickens and lengthens in the upstream direction (Smith, 1980; Beltaos, 1995). Factors controlling the disintegration of breakup jams are not fully understood, but ice conditions, river discharge, thermal effects, and ice competence are all considered important (Beltaos, 1995). The primary mechanism that triggers the release of breakup jams is hydrostatic pressure from water and ice, which build up behind the breakup jam (Smith, 1980; Beltaos, 1995).

The release of breakup jams and the flushing of surface ice from the river during breakup commonly is rapid and, in some instances, causes downstream flooding (Smith, 1980; Church, 1988; Beltaos, 1995). The sudden release of a breakup jam is similar to that of a dam failure and the severity of both the breakup ice jam and subsequent release depends on (1) the thickness, resistance, and strength of the ice jam, which determines the volume of water stored upstream from the obstruction; (2) the intensity and duration of warm weather; and (3) the intensity of runoff from snowmelt and/or rainfall, which determines the magnitude of discharge during the breakup of the ice jam (Smith, 1980; Beltaos, 1995).

Breakup ice jam floods sometimes generate water levels along the river that are considerably higher than floods produced from snowmelt or rainfall. In the United States, there were significant breakup ice jam floods along the Missouri River and its tributaries in North Dakota, South Dakota, Iowa, and Nebraska in April 1952 (United States Weather Bureau, 1954; Hoyt and Langbein, 1955). The combination of above-average snowcover, antecedent soil conditions, the lateness and intensity of spring warming, abundant breakup ice, and the thickness and competency of the ice together contributed to severe flooding on the Missouri River and its tributaries (United States Weather Bureau, 1954). Breakup ice jams formed at numerous locations along the Missouri River, causing water and ice to build up behind the jams. When the breakup jams released, tremendous surges of water and ice at unprecedented depths inundated towns and farms. The breakup ice jams systematically formed and released from the upper reaches of the Missouri River to the lower reaches of the Missouri River. The breakup ice jam floods along the Missouri River caused $170 million of damage and killed nine people (United States Weather Bureau, 1954). At many locations along the Missouri River, the recorded discharges from the 1952 flood were the largest since such data have been collected (United States Weather Bureau, 1954).

More recently, in March 1993, the combination of breakup ice jams and rapid snowmelt caused severe flooding along the lower Platte River in Nebraska (White and Kay, 1996). The floods caused an estimated $25 million of damage, which included damage to roads, bridge abutments, flood levees and dikes, agricultural areas, residential homes, and commercial and industrial buildings (White and Kay, 1996). Flooding was particularly severe where two flood levees were overtopped and breached (White and Kay, 1996). Breakup ice jams occur almost every year on the Aroostook River and the St. John River in northern Maine; however, severe flooding occurs only when conditions enable large volumes of water to accumulate upstream from the ice jam and when the ice jam breakup is rapid (Wuebben et al., 1995). In the 20-year period from 1973 to 1992, there have been six instances of severe flooding and damage from breakup ice jams (Wuebben et al., 1995). In April 1991, breakup ice jam flooding along the St. John River and the Aroostook River caused $14 million of damage to roads, bridges, and homes (Wuebben et al., 1995).

Breakup ice jam floods not only inundate large areas of land causing severe property damage and, in some cases, loss of life but can also have a substantial geomorphic effect on channel morphology by eroding channel beds and banks (Smith, 1980). When breakup ice jam floods exceed bankfull conditions, trees along the channel margin can be severely scarred by the ice (Smith, 1980). However, breakup ice jam floods may be essential to ecosystems such as the Mackenzie Delta in northwestern Canada (Marsh and Hey, 1989; Gerard and Davar, 1995). Marsh and Hey (1989) have suggested that the yearly breakup ice jam floods on channels throughout the Mackenzie Delta are critical to maintaining the hydrologic regime and biological productivity of the numerous lakes in the delta.

Organic Dams (Beaver Dams)

Beavers can alter channel processes and morphology by constructing large barriers of woody debris across streams to create ponds (Hutchinson, 1957; Butler, 1995). Beaver dams are typically constructed of logs, branches, and mud that are secured or weighted down with rocks (Woo and Waddington, 1990; Butler, 1995). Beaver dams alter channel morphology by creating a large pond upstream from the dam and a distinct step downstream from the dam. Beavers live in a variety of environments. For example, the habitat of the North American beaver (*Castor canadensis*) ranges from the subarctic wetlands in Canada to the margins of deserts in Mexico (Woo and Waddington, 1990; Butler, 1995). Beaver dams are generally limited to small channels with low discharges (Butler and Malanson, 1994; Butler, 1995). Beaver dams vary in size but typically have lengths between 15 and 70 m and widths from 1 to 2 m (Hutchinson, 1957; Butler, 1995). Beaver dams are typically stable structures that can exist for several decades or centuries when they are actively maintained by beavers (Butler and Malanson, 1994; Butler, 1995). Once the beavers abandon a dam, however, the dam gradually decomposes and collapses (Stock and Schlosser, 1991; Butler, 1995).

Beaver dams are not impervious structures and water flows through or over the surface of these structures without damaging the integrity of the dam in most situations. Although very few beaver dams fail abruptly, excessive precipitation or snowmelt can cause beaver dams to fail (Butler, 1989; Stock and Schlosser, 1991). The failure mechanisms of beaver dams are unknown, but failure probably occurs when normal seepage and overflow processes through the dam are not capable of handling the rapid rise in water level behind the dam during the precipitation event. Because beaver-dam failures commonly occur in conjunction with flooding from excessive precipitation or snowmelt, the contribution of the beaver-dam failures to flooding is often overlooked. Additionally, beavers repair their damaged dams rapidly after failure, making it difficult to establish any relationship between beaver-dam failure and subsequent flooding (Butler, 1989).

There are a few studies that have recognized the flooding caused by the failure of beaver dams (Butler, 1989; Stock and Schlosser, 1991). Butler (1989) documented five cases of beaver-dam failure and subsequent flooding in northeastern Georgia and western South Carolina, USA. Of these five, the most significant was in September 1987 when the collapse of a beaver dam on Millstone Creek, Georgia, caused a rapid increase in stream discharge and inundated a car that was attempting to ford the stream. Four of the six occupants were killed when the flood waters swept them downstream as they tried to climb to safety. This flood caused extensive erosion and deposition along the flood route; boulders up to 1 m in diameter were mobilized by the flood.

Beaver-dam failures can have a significant impact on aquatic communities immediately downstream from where the dam failed. Stock and Schlosser (1991) showed that the collapse of a beaver dam and subsequent flood caused a 90% decline in benthic insect density in a pool and riffle sequence just below the dam. After 60 days, benthic insect densities had recovered in the pool but remained 40% below predam failure levels in the riffle. Stock and Schlosser (1991) illustrated that the abundance of fish in this pool-riffle sequence declined after the flood because of the short-term reduction in the benthic insect density.

Floods from Artificial Dams

Humans have built dams for the past 5000 years for irrigation, water supply, hydroelectric power, flood control, recreation, and containment of mining waste (Smith, 1971; International Commission on Large Dams, 1973; Jansen, 1980; United States Committee on Large Dams, 1994). Every dam that is built has an inherent risk of failure. Since the 12th century there have been approximately 2000 artificial dam failures (Jansen, 1980). During the 19th and 20th centuries there have been 212 significant artificial dam failures (International Commission on Large Dams, 1973; Jansen,

Table 3.1. *Percent failure rate of various dam types in the period between 1831 and 1965 (data from International Commission of Large Dams, 1973)*

Dam type	Number of dams built	Number of dams that failed	Failure rate (%)
Embankment			
Earth	4551	121	2.66
Rock	285	13	4.56
Total	**4836**	**134**	**2.77**
Concrete			
Arch	566	7	1.24
Buttress	373	7	1.88
Gravity	2271	40	1.76
Total	**3210**	**54**	**1.68**
Combined total	**8046**	**188**	**2.34**

1980; MacDonald and Langridge-Monopolis, 1984; Costa, 1988). Thirty-three dam-failure floods have resulted in the deaths of at least 13,000 people (Jansen, 1980).

There are two basic types of dams: embankment dams and concrete or masonry dams. Embankment dams are the most common type of dam found throughout the world (Table 3.1). Embankment dams can be constructed of natural material that is similar in size, or the dam may consist of an internal impermeable barrier that is overlain by earthfill or rockfill (Wahlstrom, 1974). Concrete or masonry dams can be subdivided into gravity dams, arch dams, and buttress dams. Gravity dams depend on their weight for stability (Wahlstrom, 1974). Arch dams are thin, curved structures made of concrete reinforced with steel rods. Arch dams are designed to distribute horizontal stresses within the dam toward the abutments for stability and are typically built in narrow valleys (Wahlstrom, 1974). A buttress dam consists of a thick slab of concrete that rests on a succession of upright buttresses that hold the dam in place (Wahlstrom, 1974). Based on the number of embankment and concrete dams constructed and the number that have failed during the period between 1831 and 1965, arch dams have the lowest failure rate followed by gravity dams, buttress dams, earthfill dams, and rockfill dams (Table 3.1).

The International Commission on Large Dams (1973) reported that, for dams constructed higher than 15 m, there were 87 failures between 1900 and 1965 (Table 3.2). Of the 64 embankment dams that failed, 38% failed when the dam was overtopped by excessive flood waters or inadequate spillways; 30% failed because of piping and seepage through the fill material; 16% failed because of improper design that led to cracking, deformation, or settlement of the dam structure; 9% failed because of foundation defects such as uneven settlement, weathered and fractured rock, seepage, and excess pore pressure; and 7% failed because of surface erosion of the embankments or slope failure on oversteepened embankments (Table 3.2; International Commission on Large Dams, 1973). Of the 23 concrete dams that failed, 48% failed because of foundation defects such as uneven settlement, weathered and fractured rock, seepage, excess pore pressures, and abutment shearing along bedding planes; 30% failed when the dam was

Table 3.2. *Summary of failure mechanisms of embankment dams built and that failed in the period between 1900 and 1965 that were higher than 15 m (data from International Commission on Large Dams, 1973)*

Dam name, location	Type of dam	Date failed	Dam height (m)	Failure mechanism(s)
Briseis, Australia	earth	1929	27	overtopping
La Regadera, Colombia	earth	1937	37	foundation defects
Bila Desna, Czech Rep.	earth	1916	17	overtopping
Ahraura, India	earth	1954	26	piping and seepage
Kaddam, India	earth	1958	41	overtopping
Kaila, India	earth	1965	26	piping and seepage
Kedar Nala, India	earth	1964	20	piping and seepage
Kharagpur, India	earth	1961	24	overtopping
Pagara, India	earth	1943	30	overtopping
Panshet, India	earth	1961	49	improper design
Ashizawa, Japan	earth	1956	15	overtopping
Heiwaika, Japan	earth	1951	20	overtopping
Ogayarindo T, Japan	earth	1963	19	overtopping
Bon accord, S. Africa	earth	1937	18	improper design
Nizhne Svirskaya, USSR	earth	1935	28	improper design
Ovcar Banja, Yugoslavia	earth	1965	27	overtopping
Apishapa, USA	earth	1923	35	piping and seepage
Baldwin Hills, USA	earth	1963	80	piping and seepage
Balsam, USA	earth	1929	18	overtopping
Black Rock, USA	rock	1909	21	piping and seepage
Black Rock, USA	rock	1936	21	piping and seepage
Bully Creek, USA	rock	1925	38	overtopping
Calaveras, USA	earth	1938	67	surface erosion
Colley Lake, USA	earth	1960	19	unknown
Colley Lake, USA	earth	1963	19	overtopping
Corpus Chris, USA	earth	1930	32	piping and seepage
Fred Burr, USA	earth	1948	18	overtopping
Goodrich, USA	earth	1956	15	piping and seepage, overtopping
Graham Lake, USA	earth	1923	34	piping and seepage, overtopping
Greenlick, USA	earth	1904	19	piping and seepage
Hatchtown, USA	earth	1900	19	surface erosion
Hatchtown, USA	earth	1914	19	piping and seepage
Hatfield, USA	earth	1911	15	overtopping
Hebron, USA	earth	1914	17	foundation defects
Hebron, USA	earth	1942	17	overtopping
Jackson Bluff, USA	earth	1957	17	improper design
Jenning D3, USA	earth	1963	21	foundation defects
Jenning D16, USA	earth	1964	17	foundation defects
Julesburg, USA	earth	1910	16	surface erosion
Kenray, USA	earth	1962	16	improper design
Lake Francis, USA	earth	1935	34	piping and seepage
Lake Toxaway, USA	earth	1916	19	piping and seepage
Lake Waco, USA	earth	1947	21	overtopping
Litt Deer CR, USA	earth	1963	26	improper design
Littlefield, USA	rock	1929	37	piping and seepage
Lookoutshoal, USA	earth	1916	25	overtopping
Lyman, USA	earth	1915	20	foundation defects
Mammoth, USA	earth	1917	23	overtopping
McMahon Gul, USA	earth	1926	17	overtopping
Owen, USA	earth	1914	17	piping and seepage
Point O Rock, USA	earth	1915	26	surface erosion
Point O Rock, USA	earth	1927	26	overtopping
Schaeffer, USA	earth	1921	30	foundation defects

Table 3.2. *(Continued)*

Dam name, location	Type of dam	Date failed	Dam height (m)	Failure mechanism(s)
Sepulveda Ca, USA	earth	1914	20	overtopping
Sinker Creek, USA	earth	1943	21	improper design
Stockton, USA	earth	1950	33	improper design
Swift, USA	rock	1964	48	piping and seepage, overtopping
Terrace, USA	earth	1957	48	piping and seepage
Wachusett, ND, USA	earth	1907	25	improper design
Wahiawa, USA	rock	1921	41	overtopping
Wesl E Seale, USA	earth	1965	35	overtopping
Wiscon Dells, USA	rock	1911	18	overtopping
Woodrat Knob, USA	earth	1961	26	piping and seepage
Wyandotte, USA	earth	1937	28	improper design

overtopped by excessive flood waters or inadequate spillways; and 22% failed because of improper design, substandard construction material, or poor maintenance (Table 3.3; International Commission on Large Dams, 1973).

All artificial dam failures due to foundation defects have occurred within the first 12 years of the dam's existence, indicating inadequate consideration of geological properties below and adjacent to the dam

Table 3.3. *Summary of failure mechanisms of concrete dams built and that failed in the period between 1900 and 1965 that were higher than 15 m (data from International Commission on Large Dams, 1973)*

Dam name, location	Type of dam	Date failed	Dam height (m)	Failure mechanism(s)
Chichester, Australia	gravity	1923	41	improper design, inadequate concrete
Scott Falls, Canada	gravity	1923	15	poor construction
Malpasset, France	arch	1959	66	foundation defects
Kundli, India	gravity	1925	45	improper design, poor construction
Low Khajuri, India	gravity	1949	16	overtopping
Tigra, India	gravity	1917	28	foundation defects
Gleno, Italy	buttress	1923	49	foundation defects
Zerbino, Italy	gravity	1935	16	overtopping
Komoro, Japan	buttress	1928	16	overtopping
Vega de Tera, Spain	buttress	1959	34	improper design
Ashley Dam, USA	buttress	1909	18	foundation defects
Bayless, USA	gravity	1910	15	foundation defects
Bayless, USA	gravity	1911	15	foundation defects
Elwha, USA	buttress	1912	34	foundation defects
Gallinas, USA	arch	1957	29	poor construction, poor maintenance
Lake Hemet, USA	arch	1927	45	overtopping
Lake Lanier, USA	arch	1926	19	foundation defects
Moyie River, USA	arch	1926	16	overtopping
Owerholser, USA	buttress	1923	17	overtopping
St. Francis, USA	gravity	1928	56	foundation defects
Stony River, USA	buttress	1914	16	foundation defects
Vaighn Creek, USA	arch	1926	19	foundation defects
Wilbur, USA	gravity	1940	21	overtopping

(International Commission on Large Dams, 1973; Jansen, 1980; Costa, 1988). Additionally, many of the foundation failures were during the initial filling of the reservoir, exposing weaknesses and flaws in the foundation (Jansen, 1980; Costa, 1988). For example, the Teton Dam in Idaho, USA, failed on 5 June 1976 during its initial filling (Jansen, 1980). The principal cause of failure of this 93-m-high earthfill embankment dam was attributed to water seepage into the moderately to intensely jointed volcanic rocks along the right abutment (Jansen, 1980). During reservoir filling, water flowed rapidly through these joints to the foundation of the dam, which in turn initiated piping erosion at the base of the foundation, which then expanded and progressed through the main body of the dam until the dam completely failed (Jansen, 1980). Once the dam failed, approximately 215 million m^3 of water drained through the breach in 143 min, resulting in a peak discharge of 65,000 m^3/s 1.5 km downstream from the breach (Ray and Kjelstrom, 1978). A canyon immediately below the dam was substantially modified by the flood as nearly all the soil and vegetation was eroded from the floor of the canyon for a distance of 8 km (Ray and Kjelstrom, 1978). The failure and subsequent flooding killed 11 people and caused at least $400 million of damage.

Another example of a foundation defect that led to the failure of a dam occurred in southern France on 2 December 1959 (International Commission on Large Dams, 1973; Jansen, 1980). The Mapasset Dam, a 66-m-high concrete arch dam, collapsed when the left abutment failed (Jansen, 1980). The left abutment failed along shear surfaces in the foliated gneiss as the reservoir was filled for the first time (International Commission on Large Dams, 1973; Jansen, 1980). Approximately 22 million m^3 of water were released when the dam collapsed (Jansen, 1980). The subsequent flood had an estimated peak discharge of 28,320 m^3/s (Costa, 1988) and killed 421 people in the 11-km distance between the reservoir and the Mediterranean Sea (Jansen, 1980).

Approximately 80% of artificial dam failures attributable to overtopping occurred during the first 30 years after construction (International Commission on Large Dams, 1973; Costa, 1988). As pointed out by Costa (1988), most overtopping failures were due to underestimating the magnitude and frequency of extreme floods in the reservoir basin. For example, the Machhu II Dam in northwestern India failed on 11 August 1979 during an intense monsoonal storm when the rising waters in the reservoir overtopped the earthfill embankment despite the fact that all the spillway gates were opened to near capacity (Jansen, 1980). The rate of inflow into the reservoir was three times greater than the design inflow rate and exceeded the rate of outflow through the spillway gates by approximately 1000 to 5000 m^3/s (Jansen, 1980). Approximately 100 million m^3 of water were drained from the reservoir when the earthfill embankments were overtopped and breached (Jansen, 1980). The resulting flood inundated and damaged 68 villages along the Machhu River. The city of Morvi, located 9 km from the reservoir and inhabited by 75,000 people, was inudated by flood waters with depths of up to 6 m, resulting in the destruction of most buildings and the deaths of at least 2000 people (Jansen, 1980).

The worst artificial dam failure in the United States was in western Pennsylvania in 1889 when the South Fork Dam failed. This embankment dam failed when the water in the reservoir overtopped and breached the dam, releasing approximately 19 million m^3 of water (Jansen, 1980). The dam spillway and outlet pipe were inadequate to control the water level rise in the reservoir during an intense rainfall (Jansen, 1980). The resulting dam-failure flood destroyed a large portion of the city of Johnstown, located 14 km downstream from the dam, and killed 2209 people.

Approximately 90% of embankment dam failures due to piping and seepage occurred within 30 years of construction (International Commission on Large Dams, 1973; Costa, 1988). Piping and seepage through the embankment or at the base of the embankment may reflect (1) inadequate compaction of the embankment during construction, (2) insufficient measures to prevent or control seepage beneath or through the dam, (3) inadequate monitoring of seepage characteristics and water pressures in the embankment and its foundation to identify changes in seepage properties and conditions, and (4) not properly maintaining outlet valve structures (Jansen, 1980). The Nanaksagar Dam in northern India failed on 8 September 1967 because of uncontrolled seepage and erosion through the 16-m-high earthfill embankment (Jansen, 1980). As seepage and erosion progressed, the embankment began to settle until it was overtopped and breached (Jansen, 1980). Approximately 210 million m^3 of water were released from the reservoir, destroying 32 villages along the Deoha River and killing approximately 100 people (Jansen, 1980). Evacuation of the Deoha River Valley before the dam failure significantly reduced the loss of life (Jansen, 1980).

Another example of piping that resulted in a dam failure occurred in Rocky Mountain National Park, USA, on 15 July 1982 (Jarrett and Costa, 1986). Deterioration of lead caulking between the outlet pipe and gate valve caused seepage into the 8-m-high earthfill embankment Lawn Lake Dam (Jarrett and Costa, 1986). This seepage led to piping and erosion, which weakened the embankment and caused it to collapse, releasing 830,000 m^3 of water (Jarrett and Costa, 1986). The subsequent flood had an estimated peak discharge of 500 m^3/s at the point of release and caused extensive erosion and deposition for 7 km along the steep, mountainous channel immediately downstream from the dam (Figure 3.3) (Jarrett and Costa, 1986).

Figure 3.3. Upstream view of the Roaring River valley approximately 2 to 2.5 km downstream from the Lawn Lake Dam that failed in 1982. Note the scoured slopes on the left bank and boulder deposition across the valley bottom. Photo was taken in August 1995.

Although the discharge of the flood attenuated to 200 m^3/s at approximately 11 km downstream from the breach, the flood caused the Cascade Lake Dam, a 5-m-high concrete gravity dam, to overtop and fail. The combined flood waters had an estimated peak discharge of 450 m^3/s at the second dam failure and caused extensive flooding in a small town 10 km downstream from the Cascade Dam (Jarrett and Costa, 1986). The combined effects of these dam-failure floods resulted in the deaths of three people and $31 million in damage (Jarrett and Costa, 1986). Additionally, studies by Bathurst et al. (1986) in 1984 and 1985 and by Bathurst and Ashiq (1998) in 1995 along the upper 7 km of the flood route documented enhanced bedload transport that substantially exceeded predam failure bedload transport. Bathurst and Ashiq (1998) suggested that bedload transport rates would remain higher than predam failure conditions until the exposed unconsolidated sediment on the steep valley slopes (Figure 3.3) is revegetated and stabilized.

Mine tailings dams are classified separately from other artificial dams because they have unique design and construction characteristics (United States Committee on Large Dams, 1994). Mine tailings are solid mineral processing wastes produced during mining activities and are often used to impound water and waste from mining activity to control pollution. The primary characteristics that distinguish mine tailings dams from other artificial dams are (1) the embankments of mine tailings dams are often raised during their operational phase to increase storage capacity and (2) the integrity of the mine tailings dam embankments may be altered over time because of the changes in the properties of the mining waste that makes up the embankments (United States Committee on Large Dams, 1994). The failure of mine tailings dams that store mining waste can have a severe environmental impact on areas flooded. The United States Committee on Large Dams (1994) documented 185 dam incidents involving mine tailings throughout the world since 1917. Of these incidents, 106 have been failures that caused flooding downstream from the dam, reduced water quality, and contaminated sediment and vegetation. In some instances, the failure of these dams produced floods that caused loss of life and extensive property damage. For example, on 26 February 1972 the Buffalo Creek tailings dam in West Virginia failed, releasing 500,000 m^3 of water and coal mining waste sludge downstream (Davies, 1973; Jansen, 1980). The flood inundated the entire Buffalo Creek Valley, extensively damaged several towns causing $50 million in damage, and killed 125 people (Davies, 1973; Jansen, 1980). Soon after the failure of the Buffalo Creek tailings dam, federal and state dam safety legislation was established in the United States (United States Committee on Large Dams, 1994).

There are numerous case studies documenting the environmental impact that contaminated flood waters from the failure of mine tailings dams have had on flooded areas (see United States Committee on Large Dams, 1994, for references). For example, the failure of the Cities Service Dam at Fort Meade, Florida, caused large quantities of phosphatic clay residue to be released into the Peace River (United States Committee on Large Dams, 1994). The reasons for the failure were not determined. Although not toxic to humans, the high concentration of phosphatic clay residue in the Peace River resulted in a massive fish kill because the residue coated the gills of fish and caused them to suffocate (United States Committee on Large

Dams, 1994). Mine tailings dam failures can potentially have long-term effects on groundwater and surface-water resources. The failure of a uranium tailings dam at Church Rock, New Mexico, on 16 July 1979 resulted in the release of approximately 350,000 m^3 of contaminated water and approximately 100,000 kg of hazardous solid waste into the Rio Puerco River (United States Representatives Committee on Interior and Insular Affairs, 1980; United States Committee on Large Dams, 1994). For a distance of approximately 162 km along the Rio Puerco River, sediment and vegetation were contaminated by uranium, other radionuclides, and metals in the flood waters (Van Metre and Gray, 1992; United States Committee on Large Dams, 1994). Further discussion on the environmental impact of contaminated flood waters on channels can be found in the chapter by Finley (Chapter 7, this volume).

Peak Discharge Estimates and Discharge Hydrographs of Dam-Failure Floods

Peak Discharge Estimates of Dam-Failure Floods

Direct discharge measurements of dam-failure floods are rare because streams below the breached impoundments are typically ungaged and, even if gaged, the gaging stations are usually destroyed by the flood. In the absence of gaging stations, the peak discharge of a flood can be determined by paleoflood techniques such as regime-based paleoflood reconstructions, paleocompetence studies, or paleostage indicator estimates. Refer to Baker in Chapter 13 of this volume for a more detailed discussion of these techniques. Physically based models have been developed to determine breach formation, lake drainage, and the peak discharge at the breach based on lake volume, water depth, lake geometry, breach geometry, and the rate of erosion at the breach (Froehlich, 1987; Fread, 1989; Waythomas et al., 1996). Once the breach hydrograph is determined, flood routing models for unsteady flow can be used to evaluate and estimate the peak discharge and travel time of the flood surge as it progresses downstream (e.g., Jarrett and Costa, 1986; Fread, 1989; and O'Connor, 1993).

Regression equations have been developed from past dam failures to predict the potential peak discharge of dam-failure floods based on the volume of water drained from the lake (V), the drop in lake level depth (d), or the product of the volume of water drained and the drop in lake level (Vd) (MacDonald and Langridge-Monopolis, 1984; Costa, 1988; Costa and Schuster, 1988; Walder and Costa, 1996; Walder and O'Connor, 1997). Table 3.4 summarizes updates to the most recent regression equations for glacier dams (Walder and Costa, 1996) and landslide dams, moraine dams, and artificial dams (Walder and O'Connor, 1997). Many of the peak discharges used to develop these predictive equations were based on direct measurements of V and d and indirect measurements or estimates of peak discharge some distance downstream from the breach with no correction for flood attenuation (Walder and O'Connor, 1997). Although the regression equations in Table 3.4 provide a rapid estimate of the potential peak discharges of dam-failure floods, they usually underestimate the peak discharge at the breach and at best provide only an order-of-magnitude estimate of the probable peak discharge (Costa, 1988; Waythomas et al., 1996;

Table 3.4. *Regression equations for predicting the probable peak discharge (Q_p) at the breach for various types of dam failures*

Type of failure	Equation	r^2	s	n	Notes
Artificial	$Q_p = 2.2V^{0.46}$	0.60	0.49	35	a
Landslide	$Q_p = 3.4V^{0.46}$	0.73	0.53	19	a
Moraine	$Q_p = 0.060V^{0.69}$	0.63	0.46	10	a, b
Glacier (subglacial)	$Q_p = 0.0050V^{0.66}$	0.70		26	c
Glacier (overtopping)	$Q_p = 2.5V^{0.44}$	0.58		6	c
Artificial	$Q_p = 3.9d^{2.29}$	0.80	0.34	35	a
Landslide	$Q_p = 24d^{1.73}$	0.53	0.70	19	a
Moraine	$Q_p = 210d^{0.92}$	0.11	0.72	10	a, b
Artificial	$Q_p = 0.97(Vd)^{0.43}$	0.70	0.42	35	a
Landslide	$Q_p = 1.9(Vd)^{0.40}$	0.76	0.50	19	a
Moraine	$Q_p = 0.30(Vd)^{0.49}$	0.51	0.53	10	a, b

Symbols are as follows: Q_p, peak discharge (m^3/s); V, volume of water drained from lake (m^3); d, lake level decline (m); r^2, coefficient of correlation; s, standard error of the estimate; n, sample size.

Notes: a. This regression equation is different than the one presented by Walder and O'Connor (1997). Walder and O'Connor (1997) did not correct for the bias associated with back transformations of logarithmic coefficients to predict the peak discharge (Q_p). The predicted value of Q_p is underestimated with the uncorrected antilogged regression because the equation describes the geometric mean of the distribution and not the arithmetic mean (Miller, 1984; Ferguson, 1986; Desloges et al., 1989; Walder and Costa, 1996). With the data compiled by Walder and O'Connor (1997), the regression equations were determined and corrected for the bias associated with antilogged regression by the equation $a' = a\exp(2.65s^2)$ where a' is the corrected power law intercept, a is the uncorrected power law intercept, exp is the natural exponent, and s is the standard error of the estimate.

b. Data provided by Walder and O'Connor (1997) were modified at the two Nepal sites to reflect recent discharge data presented by Cenderelli (1998). The peak discharge of the 1977 moraine-dam failure flood was changed from 1100 m^3/s to 1850 m^3/s and the peak discharge of the 1985 moraine-dam failure flood was changed from 1600 m^3/s to 2350 m^3/s.

c. Regression equation is from Walder and Costa (1996). The equation they presented corrected for the bias associated with back transformations of logarithmic coefficients.

Walder and O'Connor, 1997). The predictive value of the regression equations in Table 3.4 is limited because they do not consider the influence of (1) the breach erosion rate on the peak discharges of landslide-, moraine-, and artificial-dam failures (Walder and O'Connor, 1997) or (2) lake temperature, creep closure, and tunnel enlargement by melting on the peak discharge for glacier-dam failures (Walder and O'Connor, 1997). In other words, factors in addition to depth and volume are important in determining the peak discharge at a breach for a given dam-failure flood.

Recent studies by Clarke (1982), Walder and Costa (1996), and Walder and O'Connor (1997) developed physically based equations to predict the potential peak discharge at the breach for various types of dam failures. For subglacial drainage of lakes, Clarke (1982) developed the following equations to predict the peak discharge (Q_p) of lakes draining through tunnels beneath the glacier:

$$Q_p = V_o(-\partial\phi/\partial s)^{0.69} N^{-0.5}(\rho_i L')^{-1.33} \tag{1}$$

$$Q_p = 0.424[k_w(\theta_1 - \theta_i)V_o(\rho_i L')^{-1}]^{0.8}[(-\partial\phi/\partial s)N^{-1}]^{0.42}\left[2\rho_{w\Pi}^{-0.5}\eta_w^{-1}\right]^{0.64} \tag{2}$$

where V_o is the initial lake volume, ϕ is fluid potential, s is distance along drainage tunnel, N is a circular tunnel constant, ρ_i is the density of ice,

L' is the effective latent heat of fusion for ice, k_w is the thermal conductivity of water, θ_l is the temperature of the lake, θ_i is the temperature of the ice, ρ_w is the density of water, and η_w is the viscosity of water. Equation 1 is used when the thermal energy of the lake does not contribute to ice melting, and equation 2 is used when the thermal energy of the lake significantly contributes to ice melting. Equations 1 and 2 consider the influence of lake volume, lake geometry, lake temperature, creep closure, and tunnel enlargement by melting on peak discharge. Using equation 2, Clarke (1982) calculated a peak discharge of 500 m^3/s at the breach for the 1978 jökulhlaup from Hazard Lake, Yukon Territory, Canada. This discharge was only 21% lower than the peak discharge predicted by the decline in lake level with time. Clarke (1982) suggested that the peak discharge at the breach was underestimated by equation 2 because the equation does not consider the effects of tunnel enlargement from the release of gravitational energy (Clarke, 1982).

Walder and Costa (1996) developed physically based equations to determine the peak discharge at the breach of glacier dams that fail when water overtops the ice impoundment:

$$Q_p = 0.27C_1(p)[f_R\rho_w V_o(\rho_i L')^{-1}]^{0.5} g h_i^{1.5} \quad (3)$$

$$Q_p = 0.32C_2(p)[k_w(\theta_l - \theta_i)V_o(\rho_i L')^{-1}]^{0.6}\left[\rho_w\eta_w^{-1}\right]^{0.48} g^{0.44} h_i^{0.52} \quad (4)$$

where the product of $C_1(p)$ describes the shape of the breach (C_1) and the shape of the lake (p) when the thermal energy of the lake is low, the product of $C_2(p)$ describes the shape of the breach (C_2) and the shape of the lake (p) when the thermal energy of the lake is high, f_R is the drag coefficient, g is gravitational acceleration, and h_i is the initial lake elevation. Equation 4 is used when the temperature of the lake is greater than 0°C. The underlying assumption of equations 3 and 4 is that breach widening is controlled by the melting rate of the ice. When h_i and V_o are unknown, Walder and Costa (1996) recommended that h_i be approximated as 90% of the glacier dam thickness and that V_i be determined by its hypsometric relation with h_i. Walder and Costa (1996) used equation 4 to estimate the peak discharges of the multiple floods that occurred from 1958 to 1966 when the glacier-dammed Lake George, USA, overtopped and breached its ice impoundment. The predicted peak discharge, using equation 4, was on average 40% less than the measured peak discharge of the Lake George floods, indicating that the physically based equation provides a reasonable estimate for predicting the peak discharge of a flood from the overtopping and breaching of a glacier dam.

For landslide, moraine, and artificial dams, Walder and O'Connor (1997) developed the following predictive equations for estimating the peak discharge at the breach:

$$Q_p = 1.51(g^{0.5}d^{2.5})^{0.06}(kVd^{-1})^{0.94} \quad \text{when } kV(gd)^{-0.5}d^{-3} < 0.6 \quad (5)$$

$$Q_p = 1.94g^{0.5}d^{2.5}(D_c d^{-1})^{0.75} \quad \text{when } kV(gd)^{-0.5}d^{-3} \gg 1 \quad (6)$$

where g is gravitational acceleration, k is the erosion rate at the breach, V is lake volume drained, d is the lake level decline during the flood, and D_c is the height of the dam crest relative to the dam base. The erosion rate at

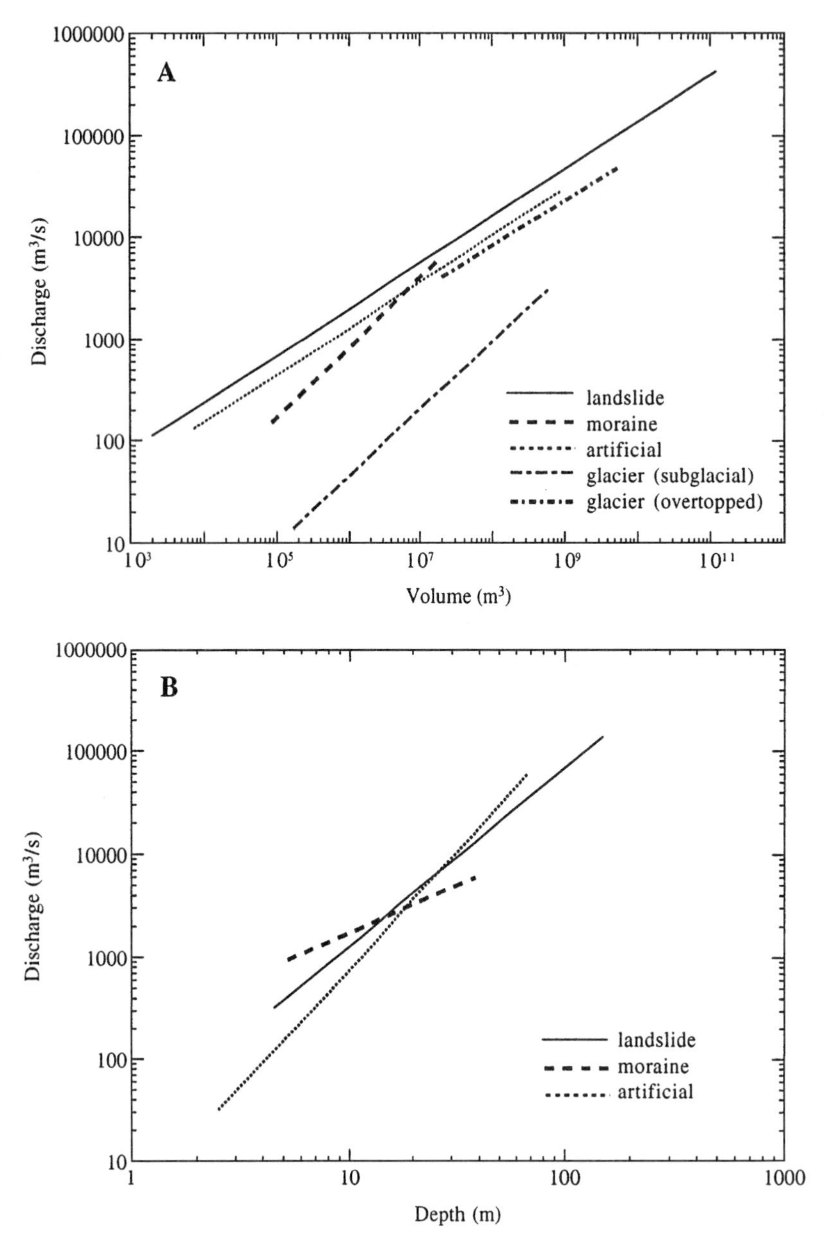

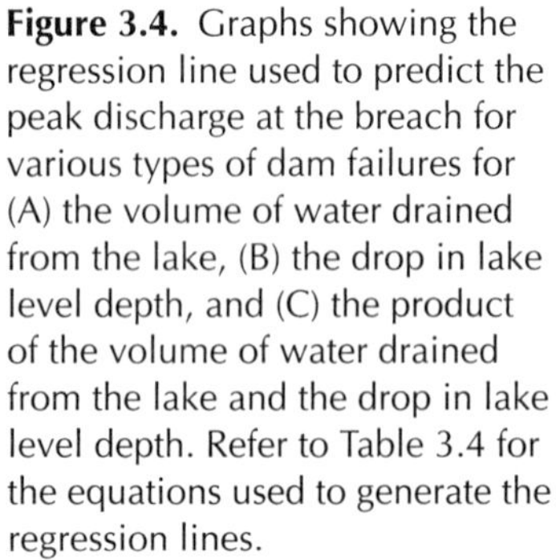

Figure 3.4. Graphs showing the regression line used to predict the peak discharge at the breach for various types of dam failures for (A) the volume of water drained from the lake, (B) the drop in lake level depth, and (C) the product of the volume of water drained from the lake and the drop in lake level depth. Refer to Table 3.4 for the equations used to generate the regression lines.

the breach is determined by

$$k = (w_i - b_f)t^{-1} \qquad (7)$$

where w_i is the initial lake level relative to the base of the dam, b_f is height to the breach relative to the base of the dam, and t is the time for the breach to form. A slow rate of breach formation or small lake volume are indicated by $kV(gd)^{-0.5}d^{-3}$ values less than 0.6, whereas values of $kV(gd)^{-0.5}d^{-3}$ greater than 1 indicate a fast rate of breach formation or a large lake volume. The limitation of equations 5 and 6 is quantifying the time for the breach to form (t) in equation 7, which determines the erosion rate of the breach (k). For example, Vuichard and Zimmermann (1987) reported that the 1985 moraine dam failure flood in Nepal took approximately 30 to 60 min to reach its maximum stage. This time is assumed to be the time it took for the breach to form. The moraine-dammed lake had a volume of 5 million m^3 and a maximum lake depth of 18 m. Using equation 5, the predicted peak discharge at the breach was 4300 m^3/s for t = 0.5 h and 2250 m^3/s for t = 1 h. At 7 km below the breached moraine, Cenderelli (1998), using a step-backwater model in conjunction with paleostage evidence, estimated that the peak discharge of the flood was 2350 m^3/s. This suggests that the time for the breach to form (t) was closer to 1 h than 30 min, but it also demonstrates the sensitivity of equation 5 to the variable t, which determines the breach erosion rate (k) in equation 7. As pointed out by Walder and O'Connor (1997), the uncertainties in determining breach erosion rates are due to the lack of detailed observations of breach formation and the variety of processes involved in breach formation, including sediment entrainment, knickpoint retreat, seepage and piping, and mass movements.

Discharge Hydrographs of Dam-Failure Floods

Despite the limitations of the regression equations discussed in the previous section, they do provide useful information on general trends in the peak discharge at a breach for various types of dam-failure floods (Figure 3.4). In general, the regression equations predicting the peak discharge at the breach for landslide, moraine, artificial, and overtopped glacier dams for a given lake volume are similar (Figure 3.4a). The regression equations for predicting the peak discharge for a given lake volume show that the slopes (the exponent in the regression equations provided in Table 3.4) for landslide-, artificial-, and overtopped glacier-dam failures are similar and that for a similar lake volume, landslide-dam failures will have a larger peak discharge than artificial- and overtopped glacier dam-failure floods (Figure 3.4a). For moraine-dam failures, the slope of the regression is steeper than the slopes for landslide-, artificial-, and overtopped glacier-dam failures. The peak discharges of moraine-dam failure floods are lower than the peak discharges of landslide-, artificial-, and overtopped glacier-dam failure floods at small lake volumes but discharge values for higher lake volumes are similar (Figure 3.4a). Glacier-dammed lakes that drain subglacially have peak discharges that are substantially lower than those for landslide, moraine, artificial, and overtopped glacier dams for a given lake volume (Figure 3.4a). This relation indicates that the processes involved in the subglacial drainage of a glacier dam are different than dams that fail from breaching of the impoundment.

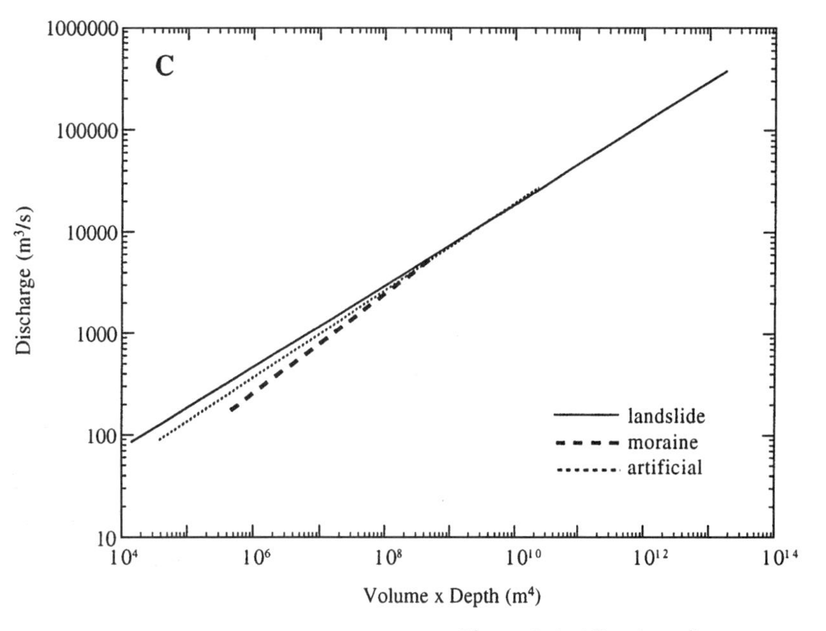

Figure 3.4. *(Continued).*

The regression equations for predicting the peak discharge based on the depth that the lake level decreased show similar trends for landslide-, artificial, and moraine-dam failures (Figure 3.4b). The relationship between peak discharge and depth for moraine-dam failures is not statistically significant. The regression equations for predicting the peak discharge from the product of the lake volume drained and the depth that the lake level dropped (Vd) show similar relations for landslide-, artificial-, and moraine-dam failures (Figure 3.4c).

Discharge hydrographs produced by dam-failure floods are a function of (1) the breach-forming mechanism, (2) the rate at which the reservoir drains, (3) distance from the point of release, and (4) the channel and valley geometry along the flood route. With the exception of jökulhlaups produced from subglacial tunnel drainage, dam-failure floods are characterized by a rapid rise in discharge (Figure 3.5). The rate at which discharge increases, or the steepness of the rising limb, is strongly controlled by the failure mechanism and the physical characteristics of the dam. Although each dam failure is unique with respect to these factors, in close proximity to the dam there is a general relationship between the shape of the discharge hydrograph and the specific type of dam (Figure 3.5).

For example, when concrete gravity dams and arch dams fail, the collapse is usually sudden because large sections or the entire structure fail

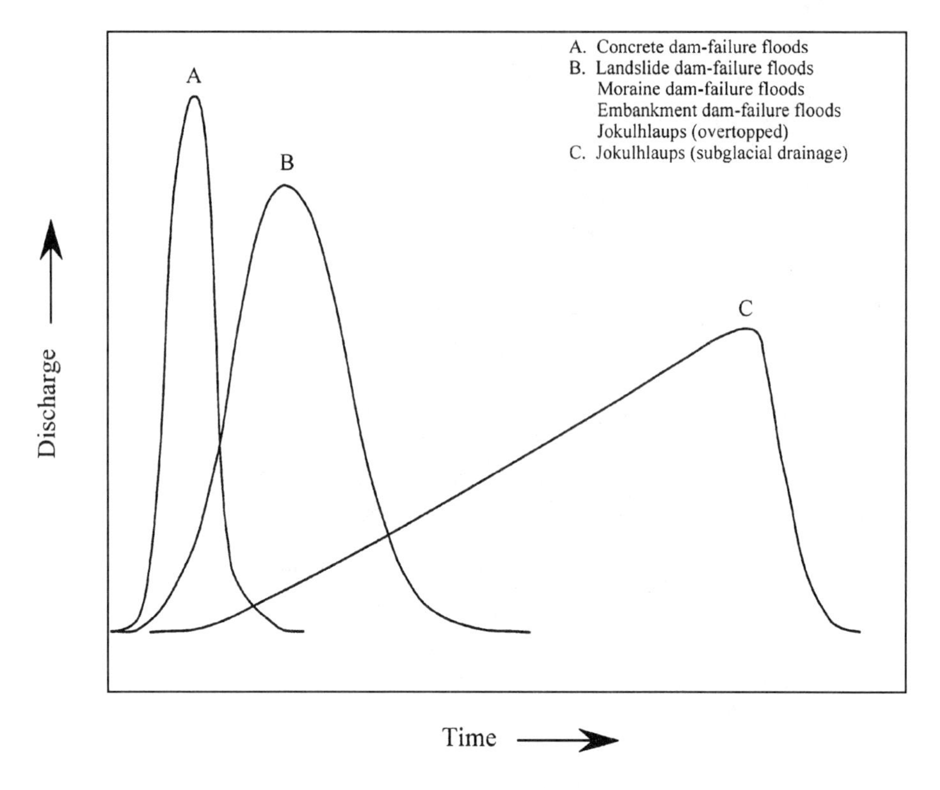

Figure 3.5. Hypothetical dam-failure flood hydrographs.

simultaneously (Jansen, 1980; MacDonald and Langridge-Monopolis, 1984). This sudden and usually complete failure causes the water stored behind the dam to drain rapidly. The discharge hydrographs produced by arch and gravity dam-failure floods are characterized by steep rising and falling limbs, sharp peak discharges, and short flood durations compared with other dam-failure floods (Figure 3.5). In contrast, artificial embankment dams, landslide dams, moraine dams, and glacier dams fail more slowly than concrete dams because they are breached by progressive erosion of the barrier by water flowing through or over the dam (Jansen, 1980; Langridge-Monopolis, 1984; Costa and Schuster, 1988). Usually the rate of breaching increases with time as the structure weakens and, consequently, the rate of drainage increases with time. The discharge hydrographs of artificial embankment-, landslide-, moraine-, and overtopped glacier-dam failure floods have less-steep rising and falling limbs, a lower and longer duration peak discharge, and a longer flood duration compared with the hydrographs of concrete dam-failure floods (Figure 3.5).

Landslide dams are usually much wider and have less-steep slopes than moraine or artificial-embankment dams (Costa and Schuster, 1988). Although there is more material to be eroded during the breaching of a landslide dam than a moraine dam or artificial-embankment dam (Costa and Schuster, 1988), a landslide-dam failure flood will have a similar peak discharge compared with floods from moraine and artificial-embankment failures with similar lake volumes and depths (Figures 3.4c and 3.5). This suggests that the material of landslide dams is more easily eroded because it undergoes minimal compaction during formation compared with moraine dams or artificial embankment dams.

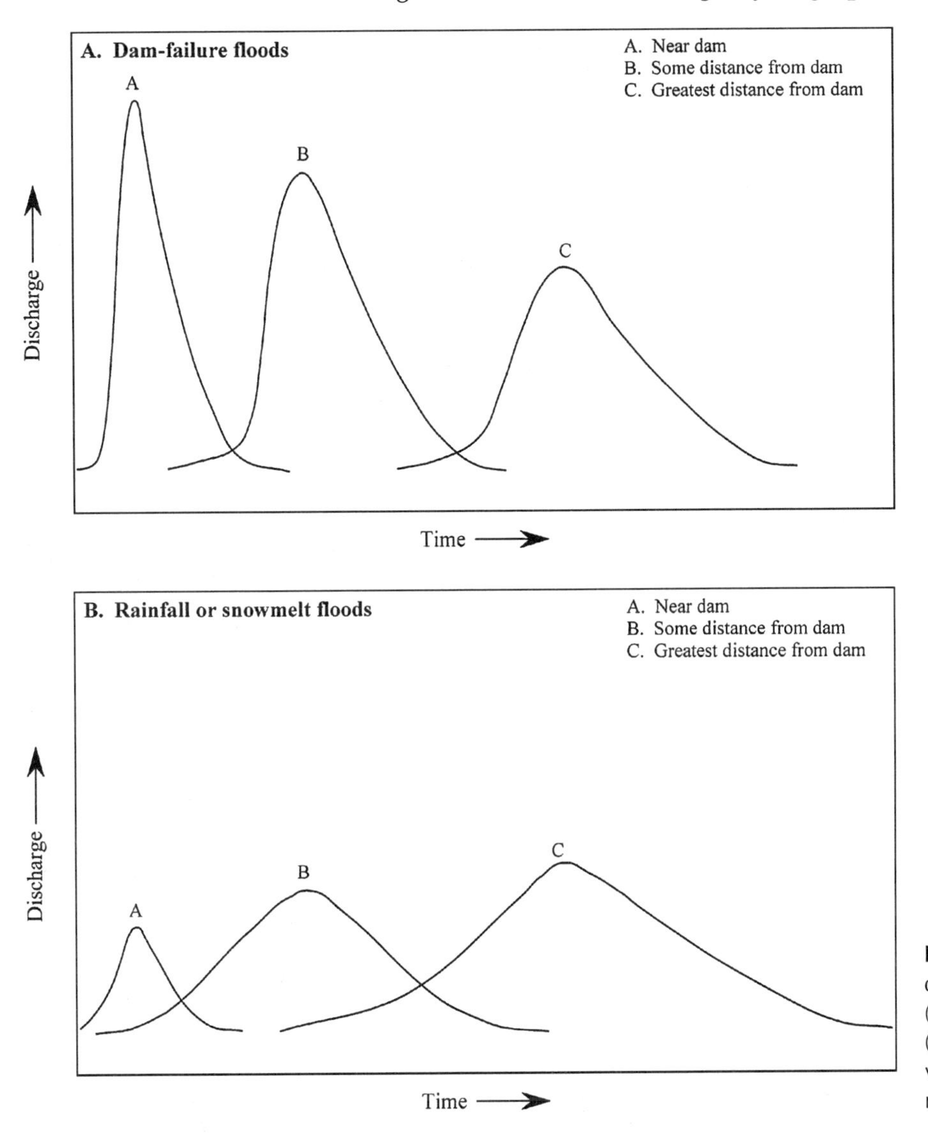

Figure 3.6. Hypothetical discharge hydrographs of (A) dam-failure floods and (B) snowmelt or rainfall floods at various distances along the flood route.

Jökulhlaups that are produced by the progressive enlargement of subglacial tunnels during the drainage of the lake will typically have discharge hydrographs that have gradual rising limbs, a sharp peak, and a steep falling limb (Figure 3.6) (Thorarinsson, 1953; Young, 1980; Clarke, 1982; Haeberli, 1983; Costa and Schuster, 1988; Walder and Costa, 1996). Compared with other artificial and natural dam failures with similar water volumes, the drainage of the lake through subglacial tunnels results in a flood with a lower peak discharge but a longer flood duration (Figures 3.4a and 3.5). There are deviations from the typical subglacial tunnel jökulhlaup. For example, the rising and falling limbs during a jökulhlaup may deviate from the overall trend in increasing or decreasing discharge with respect to time. These deviations may reflect icebergs blocking tunnels or the collapse of tunnels during the jökulhlaup (Sturm et al., 1987). Björnsson (1992) documented a gradual falling limb during several subglacial tunnel jökulhlaups in Iceland, which may reflect drainage from beneath the ice in combination with subglacial tunnel drainage.

The discharge hydrograph of a river ice jam flood is characterized by steep rising and falling limbs and a sharp peak if the release of the ice jam is sudden (e.g., United States Weather Bureau, 1954; Marsh and Hey, 1989). A less-sudden release of the ice jam will result in a flood hydrograph with more gradual rising and falling limbs, a lower peak discharge, and a longer flood duration. There is little hydrologic information on the rate at which beaver dams fail and subsequently drain. However, based on descriptions of rapid increases in flow depth downstream from a beaver-dam failure (Butler, 1989; Stock and Schlosser, 1991), one can infer that the failure of a beaver dam is probably abrupt. Thus, the discharge hydrograph of a beaver-dam failure flood would be characterized by steep rising and falling limbs and a sharp peak.

Natural and artificial dam-failure floods differ from floods produced by rainfall or snowmelt runoff because water is derived from a point source instead of the cumulative contribution of water from the entire drainage basin. As a result, discharge hydrographs of dam-failure floods along the flood route, regardless of the type of dam failure, will have different characteristics than the discharge hydrographs of rainfall or snowmelt floods (Figure 3.6). Hypothetical discharge hydrographs of a dam-failure flood at various distances from the dam failure show that (1) the peak discharge attenuates but is of longer duration with greater distance from the dam failure, (2) the rising and falling limbs of the hydrographs become less steep with greater distance from the dam failure, and (3) the duration of the flood is longer with greater distance from the dam failure (Figure 3.6a). The degree to which a dam-failure flood attenuates is strongly controlled by valley and channel geometry, distance from the dam failure, and the contribution of water from joining tributaries. For example, if the valley downstream from a dam is steep and narrow, the peak discharge of the flood will attenuate substantially less than if the valley downstream from the dam is wider and less steep. Costa (1988) gives several examples of attenuation rates of dam-failure floods for different valley geometries. In contrast, hypothetical rainfall and snowmelt hydrographs at the same stations as those used for the dam-failure flood illustrate that (1) the peak discharge increases and is of longer duration with increasing contributing drainage area, (2) the rising and falling limbs become less steep with increasing drainage area, and (3) the flood duration is longer with increasing drainage area (Figure 3.6b).

Summary

The failures of natural and artificial dams have the potential to cause considerable damage and loss of life because the sudden release of stored water generates floods of extraordinary magnitude. During the past two centuries, artificial and natural dam-failure floods have caused the deaths of at least 30,000 people, millions of dollars in property damage, and severe alteration of channels. Despite our increased awareness of the potential hazards associated with dam-failure floods in mountainous regions, human settlement in and development of these regions has increased steadily during the 20th century. As a result, the potential hazards and risks associated with natural dam-failure floods will probably increase in the future. For example, Haeberli (1983) has shown that even though the magnitude of jökulhlaups in Switzerland has decreased during the 20th century, human settlement and development in close proximity to glaciers have caused an increase in property damage and loss of life. Similarly, anticipated future flooding from a crater-rim failure or a volcanic eruption on Mount Ruapehu, New Zealand, is potentially extremely hazardous because of the increased development of the mountain for skiing (Houghton et al., 1987).

Most artificial dam failures are the result of overtopping, foundation defects, and piping and seepage (International Commission on Large Dams, 1973; Jansen, 1980; Costa, 1988). These occur because of (1) lack of understanding of the geological properties beneath and adjacent to the dam, (2) underestimating the magnitude and frequency of possible extreme floods in the reservoir basin, and (3) improper design or construction of the dam. The potential risks and hazards associated with flooding from the failure of artificial dams are steadily increasing because of increased human settlement and development in close proximity to artificial dams and the construction of dams at less than ideal sites (Jansen, 1980). In addition to severe flooding caused by artificial-dam failure, the failure of mine-tailings dams that store toxic waste can have a severe environmental impact on flooded areas.

Acknowledgments

Various drafts of this chapter benefited from helpful reviews by Sara Rathburn, Dave Merritt, Jim O'Connor, and Chris Waythomas.

References

Aitkenhead, N. (1960). Observations on the drainage of a glacier-dammed lake in Norway. *Journal of Glaciology*, **3**, 607–609.

Arnborg, L. (1955). Hydrology of the glacial river, Austurfljót. *Geografiska Annaler*, **37**, 185–201.

Baker, V.R. (1973). Paleohydrology of catastrophic Pleistocene flooding in eastern Washington. *Geological Society of America Special Paper 144*, 144 pp. Boulder, CO: Geological Society of America.

Baker, V.R., Benito, G., and Rudoy, A.N. (1993). Paleohydrology of late Pleistocene superflooding, Atlay Mountains, Siberia. *American Journal of Science*, **259**, 348–350.

Bathurst, J.C. and Ashiq, M. (1998). Dambreak flood impact on mountain stream bedload transport after 13 years. *Earth Surface Processes and Landforms*, **23**, 643–649.

Bathurst, J.C., Leeks, G.J.L., and Newson, M.D. (1986). Relationship between sediment supply and sediment transport for the Roaring River, Colorado, USA. In *Drainage Basin Sediment Delivery*, ed. R.F. Hadley, pp. 105–117. International Association of Hydrological Sciences, Publication no. 159. Wallingford, UK.

Beltaos, S. (1995). Ice jam processes. In *River Ice Jams*, ed. S. Beltaos, pp. 71–104. Highlands Ranch, Colorado: Water Resources Publications, LLC.

Benson, L. and Thompson, R.S. (1987). The physical record of lakes in the Great Basin. In *North America and Adjacent Oceans During the Last Deglaciation*, eds. W.F. Ruddiman and H.E. Wright, pp. 241–260. Boulder, Colorado: Geological Society of America, The Geology of North America, Volume K-3.

Björnsson, H. (1992). Jökulhlaups in Iceland: prediction, characteristics and simulation. *Annals of Glaciology*, **16**, 95–106.

Blown, I. and Church, M. (1985). Catastrophic lake drainage within the Homathko River basin, British Columbia. *Canadian Geotechnical Journal*, **22**, 551–563.

Bolsenga, S.J. (1968). River ice jams: A literature review. *U.S. Army Corps of Engineers Research Report 5-5*, 568 pp. Lake Survey District, Detroit, Michigan: U.S. Army Corps of Engineers.

Bretz, J.H. (1969). The Lake Missoula floods and the Channeled Scabland. *Journal of Glaciology*, **77**, 505–543.

Brown, A.S. (1969). Aiyansh lava flow, British Columbia. *Canadian Journal of Earth Sciences*, **6**, 1460–1468.

Butler, D.R. (1989). The failure of beaver dams and resulting outburst flooding: A geomorphic hazard of the southeastern Piedmont. *Geography Bulletin*, **31**, 29–38.

Butler, D.R. (1995). *Zoogeomorphology*. New York: Cambridge University Press, 231 pp.

Butler, D.R. and Malanson, G.P. (1994). Canadian landform examples – 27, beaver landforms. *The Canadian Geographer*, **38**, 76–79.

Carlson, R.F., Zarling, J.P, and Link, L.E. (1989). Cold regions engineering research – Strategic plan. *Journal of Cold Regions Engineering*, **3**, 172–190.

Cenderelli, D.A. (1998). *Glacial-Lake Outburst Floods in the Mount Everest Region of Nepal: Flow Processes, Flow Hydraulics, and Geomorphic Effects*. Ft. Collins, Colorado: Unpublished Ph.D. dissertation, Colorado State University, 247 pp.

Cenderelli, D.A. and Wohl, E.E. (1997). Hydraulics and geomorphic effects of the 1985 glacial-lake outburst flood in the Mount Everest region of Nepal. *Geological Society of America, Abstracts with Programs*, **29**, A216.

Church, M. (1988). Floods in cold climates. In *Flood Geomorphology*, eds. V.R. Baker, R.C. Kochel, and P.C. Patton, pp. 205–229. New York: John Wiley and Sons.

Clague, J.J. and Evans, S.G. (1997). The 1994 jökulhlaup at Farrow Creek, British Columbia, Canada. *Geomorphology*, **19**, 77–87.

Clarke, G.K.C. (1982). Glacier outburst floods from "Hazard Lake," Yukon Territory, and the problem of flood magnitude prediction. *Journal of Glaciology*, **28**, 3–21.

Code, J.A. and Sirhindi, S. (1986). Engineering implications of impoundment of the Indus River by an earthquake-induced landslide. In *Landslide Dams: Processes, Risk, and Mitigation*, ed. R.L. Schuster, pp. 97–110. New York: American Society of Civil Engineers, Geotechnical Special Publication no. 3.

Costa, J.E. (1988). Floods from dam failures. In *Flood Geomorphology*, eds. V.R. Baker, R.C. Kochel, and P.C. Patton, pp. 439–463. New York: John Wiley and Sons.

Costa, J.E. and Schuster, R.L. (1988). The formation and failure of natural dams. *Geological Society of America Bulletin*, **100**, 1054–1068.

Cotton, C.A. (1944). *Volcanoes as Landscape Forms*. New Zealand: Whitcombe and Tombs Limited, 416 pp.

Davies, W.E. (1973). Buffalo Creek Dam Disaster: Why it happened. *Civil Engineering*, **43**, 69–72.

Desloges, J.R. and Church, M. (1992). Geomorphic implications of glacier outburst flooding: Noeick River valley, British Columbia. *Canadian Journal of Earth Sciences*, **29**, 551–564.

Desloges, J.R., Jones, D.P., and Ricker, K.E. (1989). Estimates of peak discharge from the drainage of ice-dammed Ape Lake, British Columbia, Canada. *Journal of Glaciology*, **35**, 349–354.

Driedger, C.L. and Fountain, A.G. (1989). Glacier outburst floods at Mount Rainier, Washington State, USA. *Annals of Glaciology*, **13**, 51–55.

Elfström, A. (1987). Large boulder deposits and catastrophic floods. *Geografiska Annaler*, **69a**, 101–121.

Evans, S.G. (1986). Landslide damming in the Cordillera of western Canada. In *Landslide Dams: Processes, Risk, and Mitigation*, ed. R.L. Schuster, pp. 111–130. New York: American Society of Civil Engineers, Geotechnical Special Publication no. 3.

Evans, S.G. and Clague, J.J. (1994). Recent climatic change and catastrophic geomorphic processes in mountain environments. *Geomorphology*, **10**, 107–128.

Ferguson, R.I. (1986). River loads underestimated by rating curves. *Water Resources Research*, **22**, 74–96.

Fread, D.L. (1989). National Weather Service models to forecast dam-breach floods. In *Hydrology of Disasters*, eds. O. Starosolszky and O.M. Melder, pp. 192–211. London: James and James.

Froehlich, D.C. (1987). Embankment-dam breach parameters. In *Hydraulic Engineering*, ed. R.M. Ragan, pp. 570–575. New York: Proceedings of the 1987 national conference on hydraulic engineering, American Society of Civil Engineers.

Fuller, R.E. (1931). The aqueous chilling of basaltic lava on the Columbia River Plateau. *American Journal of Science*, **221**, 281–300.

Fushimi, H., Ikegami, K., and Higuchi, K. (1985). Nepal case study: Catastrophic floods. In *Techniques for Prediction of Runoff from Glacierized Areas*, ed. G.J. Young, pp. 125–130. IAHS pub. 149. Wallingford, UK: IAHS Pubs.

Gerard, R.L. and Davar, D.S. (1995). Introduction. In *River Ice Jams*, ed. S. Beltaos, pp. 1–28. Highlands Ranch, Colorado: Water Resources Publications, LLC.

Gilbert, G.K. (1878). The ancient outlet of Great Salt Lake. *American Journal of Science*, **15**, 256–259.

Grove, J.M. (1988). *The Little Ice Age*. London: Methuen (reprinted, 1990, New York, Routledge), 498 pp.

Gruner, E. (1963). Dam disasters. *Proceedings, Institution of Civil Engineers*, **24**, 47–60.

Haeberli, W. (1983). Frequency and characteristics of glacier floods in the Swiss Alps. *Annals of Glaciology*, **4**, 85–90.

Hamblin, W.K. (1994). Late Cenozoic Lava Dams in the Western Grand Canyon. *Geological Society of America*, **Memoir 183**, 139 pp.

Hewitt, K. (1982). Natural dams and outburst floods of the Karakoram Himalaya. In *Hydrological Aspects of Alpine and High-Mountain Areas*, ed. J.W. Glen, pp. 259–269. IAHS pub. 138. Wallingford, UK: IAHS Pubs.

Hewitt, K. (1998). Catastrophic landslides and their effects on the Upper Indus streams, Karakoram Himalaya, northern Pakistan. *Geomorphology*, **26**, 47–80.

Houghton, B.F., Latter, J.H., and Hackett, W.R. (1987). Volcanic hazard assessment for Ruapehu composite volcano, Taupo Volcanic Zone, New Zealand. *Bulletin of Volcanology*, **49**, 737–751.

Hoyt, W.G. and Langbein, W.B. (1955). *Floods*. Princeton, NJ: Princeton University Press, 467 pp.

Hutchinson, G.E. (1957). *A Treatise on Limnology*, Volume 1. New York: Wiley and Sons, 1015 pp.

International Commission on Large Dams. (1973). *Lessons from Dam Incidents*. Paris (reduced edition), 205 pp.

Ives, J.D. (1986). Glacial lake outburst floods and risk engineering in the Himalaya: A review of the Langmoche Disaster, Khumbu Himal, 4 August 1985. *International Centre for Integrated Mountain Development Occasional Paper No. 5*, 42 pp. Kathmandu, Nepal.

Jackson, L.E. (1979). A catastrophic glacial outburst flood (jökulhlaup) mechanism for debris flow generation at the Spiral Tunnels, Kicking Horse River basin, British Columbia. *Canadian Geotechnical Journal*, **16**, 806–813.

Jansen, R.B. (1980). *Dams and Public Safety*. Denver, Colorado: U.S. Department of Interior, Water and Power Resources Service, 332 pp.

Jarrett, R.D. and Costa J.E. (1986). Hydrology, geomorphology, and dam-break modeling of the July 15, 1982 Lawn Lake Dam and Cascade Lake Dam Failures, Larimer County, Colorado. *U.S. Geological Survey Professional Paper 1369*, 78 pp. Washington, DC: U.S. Government Printing Office.

Jarrett, R.D. and Malde, H.E. (1987). Paleodischarge of the late Pleistocene Bonneville flood, Snake River, Idaho, computed from new evidence. *Geological Society of America*, **99**, 127–134.

Jones, P.D., Wigley, T.M.L., and Wright, P.B. (1986). Global temperature variations between 1861 and 1984. *Nature*, **322**, 430–434.

Lliboutry, L., Arnao, B.M., Pautre, A., and Schneider, B. (1977). Glaciological problems set by the control of dangerous lakes in the Cordillera Blanca, Peru. I. Historical failures of morainic dams, their causes and prevention. *Journal of Glaciology*, **18**, 239–254.

MacDonald, T.C., and Langridge-Monopolis, J. (1984). Breaching characteristics of dam failures. *Journal of Hydraulic Engineering*, **110**, 567–586.

Malde, H.E. (1968). The catastrophic late Pleistocene Bonneville Flood in the Snake River Plain, Idaho. *U.S. Geological Survey Professional Paper 596*, 52 pp. Washington, DC: U.S. Government Printing Office.

Marcus, M.G. (1960). Periodic drainage of glacier-dammed Tulsequah Lake, British Columbia. *Geographical Review*, **50**, 89–106.

Marsh, P. and Hey, M. (1989). The flooding hydrology of Mackenzie Delta Lakes near Inuvik, N.W.T., Canada. *Arctic*, **42**, 41–49.

Mason, K. (1929). Indus floods and Shyok Glaciers. *The Himalayan Journal*, **1**, 10–29.

Mayo, L.R. (1989). Advance of Hubbard Glacier and 1986 outburst of Russell Fiord, Alaska, USA. *Annals of Glaciology*, **13**, 189–194.

Meek, N. (1989). Geomorphic and hydrologic implications of the rapid incision of Afton Canyon, Mojave Desert, California. *Geology*, **17**, 7–10.

Miller, D.M. (1984). Reducing transformation bias in curve fitting. *American Statistician*, **38**, 124–126.

Mool, P.K. (1995). Glacier-lake outburst floods in Nepal. *Journal of Nepal Geological Society. Kathmandu*, **11**, 273–280.

Nichols, R.L. and Miller, M.M. (1952). The Moreno Glacier, Lago Argentino, Patagonia. *Journal of Glaciology*, **2**, 41–50.

Nye, J.F. (1976). Water flow in glaciers: jökulhlaups, tunnels and veins. *Journal of Glaciology*, **17**, 181–207.

O'Connor, J.E. (1993). Hydrology, hydraulics, and geomorphology of the Bonneville Flood. *Geological Society of America Special Paper 274*. 83 pp.

O'Connor, J.E. and Baker, V.R. (1992). Magnitudes and implications of peak discharges from glacial Lake Missoula. *Geological Society of America Bulletin*, **104**, 267–279.

O'Connor, J.E. and Costa, J.E. (1993). Geologic and hydrologic hazards in glacierized basins in North America resulting from 19th and 20th century global warming. *Natural Hazards*, **8**, 121–140.

O'Connor, J.E., Hardison, J.E., and Costa, J.E. (1993). Debris flows from recently deglaciated areas on central Oregon Cascade Range volcanoes. *American Geophysical Union, EOS*, **74**, 314.

O'Connor, J.E., Hardison, J.E., and Costa, J.E. (1994). Breaching of lakes impounded by Neoglacial moraines in the Cascade Range, Oregon and Washington. *Geological Society of America, Abstracts with Programs*, **26**, A218–A219.

O'Connor, J.E., Hardison, J.E., and Costa, J.E. (in press). Debris flows from moraine-dammed lakes in the Three Sisters and Mt. Jefferson Wilderness areas, Oregon. *U.S. Geological Survey Water-Supply Paper*. Washington, DC: U.S. Government Printing Office.

Post, A. and Mayo, L.R. (1971). Glacier dammed lakes and outburst floods in Alaska. *U.S. Geological*

Survey Hydrologic Investigations Atlas HA-455, 3 sheets. Washington, DC: U.S. Government Printing Office.

Ray, H.A. and Kjelstrom, L.C. (1978). The flood in southeastern Idaho from the Teton Dam failure of June 5, 1976. *U.S. Geological Survey Open-File Report 77–765*, 48 pp. Washington, DC: U.S. Government Printing Office.

Richardson, D. (1968). Glacier outburst floods in the Pacific Northwest. *U.S. Geological Survey Professional Paper 600D*, pp. D79–D86. Washington, DC: U.S. Government Printing Office.

Schuster, R.L. and Costa, J.E. (1986). A perspective on landslide dams. In *Landslide Dams: Processes, Risk, and Mitigation*, ed. R.L. Schuster, pp. 1–20. New York: American Society of Civil Engineers, Geotechnical Special Publication no. 3.

Smith, D.G. (1980). River ice processes: Thresholds and geomorphologic effects in northern and mountain rivers. In *Thresholds in Geomorphology*, eds. D.R. Coates and J.D.Vitek, pp. 297–321. Boston: George Allen and Unwin.

Smith, G.I. and Street-Perrott, F.A. (1983). Pluvial lakes of the western United States. In *Late Quaternary Environments of the United States*, ed. H.E. Wright, pp. 190–212. Minneapolis: University of Minnesota Press.

Smith, N. (1971). *A History of Dams*. Secaucus, NJ: The Citadel Press, 279 pp.

Stearns, H.T., Crandall, L., and Steward, W.G. (1938). Geology and ground-water resources of the Snake River Plain in southeastern, Idaho. *U.S. Geological Survey Water-Supply Paper 774*, 268 pp. Washington, DC: U.S. Government Printing Office.

Stock, J.D. and Schlosser, I.J. (1991). Short-term effects of a catastrophic beaver dam collapse on a stream fish community. *Environmental Biology of Fishes*, **31**, 123–129.

Sturm, M. and Benson, C.S. (1985). A history of jökulhlaups from Strandline Lake, Alaska, USA. *Annals of Glaciology*, **31**, 272–280.

Sturm, M., Beget, J., and Benson, C. (1987). Observations of jökulhlaups from ice-dammed Strandline Lake, Alaska: implications for paleohydrology. In *Catastrophic Flooding*, eds. L. Mayer and D. Nash, pp. 79–94. Boston: Allen and Unwin.

Thorarinsson, S. (1939). The ice-dammed lakes of Iceland with particular reference to their values as indicators of glacier oscillations. *Geografiska Annaler*, **21**, 216–242.

Thorarinsson, S. (1953). Some new aspects of the Grímsvötn problem. *Journal of Glaciology*, **2**, 267–274.

Tianche, L., Schuster, R.L., and Wu, J. (1986). Landslide dams in south-central China. In *Landslide Dams: Processes, Risk, and Mitigation*, ed. R.L. Schuster, pp. 146–162. New York: American Society of Civil Engineers, Geotechnical Special Publication no. 3.

Tvede, A.M. (1989). Floods caused by a glacier-dammed lake at the Folgefonni ice cap, Norway. *Annals of Glaciology*, **13**, 262–264.

United States Committee on Large Dams. (1994). *Tailings Dam Incidents*. 82 pp. Denver, CO.

United States Representatives Committee on Interior and Insular Affairs. (1980). *Mill Tailings Dam Break at Church Rock, New Mexico*. Oversight hearing before the subcommittee on energy and the environment of the committee on interior and insular affairs, House of Representatives, 96th Congress, United States Government Printing Office, serial number 96-25, 232 pp.

United States Weather Burea. (1954). Floods of 1952, Upper Mississippi-Missouri-Red River of the North. *U.S. Weather Bureau Technical Paper no. 23*, U.S. Department of Commerce, Hydrologic Services Division, 93 pp. Washington, DC.

Van Metre, P.C. and Gray, J.R. (1992). Effects of uranium mining discharges on water quality in the Puerco River Basin, Arizona and New Mexico. *Hydrological Sciences Journal*, **37**, 463–480.

Vuichard, D. and Zimmermann, M. (1986). The Langmoche flash-flood, Khumbu Himal, Nepal. *Mountain Research and Development*, **6**, 90–94.

Vuichard, D. and Zimmermann, M. (1987). The 1985 catastrophic drainage of a moraine-dammed lake, Khumbu Himal, Nepal: Cause and consequences. *Mountain Research and Development*, **7**, 91–110.

Wahlstrom, E.E. (1974). *Dams, Dam Foundations, and Reservoir Sites*. New York: Elsevier Scientific Publishing Company, 278 pp.

Waitt, R.B. (1985). Case for periodic, colossal jökulhlaups from glacial Lake Missoula. *Geological Society of America Bulletin*, **95**, 1271–1286.

Walder, J.S. and Costa, J.E. (1996). Outburst floods from glacier-dammed lakes: The effect of mode of lake drainage on flood magnitude. *Earth Surface Processes and Landforms*, **21**, 701–723.

Walder, J.S. and O'Connor, J.E. (1997). Methods for predicting peak discharge of floods caused by failure of natural and constructed earthen dams. *Water Resources Research*, **33**, 2337–2348.

Watanabe, T., Ives, J.D, and Hammond, J.E. (1994). Rapid growth of a glacial lake in Khumbu Himal, Himalaya: Prospects for a catastrophic flood. *Mountain Research and Development*, **14**, 329–340.

Waythomas, C.F., Walder, J.S., McGimsey, R.G., and Neal, C.A. (1996). A catastrophic flood caused by drainage of a caldera lake at Aniakchak Volcano, Alaska, and implications for volcanic hazards assessment. *Geological Society of America Bulletin*, **108**, 861–871.

White, K.D. and Kay, R.L. (1996). Ice jam flooding and mitigation: Lower Platte River Basin, Nebraska. *U.S. Army Corps of Engineers, Cold Regions Research and Engineering Laboratory*, **96-1**, 62 pp.

Woo, M. and Waddington, J.M. (1990). Effects of beaver dams on subarctic wetland hydrology. *Arctic*, **43**, 223–230.

Wuebben, J.L., Deck, D.S., Zufelt, J.E., and Tatinclaux, J.C. (1995). *U.S. Army Corps of Engineers, Cold Regions Research and Engineering Laboratory*, **95-15**, 22 pp.

Xiangsong, Z. (1992). Investigations of glacier burst of the Yarkant River in Xinjiang, China. *Annals of Glaciology*, **16**, 135–139.

Yesenov, U.Y. and Degovets, A.S. (1979). Catastrophic mudflow on the Bol'shaya Almatinka River in 1977. *Soviet Hydrology*, **18**, 158–160.

Yongjian, D. and Jingshi, L. (1992). Glacier lake outburst flood disasters in China. *Annals of Glaciology*, **16**, 180–184.

Young, G.J. (1980). Monitoring glacier outburst floods. *Nordic Hydrology*, **11**, 285–290.

CHAPTER 4

Anthropogenic Impacts on Flood Hazards

Ellen E. Wohl
Department of Earth Resources
Colorado State University

Human activities can increase flood hazards by modifying the drainage basin characteristics that govern the movement of water and sediment from hillslopes into channels. Land uses that modify drainage basin characteristics include timber harvest and road building, grazing, crops, urbanization, and climate change. Human activities may also modify the movement of water and sediment along channels, as occurs in association with loss of wetlands and beavers, exotic vegetation, dams and flow regulation, channelization and levees, channel stabilization, and mining. Finally, humans can increase flood hazards via increasing human occupation of the floodplain. Because the processes that control floods vary with time and space, the impacts of human activities on floods and flood hazards also vary with scale. Human influences are most pronounced in small-to-medium-sized drainage basins and for small-to-moderate-magnitude floods.

Introduction

Hazard may be defined as a source of danger through loss or harm. Flood hazards are associated with the movement of water and sediment into and along channels during floods, and with changes in channel configuration during floods, as these cause loss or harm to humans or to riparian and aquatic communities. Human activities may increase the hazards associated with flooding in at least three ways: (1) by modifying the drainage basin characteristics that govern the movement of water and sediment from hillslopes into channels; (2) by modifying the movement of water and sediment along channels; and (3) by increasing human occupation of the floodplain.

Because the processes that control floods vary across spatial and temporal scales, the impacts of human activities on floods and flood hazards also vary with scale. Most research to date has focused on how human activities affect physical processes in small river basins (<10 km^2), but these effects cannot necessarily be extrapolated to larger basins. In small basins differences in land-surface characteristics (topography, soils, vegetation), precipitation intensity, and runoff-generating mechanisms may produce substantial differences in flood characteristics. In large basins the timing and magnitude of the flood peak are controlled by the spatial distribution and amount of precipitation and by the routing of flow through the channel network (Benson, 1964; Kirkby, 1976, 1988; Pitlick, 1994). The mechanisms of runoff production on individual hillslopes, and thus the impact of human activities on these mechanisms, are not as important in large as in small basins. The influence of land-surface characteristics also varies with flood magnitude, becoming less important than precipitation for larger floods.

Humans inevitably modify any landscape they occupy. The activities of hunter-gatherer communities may have only a minimal impact on vegetation cover and soil conditions within a drainage basin unless extensive burning is practiced. However, as soon as a community engages in

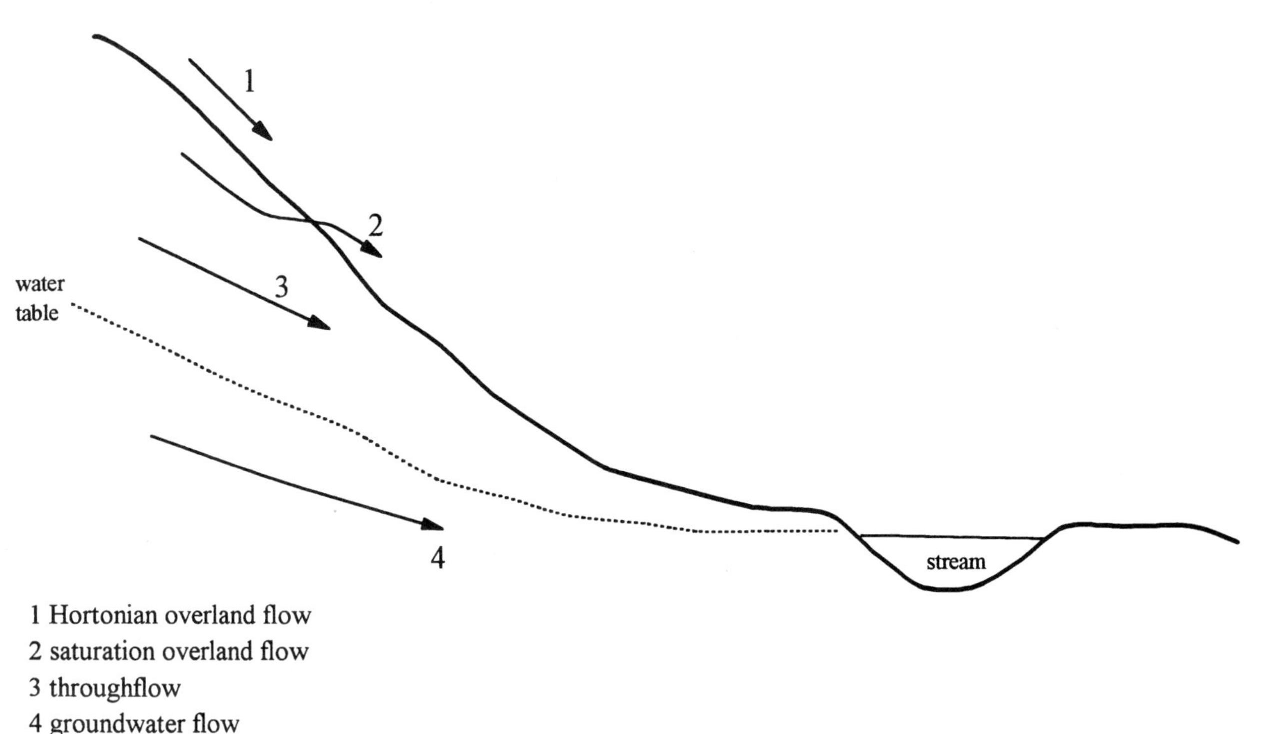

Figure 4.1. Pathways of movement of water down a hillslope.

agriculture (crops and grazing animals), the characteristics of water and sediment movement on hillslopes begin to be affected.

Water may move down a slope via either surface or subsurface pathways (Figure 4.1). Surface runoff in the form of Hortonian or saturation overland flow occurs when precipitation intensity exceeds the infiltration capacity of the ground surface. This may occur almost immediately, as in the case of intense rainfall on a surface of low permeability, or surface runoff may form after prolonged precipitation has saturated the subsurface. Precipitation intensity is governed by storm type and general climatic circulation patterns. Infiltration capacity is a function of both vegetation and slope substrate. Vegetation increases infiltration capacity by intercepting raindrops and reducing their impact and by fostering a relatively porous and permeable organic-rich surface horizon. Slope substrate characteristics control porosity and permeability of the ground surface.

Water flowing down the surface of a slope may initially move as threadflow around individual roughness elements on the slope. As runoff continues and average flow depth increases, threadflow may integrate into sheetflow that submerges individual roughness elements. Sheetflow tends to concentrate along preferential flow paths as a result of surface irregularities. As flow depth increases, the shear stress exerted on the flow boundaries also increases, and the flow erodes a defined channel, or rill, into the slope surface. Rill networks grow through self-enhancing feedback whereby continued erosion of the rill creates cross-graded side slopes into the rill, progressively capturing more of the slope's drainage.

Drainage basins with relatively high proportions of surface runoff also tend to have high rates of sediment movement along hillslopes. Both sheetflow and rillflow may be very erosive on the sparsely vegetated surfaces that are characterized by surface runoff (Dunne et al., 1995). Langbein and Schumm (1958) demonstrated that sediment yield tends to be highest in semiarid climates where rainfall is sufficient to create integrated surface flow, but vegetation is sparse.

Surface flow concentrates water relatively rapidly and may produce the peaked, short duration hydrograph of a flash flood. Subsurface flow moves water more gradually toward a channel. Throughflow occurs in the unsaturated zone above the water table and may take the form of either fairly uniform downslope movement or concentrated flow along pipes or macropores where horizons of limited permeability control deeper infiltration. In some cases, the velocity and volume of pipeflow may exceed those of saturation overland flow (Jones, 1981). Pipe systems may form at different levels in the soil, with each level activated by storms of different magnitude (Gilman and Newson, 1980).

The deepest and slowest flow path, that of groundwater flow, may be affected by precipitation associated with flooding but is commonly more important in controlling the baseflow of a channel between floods.

The relative importance of each of the four principal flow paths will govern how quickly water moves into a stream channel and thus the frequency, magnitude, and duration of floods for small drainage basins. For larger (>10 km^2) drainage basins, the travel time of runoff on a hillslope is generally very small compared with the travel time through the channel network.

The flow paths of water will also partly determine sediment introduction to channels. Impermeable, cohesive substrates such as bedrock generally have high surface runoff and low sediment yield. Low-permeability, unconsolidated sediments of the type commonly found in arid–semiarid regions commonly have high surface runoff and high sediment yield. Surfaces of high permeability generally have slower rates of water movement and lower sediment yields, unless subsurface decreases in permeability facilitate the development of either pipes or perched saturated zones that cause slope failures as debris flows or landslides.

Naturally forested drainage basins thus tend to have low peak magnitude, long duration floods, and low sediment yields relative to unforested basins. Human activities in the form of timber harvest, grazing, planting of crops, and urbanization all tend to decrease infiltration capacity by removing vegetation and compacting the surface. This in turn results in more frequent floods of higher magnitude and shorter duration. These activities may also increase sediment yield by exposing previously protected (vegetated) surfaces to direct precipitation and increased surface runoff; this effect is most pronounced for timber harvest, planting crops, and grazing. A compilation of studies around the world indicated that grazing increased suspended sediment yield up to 100 times, deforestation increased this yield up to 39 times, and deforestation combined with cultivation increased suspended sediment yield up to 4500 times (Walling and Webb, 1983). Increased sediment yields to channels may result in channel aggradation or lateral instability and a concomitant increase in overbank flooding or in bank collapse.

Human activities may also modify the movement of water and sediment along channel networks, either through interference with water and sediment supply from upstream reaches or through altering channel configuration. The geometry and stability of a natural channel are adjusted to the prevailing characteristics of water and sediment discharge. Leopold and Maddock (1953) proposed the hydraulic geometry theory of channel characteristics, in which channel width, flow depth, velocity, and channel slope vary as power functions of water discharge. Wolman and Miller (1960) defined the dominant discharge as that which transports the most sediment. Wolman and Gerson (1978) refined the definition to include the flow that shapes channel configuration. For low-gradient alluvial rivers, dominant discharge equates fairly well to the bankfull discharge that recurs approximately every 1–2 years. Subsequent studies have demonstrated that less frequent extreme floods may dominate channels with a more variable hydrologic regime (Baker, 1977; Gupta, 1988) or with greater boundary resistance (O'Connor et al., 1986; Wohl, 1992).

Attention to a single, channel-forming discharge may oversimplify the relations between discharge and channel geometry. The disparate characteristics of natural channels are commonly maintained by a range of flows (Pizzuto, 1994; Pitlick and Van Steeter, 1998). Coarse sediment deposition on midchannel bars along the Yampa River, Colorado (USA), during high discharge creates and maintains the bars for fish spawning, for example, but the recessional flood hydrograph and subsequent low flows are equally important for flushing fine sediments from the bars (Harvey et al., 1993).

Natural channels exhibit seasonal or periodic adjustment to variations in water and sediment supply by adjusting channel bedforms, width/depth ratio, planform, or slope. A channel is a dynamic system along which there is a constant exchange of water, sediment, and energy between the flow and the channel boundaries. Any human activity that interferes with this exchange may cause a change in flow or channel configuration that increases flood hazards.

Human modification of the movement of water and sediment along channels takes two basic forms. (1) Flow diversion and regulation generally reduce the temporal variations in the magnitude and frequency of water and sediment movement by reducing peak flows, increasing base flows, and trapping sediment in reservoirs. These changes in turn affect channel geometry, so that flood hazards may be increased by direct loss of flood peaks or by channel adjustment to altered discharge. Examples of hazards associated with the loss of flood peaks include a reduction in the erosion and subsequent redeposition of sediment necessary to create sites suitable for germination by cottonwood seedlings (Friedman and Auble, Chapter 8, this volume) or a reduction in winnowing of the fine sediments that may render gravel bars unsuitable as spawning habitat for salmonid fish (Wydoski and Wick, Chapter 9, this volume). Flood hazards may be increased by channel adjustment to altered discharge when channel narrowing increases the potential for overbank flooding, for example. (2) Channel alteration in the form of channelization (straightening and deepening), levees, bank stabilization, or draining of wetlands tends to produce more peaked flood hydrographs by increasing the efficiency of water conveyance along the channel and reducing the opportunities for flood flows to entrain sediment. A more peaked flood hydrograph with less sediment in transport may increase the hazards of overbank flooding or of bank collapse and channel incision.

Finally, increasing human occupation of lowlands subject to periodic flooding directly increases the flood hazards to which humans are exposed and indirectly increases flood hazards to human, riparian, and aquatic communities by encouraging channel and floodplain modifications (Merritts, Chapter 10, this volume; Hamilton and Joaquin, Chapter 18, this volume). At its worst, this cycle of accelerating hazards results in replacement of naturally functioning channels with concrete canals. Any large city in an industrialized nation presents an example: As people begin to settle along a river, they interfere with the natural flow regime to provide desired water supplies for agriculture, navigation, municipal, or industrial use. The alterations in both flow regime and in water and sediment yield resulting from land-use changes within the drainage basin may trigger channel adjustments. People settled close to the channel will not tolerate channel changes or overbank flooding, so portions of the channel bed or banks are artificially fixed in position, and flow is confined to the channel with artificial levees (Watson and Biedenharn, Chapter 14, this volume). Artificial stabilization causes further change in adjoining channel reaches, as the exchange of sediment between the flow and the channel boundaries is further disrupted, and confinement of the flow increases erosive energy. Eventually the entire channel is artificially stabilized, encouraging people to settle more densely along its margins, and thus necessitating further safeguards against overbank flooding and channel instability. The channel is now effectively a plumbing system that transports little sediment and supports very few aquatic or riparian organisms.

The remainder of this chapter describes in more detail the mechanisms by which various land-use activities affect flood hazards, with examples from the literature.

Modification of Drainage Basin Characteristics

Timber Harvest and Road Building

Timber harvest and the associated activity of road building tend to greatly increase sediment yield from hillslopes over the short term (approximately 1–10 years) and to increase water yield over longer periods (approximately 1–30 years) until vegetation becomes reestablished (Bosch and Hewlett, 1982; Roberts et al. 1994). These increases may be dramatic. Studies in Malaysia, for example, have documented 70–90% increases in suspended sediment yield (Kasran, 1988) and 470% increases in water yield (Nik, 1988) 4–6 years after logging. The magnitude of these changes will be a function of climate and slope characteristics (substrate type, steepness, length, aspect) as well as of the methods used in timber harvest and the intensity and spatial coverage of forest removal and subsequent recovery.

Mineral soils are commonly exposed during logging or site preparation activities and thereby become increasingly susceptible to surface erosion (Megahan and Kidd, 1972; Johnson and Beschta, 1980). Compaction reduces infiltration rates, often resulting in increased overland flow and sheetwash or rill erosion. Elevated soil moisture levels caused by a reduction in interception and evapotranspiration increase the weight of the soil (Megahan and Bohn, 1989). Positive pore water pressures become more likely, which can increase the likelihood of landslides (Swanston and

Swanson, 1976; Dietrich et al., 1993; Montgomery and Dietrich, 1994). Removal of vegetation results in a decrease in the tensile strength of root material within the harvest unit (Gray and Megahan, 1981), lowering the soil shear strength. Roads increase the slope angle at the cut-and-fill slope, redistribute the weight of materials on hillslopes (Figure 4.2), alter the routing of water downslope (Swanston and Swanson, 1976), and provide a major source of fine sediment (Reid and Dunne, 1984) long after vegetative cover has been reestablished. Road construction may also greatly increase sediment yield to nearby channels. A sediment budget study in Hawaii indicated that 90% of the fluvial fine-sediment load in the catchment came from highway construction (Hill et al., 1998).

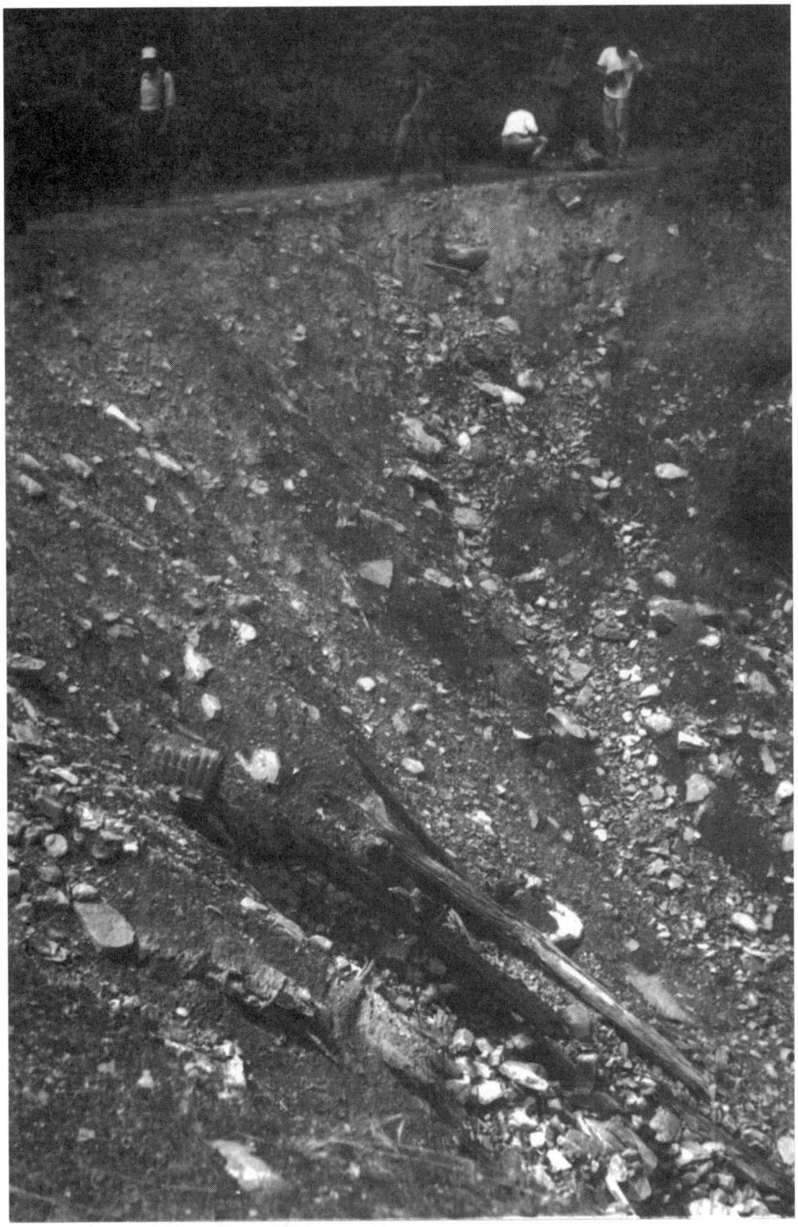

Several mechanisms contribute to increased water yields. Vegetation removal reduces evapotranspiration (Cline et al., 1977) and may increase the rate of snowmelt (Berris and Harr, 1987). Burning during site preparation may form temporary hydrophobic soil layers that reduce infiltration (DeByle, 1973). Compaction along logging roads reduces infiltration and intercepts subsurface flow. Increased water yields in turn increase peak flows (Ziemer, 1981; Harr, 1986; Cheng, 1989). The combination of increased water and sediment yields after timber harvest commonly causes channel aggradation, bank erosion, loss of pool habitat and woody debris, and increased potential for overbank flow and channel change during floods (Madsen, 1995).

Figure 4.2. Hillslope failure initiated at an unpaved road in the Kootenai National Forest, Montana. People at top of photograph are standing on a road; timbers and drainage culvert that have been carried down the slope are at lower left.

Numerous studies describe the processes by which changes in hillslope water and sediment yields trigger changes in channel geometry. Heede (1991) described a 62% increase in the peak discharge of the South Fork of Thomas Creek, Arizona (USA), a small mountain catchment, during an 8-year period after a basal area of timber stand was reduced to 28%. The increase in peak discharge caused channel cross-sectional area to increase by 10% and removed nearly half the preharvest natural control structures (log and clast steps) within the channel. Clearing of oak forests in southern Poland during the first and second centuries A.D. caused

5–6 m of aggradation along major channels such as the Vistula River (Starkel, 1988). Large-scale deforestation in the Severn Basin of the United Kingdom circa 650 B.C. left distinctive alluvial deposits along the channels (Shotton, 1978). During a 15-year study of the 10-km^2 Carnation Creek watershed in western Canada, bank erosion, channel mobility, and channel width all increased within clearcut reaches but not along uncut channel reaches (Hartman and Scrivener, 1990). Forest clearing in the Ruahine Range of New Zealand caused the width of the Tanaki River to increase from 5 to 10 m in the early 1920s, to 54 m by 1942, and to 60 m by 1976 (Mosley, 1978). Between 1851 and 1926, almost all the forest cover in the Bega River catchment of southeastern Australia was removed, resulting in channel aggradation and widening and an increase in the peakedness of the flood hydrograph (Brooks and Brierley, 1997). One of the hazards associated with channel change is that floodplain management becomes more difficult because the areas subject to flooding constantly change (Pearthree, 1982).

In addition to altered channel morphology, the increase of water yield from slopes after removal of trees may trigger increased flooding. Deforestation, in combination with minor climatic variability, caused an increase in the number of damaging floods in Switzerland during the 19th century (Vischer, 1989). The flood that occurred in the 20,000-km^2 Huaihe River basin of China during the summer of 1991 had peak discharges that averaged 70% of the peak discharges of a large flood in 1954 (Hugen and Jiaquan, 1993). But peak stage of the 1991 flood averaged 102% of the 1954 peak stage because of channel aggradation after a reduction in the basin's forest cover from 64% to $<35\%$. The 1991 flood caused far more extensive damage than the flood of 1954.

An increase in sediment yield analogous to that caused by timber harvest has been described for rock quarrying in the vicinity of the Sungai Relau (river) in Malaysia (Ismail and Rahaman, 1994). Nearly 30 m of weathered rock must be removed to reach fresh rock in this humid tropical catchment, and the disruption of slope vegetation and substrate has produced suspended sediment concentrations 1200 times greater than those under natural forest conditions.

Grazing

Grazing of domesticated animals was one of the earliest human land uses to affect slope water and sediment yields. Evidence of land-use effects on sedimentation dates to preagricultural Mesolithic people in Britain, for example (Limbrey, 1983).

In upland regions of a drainage basin, heavy grazing compacts the soil, reduces infiltration, and increases runoff to the point that the runoff regime may be transformed from a variable source area to Hortonian overland flow (Trimble and Mendel, 1995). One study in South Carolina (USA) demonstrated decreases in the infiltration rate of up to 80% as a result of upland grazing (Holtan and Kirkpatrick, 1950). Erosion and sediment yield are increased (Bari et al., 1993) as vegetation cover is reduced (Hofmann and Ries, 1991), fertility and organic matter content decrease (Trimble and Mendel, 1995), and soil aggregate stability is decreased by trampling. The effects of increased water and sediment yield on channel geometry and stability, and on flood hazards, are as described for timber harvest and road building.

Grazing within the riparian zone decreases bank resistance by reducing vegetation (Platts, 1981; Marlow and Pogacnik, 1985; Myers and Swanson, 1996b) and exposing more vulnerable substrate and by trampling that directly erodes banks (Kauffman and Krueger, 1984; Trimble and Mendel, 1995). Cows commonly create ramps along steep or wooded channel banks, and these bank irregularities create turbulence and accelerated bank erosion during high flows (Trimble, 1994). Trails along the floodplain may also become locations of enhanced erosion during overbank flows (Cooke and Reeves, 1976). Enhanced bank and overbank erosion may result in deposition of fine sediments on the channel bed, reducing pool volume and spawning habitat (Myers and Swanson, 1996a). Channels with noncohesive sand and gravel banks are most susceptible to these effects (Myers and Swanson, 1992) (Figure 4.3).

As noted by Trimble and Mendel (1995), both riverine and upland areas are commonly grazed simultaneously, so that channel erosion is increased by increased water yield and bank weakening. Channel erosion is most pronounced during flood flows, creating hazards for (i) aquatic communities by siltation along the channel bed (Myers and Swanson, 1991) and loss of bank cover by overhangs and riparian vegetation, (ii) riparian communities by bank collapse, and (iii) human communities by enhanced overbank flooding and erosion.

Crops

As in the cases of timber harvest and grazing, planting crops alters the infiltration capacity of the soil and hence the water and sediment yields from slopes (Dunne, 1979). The magnitude of these alterations depends on the spatial extent and type of crops. Klimek (1987) describes enhanced sediment yields, increased flood peaks, and a change from meandering to braided channel pattern after the rapid development of "potato plantations" in the Polish Carpathian Mountains during the second half of the 19th century. Starkel (1988) notes that increased sedimentation, channel instability, and overbank sedimentation along the Vistula River of southern Poland have been associated with the spread of agriculture since the early Neolithic (7000–6000 years ago). Adaptation of agriculture on the Loess plateau of central China about 5000 years ago resulted in increased soil erosion, a tripling of sediment load, and a dramatic increase in the frequency of damaging floods along the lower Hwang Ho (Yellow River) (Mei-e and Xianmo, 1994). In the eastern United States, sediment loads of rivers draining to the Atlantic Ocean increased four to five times after European settlement (Chorley et al., 1984), and sedimentation rates in the Chesapeake Bay increased with agricultural land clearance in the Appalachian uplands (Kearney and Stevenson, 1991). Knox (1987) describes an increase from 0.02 cm/year to up to 5 cm/year in overbank floodplain sedimentation in the upper Mississippi Valley in association with agricultural practices. Suspended sediment deposition in small basins on the Canadian prairies increased by a factor of three during the 1950s and 1960s as the result of an increase in the area under field crops and the increased use of heavy agricultural machinery. Deposition subsequently decreased as a result of soil

Figure 4.3. (a) Downstream view of a channel reach excluded from grazing, Sheep Creek, Colorado. The channel is approximately 4 m wide and densely shaded by willows. Flow is approximately 0.4 m deep. (b) Cross-channel view of Sheep Creek immediately upstream from (a). This portion of the channel is open to cattle grazing. Channel is approximately 10 m wide, with 10-cm-deep flow.

conservation measures (De Boer, 1997). Similarly, implementation of soil conservation practices in a 575-km^2 agricultural catchment in Wisconsin resulted in a decrease in flood peaks and in winter/spring flood volumes (Potter, 1991).

Increased sediment yields, surface runoff, and attendant flood damages after agricultural activities have been documented for northern Greece (Astaras, 1984), southern England (Boardman, 1995; Boardman et al., 1996), the southeastern United States (Harvey et al., 1983), and the north-central United States (Knox, 1977). In contrast, drainage of upland

Table 4.1. *Data from Wolman (1967)*

Land use	Sediment yield (metric tons km^{-2} yr^{-1})	Channel condition
Forest	<35	Stable
Agriculture	100–280	Aggrading
Urban construction phase	>35,000	Aggrading
Postconstruction	18–35	Scour and bank erosion

agricultural areas in central Wales decreased peak discharge for moderate flows by lowering the water table (Newson and Robinson, 1983).

Urbanization

The change from natural or agricultural vegetation to buildings and roads has dramatic effects on water and sediment yield from a drainage basin. During the initial, construction phase of urbanization, vegetation at the building site is completely removed, and the land surface is leveled or artificially contoured. This commonly results in extremely high sediment yields as the unconsolidated surface is exposed to rainbeat, sheet flow, and rilling (Goldman et al., 1986; Ruslan, 1995). Construction sites in the United States routinely have sediment fences (1-m-wide continuous sheet of plastic placed vertical to the ground) around the site perimeter to trap fine sediment, but these fences are not always properly maintained. Once construction is complete and ground surfaces are stabilized beneath roads, buildings, and lawns, sediment yield drops to a negligible value that is commonly lower than preurbanization values.

The longer-term result of urbanization is a substantial increase in water yield. The impervious surface area in the basin is greatly increased, so that a great percentage of precipitation falling to the surface leaves the basin as Hortonian overland flow. This increase in surface runoff may cause sheet-flooding and damage to low-lying structures such as roads and basements unless it is quickly channelized and removed from the surface. The installation of storm sewers has the effect of rapidly draining road surfaces as well as further decreasing infiltration and time for water to move from slopes and into channels. The net effect of urbanization is thus that more water reaches stream channels more rapidly. This creates flood peaks of higher magnitude and shorter duration, which may trigger channel instability and increase flood hazards.

A classic study of the effects of urbanization on channel characteristics and flood regime was conducted on channels in the Piedmont region of Maryland (USA) (Wolman, 1967; Wolman and Schick, 1967). The study documented changes in sediment yield and channel condition related to land use (Table 4.1) and recommended the reservation of floodplain lands for parks and open spaces. Numerous studies have supported these findings for a range of environments in the United States (Miller et al., 1971; Morisawa and LaFlure, 1979; Harvey et al., 1983; Urbonas and Benik, 1995), Australia (Nanson and Young, 1981), England (Park, 1977), Russia (Lvovich and Chernogaeva, 1977), Malaysia (Balamurugan, 1991; Ruslan, 1995), and other regions of the world.

Several studies of the effects of urbanization have also focused specifically on flood characteristics. Leopold (1968) suggested that the frequency of low to intermediate magnitude floods is greatly increased by urbanization. Examining data from 81 sites in the vicinity of Washington, DC (USA), Anderson (1970) found that increased urbanization reduced lag time between precipitation peak and discharge peak by a factor of 8 relative to natural channels and increased peak flow by a factor ranging from 2 to 8 because of greater runoff volumes. Espey and Winslow (1974) documented comparatively larger discharges for all return periods in urbanized basins of Texas (USA) than in nearby nonurbanized basins. Working in eastern New England (USA), Doehring and Smith (1978) noted the difficulties of floodplain zoning in basins undergoing urbanization because of changing rainfall-runoff relations and channel patterns. Summarizing research throughout Britain, Roberts (1989) documented channel enlargement as a direct and proportional response to change in flood peaks resulting from urbanization, although channel response was complicated by upstream and downstream feedbacks associated with changing sediment supply and riparian vegetation and bank resistance. She concluded that frequent small floods are more seriously affected by urbanization than are more extreme floods, presumably because a drainage basin in its natural state may respond as if impermeable when large areas are saturated by extreme storms. Simulation studies of extensive flooding along the Mississippi River basin in 1993 also indicate that the effect of land-use practices on runoff from watersheds $>1000\,km^2$ diminishes for storms with return periods greater than 25 years (SAST, 1994). The potential for storing excess runoff in wetlands is limited, particularly during very extreme (>100-year) events (Pitlick, 1997).

Recent papers examining the effects of urbanization have dealt with models such as SEDCAD, which may be used to predict changes in sediment yield (McClintock and Harbor, 1995), and with statistical and deterministic modeling of rainfall-runoff relations and urban flood frequency (Hirsch et al., 1990). Alley and Veenhuis (1983) demonstrated that the connectedness and efficiency of urban drainage networks have a more important effect on floods than does total impervious area.

Climate Change

The potential hydrologic effects of regional or global climate change are one of the least understood human influences on flood hazards. Studies of decadal to century-long weather patterns in the vicinity of large urban areas have demonstrated a "heat island effect" whereby local climate becomes warmer and drier as a result of increases in paved surface relative to natural vegetation and decreases in soil moisture (Barry and Chorley, 1987). On a global scale, records of atmospheric CO_2 levels document a steady increase in CO_2 since the Industrial Revolution, with a particularly rapid rise since the 1950s (Revelle and Suess, 1957; Siegenthaler and Oeschger, 1987). Many studies have predicted that this increase in CO_2 will enhance the atmosphere's ability to trap outgoing longwave radiation and hence cause a rise in average global temperature of 2–4°C (U.S. National Academy of Sciences, 1983). Climatologists have attempted to model how a warmer atmosphere may affect atmospheric and oceanic circulation patterns, but many variables of this complex system – such as cloud cover and oceanic uptake of CO_2 – remain poorly understood (North et al., 1981). Nevertheless, hydrologists have used the output of the climatic general circulation models (GCMs) as a starting point for rainfall-runoff models that translate climatic changes into changes in flow regime (Klemes, 1985; Nash and Gleick, 1991; Mimikou, 1995) (Figure 4.4). Various studies suggest that precipitation in climatically distinct regions may change by ±20%, and runoff may change by ±50% (Schneider et al., 1990).

Studies of the hydrological aspects of climate change have been conducted at various spatial scales. At the global scale, GCMs focus on temperature and precipitation (Airey and Hulme, 1995) with only very coarse representations of surface runoff (Eagleson, 1986; Coe, 1995; Wilby, 1995). For example, subsurface and surface flows are not differentiated (Kuhl and Miller, 1992), spatial resolution varies from 50 to 400 km (Viner, 1994), there is no lateral transfer of water between adjacent model grid cells (Kite et al., 1994), most precipitation occurs at scales smaller than that of the gridboxes of the highest resolution models, and no single currently available global dataset of precipitation fulfills all the requirements for model evaluation (Airey and Hulme, 1995; Panagoulia, 1995). Consequently, hydrologic changes are more effectively modeled at the regional or basin scales, as in the EUROFLOOD project, which is investigating the nature and extent of present and probable future flood hazards in Europe (Penning-Rowsell et al., 1992).

As reviewed by Chang et al. (1992) and Mimikou (1995), assessment of the impact of climatic changes on the hydrologic cycle at the regional or basin scale involves: (1) quantitative estimates of long-term changes in temperature, precipitation, and evapotranspiration; (2) simulation of the hydrologic cycle for the area of interest, using the scenarios developed in (1); and (3) assessment of the implications of hydrologic variations for dams and reservoirs, aquifers, flood hazards, etc. Several types of regional hydrologic models may be used in this manner (Gleick, 1989). Empirical and statistical models use empirical relationships and statistical properties of hydrologic processes (e.g., Cohen, 1987; Bardossy and Caspary, 1991). Deterministic or conceptual models use physically based mathematical descriptions of hydrologic processes (e.g., Flaschka et al., 1987; Mimikou et al., 1991). Historical analog models use either recent or geologic data sets to assess the actual impact of warming climate on hydrology and then extrapolate to potential future scenarios (e.g. Baker, 1995; Baker et al., 1995). Several studies have documented long-period (centuries to millennia) variability in flood magnitude and frequency as a result of changes in regional atmospheric circulation patterns (Ely et al., 1993; Knox, 1993; Wohl et al., 1994; Benito et al., 1996). The approaches used to assess the impact of climate change may also be used in modeling the effects of land-use changes on hydrologic systems (e.g., Ott et al., 1991), and vice versa: Elliott and Parker (1992) used reductions in discharge and sediment entrainment resulting from reservoir construction along the Gunnison River, Colorado, USA, as an indication of potential climate-induced changes in the basin.

McCabe and Hay (1995) used hypothetical climate changes to alter the contemporary time series of temperature and precipitation in the East River basin, Colorado (USA). Assuming an increase of 4°C in mean annual temperature, they predicted a 4–5% increase in annual precipitation and larger, earlier runoff peaks for this snowmelt-dominated 775-km^2 basin. Kwadijk and Rotmans (1995) coupled a model for climate assessment to a

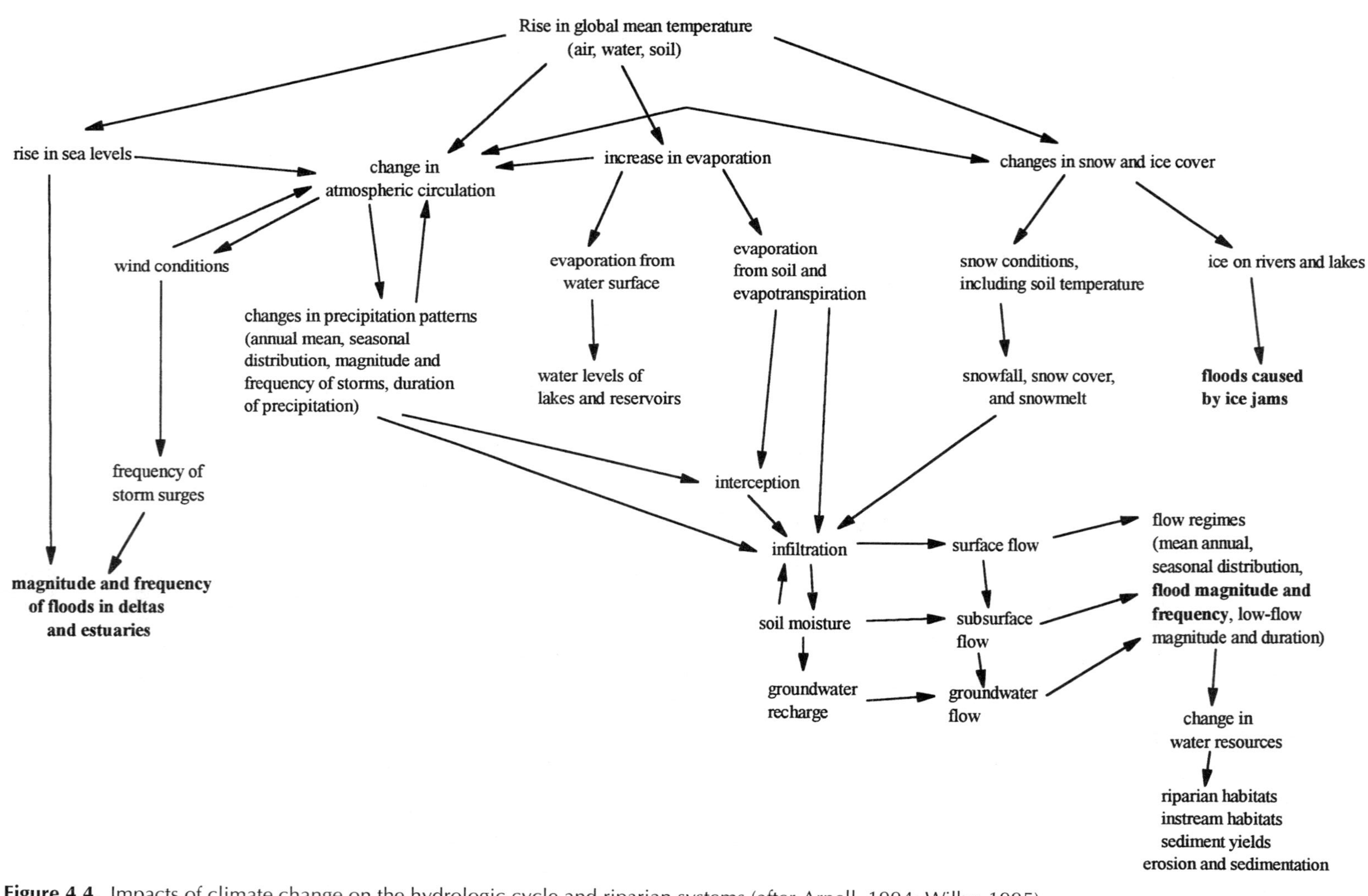

Figure 4.4. Impacts of climate change on the hydrologic cycle and riparian systems (after Arnell, 1994; Wilby, 1995).

water balance model to estimate changes in regional annual water availability and seasonal discharge in the Rhine River basin. They found that the approximately 151,000-km^2 basin changed from a combined snowmelt-rainfall-fed flow regime to a rainfall-dominated regime. The difference between the contemporary high average winter discharge and low average autumn discharge increased, and summer discharge decreased. Extreme flows (both high and low) increased in frequency in the downstream (Dutch) portion of the river. Rao (1993, 1995) used temperature, precipitation, and discharge data for the period 1901–1980 to assess the impact of climate change on the 141,600-km^2 Mahanadi River basin, India. Despite an increase in average air temperature of 1.1°C per century, there have been only insignificant decreases in precipitation regime in this catchment. However, annual river flows have gradually decreased by 40–50%, probably because of increased evaporation associated with warmer temperatures and with land-use changes.

Two potential changes in hydrologic regime are of particular concern with respect to flood hazards for human communities: changes in the magnitude and frequency of floods (Stedinger, Chapter 12, this volume), and changes in mean annual or seasonal distribution of runoff as these affect the storage capacity of flood-control reservoirs (Ford and Thornton, 1992). Physical arguments, empirical evidence, and GCMs all suggest that a global increase in average temperature will cause an increase in the frequency of heavy precipitation events because of a more intense hydrological cycle (increased convective height of clouds) and the increased water-holding capacity of a warmer atmosphere (Mearns et al., 1990; Fowler and Hennessy, 1995). In some regions this change will also affect the seasonality and type of precipitation; a modeling study of the response to global warming of the Sacramento–San Joaquin River basin in California suggested an increase in winter rain-on-snow events and annual flood maxima (Lettenmaier and Gan, 1990). However, different regions of a continent or subcontinent will respond differently to a given climate change. Warming over the Indian subcontinent, for example, will increase the surface runoff in flood-prone areas of the northeastern and central plains of India during the monsoon season but will decrease winter monsoon precipitation and surface runoff over southern peninsular India (Lal and Chander, 1993). Basins located in comparatively drier regions are more sensitive to climatic changes (Dahm and Molles, 1992), although the specifics of soil type, moisture-holding capacity, and runoff coefficient strongly influence basin response (Divya and Mehrotra, 1995). Contemporary analogs exist for the effect of atmospheric changes on discharge regime in different parts of a continent: periodic El Niño/Southern Oscillation events, which influence the seasonal latitudinal movement of the intertropical convergence zone, correlate well with low discharge on rivers of northern South America (Hastenrath, 1990) and with high discharge in southeastern Australia (Whetton and Baxter, 1989).

Climate-induced changes in flood regime are also likely to affect riparian and aquatic ecosystems. The magnitude, frequency, duration, and predictability of environmental extremes (including flow) are agents of disturbance that play an important role in regulating ecological processes and patterns (Poff, 1992). A study of Alaskan streams that assessed the impact of warming-induced increases in glacial runoff, for example, demonstrated that as flow regime, water temperature, and sediment discharge changed, diverse response variables such as channel substrate, bedforms, channel stability, leaf litter quality and quantity, and habitat complexity also changed (Oswood et al., 1992). In desert streams, changes in flood magnitude and frequency may trigger channel incision, removing the deep hyporheic sediments that support microbial communities, while changes in precipitation and runoff variability alter the availability of nitrogen, which is a limiting element in desert streams (Grimm and Fisher, 1992). In general, local community assembly and species persistence have been influenced by a characteristic disturbance regime, so that frequently disturbed communities are relatively resilient (Schlosser, 1990; Wallace, 1990). The historical variability in flow regime thus becomes important in predicting possible riparian and aquatic ecosystem responses to changing flood regime (Poff, 1992).

Human-induced climate change is a global experiment that we have unwittingly set in motion. Because of the complexity and numerous feedback mechanisms of the global hydrologic cycle (Figure 4.2), our present attempts to predict changes as specific as flood hazards in a given drainage basin are very imprecise. One of the most challenging and alarming aspects of climate change is that we have not simply changed one parameter, or process, or response; we have begun a cascade that could change everything. We may be remaking the world.

Modification of In-Channel Flow and Morphology

Loss of Wetlands and Beavers

The term wetlands has numerous legal and scientific definitions (Mitsch and Gosselink, 1993). The legal definitions arise mainly from the relatively recent recognition that the loss of wetlands in the industrialized nations of the world has had profound effects on problems as diverse as extinction of species and flood hazards. For the purposes of this discussion, a wetland is an area that is inundated or saturated by surface or groundwater at a frequency and duration sufficient to support a prevalence of vegetation typically adapted for life in saturated soil conditions (Patrick, 1995). Wetlands thus include swamps, marshes, bogs, deltas, potholes, oxbow lakes, beaver ponds, and portions of the riparian corridor such as sloughs and overflow channels. With respect to flood hazards, wetlands may be thought of as low-lying areas within the floodplain that act as rough-edged sponges: they absorb and store both surface and subsurface flow, releasing this flow more slowly than the surrounding uplands; the vegetation and irregular topography of wetlands create hydraulic roughness during overbank flows.

Three general processes are predominantly responsible for forming riparian wetlands. (1) Wetlands may result from dams built by beavers or other animals or from the accumulation of woody debris in logjams (Patrick, 1995). (2) Abandoned river courses serve as a site for riparian wetlands. (3) Formation of natural levees and local base levels also enhance riparian wetlands.

Beavers are herbivorous rodents that build dams and canals along waterways. Along streams with a suitable habitat of permanent and relatively

Figure 4.5. Downstream view of a series of beaver dams (approximately 1.5 m high) and ponds along a channel in Colorado.

constant water flow, valley widths of approximately 45 m, channel gradients less than 15%, and aspen or willow growing nearby, beaver density averages two or three colonies per kilometer (Allen, 1983; DeByle, 1985). Estimates of the beaver population of North America before the coming of the Europeans range from 60 to 400 million, or 6 to 40 animals per kilometer of stream (Naiman et al., 1986). Beaver were also historically present throughout Europe (Gurnell, 1998).

Channels occupied by beavers have a stepped appearance, with water ponded behind the beaver dams creating reaches of low gradient that are punctuated by abrupt drops of 1–2 m downstream from the dams (Figure 4.5). These ponds decrease flow velocity, trap sediment, reduce channel bed and bank erosion, promote more uniform stream flows during periods of high and low discharge, and diversify aquatic and riparian habitat (Olson and Hubert, 1994). As the beaver pond gradually fills with sediment, meadow grasses and riparian thickets grow across the site, providing a broad, stable floodplain that continues to slow the passage of floodwaters.

Logjams have the same effects as beaver dams on water and sediment movement along channels. Veatch (1906) described more than 20 lakes formed by logjam impoundments along a 260-km stretch of the Red River, Louisiana (USA). The whole structure, called the Great Raft, constantly migrated upstream as new woody debris was added to the upper reaches and debris about 200 years older decayed and dispersed at the downstream end. Similarly massive jogjams are described in historic records for the Mississippi River, where 800,000 snags were removed along a 16-km reach of the lower river from 1870 to 1920 (Sedell et al., 1982) as well as for rivers of the Pacific Northwest (USA) (Sedell and Froggatt, 1984) and for Europe (Piégay and Gurnell, 1997).

Beaver trapping for the fur trade nearly eradicated the animal in North America. The trappers moved progressively westward, trapping out the eastern portions of the continent beginning in 1604 and the interior and western regions by the 1830s (Sandoz, 1964). Current population estimates are 15–20% of those before trapping (Naiman et al., 1986). In the upper Mississippi River basin, 99% of the beaver ponds present in A.D. 1600 were gone by 1990. Hey and Philippi (1995) estimated that, at a depth of 1 m, the original area ponded by beaver dams in the upper Mississippi basin (21 million ha) could have stored more than three floods the size of the destructive summer 1993 flood on the Mississippi River. Other investigators, while acknowledging the importance of wetlands in controlling flood magnitude and duration for smaller basins or smaller floods, question whether reductions in wetland extent substantially affected the 1993 flood characteristics (Pitlick, 1997).

The great logjams were also being destroyed during the 18th and 19th centuries. The logs were cleared from the Red River in 1873, and the lakes began draining (Veatch, 1906). The process continued as the river incised in response to increased velocity.

The logjams were often cleared for purposes of navigation or flood control in the form of reduced overbank flooding. Unforeseen consequences of both removal of logjams and beaver trapping included channel incision, bank erosion, increased channel gradient, and more peaked flood hydrographs. The magnitude of these effects was surprising. When beavers were reestablished along Currant Creek, Wyoming (USA), for example, daily sediment transport decreased from 30,000 to 3,600 kg (Brayton, 1984; Parker, 1986).

In addition to wetlands formed in association with beaver dams or logjams, riparian wetlands may be formed along abandoned river courses. As meander cutoff or channel avulsion isolates meander bends or former channel reaches, these areas scoured below the surrounding floodplain may be maintained as wetlands by precipitation, overbank flooding of the main channel, and groundwater flow (Patrick, 1995). Ohmart et al. (1975) describe such wetlands historically present along the lower Colorado River (USA) where it flows through the Mohave Desert. These lakes had surface areas of 10–530 ha, and they typically lasted 50–70 years in this arid environment. Only 4% of the arid lower Colorado River basin is currently wetlands, compared with 21% of the humid lower Mississippi River basin (Patrick, 1995).

Finally, the formation of natural levees and local base levels may enhance riparian wetlands. The levees slow the movement of floodwaters and springwaters between the main channel and the floodplain, so that water accumulates in depressions to form swamps and ponds. Darlington (1943) described how the levees of the ancestral meandering Cranberry River, West Virginia (USA), formed a broad bog in the river's headwaters toward the end of the last glacial interval. The river subsequently incised through a resistant conglomerate that had formed a local base level, and the channel straightened and cut through its levees, draining the bog. Similar situations have been described for channels in the Georgia (USA) piedmont, the Mississippi River, the Arkansas River (Patrick, 1995), and southern Indiana (Potzger, 1934).

Because wetlands traditionally have been perceived as unhealthy regions where land is wasted by flooding, the trend throughout much of the world during the past four centuries has been the artificial draining and loss of wetlands. Land drainage may occur via: (1) installation of tile drains under agricultural fields, typically in areas of low land-surface gradients, in low-lying coastal areas, or in areas of heavy, minimally porous soils; (2) surface ditching of features, such as prairie potholes or coastal marshes; and (3) reshaping the land surface, particularly in coastal areas, by transforming partially submerged areas into high ground (Hirsch et al., 1990). In the United States, a net 54% loss of wetlands from the mid-1950s to the mid-1970s occurred primarily as a result of agricultural development (87% of the loss) and to a much lesser extent from urbanization (8%) (Tiner, 1984). The contribution of urbanization has probably increased during the past two decades. In many states, the riparian vegetation often associated with wetlands has been reduced in area by more than 80% since the 1700s (Swift, 1984). In the upper Mississippi River basin, 59% of the wetlands present in 1780 had been drained by 1980, and the 10 million ha of drained wetlands are hypothesized to have had sufficient area to store two times the amount of floodwater that passed the city of St. Louis during the summer 1993 flood on the Mississippi River (Hey and Philippi, 1995). Major draining works were complete on most European rivers by 1880 (Petts, 1990). For example, the regulation of the River Theisz (Hungary), beginning in 1845, drained 12.5×10^6 ha of floodplain marsh and shortened the river course by 340 km (Petts, 1990). Piégay and Salvador (1997) note that, except for riparian forests located on international boundaries such as the Rhine or the Danube River, the riparian vegetation of most western European rivers was cut during the Middle Ages.

The effects of wetland losses on flood peaks vary. The increased thickness of the unsaturated zone after drainage means that greater precipitation is required to fill the available pore spaces and produce runoff (Hirsch et al., 1990). On the other hand, artificial drainage increases the density and efficiency of the channel network; surface or ditch drainage is most effective at increasing peak discharges (National Hydrology Research Institute, 1982). Artificial drainage of wetlands in the lower portion of a river basin may reduce peak discharge at the basin outlet because water from the lower basin will be removed before the arrival of flood peaks from the upper basin (Wisler and Brater, 1965).

By the mid-19th century many people in both Europe (Hermann et al., 1890) and the United States (Ellett, 1853) believed that an observed increase in flood frequency and magnitude was caused by upland deforestation, land drainage, and artificial levees and cutoffs. These observations have been supported by studies such as that of Mustonen and Seuna (1971) in Finland, where by 1970 about one-third of the country's peatland had been drained, leading to a 40% increase in annual mean runoff and an increase in flooding. The increase in flood hazards in turn led to intensified channel engineering that sometimes exacerbated the problem (Petts, 1990). Additional examples of flood hazards exacerbated by floodplain and wetland loss come from several African nations (Acreman, 1994) as well as from the United States (Hollis and Acreman,1994).

Exotic Vegetation

Exotic vegetation refers to vegetation that is not native to a region but is introduced either intentionally or unintentionally by humans. In the absence of coevolved parasites or predators, exotic vegetation may outcompete native species and come to dominate a region. Away from the Eurasian continent, numerous examples exist on islands and on relatively isolated continents such as Australia and North America. When exotic vegetation becomes established along the riparian corridor, it may alter channel-bank resistance and stability, overbank hydraulic roughness and sedimentation, and channel width and planform. Burkham (1976), for example, noted that changes in riparian vegetation along the Gila River, Arizona, between 1965 and 1972 caused significant differences in stage, mean cross-sectional velocity and depth, and boundary roughness at peak discharges during three major floods. During an exceptional 400-year flood along the Ouveze River, France, riparian vegetation reduced the flow capacity of the floodplain as well as the lateral mineral fluxes and organic deposition (Piégay and Bravard, 1997). Examples of widespread exotic vegetation along riparian corridors in the United States include kudzu in the Southeast, tamarisk in the arid West, and Russian olive in the semiarid prairies.

The influence of exotic vegetation commonly occurs in conjunction with other anthropogenic modifications of the drainage basin, as demonstrated by the example of the Green River. The Green River drains 115,800 km^2 of the western United States. The portion of the river in Canyonlands National Park, Utah, flows through sandstone canyons up to 330 m deep. The rock walls and tributary channels provide a steady source of sand that accumulates as islands and bars in the main channel. Tamarisk (*Tamarix chinensis*) was introduced into the American Southwest in the late 1800s. The plant reached the Green River between 1925 and 1931, spreading upstream at about 20 km/year as it colonized what had been sparsely vegetated channel banks and bars (Graf, 1978). The dense tamarisk thickets trapped and stabilized sediment, causing an average reduction in channel width of 27% along the 60 to 140-m-wide channel.

Exotic vegetation may thus have at least three effects on flood hazards: (1) By increasing bank stability, the vegetation can reduce bank collapse during floods. (2) By increasing channel-boundary flow resistance and decreasing channel discharge capacity, vegetation may increase overbank flooding. This effect is significant only for channels with a width/depth ratio less than 16 (Masterman and Thorne, 1992). (3) Vegetation growing on the channel bed greatly increases flow resistance, causing channel widening and reduced flow velocity (Huang and Nanson, 1997) and thus altering overbank flood frequency.

Dams and Flow Regulation

The construction of dams and the regulation of river flow by reservoirs and diversion have been undertaken for purposes of hydroelectric power generation, agricultural and municipal water supplies, navigation, and flood control. The earliest-known dam was built circa 2800 B.C. in Egypt (Smith, 1971), but the construction of dams greater than 15 m tall has accelerated greatly since the 1950s (Goldsmith and Hildyard, 1984; Petts, 1984). As of 1989, large dams greater than 150 m high were being completed at a rate of one every 1.65 years in Europe alone (Petts, 1989).

The effect of a dam or of flow regulation on flood hazards will be a function of the nature of the alteration in flow regime and of the characteristics of the channel. Dam construction commonly has four primary effects on flow regime: (1) a reduction in the mean and the coefficient of variation of annual peak flow; (2) an increase in minimum flows; (3) a shift in seasonal variability; and (4) particularly for hydroelectric power generating dams, a dramatic increase in diurnal fluctuations. Williams and Wolman (1984) documented average annual floods downstream from 29 dams in the central and western United States. These floods ranged from 3% to 90% of predam values, with an average of 40%. Most dams operate under minimum-release requirements that increase the magnitude of low flows occurring with a given recurrence interval. Annual 7-day low flows on the Columbia River in southwestern Canada nearly doubled after reservoir construction (Hirsch et al., 1990).

The Colorado River basin has been more profoundly modified by dams and flow diversions than any other large basin in North America. The lower portion of the basin has a 3.34 ratio of reservoir storage to annual supply, compared with values of 0 to 1.45 for the rest of the continent (Hirsch et al., 1990). Before 1935 there was virtually no flow regulation in the Colorado River basin. Completion of Hoover Dam in 1935, Glen Canyon Dam in 1963, and several smaller dams reduced average annual peak flow and increased minimum flow (Graf, 1996). Before regulation, the annual hydrograph was dominated by the May–June snowmelt flood. Now the river has a broad peak from April to September, corresponding to maximum irrigation and municipal demands, with less than half the predam seasonal variability. However, an experimental flood release of 1270 m^3/s (approximately one-third to one-quarter of predam large floods) through the Grand Canyon in March 1996 that was designed for evaluation of flood-related changes in sediment transport and channel morphology may be repeated regularly in the future to help maintain channel characteristics (Collier et al., 1997).

The high flows released during peak production times (approximately 6 h/day, 5 days/week) from hydroelectric dams may be nearly as large as the mean annual flood that occurred in predam conditions (Hirsch et al., 1990). During evening and weekend low production, flow release may drop precipitously.

These changes in flow regime affect flood hazards in at least two ways. First, the channel downstream of a dam often adjusts its morphology to the altered flow regime and to the reduction in sediment transport caused by the sediment-trapping properties of the dam. The changes in sediment transport may be formidable. Three major Missouri River (USA) dams were finished between 1952 and 1955. The postdam annual suspended sediment load 8 km downstream from the lowest dam was <1% of the predam load. Twelve hundred kilometers downstream, the load was only 30% of predam load (Williams and Wolman, 1985).

Channel adjustment may take the form of incision. Sand-bed rivers commonly incise until a stable, gravel-armored bed is created (Jiongxin, 1996) or until the slope is reduced to the point that sediment removal ceases (Kellerhals, 1982). Bed incision exceeded 7 m downstream of Hoover Dam on the Colorado River. Such effects may continue far downstream. Incision occurred up to 300 km downstream of the Sariyar Dam in Turkey (Galay, 1983). Observations at 111 sites on rivers across the United States demonstrated that most of the channel incision occurs during the first few years after dam construction. Channel incision extended 480 km downstream during the first 4 years after dam completion on the Hanjiang River, China, and 670 km downstream during the next 10 years (Han and Tong, 1982).

Conversely, the reduction in peak flows downstream from a dam may reduce sediment transport capacity so that tributary inputs cause channel aggradation or narrowing, as along the Brazos River, Texas (USA) (Allen et al., 1989), or the River Derwent in England (Petts, 1977). Flushing flows may be released periodically to alleviate these problems (Kondolf and Wilcock, 1996; Wilcock et al., 1996). Dam-related reduction in flood scouring may also enhance riparian vegetation establishment, stabilizing bars, islands, and channel banks and causing channel narrowing. The severity of ice jams during spring breakup has increased after damming of some Canadian rivers, so that terraces too high to be flooded before dam construction are now inundated (Kellerhals, 1982).

Channel adjustment to a dam commonly varies with distance downstream from the dam (e.g., Erskine, 1985), as illustrated by the Green River, a tributary of the Colorado River (Andrews, 1986). Since completion of the Flaming Gorge dam in 1962, mean annual sediment discharge has decreased by 45%. Immediately downstream from Flaming Gorge Reservoir, the capacity of the river to transport sediment is greater than the sediment supply, and the channel bed has incised along a 35-km reach. Approximately 170 km downstream, a 100-km reach is in quasi-equilibrium with respect to sediment, although channel width has narrowed by approximately 10% because of reduced peak flow. Four hundred and sixty kilometers downstream of the reservoir, transport capacity has decreased below the sediment supply from small tributaries, and the reach is aggrading. Similarly variable responses downstream have been documented for dam construction on the Rio Grande (USA) (Lagasse, 1981). These channel adjustments to altered water and sediment discharge may exacerbate flooding hazards because of increased bank collapse in incising reaches (Allen et al., 1989) or overbank flooding in aggrading reaches (Lagasse, 1981; Williams and Wolman, 1985).

A second type of increase in flood hazards involves the changes imposed on riparian and aquatic communities by changes in flood regime. Dams interfere not only with downstream sediment movement but also with the movement of nutrients and the coarse woody debris that enhance pool formation and the diversity of aquatic habitat, with the migration of fish, and with water temperature regime and water chemistry (Baxter, 1977; Brooker, 1981). When channel morphology adjusts to altered water and sediment flows, riparian vegetation may lose the freshly-scoured surfaces necessary for seedling establishment, seedlings may be killed by prolonged submersion, or changing grain-size distribution along the banks and bars may inhibit seedling establishment (Pearlstine et al., 1985; Friedman and Auble, Chapter 8, this volume). Tree growth of all floodplain species along the Missouri River downstream of Garrison Dam declined after dam construction, whereas trees along undammed reaches were unaffected (Reily and Johnson, 1982). The dam shifted peak flow from early in the growing season to a later time that is out of phase with the vernal growth pattern typical of floodplain trees. Numerous studies document lower species-richness and percentage cover of vegetation along regulated rivers relative to unregulated rivers (Nilsson et al., 1991; Dynesius and Nilsson, 1994). Along channels in arid and semiarid regions, changes in the water table beneath the floodplain as a result of changed flow regime may cause soil salinization that kills riparian vegetation (Jolly, 1996).

Fish and macroinvertebrates are also affected by dam-induced changes in flood regime (Ligon et al., 1995). The Colorado squawfish (*Ptychocheilus lucius*), an endangered species that inhabits the Colorado River basin, spawns at a limited number of gravel bars during the recessional limb of the annual snowmelt hydrograph (Harvey et al., 1993). Sediment deposition and bar formation occur at discharges greater than 280 m^3/s, but spawning habitat is formed by bar dissection and erosion of fine sediments at flows between 10 and 140 m^3/s. Thus, the fish cannot successfully spawn if the historical May–June peak flow is prolonged too late into the season by dam releases. This has been a problem since construction of an upstream dam in 1962 and it has contributed to the squawfish's endangered status (Wydoski and Wick, Chapter 9, this volume).

The squawfish and other endangered fish species are also affected by channel narrowing and a loss of channel complexity as backwater habitat is lost (Van Steeter and Pitlick, 1998). Dam operations have caused annual peak discharges on the Colorado River near Grand Junction, Colorado, to decrease by 29–38% since 1950, while base flows have been augmented. Average annual suspended sediment load has decreased approximately 40% during this period, with much of the sediment apparently being stored in former side channels that provided a nursery habitat for native fish species.

The Flathead River basin of Montana (USA) contains both pristine channel reaches and three dams operated for flood control and hydroelectric power production. Along the unregulated channels, wetting and subsequent drying of the channel edges and floodplain (the varial zone) occur once a year during spring snowmelt floods. Along regulated reaches, the varial zone is unpredictably flooded and dried so that aquatic and riparian biota have little chance of naturally colonizing new areas as the stage rises or emigrating when the stage falls (Stanford and Hauer, 1992). Aquatic biodiversity is drastically reduced because of these changes in the flood regime.

Fish also may be severely affected by the loss of floodplain spawning and nursery habitat when reduction of peak flows curtails overbank flooding. Declines in number of fish species using floodplains have been attributed to dam construction along the Murray River of Australia (Lake, 1975), the Pongolo River of South Africa (Jubb, 1972), the Rivers Dnieper and Volga of Russia and Ukraine (Zalumi, 1970; Chikova, 1974), and the Missouri River of the United States (Whitley and Campbell, 1974). Stream biota of all types may be adversely affected by sediment releases from reservoirs (Gray and Ward, 1982). Biota may also be affected by artificial accumulation of organic materials. Richardson (1981) documented several fish kills along the Belmore River of southeastern Australia. These fish kills occurred when rapid release of large masses of organic matter trapped behind floodgates caused rapid deoxygenation of river water.

Flow diversion has many of the same effects on flood hazards as do dams and reservoirs. Diversion for irrigation or for flood control commonly reduces both base flows and flood peaks (Schleusener et al., 1962; Richards and Wood, 1977) and thus affects channel morphology (Ryan, 1997) and riparian and aquatic communities (Johnson, 1978). Along many channels, it has been found necessary to provide a minimum flow regime to maintain channel morphology (channel maintenance flow) or biotic communities (instream flow; Stromberg and Patten, 1989, 1990; Stalnaker and Wick, Chapter 16, this volume).

Finally, dams may create catastrophic flood hazards when they fail (Cenderelli, Chapter 3, this volume). Dam failures presumably have been occurring for as long as dams have been built. Some of the more devastating failures include the Puentes Dam in Spain (1802; 600 people killed), the Dale Dike in Britain (1864; 244 people killed), and the Johnstown Dam in the United States (1889; 2200 people killed) (Dietrich, 1995). Approximately 1% of the small dams (<150 m high) around the world fail each year (Goldsmith and Hildyard, 1984).

Channelization and Levees

Channelization involves removal of instream obstructions and artificial widening, deepening, and/or straightening of a channel, usually by meander cutoff, in order to increase channel capacity and conveyance of floodwaters. Levees are linear ridges parallel to flow that form along the channel banks during overbank flow when the increase in cross-sectional area and decrease in flow velocity promote sediment deposition. Levees are also commonly artificially enhanced, and the following discussion refers to artificial levees (Watson and Biedenharn, Chapter 14, this volume).

Channelization has been practiced in Europe since about A.D. 1750, when the Rio Guadalquivir (Spain) was reduced in length by 40% (Petts, 1989). During the past 150 years, more than 340,000 km of channels have been modified in the United States, primarily to reduce flooding, to drain land for agricultural use, and to enhance navigation (Schoof, 1980) (Figure 4.6). In British river basins, the percentage of the main river that is channelized varies from 41% in London to 12% in less densely populated regions (Brookes, 1988). These alterations have had many unforeseen consequences (Schoof, 1980; Swales, 1982; Brookes, 1988; Wyzga, 1996): channels in noncohesive sediments are prone to incision or bank collapse during floods as the channel adjusts to its new slope, increasing sediment loads downstream; flood waves may become more peaked as channels incise; channelization alters flow velocities, which in turn alter substrate particle size distributions, affecting aquatic macroinvertebrates and fish; removal of riparian vegetation during channelization reduces organic matter input and nutrient supply, as well as habitat diversity and cover, and may increase water temperature; and lowering of tributary base levels when a larger channel is straightened may initiate massive incision along the tributaries. An example of the latter response comes from the 110-km^2 Oaklimiter Creek watershed in Mississippi (USA) (Harvey et al., 1983). Channel aggradation and increased overbank flooding after European occupation of the region led to a channelization program in the 1960s. Straightening of the naturally sinuous (1.3–2.5) channel resulted in steeper gradients, and implementation of soil conservation measures reduced sediment yield but not runoff. Retention of larger flows within the channel increased the sediment transport capacity by a factor of approximately 50, causing severe channel incision and widening. Fifteen years after channelization, channel capacity had increased by a factor of 10. Similar examples come from Big Pine Creek, Indiana (USA) (Barnard and Melhorn, 1982), Santa Rosa Wash, Arizona (USA)

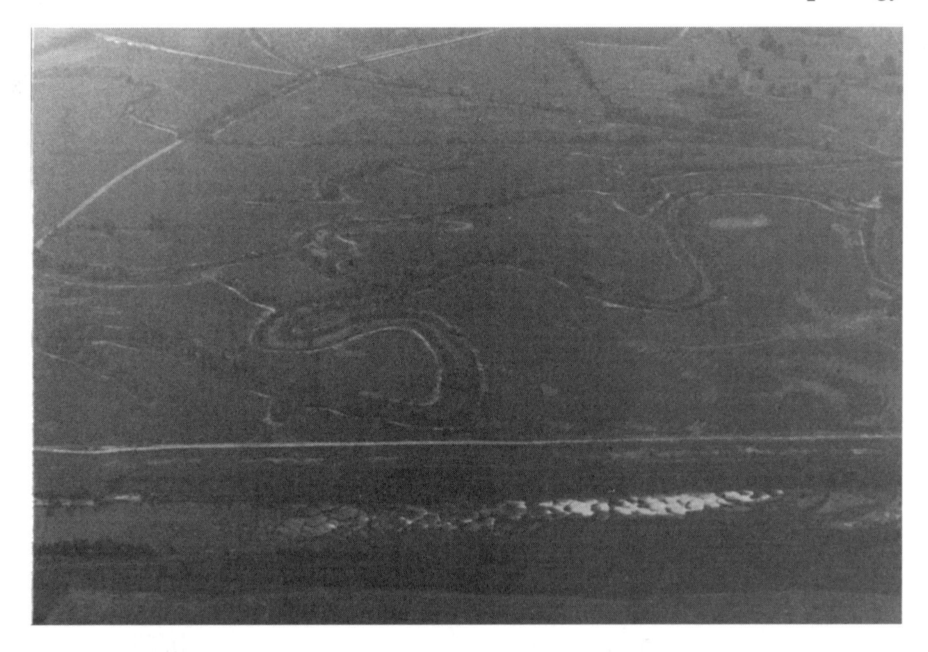

Figure 4.6. Aerial view of Yalobusha River, Mississippi, showing natural meandering channel (center of photograph) and straightened channel (base of photograph).

(Rhoads, 1990), numerous alluvial rivers in Japan, where rivers have been channelized since the 17th century (Fujita and Yamamoto, 1996), the Blackwater River, Missouri, USA (Emerson, 1971), the River Foyle of Ireland (Elson, 1975), the River Perry of England (Swales, 1982), the East and West Prairie Rivers of Alberta, Canada (Parker and Andres, 1976), the Mississippi River, USA (Winkley, 1994), and the South Fork of the Forked Deer River, Tennessee, USA (Hirsch et al., 1990). Channelization along the Raba River, Poland, during the late 1950s caused channel incision and an increase in the magnitude of flood waves and flood hazards (Wyzga, 1996).

Examples of the biological effects of floods along channelized rivers come from (1) northern Japan, where channelization of the meandering Kuchoro River caused increased suspended sediment to be delivered to the Kushiro Marsh, altering edaphic conditions and wetland vegetation (Nakamura et al., 1997); (2) the Bunyip River of southeastern Australia, where total biomass of fish and species richness were reduced by channelization (Hortle and Lake, 1982); and (3) northern Europe, where studies have reported adverse impacts on fish populations in France (Bouchard et al., 1979), Germany (Tesch and Albrecht, 1961), Scandinavia (Müller, 1962), and the United Kingdom (Swales, 1980).

Two of the most famous examples of large rivers with extensive artificial levees are the Mississippi River and China's Hwang Ho (Yellow River). Levees are designed to contain floodwaters within a channel. They are commonly effective flood control measures, because they are relatively inexpensive and easy to build. However, levees may: (1) negatively affect the flow regime both upstream and downstream and exacerbate flooding in other places as water accumulates upstream of the levee system, raising flood stage, and as accelerated water leaves the confinement of the levees downstream; (2) confine floodwaters so that increased stage and velocity cause channel erosion; (3) provide a misleading sense of security that encourages human occupation of floodplains; and (4) disrupt ecological exchanges between channels and floodplains.

Levee construction along the Mississippi River began in 1717 at New Orleans (NHRAIC, 1992). Levee building along the river increased through the 20th century, when a series of Flood Control Acts between 1917 and 1944 gave new impetus to flood protection (Tobin, 1995). In the upper Mississippi River basin there are currently approximately 13,000 km of levees. During the summer 1993 flood, about 70% of these levees failed, primarily because of technical weaknesses, which caused locally enhanced flood hazards (Tobin, 1995). These failures prompted a reexamination of the more than 42,000 km of levees and floodwalls in the United States, which protect approximately 5.5% of floodplain human communities (NHRAIC, 1992). Two of the realizations of this process were the importance of an integrated, basinwide approach to flood hazard mitigation and the necessity of maintaining floodplain-channel exchanges of energy to preserve riparian and aquatic communities.

Levee construction and repair has a much longer history along China's 745,000-km^2 Hwang Ho drainage basin. During the past 3500 years, the river has breached its levees some 1500 times (Clark, 1982). These levees are 30 m wide at the base, 15 m wide at the crest, and 9 m tall. The levees are so effective at confining this silt-laden river that, in some areas, the Hwang Ho runs almost 10 m above the surrounding North China Plain (Anonymous, 1972). When the Hwang Ho breaches its levees, the resulting floods are immense: the 1887–89 flood killed between 900,000 and 2.5 million people, and the 1938 flood covered more than 23,000 km^2 and killed 500,000 people (Clark, 1982).

Widespread levee systems have also been constructed in India (Goldsmith and Hildyard, 1984) and in most European countries (Petts, 1984). The density of channelized rivers in England and Wales was 0.06 km/km^2 as of 1980, compared with 0.003 km/km^2 in the USA (Gregory, 1987).

Channel Stabilization

Channel stabilization refers to the artificial strengthening of channel bed or banks against erosion. This may be accomplished by grade-control structures or check dams for the bed and riprap, soil cement, gabions, revetments, training structures, or riparian vegetation for the banks (Coppin and Richards, 1990). The purpose of all these approaches is to reduce bed and bank erosion, particularly during floods, and to reduce downstream sediment yields. However, although individual structures may fulfill their design purpose at a specific location, they commonly trigger upstream and downstream channel adjustments that may increase flood hazards associated with channel change and sediment transport. During the October 1983 flood along the Santa Cruz River, Arizona (USA), severe bank erosion occurred immediately downstream from every portion of the channel protected by soil cement (Kresan, 1988).

Mining

In-channel mining for sand and gravel and for placer deposits of precious metals has been occurring for millennia in some regions of the world.

However, systematic studies of the effects of these activities on river channels were not undertaken until the early 20th century (Gilbert, 1917). It is now widely recognized that mining disrupts channel flow and sediment transport, causing various channel adjustments in response. If mining occurs in the form of a large, localized excavation, a knickpoint can migrate upstream from the excavation, and both bed incision and bank erosion may be accelerated downstream from the excavation as downstream sediment transport is reduced. If mining involves disruption of the channel surface, as in placer mining, increased sediment mobility and preferential transport of fine sediments may cause downstream aggradation and bank instability, leading to channel braiding as well as adversely affecting aquatic organisms.

Examples of mining-induced channel change and attendant flood hazards come from around the world. The American River of California aggraded substantially (1.5–9 m) during the primary hydraulic mining period (1861–1884), causing channel avulsions (James, 1994) and overbank flooding and deposition (Fischer and Harvey, 1991), and later degraded as sediment input decreased (Gilbert, 1917). It is now possible that the river will undergo a period of channel enlargement by bank erosion and lateral migration, increasing flood hazards to communities that have developed since mining ceased (CFCA, 1995). Similar hazardous instability has been documented along the neighboring Bear and Yuba Rivers (James, 1991). Placer tin mining along the Ringarooma River, Tasmania, caused >5 m of aggradation, increases in channel width of up to 300%, and enhanced braiding, so that bridges had to be frequently replaced after floods (Knighton, 1989). A 50-year flood along Tujunga Wash, California, triggered knickpoint migration from a 300 × 460 × 15–23 m deep gravel pit (Bull and Scott, 1974). The knickpoint caused the failure of three major highway bridges, and lateral scour downstream from the pit destroyed seven homes and a long section of a major highway. Other examples of channel aggradation, braiding, or channel instability induced by placer or sand and gravel mining include Birch Creek, Alaska, USA (Van Haveren, 1991), the lower Mississippi River (Lagasse et al., 1980), San Juan Creek, California, USA (Chang, 1987), and the South Platte River, Colorado, USA (Hilmes and Wohl, 1995). Where the mining sediments transported by floods include toxic metals, as along the River Ystwyth of Wales (Lewin et al., 1977), the Guadiamar River of Spain (Van Geen and Chase, 1998), and the Knabeåna-Kvina Rivers of Norway (Langedal, 1997), flood hazards take on a whole new dimension (Finley, Chapter 7, this volume). Studies documenting adverse effects of mining on aquatic organisms include those on the Loire, Allier, and Tarn Rivers, France (Swales, 1982), the Yukon River, Canada (McLeay et al., 1987), and subarctic streams of Alaska, USA (Van Nieuwenhuyse and LaPerriere, 1986).

Settlement within the Flood Zone

A third category of human activity that increases flood hazards is settlement along channels and floodplains. Humans have settled along river corridors for access to food, water, and transportation since the beginning of human history. However, many human communities were built beyond the reach of all but the most extreme floods because of the hazards and expense of rebuilding after each flood. During the 19th and 20th centuries, however, increasing human population density and rising land costs as well as a sense of security from floods as a result of flood control works (levees, detention basins, reservoirs) have encouraged denser human settlement within flood zones (Laituri, Chapter 17, this volume). A recent example comes from the summer 1993 floods along the upper Mississippi River, which caused approximately $10 billion in damage and killed 38 people (IFMRC, 1994). Seventy percent of the land flooded in 1993 is used for agriculture (only 5% is urban), and development of the floodplain for agriculture has steadily increased during the past century. Parker (1995) describes an escalator effect for floodplain settlement in England and Wales, whereby progressively greater flood defense in response to progressively increasing flood damage encourages further floodplain settlement. Data from six case studies showed increases in 100-year floodplain properties of 108–476% after flood control projects (Parker, 1995). Similar trends exist in the United States (White et al., 1958) and along the Ganges River in India (Ramachandran and Thakur, 1974). In addition to exposing more people and structures to flood damage, encroachment into the floodplain with structures results in a loss of flood storage capacity. This can lead to increased flood elevation and velocities within the encroached reach and increased downstream peak flows (Carlton et al., 1989).

Summary

This chapter has detailed numerous examples from around the world of how human activities, by altering water and sediment from hillslopes and along channels, have increased flood hazards for human, riparian, and aquatic communities. The types of studies summarized in this chapter have led to a consensus among the scientific community that human effects on river systems have commonly been insufficiently planned and thus unnecessarily damaging (Hamilton and Joaquin, Chapter 18, this volume). This consensus has emerged none too soon. As global human population continues to double at progressively shorter time intervals and as nonindustrialized countries rapidly exploit their natural resources to attain the socioeconomic standards of the industrialized nations, human influences on river corridors will increase dramatically. The sheer scale of these effects has reached a new level during the past few decades: for example, Turkey's Southeast Anatolia Project calls for construction of 80 dams, 66 hydroelectric power stations, and 68 irrigation projects covering up to 2 million ha (Hillel, 1994). As of 1980, 16 interbasin water transfers greater than 25 m^3/s existed in Canada, with more planned (Hirsch et al., 1990). In Brazil, plans to dam the Amazon River and its tributaries will flood approximately 350,000 km^2 (Goldsmith and Hildyard, 1984). These massive projects will necessarily affect every aspect of the drainage basins in which they are built.

If negative impacts of flooding are to be minimized, then we must consider human activities at the scale of a drainage basin, rather than a localized reach of channel, and balance the needs of riparian and aquatic communities with the short-term needs of human communities. A basin scale approach to human activities recognizes the connections between hillslopes and channel, channel and floodplains, and upstream and downstream. Recognition that gravel mining may cause scour both upstream and downstream, for example, can be used to design mining operations to minimize the disruption of sediment transport. The scale and rate of gravel extraction per year could be reduced and extended over a longer time period to minimize channel disruption, for example, instead of concentrated to maximize the profit to an individual gravel mining company.

Balancing environmental concerns against short-term human needs will involve using alternative solutions to problems of overbank flooding, channel instability, and resource use. Reservation of floodplain lands as open-space corridors and wildlife habitat, bank stabilization by replanting with native riparian species, and bed stabilization by restoration of a pool-riffle sequence are all examples of nonstructural approaches to flood hazards that benefit river ecosystems and, in the long run, are economically more viable than traditional river engineering (Hey, 1994; Hey and Philippi, 1995; Gruntfest, Chapter 15, this volume). As the annual damages from floods continue to rise, it becomes imperative that we not continue with business as usual.

Acknowledgments

This chapter benefited greatly from constructive reviews by John Pitlick and Lisa Ely.

References

Acreman, M. (1994). The role of artificial flooding in the integrated development of river basins in Africa. In *Integrated River Basin Development*, eds. C. Kirby and W.R. White, pp. 35–44. Chichester: Wiley.

Airey, M. and Hulme, M. (1995). Evaluating climate model simulations of precipitation: methods, problems and performance. *Progress in Physical Geography*, **19**, 427–448.

Allen, A.W. (1983). *Habitat Suitability Index Models: Beaver*. Ft. Collins, Colorado: U.S. Fish and Wildlife Services, 20 pp.

Allen, P.M., Hobbs, R., and Maier, N.D. (1989). Downstream impacts of a dam on a bedrock fluvial system, Brazos River, central Texas. *Bulletin of the Association of Engineering Geologists*, **26**, 165–189.

Alley, W.M. and Veenhuis, J.E. (1983). Effective impervious area in runoff modeling. *American Society Civil Engineering Journal of Hydraulic Engineering*, **109**, 313–319.

Anderson, D.G. (1970). Effects of urban development on floods in northern Virginia. *U.S. Geological Survey Water Supply Paper 2001-C*, 22 pp. U.S. Government Printing Office, Washington, DC.

Andrews, E.D. (1986). Downstream effects of Flaming Gorge Reservoir on the Green River, Colorado and Utah. *Geological Society of America Bulletin*, **97**, 1012–1023.

Anonymous. (1972). *China Tames Her Rivers*. Beijing: Foreign Languages Press, 51 pp.

Arnell, N.W. (1994). Hydrology and climate change. In *The Rivers Handbook: Hydrological and Ecological Principles*, eds. P. Calow and G.E. Petts, **2**, 173–186. Oxford: Blackwell Scientific Publications.

Astaras, T. (1984). Drainage basins as process-response systems: an example from central Macedonia, North Greece. *Earth Surface Processes and Landforms*, **9**, 333–341.

Baker, V.R. (1977). Stream-channel response to floods, with examples from central Texas. *Geological Society of America Bulletin*, **88**, 1057–1071.

Baker, V.R. (1995). Global palaeohydrological change. *Quaestiones Geographicae*, Special Issue 4, 27–35.

Baker, V.R., Ely, L.L., Enzel, Y., and Kale, V.S. (1995). Understanding India's rivers: late Quaternary palaeofloods, hazard assessment and global change. *Memoir 32, Geological Society of India*, pp. 61–77.

Balamurugan, G. (1991). Some characteristics of sediment transport in the Sungai Kelang Basin, Malaysia. *Journal of the Institution of Engineers, Malaysia*, **48**, 31–52.

Bardossy, A. and Caspary, H.J. (1991). Conceptual model for the calculation of the regional hydrologic effects of climate change. In *Hydrology for the Water Management of Large River Basins*, eds. F.H.M. van de

Ven, D. Gutknecht, D.P. Loucks, and K.A. Salewicz, pp. 73–82. IAHS Publ. no. 201. Wallingford UK.

Bari, F., Wood, M.K., and Murray, L. (1993). Livestock grazing impacts on infiltration rates in a temperate range of Pakistan. *Journal of Range Management*, **46**, 367–372.

Barnard, R.S. and Melhorn, W.N. (1982). Morphologic and morphometric response to channelization: the case history of Big Pine Creek Ditch, Benton County, Indiana. In *Applied Geomorphology*, eds. R.G. Craig and J.L. Craft, pp. 224–239. London: Allen and Unwin.

Barry, R.G. and Chorley, R.J. (1987). *Atmosphere, Weather and Climate*, 5th ed. New York: Methuen, 460 pp.

Baxter, R.M. (1977). Environmental effects of dams and impoundments. *Annual Reviews of Ecological Systems*, **8**, 25–283.

Benito, G., Machado, M.J., and Perez-Gonzalez, A. (1996). Climate change and flood sensitivity in Spain. In *Global Continental Changes: The Context of Palaeohydrology*, eds. J. Branson, A.G. Brown, and K.J. Gregory, pp. 85–98. Geological Society Special Publication no. 115. London.

Benson, M.A. (1964). Factors affecting the occurrence of floods in the Southwest. *U.S. Geological Survey Water-Supply Paper 1580-D*, 72 pp. U.S. Govt. Printing Office, Washington DC.

Berris, S. and Harr, R.D. (1987). Comparative snow accumulation and melt during rainfall in forested and clearcut plots in the western Cascades of Oregon. *Water Resources Research*, **23**, 135–142.

Boardman, J. (1995). Damage to property by runoff from agricultural land, South Downs, southern England, 1976-1993. *The Geographical Journal*, **161**, 177–191.

Boardman, J., Burt, T.P., Evans, R., Slattery, M.C., and Shuttleworth, H. (1996). Soil erosion and flooding as a result of a summer thunderstorm in Oxfordshire and Berkshire, May 1993. *Applied Geography*, **16**, 21–34.

Bosch, J.M. and Hewlett, J.D. (1982). A review of catchment experiments to determine the effects of vegetation changes on water yield and evapotranspiration. *Journal of Hydrology*, **55**, 3–23.

Bouchard, B., Clavel, P., Hamon, Y., and Romaneix, C. (1979). Ecological incidences of dredging up alluvial materials and regulation of streams on the aquatic ecosystem. *Bulletin Français Pisciculture*, **273**, 137–156 (in Brookes, 1988).

Brayton, S.D. (1984). The beaver and the stream. *Journal of Soil and Water Conservation*, **39**, 108–109.

Brooker, M.P. (1981). The impact of impoundments on the downstream fisheries and general ecology of rivers. In *Advances in Applied Biology*, ed. T.H. Coaker, **6**, 91–152. London: Academic Press.

Brookes, A. (1988). *Channelized Rivers: Perspectives for Environmental Management*. Chichester: John Wiley and Sons, 326 pp.

Brooks, A.P. and Brierley, G.J. (1997). Geomorphic responses of lower Bega River to catchment disturbance, 1851–1926. *Geomorphology*, **18**, 291–304.

Bull, W.B. and Scott, K.M. (1974). Impact of mining gravel from urban stream beds in the southwestern US. *Geology*, **2**, 171–174.

Burkham, D.E. (1976). Hydraulic effects of changes in bottom-land vegetation on three major floods, Gila River in southeastern Arizona. *U.S. Geological Survey Professional Paper 655-J*, 14 pp.

Carlton, D.K., Barker, B., Nelson, R., and Stypula, J. (1989). Effect of lost floodplain storage on flood peaks. In *Partnerships: Effective Flood Hazard Management*. Proc., 13th Annual Conference of the Association of State Floodplain Managers, May 22–27, 1989, Scottsdale, Arizona, USA, pp. 237–240.

CFCA (Committee on Flood Control Alternatives in the American River Basin). (1995). *Flood Risk Management and the American River Basin: An Evaluation*. Washington, DC: National Academy Press.

Chang, H.H. (1987). Modelling fluvial processes in streams with gravel mining. In *Sediment Transport in Gravel-Bed Rivers*, eds. C.R. Thorne, J.C. Bathurst, and R.D. Hey, pp. 977–988. Chichester: Wiley.

Chang, L.H., Hunsaker, C.T., and Draves, J.D. (1992). Recent research on effects of climate change on water resources. *Water Resources Bulletin*, **28**, 273–286.

Cheng, J.D. (1989). Streamflow changes after clearcut logging of a pine-beetle infested watershed in southern British Columbia, Canada. *Water Resources Research*, **25**, 449–456.

Chikova, V.M. (1974). Species and age composition of fishes in the lower reach (downstream) of the V.I. Lenin Hydroelectric Station. In *Biological and Hydrological Factors of Local Movements of Fish in Reservoirs*, ed. B.S. Kuzin, pp. 185–192. New Delhi, India: Amerind Publishing Company.

Chorley, R.J., Schumm, S.A., and Sugden, D.E. (1984). *Geomorphology*. London: Methuen.

Clark, C. (1982). *Flood*. Alexandria, Virginia: Time-Life Books, 176 pp.

Cline, R.G., Haupt, H., and Campbell, G. (1977). Potential water yield response following clearcut harvesting on north and south slopes in northern Idaho. *USDA Forest Service Research Paper INT-191*. U.S. Govt. Printing Office, Washington, DC.

Coe, M.T. (1995). The hydrologic cycle of major continental drainage and ocean basins: a simulation of the modern and mid-Holocene conditions and a comparison with observations. *Journal of Climate*, **8**, 535–543.

Cohen, S.J. (1987). Projected increases in municipal water use in the Great Lakes due to CO_2-induced climate change. *Water Resources Bulletin*, **23**, 91–101.

Collier, M.P., Webb, R.H., and Andrews, E.D. (1997). Experimental flooding in Grand Canyon. *Scientific American*, **276**, 66–73.

Cooke, R.U. and Reeves, R.W. (1976). *Arroyos and Environmental Change in the American South-West.* Oxford: Clarendon Press.

Coppin, N.J. and Richards, I.G. (1990). *Use of Vegetation in Civil Engineering.* London: Butterworths, 292 pp.

Dahm, C.N. and Molles, M.C., Jr. (1992). Streams in semi-arid regions as sensitive indicators of global climate change. In *Global Climate Change and Freshwater Ecosystems*, eds. P. Firth and S.G. Fisher, pp. 250–260. New York: Springer-Verlag.

Darlington, H.C. (1943). Vegetation and substrate of Cranberry Glades, West Virginia. *Botanical Gazette*, **104**, 371–393.

De Boer, D.H. (1997). Changing contributions of suspended sediment sources in small basins resulting from European settlement on the Canadian Prairies. *Earth Surface Processes and Landforms*, **22**, 623–639.

DeByle, N.V. (1973). Broadcast burning of logging residues and the water repellency of soils. *Northwest Science*, **47**, 77–87.

DeByle, N.V. (1985). Wildlife. In *Aspen: Ecology and Management in the Western US*, eds. N.V. DeByle and R.P. Winokur, pp. 135–152. Ft. Collins, Colorado: U.S. Fish and Wildlife Service, Rocky Mountain Forest and Range Experiment Station.

Dietrich, W. (1995). *Northwest Passage: The Great Columbia River.* Seattle: University of Washington Press, 448 pp.

Dietrich, W.E., Wilson, C.J., Montgomery, D.R., and McKean, J. (1993). Analysis of erosion thresholds, channel networks, and landscape morphology using a digital terrain model. *Journal of Geology*, **101**, 259–278.

Divya and Mehrotra, R. (1995). Climate change and hydrology with emphasis on the Indian subcontinent. *Hydrological Sciences Journal*, **40**, 231–242.

Doehring, D.O. and Smith, M.E. (1978). *Modelling the Dynamic Response of Floodplains to Urbanization in Eastern New England.* Environmental Resources Center, Completion Report No. 83, 95 pp. Colorado State University, Ft. Collins, Co.

Dunne, T. (1979). Sediment yield and land use in tropical catchments. *Journal of Hydrology*, **42**, 281–300.

Dunne, T., Whipple, K.X., and Aubry, B.F. (1995). Microtopography of hillslopes and initiation of channels by Horton overland flow. In *Natural and Anthropogenic Influences in Fluvial Geomorphology*, eds. J.E. Costa, A.J. Miller, K.W. Potter, and P.R. Wilcock, pp. 27–44. Washington, DC: American Geophysical Union Geophysical Monograph 89.

Dynesius, M. and Nilsson, C. (1994). Fragmentation and flow regulation of river systems in the northern third of the world. *Science*, **266**, 753–762.

Eagleson, P.S. (1986). The emergence of global scale hydrology. *Water Resources Research*, **22**, 6S–14S.

Ellett, C. (1853). *The Mississippi and Ohio Rivers.* Philadelphia: Lippincott, Grambo and Company.

Elliott, J.G. and Parker, R.S. (1992). Potential climate-change effects on bed-material entrainment, the Gunnison Gorge, Colorado. In *Managing Water Resources During Global Change.* American Water Resources Association, pp. 751–759. Bethesda, Maryland.

Elson, P.F. (1975). In *The Foyle Fisheries – New Basis for Rational Management*, eds. P.F. Elson and A.L.W. Twomi, pp. 59–64 and 142–148. Northern Ireland: L.M. Press, Ltd. Belfast.

Ely, L.L., Enzel, Y., Baker, V.R., and Cayan, D.R. (1993). A 5000-year record of extreme floods and climate change in the southwestern United States. *Science*, **262**, 410–412.

Emerson, J.W. (1971). Channelisation: a case study. *Science*, **173**, 325–326.

Erskine, W.D. (1985). Downstream geomorphic impacts of large dams: the case of Glenbawn Dam, NSW. *Applied Geography*, **5**, 195–210.

Espey, W.H. Jr. and Winslow, D.E. (1974). Urban flood frequency characteristics. *American Society of Civil Engineering Proceedings, Hydraulics Division*, HY 2, 279–294.

Fischer, K.J. and Harvey, M.D. (1991). Geomorphic response of lower Feather River to 19th century hydraulic mining operations. In *Inspiration: Come to the Headwaters.* Proc., 15th Annual Conference Association of State Floodplain Managers, June 10–14, 1991, Denver, Colorado, USA, pp. 128–132.

Flaschka, I.M., Stockton, C.W., and Boggers, W.R. (1987). Climatic variation and surface water resources in the Great Basin region. *Water Resources Bulletin*, **23**, 47–57.

Ford, D.E. and Thornton, K.W. (1992). Water resources in a changing climate. In *Global Climate Change and Freshwater Ecosystems*, eds. P. Firth and S.G. Fisher, pp. 26–47. New York: Springer-Verlag.

Fowler, A.M. and Hennessy, K.J. (1995). Potential impacts of global warming on the frequency and magnitude of heavy precipitation. *Natural Hazards*, **11**, 283–303.

Fujita, K. and Yamamoto, K. (1996). Response of alluvial rivers to human activities in Japan. In *Proceedings of the Workshop on the Effects of Global Climate Change on Hydrology and Water Resources at the Catchment Scale*, Feb. 3–6, 1996, Tsukuba,

Japan. Japan-US Committee on Hydrology, Water Resources, and Global Climate Change, Publication No. 1, pp. 445–457.

Galay, V.J. (1983). Causes of river bed degradation. *Water Resources Research*, **19**, 1057–1090.

Gilbert, G.K. (1917). Hydraulic-mining debris in the Sierra Nevada. *U.S. Geological Survey Professional Paper 105*.

Gilman, K. and Newson, M.D. (1980). *Soil Pipes and Pipeflow – A Hydrological Study in Upland Wales.* Norwich: Geobooks.

Gleick, P.H. (1989). Climate change, hydrology, and water resources. *Reviews of Geophysics*, **27**, 329–344.

Goldman, S., Jackson, K., and Bursktynsky, T. (1986). *Erosion and Sediment Control Handbook.* New York: McGraw Hill.

Goldsmith, E. and Hildyard, N. (1984). *The Social and Environmental Effects of Large Dams.* San Francisco: Sierra Club Books, 404 pp.

Graf, W.L. (1978). Fluvial adjustments to the spread of tamarisk in the Colorado Plateau region. *Geological Society of America Bulletin*, **89**, 1491–1501.

Graf, W.L. (1996). Geomorphology and policy for restoration of impounded American rivers: What is 'natural'? In *The Scientific Nature of Geomorphology*, eds. B.L. Rhoads and C.E. Thorn, pp. 443–473. New York: Wiley.

Gray, D.H. and Megahan, W.F. (1981). Forest vegetation removal and slope stability in the Idaho batholith. *USDA Forest Service Research Paper INT-271*, 23 pp. U.S. Govt. Printing Office, Washington, DC.

Gray, L.J. and Ward, J.V. (1982). *Effects of Releases of Sediment from Reservoirs on Stream Biota.* Ft. Collins, Colorado: Colorado Water Resources Research Institute Completion Report No. 116, Colorado State University, 82 pp.

Gregory, K.J. (1987). River channels. In *Human Activity and Environmental Processes*, eds. K.J. Gregory and D.E. Walling, pp. 207–235. Chichester: John Wiley and Sons.

Grimm, N.B. and Fisher, S.G. (1992). Responses of arid-land streams to changing climate. In *Global Climate Change and Freshwater Ecosystems*, eds. P. Firth and S.G. Fisher, pp. 211–233. New York: Springer-Verlag.

Gupta, A. (1988). Large floods as geomorphic events in the humid tropics. In *Flood Geomorphology*, eds. V.R. Baker, R.C. Kochel, and P.C. Patton, pp. 301–320. New York: John Wiley and Sons.

Gurnell, A.M. (1998). The hydrogeomorphological effects of beaver dam-building activity. *Progress in Physical Geography*, **22**, 167–189.

Han, Q.W. and Tong, Z.J. (1982). The impact of Danjiangkou Reservoir on the downstream river channel and the environment. *Fourteenth International Congress on Large Dams*, v. 3, Question 54, pp. 189–200. Rio de Janeiro, Brazil: Transactions. International Society of Large Dams.

Hartman, G.F. and Scrivener, J.C. (1990). Impacts of forestry practices on a coastal stream ecosystem, Carnation Creek, British Columbia. *Canadian Bulletin of Fisheries and Aquatic Sciences*, **223**, 148 pp.

Harr, R.D. (1986). Effects of clearcutting on rain on snow runoff in western Oregon: a new look at old studies. *Water Resources Research*, **22**, 1095–1100.

Harvey, M.D., Mussetter, R.A., and Wick, E.J. (1993). A physical process-biological response model for spawning habitat formation for the endangered Colorado squawfish. *Rivers*, **4**, 114–131.

Harvey, M.D., Watson, C.C., and Schumm, S.A. (1983). Channelized streams: an analog for the effects of urbanization. *1983 International Symposium on Urban Hydrology, Hydraulics and Sediment Control.* University of Kentucky, pp. 401–409. Lexington, Kentucky.

Hastenrath, S. (1990). Predictability of anomalous river discharge in Guyana. *Nature*, **345**, 53–54.

Heede, B.H. (1991). Increased flows after timber harvest accelerate stream disequilibrium. In *Erosion Control: A Global Perspective*, pp. 449–454. Orlando, FL: Proceedings of the Conference XXII, International Erosion Control Association.

Hermann, Klein, J. and Thome, (1890). *Land, Sea and Sky* (transl. by J. Minshull), London: Ward, Lock and Co.

Hey, D.L. and Philippi, N.S. (1995). Flood reduction through wetland restoration: the upper Mississippi River basin as a case history. *Restoration Ecology*, **3**, 4–17.

Hey, R.D. (1994). Environmentally sensitive river engineering. In *The Rivers Handbook: Hydrologic and Ecologic Principles*, eds. P. Calow and G.E. Petts, **2**, 337–362. Oxford: Blackwell Scientific Publications.

Hill, B.R., Decarlo, E.H., Fuller, C.C., and Wong, M.F. (1998). Using sediment 'fingerprints' to assess sediment budget errors, North Halawa Valley, Oahu, Hawaii, 1991–92. *Earth Surface Processes and Landforms*, **23**, 493–508.

Hillel, D. (1994). *Rivers of Eden: The Struggle for Water and the Quest for Peace in the Middle East.* New York: Oxford University Press, 355 pp.

Hilmes, M.M. and Wohl, E.E. (1995). Changes in channel morphology associated with placer mining. *Physical Geography*, **16**, 223–242.

Hirsch, R.M., Walker, J.F., Day, J.C., and Kallio, R. (1990). The influence of man on hydrologic systems. In *Surface Water Hydrology*, eds. M.G. Wolman and H.C. Riggs, pp. 329–359. Boulder, Colorado: Geological Society of America.

Hofmann, L. and Ries, R.E. (1991). Relationship of soil and plant characteristics to erosion and runoff on pasture and range. *Journal of Soil and Water Conservation*, **46**, 143–147.

Hollis, G.E. and Acreman, M.C. (1994). The functions of wetlands within integrated river basin development: international perspectives. In *Integrated River Basin Development*, eds. C. Kirby and W.R. White, pp. 351–365. Chichester: Wiley.

Holtan, H.N. and Kirkpatrick, M.H. (1950). Rainfall infiltration and hydraulics of flow in runoff computation. *Transactions American Geophysical Union*, **31**, 771–779.

Hortle, K.G. and Lake, P.S. (1982). Macroinvertebrate assemblages in channelized and unchannelized sections of the Bunyip River, Victoria. *Australian Journal of Marine and Freshwater Research*, **35**, 1071–1082.

Huang, H.Q. and Nanson, G.C. (1997). Vegetation and channel variation; a case study of four small streams in southeastern Australia. *Geomorphology*, **18**, 237–249.

Hugen, Z. and Jiaquan, W. (1993). The effect of forests on flood control – some comments on the flood disaster in the Huaihe River basin of Anhui Province in 1991. *Journal of Environmental Hydrology*, **1**, 38–43.

IFMRC (Interagency Floodplain Management Review Committee). (1994). *Sharing the Challenge: Floodplain Management into the 21st Century.* Washington, DC: Administration Floodplain Management Task Force, 190 pp.

Ismail, W.R. and Rahaman, Z.A. (1994). The impact of quarrying activity on suspended sediment concentration and sediment load of Sungai Relau, Pulau Pinang, Malaysia. *Malaysian Journal of Tropical Geography*, **25**, 45–57.

James, L.A. (1991). Incision and morphologic evolution of an alluvial channel recovering from hydraulic mining sediment. *Geological Society of America Bulletin*, **103**, 723–736.

James, L.A. (1994). Channel changes wrought by gold mining: northern Sierra Nevada, California. In *Effects of Human-Induced Changes on Hydrologic Systems*. American Water Resources Association, pp. 629–638. Bethesda, Maryland.

Jiongxin, X. (1996). Underlying gravel layers in a large sand bed river and their influence on downstream-dam channel adjustment. *Geomorphology*, **17**, 351–359.

Johnson, M.G. and Beschta, R.L. (1980). Logging, infiltration and surface erodibility in western Oregon. *Journal of Forestry*, **78**, 334–337.

Johnson, R.R. (1978). The lower Colorado River: a western system. In *Strategies for Protection and Management of Floodplain Wetlands and Other Riparian Ecosystems*. USDA Forest Service General Technical Report WO-12, pp. 41–55. Washington, DC.

Jolly, I.D. (1996). The effects of river management on the hydrology and hydroecology of arid and semi-arid floodplains. In *Floodplain Processes*, eds. M.G. Anderson, D.E. Walling, and P.D. Bates, pp. 577–609. Chichester: Wiley.

Jones, J.A.A. (1981). *The Nature of Soil Piping: A Review of Research*. Norwich: Geobooks.

Jubb, R.A. (1972). The J.G. Strydom Dam: Pongolo River: northern Zululand. The importance of floodplains below it. *Piscator*, **86**, 104–109.

Kasran, B. (1988). Effect of logging on sediment yield in a hill Dipterocarp forest in Peninsular Malaysia. *Journal of Tropical Forest Science*, **1**, 56–66.

Kauffman, J.B. and Krueger, W.C. (1984). Livestock impacts on riparian ecosystems and streamside management implications . . . a review. *Journal of Range Management*, **37**, 430–438.

Kearney, M.S. and Stevenson, J.C. (1991). Island land loss and marsh vertical accretion rate evidence for historical sea-level changes in Chesapeake Bay. *Journal of Coastal Research*, **7**, 403–415.

Kellerhals, R. (1982). Effects of river regulation on channel stability in gravel bed rivers. In *Gravel Bed Rivers*, eds. R.D. Hay, J.C. Bathurst, and C.R. Thorne, pp. 685–715. New York: John Wiley and Sons.

Kirkby, M.J. (1976). Tests of the random network model, and its application to basin hydrology. *Earth Surface Processes*, **1**, 197–212.

Kirkby, M.J. (1988). Hillslope runoff processes and models. *Journal of Hydrology*, **100**, 315–339.

Kite, G.W., Dalton, A., and Dion, K. (1994). Simulation of streamflow in a macroscale watershed using general circulation model data. *Water Resources Research*, **30**, 1547–1559.

Klemes, V. (1985). *Sensitivity of Water Resource Systems to Climate Variations*. Geneva: World Climatology Applications Programme WCP-98, World Meteorological Organization, 17 pp.

Klimek, K. (1987). Man's impact on fluvial processes in the Polish Western Carpathians. *Geografiska Annaler*, **69A**, 221–225.

Knighton, A.D. (1989). River adjustment to changes in sediment load: the effects of tin mining on the Ringarooma River, Tasmania, 1875–1984. *Earth Surface Processes and Landforms*, **14**, 333–359.

Knox, J.C. (1977). Human impacts on Wisconsin stream channels. *Annals of the Association of American Geographers*, **67**, 323–342.

Knox, J.C. (1987). Historical valley floor sedimentation in the Upper Mississippi Valley. *Annals of the Association of American Geographers*, **77**, 224–244.

Knox, J.C. (1993). Large increases in flood magnitude in response to modest changes in climate. *Nature*, **361**, 430–432.

Kondolf, G.M. and Wilcock, P.R. (1996). The flushing flow problem: defining and evaluating objectives. *Water Resources Research*, **32**, 2589–2599.

Kresan, P.L. (1988). The Tucson, Arizona, flood of October 1983. In *Flood Geomorphology*, eds. V.R. Baker, R.C. Kochel, and P.C. Patton, pp. 465–489. New York: John Wiley and Sons.

Kuhl, S.C. and Miller, J.R. (1992). Seasonal river runoff calculated from a global atmospheric model. *Water Resources Research*, **28**, 2029–2039.

Kwadijk, J. and Rotmans, J. (1995). The impact of climate change on the River Rhine: a scenario study. *Climatic Change*, **30**, 397–425.

Lagasse, P.F. (1981). Geomorphic response of the Rio Grande to dam construction. *New Mexico Geological Society, Special Publication No. 10*, pp. 27–46.

Lagasse, P.F., Winkley, B.R., and Simons, D.B. (1980). Impact of gravel mining on river system stability. *ASCE Journal of the Waterway, Port, Coastal, and Ocean Division*, **106**, 389–404.

Lake, J.S. (1975). Fish of the Murray River. In *The Book of the Murray*, eds. G.C. Lawrence and G.K. Smith, pp. 213–224. Adelaide, Australia: Rigby.

Lal, M. and Chander, S. (1993). Potential impacts of greenhouse warming on the water resources of the Indian subcontinent. *Journal of Environmental Hydrology*, **1**, 3–13.

Langbein, W.B. and Schumm, S.A. (1958). Yield of sediment in relation to mean annual precipitation. *Transactions, American Geophysical Union*, **39**, 1076–1084.

Langedal, M. (1997). The influence of a large anthropogenic sediment source on the fluvial geomorphology of the Knabeåna-Kvina rivers, Norway. *Geomorphology*, **19**, 117–132.

Leopold, L.B. (1968). Hydrology for urban land planning – a guidebook on the hydrologic effects of urban land use. *U.S. Geological Survey Circular No. 554*, 18 pp.

Leopold, L.B. and Maddock, T.W. Jr. (1953). The hydraulic geometry of stream channels and some physiographic implications. *U.S. Geological Survey Professional Paper 252*, 57 pp.

Lettenmaier, D.P. and Gan, T.Y. (1990). Hydrologic sensitivities of the Sacramento-San Joaquin River basin, California, to global warming. *Water Resources Research*, **26**, 69–86.

Lewin, J., Davies, B.E., and Wolfenden, P.J. (1977). Interactions between channel change and historic mining sediments. In *River Channel Changes*, ed. K.J. Gregory, pp. 353–367. Chichester: John Wiley and Sons.

Ligon, F.K., Dietrich, W.E., and Trush, W.J. (1995). Downstream ecological effects of dams. *BioScience*, **45**, 183–192.

Limbrey, S. (1983). Archaeology and palaeohydrology. In *Background to Palaeohydrology: A Perspective*, ed. K.J. Gregory, pp. 189–212. Chichester: John Wiley and Sons.

Lvovich, M.I. and Chernogaeva, G.M. (1977). The water balance of Moscow. In *Effects of Urbanization and Industrialization on the Hydrological Regime and on Water Quality*. IAHS Publ. 123, pp. 48–51. Wallingford, UK.

Madsen, S.W. (1995). *Channel Response Associated with Predicted Water and Sediment Yield Increases in Northwest Montana*. Ft. Collins, Colorado: Unpublished MS thesis, Colorado State University, 230 pp.

Marlow, C.B. and Pogacnik, T.M. (1985). Time of grazing and cattle-induced damage to streambanks. In *Riparian Ecosystems and Their Management; Reconciling Conflicting Uses*. First North American Riparian Conference, Tucson, Arizona, USDA Forest Service General Technical Report RM-120, pp. 279–284.

Masterman, R. and Thorne, C.R. (1992). Predicting influence of bank vegetation on channel capacity. *Journal of Hydraulic Engineering*, **118**, 1052–1058.

McCabe, G.J. Jr. and Hay, L.E. (1995). Hydrological effects of hypothetical climate change in the East River basin, Colorado, USA. *Hydrological Sciences Journal*, **40**, 303–318.

McClintock, K. and Harbor, J.M. (1995). Modeling potential impacts of land development on sediment yields. *Physical Geography*, **16**, 359–370.

McLeay, D.J., Birtwell, I.K., Hartman, G.F., and Ennis, G.L. (1987). Responses of Arctic grayling (*Thymallus arcticus*) to acute and prolonged exposure to Yukon placer mining sediment. *Canadian Journal of Fisheries and Aquatic Sciences*, **44**, 658–673.

Megahan, W.F. and Bohn, C.C. (1989). Progressive, long-term slope failure following road construction and logging on noncohesive, granitic soils of the Idaho Batholith. In *Headwaters Hydrology*, pp. 501–510. Bethesda, Maryland: American Water Resources Association.

Megahan, W.F. and Kidd, W.J. (1972). Effects of logging and logging roads on erosion and sediment deposition from steep terrain. *Journal of Forestry*, **70**, 136–141.

Mei-e, R. and Xianmo, Z. (1994). Anthropogenic influences on changes in the sediment load of the Yellow River, China, during the Holocene. *The Holocene*, **4**, 314–320.

Miller, R.A., Troxell, J., and Leopold, L.B. (1971). Hydrology of two small river basins in Pennsylvania before urbanization. *U.S. Geological Survey Professional*

Paper 701-A, 57 pp. U.S. Government Printing Office, Washington, DC.

Mimikou, M.A. (1995). Climatic change. In *Environmental Hydrology*, ed. V.P. Singh, pp. 69–106. The Netherlands: Kluwer Academic Publishers.

Mimikou, M.A., Hadjisavva, P.S., and Kouvopoulos, Y.S. (1991). Regional effects of climate change on water resources systems. In *Hydrology for the Water Management of Large River Basins*, eds. F.H.M. van de Ven, D. Gutknecht, D.P. Loucks, and K.A. Salewicz, pp. 173–182. IAHS Publ. no. 201. Wallingford, UK.

Mitsch, W.J. and Gosselink, J.G. (1993). *Wetlands*, 2nd ed. New York: Reinhold, 722 pp.

Montgomery, D.R. and Dietrich, W.E. (1994). A physically based model for the topographic control on shallow landsliding. *Water Resources Research*, **30**, 1153–1171.

Morisawa, M.E. and LaFlure, E. (1979). Hydraulic geometry, stream equilibrium and urbanization. In *Adjustments of the Fluvial System*, eds. D.D. Rhodes and G.P. Williams, pp. 333–350. Allen and Unwin. Boston.

Mosley, M.P. (1978). Erosion in the south-east Ruahine Range: its implications for downstream river control. *New Zealand Journal of Forestry*, **23**, 21–48.

Müller, K. (1962). Limnologisch-Fischereibiologische Untersuchungen in regulierten Gewässern Schwedisch Lapplands. *Oikos*, **13**, 125–154. (In Brookes, 1988).

Mustonen, S.E. and Seuna, P. (1971). Metsaojituksen vaikutuksesta suon hydrologiaan. Vesientutkimuslaitoksen Julkaisuja 2: In, D.E. Walling, (1987), Hydrological processes, in *Human Activity and Environmental Processes*, eds. K.J. Gregory and D.E. Walling, pp. 53–85. Chichester: John Wiley and Sons.

Myers, T.J. and Swanson, S. (1991). Aquatic habitat condition index, stream type, and livestock bank damage in northern Nevada. *Water Resources Bulletin*, **27**, 667–677.

Myers, T.J. and Swanson, S. (1992). Variation of stream stability with stream type and livestock bank damage in northern Nevada. *Water Resources Bulletin*, **28**, 743–754.

Myers, T.J. and Swanson, S. (1996a). Long-term aquatic habitat restoration: Mahogany Creek, Nevada, as a case study. *Water Resources Bulletin*, **32**, 241–252.

Myers, T.J. and Swanson, S. (1996b). Temporal and geomorphic variations of stream stability and morphology: Mahogany Creek, Nevada. *Water Resources Bulletin*, **32**, 253–265.

Naiman, R.J., Melillo, J.M., and Hobbie, J.E. (1986). Ecosystem alteration of boreal forest streams by beaver (*Castor canadensis). Ecology*, **67**, 1254–1269.

Nakamura, F., Sudo, T., Kameyama, S., and Jitsu, M. (1997). Influences of channelization on discharge of suspended sediment and wetland vegetation in Kushiro Marsh, northern Japan. *Geomorphology*, **18**, 279–289.

Nanson, G.C. and Young, R.W. (1981). Downstream reduction of rural channel size with contrasting urban effects in small coastal streams of southeastern Australia. *Journal of Hydrology*, **52**, 239–255.

Nash, L.L. and Gleick, P.H. (1991). Sensitivity of streamflow in the Colorado Basin to climatic changes. *Journal of Hydrology*, **125**, 221–241.

National Hydrology Research Institute. (1982). *NHRI Research Project: Agricultural Land Drainage*. Ottawa: Environment Canada, 38 pp.

Newson, M.D. and Robinson, M. (1983). Effects of agricultural drainage on upland streamflow: case studies in mid-Wales. *Journal of Environmental Management*, **17**, 333–348.

NHRAIC (Natural Hazards Research and Applications Information Center). (1992). *Floodplain Management in the US: An Assessment Report*. Summary Report, Federal Interagency Floodplain Management Task Force, **1**, 69 pp. Washington, DC.

Nik, A.R. (1988). Water yield changes after forest conversion to agricultural landuse in Peninsular Malaysia. *Journal of Tropical Forest Science*, **1**, 67–84.

Nilsson, C., Ekblad, A., Gardfjell, M., and Carlberg, B. (1991). Long-term effects of river regulation on river margin vegetation. *Journal of Applied Ecology*, **28**, 963–987.

North, G.R., Calahan, R.F., and Coakley, J.A. (1981). Energy balance climate models. *Reviews of Geophysics and Space Physics*, **19**, 19–121.

O'Connor, J.E., Baker, V.R., and Webb, R.H. (1986). Paleohydrology of pool and riffle pattern development, Boulder Creek, Utah. *Geological Society of America Bulletin*, **97**, 410–420.

Olson, R. and Hubert, W.A. (1994). *Beaver: Water Resources and Riparian Habitat Manager*. Laramie: University of Wyoming, 48 pp.

Oswood, M.W., Milner, A.M., and Irons, J.G., III. (1992). Climate change and Alaskan rivers and streams. In *Global Climate Change and Freshwater Ecosystems*, eds. P. Firth and S.G. Fisher, pp. 192–210. New York: Springer-Verlag.

Ott, M., Su, Z., Schumann, A.H., and Schultz, G.A. (1991). Development of a distributed hydrological model for flood forecasting and impact assessment of land-use change in the international Mosel River basin. In *Hydrology for the Water Management of Large River Basins*, eds. F.H.M. van de Ven, D. Gutknecht, D.P. Loucks, and K.A. Salewicz, pp. 183–194. IAHS Publ. no. 201. Wallingford, UK.

Panagoulia, D. (1995). Assessment of daily catchment precipitation in mountainous regions for climate change interpretation. *Hydrological Sciences Journal*, **40**, 331–350.

Park, C.C. (1977). Man-induced changes in stream channel capacity. In *River Channel Changes*, ed. K.J. Gregory, pp. 121–144. Chichester: John Wiley and Sons.

Parker, D.J. (1995). Floodplain development policy in England and Wales. *Applied Geography*, **15**, 341–363.

Parker, G. and Andres, D. (1976). Detrimental effects of river channelization. *Rivers '76*, pp. 1248–1266. American Society of Civil Engineering. New York, N.Y.

Parker, M. (1986). *Beaver, Water Quality, and Riparian Systems*, Larmie: Wyoming Water and Streamside Zone Conferences, Wyoming Water Research Center, University of Wyoming.

Patrick, R. (1995). *Rivers of the United States*, vol. 2, *Chemical and Physical Characteristics*, chp. 6, Wetlands, pp. 142–194. New York: John Wiley and Sons.

Pearlstine, L., McKellar, H., and Kitchens, W. (1985). Modelling the impacts of a river diversion on bottomland forest communities in the Santee River floodplain, South Carolina. *Ecological Modelling*, **29**, 283–302.

Pearthree, M.S. (1982). *Channel Change in the Rillito Creek System, Southeastern Arizona: Implications for Floodplain Management*. Tucson, Arizona: Unpublished MS thesis, University of Arizona, 128 pp.

Penning-Rowsell, E., Peerbolte, B., Correia, F.N., Fordham, M., Green, C., Pflüger, W., Rocha, J., da-Graca Saraiva, M., Schmidtke, R., Torterotot, J.P., and Van der Veen, A. (1992). Flood vulnerability analysis and climate change: towards a European methodology. In *Floods and Flood Management*, ed. A.J. Saul, pp. 343–361. Dordrecht: Kluwer.

Petts, G.E. (1977). Channel response to flow regulation: the case of the River Derwent, Derbyshire. In *River Channel Changes*, ed. K.J. Gregory, pp. 145–164. Chichester: John Wiley and Sons.

Petts, G.E. (1984). *Impounded Rivers: Perspectives for Ecological Management*. Chichester: John Wiley and Sons, 326 pp.

Petts, G. (1989). Historical analysis of fluvial hydrosystems. In *Historical Change of Large Alluvial Rivers: Western Europe*, ed. G.E. Petts, pp. 1–18. Chichester: John Wiley and Sons.

Petts, G.E. (1990). Forested river corridors. In *Water, Engineering and Landscape*, eds. D. Cosgrove and G. Petts, pp. 188–208. London: Belhaven Press.

Piégay, H. and Bravard, J.-P. (1997). Response of a Mediterranean riparian forest to a 1 in 400 year flood, Ouveze River, Drome-Vaucluse, France. *Earth Surface Processes and Landforms*, **22**, 31–43.

Piégay, H. and Gurnell, A.M. (1997). Large woody debris and river geomorphological pattern: examples from S.E. France and S. England. *Geomorphology*, **19**, 99–116.

Piégay, H. and Salvador, P.-G. (1997). Contemporary floodplain forest evolution along the middle Ubaye River, southern Alps, France. *Global Ecology and Biogeography Letters*, **6**, 1–10.

Pitlick, J. (1994). Relation between peak flows, precipitation, and physiography for five mountainous regions in the western USA. *Journal of Hydrology*, **158**, 219–240.

Pitlick, J. (1997). A regional perspective of the hydrology of the 1993 Mississippi River basin floods. *Annals of the Association of American Geographers*, **87**, 135–151.

Pitlick, J. and Van Steeter, M. (1998). Geomorphology and endangered fish habitats of the upper Colorado River. 2. Linking sediment transport to habitat maintenance. *Water Resources Research*, **34**, 303–316.

Pizzuto, J.E. (1994). Channel adjustments to changing discharges, Powder River, Montana. *Geological Society of America Bulletin*, **106**, 1494–1501.

Poff, N.L. (1992). Regional hydrologic response to climate change: an ecological perspective. In *Global Climate Change and Freshwater Ecosystems*, eds. P. Firth and S.G. Fisher, pp. 88–115. New York: Springer-Verlag.

Potter, K.W. (1991). Hydrological impacts of changing land management practices in a moderate-sized agricultural catchment. *Water Resources Research*, **27**, 845–855.

Potzger, J.E. (1934). A notable case of bog formation. *American Midland Naturalist*, **15**, 567–580.

Ramachandran, R. and Thakur, S.C. (1974). India and the Ganga floodplain. In *Natural Hazards: Local, National, Global*, ed. G.F. White, pp. 36–43. New York: Oxford University Press.

Rao, P.G. (1993). Climatic changes and trends over a major river basin in India. *Climate Research*, **2**, 215–223.

Rao, P.G. (1995). Effect of climate change on streamflows in the Mahanadi River basin, India. *Water International*, **20**, 205–212.

Reid, L.M. and Dunne, T. (1984). Sediment production from forest road surfaces. *Water Resources Research*, **20**, 1753–1761.

Reily, P.W. and Johnson, W.C. (1982). The effects of altered hydrologic regime on tree growth along the Missouri River in North Dakota. *Canadian Journal of Botany*, **60**, 2410–2423.

Revelle, R.R. and Suess, H.E. (1957). Carbon dioxide exchange between atmosphere and ocean and the question of an increase of atmospheric CO_2 during the past decades. *Tellus*, **9**, 18–27.

Rhoads, B.L. (1990). The impact of stream channelization on the geomorphic stability of an arid-region river. *National Geographic Research*, **6**, 157–177.

Richards, K.S. and Wood, R. (1977). Urbanization, water redistribution, and their effect on channel processes. In *River Channel Changes*, ed. K.J. Gregory, pp. 369–388. Chichester: John Wiley and Sons.

Richardson, B.A. (1981). Fish kill in the Belmore River, Macleay River drainage, NSW, and the possible influence of flood mitigation works. *Proc., Floodplain Management Conference Australian Water Resources Council*, Conf. Series No. 4, Canberra: Australian Government Public Service, pp. 50–60.

Roberts, C.R. (1989). Flood frequency and urban-induced channel change: some British examples. In *Floods: Hydrological, Sedimentological and Geomorphological Implications*, eds. K. Beven and P. Carling, pp. 57–82. Chichester: John Wiley and Sons.

Roberts, G., Hudson, J., Leeks, G., and Neal, C. (1994). The hydrological effects of clear-felling established coniferous forestry in an upland area of mid-Wales. In *Integrated River Basin Development*, eds. C. Kirby and W.R. White, pp. 187–199. Chichester: Wiley.

Ruslan, I.W. (1995). Impact of urbanisation and uphill land clearances on the sediment yield of an urbanising catchment of Pulau Pinang, Malaysia. In *Postgraduate Research in Geomorphology; Selected Papers from the 17th BGRG Postgraduate Symposium*, eds. S.J. McLelland, A.R. Skellern, and P.R. Porter, pp. 28–33. University of Leeds: British Geomorphological Research Group, School of Geography.

Ryan, S. (1997). Morphologic response of subalpine streams to transbasin flow diversion. *Journal of the American Water Resources Association*, **33**, 839–854.

Sandoz, M. (1964). *The Beaver Men: Spearheads of Empire*. Lincoln, Nebraska: University of Nebraska Press, 335 pp.

SAST (Scientific Assessment and Strategy Team). (1994). Science for *Floodplain Management into the 21st Century*. Preliminary report. Washington, DC: U.S. Government Printing Office.

Schleusener, R.A., Smith, G.L., and Chen, M.C. (1962). Effect of flow diversion for irrigation on peak rates of runoff from watersheds in and near the Rocky Mountain foothills of Colorado. *International Association of Hydrologists Bulletin*, **7**, 53–61.

Schlosser, I.J. (1990). Environmental variation, life history attributes, and community structure in stream fishes: implications for environmental management assessment. *Environmental Management*, **14**, 621–628.

Schneider, S.H., Gleick, P.H., and Mearns, L.O. (1990). Prospects for climate change. In *Climate Change and U.S. Water Resources*, ed. P.E. Waggonner, pp. 41–73. New York: John Wiley and Sons.

Schoof, R. (1980). Environmental impact of channel modification. *Water Resources Bulletin*, **16**, 697–701.

Sedell, J.R., Everest, F.H., and Swanson, F.J. (1982). Fish habitat and streamside management: past and present. In *Proceedings of the Society of American Foresters*, Annual Meeting, pp. 244–255.

Sedell, J.R. and Froggatt, J.L. (1984). Importance of streamside forests to large rivers: the isolation of the Willamette River, Oregon, USA, from its floodplain by snagging and streamside forest removal. *Verhandlungen International Verein. Limnol.*, **22**, 1828–1834.

Shotton, F.W. (1978). Archaeological inferences from the study of alluvium in the lower Severn-Avon Valleys. In *The Effect of Man on the Landscape: The Lowland Zone*, eds. S. Limbrey and I.G. Evans, pp. 27–32. Council for British Archaeology, London.

Siegenthaler, U. and Oeschger, H. (1987). Biospheric CO_2 emissions during the last 200 years reconstructed by the deconvolution of ice core data. *Tellus*, **398**, 140–154.

Smith, N. (1971). *A History of Dams*. London: Peter Davies.

Stanford, J.A. and Hauer, F.R. (1992). Mitigating the impacts of stream and lake regulation in the Flathead River catchment, Montana, USA: an ecosystem perspective. *Aquatic Conservation: Marine and Freshwater Ecosystems*, **2**, 35–63.

Starkel, L. (1988). Tectonic, anthropogenic and climatic factors in the history of the Vistula River valley downstream of Cracow. In *Lake, Mire and River Environments During the Last 15,000 Years*, eds. G. Lang and C. Schluchter, pp. 161–170. Rotterdam: A.A. Balkema.

Stromberg, J.C. and Patten, D.T. (1989). Instream flow requirements for riparian vegetation. *ASCE Proceedings on Legal, Institutional, Financial, and Environmental Aspects of Water Issues*, July 1989, University of Delaware, pp. 123–130. New York, N.Y.

Stromberg, J.C. and Patten, D.T. (1990). Riparian vegetation in stream flow requirements: a case study from a diverted stream in the eastern Sierra Nevada, California, USA. *Environmental Management*, **14**, 185–194.

Swales, S. (1980). *Investigations of the Effects of River Channel Works on the Ecology of Fish Populations*. Unpublished PhD dissertation, University of Liverpool, United Kingdom (in Brookes, 1988).

Swales, S. (1982). Environmental effects of river channel works used in land drainage improvement. *Journal of Environmental Management*, **14**, 103–126.

Swanston, D.N. and Swanson, F.J. (1976). Timber harvesting, mass erosion and steepland forest geomorphology in the Pacific Northwest. In *Geomorphology and Engineering*, ed. D.R. Coates, pp. 199–221. Stroudsburg, Pennsylvania: Dowden, Hutchinson, and Ross.

Swift, B.L. (1984). Status of riparian ecosystems in the United States. *Water Resources Bulletin*, **20**, 223–228.

Tesch, F.W. and Albrecht, M.L. (1961). Uber den Einfluss verschiedener Umweltfaktoren auf Wachstum und Bestand der Bachforelle (*Salmo trutta fario* L.) in Mittelgebirgsgewasser. *Verhandlungen der internationalen Vereinigung für theoretische und angewandte Limnologie*, **14**, 763–768.

Tiner, R.W. Jr (1984). *Wetlands of the US: Current Status and Recent Trends*. U.S. Fish and Wildlife Service, 59 pp. Washington, DC.

Tobin, G.A. (1995). The levee love affair: a stormy relationship. *Water Resources Bulletin*, **31**, 359–367.

Trimble, S.W. (1994). Erosional effects of cattle on streambanks in Tennessee, USA. *Earth Surface Processes and Landforms*, **19**, 451–464.

Trimble, S.W. and Mendel, A.C. (1995). The cow as a geomorphic agent – a critical review. *Geomorphology*, **13**, 233–253.

Urbonas, B. and Benik, B. (1995). Stream stability under a changing environment. In *Stormwater Runoff and Receiving Systems: Impact, Monitoring, and Assessment*, ed. E.E. Herricks, pp. 77–101. Boca Raton, Florida: Lewis Publishers.

U.S. National Academy of Sciences. (1983). *Changing Climate*. Washington, DC: National Academy Press.

Van Geen, A. and Chase, Z. (1998). Recent mine spill adds to contamination of southern Spain. *EOS, Transactions, American Geophysical Union*, **79**, 449–455.

Van Haveren, B.P. (1991). Placer mining and sediment problems in interior Alaska. In *Proceedings of the 5th Federal Interagency Sedimentation Conference*, eds. S.S. Fan and Y.H. Kuo, **2**, 10-69–10-73. Los Vegas, Nevada.

Van Nieuwenhuyse, E.E. and LaPerriere, J.D. (1986). Effects of placer gold mining on primary production in subarctic streams of Alaska. *Water Resources Bulletin*, **22**, 91–99.

Van Steeter, M.M. and Pitlick, J. (1998). Geomorphology and endangered fish habitats of the upper Colorado River. 1. Historic changes in streamflow, sediment load, and channel morphology. *Water Resources Research*, **34**, 287–302.

Veatch, A.C. (1906). Geology and underground water resources of northern Louisiana and southern Arkansas. *U.S. Geological Survey Professional Paper 46*, 422 pp. U.S. Government Printing Office, Washington, DC.

Viner, D. (1994). Climate change modelling and climate change scenario construction methods for impacts assessment. In *Climate Impacts LINK Symposium Proceedings 19–21 September*, Norwich: Climatic Research Unit, University of East Anglia.

Vischer, D. (1989). Impact of 18th and 19th century river training works: three case studies from Switzerland. In *Historical Change of Large Alluvial Rivers: Western Europe*, eds. G.E. Petts, H. Moller, and A.L. Roux, pp. 19–40. Chichester: John Wiley and Sons.

Wallace, J.B. (1990). Recovery of lotic macroinvertebrate communities from disturbance. *Environmental Management*, **14**, 605–620.

Walling, D.E. and Webb, B.W. (1983). Patterns of sediment yield. In *Background to Palaeohydrology*, ed. K.J. Gregory, pp. 69–100. Chichester: John Wiley and Sons.

Whetton, P.H. and Baxter, J.T. (1989). The Southern Oscillation and river behavior in south-east Australia. In *Climanz 3*, eds. T.H. Donnelly and R.J. Wasson, pp. 67–69. Canberra, Australia: CSIRO Division of Water Resources.

White, G.F., Calef, W.C., Hudson, J.W., Mayer, H.M., Sheaffer, J.R., and Volk, D.J. (1958). *Changes in Urban Occupance of Flood Plains in the United States*. Research Paper 57, Chicago: Department of Geography, University of Chicago.

Whitley, J.R. and Campbell, R.S. (1974). Some aspects of water quality and biology of the Missouri River. *Transactions Missouri Academy of Science*, **8**, 60–72.

Wilby, R.L. (1995). Greenhouse hydrology. *Progress in Physical Geography*, **19**, 351–369.

Wilcock, P.R., Kondolf, G.M., Graham Matthews, W.V., and Barta, A.F. (1996). Specification of sediment maintenance flows for a large gravel-bed river. *Water Resources Research*, **32**, 2911–2921.

Williams, G.P. and Wolman, M.G. (1984). Effects of dams and reservoirs on surface-water hydrology; changes in rivers downstream from dams. *U.S. Geological Survey Professional Paper 1286*, 83 pp. U.S. Government Printing Office, Washington, DC.

Williams, G.P. and Wolman, M.G. (1985). Effects of dams and reservoirs on surface-water hydrology – changes in rivers downstream from dams. *U.S. Geological Survey, National Water Summary 1985*, pp. 83–88.

Winkley, B.R. (1994). Response of the lower Mississippi River to flood control and navigation improvements. In *The Variability of Large Alluvial Rivers*, eds. S.A. Schumm and B.R. Winkley, pp. 45–74. New York: ASCE Press.

Wisler, C.O. and Brater, E.F. (1965). *Hydrology*. New York: John Wiley and Sons, pp. 55–56.

Wohl, E.E. (1992). Bedrock benches and boulder bars: Floods in the Burdekin Gorge of Australia. *Geological Society of America Bulletin*, **104**, 770–778.

Wohl, E.E., Webb, R.H., Baker, V.R., and Pickup, G. (1994). *Sedimentary Flood Records in the Bedrock Canyons of Rivers in the Monsoonal Region of Australia*. Ft. Collins, Colorado: Colorado State University Water Resources Paper 107, 102 pp.

Wolman, M.G. (1967). A cycle of sedimentation and erosion in urban river channels. *Geografiska Annaler*, **49A**, 385–395.

Wolman, M.G. and Gerson, R. (1978). Relative scales of time and effectiveness of climate in watershed geomorphology. *Earth Surface Processes*, **3**, 189–208.

Wolman, M.G. and Miller, J.P. (1960). Magnitude and frequency of forces in geomorphic processes. *Journal of Geology*, **68**, 54–74.

Wolman, M.G. and Schick, A.P. (1967). Effects of construction on fluvial sediment, urban and suburban areas of Maryland. *Water Resources Research*, **3**, 451–464.

Wyzga, B. (1996). Changes in the magnitude and transformation of flood waves subsequent to the channelization of the Raba River, Polish Carpathians. *Earth Surface Processes and Landforms*, **21**, 749–763.

Zalumi, S.G. (1970). The fish fauna of the lower reaches of the Dnieper: its present composition and some features of its formation under conditions of regulated and reduced river discharge. *Journal of Ichthyology*, **10**, 587–596.

Ziemer, R.R. (1981). Storm flow response to road building and partial cutting in small streams of northern California. *Water Resources Research*, **17**, 907–917.

FLOOD PROCESSES AND EFFECTS

CHAPTER 5

Inundation Hydrology

Leal A.K. Mertes
Department of Geography
University of California
at Santa Barbara

Inundation hydrology is the study of the water sources that contribute to inundation of a floodplain at the local scale of a river reach. Production of runoff from upstream watersheds is only one of many components that contribute to inundation hydrology. Interior flooding, in which the floodplain surface is inundated before overbank flooding from the river, may result from local groundwater rise, from tributary inputs, or from precipitation directly onto a fully saturated floodplain surface. Inundation hydrology varies spatially along a river depending on the geomorphology and temporally depending on the magnitude and timing of the flood. For global comparisons among channel-floodplain systems it is possible to quantify the relative influence of different hydrologic and geomorphic processes on inundation hydrology by considering flood hydroclimatology, the landscape characteristics that produce or regulate flood-producing runoff, and the human impacts that may exacerbate or ameliorate flood potential. This type of work requires an array of data types and modeling techniques. The Amazon River in Brazil and the Altamaha River in Georgia (USA) are used as reference sites to illustrate the conditions that promote development of mixing zones on inundated floodplains. A theoretical framework for analysis of the patterns of inundation hydrology is presented, by using a Geographic Information System-based method applied to the Altamaha, Tone (Japan), Ob'Irtysh (Russia), Brahmaputra (Bangladesh), and Jubba and Shebele (Somalia) Rivers. The patterns of inundation hydrology have implications for the distribution of sediment, spread of contaminants, floodplain ecology, and flood hazards.

Introduction

Floods are one of the greatest natural hazards that affect humans. How is it that, after centuries of study and hundreds of billions of dollars invested in flood control, the seemingly simple process of river flooding continues to inflict damage with floods of unexpected intensity, extent, and duration? The answer lies in the fact that flooding is a phenomenon that depends on complex interactions among climate, landscape, and human intervention. The focus of this chapter is to elucidate the components of inundation hydrology in order to better understand river flooding and the landscapes it produces. Inundation hydrology is defined here as the study of the water sources that contribute to the inundation of a floodplain at the local scale of a river reach.

Recent extreme floods have focused new attention on this very old problem. For example, the 1997 flooding in southern Somalia was astounding in its extent and persistence, just as were the 1993 floods on the Mississippi–Missouri Rivers, the 1996–1997 floods on the Sacramento–San Joaquin Rivers, the 1997 floods on the Red River in the United States, and the 1999 floods in Hungary. Newspaper reports from March 1999 described flooding in Hungary that caused the evacuation of hundreds of people and loss of several homes. The flooding was described as specifically due not to rivers overflowing but instead to excess saturation of the ground, with an expected duration of the flood conditions for 3–5 weeks.

Wherever they occur, extreme floods shake the public's confidence in government and in the engineering solutions that were installed to protect people from the dangers of flooding rivers. Current discussions in

California about constructing floodways with levees set back from the river channels on the Sacramento–San Joaquin delta suggest that there is a realization in the United States that floods will occur despite massive engineering efforts (Haeuber and Michener, 1998). Added to this new distrust of engineering solutions is the public's growing perception that climate changes could result in seemingly unpredictable episodes of massive flooding or drought, for which there has been no planning.

The failure of engineering systems to prevent floods is in part the result of an incomplete analysis of the factors that cause floods. The traditional engineering view of floods is that water runs off hillslopes into streams and converges downstream; a flood occurs if the capacity of the channel system is exceeded. New insight into the processes that control flood extent and persistence suggests that production of runoff from upstream watersheds is only one of many components that contribute to inundation hydrology (e.g., Mertes, 1997; Burt and Haycock, 1996). In their analysis of the 1993 Mississippi River floods, the Scientific Assessment and Strategy Team (SAST 1994, p. 44) referred to interior flooding, in which the floodplain surface was inundated before overbank flooding from the river. This interior flooding may have been the result of local groundwater rise or precipitation directly onto a fully saturated floodplain surface as experienced during the 1999 floods in Hungary. Traditional engineering models are not yet effective for describing such a phenomenon in which the floodplain is inundated before the river overtops its banks. Indeed, current models underestimate the extent and duration of inundation under this type of scenario.

Despite the hazards associated with floods, they also serve essential ecological, societal, and geomorphological functions. Floodplains not only function to support habitat for rich species diversity but also retard flood waves and store sediment. The hydrologic and mass transfer properties of a floodplain determine the degree to which biological and chemical processing imparts a chemical signal on the water, sediment, and dissolved materials as they pass across the floodplain into the river channel. Spatial heterogeneity of the landforms that control water and mass transfer is expected on hydrologically and geomorphically distinct floodplains. Yet, our knowledge of the dynamic properties of floodplains is limited by a tendency to study floodplains as separate from the channel instead of as a component of a channel–floodplain complex.

River channels and their floodplains share water. Bhowmik and Demissie (1982) described the spatial variability in the relative proportion of the total water flow carried by channels versus floodplains for several rivers based on field measurements of cross-sectional area and flow velocities during floods. They showed that the proportion of water or "carrying capacity" of the floodplains varied both with the magnitude as well as with the location of the flood along the stream network. Woltemade and Potter (1994) demonstrated with watershed models that even modest changes in the floodplain geomorphology could significantly change the amount of flow crossing a floodplain given a constant upstream watershed input. Mertes et al. (1995) showed, for the Amazon River, that the inundation hydrology varies spatially along the river depending on the geomorphology (Mertes et al., 1996) and temporally depending on the magnitude and timing of the annual flood (Richey et al., 1989; Mertes et al., 1993; Mertes, 1994).

For global comparisons among channel–floodplain systems it is possible to quantify the relative influence of different hydrologic and geomorphic processes on inundation hydrology by considering

- the hydroclimatology that produces a flood,
- the landscape characteristics that produce or regulate flood-producing runoff, and
- the human impacts that may exacerbate or ameliorate flood potential.

A new understanding arises from this type of analysis – namely, that from the Arctic to the Amazon large rivers often may only partially inundate their floodplains with river water during floods (Mertes 1997), and other nonriver sources of water can contribute substantially to inundation. In addition, the synoptic view of flooding in long river reaches (>100 km) provided by remote sensing data indicates that there is extensive mixing of different water types (e.g., river water, groundwater, and direct precipitation) on floodplains during flooding and that this mixing or *perirheic* zone (Mertes, 1997) may encompass a previously unidentified and significant floodplain ecotone.

Studying floods can be a challenging endeavor: floods are extreme events that can rise and subside rapidly. During floods, transportation systems are often incapacitated making timely observations difficult to achieve. Monitoring instruments are frequently destroyed or swept away during intense floods. In addition, the scale of large river systems is so vast as to try the abilities of the earth-bound scientist. For these reasons, research on patterns of inundation hydrology requires an array of data types and modeling techniques. Using hydrologic, precipitation, landscape, land-use, remote sensing, and other geographic data for rivers that experience different types of river flooding, it is possible to systematically examine the relevant patterns within a theoretical and empirical framework (Figure 5.1). Although field data on flow velocity, depth, and extent of flooding are critical (Wolman and Leopold, 1957; Velikanova and Yarnykh, 1970),

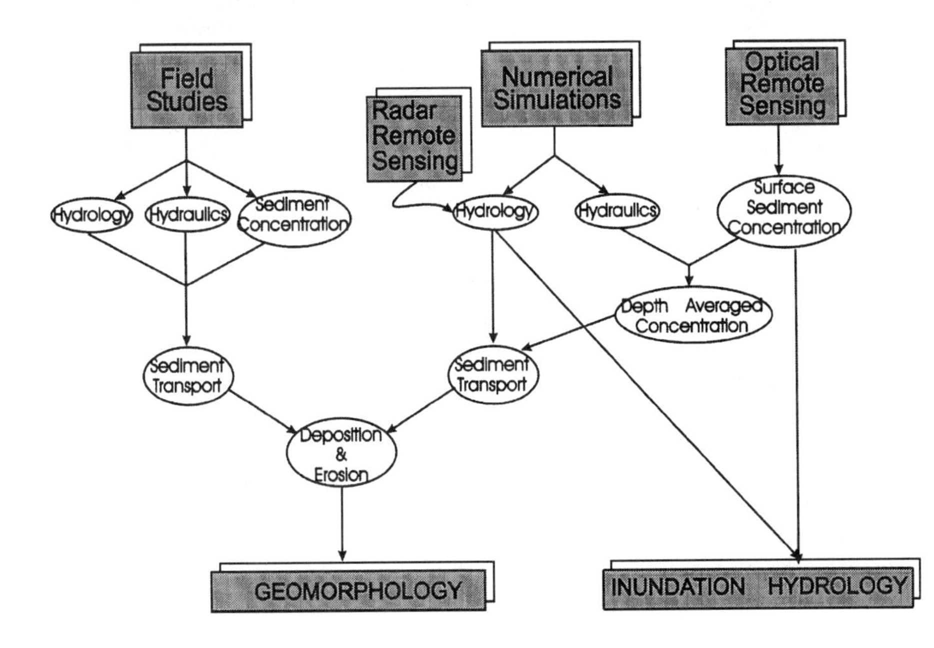

Figure 5.1. Data types and GIS analytical framework for the study of inundation hydrology and floodplain geomorphology.

use of remote sensing data and numerical simulations is rapidly changing our ability to examine patterns of inundation hydrology. Complementary to the use of optical remote sensing data for water quality (reviewed in this chapter) are microwave data for patterns of inundation extent and water surface height. A review of the use of microwave data for flood mapping is not included in this chapter, because excellent reviews by Hess et al. (1990; 1999), Smith (1997), and Townsend and Walsh (1998) cover the topic well. Recent, new analyses of microwave data include measuring large-scale changes in wetness (Sippel et al., 1998; Hamilton et al., 1996), mapping vegetation patterns keyed to a temporal inundation pattern (Hess, 1999), estimating relative heights of water surfaces (Koblinsky et al., 1993; Brakenridge et al., 1998), and developing maps of temporal changes in connectivity of water bodies on floodplains (Smith and Alsdorf, 1998).

By combining these different types of data within a Geographic Information System (GIS) it is possible to constrain interpretation of the results from one technique or data type with results from other types. For example, with respect to hydrodynamic modeling alone, Bates et al. (1998) demonstrated the impact of different degrees of topographic roughness as derived from fractal-generated topography on the local hydraulics of flooding water and duration of inundation based on results from a finite-element model of floodplain inundation. Mertes (1990, 1994) successfully compared rates of sediment deposition estimated from combining Landsat-derived sediment concentrations and computed water discharge from a two-dimensional hydrodynamic finite-difference model in a GIS to field measurements of deposition rates on the central Amazon floodplain. Within another GIS framework, inundation maps derived from microwave data were combined with maps of vegetation and flood elevations on the Roanoke River floodplain to characterize the potential effects of changing hydrology due to dam construction on duration of inundation and vegetation patterns (Townsend and Walsh, 1998).

In the first part of this chapter, the components of inundation hydrology are discussed with respect to the sources and types of water that contribute to inundation. Data from two rivers are shown as examples of reference sites for the conditions that promote development of mixing zones on inundated floodplains. A theoretical framework for analysis of the patterns of inundation hydrology then is discussed, with emphasis on a new GIS-based method that was applied to five river systems to examine potential water sources for inundation. Although considered essential to fully understanding inundation hydrology, the analysis presented in this chapter does not yet explicitly incorporate subsurface flows.

The significance of the analysis lies in the fact that floods will continue to occur and, if history is any lesson, floods will continue to have both a devastating and a nurturing effect. By ensuring first that the descriptions of flood-generating mechanisms are complete and then by systematizing our understanding of the patterns of inundation hydrology, it will be possible to more successfully predict the negative effects of floods such as potential loss of human life, damage to farmlands, and transport and deposition of contaminants. Simultaneously, by basing this understanding on the physics of water movement, it will be possible to determine the conditions of flooding critical for long-term sustainability of landscape heterogeneity, biodiversity, and ecosystem resilience in floodplain landscapes.

Inundation Hydrology

Sources of Water

Inundation of a floodplain is the result of flooding from different water sources and may occur before a river overtops its banks or levees. The components of inundation hydrology can be expressed in terms of the mass balance of water in a channel–floodplain reach (Figure 5.2), where the mass balance is expressed as a function of the volumetric change in storage of water in the reach (ΔS_r) with

$$\Delta S_r = \text{time} \times [Q_{in} + Q_{tr} + Q_l + Q_p + Q_{gw} - (Q_{out} + Q_e + Q_{gw})]. \quad (1)$$

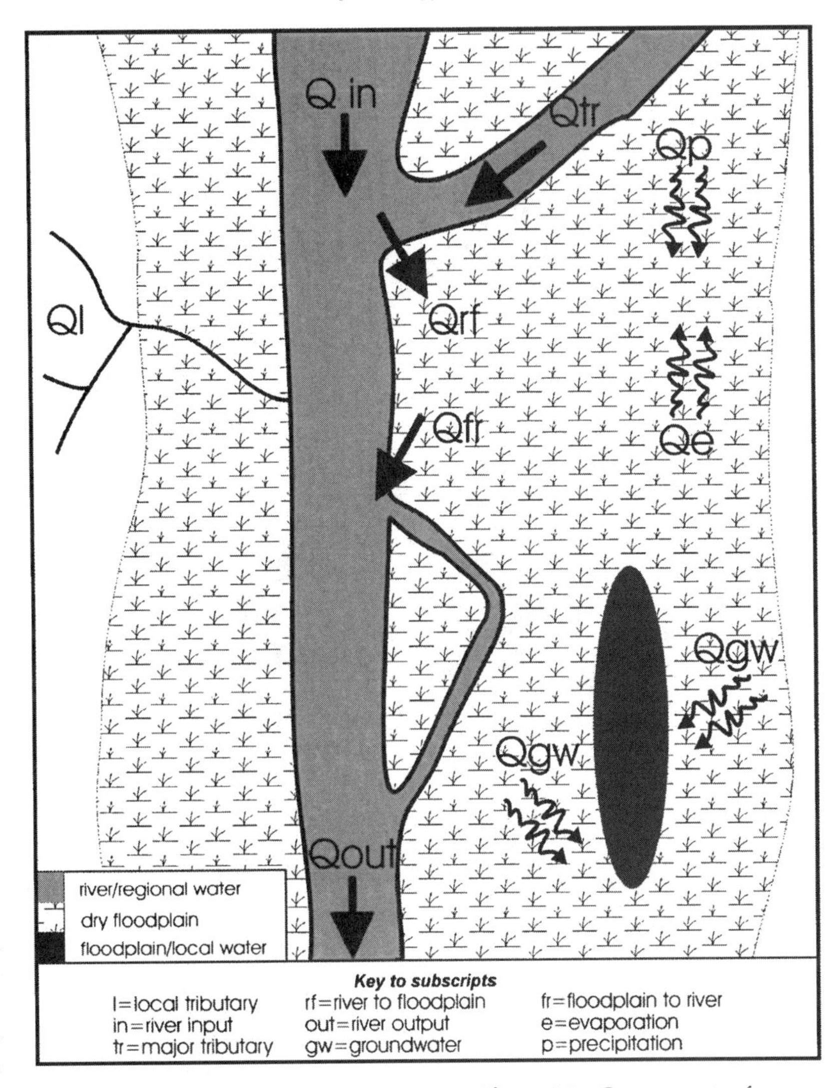

Figure 5.2. Components of inundation hydrology. Discharges for mass balance (Equations 1 and 2) are represented as Q with subscripts used to discriminate among the terms (after Mertes 1997).

Subscripted Q terms represent discharge as volume per unit time that are defined and illustrated in Figure 5.2. Positive terms are inputs and negative terms are outputs. Water can enter the reach from large and small tributaries, groundwater and precipitation onto the channel, floodplain, or slopes that drain directly onto the floodplain surface (e.g., Burt and Haycock, 1996). Evaporation from the channel and floodplain completes the mass balance for the reach.

If the water balance for just the floodplain were considered in terms of the volumetric change in storage (ΔS_f), then the exchange terms between the river and floodplain need to be included according to

$$\Delta S_f = \text{time} \times [Q_l + Q_p + Q_{rf} + Q_{gw} - (Q_{fr} + Q_e + Q_{gw})]. \quad (2)$$

Water travels to (Q_{rf}) and from (Q_{fr}) the main river channel through channelized and diffuse subsurface and surface flows. The mixing of river water and groundwater in the subsurface occurs in an area known as the hyporheic zone (Dahm and Valett, 1996). Hyporheic water can flow to or from the floodplain, thus contributing to both Q_{rf} and Q_{fr}. Surface flows from a channel to its floodplain occur during floods as nonchannelized

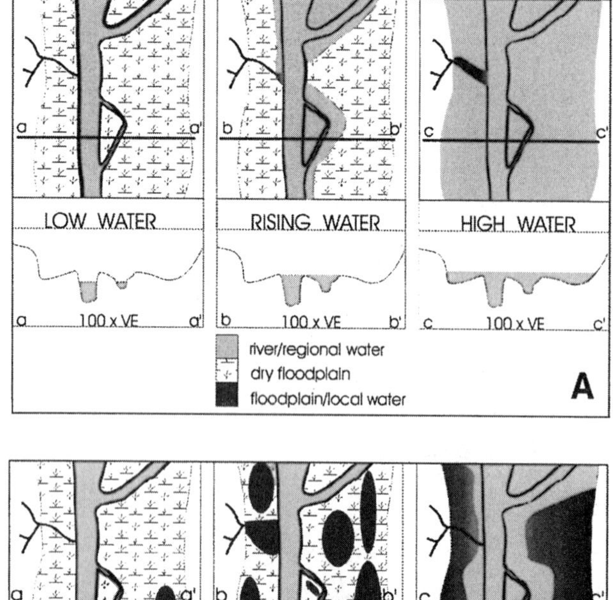

Figure 5.3. (A) Inundation pattern for a dry floodplain. Upper three graphs are planform views of a typical river, with cross sections shown in the lower three graphs with vertical exaggeration (VE) on the order of 100 times. Conditions for three water levels are illustrated, with mixing of water types shown for surface waters only. Subsurface waters are not illustrated (after Mertes 1997). (B) Inundation pattern for a saturated floodplain. Upper three graphs are planform views of a typical river, with cross sections shown in the lower three graphs with vertical exaggeration (VE) on the order of 100 times. Conditions for three water levels are illustrated, with mixing of water types shown for surface waters only. Subsurface waters are not illustrated (after Mertes 1997).

overbank flows and flow into channels on the floodplain surface. Draining of a floodplain during recession of a flood augments Q_{fr} (Hughes, 1978, 1980; Lewin and Hughes, 1980).

Patterns and Mechanisms of Flooding

The typical pattern described for movement of water between channels and their floodplains is embodied in the flood-pulse concept expressed by Junk et al. (1989) and illustrated in Figure 5.3A. Junk et al. (1989) described the flood-pulse as a moving zone of ever-increasing inundation on a floodplain. This pattern is common when the floodplain is essentially dry before overbank flooding from the river channel occurs. As depicted in Figure 5.3A, when the water rises the zone of flooding expands parallel to the river as levees are overtopped along the main channel and floodplain channels. Eventually, the entire floodplain is filled with increasingly deep river water, with the possible exception of areas near local tributaries where water from the main channel might mix with water from the tributary.

However, during floods the incursion of river water across the surface of a floodplain (Q_{rf}) may be resisted by water already present on the floodplain because of rising groundwater, hyporheic water, flooding of local tributaries, runoff from surrounding slopes, direct precipitation, water from melted snow, and antecedent water from prior floods. The saturation of the floodplain through these sources may occur before the crest of the floodwave and the onset of overbank flooding from the river channel and may produce the pattern of flooding illustrated schematically in Figure 5.3B.

The channel–floodplain system shown in Figure 5.3B is similar to that shown in Figure 5.3A, except for the presence of saturated soils or ponded water on the floodplain surface perhaps due to local soil differences and recent intense rainfall. As the water rises, but does not overtop its banks, rain may continue to fall directly on the floodplain and the floodplain will continue filling with local water. Although the river remains within its banks during the rising water period, large areas of saturated/ponded floodplain may form far from the river channel. By the time the river floods its banks, there is sufficient water on the floodplain so that the valley is only partially flooded by river water and mixing of the water types occurs.

The mixing zone is shown as parallel bands in Figure 5.3B. The particular pattern of mixing of surface waters will be a function of the composition of, and pressure distributions across, water bodies on the floodplain. The pressure distributions vary as a function of the water balances described by Equations 1 and 2. For example, a large floodplain lake with sufficient runoff from its local drainage basin could have sufficient head to completely prevent the entry of river water into the floodplain. This type of pressure balance could produce a well-defined boundary between the river and lake water. In contrast, a more heterogeneous boundary could be produced if intense rainfall occurs on the floodplain before the onset of overbank flow. As a result of the rain, patches of ponded water of multiple sizes could be distributed across the floodplain (Figure 5.3B). Overbank flow from the river may absorb the small patches of water. At the other extreme, the large patches may be surrounded by river water and may never be completely absorbed, thus producing a mixing zone that conforms approximately to the shape of the topography controlling the location of the ponding.

Mertes (1997) showed that the longitudinal zone resulting from surficial lateral gradients in turbidity and water chemistry on a floodplain is often extensive. This perirheic zone – i.e., zone of mixing surrounding (*peri*) the flowing (*rheo*) river water – is comparable to the hyporheic zone. The longitudinal perirheic zone includes the elements requisite for an ecotone – i.e., boundary, in the sense of Holland (1996) – in that it derives resources from multiple homogeneous patches and provides a temporally and spatially dynamic boundary between these resource patches. Examples of this type of mixing were recently provided by Sparks et al. (1998) for the upper Mississippi River and by Dietrich et al. (1999) for the Fly River, Papua New Guinea.

Remote Sensing Methods for Identification of Perirheic Zones. It is often possible to identify the zones influenced by river water in contrast to local water, because the water types may differ significantly in the quantity of suspended materials. Optical remote sensing data of resolution fine enough to involve at least four or five pixels across the flooded area provide an opportunity to document and characterize the perirheic zone across large floodplain areas. Optical data are ideally suited for analysis of relative concentrations of suspended materials (Dekker et al., 1995), because the upwelling radiance, which is the physical property measured by remote sensing detectors (Kirk, 1986), increases monotonically with an increase in the number of particles in suspension. In addition to the scattering of light by particles, suspended and dissolved materials and the water absorb light. All these effects combine to produce a nonlinear but monotonically increasing relationship between particle concentration and detected radiance.

Mertes et al. (1993) reviewed the use of optical remote sensing data for examining patterns of suspended sediment concentration and developed an application of spectral mixture analysis to estimate the concentration of suspended sediment in surface waters of rivers and flooded floodplains using the Amazon River as a type example. Spectral mixture analysis relies on least-squares techniques to determine meaningful gradients in image data (Smith et al., 1990; Mertes et al., 1993). The gradients or mixing lines are defined in terms of the spectra of selected end members. End members are defined as the purest representation of a material in the laboratory, in the field, or on an image. If the image can be corrected for instrument and atmospheric perturbations then absolute sediment concentration in milligrams per liter can be derived based on laboratory data (Mertes et al., 1993; Gomez et al., 1995). If the image cannot be corrected, then relative sediment concentrations can be determined (Mertes et al., 1995; Mertes 1997).

Both the calibrated and uncalibrated methods were used for the images shown in this chapter. An image from the Amazon was corrected to a surface reflectance (Figure 5.4A, see color section following p. 166.). The Amazon image was converted with coefficients calculated by Mertes et al. (1993). After correction to surface reflectance, suspended sediment concentrations were estimated for each pixel by the spectral mixture analysis technique. Reference spectral end members were based on laboratory reflectance data for sediment-water mixtures that included resuspended Mississippi River sediment (Witte et al., 1981), which was shown to be a reasonable surrogate for turbid Amazon waters (Mertes et al., 1993). Reflectance end-member fractions were related to absolute sediment concentrations by a nonlinear calibration curve (Mertes et al., 1993). The estimated error is ± 20 mg/L, and results were corroborated with field data from the Amazon (Mertes et al., 1993; Mertes, 1994).

When it is not possible to accurately estimate the appropriate radiometric and atmospheric corrections to produce images of surface reflectance, it is not possible to apply calibrations associated with laboratory spectra. For the Ob'-Irtysh and Somalian Rivers, described later in this chapter, it was not possible to produce surface reflectance images. Therefore, instead of relying on the laboratory end members, image end members for turbid and clear water were selected for these images. The relative amount of turbidity was then determined through application of spectral mixture analysis.

Field Methods for Identification of the Perirheic Zone. Field measurements of magnitude and direction of currents and suspended sediment concentrations were collected on the Amazon and Altamaha Rivers before and during flooding. Procedures described by Mertes (1994) and by Mertes et al. (1993) involve collection of a bottle sample for suspended sediment that was subsampled for filtering on tared filters with a pore size of 0.45 μm. Velocity measurements were made with a hand-held current meter over 2-min intervals.

Perirheic Reference Sites

Illustration of the types of mixing conditions depicted in Figure 5.3B is provided by examples from different types of data for two river systems, the Amazon River in Brazil and the Altamaha River in Georgia, USA (Figure 5.4). These two cases represent the type sites for perirheic conditions on inundated floodplains. The complexity of mixing of water types on the Amazon River floodplain has long been recognized (e.g., Sioli, 1957). The characterization of this mixing for the Amazon floodplain has usually been cast as the result of the unique functioning of a large river that is situated along the equator. The lagged inputs of water from the northern and southern hemispheres is seen as a primary cause for the complexity of Amazon hydrology (Richey et al., 1989; Vorosmarty et al., 1996). Additionally, lake systems and their watersheds contribute enormously to the presence of water from local sources on the floodplain (Forsberg et al., 1988; Lesack and Melack, 1995; Mertes et al., 1995). The discovery of perirheic conditions in the field on the much smaller Altamaha River led to the synthesis and framework for interpreting patterns of inundation hydrology described in this chapter.

Reference Site – Amazon River, Brazil. Analysis of Landsat images showed that the inundation hydrology of the Amazon River varies seasonally (Mertes et al., 1993; Mertes, 1994) and spatially (Mertes et al., 1995), confirming modeling results reported by Richey et al. (1989). The central Amazon River is shown in Figure 5.4A during a very high flood that occurred in summer 1989. Most natural river levees throughout this reach had been overtopped at the time this image was recorded (Mertes, 1990); yet the most striking feature of the image is the narrowness of the band of rainbow-colored sediment-laden river water (60–180 mg/L) relative to the broad expanses of clear floodplain water in blues (<40 mg/L). Other images examined for this reach (Mertes et al., 1993; Mertes, 1994) and field measurements of sediment concentration and alkalinity throughout the region for floods in 1986 and 1991 (Mertes, 1990; Dunne et al., 1998, and unpublished data) confirm that most of the sediment-rich water is confined to the narrow zone near the river channel and that a perirheic zone of diluted river water covers large areas of the floodplain (compare Figures 5.4A and 5.3B). In addition, a simulation of this flood in this reach computed from a two-dimensional hydrodynamic, finite-difference model (Mertes, 1990, 1994) showed nearly stagnant flow along the distal edges of the floodplain, suggesting that these areas are covered with clear local water as shown in Figure 5.4A.

Mertes et al. (1995, 1996) showed that the spatial variation of the inundation hydrology along the river is partially a function of the geomorphology. Upstream reaches are characterized by a high density of floodplain channels that confine flooding river water and therefore reduce the potential for mixing of river water with locally derived water during flooding. Middle reaches of the river (Figure 5.4A) are characterized by overbank flows across levees that result in greater mixing of water types in the more open floodplain areas. In the downstream reaches, many large lakes have sufficient hydraulic head to prevent river water from entering large sections of the floodplain. The lateral extent of incursion of river water into the floodplain varies from approximately 5 km in the upstream reaches to 5 to 15 km in the reach shown in Figure 5.4A.

Mertes et al. (1995) used two landscape metrics derived from classified remote sensing data – percent cover and semivariance – to describe the spatial distribution and heterogeneity of wetland classes in geomorphically representative reaches of the Amazon River. The spatial heterogeneity of the vegetation communities on similar landforms was similar and was greatest where the landforms were most diverse. Landforms were the most diverse where the perirheic zone was extensive, especially in the middle reaches of the river (Figure 5.4A), which exemplifies the subparallel banding pattern of the perirheos depicted in Figure 5.3B.

Reference Site – Altamaha River, Georgia, USA. Based on field observations this channel–floodplain system is especially susceptible to saturation of the floodplain before arrival of the flood wave (Mertes, 1997). In early October 1994, mid-November 1995, and early November 1997, conditions on the floodplain were documented through ground measurements and aerial photography. In all cases, the floodplain was nearly covered with water carrying essentially no sediment (suggesting rising groundwater), whereas the river was still rising but was below bankfull level. In particular, locally intense rainfall before the crest of the flood resulted in deep ponding on the floodplain before the October 1994 flood and presumably influenced the ability of river water to enter the floodplain. Several cross sections across the floodplain indicated that patches of ponded water on the floodplain (not directly connected to the surface flow of the river during rising water) were often decimeters higher or lower in elevation than the river water surface. The relative elevation difference immediately adjusted to zero once the connection to the river was made as higher flows topped intervening barriers. Hence, in some locations on the floodplain when the surface connection to the river flow occurred, the water level increased more than a meter, whereas the river water surface had only risen a few decimeters.

Overbank flow conditions were sampled during mid-November 1994 and mid-February 1995. Velocity measurements of flow magnitude and direction show that the river flow typically enters the floodplain at the inflection point of river bends and flows obliquely onto the floodplain, and the flow vector retains a significant downvalley component (Figure 5.4B). Therefore, the river water flows down the floodplain in a relatively straight line between river bends. Without a significant cross-valley component to the flow vector, the river water may cover only a narrow zone of the floodplain that is on the scale of the amplitude of the largest river bends (approximately 2–4 km) in a floodplain up to 10 km wide. The combination of bend geometry and barriers to surface connectivity, such as levees, control the pattern of inundation hydrology along the Altamaha River.

Theoretical Framework for Analysis of Patterns of Inundation Hydrology

Variations in the spatial and temporal pattern of inundation hydrology and, as a corollary, the extent of the perirheic zone, are a function of the relative timing and magnitude of the contributions of water from different water sources and the geomorphic characteristics of the river reach. In this section, the hydrographs that describe the rate and timing of flows that contribute to flooding of a river reach (Figure 5.2) are characterized and their influences on inundation hydrology are considered. A conceptual model is then described that forms the basis for hypotheses about the patterns of inundation hydrology expected for different river environments. Finally, new data on watershed structure related to inundation hydrology are analyzed for five different types of rivers.

Controls on the Magnitude and Duration of Floods

Hydroclimatology. The hydroclimatology of flooding (Figure 5.5A) includes considering the effects of the size of the meteorological events providing precipitation for flood events (e.g., Hayden, 1988; Hirschboeck,

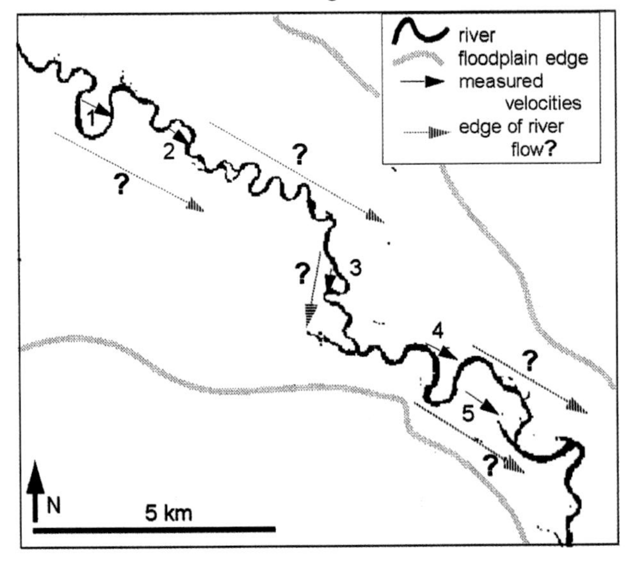

Figure 5.4. (A) Perirheic zone on Amazon River floodplain. Processed remote sensing data for suspended sediment concentrations that show the boundary for limited incursion of river water onto the Amazon River floodplain. The fully calibrated image (Landsat 5, August 2, 1989) shows concentrations ranging from 20 to 180 mg/L. For clarity, areas unaffected by flooding or areas where vegetation screened the water surface are masked. White arrow indicates proposed boundary of the incursion of river water and the perirheic zone (after Mertes 1997). (B) Perirheic zone on Altamaha River floodplain. This image shows a black and white rendering of the muddy water fraction image for the Altamaha River. Approximate boundary of the floodplain was estimated from aerial photographs, remote sensing data, and topographic maps. Solid, numbered arrows mark river bends for which velocity measurements were made during flooding in November 1994 and February 1995 (see Figure 5.7A for location of river reach where velocities were measured). These numbered arrows indicate the direction of flow across the area of the river bends. The range of (surface) and [depth-averaged] velocities in meters per second measured at these river bends include: 1, (0.08–0.40) [0.02–0.41]; 2, (0.11–0.24) [0.04–0.12]; 3, (0.05–0.21) [0.00–0.16]; 4, (0.11–1.03) [0.02–0.76]; 5, (0.06–0.62) [0.03–0.34] over depths ranging from 0.5 to 2 m. Patterned arrows indicate the proposed boundary of the incursion of river water and the perirheic zone (after Mertes 1997).

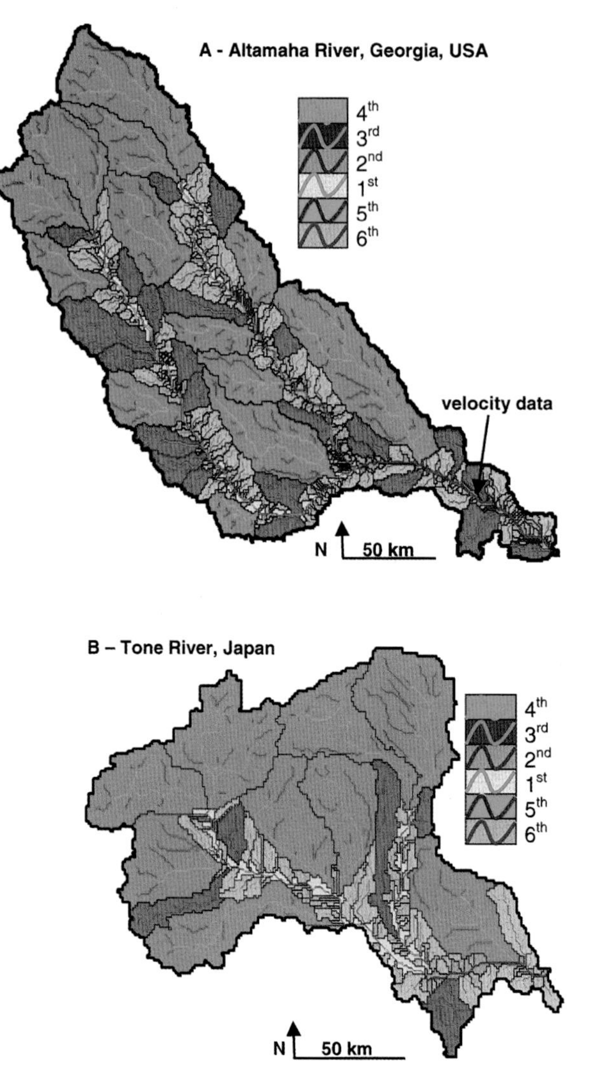

Figure 5.7. (A) Watershed morphometry of the Altamaha River, Georgia, USA. The watershed structure for the Altamaha River is shown as derived from the GTOPO30 digital elevation data set available from the EROS data center, based on cell-based flow modeling in Arc/Info/Grid. Picture emphasizes the local-scale subwatersheds (first to fourth order) that contribute directly to the larger stream channels. The watershed and stream network colors are shown in the legend. Arrow indicates the river reach where the velocity data shown in Figure 5.4B were measured. (B) Watershed morphometry of the Tone River, Japan. The watershed structure for the Tone River is shown as derived from the GTOPO30 digital elevation data set available from the EROS data center, based on cell-based flow modeling in Arc/Info/Grid. Picture emphasizes the local-scale subwatersheds (first to fourth order) that contribute directly to the larger stream channels. The watershed and stream network colors are shown in the legend.

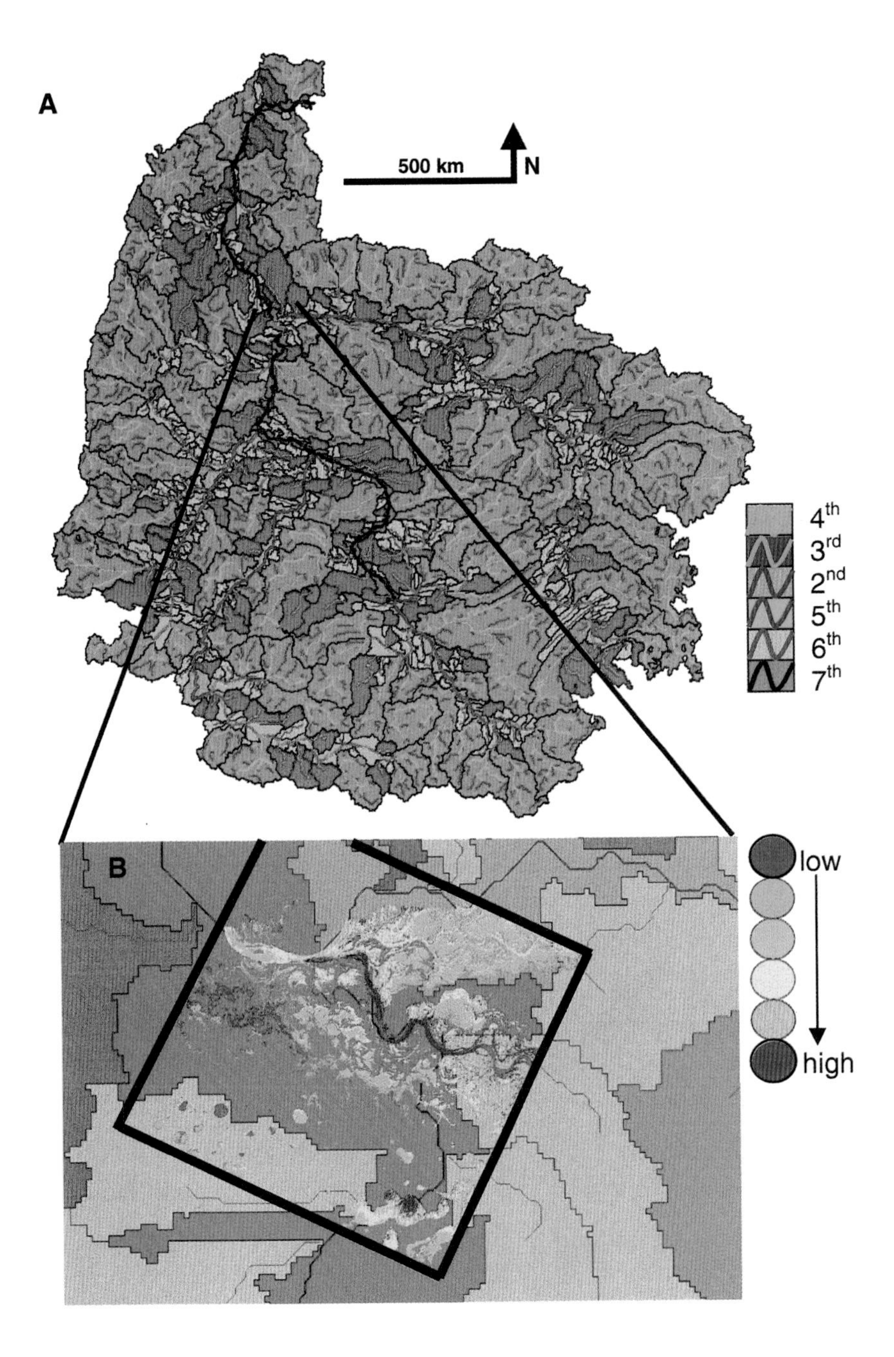

Figure 5.9. Watersheds and flood pattern of the Ob′-Irtysh River, Siberia. The watershed structure for the Ob′-Irtysh River is shown as derived from the GTOPO30 digital elevation data set available from the EROS data center, based on cell-based flow modeling in Arc/Info/Grid. Picture emphasizes the local-scale subwatersheds (first to fourth order) that contribute directly to the larger stream channels. Watershed and stream network colors are shown in the legend. Detail of the watersheds for the area near the confluence of the two rivers is shown in the lower graph. Overlaid on the detailed watershed map is an image representing the surface sediment concentration during the summer flood of 1993. This image of a color-density slice of a muddy water fraction from SPOT data (SPOT 2, July 23, 1993) shows relative turbidity with low sediment concentrations in blue and increasing concentrations progressing through green, yellow, and to the highest concentrations in red. For clarity, areas unaffected by flooding or areas where vegetation screened the water surface are masked. Full calibration to absolute sediment concentrations was not possible because of atmospheric inconsistencies.

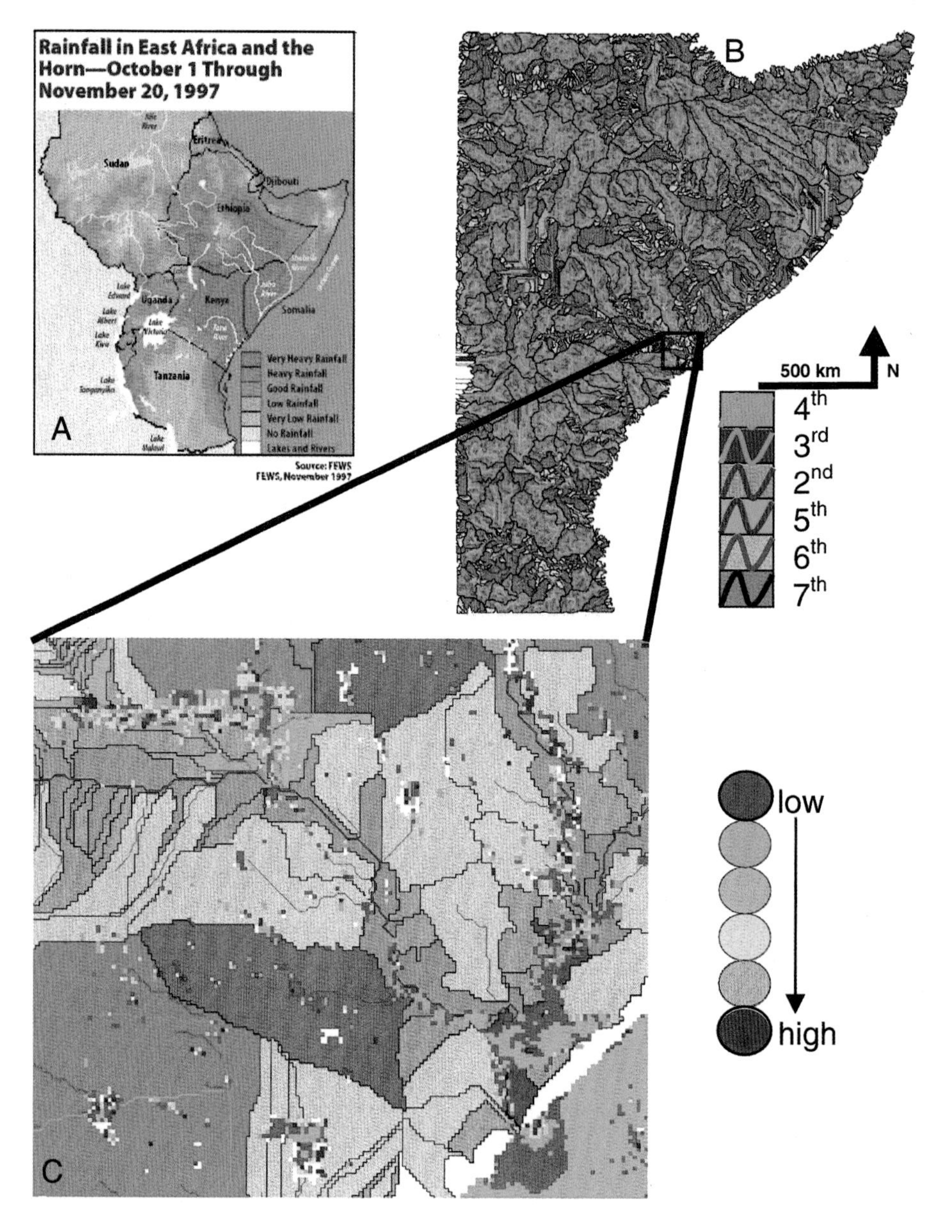

Figure 5.10. (A) Rainfall in East Africa and the Horn, October 1 through November 20, 1997. Zones of very heavy to no rainfall are shown in this image according to data from USAID/FEWS (ReliefWeb 1998). (B) Watershed morphometry of the rivers of southern Somalia, Africa. The watershed structure for the Shebele, Jubba, and smaller rivers is shown as derived from the GTOPO30 digital elevation data set available from the EROS data center, based on cell-based flow modeling in Arc/Info/Grid. Picture emphasizes the local-scale subwatersheds (first to fourth order) that contribute directly to the larger stream channels. Watershed and stream network colors are shown in the legend. (C) Detail of the watersheds for the area near the confluence of the Shebele and Jubba Rivers, Somalia. Overlaid on the detailed watershed map is an image representing the surface sediment concentration during the autumn flooding of 1997. An image of a color-density slice of a muddy water fraction from AVHRR data (November 11, 1997) shows relative turbidity with low sediment concentrations in blue and increasing concentrations progressing through green, yellow, and to the highest concentrations in red. The rainbow variation representing the sediment concentration gradient can be seen in the cone-shaped river plume entering the ocean in the lower right corner of the image. For clarity, areas unaffected by flooding, the surf zone along the coast, and areas where vegetation screened the water surface are masked. Spatial resolution of the AVHRR data is 1.1 km. Therefore, the flood pattern has a more blocky appearance than the SPOT (20-m resolution) and Landsat (30-m resolution) data shown in Figures 5.4A and 5.9. Full calibration to absolute sediment concentrations was not possible because of atmospheric inconsistencies.

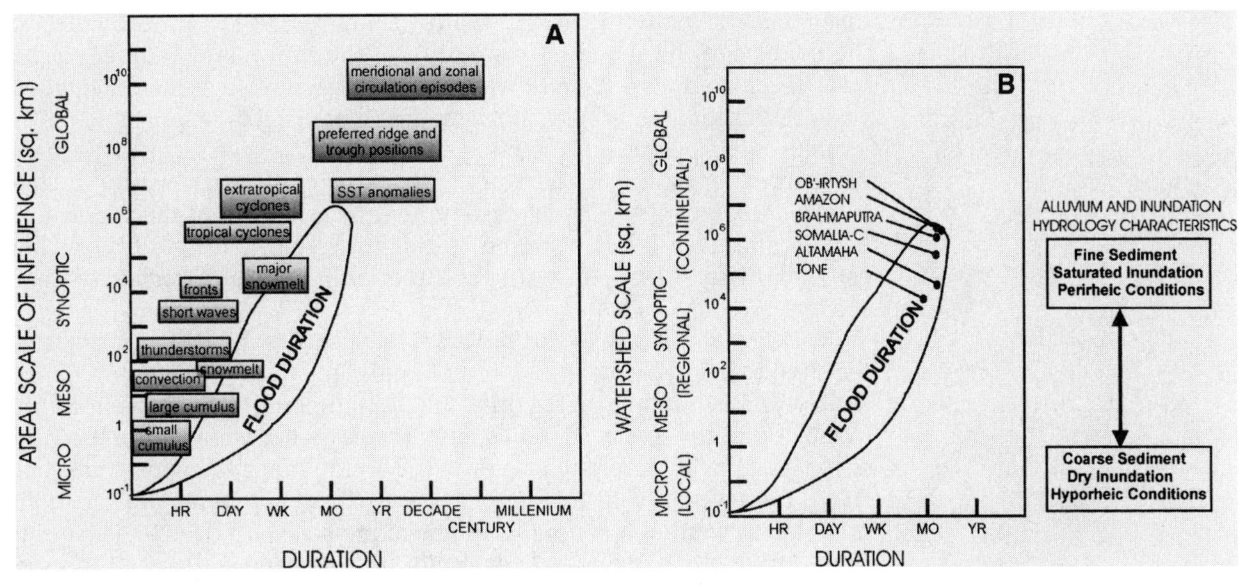

Figure 5.5. (A) Space-time domain for flood hydroclimatology (after Hirschboeck 1988). (B) Space-time domain for inundation hydrology and associated floodplain characteristics. The rivers for which data are reported are shown with respect to the size of the contributing watershed to the main channel and the estimated duration of the annual flood. The relative scale at the right for floodplains represents the hypothesized relations among sediment texture (fine versus coarse) of the alluvium, inundation hydrology pattern (Figure 5.3), and tendency for the development of hyporheic versus perirheic conditions during inundation.

1988; SAST, 1994; Magilligan and Graber, 1996; Hirschboeck et al., Chapter 2, this volume). In Figure 5.5A the space-time domains for meteorology, climatology, and flooding are depicted. As described by Hirschboeck, the spatial domain is the areal scale of influence, which could be cloud size or drainage basin size. The temporal domain represents the typical duration of a given event. The pattern of increasing time scales with increasing spatial scales holds for both the climatological elements as well as for floods.

The meteorological cell size may vary from the mesoscale (named local in this chapter for watersheds) of a thunderstorm (10^2 km^2) to the synoptic scale (named regional in this chapter for watersheds) of tropical and extratropical cyclones (10^4–10^6 km^2) according to Hirschboeck (1988, Figure 1, p. 28). The intersection of precipitation events in a given watershed will control the amount of water available from the different sources depicted in Figure 5.2 and Equations 1 and 2 and therefore the amount of the different water types available for inundation of floodplains.

Analysis of hydroclimatology provides an indication of the size of the cells providing the precipitation for floods. On the largest rivers, with continental-scale watersheds, flooding is caused primarily by synoptic-scale climatology. For example, the movement of the Intertropical Convergence Zone influences the rainfall pattern over the entire Amazon basin. However, it is important to remember that inside large watersheds are local-scale watersheds that perhaps receive locally intense rainfall. Therefore, in a large watershed the mesoscale and synoptic-scale events that trigger floods may be temporally out of phase or floods may be triggered by multiple mesoscale events that occur rapidly in succession.

Landscape Characteristics and Watershed Position. After considering the effects of climate on inundation hydrology, the next step toward building a conceptual model for patterns of inundation hydrology is to examine the intersection of the climate with the landscape. Minshall (1988) discussed the spatial-temporal scales of interest to stream ecologists in the context of the potential variability of the flows and inundation patterns associated with global river systems. Minshall suggested that, at the scale of continental river systems, the principal factors responsible for the structure and operation of stream ecosystems are river width, gradient, discharge, change in temperature, change in particulate organic matter, and water chemistry. The average annual temperature of the water, flow regime, drainage density, climax vegetation type, basin erodibility, and geologic history also constrain the functioning of the associated stream ecosystems.

These points made by Minshall are similar to the types of considerations required to evaluate the influences of landscape characteristics on hydrologic output. In summarizing this type of spatiotemporal analysis, Holling (1992) identified three different spatial scales that operate within the landscape, forming a hierarchy of influences on the functioning of ecosytems. At the finest scales vegetative processes, such as evapotranspiration, dominate from "centimeters to tens of meters in space and days to decades in time. At the other, macroscale extreme, slow geomorphological processes dominate the formation of a topographic and edaphic [i.e., soil] structure at large scales of hundreds to thousands of kilometers and centuries to millennia. At the mesoscales in between, contagious disturbance processes such as fire and water flow dominate the formation of patterns over spatial scales of hundreds of meters to hundreds of kilometers" (Holling, 1992, pp. 447–448).

From the perspective of this type of spatiotemporal analysis a flood can be described as a disturbance process. Within this paradigm, flooding is controlled by runoff processes dominated at the microscale by vegetation and soil structure. These fine-scale structures are integrated into watershed-scale networks formed on the backbone of the topography and underlying geology of the landscape. Inherent to the properties of a single flood event is the spatial and temporal variability of inundation as water is collected and transferred through the river system.

Conceptual Model for Patterns of Inundation Hydrology

In summary, inundation hydrology varies spatially and temporally under different hydroclimatological regimes (Figure 5.5A). Temporal variation in the spatial patterns of precipitation cells generating runoff for floods will cause variation in the relative arrival times to the floodplain of local versus regional runoff. The differences in timing of the hydrological events will vary partially with the size of the drainage basin and are expected to be most severe in the largest basins where regional storms produce slowly translated flood waves and, simultaneously, locally intense rainfall saturates floodplains. In the most extreme cases, sufficient local water could saturate a floodplain so that river water cannot enter the floodplain during a flood. In addition, there is a general trend for the sediment on floodplains to become finer (Figure 5.5B) as one moves downstream through a watershed (Wolman and Leopold 1957). In these downstream sections, the finer sediment in floodplains will tend to limit the development of a hyporheos and accentuate the development of a perirheos by reducing the rate of flow through the subsurface and increasing the potential for surface ponding (Mertes, 1997).

Combining the reach-scale perspective described by Equations 1 and 2 with the watershed perspective of Figure 5.5B results in a theoretical framework for examining the patterns of inundation hydrology for rivers with different climate, landscape, and human impact. This framework is illustrated schematically in Figure 5.6 where the input components of the mass balance for inundation hydrology are illustrated with representative curves. The hyetograph for precipitation (p) shows the most peaked distribution and the shortest time lag for contribution of water to storage in this reach of the river. The local tributary hydrograph (l) is also relatively peaked and has a short lag to peak because of the relatively small size of the drainage basin. The large tributary (tr) has a more attenuated hydrograph and a longer lag to peak as the result of flows converging from a larger watershed than l. The main river channel input (in) is represented by a very smooth hydrograph with the longest lag to peak of all the components. The rise and fall of the groundwater (gw) is represented by a curve with a slow rise, long constant flow rate, and slow fall. The limit on the flow rate is that the elevation of the ground surface is a constant, and therefore the flow rate cannot increase substantially once the system is fully saturated.

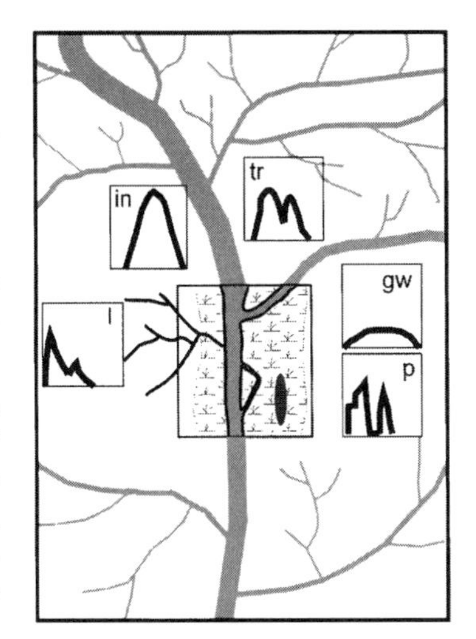

Figure 5.6. Watershed perspective for components contributing to inundation hydrology. Typical hydrographs (time versus discharge) are shown for the input components illustrated in Figure 5.2. l is a local tributary, is in the main channel input, is a larger tributary, and gw represents groundwater. Precipitation, p, is shown as a hyetograph. Magnitude, duration, and lag of the hydrographs are a function of climate, watershed characteristics, and human impacts.

These curves are generic representations of the suite of components that may contribute to the inundation hydrology of any channel–floodplain reach. Differences in the patterns of inundation hydrology among rivers or reaches on a single river are the result of spatially and temporally varying combinations of the hydrographs of the components and texture of the alluvium. For instance, it may be known when the storage capacity of the channel is exceeded and flooding occurs. After this type of overbank flooding, it may rain on the floodplain. This precipitation event may be relatively small in total water volume, but the flood depth may be increased substantially, thus creating the potential to increase hazards dramatically.

The conceptual model described here helps to define the potential patterns of inundation hydrology for different climates, landscapes, and types of human impact. By characterizing river reaches with respect to these variables, calculations of the potential hydrographs for the contributing components to the inundation hydrology can be made. After the shapes and magnitudes of the hydrographs have been determined, it is then possible to examine the effects on the patterns of flooding if the lag times of the hydrographs are varied. The key to this examination is establishing reasonable constraints on the variability of the lag times for the different components. These constraints are based primarily on the inherent variability of the climate patterns at both local and regional scales and the potential variability due to changes in land use.

To begin tests of the conceptual model within this theoretical framework, a new method for analyzing the local watershed structure surrounding a floodplain reach was developed. This method is described in the next section and was applied to five different rivers. The sizes of the watersheds and approximate duration of the average floods are plotted in Figure 5.5B for comparison among the case studies.

GIS Method for Globally Consistent Morphometric Analysis of Local Watershed Structure

GIS tools allow data sets to be combined and can be used for modeling. As illustrated in Figure 5.2, it is important to document the characteristics of the channel and the floodplain surface in the selected channel–floodplain reaches. As illustrated in Figure 5.6, it is also necessary to characterize the structure of the watersheds that have the potential to contribute to the selected channel–floodplain reach. Advances in GIS modeling tools and availability of globally consistent digital elevation data sets (e.g., GTOPO30 from the EROS Data Center) provide an opportunity to develop new types of data for the study of inundation hydrology. A morphometric analysis of watershed structure can be based on cell-based modeling techniques described for ARC/INFO/GRID (AIG) (Maidment, 1995). These GIS-based techniques involve dividing the river watershed into progressively smaller subwatershed units. AIG methods were applied to GTOPO30 data for the case studies described in this chapter.

The key to computing the morphometric structure of the watersheds, in this case based on Strahler stream ordering (Strahler, 1964), is that the subwatersheds can be derived according to the scale of a threshold function. The threshold function restricts the size of the smallest subwatershed and is based on the total area contributing to flow at a point in the stream network. In the examples described in the case studies, the minimum subwatershed size (i.e., the first-order watersheds) for the regional-scale Altamaha River and Tone River was selected to be 5 km^2. The continental-scale Ob'-Irtysh Rivers in Siberia and Brahmaputra River in south-central Asia had a minimum subwatershed size of 100 km^2. The regional-scale river systems that flooded in 1997 in southern Somalia, the Jubba, Shebele, and a few smaller rivers, have a combined drainage basin area of over 10^6 km^2, and the minimum size selected for subwatersheds was 50 km^2.

In Figures 5.7, 5.9, and 5.10, the subwatersheds are represented by different colors and are layered to emphasize the subwatersheds that locally contribute to the floodplain. If the topology of the network were perfectly integrated then smaller watersheds, first to fourth order, would not be seen along the major stream channels, usually fifth to seventh order. In a sense, these map views depict the areas where the network is not fully integrated and low-order channels feed into much higher order channels. From the perspective of topology, these nonintegrated subwatersheds have been previously named interior basins (Verdin, 1997).

When investigating a selected channel-floodplain reach (Figure 5.6) the subwatersheds contributing directly to that reach can be accounted for from the GIS-based morphometry by counting the number and measuring the area of each subwatershed. The number and area statistics for the watersheds that affect selected reaches can be compared along a river network or between rivers if the minimum size threshold is held constant. In the case studies, the morphometry of the regional-scale rivers is compared for the lower reaches (approximately 150 km) of the Altamaha and Tone Rivers. For the larger rivers, the statistics for two 200-km reaches for each river were collected. For the Somalia rivers (Somalia-C in Figure 5.5B), the morphometry of the subwatersheds at the confluence of the rivers near the coast was evaluated.

Case Studies

Altamaha and Tone Rivers

The Altamaha (37,000 km^2) and Tone (19,000 km^2) Rivers represent regional-scale rivers (Figure 5.5B) with different types of terrain. The morphometry of the watersheds is illustrated in Figure 5.7 (see color section following p. 166.). The layering of the subwatersheds shows that, for a substantial length along both stream networks, the topology is not well integrated. In addition, the patterns of the interior basins that contribute to the channel–floodplain reaches vary along the two networks.

A comparison of the morphometric statistics (Figure 5.8A,B) for the subwatersheds contributing directly to the lower reach of each river shows that the number of first-order subwatersheds contributing directly to the floodplain is nearly three times larger for the Tone than for the Altamaha River. On the other hand, the total area of the first-order subwatersheds is similar, which suggests that the terrain along the valley near the Tone River is steeper and the subwatersheds are smaller compared with the gentler terrain along the Altamaha River on the coastal plain of the southeastern United States. A reference scale for comparison of these subwatershed characteristics would be the floodplain carrying capacity (sensu Bhowmik and Demissie, 1982) along the reach. Without detailed flood measurements, a surrogate for the carrying capacity could be the floodplain area, which is approximately 400 km^2 for both river reaches. With larger local tributaries contributing directly to the Altamaha floodplain, it is not surprising that a substantial amount of the water flooding the Altamaha floodplain is from local sources (Mertes, 1997; Mertes, unpublished data). In contrast, one might predict that, although there are more first-order subwatersheds along the Tone, because they are smaller their individual effect would be relatively less with respect to influencing flooding on any reach of the Tone floodplain and flooding would be dominated by water converging from upstream. Remote sensing and field observations could be used to test the predicted pattern of inundation hydrology for the Tone floodplain based on these morphometric and floodplain statistics.

Ob'-Irtysh and Brahmaputra Rivers

The Ob'-Irtysh and Brahmaputra Rivers represent continental-scale rivers (see Figure 5.5B for scale) situated in different climatic regimes and in different terrains. According to Hayden (1988) the Ob'-Irtysh Rivers (Figure 5.9, see color section following p. 166.) are subjected to perennially baroclinic conditions with moderate rainfall due to an inadequate supply of atmospheric moisture. Local water sources for flooding may be melting snow and groundwater saturation. In contrast, the Brahmaputra River is subjected to snow-melt runoff from the Himalayas and monsoons and cyclones across the basin.

Comparison of the morphometric statistics (Figure 5.8C,D) shows that not only are the three rivers different, but the upstream and downstream reaches differ as well. The upper reach of the Brahmaputra (br-u on Figure 5.8C,D) was selected along one of the major gorges in the Himalayas. The topology of the network in this area is dominated by the high frequency of

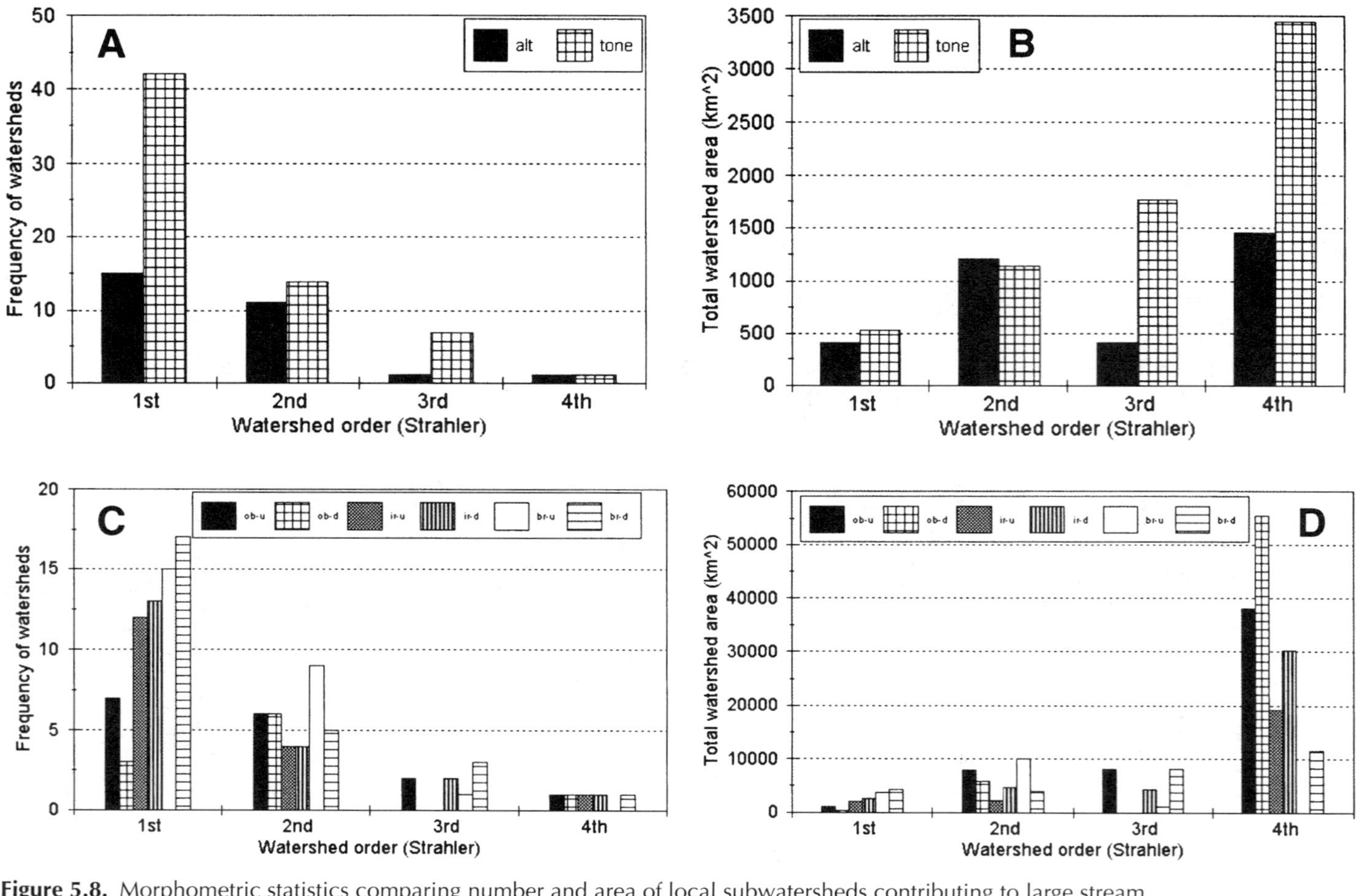

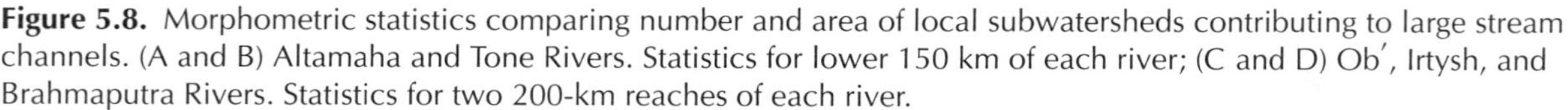
Figure 5.8. Morphometric statistics comparing number and area of local subwatersheds contributing to large stream channels. (A and B) Altamaha and Tone Rivers. Statistics for lower 150 km of each river; (C and D) Ob′, Irtysh, and Brahmaputra Rivers. Statistics for two 200-km reaches of each river.

occurrence and large area of the first- and second-order watersheds. These patterns indicate that steep, local watersheds are contributing locally to narrow, channel–floodplain reaches and that these sections of the main-channel floodplain may experience short-lived floods due to high flow from these local mountain watersheds.

The Ob'-Irtysh channel–floodplain system would be expected to behave differently because of the large size of the local tributaries contributing to the floodplain in different reaches and also because of the potential for water from snow melt to saturate the floodplain surface. By combining the morphometric analysis of the watershed structure with a remote sensing analysis for surface sediment concentration, it is possible to examine in detail the potential for the development of a perirheic zone in the Ob'-Irtysh complex (Figure 5.9). In Figure 5.9B, a SPOT image is shown that was analyzed for the sediment concentration gradient. The image is overlaid onto the map of the subwatershed structure. The Irtysh River (red at the east-central part of the image) at this time apparently supported higher sediment concentrations than the northern Ob'. The floodplain between the two rivers (Yasuoka et al., 1994) appears to be flooded with water containing sediment concentrations similar to that carried in the Ob' River. The cyan and dark blue water in the west-central part of the image is at the distal edge of the floodplain (dark green) and may represent water from *in situ* snow melt that was prevented from draining by the high water in the two river channels.

Somalia Floods, Fall 1997

In fall 1997, significant rainfall (Figure 5.10A, see color section following p. 166.) fell for approximately 2 months on the mountains and coastal areas (Figure 5.10B) of northeastern Africa (ReliefWeb, 1998). The result of this precipitation was extensive flooding throughout the region. The rainfall resulted in complete saturation of the watersheds of the Jubba and Shebele Rivers. The flooding at the confluence of these rivers is shown with a processed image acquired by the Advanced Very High Resolution Radiometer (AVHRR) satellite for November 25, 1997, that is overlaid onto a map of the subwatersheds in Figure 5.10C. The large, irregular blue patch in the southeastern corner of the image is a flooded zone. It is informative to recognize that the sediment concentrations in this flooded area near the coast are relatively lower than the flooded areas that can be seen in the northwestern corner of the image. Based on the watershed structure and reported precipitation pattern, it may be that the lower floodplain is covered by a mixture of river water contributed from the entire upstream watershed and locally derived rain water due to intense precipitation during fully saturated conditions.

It is possible to make an order-of-magnitude estimate of the potential contribution to flooding from direct precipitation and the first- to third-order subwatersheds that feed directly into the flooded area at the confluence of the rivers. In the vicinity of the confluence there are seven first-order subwatersheds with a total area of 800 km^2, three second-order subwatersheds with a total area of 1300 km^2, and one third-order subwatershed with an area of 1700 km^2. The results shown in Figure 5.11 were calculated by assuming saturated conditions on the floodplain and a 15-cm rainstorm in 24 h.

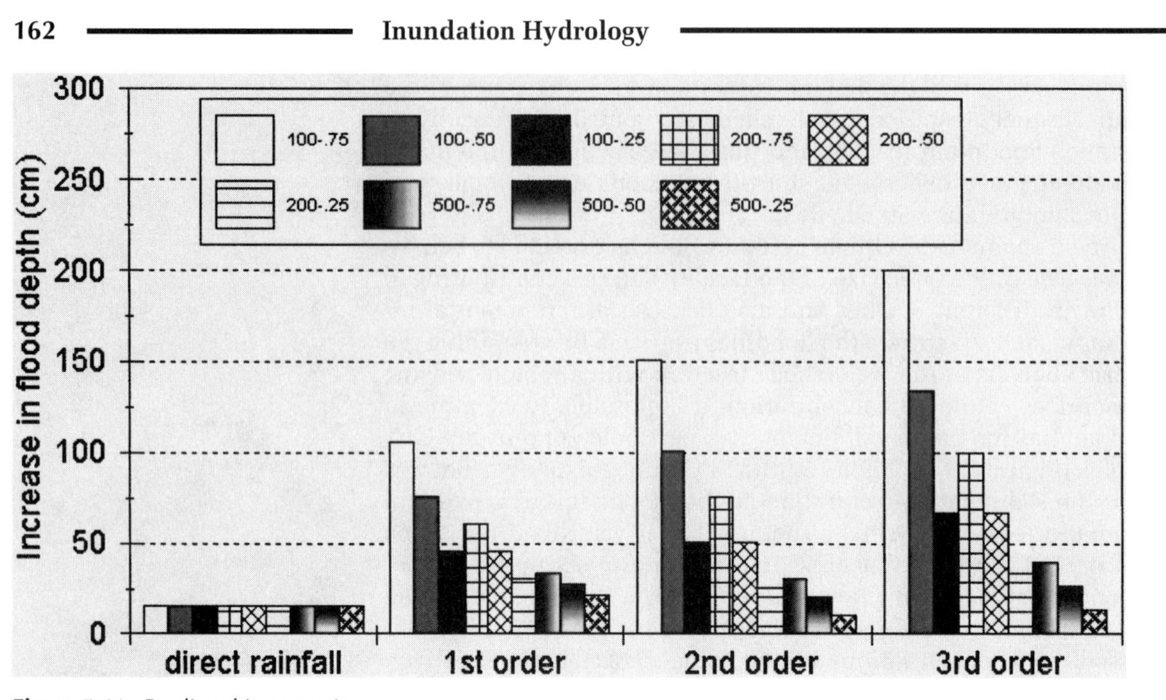

Figure 5.11. Predicted increase in flood depth due to runoff from local subwatersheds. Results shown in this graph are based on an example for a 15-cm rainfall event over all the local watersheds (first to third order) that contribute directly to the channel–floodplain system near the confluence of the Jubba and Shebele Rivers (Figure 5.10C). Coefficients of runoff (0.25, 0.50, and 0.75) and the floodplain area over which the water could spread (100, 200, and 500 km^2) were varied.

The potential increase in flood depth is shown for different runoff coefficients (0.25, 0.50, and 0.75) and different floodplain areas for storage (100, 200, and 500 km^2) as a function of the potential contribution from direct rainfall or the individual contributions from the subwatersheds grouped by order. The immediate effect of a 15-cm rainstorm is that the flood depth will increase by 15 cm. Depending on the lag time for arrival of the water from the different subwatersheds and the floodplain area over which the water spreads, the increase in flood depth due to the local runoff ranges from 10 to 200 cm.

These calculations scale the magnitude of the potential effect on flood depth resulting from local precipitation events. Based on this order-of-magnitude calculation, the remarkable extent of inundation by clear water in the lower sections of the rivers in Somalia in 1997 was probably as much due to the continuous intense rainfall on the lower reaches of the river system as to the arrival of the flood wave from upstream.

Summary

As described by Ramirez (Chapter 11, this volume) watershed-scale evaluation of flooding potential is effective for predicting the probability that a certain water discharge will be the result of a particular precipitation pattern (Hirschboeck et al., Chapter 2, this volume). Combining this discharge information with information on storage capacity of the channel through measurement of cross-sectional area and flow velocities allows for prediction of the potential for flooding along river reaches. Nevertheless, a reach-scale examination of the components of inundation hydrology shows that, for some channel–floodplain complexes, inundation of the floodplain may occur due to rising groundwater, hyporheic water, flooding of local tributaries, runoff from surrounding slopes, direct precipitation, water from melted snow, and antecedent water from prior floods (Mertes, 1997). A complete model for flood potential would therefore have to include components that simulated the development of saturated conditions on the floodplain surface both in response to a rising water table as the water surface in the channel rises and to water arriving from more diverse sources than just upstream watersheds.

One of the results of flooding from diverse water sources is that a floodplain, at any one time, may have water bodies perched at different elevations or with different pressure distributions due to variation in the source of water. As a flood develops, these disconnected water bodies may be washed away by relatively more substantial overbank river water, remain as discrete disconnected bodies of local water surrounded by overbank river water, or may coalesce into a separate body of flood water composed of local water that partially or wholly blocks river water from directly entering the floodplain. Where substantial amounts of local water inundate a floodplain, there is the potential for development of a mixing zone between local water and river water on the floodplain. The extent and pattern of this mixing zone, named the perirheic zone (Mertes, 1997), can also be used to evaluate potential sources of water contributing to inundation.

A conceptual model developed in this chapter is that the spatial and temporal variation of inundation patterns should be predictable based on the hydroclimatology, landscape characteristics of both the watershed and floodplain, and human impacts on the floodplain surface. Several approaches for studying the predicted patterns of inundation hydrology were described in this chapter and a flowchart showing an analysis structure within a GIS framework was presented. Although field data on flow velocity, depth, and extent of flooding are critical, use of remote sensing data and numerical simulations is rapidly changing our ability to examine patterns of inundation hydrology. For example, as reviewed in this chapter, optical remote sensing data of resolution fine enough to involve at least four or five pixels across the flooded area provide an opportunity to document and characterize the perirheic zone across large floodplain areas (Mertes, 1997). Complementary to optical data used to measure water quality are microwave data used to measure inundation extent and water surface height. GIS analyses of local watershed structure based on a new technique described in this chapter provide additional data for estimating the potential for local water to contribute to inundation of floodplains at the reach scale.

Although the theoretical framework, conceptual model, and techniques described in this chapter for analyzing inundation patterns are new and not yet proven, analysis of the results for several rivers suggests that this framework is a reasonable approach. For example, the Amazon River is subjected to a hydroclimatology that allows for substantial local flooding in the Brazilian reaches of the system before inundation due to overbank flooding. In a similar fashion, large local watersheds along the Altamaha River floodplain in Georgia were shown to be responsible for contributing to the presence of perirheic conditions during most floods on this coastal plain floodplain. In contrast, it was predicted from watershed data that the impact of local tributaries on flooding along the Tone River in Japan may be less significant than on the Altamaha River, because the local tributaries are relatively smaller than the floodplain area along the main river channel of the Tone. At the other extreme, the narrow gorge of the Brahmaputra River in the Himalayas limits the size of the floodplain along the main channel so that the steep, local first-order watersheds may produce reach-scale flash floods along the main channel that may not be synchronous with flooding due to water arriving from upstream watersheds. Finally, the 1997 Somalia floods started with the convergence of water from regional-scale watersheds in the mountains. The persistence of the flood was due to continuous rainfall that occurred on saturated portions of the floodplain and local watersheds. An order-of-magnitude calculation of the potential increase in flood depth due to a 15-cm rainstorm on the saturated, local watersheds that drained directly into the flooded area was made, and it showed that the flood depth could have increased by up to 2 m just from the local rainfall.

The implications for the effect of these patterns of inundation hydrology on the distribution of sediment, spread of contaminants, floodplain ecology, and flood hazards are many. Dietrich et al. (1999) described variations in the geomorphology of the Fly River floodplain that are related to variations in the patterns of flooding due to the presence or absence of local water. Finley (Chapter 7, this volume) describes a variety of types of contaminants that may be transferred to river floodplains during inundation. However, the pattern of contamination will depend in part on how the water is distributed on the floodplain. For example, if local water blocks river water during inundation, the patterns of contamination will reflect this distribution. E. Fustec (personal communication) reported this type of situation for the Aube River floodplain near Paris, where floodplain areas that receive only local water draining from agricultural fields have significantly higher nutrient levels in the soils than areas that receive river water during floods. Even for a heavily engineered floodplain, Sparks et al. (1998) described the need for understanding in detail the connectivity of the upper Mississippi River channel to different parts of its floodplain in order to ensure that the floodplain ecology of areas that have received only local waters are not damaged by engineered floodplain restoration that mistakenly allows river water to flood them.

Acknowledgments

This research was supported by funds from NASA (NAG5-3498) and NSF (EAR 94-05757). The raw Amazon image was obtained from R. Almeida Filho of the Instituto Nacional Pesquisas Especiais of Brazil (INPE). The Ob'-Irtysh River image was developed in collaboration with Y. Yasuoka and M. Tamura of the National Institute of Environmental Studies of Japan. The raw Somalia image was provided by G.R. Brakenridge of the Department of Geography, Dartmouth College.

References

Bates, P.D., Horritt, M., and Hervouet, J.-M. (1998). Investigating two-dimensional, finite element predictions of floodplain inundation using fractal generated topography. *Hydrological Processes*, **12**, 1257–1277.

Bhowmik, N.G. and Demissie, M. (1982). Carrying capacity of flood plains. *Proceedings ASCE Hydraulics Division*, **108**, 443–451.

Brakenridge, G.R., Tracy, G.R., and Knox, J.C. (1998). Orbital SAR remote sensing of a river flood wave. *International Journal of Remote Sensing*, **19**, 1439–1445.

Burt, T.P. and Haycock, N.E. (1996). Linking hillslopes to floodplains. In *Floodplain Processes*, eds. M.G. Anderson, D.E. Walling, and P.D. Bates, pp. 461–492. New York: John Wiley & Sons.

Dahm, C.N. and Valett, H.M. (1996). Hyporheic zones. In *Methods in Stream Ecology*, eds. F.R. Hauer and G.A. Lamberti, pp. 107–119. New York: Academic Press, Inc.

Dekker, A.G., Malthius, T.J., and Hoogenboom, H.J. (1995). The remote sensing of inland water quality. In *Advances in Environmental Remote Sensing*, eds. F.M. Danson and S.E. Plummer, pp. 123–142. New York: John Wiley & Sons.

Dietrich, W.E., Day, G., and Parker, G. (1999). The Fly River, Papua New Guinea: Inferences about river dynamics, floodplain sedimentation and fate of sediment. In *Varieties of Fluvial Form*, eds. A.J. Miller and A. Gupta, pp. 345–376. New York: John Wiley & Sons.

Dunne, T., Mertes, L.A.K., Meade, R.H., Richey, J.E., and Forsberg, B.R. (1998). Exchanges of sediment between the flood plain and channel of the Amazon River in Brazil. *Geological Society of America Bulletin*, **110**, 450–467.

Forsberg, B.R., Devol, A.H., Richey, J.E., Martinelli, L.A., and dos Santos, H. (1988). Factors controlling nutrient concentrations in Amazon floodplain lakes. *Limnology & Oceanography*, **33**, 41–56.

Gomez, B., Mertes, L.A.K., Phillips, J.D., Magilligan, F.J., and James, L.A. (1995). Sediment characteristics of an extreme flood, 1993 upper Mississippi River valley. *Geology*, **23**, 963–966.

Hamilton, S.K., Sippel, S.J., and Melack, J.M. (1996). Inundation patterns in Pantanal wetland of South America determined from passive microwave remote sensing. *Archives of Hydrobiologie*, **137**, 1–23.

Haeuber, R.A. and Michener, W.K. (1998). Flooding natural and managed disturbances – A special issue of BioScience devoted to flooding as a disturbance. *BioScience*, **48**, 677–680.

Hayden, B.P. (1988). Flood climates. In *Flood Geomorphology*, eds. V.R. Baker, R.C. Kochel, and P.C. Patton, pp. 13–26. New York: John Wiley & Sons.

Hess, L.L. (1999). *Monitoring Flooding and Vegetation on Seasonally Inundated Floodplains with Multifrequency Polarimetric Synthetic Aperture Radar*. Ph.D. dissertation, University of California, Santa Barbara.

Hess L.L., Melack, J.M., and Simonett, D.S. (1990). Radar detection of flooding beneath the forest canopy – a review. *International Journal of Remote Sensing*, **11**, 1313–1325.

Hirschboeck, K.K. (1988). Flood hydroclimatology. In *Flood Geomorphology*, eds. V.R. Baker, R.C. Kochel, and P.C. Patton, pp. 27–49. New York: John Wiley & Sons.

Holland, M.M. (1996). Wetlands and environmental gradients. In *Wetlands: Environmental Gradients, Boundaries, and Buffers*, eds. G. Mulamoottil, B.G. Warner, and E.A. McBean, pp. 19–43. New York: Lewis Publishers.

Holling, C.S. (1992). Cross-scale morphology, geometry, and dynamics of ecosystems. *Ecological Monographs*, **62**, 447–502.

Hughes, D.A. (1978). *Flooding and Floodplain Inundation*. Ph.D. dissertation, University College of Wales, Aberystwyth, 252 pp.

Hughes, D.A. (1980). Floodplain inundation: Processes and relationships with channel discharge. *Earth Surface Processes*, **5**, 297–304.

Junk, W.J., Bayley, P.B., and Sparks, R.E. (1989). The flood pulse concept in river-floodplain systems. In *Proceedings International Large River Symposium*, ed. D.P. Doge, pp. 110–127. Canadian Fisheries & Aquatic Sciences Special Publication, Ottawa, Canada.

Kirk, J.T.O. (1986). *Light and Photosynthesis in Aquatic Ecosystems*. 401 pp., Cambridge: Cambridge University Press.

Koblinsky, C.J., Clarke, R.T., Brenner, A.C., and Frey, H. (1993). Measurement of river level variations with satellite altimetry. *Water Resources Research*, **29**, 1839–1848.

Lesack, L.F.W. and Melack, J.M. (1995). Flooding hydrology and mixture dynamics of lake water derived from multiple sources in an Amazon floodplain lake. *Water Resources Research*, **31**, 329–346.

Lewin, J. and Hughes, D.A. (1980). Welsh floodplain studies II. Application of a qualitative inundation model. *Journal of Hydrology*, **46**, 35–49.

Magilligan, F.J. and Graber, B.E. (1996). Hydroclimatological and geomorphic controls on the timing and spatial variability of floods in New England, USA. *Journal of Hydrology*, **178**, 159–180.

Maidment, D.R. (1995). *GIS and Hydrology*. User Notes for 15th Annual ESRI User Conference, Palm Springs, CA.

Mertes, L.A.K. (1990). *Hydrology, Hydraulics, Sediment Transport, and Geomorphology of the Central Amazon Floodplain*. Ph.D. dissertation, University of Washington, Seattle, 225 pp.

Mertes, L.A.K. (1994). Rates of flood-plain sedimentation on the central Amazon River. *Geology*, **22**, 171–174.

Mertes, L.A.K. (1997). Documentation and significance of the perirheic zone on inundated floodplains. *Water Resources Research*, **33**, 1749–1762.

Mertes, L.A.K., Daniel, D.L., Melack, J.M., Nelson, B., Martinelli, L.A., and Forsberg, B.R. (1995). Spatial patterns of hydrology, geomorphology, and vegetation on the floodplain of the Amazon River in Brazil from a remote sensing perspective. *Geomorphology*, **13**, 215–232.

Mertes, L.A.K., Dunne, T., and Martinelli, L.A. (1996). Channel-floodplain geomorphology along the Solimões-Amazon River, Brazil. *Geological Society of America Bulletin*, **108**, 1089–1107.

Mertes, L.A.K., Smith, M.O., and Adams, J.B. (1993). Estimating suspended sediment concentrations in surface waters of the Amazon River wetlands from Landsat images. *Remote Sensing Environment*, **43**, 281–301.

Minshall, G.W. (1988). Stream ecosystem theory: a global perspective. *Journal of the North American Benthological Society*, **7**, 263–288.

ReliefWeb. (1998). Rainfall in East Africa and the Horn – October 1 through November 20, 1997, United Nations Office for the Coordination of Humanitarian Affairs (http://www.reliefweb.int), US Agency for International Development/FEWS. Washington, DC.

Richey, J.E., Mertes, L.A.K., Dunne, T., Victoria, R., Forsberg, B.R., Tancredi, A., and Oliveira, E. (1989). Sources and routing of the Amazon River flood wave. *Global Biogeochemical Cycles*, **3**, pp. 191–204.

SAST. (1994). *Science for Floodplain Management into the 21st Century: A Blueprint for Change – Part V.* Washington, DC.: Report of the Interagency Floodplain Management Review Committee, 272 pp.

Sioli, H. (1957). Sedimentation im Amazonasgebiet. *Geologicsche Rundschau*, **45**, 608–633.

Sippel, S.J., Hamilton, S.K., Melack, J.M., and Novo, E.M.M. (1998). Passive microwave observations of inundation area and the area/stage relation in the Amazon River floodplain. *International Journal of Remote Sensing*, **19**, 3055–3074.

Smith, L. (1997). Satellite remote sensing of river inundation area, stage, and discharge-a review. *Hydrological Processes*, **11**, 1427–1439.

Smith, L.C. and Alsdorf, D.E. (1998). Control on sediment and organic carbon delivery to the Arctic Ocean revealed with space-borne synthetic aperture radar: Ob' River, Siberia. *Geology*, **26**, 395–398.

Smith, M.O., Ustin, S.L., Adams, J.B., and Gillespie, A.R. (1990). Vegetation in deserts I. A regional measure of abundance from multispectral images. *Remote Sensing of Environment*, **31**, 1–26.

Sparks, R.E., Nelson, J.C., and Yin, Y. (1998). Naturalization of the flood regime in regulated rivers: the case of the upper Mississippi River. *BioScience*, **48**, 706–720.

Strahler, A.N. (1964). Quantitative geomorphology of drainage basins and channel networks. In *Handbook of Applied Hydrology*, ed. V.T. Chow, pp. 75–94. New York: McGraw-Hill.

Townsend, P.A. and Walsh, S.J. (1998). Modeling floodplain inundation using an integrated GIS with radar and optical remote sensing. *Geomorphology*, **21**, 295–312.

Velikanova, Z.M., and Yarnykh, N.A. (1970). Field investigations of the hydraulics of a floodplain during a high flood. *Soviet Hydrology: Selected Papers*, **5**, 33–53.

Verdin, K.L. (1997). A system for topologically coding global drainage basins and stream networks. In *Proceedings 17th Annual ESRI Users Conference*, San Diego, CA, July.

Vorosmarty, C.J., Willmott, C.J., Choudhury, B.J., Schloss, A.L., Stearns, T.K., Robeson, S.M., and Dorman, T.J. (1996). Analyzing the discharge regime of a large tropical river through remote sensing, ground-based climatic data and modeling. *Water Resources Research*, **32**, 3137–3150.

Witte, W.G., Whitlock, C.H., Usry, J.W., Morris, W.D., and Gurganus, E. (1981). Laboratory measurements of physical, chemical, and optical characteristics of Lake Chicot sediment waters. *NASA Technical Paper 1941*, **27** pp. Washington, DC.

Wolman, M.G. and Leopold, L.B. (1957). River flood plains: some observations on their formation. *USGS Professional Paper 282-C*, 81–107. U.S. Government Printing Office, Washington, DC.

Woltemade, C.J. and Potter, K.W. (1994). A watershed modeling analysis of fluvial geomorphological influence on flood peak attenuation. *Water Resources Research*, **30**, 1933–1942.

Yasuoka, Y., Tamura, M., and Yamagata, Y. (1994). Application of remote sensing to environmental monitoring – global wetland monitoring. In *Optical Methods in Biomedical and Environmental Sciences*, eds. H. Ohzue and S. Komatsu, pp. 269–272. New York: Elsevier Science.

CHAPTER 6

Geomorphic Effects of Floods

Ellen E. Wohl
Department of Earth Resources
Colorado State University

Floods create geomorphic hazards via changes in channel configuration and sediment transport. These changes may occur during a single flood or over a period of time because of changes in flood magnitude and frequency. The rate and magnitude of channel change will be influenced by flood hydrology and hydraulics, channel boundary resistance, sediment supply, and flow history. Erosion dominates where velocity and shear stress are higher than adjacent reaches and may occur as bank, bed, or overbank erosion. Deposition may occur in the channel as the growth of bars or islands, on tributary fans as boulder berms, or as lateral or vertical floodplain accretion. Lateral channel movement may occur as meander migration, meander cutoff, or avulsion. Channel planform may alternate regularly in association with seasonal flooding, may change during a single large flood, or may change gradually in response to a change in flood regime. Channels shaped predominantly by large, infrequent floods have flashy hydrographs; high channel gradient; abundant coarse bedload; low bank cohesion; and deep, narrow cross-sectional geometry. A threshold of flood-induced channel modification may be defined as a function of unit stream power and drainage area. Changes in flood regime associated with flow regulation can alter sedimentation patterns and channel characteristics and put human and nonhuman structures at risk.

Introduction

Analyzing the impact of floods on channel morphology and sediment transport has been a primary focus of fluvial geomorphology. A flood may cause dramatic changes along some reaches of a channel and have relatively little effect on other reaches. Similarly, a flood that occurs on average once every hundred years may create erosional and depositional forms that are completely reworked within 10 years along one channel but that persist for decades along a neighboring channel. Investigations of the geomorphic effects of floods have sought to explain these differences among channel responses to floods by identifying patterns between flood magnitude–frequency relationships and channel responses and by identifying the influences of channel-boundary resistance, thresholds of channel change, and recovery and relaxation times of specific channel features. These patterns may be used to understand the magnitude and location of geomorphic changes that occur during floods, and thus the hazards to human, riparian, and aquatic communities that may result from these geomorphic changes.

Hazardous geomorphic effects associated with flooding take the form of changes in channel configuration and changes in sediment transport. Changes in channel geometry may occur very rapidly during a single flood, or they may occur over a period of years to decades as progressive, small alterations resulting from a shift in the magnitude and frequency of floods. Changes in sediment–transport relations may also occur during a single flood or over an extended period of time that includes multiple floods.

Changes in channel geometry during a single flood are a response to the abrupt change in water and sediment discharges that accompany passage of a flood wave. Leopold and Maddock (1953) proposed that natural channels are adjusted to water discharge so that channel width, depth, slope, and flow velocity are all proportional to discharge. As implied in the continuity equation

$$Q = wdv$$

(where Q is discharge, w is width, d is depth, and v is velocity), a change in discharge will cause a proportional change in width, depth, or velocity. The increased discharge of a flood commonly causes an increase in flow velocity. This faster flow exerts greater shear stress on the channel boundaries, which may cause channel boundary erosion and an increase in width and/or depth. Even in resistant boundary channels, depth and velocity will increase with discharge, generating increased shear stress that may mobilize bed alluvium, which remains stationary during normal flows.

The rate and magnitude of channel change during a flood (Table 6.1) will be a function of (1) flood hydrology and hydraulics, (2) channel boundary resistance, (3) sediment supply, and (4) flow history.

Flood Hydrology and Hydraulics

Flood hydrology refers to the rate of rise and fall, the magnitude and duration of the flood peak, and the ratio of the flood peak to base flow or to regularly recurring peak flows. A flood that is very large in relation to base flows or to regularly recurring peak flows is more likely to cause substantial channel change because of the adjustments necessary to accomodate the increased discharge. Floods with a very steep rising limb and very short duration tend to generate large peak stream power per unit area but low total energy expenditure (expressed as the product of stream power and flood duration). These floods may be ineffective agents of geomorphic change (Costa and O'Connor, 1995). Floods of long duration, moderate-to-large energy expenditure, and low peak stream power may also be geomorphically ineffective. Floods of medium to long duration, with medium-to-large total energy expenditure and large peak stream power, are likely to be the most geomorphically effective because they exert large stream power against the channel boundaries for a sufficient length of time to substantially alter those boundaries (Costa and O'Connor, 1995).

Flood hydraulics refer to the distribution of energy expenditure by the flood flow across time and space. Temporal distribution of flood hydraulics depends on rate of discharge rise and recession as well as duration. Prestegaard et al. (1994) attributed minimal erosion from the summer 1993 floods along the upper portions of Raccoon River, Iowa (USA), in part to relatively short time of flooding. Downstream reaches in the basin had more prolonged flood peaks and severe erosion. Costa and O'Connor (1995) describe two damburst floods in Washington and Oregon that caused minimal geomorphic channel change. They attributed this lack of geomorphic change to the short duration of the floods (6 and 16 min, respectively).

Channel geometry plays a vital role in controlling spatial characteristics of energy expenditure. Flood flows along a confined (deep, narrow) channel may generate extremely large values of boundary shear stress and of stream power per unit area (Rathburn, 1993). Flood changes in flow stage and hydraulics may be particularly dramatic along river reaches with inner channels. The 1870 flood along the Chang Jiang River produced more than 80 m of stage change, generating stream powers of 3×10^3 W/m^2

Table 6.1. *Examples of geomorphic effects of floods*

River	Flood recurrence interval (yr)	Geomorphic effect	Reference
Chilliwack River, Canada	20	transported 4 times average annual bed-material yield	Church, 1988
Genoa River, Australia	>100	>4 m of bed fill in 8- to 10-m-deep channel	Erskine, 1993
Twenty Mile Creek, Canada	–	plucking of sandstone slabs (1.2 × 0.4 × 0.1 m) by flows 0.5–1.8 m deep	Tinkler, 1993
Colorado River–Glen Canyon Dam spillway	–	cavitation holes up to 10.6 m deep and 15 m wide in concrete spillway	Eckley and Hinchcliff, 1996
Patuxent River, USA	100	up to 6 m of channel widening via slumping along 12-m-wide channel	Gupta and Fox, 1974
Bas Glacier d'Arolla, Switzerland	–	≤6.5-m bank recession along 4- to 7-m-wide proglacial channel	Warburton, 1994
Upper Mississippi River, USA	>100	<1–4 mm suspended sediment deposition	Gomez et al., 1997
Connestoga River, USA	–	21 times suspended sediment transport for rest of the year	Kochel, 1988
Eel River, USA	–	51% of suspended sediment discharge for 10-yr period	Helley and LaMarche, 1973
Hwang Ho, China	annual	80% of annual sediment deposition during flood season	Zhou and Pan, 1994
Plum Creek, USA	–	average channel width increased from 26 to 68 m	Osterkamp and Costa, 1987
Hwang Ho, China	annual	180-m-wide channel scoured 3.6–4.6 m to bedrock	Leopold et al., 1964
Gasconade River, USA	–	scour hole 160 m long, 60 m wide, 7 m deep at bridge pier	Ritter, 1988
Orange River, South Africa	–	linear scours 50 m wide, 1 km long, 1–3 m deep on floodplain	Smith and Zawada, 1988
Rubicon River, USA	–	boulder fronts up to 2 m high and 76 m long; 38 m of lateral movement along 114-m-wide channel	Scott and Gravlee, 1968
Macdonald River, Australia	–	average 0.3 m of bed fill	Erskine and Melville, 1983
Roaring River, USA	–	280,000 m^3 debris fan with boulders up to 3 m diameter	Jarrett and Costa, 1986
Mississippi River, USA	–	average rates of meander migration 0.6–300 m/year	Schumm, 1977
Waal River, The Netherlands	approx. 70	169,000–217,000 m^3 of overbank deposition, and sand output of 19,000–67,000 m^3 (average yearly transport of 500,000 m^3)	Ten Brinke et al., 1998
Ha!Ha! River, Canada	8 times the 100-yr flood (damburst)	new channel 2 km long, 90 m wide, and 15 m deep eroded, and 6×10^6 m^3 transported to lower gradient reaches downstream; more than 9 million m^3 sediment exported to the ocean	Lapointe et al., 1998

(Baker and Kochel, 1988). A comparable flood along the Amazon River would have a stage change of 15 m and would generate unit stream power of 12 W/m^2 (Baker and Kochel, 1988). In contrast, flood flows along a shallow channel with a broad floodplain will not greatly increase shear stress or stream power relative to frequent high flows (Baker and Costa, 1987) because flow depths are limited and much of the flow energy will be dissipated across the floodplain. Local changes in channel geometry often control reach-length patterns of flood hydraulics and associated channel change. Wolman and Eiler (1958) demonstrated that valley width and stream gradient were closely associated with local erosion and sedimentation during the 1955 floods in Connecticut (USA). Patton (1988) noted similar patterns across New England; floodplain surfaces immediately downstream of bedrock valley constrictions are narrower and more scoured by overflow channels than floodplains upstream of the constrictions. Miller (1990) found that severe erosion downstream of a bedrock constriction is most common where the channel bends and flood waters continue straight downvalley across the floodplain.

Channel Boundary Resistance

Channel boundary resistance relative to flood erosive force will determine how much flow energy is expended in modifying the channel boundaries. Because of relatively low rates of sediment transport, a large proportion of flow energy may be expended in turbulence for resistant boundary channels (Baker, 1988). In extremely resistant channels, macroturbulence occurs. Macroturbulence results from (1) a steep energy gradient, (2) a low ratio of actual to potential sediment transport, and (3) an irregular, rough boundary capable of generating flow separation (Matthes, 1947). At a peak discharge of approximately 3×10^4 m^3/s, the bedrock channel of the Pecos River in Texas (30 m deep; slope, 3×10^{-3}) has much greater flow energy (9×10^{12} N/m^2 shear stress, 1×10^4 W/m^2 for stream power per unit area) than does the alluvial Mississippi River (12 m deep; slope, 5×10^{-5}; 6 N/m^2 shear stress, 12 W/m^2 stream power) (Baker and Kochel, 1988).

Alluvial channels with erodible boundaries may be subject to substantial erosion as flow energy is expended in sediment entrainment and transport. On the Chilliwack River, Canada, a December 1975 flood with a recurrence interval of about 20 years flushed approximately four times the average annual bed-material yield from the catchment (Church, 1988). Over 4 m of bed fill occurred along reaches of the 8- to 10-m deep Genoa River in Victoria, Australia, during a 1971 flood with a return period in excess of 100 years (Erskine, 1993).

Channel boundary resistance also controls the relative importance of the various erosive processes. Cohesive boundary channels may be eroded through corrosion, corrasion, and cavitation. Corrosion is the chemical weathering and dissolution of cohesive boundaries. Corrosion may weaken cohesive boundaries and render them more susceptible to corrasion and cavitation. On the time scale of a single flood, however, corrosion is most effective at eroding carbonate-rich boundaries. There are few published descriptions of corrosion rates for a single flood. Smith et al. (1995) list rates of 0.02 to 0.20 mm/year for limestone-bounded rivers in southeastern Australia.

Corrasion is the physical abrasion of cohesive boundaries by sediment in transport during a flood. The abundance and abrasive resistance of sediment in transport relative to the abrasive resistance of the channel boundaries determine the effectiveness of corrasion. As for corrosion, there are few published rates of corrasion for a single flood. Tinkler (1993) describes flaking of horizontal slabs of sandstone 5–11 cm thick (1.2×1.45 m surface dimensions) during flows 0.5–1.8 m deep with velocities of 2–3 m/s along Twenty Mile Creek, Ontario (Canada). Tinkler and Wohl (1998) compiled long-term rates of bedrock channel incision that ranged from 7 cm per thousand years in a basin with resistant rocks and dry climate to 130 cm per thousand years in basins with sedimentary rocks and tectonic uplift. Most of this corrasion may be assumed to occur during floods, particularly in arid climates.

Cavitation is the implosion of vapor bubbles as a result of pressure fluctuations in a flow. The implosion generates shock waves that weaken cohesive boundaries. Cavitation has been an especially severe process in the sediment-free flows on dam spillways (Falvey, 1982; Kells and Smith, 1991). During the 1983 flood on the Colorado River (USA), engineers at Glen Canyon Dam were forced to release discharges up to 900 m^3/s down the dam's 12.5-m-diameter concrete-lined spillways. Cavitation during these flows eroded holes up to 10.6 m deep and 15 m wide into the spillway tunnels, creating an enormous step-pool sequence (Eckley and Hinchcliff, 1986). Cavitating flow conditions also may cause the intense bedrock scouring that produces flute marks, polished surfaces, and, at larger scales, scablands. Such flows have been described for the Narmada River, India (Ely et al., 1996), and for the Tapi River, India, where a July 1991 flood produced channel velocities of 5–11 m/s that tore rock slabs 1–3 m long and 0.5–1.5 m wide from a bridge (Kale et al., 1994).

Other hydrodynamic forces may also weaken cohesive boundaries. Pressure fluctuations associated with flow turbulence affect the channel boundaries. And the vertical velocity gradient of the flow can generate a lift force capable of plucking bedrock along joints or bedding planes. Baker (1978) attributed the formation of potholes and rock basins in the channeled scablands of Washington (USA) to plucking by large vertical vortices. Discharges as great as 21.3×10^6 m^3/s were conveyed through the channeled scabland during the draining of Glacial Lake Missoula. These floods created erosional features as large as the 11-km-long, 30-m-deep Rock Lake (Baker, 1978).

Noncohesive channel boundaries may be eroded by entrainment of individual clasts from the bed and banks. Channel erosion also occurs through bank collapse caused by fluvial erosion of the bank toe, rotational or planar slipping along failure surfaces, frost action, sapping or piping, and undercutting by ice blocks in transport (Russell et al., 1995). Gupta and Fox (1974) describe channel widening of up to 6 m (along a 12 m wide channel) via slumping along the Patuxent River, Maryland (USA), during the 100-year recurrence interval 1971 flood. Up to 6.5 m of bank recession occurred along the 4- to 7-m-wide proglacial channel of the Bas Glacier d'Arolla, Switzerland, during a flood in July 1986 (Warburton, 1994). Bank erosion along many channels increases proportionally with discharge (e.g., Pickup and Warner, 1976).

Sediment Supply

Sediment supply influences the erosive forces exerted on the channel boundaries during a flood. Large suspended-sediment concentrations may increase flow viscosity to the level that velocity increases, and turbulence and channel boundary erosion are reduced (Chang, 1988). Suspended-sediment transport commonly increases dramatically during a flood (Inbar, 1987; Conesa García, 1995). Three days of flooding on Pennsylvania's Connestoga River basin in June 1972 transported up to 21 times the suspended sediment moved through the basin during the rest of the year (Kochel, 1988). The December 1964 flood on the Eel River, California (USA), carried 51% of the channel's total suspended sediment discharge during the period 1957–1967 (Helley and LaMarche, 1973). Leopold and Maddock (1953) showed that the highest discharges along the Powder River in Wyoming transported approximately 1000 times the suspended sediment carried by average flow discharges. However, the summer 1993 Mississippi River flood was notable for its high-magnitude, long-duration, and low suspended-sediment discharge. In early January 1993, suspended sediment concentrations >500 mg/L were associated with average daily discharges between 4670 and 6030 m^3/s. In early August 1993, suspended-sediment concentration was 42 mg/L and peak flow was 16,900 m^3/s (Gomez et al., 1995).

Along seasonally flooded rivers, suspended-sediment discharge may show dramatic seasonal contrasts. More than 95% of the Brahmaputra River's annual average suspended load of 345 million metric tons is carried during the May–October monsoon floods (Coleman, 1969; Goswami, 1982). The greatest volume of this sediment is moved by a discharge of 38,000 m^3/s, which is equaled or exceeded 18% of the time (Thorne et al., 1993).

Peak suspended-sediment concentrations may correspond to peak discharge in the upper portion of a basin but lag increasingly behind the peak discharge in a downstream direction as flood waves move downstream faster than the flow velocity (Marcus, 1989). However, Kinoshita (1982) found that maximum near-surface suspended-sediment concentration preceded maximum discharge by 5–8 h during a large flood on the lower Tone River, Japan. Mertes (1990) noted a similar pattern of peak suspended-sediment concentration preceding maximum water discharge on the East Fork River, Wyoming (USA), the Kyronjoki River (Finland), and the Amazon River (Brazil) (drainage areas of 800, 5000, and 2 million km^2, respectively). Peak suspended-sediment concentrations may precede peak discharge in small basins where suspended sediment is entrained from the channel bed (Kurashige, 1994). Where suspended sediment enters the channel mainly via runoff from hillslopes or from bank failures, sediment peak may coincide with or follow discharge peak (Kurashige, 1994). Bull et al. (1995) reported waves of suspended sediment moving down the River Severn (United Kingdom) during floods and interpreted these waves as resulting from changes in the amount of sediment moving into and out of storage.

Coarser sediment moving as bed material may abrade even resistant channel boundaries. As for suspended load, bed-material load commonly increases during a flood. The 5-year recurrence interval discharge on the Alabama River (USA) has a water discharge 5.8 times greater than the bankfull flow, but sand discharge for the 5-year flood is 102 times greater than that for the bankfull discharge (Harvey and Schumm, 1994). Sand discharge during the 100-year recurrence interval flood is 500 times greater than the bankfull sand discharge (Harvey and Schumm, 1994). Approximately 80% of the annual sediment deposition along the Hwang Ho (Yellow River) of China occurs during the flood season, and 82% of this sediment is bed-material load (Zhou and Pan, 1994).

A flood flow with limited sediment supply and limited energy expenditure in sediment transport may generate extreme turbulence that enhances boundary erosion. This effect is commonly observed downstream of dams, culverts, or reaches of artificially stabilized channel. The interrupted sediment exchange between the flow and the channel boundaries along the stabilized reach results in abrupt channel widening or incision where the stabilization ends. During the October 1983 flood along the Santa Cruz River in Tucson, Arizona (USA), severe bank erosion (a doubling of channel width) occurred immediately downstream of every channel reach with banks protected by soil cement (Kresan, 1988).

Flow History

The sequence of flows during the period preceding a flood determines how much adjustment is required for the channel to convey the flood flow. Brunsden and Thornes (1979) defined a transient form ratio for geomorphic features produced by extreme events, where $TF =$ (mean relaxation time)/(mean recurrence time of events). Relaxation time here refers to the time that elapses between the time of a change in the energy input to the system (e.g., a flood) and the time when the system reaches a new, stable morphology (Allen, 1974). This may be applied to channel changes associated with floods. For $TF > 1$, the recurrence time of floods capable of producing change is shorter than the time for the channel to recover, and the channel will be adjusted to flood discharges. Where $TF < 1$, the channel recovers its preflood characteristics before the next flood occurs, and flood features are transient. As an example, Gupta and Fox (1974) found that a 100-year recurrence interval flood in September 1971 along the Patuxent River, Maryland, caused substantial channel widening. However, a record flood along the same channel in June 1972 caused relatively minor channel widening, presumably because the channel had already been widened by the September 1971 flood to accomodate such high flows. Similar effects have been described for the Dudh Kosi region of Nepal in association with glacier-lake outburst floods in 1977 and 1985 (Cenderelli and Wohl, 1997; Cenderelli, 1998).

A single flood or a series of floods substantially larger or smaller than preceding floods may also cross a geomorphic threshold, so that channel form begins to evolve in a new direction. Webb (1987) describes the initiation of channel incision during a 1909 flood on the Escalante River basin, Utah (USA). Floods during the next 30 years created a channel 10 times as wide and 9 times as deep as the pre-1909 channel. Schumm and Lichty

(1963) describe a 30-year phase of channel widening along the Cimarron River in Kansas (USA) that was initiated by a major flood in 1914. Subsequent moderate floods continued to cause channel widening because the 1914 flood had destabilized the channel.

Rate, magnitude, and nature of channel change during a single flood may vary widely among channels and along a single channel (e.g., Interagency Floodplain Management Review Committee, 1994). Boundary change may occur as primarily erosion, deposition, lateral movement, or a change of channel planform. Either erosion or deposition may occur predominantly along the channel bed or banks or along both.

Erosional Patterns

Erosion dominates along reaches of channel where flow velocity, shear stress, or stream power per unit area are higher than along adjacent reaches. Such an increase is generally associated with steeper or narrower channel configurations (e.g., Harvey, 1984). During the July 1976 flood on the Big Thompson River, Colorado (USA), channel reaches <40 m wide and with gradients of 0.02–0.04 had intensive scour, whereas reaches >80 m wide and with gradients <0.02 had limited scour and widespread deposition (Shroba et al., 1979).

Substantial amounts of sediment may be moved by flood erosion. A 1985 glacier-lake outburst flood along Langmoche Khola, Nepal, eroded an estimated 2.6 million m^3 of sediment from 26 km of a valley that ranged from 10 to 200 m wide (Vuichard and Zimmermann, 1987) (Figure 6.1). Erosion from channel scour and hillslope failure reached 34,600–50,000 m^3/km^2 during the August 1969 flood in Virginia (USA) (Williams and Guy, 1973).

Figure 6.1. Erosion along Langmoche Khola, Nepal, after the 1985 glacier-lake outburst flood (photograph taken in 1995). View is upstream along a portion of the valley widened by erosion of fluvioglacial terraces. Active channel is approximately 10 m wide. (Photograph courtesy of Daniel Cenderelli)

Bank Erosion

Channel erosion may take many different forms. Bank erosion can cause bank collapse and channel widening during a flood. During the October 1983 flood along the Santa Cruz River, Arizona, land outside the designated 100- to 500-year floodplain collapsed into the channel when channel width quadrupled (Kresan, 1988). The June 1965 flood along Plum Creek, Colorado (USA), increased average channel width from 26 to 68 m (Osterkamp and Costa, 1987). Between 1949 and 1958, floods along the Hwang Ho of China caused an average bank-caving rate of 1.5 ha km^{-1} $year^{-1}$ along the central, braided portion of the river (Zhou and Pan, 1994). A 1984 glacier-lake outburst flood along the Noeick River, Canada, eroded tributary debris and avalanche fans, widening the channel 14 km below the lake from a preflood average of 75 m to almost 200 m (Desloges and Church, 1992). The magnitude and processes of bank erosion tend to be highly variable across time and space as a result of variability in flow hydraulics, bank composition, bank preconditioning, and other factors (Lawler, 1994; Couperthwaite et al., 1996).

Bed Erosion

Erosion of the bed may occur across an entire channel cross section or along an entire channel reach. During the December 1964 flood on the Eel River basin, California (USA), the channel degraded 1.8–2.4 m over reaches several kilometers long (Helley and LaMarche, 1973). Along some channels, erosion of the bed is limited to removal of alluvium overlying resistant bedrock, as occurred along reaches of the Rubicon River, California (USA), during a December 1964 flood surge (Scott and Gravlee, 1968). Annual floods on the Hwang Ho regularly scour the 180-m-wide channel 3.6–4.6 m, down to bedrock (Leopold et al., 1964). Bed erosion also may erode the bedrock itself, as occurred along reaches of the Genoa River, Australia, during a February 1971 flood (Erskine, 1993).

Erosion is commonly discontinuous along a channel and among neighboring channels because of large spatial and temporal variability in both flood hydraulics and channel-boundary resistance and configuration. Longitudinal variations in valley width and channel orientation are potentially more important than average width in determining location and severity of flood impacts, for example (Miller, 1995). Williams and Guy (1973) describe scour 3–6 m wide and 1.5–3 m deep along small alluvial channels in Virginia (USA) flooded in August 1969. Adjacent channels 400 m away had little or no erosion.

Erosion of a resistant channel bed also may occur selectively, so that distinctive erosional bedforms are created. Potholes are cylindrical to irregularly shaped depressions worn into the channel by the combined hydraulic and abrasive forces of flood flows. Potholes may be minor features a few centimeters wide and deep, or they may be substantial cavities 11 m wide and 10 m deep (Sato et al., 1987). Longitudinal grooves are deep, narrow grooves parallel to flow that may reach dimensions of tens of meters in length, and nearly a meter in width and depth (Wohl, 1993). Prolonged bed erosion may cause either potholes or longitudinal grooves to coalesce into a deep, narrow inner channel. Inner channels are particularly characteristic of large bedrock rivers, such as the Colorado (USA), the Columbia (USA), the Narmada (India), and the Chang Jiang (China) Rivers. The inner channels produced by the Missoula flood in the channeled scabland are more than 500 m wide and 200 m deep (Baker, 1978).

Along both bedrock and alluvial channels, bed erosion may occur predominantly in the pools of a pool-riffle bedform sequence. Working in the bedrock gorge of the Katherine River, Australia, Baker and Pickup (1987) estimated flood values of stream power per unit area in pools more than three times greater than the values for adjacent boulder riffles (3240 and 980 W/m^2, respectively). Baker and Pickup (1987) hypothesized that pool locations were dictated by structural control, with scour resulting from vortices localized at joint intersections. Pools along the sand-bed channel of the Genoa River, Australia, scour by approximately 3 m (bankfull channel width is approximately 120 m) dı ring moderate to large floods and then infill by delta progradation during low flows (Erskine, 1993).

Bed scour may occur at the base of immobile obstacles, such as large boulders, or bridge piers (e.g., Harvey, 1984). Bridge scour is one of the most costly forms of flood damage and accounts for most bridge failures in the United States (Landers et al., 1994). During a December 1982 flood in the 9300-km^2 Gasconade River basin of Missouri (USA), scour initiated by macroturbulence around bridge piers produced a hole 160 m long, 60 m wide, and up to 7 m deep (Ritter, 1988). The August 1969 flood in Virginia (USA) produced numerous examples of severe scour next to bridges and railway trestles (Williams and Guy, 1973). Up to 6 m of channel bank was eroded on each side of the channel at some bridges, particularly where the bridge trapped debris. Where the debris jams grew large enough to divert the flow, a new channel was scoured. In other cases, the bridge became a dam with ponded water upstream. Sediment wedges up to 200 m wide and 4 m deep built up to the level of the bridge, and sediment transport continued across the bridge (Williams and Guy, 1973). Prestegaard et al. (1994) found that bridge failure in the Raccoon River, Iowa, during the summer 1993 floods was associated not with sites of particularly high shear stress but with high channel velocities and low bank resistance.

Overbank Erosion

Channel erosion may also affect overbank flow areas. During the November 1985 flood in the South Branch Potomac River Basin of West Virginia, widespread erosion of bottomlands occurred via both longitudinal grooves greater than 1 m wide and deep and stripping of the valley floor through the formation of scour marks tens to hundreds of meters long and tens of meters wide (Miller and Parkinson, 1993). Such overbank erosion may be initiated by turbulence associated with topographic irregularities on the floodplains (Collins and Schalk, 1937). Linear scours 50 m wide, 1 km long, and 1–3 m deep formed along the Orange River (South Africa) floodplain during a 1988 flood (Smith and Zawada, 1988). Inbar (1987) describes stripping to a depth of 1 m across a floodplain 100–400 m wide and 2000 m long during a 1969 flood along the Jordan River, Israel. A unique form of overbank channel erosion that may occur in carbonate terrains is formation of sinkholes as floodwaters saturate and liquefy overburden, causing soil arches to collapse and flow into cavities in bedrock (Hyatt and Jacobs, 1996). These examples of episodic overbank erosion during floods are localized or of limited depth. In contrast, Nanson (1986) describes catastrophic stripping of overbank areas along high-energy, laterally stable channels in southeastern Australia. Gradual overbank deposition progressively concentrates flow in the main channel, concentrating erosional energy, until a single large flood or a series of more moderate floods strips the floodplain to a basal lag deposit. The dynamic nature of processes that operate between the channel and overbank areas is indicated by the findings of Dunne et al. (1998) for the Amazon River. They found that exchange between the channel and the floodplain in each direction exceed the annual flux of sediment out of the river.

For many channels, the initiation of bed erosion involves a threshold effect. The coarse surface layer commonly present along high-gradient (>0.02) channels remains stable at lower flows (Grant and Swanson, 1995). During a flood, entrainment of the coarse surface layer may be followed by rapid bed erosion as the underlying finer sediments are readily entrained (Pickup and Warner, 1976). The channel-bed configuration set during large flows may also strongly control deposition during subsequent moderate and more frequent flows. The November 1985 flood on the Cheat River, West Virginia (USA), transported boulders up to 4 m in diameter along Cheat Canyon (Kite and Linton, 1993). This was the only 20th century flood capable of transporting these boulders, which control valley-bottom morphology and flood hydraulics in the canyon.

Sediment entrainment and associated channel erosion are commonly greatest during the rising limb of a flood and less during the recessional limb of the hydrograph. This dichotomy is reflected by both the hysteresis of suspended sediment and scour and fill of the channel bed. Various studies have shown suspended sediment concentrations that are (1) more than 3 times higher on the rising limb than on the falling limb for equivalent flow stage on the River Culm, England (Nicholas and Walling, 1995); (2) 10 times higher on the rising limb for the San Juan River, Utah (USA) (Leopold et al., 1964); and (3) nearly 3 times higher on the rising limb for the Colorado River, Arizona (USA) (Leopold and Maddock, 1953). Analogous values for channel-bed scour and fill during a flood are (1) more than 1 m along the Fraser River, Canada, which averages 220 m wide and 7.1 m deep, with a sandy gravel bed that is seasonally mobile in response to snowmelt-induced sustained high flows (Hickin, 1995); (2) a maximum of 38 m along the gravel and cobble bed of the Black Canyon of the Gunnison River, Colorado (USA), for a flow depth of 10.6 m (Leopold et al., 1964); and (3) 2.4 m along the sand bed of the Colorado River (USA) for a flow depth of 6.7 m (Leopold et al., 1964).

The magnitude and pattern of channel erosion during a flood will depend partly on channel planform. Erosion along meandering channels may occur as meander migration along the outside of individual bends, facilitated by changes in stage or by freezing and thawing of ice (Wolman, 1959; Lawler, 1993; Gatto, 1995), or as meander cutoff. Braided channels are characterized by banks with low cohesion. These channels commonly have less predictable patterns of erosion and higher magnitudes of erosion during a flood than do meandering channels (Simon, 1992). Straight channels with well-developed pool-riffle or step-pool bedforms may be dominated by pool scour during a flood. However, if the flood has a large sediment concentration, preferential deposition in the pools may obscure the bedform sequence.

Depositional Patterns

Transport of sediment may occur as dissolved load (material carried in solution), as wash load (suspended sediment finer than the particles usually found in the channel bed), as suspended load (suspended sediment of particle sizes found in appreciable quantity in the channel bed), and as bed load (sediment transported by rolling, sliding, or saltating along the channel bed) (Knighton, 1984). Wash load, suspended load, and bedload transport all generally increase during a flood. The movement of bedload is especially difficult to predict, as it commonly occurs via episodic waves or pulses (e.g., Carey, 1985; Dinehart, 1989, 1992). Erskine (1993) described large amplitude (>4 m) sand slugs or bedload waves moving along the Genoa River, Australia. Boulder fronts up to 2 m high and 76 m long moved as slugs along the Rubicon River, California, during a December 1964 flood (Scott and Gravlee, 1968).

Sediment deposition during a flood may occur as (1) in-channel deposition that raises the level of the channel bed, (2) growth of bars or islands within a channel, (3) formation of an alluvial fan where a steep, confined channel joins a broader valley, (4) lateral boulder berms beside the low-flow channel, (5) lateral accretion of the floodplain through construction of point bars as a meander bend migrates, and (6) vertical accretion of the floodplain as sediment settles from suspension in overbank flows. Magnitudes and patterns of deposition will depend on sediment supply, spatial and temporal characteristics of flow hydraulics, and channel planform.

In-Channel Deposition

Deposition occurs where sediment supply overwhelms flow transport capacity. This may occur along the entire length of a channel or only in localized reaches where a change in channel geometry causes flow velocity to decrease. Nolan and Marron (1985) documented differences in channel response to flooding as a function of sediment supply. Floods along intermediate- and high-order channels in northwestern California (USA) result in long-lasting widening and filling along the length of these channels because of the large volumes of sediment supplied by streambank erosion and failure of streamside hillslopes. In contrast, floods along similar channels in central California result in only local deposition and channel change because these channels have widespread bedrock control and relatively few slope failures. Erskine and Melville (1983) described fill and subsequent erosion along the lower Macdonald River in southeastern Australia. An average of 0.3 m of bed fill occurred during a moderate flood in March 1978, with large volumes of sediment supplied during the flood by bank erosion. A smaller flood in June 1978 did not cause bank erosion but resulted in 0.1 m of bed scour.

Lisle (1982) described an example of fairly continuous deposition during the December 1964 flood in northern California. This rainfall-generated flood mobilized large quantities of sediment from recently logged

Figure 6.2. Deposition along Langmoche Khola, Nepal, after the 1985 glacier-lake outburst flood (photograph taken in 1995). View is downstream, and shows channel reach located immediately upstream from the view in Figure 6.1. Active channel is approximately 10 m wide. (Photograph courtesy of Daniel Cenderelli)

hillslopes. The sediment caused up to 4 m of channel aggradation along some gravel-bed channels in northwestern California, with maximum deposition occurring in the pools. More localized deposition during a flood may result from a decrease in channel bed gradient, an increase in channel width, or other lower velocity zones such as the inside of channel bends. O'Connor et al. (1986) describe boulder deposition at bends along Boulder Creek, Utah (USA), during large floods. The 1985 Langmoche Khola outburst-flood in Nepal deposited 2,580,000 m^3 of sediment along 26 km of channel, with deposition occurring primarily along wider, lower gradient reaches (Vuichard and Zimmermann, 1987) (Figure 6.2).

Growth of Bars and Islands

Bars or islands within a channel may grow by vertical and lateral accretion during a flood. The development of bars and islands is facilitated by (1) locally erodible banks that increase channel width, redirect flow away from the bars, and introduce sediment to the flow; (2) in-channel obstacles such as large, immobile boulders downstream from which flow separation develops (Baker, 1978); and (3) boundary roughness such as vegetation. Scott and Gravlee (1968) noted formation of lateral bars of locally derived stream alluvium during the December 1964 flood along the Rubicon River, California.

Tributary Fans

Flood deposition in the form of alluvial fans has been described for several floods occurring in hilly or mountainous regions. Williams and Guy (1973) described fans ranging in length from a few meters to nearly 700 m as a result of the August 1969 flood in central Virginia. These fans had up to 8500 m^3 of sand- to boulder-sized sediments. The July 1982 Lawn Lake

Figure 6.3. Lawn Lake flood debris fan. View across the Fall River valley to the junction with the Roaring River, which comes down from the mountains at the back of this view.

flood in Colorado created a 280,000 m^3 debris fan with boulders up to 3 m in diameter (Jarrett and Costa, 1986) (Figure 6.3). A January 1971 flood in the Sinai Desert (Egypt) created a 6200 m^3 fan where the 13-km^2 Wadi Mikeimin entered a larger channel (Schick and Lekach, 1987). Debris fans formed at channel junctions may occasionally dam the receiving channel, creating the potential for a damburst flood. An intense rainstorm in December 1980 triggered a debris flow at the head of Polallie Creek, Oregon (USA). When this flow reached the debris fan at the confluence with the East Fork Hood River, the flow deposited approximately 76,400 m^3 of debris that formed a temporary dam and sent a 560-m^3/s flood wave down the East Fork Hood River (Gallino and Pierson, 1985). This flood wave severely damaged a highway located downstream.

Boulder Berms

Boulder berms are linear features parallel to the channel boundaries. Scott and Gravlee (1968) measured berm surfaces as much as 8.5 m above the post-flood channel thalweg after the December 1964 flood along the Rubicon River, California. Failure of a small moraine-dam in England in 1927 caused an outburst flood that deposited boulder berms approximately 1 m deep and 30 m wide at major expansions in valley width (Carling and Glaister, 1987). A glacier-lake outburst flood in Nepal created multiple linear and curvilinear longitudinal cobble and boulder bars up to 2.5 m high, 20 m wide, and 100 m long that are parallel or subparallel to valley alignment (Cenderelli and Wohl, 1996; Cenderelli and Cluer, 1998).

Lateral Floodplain Accretion

Lateral floodplain accretion via point bar migration occurs when a meander bend migrates laterally. Meander migration rates are strongly controlled by (1) bend curvature, with maximum migration rates at radius of curvature/channel width ratios of 2–3; (2) bank strength (migration rate decreases as bank strength increases); and (3) stream power (migration rate increases with increasing stream power) (Hickin and Nanson, 1984). Erosion along the outside of the bend and point-bar deposition along the inside of the bend may move an individual bend tens of meters across a floodplain during a single flood. The Bhagirathi River of West Bengal, India (approximately 300 m wide), had erosion on the order of 0.9×10^6 m^3 on the concave side of a single meander loop during 1987–1988 (Chakraborty, 1993). Simon and Darby (1997) report 2 m of migration in 9 months along an incised meander of Goodwin Creek, Mississippi (USA). Relatively few studies report event-based rates of meander migration; most reported rates represent an average over a period of decades. Such average rates may vary from 0.6 to 300 m/year along different portions of the same channel – in this case, the Mississippi River (USA) (Schumm, 1977).

Vertical Floodplain Accretion

Vertical floodplain accretion during a flood is enhanced by a high suspended-sediment load and by a wide and hydraulically rough floodplain. Mertes (1990) noted that the rate of sediment deposition on the floodplain surface is likely to reach a maximum during the peak stages of inundation even though the concentrations of sediment are decreasing. Deposition may occur as a layer of fine sediments that settle from suspension to blanket preexisting topography or as discrete sheets, lobes (Scott and Gravlee, 1968), or splays. A December 1982 flood in southeastern Missouri (USA) left thin (<0.5 m) sheetlike layers of coarse gravel and sand spread evenly across the Gasconade River floodplains as well as a sand and gravel lobe 30–100 m wide, 1 km long, and 1–2 m thick (Ritter, 1988). The 1965 Plum Creek flood (Colorado, USA) deposited up to 2 m of sediment on the floodplain (Osterkamp and Costa, 1987). Along sinuous channels such as the River Waal in The Netherlands, helicoidal currents may induce erosion of sediments in the concave (outer) bank, which are subsequently transported to and deposited in the convex (inner) bank (Ten Brinke et al., 1998).

A number of factors may influence floodplain spatial and temporal patterns of vertical accretion. Deposition from suspension typically grows thinner and finer grained with distance from the channel (Gomez et al., 1997; Allison et al., 1998), which suggests that sediment is being transported by turbulent diffusion (Allen, 1985; Pizzuto, 1987). Where water flow over the floodplain has a component perpendicular to the channel, rather than being parallel to the main channel, convection may be the dominant mechanism of sediment transport (Middelkoop and Asselman, 1998). Deposition may also vary as a function of flood magnitude. Along the Rhine and Meuse Rivers in The Netherlands, deposition of sand on natural levees occurs mainly during high-magnitude floods, whereas low-magnitude floods and slowly receding floods are important for deposition of silt and clay in floodplain depressions farther from the main channel (Asselman and Middelkoop, 1998). Gomez et al. (1998) documented vertical accretion rates of 14–18 mm/h during floods with recurrence intervals of 5–30 years in the Waipaoa River basin of New Zealand, whereas vertical accretion averaged only 6 mm/h during a larger flood with higher flow velocities across the floodplain. The source(s) of floodplain water may control suspended sediment concentrations and hence depositional patterns (Mertes, 1997; Mertes, Chapter 5, this volume). Finally, deposition may vary with time in association with climatic shifts (Knox, 1972) or land-use changes (Gomez et al., 1998) that affect sediment production and transport.

Floods along the Connecticut River (USA) in 1936 and 1938 deposited up to 2 m of sediment near the channel but only a thin veneer of sediment along the floodplain margins; an overall average of 9 mm was added to the floodplain surface (Jahns, 1947). Mertes (1990) reported a vertical deposition rate of 7.6 cm/year within 500 m of the channel on the central Amazon River floodplain and a rate of 0.8 cm/year across the entire floodplain surface. Mean annual deposition rates for floodplains along British rivers are commonly within the range of 0–10 mm/year (Nicholas and Walling, 1997). Long-term rates such as these may be constrained by a ^{14}C or ^{137}Cs chronology (Collins et al., 1997; Allison et al., 1998). Knox (1987) reported decadal-scale average historical rates of overbank floodplain sedimentation ranging from 0.3 to 5 cm/year in the upper Mississippi Valley.

The storage of overbank sediments after a flood varies as a function of channel characteristics. Working on channels in southern Wisconsin, USA, Lecce (1997) found that alluvial storage increased with valley width and decreased with cross-sectional stream power. The development of meander belts also influenced overbank sediment storage.

Erosional and depositional responses to a flood commonly continue for a period of years and may be asynchronous along different reaches of a channel. A dam failure on Lawn Lake, Colorado (USA) in July 1982 caused an outburst flood that peaked at 500 m^3/s on the Roaring River, about 30 times the 500-year flood for the Roaring River (Jarrett and Costa, 1986). The flood cut a channel as much as 15 m deep along the steeper reaches of the Roaring River (Pitlick, 1993). Where the Roaring River joins the wider, lower gradient valley of the Fall River, the flood deposited a large (280,000 m^3) debris fan. During the 1983 snowmelt runoff, approximately 15.5 million kg of bedload sediment was eroded from the fan and adjacent channel reaches. This sediment was carried downstream to a highly sinuous portion of the Fall River that had not been much affected by the Lawn Lake flood. Sedimentation during 1983 aggraded the 2-m-deep channel to the level of the floodplain, forming a 2.3-km-long continuous zone of deposition. This sediment was gradually transported downstream during the next decade (Bathurst and Ashiq, 1998). Pitlick (1993) noted that different reaches of the Fall River were never in the same phase of adjustment; erosion in the debris-fan reach coincided with deposition along the sinuous reach. Similar prolonged and asynchronous channel adjustment to flooding has been described for other rivers (Kelsey, 1980).

Lateral Channel Movement and Channel Planform Change

Lateral channel movement during a single flood may take the form of predictable, albeit rapid, mobility, such as meander migration, or it may take the form of less predictable meander cutoff or of avulsion along any type of channel. Meander cutoff is a symptom of channel instability (Knighton, 1984). When the channel can no longer maintain an increasing sinuosity, erosion during a flood(s) will cut a new channel across the base of a meander bend (Gay et al., 1998). The lateral shift of flow from the abandoned meander to the new straight channel may involve a distance of 1–2 km, as documented along the lower Hunter River, Australia, during floods since 1870 (Erskine et al., 1992). Along the meandering rivers of Bangladesh, tectonic movements, catastrophic floods, high sediment loads, and human modifications all contribute significantly to the large, rapid lateral movements that characterize these channels. Between 1973 and 1988 a 35-km-long reach of the Arial Khan River, for example, developed three major meander cutoffs (each shortened the channel by approximately 6 km) and two new meanders (Winkley et al., 1994). The summer 1993 flooding on the Mississippi River (USA) created a large cutoff; during the peak discharge, 25% of the total flow was contained in the incipient cutoff channel, which scoured to depths of 21 m and extended 2 km downstream from the levee break (Jacobson et al., 1993).

Avulsion during a flood has been described for various types of channels. Flood flows along sinuous channels may scour a straight channel directly across meander bends of the preflood channel, as along the Santa Cruz River, Arizona, during the 1983 flood (Kresan, 1988), or Plum Creek, Colorado, during the 1965 flood, when avulsions moved the channel 200 m laterally (Osterkamp and Costa, 1987). Scott and Gravlee (1968) measured 38 m of lateral movement of the thalweg along portions of the 114-m-wide Rubicon River, California, during a December 1964 flood surge. During an August 1954 flood on the Hwang Ho, a braided portion of the river moved 6 km back and forth across the floodplain within 24 h (Zhou and Pan, 1994).

In addition to straightening of sinuous channels, channel planform changes during a flood most commonly take the form of an increase in width to the point where a single channel becomes braided (Virginia Division of Mineral Resources, 1969). Meltwater flooding in the valley train of the Bas Glacier d'Arolla, Switzerland, during 1986 and 1987 changed a low sinuosity, single-thread channel to a braided channel several times wider (Warburton, 1994).

Channel planform changes may also occur as regular alternations tied to seasonal flooding. Nanson et al. (1986) describe a seasonal alternation between braided and anastomosing channel patterns on Cooper Creek, Australia. During floods, a low sinuosity braided system develops, with clay-rich sediment that has hardened during dessication of the floodplain being transported as sand-sized aggregates. As the clay gradually disaggregates, an anastomosing pattern of channels with lower width/depth ratios develops. This anastomosing pattern is maintained during moderate flows. The Auranga River, India, alternates between a braided pattern during the dry season and a meandering pattern during the wet monsoon (Gupta and Dutt, 1989). Rivers of the seasonal tropics commonly display a channel-in-channel morphology of a small inner channel that carries high flows of the wet season and a larger channel maintained by less frequent, high-magnitude floods (Gupta, 1995). Fahnestock (1963) describes a regular seasonal change along the proglacial White River (19 km^2) draining the flank of Mt. Rainier, Washington. The channel pattern changes from meandering during low flows to braided during high summer flows.

Gradual Channel Changes

Changes in channel planform, and in patterns of erosion and deposition, also may occur over a period of years to decades as a result of changes in the flood regime. The alluvial channels of the western Great Plains of the United States provide an example. During the past century (approximately 1870–1970), the South Platte, North Platte, and other channels of the western Great Plains have narrowed dramatically. During the late 19th century, these were broad, braided, relatively straight channels with reaches of intermittent flow. Flood peaks were driven by spring snowmelt in the Rocky Mountains and by rainfall from summer thunderstorms. With the development of irrigated agriculture from the 1870s onward, peak flows decreased substantially; average annual peak flow for 1957–1970 was 0.1–0.3 the average yearly peak for pre-1909 on the North Platte and Platte Rivers (Williams, 1978). Infiltration of irrigation water raised water tables. New riparian forests became established along the previously sparsely vegetated channels, and this vegetation increased overbank roughness and enhanced floodplain sedimentation (Eschner et al., 1983; Johnson, 1994). Channels that had been 450 m wide in the 1800s had narrowed to 100-m-wide sinuous courses by 1970 (Nadler and Schumm, 1981).

The Gila River of southern Arizona (USA) also changed during the period 1846 to 1970 as a result of changes in flood characteristics (Burkham, 1972). Before 1905 the channel was sinuous, relatively stable, and narrow (average width 90 m). Between 1905 and 1917 several large winter floods with low sediment loads eroded the floodplain and widened the channel to approximately 600 m. Much of this widening occurred during floods in 1905 and 1906. Since 1918 the floodplain has been rebuilt by smaller floods with high sediment loads. By 1964 the Gila River had again developed a meandering pattern, with an average channel width of 60 m.

Changes in flood regime associated with reservoirs and dam operations are becoming increasingly common. Depending on the characteristics of the channel and the dam, channels may incise, aggrade, change their patterns, develop coarser or finer-grained beds, widen or narrow, or increase or decrease lateral migration (Ligon et al., 1995). Sediment load along the Snake River, Idaho, the most extensively dammed river in the western United States, has decreased substantially since the construction of dams began in the 1940s (Collier et al., 1996). As a result, beaches along the Hells Canyon portion of the Snake River shrank by 75% between 1955 and 1982 (Grams, 1991). Construction of numerous dams along the Rio Grande, New Mexico (USA), began in 1916. Channel response ranged from nearly 1 m of bed scour downstream from some dams (Lagasse, 1980) to more than 4 m of channel aggradation where decreased main channel flows could not rework tributary sediment inputs (Everitt, 1993). Summarizing channel-bed responses downstream from dams for 23 rivers in the western United States, Williams and Wolman (1984) found that maximum streambed lowering ranged from a negligible amount to as much as 7.6 m. After completion of the Sanmenxia Reservoir in 1960, the sediment concentration on the Hwang Ho dropped to approximately 21% of predam values, and some 300 km of channel downstream from the dam incised by an average of 1 m (Zhou and Pan, 1994).

Geomorphic Importance of Floods

Numerous case studies of the geomorphic effects of individual floods have demonstrated that some channel systems are predominantly shaped by fairly small, frequent floods, whereas other channels are controlled by large, infrequent floods (e.g., Gupta, 1983; Lewin, 1989; Wohl et al., 1994). Kochel (1988) concluded that streams experiencing major geomorphic changes during large floods are characterized by (1) flashy hydrographs, with a high Q_{max}/Q_{mean} ratio, (2) high channel gradient, (3) abundant coarse bedload, (4) relatively low bank cohesion, and (5) relatively deep, narrow cross-sectional geometry, which facilitates high shear stresses during flood flows. Patton's (1988) summary of the geomorphic significance of floods in southern New England provides conclusions that support those of Kochel. Patton (1988) found that highland drainages have floodplains dominated by sedimentation during large floods. Erosion during large floods along lowland streams, however, is no greater than the erosion caused by a succession of more moderate flows. The relative magnitude of channel change during a flood may thus be estimated as a function of flood hydrology and channel characteristics. Generalizing from a dataset that included several floods, Magilligan (1992) defined a threshold unit stream power of approximately 300 W/m^2 for causing major morphological adjustments (primarily via erosion) along low-gradient alluvial channels in humid to subhumid environments. Miller (1990) also identified a stream power threshold of 300 W/m^2 for channels in the central Appalachians. Brookes (1990) found that channelized rivers in the United Kingdom with gradients of 0.001–0.01 tended to be erosionally modified when stream power exceeded 35 W/m^2. Wohl et al. (in press) defined a threshold of substantial erosional or depositional modification during floods along bedrock canyon rivers that varied from approximately 100 to 2000 W/m^2, as a function of drainage area. This can be expressed in the form $y = 21x^{0.36}$, where y is stream power per unit area (W/m^2) and x is drainage area (km^2). Ultimately, the geomorphic effectiveness of a flood is linked directly not so much to magnitude or frequency as to shear stress and stream power per unit area relative to channel-boundary resistance to erosion (Baker and Costa, 1987).

Geomorphic Hazards Associated with Flooding

Enhanced erosion, deposition, lateral channel movement, or changes in channel planform during a flood present obvious hazards to both humans and aquatic and riparian communities. Human use of the channel and floodplain is commonly based on the assumption that these features will remain relatively static. Human-built structures are particularly vulnerable to geomorphic hazards during floods (Merritts, Chapter 10, this volume). Channel-bed scour may undermine bridges, dams, and channel-stabilization structures. Bank collapse and lateral channel movement may destroy buildings and roads. In-channel deposition may clog irrigation intake structures, increase flooding of the floodplain, and hinder the drainage of backswamps. Overbank deposition may damage buildings.

The prolonged changes in flood regime associated with flow regulation may also alter sedimentation patterns and put human structures at risk. The 1963 completion of Glen Canyon Dam along the Colorado River (USA) resulted in substantially reduced flood peaks and suspended sediment concentrations downstream from the dam, in Grand Canyon National Park (Collier et al., 1996). Before dam construction, spring floods on the Colorado River annually scoured and filled the channel bed below the dam by 6–9 m. After a June 1965 flood, the bed elevation fell by approximately 4.5 m and did not recover its previous elevation. Present flow releases from the dam have little effect on the armored bed (Collier et al., 1996). High beaches along the channel margins, which used to be replenished by sediment deposition from suspension during large floods, have been gradually eroding for the past three decades (Kearsley et al., 1994). Beach erosion has both undermined adjacent alluvial terraces that contain archeological sites and reduced recreational camping sites along the Colorado River. Closure of Hoover Dam approximately 450 km downstream on the Colorado River caused channel-bed erosion for more than 500 km downstream. Intake structures for municipal supply and irrigation were rendered useless and bridge foundations were undermined. The channel bed was lowered 2.8 m at Yuma, Arizona, 560 km downstream from Hoover Dam (Dunne, 1988).

Although riparian and aquatic species are generally adapted to flooding and the associated geomorphic changes, human activities may make it difficult for these species to recover after a flood. Recolonization of a channel disturbed by flooding commonly occurs from nearby undisturbed reaches, such as tributaries or adjacent drainage basins, or from temporary refugia such as the floodplain (Gregory et al., 1991). Along heavily developed channels where only one drainage basin or a small segment of a basin still supports certain species or where the floodplain is effectively innaccessible to aquatic and riparian organisms, successful recolonization may not occur (Wydoski and Wick, Chapter 9, this volume). Where irrigation diversions, reservoir operation, or other human activities have altered both hydrologic and geomorphic patterns associated with flooding, the existing aquatic and riparian species may be incapable of adapting to the change (Brinson et al., 1981; Bayley, 1991; Malanson, 1993; Power et al., 1995; Friedman and Auble, Chapter 8, this volume). Johnson (1992) predicted a decline in the areal extent of pioneer riparian forests (cottonwood, willow) along the upper Missouri River, North Dakota (USA), as a result of reduced rates of river meandering associated with reduced peak flows downstream from dams. Ligon et al. (1995) described three examples of geomorphic changes resulting from flow alteration that stressed riparian and aquatic biota: (1) Damming of the McKenzie River, Oregon (USA), reduced peak flows and channel avulsion, thereby eliminating the recruitment of coarse sediment necessary for salmon spawning; (2) damming of the Oconee River, Georgia (USA), caused downstream channel incision, a reduction in overbank flooding, and a reduction in densities of fish; and (3) damming of the braided Waitaki River (New Zealand) stabilized the channel and allowed riparian vegetation to colonize the formerly exposed bars that provide nesting habitat for the endangered black stilt (*Himantopus novaezealandiae*). As a result of local or basin-wide human alterations in flood-related geomorphic processes, geomorphic flood hazards to riparian and aquatic species may increase to the point that individual species or communities can no longer survive along a river.

Conclusions

Floods are commonly associated with changes in the movement of both water and sediment along a channel and floodplain system. These changes cause channel adjustments in the form of erosion, deposition, lateral channel movement, and altered channel pattern. If the channel adjustments are of a magnitude that cannot be withstood by structures or adapted to by riparian and aquatic communities, then hazardous conditions are created. Using threshold values of stream power defined by several investigators, in combination with flow and sediment transport models, the general trends of likely channel adjustment in response to a specified flood scenario may be estimated as a function of flood hydrology, channel geometry, sediment supply, and flow history. An example of this approach is provided by investigations along the Colorado River in the Grand Canyon. Sediment retention behind Glen Canyon Dam upstream from the Grand Canyon has created concerns about bar erosion within Grand Canyon. After a tributary flood in 1993 that introduced substantial volumes of sand (approximately 4.2 Tg) into the Colorado River, Wiele et al. (1996) compared measured changes in cross-sectional area along the Colorado River as a result of sand deposition to changes computed by a two-dimensional flow and sediment transport model. The total difference between measured and computed changes in cross-sectional area was 6%. Despite the success in this example, the specific characteristics of channel change in response to flooding along many channels are commonly too stochastic to be reliably predicted for use in engineering design.

Acknowledgments

Andrew Simon and Frank Magilligan provided helpful reviews of this chapter.

References

Allen, J.R.L. (1974). Reaction, relaxation and lag in natural sedimentary systems: general principles, examples and lessons. *Earth-Science Reviews*, **10**, 263–342.

Allen, J.R.L. (1985). Principles of physical sedimentology. London: George Allen and Unwin.

Allison, M.A., Kuehl, S.A., Martin, T.C., and Hassan, A. (1998). Importance of flood-plain sedimentation for river sediment budgets and terrigenous input to the oceans: insights from the Brahmaputra-Jamuna River. *Geology*, **26**, 175–178.

Asselman, N.E.M. and Middelkoop, H. (1998). Temporal variability of contemporary floodplain sedimentation in the Rhine-Meuse Delta, The Netherlands. *Earth Surface Processes and Landforms*, **23**, 595–609.

Baker, V.R. (1978). Large-scale erosional and depositional features of the Channeled Scabland. In *The channeled scabland*, eds. V.R. Baker and D. Nummedal, pp. 81–115. Washington, DC: National Aeronautics and Space Administration.

Baker, V.R. (1988). Flood erosion. In *Flood geomorphology*, eds. V.R. Baker, R.C. Kochel, and P.C. Patton, pp. 81–95. New York: John Wiley and Sons.

Baker, V.R. and Costa, J.E. (1987). Flood power. In *Catastrophic flooding*, eds. L. Mayer and D. Nash, pp. 1–21. Boston: Allen and Unwin.

Baker, V.R. and Kochel, R.C. (1988). Flood sedimentation in bedrock fluvial systems. In *Flood geomorphology*, eds. V.R. Baker, R.C. Kochel, and P.C. Patton, pp. 123–137. New York: John Wiley and Sons.

Baker, V.R. and Pickup, G. (1987). Flood geomorphology of the Katherine Gorge, Northern Territory, Australia. *Geological Society of America Bulletin*, **98**, 635–646.

Bathurst, J.C. and Ashiq, M. (1998). Dambreak flood impact on mountain stream bedload transport after 13 years. *Earth Surface Processes and Landforms*, **23**, 643–649.

Bayley, P.R. (1991). The flood pulse advantage and the

restoration of river-floodplain systems. *Regulated Rivers: Research and Management*, **6**, 75–86.

Brinson, M.M., Swift, B.L., Plantico, R.C., and Barclay, J.S. (1981). *Riparian ecosystems: their ecology and status*. U.S. Fish and Wildlife Service, FWS/OBS-81/17, 154 pp.

Brookes, A. (1990). Restoration and enhancement of engineered river channels: some European experiences. *Regulated Rivers: Research and Management*, **5**, 45–56.

Brunsden, D. and Thornes, J.B. (1979). Landscape sensitivity and change. *Transactions of the Institute of British Geographers New Series*, **4**, 463–484.

Bull, L.J., Lawler, D.M., Leeks, G.J.L., and Marks, S. (1995). Downstream changes in suspended sediment fluxes in the River Severn, United Kingdom. In *Effects of scale on interpretation and management of sediment and water quality*, pp. 27–37. Proceedings, Boulder Symposium, July 1995. IAHS Publ. no. 226. Wallingford, UK: IAHS Pubs.

Burkham, D.E. (1972). Channel changes of the Gila River in Safford Valley, Arizona, 1846–1970. *U.S. Geological Survey Professional Paper 655G*, 24 pp. Washington, DC: U.S. Geological Survey Pubs.

Carey, W.P. (1985). Variability in measured bedload-transport rates. *Water Resources Bulletin*, **21**, 39–48.

Carling, P.E. and Glaister, M.S. (1987). Reconstruction of a flood resulting from a moraine-dam failure using geomorphological evidence and dam-break modeling. In *Catastrophic flooding*, eds. L. Mayer and D. Nash, pp. 181–200. Boston: Allen and Unwin.

Cenderelli, D.A. (1998). *Glacial-lake outburst floods in the Mount Everest region of Nepal: flow processes, flow hydraulics, and geomorphic effects*. Unpublished PhD dissertation, Colorado State University, Ft. Collins, Colorado, 247 pp.

Cenderelli, D.A. and Cluer, B.L. (1998). Depositional processes and sediment supply in resistant-boundary channels: examples from two case studies. In *Rivers over rock: fluvial processes in bedrock channels*, eds. K.J. Tinkler and E.E. Wohl, pp. 105–131. Washington, DC: American Geophysical Union Geophysical Monograph 107.

Cenderelli, D.A. and Wohl, E.E. (1996). Sedimentology and clast orientation of large bar complexes generated by glacial-lake outburst floods in the Mt. Everest region of Nepal. *Geological Society of America Abstracts with Programs*, **28**, A110.

Cenderelli, D.A. and Wohl, E.E. (1997). Hydraulics and geomorphic effects of the 1985 glacial-lake outburst flood in the Mount Everest region of Nepal. *Geological Society of America Abstracts with Programs*, **29**, A-216.

Chakraborty, K. (1993). Some geo-technical aspects of meander cutoff in alluvial environment. *Journal of the Geological Society of India*, **42**, 231–242.

Chang, H.H. (1988). *Fluvial processes in river engineering*. New York: John Wiley and Sons, 432 pp.

Church, M. (1988). Floods in cold climates. In *Flood geomorphology*, eds. V.R. Baker, R.C. Kochel, and P.C. Patton, pp. 205–229. New York: John Wiley and Sons.

Coleman, J.M. (1969). Brahmaputra River: channel processes and sedimentation. *Sedimentary Geology*, **3**, 129–239.

Collier, M., Webb, R.H., and Schmidt, J.C. (1996). Dams and rivers: a primer on the downstream effects of dams. *U.S. Geological Survey Circular 1126*, 94 pp. Washington, DC: U.S. Geological Survey Pubs.

Collins, A.L., Walling, D.E., and Leeks, G.J.L. (1997). Use of the geochemical record preserved in floodplain deposits to reconstruct recent changes in river basin sediment sources. *Geomorphology*, **19**, 151–167.

Collins, R.F. and Schalk, M. (1937). Torrential flood erosion in the Connecticut Valley, March 1936. *American Journal of Science*, **34**, 293–307.

Conesa García, C. (1995). Torrential flow frequency and morphological adjustments of ephemeral channels in south-east Spain. In *River morphology*, ed. E.J. Hickin, pp. 169–192. Chichester: Wiley.

Costa, J.E. and O'Connor, J.E. (1995). Geomorphically effective floods. In *Natural and anthropogenic influences in fluvial geomorphology*, eds. J.E. Costa, A.J. Miller, K.W. Potter, and P.R. Wilcock, pp. 45–56. Washington, DC: American Geophysical Union Geophysical Monograph 89.

Couperthwaite, J.S., Bull, L.J., Lawler, D.M., and Harris, N.M. (1996). Downstream change in channel hydraulics and river bank erosion rates in the Upper Severn, United Kingdom. In *Hydrologie dans les pays celtiques*, pp. 93–100. Rennes, France: INRA.

Desloges, J.R. and Church, M. (1992). Geomorphic implications of glacier outburst flooding: Noeick River valley, British Columbia. *Canadian Journal of Earth Sciences*, **29**, 551–564.

Dinehart, R.L. (1989). Dune migration in a steep, coarse-bedded stream. *Water Resources Research*, **25**, 911–923.

Dinehart, R.L. (1992). Gravel-bed deposition and erosion by bedform migration observed ultrasonically during storm flow, North Fork Toutle River, Washington. *Journal of Hydrology*, **136**, 51–71.

Dunne, T. (1988). Geomorphologic contributions to flood control planning. In *Flood geomorphology*, eds. V.R. Baker, R.C. Kochel, and P.C. Patton, pp. 421–438. New York: John Wiley and Sons.

Dunne, T., Mertes, L.A.K., Meade, R.H., Richey, J.E., and Forsberg, B.R. (1998). Exchanges of sediment between the flood plain and channel of the Amazon

River in Brazil. *Geological Society of America Bulletin*, **110**, 450–467.

Eckley, M.S. and Hinchliff, D.L. (1986). Glen Canyon Dam's quick fix. *Civil Engineering*, **56**, 46–48.

Ely, L.L., Enzel, Y., Baker, V.R., Kale, V.S., and Mishra, S. (1996). Changes in the magnitude and frequency of late Holocene monsoon floods on the Narmada River, central India. *Geological Society of America Bulletin*, **108**, 1134–1148.

Erskine, W. (1993). Erosion and deposition produced by a catastrophic flood on the Genoa River, Victoria. *Australian Journal of Soil and Water Conservation*, **6**, 35–43.

Erskine, W., McFadden, C., and Bishop, P. (1992). Alluvial cutoffs as indicators of former channel conditions. *Earth Surface Processes and Landforms*, **17**, 23–37.

Erskine, W. and Melville, M.D. (1983). Impact of the 1978 floods on the channel and floodplain of the lower Macdonald River, N.S.W. *Australian Geographer*, **15**, 284–292.

Eschner, T.R., Hadley, R.F., and Cowley, K.D. (1983). Hydrologic and morphologic changes in channels of the Platte River Basin in Colorado, Wyoming, and Nebraska: a historical perspective. *U.S. Geological Survey Professional Paper 1277-A*, 39 pp. Washington, DC: U.S. Geological Survey Pubs.

Everitt, B.L. (1993). Channel responses to declining flow on the Rio Grande between Ft. Quitman and Presidio, Texas. *Geomorphology*, **6**, 225–242.

Fahnestock, R.K. (1963). Morphology and hydrology of a glacial stream – White River, Mt. Rainier, Washington. *U.S. Geological Survey Professional Paper 422A*, 61 pp. Washington, DC: U.S. Geological Survey Pubs.

Falvey, H.T. (1982). Predicting cavitation in tunnel spillways. *Water Power and Dam Construction*, **34**, 13–15.

Gallino, G.L. and Pierson, T.C. (1985). Polallie Creek debris flow and subsequent dam-break flood of 1980, East Fork Hood River basin, Oregon. *U.S. Geological Survey Water-Supply Paper 2273*, 22 pp. Washington, DC: U.S. Geological Survey Pubs.

Gatto, L.W. (1995). Soil freeze-thaw effects on bank erodibility and stability. *Cold Regions Research and Engineering Laboratory, Special Report 95-24*, 17 pp. Hanover, New Hampshire: Cold Regions Research and Engineering Laboratory.

Gay, G.R., Gay, H.H., Gay, W.H., Martinson, H.A., Meade, R.H., and Moody, J.A. (1998). Evolution of cutoffs across meander necks in Powder River, Montana, USA. *Earth Surface Processes and Landforms*, **23**, 651–662.

Gomez, B., Eden, D.N., Peacock, D.H., and Pinkney, E.J. (1998). Floodplain construction by recent, rapid vertical accretion: Waipaoa River, New Zealand. *Earth Surface Processes and Landforms*, **23**, 405–413.

Gomez, B., Mertes, L.A.K., Phillips, J.D., Magilligan, F.J., and James, L.A. (1995). Sediment characteristics of an extreme flood: 1993 upper Mississippi River valley. *Geology*, **23**, 963–966.

Gomez, B., Phillips, J.D., Magilligan, F.J., and James, L.A. (1997). Floodplain sedimentation and sensitivity: summer 1993 flood, Upper Mississippi River valley. *Earth Surface Processes and Landforms*, **22**, 923–936.

Goswami, D.C. (1982). *Brahmaputra River, Assam (India): Suspended sediment transport, valley aggradation and basin denudation.* Unpublished PhD dissertation, Johns Hopkins University, Baltimore, Maryland. Cited in A. Gupta, (1988), Large floods as geomorphic events in the humid tropics. In *Flood geomorphology*, eds. V.R. Baker, R.C. Kochel, and P.C. Patton, pp. 301–315. New York: John Wiley and Sons.

Grams, P.E. (1991). *Degradation of alluvial sand bars along the Snake River below Hells Canyon Dam, Hells Canyon National Recreation Area, Idaho.* Middlebury College, Middlebury, Vermont, unpublished Senior Paper, 98 pp. Cited in Collier et al., 1996.

Grant, G.E. and Swanson, F.J. (1995). Morphology and processes of valley floors in mountain streams, western Cascades, Oregon. In *Natural and anthropogenic influences in fluvial geomorphology*, eds. J.E. Costa, A.J. Miller, K.W. Potter, and P.R. Wilcock, pp. 83–101. Washington, DC: American Geophysical Union Geophysical Monograph 89.

Gregory, S.V., Swanson, F.J., McKee, W.A., and Cummins, K.W. (1991). An ecosystem perspective of riparian zones. *BioScience*, **41**, 540–551.

Gupta, A. (1983). High-magnitude floods and stream channel response. *Special Publications International Association of Sedimentologists*, **6**, 219–227.

Gupta, A. (1995). Magnitude, frequency, and special factors affecting channel form and processes in the seasonal tropics. In *Natural and anthropogenic influences in fluvial geomorphology*, eds. J.E. Costa, A.J. Miller, K.W. Potter, and P.R. Wilcock, pp. 125–136. Washington, DC: American Geophysical Union Geophysical Monograph 89.

Gupta, A. and Dutt, A. (1989). The Auranga: description of a tropical monsoon river. *Zeitschrift fur Geomorphologie*, **33**, 73–92.

Gupta, A. and Fox, H. (1974). Effects of high-magnitude floods on channel form: a case study in Maryland piedmont. *Water Resources Research*, **10**, 499–509.

Harvey, A.M. (1984). Geomorphological response to an extreme flood: a case from southeastern Spain. *Earth Surface Processes and Landforms*, **9**, 267–279.

Harvey, M.D. and Schumm, S.A. (1994). Alabama River: variability of overbank flooding and deposition. In *The variability of large alluvial rivers*, eds. S.A. Schumm and B.R. Winkley, pp. 313–337. New York: American Society of Civil Engineering Press.

Helley, E.J. and LaMarche, V.C. Jr. (1973). Historic flood information for nothern California streams from geological and botanical evidence. *U.S. Geological Survey Professional Paper 485-E*, 16 pp. Washington, DC: U.S. Geological Survey Pubs.

Hickin, E.J. (1995). Hydraulic geometry and channel scour, Fraser River, British Columbia, Canada. In *River geomorphology*, ed. E.J. Hickin, pp. 155–167. Chichester: Wiley.

Hickin, E.J. and Nanson, G.C. (1984). Lateral migration rates of river bends. *Journal of Hydraulic Engineering*, **110**, 1557–1567.

Hyatt, J.A. and Jacobs, P.M. (1996). Distribution and morphology of sinkholes triggered by flooding following Tropical Storm Alberto at Albany, Georgia, USA. *Geomorphology*, **17**, 305–316.

Inbar, M. (1987). Effects of a high magnitude flood in a Mediterranean climate: A case study in the Jordan River basin. In *Catastrophic flooding*, eds. L. Mayer and D. Nash, pp. 333–353. Boston: Allen and Unwin.

Interagency Floodplain Management Review Committee. (1994). Floodplain geomorphology, Chapter 5. In *Sharing the challenge: floodplain management into the 21st century*, 190 pp. Washington, DC: U.S. Government Printing Office.

Jacobson, R.B., Oberg, K.A., and Westphal, J.A. (1993). The Miller City, Illinois, levee break and incipient meander cutoff – an example of geomorphic change accompanying the upper Mississippi River flood, 1993. *American Geophysical Union, Abstracts 1993 Fall Meeting*, 61 pp.

Jahns, R.H. (1947). Geologic features of the Connecticut Valley, Massachusetts as related to recent floods. *U.S. Geological Survey Water-Supply Paper 996*, pp. 1–158. Washington, DC: U.S. Geological Survey Pubs.

Jarrett, R.D. and Costa, J.E. (1986). Hydrology, geomorphology, and dam-break modeling of the July 15, 1982 Lawn Lake Dam and Cascade Lake Dam failures, Larimer County, Colorado. *U.S. Geological Survey Professional Paper 1369*, 78 pp. Washington, DC: U.S. Geological Survey Pubs.

Johnson, W.C. (1992). Dams and riparian forests: case study from the upper Missouri River. *Rivers*, **3**, 229–242.

Johnson, W.C. (1994). Woodland expansion in the Platte River, Nebraska: patterns and causes. *Ecological Monographs*, **64**, 45–84.

Kale, V.S., Ely, L.L., Enzel, Y., and Baker, V.R. (1994). Geomorphic and hydrologic aspects of monsoon floods on the Narmada and Tapi Rivers in central India. *Geomorphology*, **10**, 157–168.

Kearsley, L.H., Schmidt, J.C., and Warren, K.D. (1994). Effects of Glen Canyon Dam on Colorado River sand deposits used as campsites in Grand Canyon National Park, USA. *Regulated Rivers: Research and Management*, **9**, 137–149.

Kells, J.A. and Smith, C.D. (1991). Reduction of cavitation on spillways by induced air entrainment. *Canadian Journal of Civil Engineering*, **18**, 358–377.

Kelsey, H.M. (1980). A sediment budget and an analysis of geomorphic process in the Van Duzen River basin, north coastal California, 1941–1975. *Geological Society of America Bulletin*, **91**, 1119–1216.

Kinoshita, R. (1982). *Study of the flood deposits in the lower Ishikari River.* Unpublished report, Civil Engineering Research Institute, Hokkaido Development Bureau, 38 pp. (in Japanese): from F. Iseya. (1984). *An experimental study of dune development and its effect on sediment suspension.* University of Tsukuba (Japan), Environmental Research Center Paper No. 5, 56 pp.

Kite, J.S. and Linton, R.C. (1993). Depositional aspects of the November 1985 flood on Cheat River and Black Fork, West Virginia. In *Geomorphic studies of the storm and flood of Nov. 3–5, 1985, in the upper Potomac and Cheat River basins in West Virginia and Virginia*, ed. R.B. Jacobson, pp. D1–D24. U.S. Geological Survey Bulletin 1981. Washington, DC.

Knighton, D. (1984). *Fluvial forms and processes.* London: Edward Arnold, 218 pp.

Knox, J.C. (1972). Valley alluviation in southwestern Wisconsin. *Annals of the Association of American Geographers*, **62**, 401–410.

Knox, J.C. (1987). Historical valley floor sedimentation in the upper Mississippi Valley. *Annals of the Association of American Geographers*, **77**, 224–244.

Kochel, R.C. (1988). Geomorphic impacts of large floods: review and new perspectives on magnitude and frequency. In *Flood geomorphology*, eds. V.R. Baker, R.C. Kochel, and P.C. Patton, pp. 169–187. New York: John Wiley and Sons.

Kresan, P.L. (1988). The Tucson, Arizona, flood of October 1983: implications for land management along alluvial river channels. In *Flood geomorphology*, eds. V.R. Baker, R.C. Kochel, and P.C. Patton, pp. 465–489. New York: John Wiley and Sons.

Kurashige, Y. (1994). Mechanisms of suspended sediment supply to headwater rivers. *Transactions, Japanese Geomorphological Union*, **15A**, 109–129.

Lagasse, P.F. (1980). *An assessment of the response of the Rio Grande to dam construction – Cochiti to Isleta reach, Albuquerque, New Mexico.* U.S. Army

Corps of Engineers, Technical Report, 133 pp. Albuquerque, New Mexico.

Landers, M.N., Jones, J.S., and Trent, R.E. (1994). Brief summary of national bridge scour data base. In *Hydraulic engineering '94*, eds. G.V. Cotroneo and R.R. Rumer, pp. 41–45. Proceedings of the American Society Civil Engineers Conference. New York, NY.

Lapointe, M.F., Secretan, Y., Driscoll, S.N., Bergeron, N., and Leclerc, M. (1998). Response of the Ha!Ha! River to the flood of July 1996 in the Saguenay region of Quebec: large-scale avulsion in a glaciated valley. *Water Resources Research*, **34**, 2383–2392.

Lawler, D.M. (1993). Needle ice processes and sediment mobilization on river banks: the River Ilston, West Glamorgan, United Kingdom. *Journal of Hydrology*, **150**, 81–114.

Lawler, D.M. (1994). Temporal variability in streambank response to individual flow events: the River Arrow, Warwickshire, United Kingdom. In *Variability in stream erosion and sediment transport*. Proceedings, Canberra Symposium, IAHS Publication No. 224, pp. 171–180. Wallingford, UK.

Lecce, S.A. (1997). Spatial patterns of historical overbank sedimentation and floodplain evolution, Blue River, Wisconsin. *Geomorphology*, **18**, 265–277.

Leopold, L.B. and Maddock, T., Jr. (1953). The hydraulic geometry of stream channels and some physiographic implications. *U.S. Geological Survey Professional Paper 252*, 56 pp. Washington, DC: U.S. Geological Survey Pubs.

Leopold, L.B., Wolman, M.G., and Miller, J.P. (1964). *Fluvial processes in geomorphology*. San Francisco: W.H. Freeman and Company, 522 pp.

Lewin, J. (1989). Floods in fluvial geomorphology. In *Floods: hydrological, sedimentological and geomorphological implications*, eds. K. Beven and P. Carling, pp. 265–284. Chichester: John Wiley and Sons.

Ligon, F.K., Dietrich, W.E., and Trush, W.J. (1995). Downstream ecological effects of dams. *BioScience*, **45**, 183–192.

Lisle, T.E. (1982). Effects of aggradation and degradation on riffle-pool morphology in natural gravel channels, northwestern California. *Water Resources Research*, **18**, 1643–1651.

Magilligan, F.J. (1992). Thresholds and the spatial variability of flood power during extreme floods. *Geomorphology*, **5**, 373–390.

Malanson, G.P. (1993). *Riparian landscapes*. Cambridge, United Kingdom: Cambridge University Press, 296 pp.

Marcus, W.A. (1989). Lag-time routing of suspended sediment concentrations during unsteady flow. *Geological Society of America Bulletin*, **101**, 644–651.

Matthes, G.H. (1947). Macroturbulence in natural stream flow. *Transactions American Geophysical Union*, **28**, 255–262.

Mertes, L.A.K. (1990). *Hydrology, hydraulics, sediment transport and geomorphology of the central Amazon floodplain*. Unpublished PhD dissertation, University of Washington, Seattle, 225 pp.

Mertes, L.A.K. (1997). Documentation and significance of the perirheic zone on inundated floodplains. *Water Resources Research*, **33**, 1749–1762.

Middelkoop, H. and Asselman, N.E.M. (1998). Spatial variability of floodplain sedimentation at the event scale in the Rhine-Meuse delta, The Netherlands. *Earth Surface Processes and Landforms*, **23**, 561–573.

Miller, A.J. (1990). Flood hydrology and geomorphic effectiveness in the central Appalachians. *Earth Surface Processes and Landforms*, **15**, 119–134.

Miller, A.J. (1995). Valley morphology and boundary conditions influencing spatial patterns of flood flow. In *Natural and anthropogenic influences in fluvial geomorphology*, eds. J.E. Costa, A.J. Miller, K.W. Potter, and P.R. Wilcock, pp. 57–81. Washington, DC: American Geophysical Union Geophysical Monograph 89.

Miller, A.J. and Parkinson, D.J. (1993). Flood hydrology and geomorphic effects on river channels and flood plains: the flood of November 4–5, 1985, in the South Branch Potomac River Basin of West Virginia. In *Geomorphic studies of the storm and flood of Nov. 3–5, 1985, in the upper Potomac and Cheat River basins in West Virginia and Virginia*, ed. R.B. Jacobson, pp. E1–E96. U.S. Geological Survey Bulletin 1981. Washington, DC.

Nadler, C.T. and Schumm, S.A. (1981). Metamorphosis of South Platte and Arkansas Rivers, eastern Colorado. *Physical Geography*, **2**, 95–115.

Nanson, G.C. (1986). Episodes of vertical accretion and catastrophic stripping: a model of disequilibrium flood-plain development. *Geological Society of America Bulletin*, **97**, 1467–1475.

Nanson, G.C., Rust, B.R., and Taylor, G. (1986). Coexistent mud braids and anastomosing channels in an arid-zone river: Cooper Creek, central Australia. *Geology*, **14**, 175–178.

Nicholas, A.P. and Walling, D.E. (1995). Modelling contemporary overbank sedimentation on floodplains: some preliminary results. In *River geomorphology*, ed. E.J. Hickin, pp. 131–153. Chichester: Wiley and Sons.

Nicholas, A.P. and Walling, D.E. (1997). Investigating spatial patterns of medium-term overbank sedimentation on floodplains: a combined numerical modelling and radiocaesium-based approach. *Geomorphology*, **19**, 133–150.

Nolan, K.M. and Marron, D.C. (1985). Contrast in stream-channel response to major storms in two mountainous areas of California. *Geology*, **13**, 135–138.

O'Connor, J.E., Webb, R.H., and Baker, V.R. (1986). Paleohydrology of pool-and-riffle pattern development: Boulder Creek, Utah. *Geological Society of America Bulletin*, **97**, 410–420.

Osterkamp, W.R. and Costa, J.E. (1987). Changes accompanying an extraordinary flood on a sand-bed stream. In *Catastrophic flooding*, eds. L. Mayer and D. Nash, pp. 201–224. Boston: Allen and Unwin.

Patton, P.C. (1988). Geomorphic response of streams to floods in the glaciated terrain of southern New England. In *Flood geomorphology*, eds. V.R. Baker, R.C. Kochel, and P.C. Patton, pp. 261–277. New York: John Wiley and Sons.

Pickup, G. and Warner, R.F. (1976). Effects of hydrologic regime on magnitude and frequency of dominant discharge. *Journal of Hydrology*, **29**, 51–75.

Pitlick, J. (1993). Response and recovery of a subalpine stream following a catastrophic flood. *Geological Society of America Bulletin*, **105**, 657–670.

Pizzuto, J.E. (1987). Sediment diffusion during overbank flows. *Sedimentology*, **34**, 301–317.

Power, M.E., Parker, G., Dietrich, W.E., and Sun, A. (1995). How does floodplain width affect floodplain river ecology? A preliminary exploration using simulations. *Geomorphology*, **13**, 301–317.

Prestegaard, K.L., Matherne, A.M., Shane, B., Houghton, K., O'Connell, M., and Katyl, N. (1994). Spatial variations in the magnitude of the 1993 floods, Raccoon River basin, Iowa. *Geomorphology*, **10**, 169–182.

Rathburn, S.L. (1993). Pleistocene cataclysmic flooding along the Big Lost River, east central Idaho. *Geomorphology*, **8**, 305–319.

Ritter, D.F. (1988). Floodplain erosion and deposition during the December 1982 floods in southeast Missouri. In *Flood geomorphology*, eds. V.R. Baker, R.C. Kochel, and P.C. Patton, pp. 243–259. New York: John Wiley and Sons.

Russell, A.J., Van Tatenhove, F.G.M., and Van De Wal, R.S.W. (1995). Effects of ice-front collapse and flood generation on a proglacial river channel near Kangerlussuaq (Sondre Stromfjord), West Greenland. *Hydrological Processes*, **9**, 213–226.

Sato, S., Matsuura, H., and Miyazaki, M. (1987). *Potholes in Shikoku, Japan, Part I. Potholes and their hydrodynamics in the Kurokawa River, Ehime*. Memoirs of the Faculty of Education, Ehime University (Japan), Series III, **7**, 127–190.

Schick, A.P. and Lekach, J. (1987). A high magnitude flood in the Sinai Desert. In *Catastrophic flooding*, eds. L. Mayer and D. Nash, pp. 381–410. Boston: Allen and Unwin.

Schumm, S.A. (1977). *The fluvial system*. New York: John Wiley and Sons, 338 pp.

Schumm, S.A. and Lichty, R.W. (1963). Channel widening and flood-plain construction along Cimarron River in southwestern Kansas. *U.S. Geological Survey Professional Paper 352-D*, pp. 71–88. Washington, DC: U.S. Geological Survey Pubs.

Scott, K.M. and Gravlee, G.C., Jr. (1968). Flood surge on the Rubicon River, California – hydrology, hydraulics, and boulder transport. *U.S. Geological Survey Professional Paper 422-M*, 38 pp. Washington, DC: U.S. Geological Survey Pubs.

Shroba, R.R., Schmidt, P.W., Crosby, E.J., Hansen, W.R., and Soule, J.M. (1979). Storm and flood of July 31–August 1, 1976, in the Big Thompson River and Cache la Poudre River basins, Larimer and Weld Counties, Colorado. Part B. Geologic and geomorphic effects in the Big Thompson Canyon area, Larimer County. *U.S. Geological Survey Professional Paper 1115*, pp. 87–152. Washington, DC: U.S. Geological Survey Pubs.

Simon, A. (1992). Energy, time, and channel evolution in catastrophically disturbed fluvial systems. *Geomorphology*, **5**, 345–372.

Simon, A. and Darby, S. (1997). Process-form interactions in unstable sand-bed river channels: a numerical modeling approach. *Geomorphology*, **21**, 85–106.

Smith, A.M. and Zawada, P.K. (1988). The role of the geologist in flood contingency planning. *South African Geological Survey, Paper 7.3*, 9 pp.

Smith, D.I., Greenaway, M.A., Moses, C., and Spate, A.P. (1995). Limestone weathering in eastern Australia. Part 1: erosion rates. *Earth Surface Processes and Landforms*, **20**, 451–463.

Ten Brinke, W.B.M., Schoor, M.M., Sorber, A.M., and Berendsen, H.J.A. (1998). Overbank sand deposition in relation to transport volumes during large-magnitude floods in the Dutch sand-bed Rhine River system. *Earth Surface Processes and Landforms*, **23**, 809–824.

Thorne, C.R., Russell, A.P.G., and Alam, M.K. (1993). Planform pattern and channel evolution of the Brahmaputra River, Bangladesh. In *Braided rivers*, eds. J.L. Best and C.S. Bristow, pp. 257–276. Geological Society (London) Special Publ. No. 75, pp. 257–276.

Tinkler, K.J. (1993). Fluvially sculpted rock bedforms in Twenty Mile Creek, Niagara Peninsula, Ontario. *Canadian Journal of Earth Sciences*, **30**, 945–953.

Tinkler, K.J. and Wohl, E.E. (1998). A primer on bedrock channels. In *Rivers over rock: fluvial processes in bedrock channels*, pp. 1–18. Washington, DC: American Geophysical Union Geophysical Monograph 107.

Virginia Division of Mineral Resources. (1969). *Natural features caused by a catastrophic storm in Nelson and Amherst Counties, Virginia*. Commonwealth of Virginia, 20 pp. Richmond, Virginia.

Vuichard, D. and Zimmermann, M. (1987). The 1985 catastrophic drainage of a moraine-dammed lake, Khumbu Himal, Nepal. *Mountain Research and Development*, **7**, 91–110.

Warburton, J. (1994). Channel change in relation to meltwater flooding, Bas Glacier d'Arolla, Switzerland. *Geomorphology*, **11**, 141–149.

Webb, R.H. (1987). Occurrence and geomorphic effects of streamflow and debris flow floods in northern Arizona and southern Utah. In *Catastrophic flooding*, eds. L. Mayer and D. Nash, pp. 247–265. Boston: Allen and Unwin.

Wiele, S.M., Graf, J.B., and Smith, J.D. (1996). Sand deposition in the Colorado River in the Grand Canyon from flooding of the Little Colorado River. *Water Resources Research*, **32**, 3579–3596.

Williams, G.P. (1978). The case of the shrinking channels – the North Platte and Platte Rivers in Nebraska. *U.S. Geological Survey Circular 781*, 48 pp. Washington, DC: U.S. Geological Survey Pubs.

Williams, G.P. and Guy, H.P. (1973). Erosional and depositional aspects of Hurricane Camille in Virginia, 1969. *U.S. Geological Survey Professional Paper 804*, 80 pp. Washington, DC: U.S. Geological Survey Pubs.

Williams, G.P. and Wolman, M.G. (1984). Downstream effects of dams on alluvial rivers. *U.S. Geological Survey Professional Paper 1286*, 83 pp. Washington, DC: U.S. Geological Survey Pubs.

Winkley, B.R., Lesleighter, E.J., and Cooney, J.R. (1994). Instability problems of the Arial Khan River, Bangladesh. In *The variability of large alluvial rivers*, eds. S.A. Schumm and B.R. Winkley, pp. 269–284. New York: American Society of Civil Engineering Press.

Wohl, E.E. (1993). Bedrock channel incision along Piccaninny Creek, Australia. *Journal of Geology*, **101**, 749–761.

Wohl, E.E., Cenderelli, D., and Mejia-Navarro, M. (in press). Geomorphic hazards from extreme floods in canyon rivers. In *The Schumm volume*, eds. M.D. Harvey and D.J. Anthony. Littleton, Colorado: Water Resources Publications.

Wohl, E.E., Greenbaum, N., Schick, A.P., and Baker, V.R. (1994). Controls on bedrock channel incision along Nahal Paran, Israel. *Earth Surface Processes and Landforms*, **19**, 1–13.

Wolman, M.G. (1959). Factors influencing erosion of a cohesive river bank. *American Journal of Science*, **257**, 204–216.

Wolman, M.G. and Eiler, J.P. (1958). Reconnaissance study of erosion and deposition produced by the flood of August 1955 in Connecticut, USA. *Transactions, American Geophysical Union*, **39**, 1–14.

Zhou, Z. and Pan, X. (1994). Lower Yellow River. In *The variability of large alluvial rivers*, eds. S.A. Schumm and B.R. Winkley, pp. 363–393. New York: American Society of Civil Engineering Press.

CHAPTER 7

Contaminant Transport Hazards during Flooding

Jim B. Finley
Shepherd Miller, Inc.

The nature and extent of the contaminant hazard associated with floods are time dependent. On the time scale of the flood, inorganic geochemical processes can affect the distribution of metals among the possible forms (free ion, sediment bound, precipitated, or complexed). Long-term effects to the riparian environment also arise from the flood transport and deposition of metal-bearing sediments, as a combination of geochemical reactions and geomorphological evolution of floodplain sediments act to mobilize deleterious concentrations of metals to the active fluvial ecosystem. In contrast to the inorganic contaminants, synthetic organic compounds typically persist on time scales that are longer than the duration of most floods. Redistribution of sediment-bound organic compounds may represent a long-term hazard in much the same manner as do metal-bearing sediments deposited in the floodplain; the major difference is that synthetic organic compounds normally do not concentrate in sedimentary deposits. Inorganic chemical reactions that affect contaminants in river systems include (1) oxidation–reduction, (2) mineral solubility, (3) surface complexation, and (4) cation exchange. Most of the chemical reactions that affect synthetic organic compounds in the riparian environment are influenced by the presence of natural organic matter and microbially mediated reactions. Biogeochemical reactions that modify synthetic organic compounds include hydrophobic partitioning, biodegradation, and volatilization.

Introduction

Humans have interacted with the fluvial system since early evolution and, as such, have become an integral part of the response of the fluvial system to floods at all scales. One result of the formalization of human civilization, beginning with development of the agrarian society, has been the need for disposal of waste as a natural consequence of developing technologies. Originally the waste probably consisted primarily of human waste, and because the total population was relatively small, the effect of that form of sewage disposal on the surrounding environment was small. As human civilization changed through time, the nature of the waste, described succinctly by Gore (1993) in an economic sense as "externalities," introduced into the fluvial system also changed, increasing in sophistication first with the discovery of metals and smelting (gold ca. 6000 BC) and ultimately followed by the complex, synthetic organic compounds developed in the early 20th century as by-products of petroleum refining.

Waste materials or by-products of evolving technologies arise from two basic causes regardless of the nature of the materials. In the first case, resources must be gathered and concentrated until a sufficient quantity of resource exists for practicable exploitation. This is the case when the resource has been identified, but the natural abundance of the resource is too low for direct extraction and use. Thus, technologies were developed that allowed removal of the target resource from its associated materials, the latter of which is treated as a waste (externality) and most often disposed of in the surrounding environment. The other case evolved when the population of humans increased beyond the capacity of the immediately surrounding environment to buffer the discharge of wastes from the concentrated population, be the wastes liquid (e.g., sewage), gaseous (e.g., sulfur dioxide associated with burning coal), or solid (e.g., landfill solids or coal-fly ash from coal-fired power plants).

Modern human civilization evolved in close association with the fluvial system, primarily because rivers and waterways provided the most readily accessible form of transportation (e.g., De Voto, 1952). As such, early settlements were almost always established in close proximity with some form of waterway; a trend followed throughout history to the present. Thus, the components of human society that define civilization, from domicile to industry, were established within the active fluvial system.

This volume has been assembled to discuss and explore hazards associated with flooding, linking physical processes (e.g., Hirschboeck et al., Chapter 2, this volume) and biological processes and responses (e.g., Friedman and Auble, Chapter 8, this volume; Wydoski and Wick, Chapter 9, this volume) with human responses (e.g., Merritts, Chapter 10, this volume; Laituri, Chapter 17, this volume). Floods involve transport of both water and solids, with flood waters often reaching portions of the river valley that are only rarely inundated. Because waste materials associated with human activities are also stored within the fluvial system, the waste also may be transported during floods along with naturally occurring materials.

The complexity of response to flooding scales with the size of the drainage basin and the size of the flood. As the drainage basin size increases, the size of the floodplain may increase as well. A larger floodplain introduces secondary processes that may lead to flooding, not from overbank flow from the river but from localized flooding in the perirheic zone (Mertes, Chapter 5, this volume). In addition to the complexity of response added as the floodplain size increases, the longevity of the response to flooding may also increase. Thus, in large river systems, there may be portions of the floodplain that remain flooded for a period of time beyond the passage of the main flood. The potential range of biogeochemical reactions, and potential sources of contamination associated with flooding, that may occur under these conditions are large. The focus of this chapter is to consider the time scale of the major portion of the flooding, but there may well be contaminant effects from longer-term retention of water in isolated zones of the floodplain.

This chapter explores the chemical processes involved with the flood transport of wastes that may prove hazardous because of the chemical reactivity of constituents contained in the waste. Rather than providing a complete historical overview of the literature produced in developing the current level of understanding, the primary objective of this chapter is to describe the various chemical processes that may affect the fate and transport of chemical compounds in the context of the contaminant hazard posed by transport of chemical compounds during flooding. The first portion of the chapter considers the range of biogeochemical processes that affect the fate and transport of contaminants (both inorganic and organic), especially the combined effects of physical and chemical conditions that may be present only during flooding. Three case studies are then presented that provide examples of the chemical processes that contribute to hazards that accompany the transport of contaminants during floods.

Biogeochemistry of Contaminants in the Fluvial System

The nature and extent to which a given contaminant, heavy metal or synthetic organic, presents a hazard in the fluvial system often depends on the physical and chemical form in which the contaminant exists. Thus, knowledge of the fundamental biogeochemical processes that influence the partitioning of contaminants in the fluvial system is prerequisite to building an understanding of how contaminants are transported during flooding and of the nature of the hazard presented by the transport of contaminants.

Inorganic Contaminants

Inorganic contaminants in the environment most often consist of transition metals (iron, zinc, copper, cadmium, lead, silver, gold) and metalloids (arsenic and selenium). Sources of these chemical constituents include (1) naturally occurring releases of metal-bearing water from mineralized rocks (e.g., Plumlee et al., 1992), (2) acidic, metal-bearing waters associated with mine workings (e.g., Nordstrom, 1982), (3) waters discharging from urban areas (e.g., Imhoff et al., 1980; Rang and Schouten, 1989; Gonçalves et al., 1994; Garban et al., 1996; Ciszewski, 1997; Singh et al., 1997), and (4) mining and milling tailings released into the fluvial system that are subsequently distributed within the river valley along with non-metal-bearing sediments (e.g., Marron, 1989; Axtmann and Louma, 1991). The geochemical processes discussed in the following sections are specific to the chemical transformation of mining-related sulfide minerals in the fluvial system with the subsequent release of metals.

The extraction of metals from ore rocks has contributed significantly to the rise of civilization, and one legacy of the extraction process has been the introduction of large quantities of metal-bearing sediments into the surrounding environment. For example, 100 million metric tons of tailings were released into the Clark Fork River, Montana, USA, between 1880 and 1950 (Moore and Louma, 1990; Axtmann and Luoma, 1991). Releases of mining-related sediments have been used by geomorphologists to elucidate many processes associated with sediment transport and storage in the fluvial system (e.g., Wolfenden and Lewin, 1977; Bradley, 1984; Lewin and Macklin, 1987; Marcus, 1987; Macklin and Dowsett, 1989; Macklin and Lewin, 1989; Bradley and Cox, 1990; Marron, 1989, 1992; Macklin et al., 1992, 1994; Graf, 1994; Macklin, 1996; Taylor, 1996).

Oxidation–Reduction. Metals introduced into the environment from the mining and milling of ore typically constitute that fraction of the mineralized rock that has minimal economic value (e.g., iron, arsenic, selenium). Additionally, some metals are present in the mineralized rock in a low enough concentration that there is no economic benefit to extracting the metal (e.g., cadmium, aluminum, nickel, chromium, antimony, zinc, lead, copper). The most common metal-bearing minerals that ultimately end up as part of the particulate load in the fluvial system are sulfides. For example, pyrite (FeS_2), arsenopyrite (FeAsS), galena (PbS), and sphalerite (ZnS) are all relatively common minerals in ore and may actually compose the most abundant mineral phase by weight (especially pyrite) (e.g., Macklin and Dowsett, 1989; Axtmann and Louma, 1991; Marron, 1992). Although the chemical formulae of the sulfide minerals are well constrained theoretically, the actual chemical composition is rarely the pure phase. More commonly, other metals are incorporated into the mineral when the "parent" sulfide mineral first forms during mineralization.

Once exposed to the atmosphere, sulfide minerals undergo a chemical reaction that causes dissolution of the original mineral and release of the associated trace metals into solution. The chemical reaction is referred to as oxidation and may proceed by consumption of oxygen or ferric iron [Fe(III)] or may be catalyzed by bacteria (Stumm and Morgan, 1981; Nordstrom, 1982; Moses et al., 1987). The exact geochemical mechanism is not important for this discussion; rather, the important point is to recognize that sulfide minerals constitute the ultimate source of most metals in areas affected by historical mining.

Given the historical propensity for storing mine tailings adjacent to river systems, it naturally follows that flooding will mobilize sulfide-bearing sediments from mine workings, transporting and depositing the sediments within the fluvial system as a function of the magnitude of the event relative to the physical characteristics of the channel (e.g., incised or braided). Thereafter, the sulfide minerals oxidize and release metals into the surrounding environment or may be reentrained as the river channel migrates through metal-bearing floodplain sediments (Marron, 1989; Moore et al., 1989).

Mineral Solubility. The fate and mobility of metals released from oxidized sulfide minerals depend strongly on the mineralogy of surrounding sediments. Oxidation results not only in release of metals but also in production of hydrogen ions, which may translate directly into low pH. Thus, if the surrounding sediments, either intercalated with tailing sediments in a fluvial deposit or in the sediments bounding the tailing deposit, do not have geochemically reactive buffering minerals (e.g., calcite), then metals may remain in solution, potentially contaminating both surface and groundwater. On the other hand, calcite is a fairly abundant mineral and readily reacts with low-pH waters to increase pH by buffering the acid released during oxidation of sulfide minerals.

A variety of metal oxide minerals precipitate at circumneutral pH, and the most common are iron and manganese oxyhydroxide minerals (e.g., Cherry et al., 1986; Drever, 1988; Appelo and Postma, 1993). Solubility minima for metal oxyhydroxide minerals that are important in controlling the mobility of trace metals in fluvial systems occur in the pH range from approximately 6 to 7.5 (Figure 7.1). Most natural waters have pH values within the range of 5.5 to approximately 8 (Drever, 1988). Thus, when iron-bearing sulfide minerals oxidize upon exposure to the atmosphere, the iron released (as ferric iron, Fe^{3+}) rapidly precipitates as iron oxyhydroxide (Stumm and Morgan, 1981; Drever, 1988; Langmuir, 1997). The metal oxyhydroxide minerals that precipitate exert significant control on the mobility of trace metals that are also released during oxidation of sulfide minerals (Cherry et al., 1986; Moore et al., 1989; Axtmann and Louma, 1991).

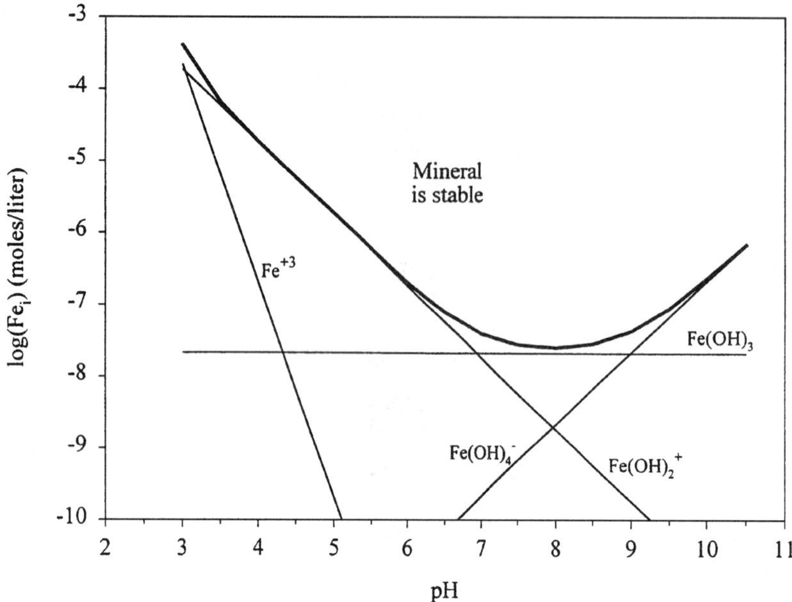

Figure 7.1. Solubility of iron oxyhydroxide as a function of pH, showing the solubility minimum that is characteristic of most metal oxyhydroxide minerals. The mineral is stable for conditions above the thick solid line. Figure was created with the geochemical computer model PHREEQC (Parkhurst, 1995).

Evaporative concentration of metal-bearing waters may also result in the precipitation of efflorescent salts at the surface of floodplain deposits (e.g., Filipeck et al., 1987; Nimick and Moore, 1991). Precipitation of salts (metal-bearing or otherwise) most often occurs in semiarid or arid environments where evaporation rates are high. Efflorescent salts form as metal-bearing moisture is drawn upward toward the sediment–air interface. The moisture transports the oxidation products (metals and sulfate) to the sediment–air interface, or just beneath the interface, where the moisture vaporizes, causing the chemical constituents to become concentrated so that metal-bearing salt minerals precipitate. Efflorescent salts are very soluble, and subsequent wetting, either from precipitation or during flooding, dissolves the salts, releasing elevated concentrations of metals to the runoff water (Nordstrom, 1982; Nimick and Moore, 1991).

There may also be natural accumulations of soluble minerals in the geologic formations that comprise river basins. For example, in the upper Colorado River basin metals and metalloids (e.g., selenium) occur naturally in the bedrock and geologic formations (Graf, 1985). Because of the evolution of land-use practices (e.g., concentrated drainage from irrigated agricultural land and mining), both in the floodplain and in the surrounding drainage basin, the soluble metals and metalloids become mobilized and introduced into the river system, especially during floods. Irrigation return flows to the Gunnison and Uncompahgre Rivers, which are tributary to the Colorado River, contribute selenium, boron, and uranium to the upper Colorado River (Butler et al., 1991). Inundation of the floodplain may also enhance the process of flushing soluble metal compounds from geologic materials that have natural accumulations of metal and metalloid compounds.

Solid-Phase Associations. Metals released during oxidation of sulfide minerals can react geochemically in a variety of ways with materials in the fluvial deposit. Dissolved metals are affected by: (1) adsorption reactions with the electrostatically charged surfaces of minerals (surface complexation reactions), (2) adsorption on particulate organic matter, and (3) exchange reactions that consist of the metal exchanging with another chemical species already associated with a mineral phase (cation exchange) (Stumm and Morgan, 1981; Appelo and Postma, 1993).

Surface Complexation Reactions. Iron oxyhydroxide minerals are abundant in sedimentary deposits, especially in areas affected by historical mining, because the most common sulfide mineral in many ore deposits is pyrite (FeS_2). The physicochemical characteristics of iron and manganese oxyhydroxide minerals produce conditions that are favorable for the adsorption of dissolved metals. Adsorption on iron and manganese oxyhydroxide minerals is an important process controlling metal mobility in fluvial systems (Cherry et al., 1986; Macklin and Dowsett, 1989; Marron, 1989).

At the microscopic level, the surface of metal oxyhydroxide minerals interacts with water, resulting in a variation in the overall electrostatic charge on the mineral (e.g., Davis and Kent, 1990; Parks, 1990). The basic chemical representation of the process is:

$$\equiv SOH^{o} + H^{+} = \equiv SOH_2^{+}$$

and

$$\equiv SOH^{o} = \equiv SO^{-} + H^{+}$$

where $\equiv SOH^{o}$ represents the uncharged metal oxyhydroxide surface emphasizing the hydroxide that interacts at the mineral-water interface (e.g., Parks and de Bruyn, 1962). Protons are exchanged between the water and the mineral surface with the result that, under low-pH conditions, the surface of the mineral carries a positive charge and at higher pH the surface is negatively charged.

The relative affinity of trace metals for interactions (adsorption) with the charged surface of oxyhydroxide minerals varies as a function of pH (Davis and Leckie, 1978; Stumm and Morgan, 1981; Davis and Kent, 1990; Dzombak and Morel, 1990). Thus, the stability of the electrostatic bond between a trace metal, released during oxidation of a sulfide mineral, and an iron or manganese oxyhydroxide varies with pH. As a consequence of the pH dependence on the amount of trace metal adsorbed, trace metals can be released from the mineral surface and adsorbed again as the pH of the water changes. The pH dependence of adsorption for a variety of trace metals is shown in Figure 7.2. As noted earlier, most natural waters have pH values that range from approximately 5.5 to 8; thus the concentration of most trace metals associated with an iron or manganese oxyhydroxide surface is maximized, and the potential for the mineral to act as a source of trace metals depends strongly on the pH of the water. Release of trace metals from oxyhydroxide mineral surfaces during flooding occurs only if the pH of flood waters varies over the course of the event.

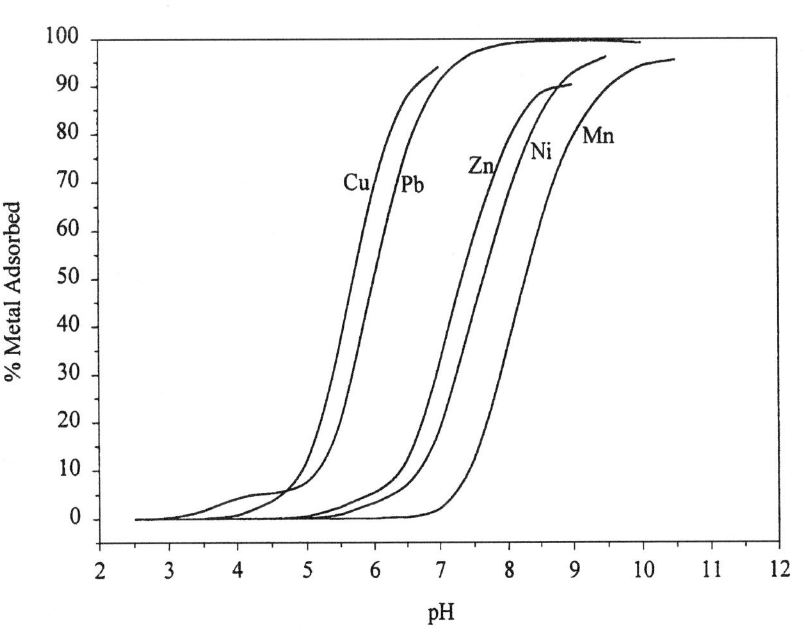

Figure 7.2. Adsorption of various dissolved metals on amorphous iron oxyhydroxide, showing the pH dependence of the process. Curves were calculated with the geochemcial computer model PHREEQC (Parkhurst, 1995) and the surface complexation model constants of Dzombak and Morel (1990).

Cation Exchange Reactions. Cation exchange is another type of mineral–water interaction that is derived from the charges on the surface of the mineral and from charged ions dissolved in the water. In this type of interaction, a dissolved, charged ion (e.g., Ca^{2+}) exchanges with a charged ion (e.g., Na^{+}) associated with the mineral surface, thus the exchange part of the name. Unlike the variable charge surfaces discussed earlier (most often iron and manganese oxyhydroxide minerals), cation exchange occurs most frequently in association with minerals that have a permanent charge imbalance typically caused by isomorphous substitution of, for example, aluminum for silica in the crystal structure of a mineral (Drever, 1988).

Cation exchange reactions are based on the concept of chemical equilibrium, but, unlike a thermodynamic equilibrium constant, the exchange coefficient is solution and mineral specific. Thus, in natural systems, defining the characteristics of cation exchange is a nontrivial task. However, there are certain principles that hold regardless of the nature of a given system. Whereas surface complexation reactions are pH dependent, cation exchange reactions and the relative selectivity of the mineral for a given charged ion depend more on the total concentration of the solution and on the selectivity of the mineral for each individual charged ion (Drever, 1988; Appelo and Postma, 1993). For example, divalent cations (e.g., Ca^{2+}, Cu^{2+}, and Zn^{2+}) are preferentially associated with cation exchange minerals primarily as a consequence of the relative size of the dissolved cation. But, because most potentially harmful metals occur in relatively low concentrations, except in the most contaminated settings, calcium is the dominant ion associated with cation exchange minerals in freshwater systems such as fluvial environments. In considering the environmental conditions in the floodplain, where metal-bearing sulfide minerals may be deposited, subsequently oxidized, and then release trace concentrations of metals, cation exchange is much less likely to be important as a control on the mobility of the trace metals because the concentrations of metals are too low. There are exceptions to this generalization; for example, lead is strongly bound to cation exchange minerals because of its physical and chemical properties, and lead is virtually immobilized by cation exchange reactions (e.g., Schnoor et al., 1987).

Organic Contaminants

Synthetic organic compounds, as well as other nutrient compounds that enhance the productivity of biotic life (e.g., nitrogen and phosphorous compounds), have become an integral part of human society. For example, the current crop yields produced by farmers worldwide have resulted from the use of large quantities of both synthetic organic compounds (herbicides, pesticides, fungicides, and others) and nitrogen and phosphorous fertilizers. The amounts of synthetic organic compounds applied annually are large with the greatest application, on a per acre basis, in the United States and Western Europe (2–3 kg) followed by the USSR (~1 kg) (Ananyeva et al., 1992). Nitrogen fertilizers are also used extensively to enhance the productivity of cropland. For example, approximately 6.3 million tons of nitrogen fertilizer were applied to cropland in the Mississippi River basin in 1991 (Goolsby et al., 1993). Industrial applications of a broad variety of synthetic organic compounds are an additional source. Large floods have the potential to mobilize very large amounts of organic compounds and constitute potential hazards to the use of water, especially drinking water supplies. This section considers the biogeochemical processes that control the mobility and potential toxicity of organic compounds once the compounds are released into the fluvial system. The discussion is confined necessarily to those biogeochemical processes that may occur on timescales similar to that of floods (days to months).

Role of Natural Organic Matter. Microbes in natural environments exist in a matrix composed of geologic minerals. The natural life cycle of organisms includes recycling of elements incorporated into organic compounds through microbially mediated degradation reactions that result in partial decomposition of the original organic structure. Residual organic material constitutes the single most important substrate for controlling the fate of organic compounds in natural systems (Schwarzenbach et al., 1993; Appelo and Postma, 1993). Most nondegradable organic material is composed of woody terrestrial plants, and this class of material is generally referred to as natural organic matter (Schnitzer and Khan, 1972; Liao et al., 1982). Natural organic matter is a complex mix of organic substances that generally can be classified as: (1) humic substances if the material is soluble in water, or (2) kerogen or humin if the natural organic matter is insoluble in water (Schwarzenbach et al., 1993). The first category of natural organic matter is the more important material affecting the fate and mobility of synthetic organic compounds in natural systems.

Humic substances are further subdivided depending on the pH-dependent solubility. Fulvic acids are soluble in liquid water within the pH range of natural waters, whereas humic acids are soluble only in high-pH solutions. Humic substances exhibit a combination of chemical characteristics that affect the distribution of synthetic organic compounds in natural systems. The chemical composition of humic substances is very complex and variable but consists primarily of carbon (40–50%), oxygen, nitrogen, sulfur, and hydrogen (Table 7.1) (Schwarzenbach et al., 1993).

Table 7.1. *Compounds composing natural organic matter and chemical properties influencing sorption (from Environmental Organic Chemistry, Schwarzenbach et al., Copyright ©1993. Reprinted by permission of John Wiley & Sons, Inc.)*

	Mole ratios of functional groups				Total moles of H-bonding
Compound	Carbon	Hydrogen	Nitrogen	Oxygen	groups per kg
Proteins	10	15	4	2	–
Cellulose (linear polymer)	10	16	0	8	3.1
Lignin	10	11	0.1	3	–
Fulvic acids					
Dissolved (river)	10	12	0.2	6	17
Soils	10	8	0.1	7	19.2
Humic acids					
Dissolved (river)	10	11	0.3	6	4.3
Sedimentary (lakes)	10	13	0.9	5	–
Soils	10	12	0.6	4	14.1
Humin	10	19	0.5	11	–

Thus, the overall structure is more polar than many of the synthetic organic compounds but not as polar as water. As a consequence, humic substances act to provide a more energetically favorable environment in which synthetic organic compounds can exist, and they are capable of interfacing with the polar structure of water and sediments.

Weathered minerals are often coated by natural organic matter. The fraction of the organic carbon (f_{oc}) present in sediments is a measure of the amount of natural organic matter and often determines the extent of sorption of synthetic organic compounds (Thurman, 1985; Schwarzenbach et al., 1993; Appelo and Postma, 1993). Thus, the distribution and transport of synthetic organic compounds in fluvial systems depend primarily on the abundance of natural organic matter, both as suspended colloids and as coatings on sediments.

Hydrophobic Partitioning. Many of the first synthetic organic compounds used in both industrial and agricultural processes were created without due consideration of how the chemical compounds would react or interact in the larger environmental system. For example, dichlorodiphenyltrichloroethane (DDT) proved very effective at controlling mosquito populations, but there was no inkling of the decimating effect the chemical would have on osprey reproduction (Carson, 1962). A characteristic of the early synthetic organic compounds that contributed to a marked persistence in the environment was their overall low solubility in water. Thus, the initially developed synthetic organic compounds preferentially partitioned from water to sediments, producing long-term sources.

When released into the environment, either as industrial discharges or flushed from agricultural fields by precipitation or irrigation return flow, hydrophobic organic compounds are preferentially partitioned to particulate matter, not so much because the particulate material (sediments, organic matter, etc.) has a greater affinity for the organic compound but because the organic molecule is essentially excluded from the highly polar structure of water (Schwarzenbach et al., 1993). In natural systems, particulate material is often coated, or partially coated, with a layer of organic matter (Thurman, 1985; Schwarzenbach et al., 1993). A detailed description of the mechanism encompassed by hydrophobic partitioning of synthetic organic compounds is beyond the scope of this paper (see Elzerman and Coates, 1987, for a thorough review), but the important point is that hydrophobic organic compounds preferentially partition from a dissolved state to particulate material. One potential hazard associated with such contaminated sediments arises when large floods mobilize the sediment, effectively redistributing and reexposing the fluvial environment to elevated concentrations of potentially toxic organic compounds.

As a consequence of the broad-scale environmental degradation wrought by the use of recalcitrant synthetic organic compounds, a large effort has been put forth to develop agricultural and industrial chemicals that are more readily biodegradable and more hydrophilic (greater solubility) (Elzerman and Coates, 1987). The increase in solubility arises because the basic carbon and hydrogen structure of the organic compound has been altered to include a variety of functional groups that serve as an interface between the core of the organic compound (hydrophobic) and water molecules. Many of the functional groups that are important components of synthetic organic compounds of environmental interest have acid-base properties with acid dissociation constants within the range of pH in natural waters (Figure 7.3) (Schwarzenbach et al., 1993). Thus, hydrolyzable functional groups on synthetic organic compounds can act chemically in a manner that is similar to iron and manganese oxyhydroxide minerals, with the interface between molecule and solution carrying positive and negative charge as a function of the pH of the bulk solution.

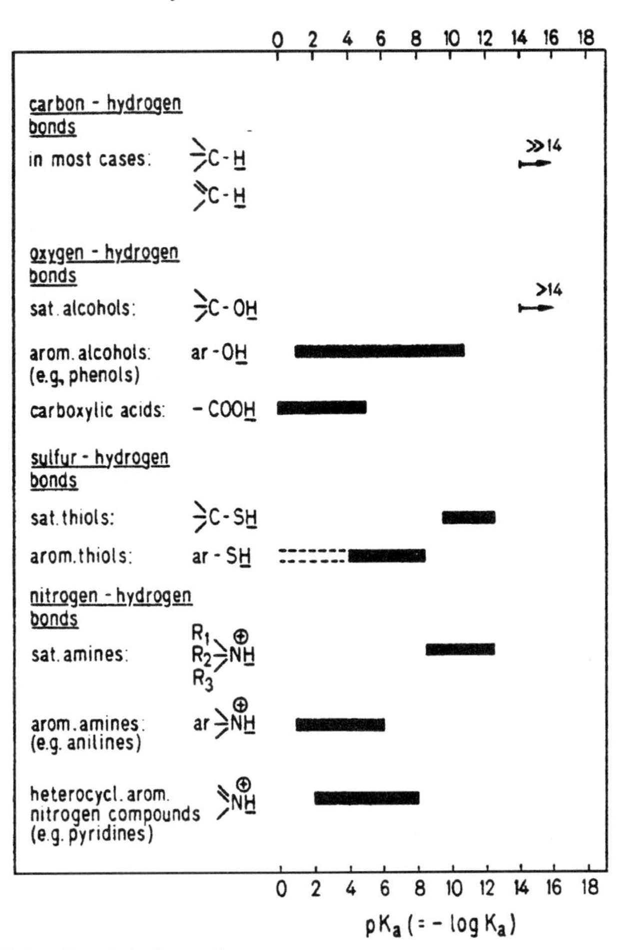

Figure 7.3. Ranges of acidity constants for hydrogen bound to different functional groups in organic compounds (from *Environmental Organic Chemistry*, Schwarzenbach et al., Copyright © 1993. Reprinted by permission of John Wiley & Sons, Inc.).

Partitioning of synthetic organic compounds between the dissolved and sorbed state dictates the potential magnitude of the environmental hazard associated with release of the organic compound to natural waters. Particulate matter (mineral and biological) has a greater affinity for organic compounds (both hydrophobic and hydrophylic) than does liquid water. Usually, the limiting factor determining the extent of partitioning of organic compounds to particulate matter is the concentration of suspended solids. During a flood, sediment transport is high, which can correlate with high concentrations of suspended solids. Therefore, adsorption of synthetic organic compounds introduced into the fluvial system during flooding may be much greater than at other times when the concentration of suspended solids is lower. As a consequence of the preferential partitioning of organic compounds to suspended solids, fluvial deposits that develop in response to flooding may act to concentrate sediment-bound organic compounds much as fluvial deposition concentrates metal contaminants.

Transformations of Organic Contaminants. Once introduced into the natural environment, synthetic organic compounds undergo a variety of secondary transformation reactions that may result in complete alteration of the original compound or may simply alter the chemistry (and reactivity) of the original compound. The transformation processes may affect the nature of the hazard posed by the synthetic organic compound throughout its life span in the natural environment. Secondary transformation processes that will be discussed include biodegradation, abiotic degradation, and volatilization. Of special significance in this discussion is the relative rate of reaction for a secondary transformation process compared with the potential rate of transport of a synthetic organic compound during a flood. The time scales over which biodegradation and abiotic degradation operate are generally longer than the typical time scale for transport during a flood. Thus, volatilization may be the most important secondary transformation process that could minimize the contamination potential of a synthetic organic compound during transport by floodwater.

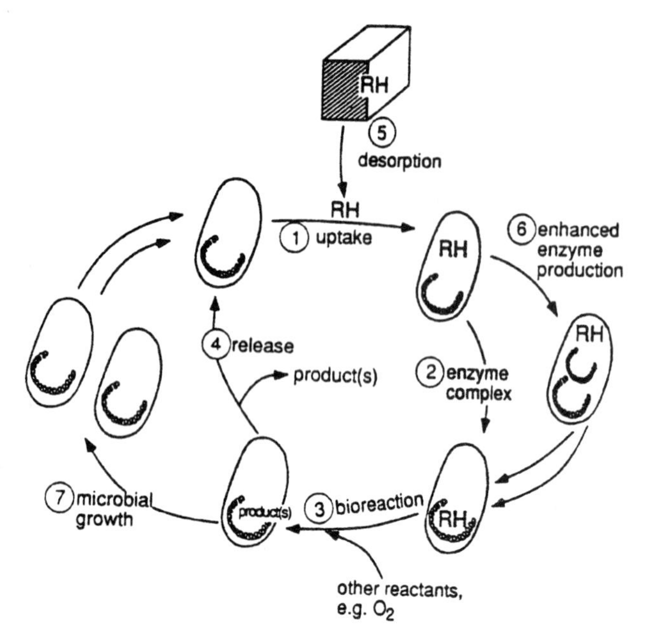

Figure 7.4. Summary of processes that occur during biotransformation of an organic compound (designated as RH) by enzymatic complexation (dark wavy line is the enzyme) (from *Environmental Organic Chemistry,* Schwarzenbach et al., Copyright © 1993. Reprinted by permission of John Wiley & Sons, Inc.).

Biodegradation of Organic Contaminants. Microorganisms exist in all natural environments and contribute fundamentally to biologic processes at the surface of the earth. Microorganisms utilize naturally occurring organic matter as a means of obtaining the energy necessary to sustain their existence. As such, microorganisms play a central role in the degradation of natural organic matter and, concomitantly, in the degradation of synthetic organic compounds. A complete discussion of the processes involved in biodegradation is beyond the scope of this chapter, and the interested reader is referred to the work of Schwarzenbach et al. (1993), and references therein, for a thorough review of the microbially mediated degradation processes.

The essence of biodegradation is that microorganisms employ similar reactive pathways and mechanisms to alter the chemical structure of synthetic organic compounds in the same way that natural organic matter is altered. Three principal classes of biochemical reactions are mediated by organisms: oxidation, reduction, and hydrolysis (Schwarzenbach et al., 1993). In all cases, the chemical reaction is enzymatically catalyzed, a process that lowers the overall energy required to complete the degradation reaction, usually by the formation of an intermediate activated complex that decreases the overall free energy of the reaction and increases the reactivity of the intermediate species (Schwarzenbach et al., 1993). The suite of processes operating during biodegradation is shown in Figure 7.4, which illustrates the variety of biochemical processes that may lead to degradation of synthetic organic compounds.

There is an inevitable lag in time between when a synthetic organic compound enters the natural system and when microorganisms begin degradation of the compound. The lag time varies as a function of the chemical complexity of the organic compound, and it arises fundamentally because the nature of the enzymatic reaction may require modification from the mechanism used to degrade natural organic compounds. There are other factors that affect the time lag, or whether biotic degradation can even occur, such as the presence of other necessary reactants (O_2 or H_2), the ability of the synthetic organic compound to transfer across the cell membrane, or the presence of a suitable consortia of microorganisms to degrade a particular compound (Schwarzenbach et al., 1993). Therefore, biodegradation, though an important mechanism in determining the long-term fate of synthetic organic compounds in the environment, is probably not an important process over the time scale of floods.

Abiotic Transformations of Organic Contaminants. Although the influence of biologically mediated transformations is difficult to distinguish from abiotic transformations in natural systems, abiotic processes do occur and may affect the overall chemical composition and reactivity of synthetic organic compounds (Wolfe, 1992; Schwarzenbach et al., 1993). The type of abiotic transformation occurring is generally differentiated by the presence

Table 7.2. *Examples of environmentally relevant chemical reactions (modified from Environmental Organic Chemistry, Schwarzenbach et al., Copyright © 1993. Reprinted by permission of John Wiley & Sons, Inc.)*

Reactants		Products
Nucleophilic substitution		
$CH_3Br + H_2O$ Methyl bromide	$\longrightarrow$	$CH_3OH + H^+ + Br^-$ Methanol
Elimination		
$Cl_2HC—CHCl_2 + OH^-$ 1,1,2,2-Tetrachloroethane	$\longrightarrow$	$ClHC{=}CCl_2 + Cl^- + H_2O$ Trichloroethene
Ester hydrolysis		
Disbutyl phthalate $+2OH^-$	$\longrightarrow$	Phthalate $+ 2HO—C_4H_9$
Oxidation		
$2CH_3SH + 1/2O_2$ Methyl mercaptan	$\longrightarrow$	$H_3C—S—S—CH_3 + H_2O$ Dimethyl disulfide
Reduction		
$C_6H_5—NO_2$ + "reduced species" $+6H^+$	$\longrightarrow$	$C_6H_5—NH_2$ + "oxidized species" $+ 2H_2O$

or absence of O_2 and sunlight, both of which contribute to the formation of intermediate chemical compounds that are more reactive than the original synthetic organic compound. Major categories of abiotic transformation reactions are shown in Table 7.2.

Unlike many of the inorganic chemical reactions that occur in solution, the abiotic transformation reactions listed in Table 7.2 exert a limited influence on the abundance of synthetic organic compounds during floods. The primary reason is that, in addition to energy constraints associated with the formation of activated complexes, abiotic transformation reactions are also rate limited – that is, the reactions occur at a slow enough rate that the relative importance of the reaction is small in the context of the time scale of a flood. A sense of the time aspect of the abiotic transformations is captured in the half-life ($t_{1/2}$), or the time span over which the initial concentration of the synthetic organic compound is decreased by a factor of 2 by a given abiotic transformation reaction (Table 7.3). The $t_{1/2}$ of an abiotic transformation reaction is compound specific and depends strongly on the structure of the synthetic organic compound; thus, the range of $t_{1/2}$ listed can vary by several orders of magnitude. Most of the abiotic transformation reactions are too slow to be important over the time scale of a single flood.

Oxidation–reduction reactions can also affect the abundance of synthetic organic compounds. For the most part, oxidation is important only for a restricted class of synthetic organic compounds for the simple reason that the organic molecules are designed to be stable in an aerobic

Table 7.3. *Half-lives of selected organic compounds in aqueous solutions (modified from Environmental Organic Chemistry, Schwarzenbach et al., Copyright ©1993. Reprinted by permission of John Wiley & Sons, Inc.)*

Compound	Range in half-life	Mechanism*
Monohalogenated hydrocarbons	23 s to 30 yr	Nucleophilic substitution
Polyhalogenated hydrocarbons	40 d to 3,500 yr	Nucleophilic substitution and nonreductive elimination
Carboxylic acid esters	4 min to 140 yr	Hydrolysis
Simple amides	1.5 yr to 40,000 yr	Hydrolysis
Simple carbamates	25 s to 55,000 yr	Hydrolysis
Phosphoric and thiophosphoric acid triesters	72 d to 5 yr	Hydrolysis

*Examples of the mechanisms are listed in Table 7.2.

environment (Schwarzenbach et al., 1993). In contrast, a broader array of synthetic organic compounds are susceptible to abiotic transformations under anaerobic conditions (Macalady et al., 1986; Jafvert and Wolfe, 1987; Wolfe, 1992). The difficulty in truly assessing abiotic reduction is that the overall oxidation–reduction status of most natural systems is controlled by microbiological processes (Stumm and Morgan, 1981; Drever, 1988). The distinction is less of a concern for this discussion because abiotic reduction is not likely to be an important process during a flood because turbulent mixing will probably maintain floodwaters close to oxygen saturation. There is the possibility, as a function of the magnitude of the flood, for the products of abiotic reduction of synthetic organic compounds to be released into the fluvial system.

Synthetic organic compounds are also susceptible to transformation reactions driven by absorption of light; a process referred to as photolysis (Schwarzenbach et al., 1993). Again, the details of the mechanism are beyond the scope of this chapter and the interested reader is directed to the text of Schwarzenbach et al. (1993). The essence of the mechanism is that absorbed light alters the energetic state of the synthetic organic compound in a manner analogous to the activated species state in biodegradation processes, resulting in an increase in the overall reactivity and ultimate transformation of the organic molecule. Photolytic transformation rates occur roughly on the same time scale as do floods and therefore may affect the concentration of synthetic organic compounds transported by floodwaters. Currently, the magnitude of the potential impact cannot be assessed, but photolytic transformation can be considered in the context of the physical conditions of river water during flooding.

The effectiveness of the photolytic transformation reactions depends on many factors but, in the context of assessing the potential magnitude of photolysis (either direct or indirect) during a flood, the depth of light penetration is probably the key factor. Rivers under flood conditions typically carry large suspended sediment loads, with the result that the waters are turbid. As a consequence of the increase in turbidity, the absorptive capacity and the depth of light penetration in floodwaters is probably less than when the river does not have an excess of energy supporting the large increase in sediment transport capacity (see, for example, Wohl, Chapter 6, this volume). However, if there is development of a perirheic zone (Mertes, Chapter 5, this volume) where the water on the floodplain may be relatively clear because of mixing with less turbid local water, photolytic activity may be more pronounced.

Volatilization of Organic Contaminants. Boiling temperatures for synthetic organic compounds vary widely relative to that of liquid water, but for some classes of organic compounds (e.g., gasoline) the vapor phase is the most stable at earth surface temperature and pressure. Other classes of synthetic organic compounds may partition preferentially to the vapor phase not because of a low boiling temperature but because of very low solubility in natural waters (Schwarzenbach et al., 1993). The physical nature of the two boundary layers – the stagnant zones just below and just above the water–air interface – determines the mass transfer flux from the dissolved to the vapor state for any given compound (including water).

In all cases, the rate of mass transfer due to volatilization varies inversely with the thickness of the boundary layers. Wind speed affects the thickness of the air boundary layer, and the magnitude of turbulence in the water influences the thickness of the water boundary layer. Thus, one might reasonably expect that the water boundary layer will be relatively thin during a flood because of the large amount of turbulence caused by the wide variety of flow conditions present in the active floodway. As a consequence, then, volatilization of synthetic organic compounds may be greatly enhanced by the accentuated hydraulics of flow that occur during flooding.

Schwarzenbach et al. (1993) present a calculation that defines a critical value for the ratio of current velocity to depth above which the motion of the fluid contributes significantly to the volatilization of organic compounds. The mass transfer velocity across the liquid boundary layer can be approximated from the empirical relationship developed by O'Connor and Dobbins (1958), who studied the mass transfer of O_2 in streams. The basic relationships are as follow:

$$r_W \approx \frac{u_W}{d_W}$$

where r_W is the rate of renewal of the water boundary layer (surface renewal rate), u_W is the water velocity, and d_W is the depth of water (O'Connor and Dobbins, 1958). When combined with the approximation for the mass transfer velocity across the water boundary layer based on r_W and the diffusion coefficient in water (D_W), the critical value for the ratio of u_W to d_W can be determined as shown by Schwarzenbach et al. (1993):

$$\begin{aligned} v_W &\approx (D_W r_W)^{1/2} \\ &\approx \left(\frac{(D_W u_W)}{d_W}\right)^{1/2} \end{aligned}$$

Turbulent effects on volatilization of chemical compounds from rivers should be considered when the term on the right is greater than the critical value of the water-velocity-to-depth ratio as presented by Schwarzenbach et al. (1993). Values of v_W are generally greater than approximately 5×10^{-4} cm s^{-1}, and the diffusion coefficient of a broad array of organic compounds in water is approximately 10^{-5} cm^2 s^{-1}. Thus, the critical ratio of current velocity to depth is estimated by

$$5 \times 10^{-4}\,\mathrm{cm\,s^{-1}} < \left(\frac{10^{-5}(\mathrm{cm^2\,s^{-1}})u_{\mathrm{W}}}{d_{\mathrm{W}}}\right)^{1/2}$$

$$\left(\frac{u_{\mathrm{W}}}{d_{\mathrm{W}}}\right) > 0.03\,\mathrm{s^{-1}}$$

Thus, for a water depth of 2 m, the water velocity must exceed approximately 6 cm s^{-1} for the turbulent motion of flow to affect the volatilization of synthetic organic compounds, a condition readily met under flood conditions. Therefore, volatilization of synthetic organic compounds is probably an important process that influences the fate and transport of organic compounds during flooding and may decrease the hazard posed by the presence of synthetic organic compounds in river systems.

Case Studies

Three case studies are presented as examples of how the combined biogeochemical processes presented in the preceding sections contribute to the hazard posed by the presence of inorganic and organic contaminants in waters discharged during flooding. In the first case study, the transport and deposition of mine wastes on the floodplain of the River Tyne (United Kingdom) are used to elucidate geomorphologic processes that occur during flooding. The second case study reviews the effects of a large flood on the Mississippi River (USA) on the transport and distribution of synthetic organic compounds and nutrients in the river system. In the third case study, the historical and current effects of metal transport in the Clark Fork River (USA) are described in relation to how floodplain deposition of contaminated mill wastes and subsequent geochemical processes produce markedly deleterious effects on the aquatic ecosystem of the river. General characteristics of each river system are listed in Table 7.4.

Metal Contaminants in the River Tyne, United Kingdom

Metal mining in the southern portion of the River Tyne drainage basin has occurred since the 16th century, and for a period of time the mining district was the most productive in the United Kingdom in terms of exploiting deposits of lead, such as galena (PbS), and zinc, such as sphalerite (ZnS) (Macklin and Lewin, 1989). Fine-grained, metal-bearing sediments from the milling processes were subsequently released into the fluvial system (Table 7.4) (Macklin et al., 1992). The fine-grained, metal-bearing sediments were then transported throughout the river system, and overbank flooding produced floodplain sedimentation that incorporated the metal-bearing sediments. Macklin and his colleagues have investigated the distribution of metal-bearing sediments throughout the River Tyne valley as a means of better understanding geomorphologic processes associated with rates of sediment transport and deposition (Lewin et al., 1977; Lewin and Macklin, 1987; Macklin and Dowsett, 1989; Macklin and Lewin, 1989; Macklin et al., 1992).

The major processes controlling the rate of transport and distribution of metal-bearing sediments in the River Tyne were summarized by Lewin and

Table 7.4. *General characteristics of river systems with contaminant hazards during flooding*

River system	Drainage basin size/mean daily discharge	Source of contaminants	Common contaminants	Biogeochemical processes
River Tyne, UK, northeastern England	2198 km^2 44 m^{3}s^{-1} (mean annual discharge)	Historical mining of lead and zinc; flooding deposited metal-laden sediments in the floodplain	Copper, lead, zinc, and cadmium	Physical transport of sulfide minerals; oxidation of sulfide minerals; precipitation of iron oxyhydroxide minerals; adsorption/coprecipitation of metals; adsorption by cation exchange on clay minerals; retention of metals in floodplain deposits
Mississippi River, USA, at Thebes, IL	~1.8 × 10^{6}km^2 4587 m^3 s^{-1}	Agricultural runoff and urban discharge	Herbicides (e.g., atrazine, alachlor, cyanazine, and metolachlor) and nitrate	Flushing of complex synthetic organic compounds and nutrients with transport through the system
Clark Fork River, USA, northern Rocky Mountains	~57,400 km^2 634 m^3 s^{-1}	Historical mining and milling in the flood plain; major historical flooding deposited metal-laden sediments in flood plain deposits	Arsenic, cadmium, copper, lead, silver, and zinc	Physical transport of sulfide minerals; oxidation of sulfide minerals either at the source of contamination or in situ in flood plain deposits; precipitation of iron oxyhydroxide minerals; adsorption/coprecipitation of metals; precipitation and dissolution of efflorescent metal-bearing salts in floodplain deposits; physical erosion of channel bank deposits releasing metal-bearing sediments into the river

Macklin (1987) and include (1) hydraulic sorting as a function of particle density (for the sulfide minerals) and size, (2) adsorption–desorption of metals on secondary iron and manganese oxyhydroxide minerals, (3) adsorption to natural organic matter, (4) dilution or concentration of metal-bearing sediments in the main river channel by the addition of uncontaminated sediments from tributary inflow or bank erosion, and (5) floodplain deposition and long-term storage. From the standpoint of biogeochemical processes, the distribution of mining-related metals in the River Tyne drainage basin is affected primarily by (1) oxidation of the sulfide minerals, (2) precipitation of iron and manganese oxyhydroxide minerals, (3) adsorption of metals by the iron and manganese oxyhydroxide minerals, and (4) some complexation of metals by organic matter in the floodplain sediments (Macklin, 1996).

Major findings regarding the distribution of metals in fluvial sediment of the River Tyne are (1) the concentration of original sulfide minerals in overbank floodplain deposits is small, which means that sulfide minerals oxidize fairly rapidly, (2) the greatest percentage of metal mass in floodplain sediments was found to be associated with the iron and manganese oxyhydroxide mineral fraction, (3) contrary to previous observations, metals associated with the silt-clay size class composed only a small fraction of the total mass of sediment-bound metals, and (4) although the bulk concentration of sediment-bound metals decreased downstream of the source area, the fraction of metals in the exchangeable fraction increased, potentially increasing the hazard posed by the metals (Macklin and Dowset,

1989). Thus, the flood-related transport, and subsequent deposition, of metal-bearing sediments represents a hazard not only within the river channel during the flood but also from flood deposits in the floodplain that may affect long-term floodplain stability due to phytotoxicity of metal-bearing sediments (Friedman and Auble, Chapter 8, this volume; Lewin et al., 1983).

Transport of Agricultural Chemicals during the 1993 Mississippi River Flood, USA

A unique set of atmospheric conditions in the upper Mississippi River basin during spring 1993 resulted in rainfall amounts that exceeded average quantities by up to 200% (Wahl et al., 1993). The bulk of the precipitation occurred shortly after crops were planted, which is usually accompanied by the application of various agricultural chemicals (e.g., herbicides and fertilizers) (Goolsby et al., 1993). Widespread flooding in the upper Mississippi River basin began in mid-June and lasted through August 1993. The flooding was caused primarily by surface runoff of rainfall, and the overland flow of rainwater acted to flush a large portion of the mass of agricultural chemicals applied to the cropland that spring.

The 1993 flood in the upper Mississippi River basin resulted in the transport of large quantities of herbicides (synthetic organic compounds) and nutrients (mostly nitrate) through the river system and into the Gulf of Mexico (Table 7.4) (Goolsby et al., 1993). The primary issue with regard to the hazard posed by the flushing of agricultural chemicals is whether concentrations of specific compounds exceed drinking-water-quality standards. Drinking-water-quality standards, or maximum concentration levels (MCLs), for herbicides are based on the average of four samples collected over a year (Goolsby et al., 1993). Thus, even though the concentrations of selected herbicides exceeded the MCL at some time during the flood, the average concentration for the year probably did not. In contrast, the concentration of nitrate (a nutrient added to soils as fertilizers) is governed by the Safe Drinking Water Act, which dictates that the concentration of nitrate in any given sample of water must be less than the MCL in order to be in compliance with the law.

One hazard caused by the large-scale flooding in the upper Mississippi River basin was the contamination of drinking water supplies. Flood waters inundated much of the floodplain, which is where most industry and sewage treatment facilities are located. Chemical compounds flushed from the agricultural land during overland flow combined with a broad spectrum of inorganic and organic chemicals mobilized from urban areas, resulting in widespread contamination of municipal water supplies. Large-scale flooding during the 1993 event in the upper Mississippi River basin mobilized contaminants from almost all possible sources, including animal wastes, domestic sewage, and industrial sites. Therefore, the magnitude of the potential hazard from contaminants during flooding is proportional to the magnitude of the flood simply because the probability of mobilizing contaminants from the floodplain increases as the proportion of the floodplain inundated increases.

In all likelihood, the extent of chemical transformations affecting the concentration of contaminants being transported by floodwater was probably minimal. However, large quantities of sediment were redistributed throughout the floodplain and may serve to concentrate contaminated sediment, resulting in a long-term hazard. Measurements of constituent concentrations taken during the flood demonstrate that a large portion of the mass of contaminants, especially agricultural chemicals (herbicides and nutrients), were transported through the Mississippi River system and discharged into the Gulf of Mexico (Goolsby et al., 1993). Therefore, the contaminant hazard during the 1993 Mississippi River flood may have been extreme during the event but generally receded with the floodwaters.

Metal Toxicity Effects during Flooding, Clark Fork River, USA

During the period 1880 to 1972 over 400 million metric tons of metal-bearing ore was mined from the Butte district in west central Montana, USA (Figure 7.1) (Moore and Louma, 1990; Axtmann and Louma, 1991; Nimick and Moore, 1991). Before the 1950s, mill tailings and rock waste were disposed of in large piles adjacent to the river channel so that metal-bearing sediments were steadily added to the Clark Fork River system. Periodic large fluxes of metal-contaminated sediment were injected into the river system during five large floods: four floods that occurred in the 1890s and the largest flood on record, which occurred in 1908 (Nimick and Moore, 1991). The metal-bearing sediments have been distributed throughout the floodplain of the Clark Fork River over a distance of approximately 200 km (Moore et al., 1989; Moore and Louma, 1990).

The ore mined in the upper Clark Fork River drainage is composed primarily of sulfide minerals of copper, silver, lead, zinc, iron, and arsenic (Nimick and Moore, 1991). Contaminated floodplain sediments contain concentrations of silver, gold, cadmium, copper, iron, manganese, nickel, lead, and zinc that are significantly elevated above concentrations observed in tributary channels, which do not drain historically mined areas (Brook and Moore, 1988; Moore et al., 1989). The metal-rich floodplain deposits have significantly affected the ecosystem of the Clark Fork River, as evidenced by periodic fish kills (Kemble et al., 1994; Phillips and Lipton, 1995) and a decrease in the productivity of floodplain vegetation communities (Rice and Ray, 1985).

The Clark Fork is a high-gradient river system such that sediments in the channel and floodplain are relatively coarse grained (Moore et al., 1989). Transport of large volumes of contaminated tailing sediment during historic floods resulted in widespread distribution of sulfide minerals throughout the Clark Fork River system (Andrews, 1987; Axtmann and Louma, 1991; Nimick and Moore, 1991), which contrasts with the way metals are distributed in the River Tyne as described earlier. The presence of sulfide minerals in the floodplain deposits of tailings means that there is a very large original source of metals throughout the Clark Fork River system. Inorganic geochemical processes exert the greatest control on the time-dependent distribution and mobility of metals released from the sulfide minerals.

Oxidation of the sulfide minerals is the basic geochemical reaction that initiates all subsequent geochemical processes that determine the distribution of metals in the floodplain. The geomorphologic response of the Clark Fork River to the introduction of large quantities of contaminated sediments was widespread aggradation, and after installation of sediment control in the source area in the 1950s the river system has undergone incision (Moore et al., 1989). Thus, the sulfide-bearing tailings deposited during flooding have existed in an aerobic environment throughout the history of the deposits. Oxidation of the sulfide minerals was followed by precipitation of iron and manganese oxyhydroxide minerals that adsorbed some of the associated metals released during the oxidation reaction. In general, the pH of pore waters associated with the secondary iron and manganese oxyhydroxide minerals is generally too low for optimum adsorption of metals, with the result being that metals released during oxidation migrate downward and ultimately precipitate as diagenetic sulfides (anaerobic conditions) or are complexed by organic matter (Nimick and Moore, 1991).

A second, and important, geochemical process contributing to the long-term hazard associated with the accumulation of metal-bearing sediments in the Clark Fork River floodplain is the precipitation of efflorescent salts. The salts precipitate as a consequence of strong evaporation gradients causing upward migration of metal-bearing water (with elevated concentrations of sulfate) until the water vaporizes. Constituents originally in the water become concentrated in the residual solution until salts, most commonly sulfate salts, precipitate just below the sediment–air interface (Nimick and Moore, 1991). Efflorescent salts are very soluble and rainfall runoff dissolves the salts before discharging into the river. The rainfall runoff water is a low-pH solution containing elevated concentrations of metals (copper and zinc) that subsequently contaminate the Clark Fork River and cause fish kills mainly because of the extreme toxicity of copper (Phillips and Lipton, 1995).

Another example of the long-term hazards associated with flood deposits of metal-bearing tailings is the introduction of metal-bearing sediments to the river during active bank erosion (Andrews, 1987). The Clark Fork River is actively meandering, and the erosion of metal-bearing sediments into the river effectively maintains the total metal concentrations in the river above the minimum values at which fish populations can survive.

The environmental situation along the Clark Fork River is a good example of the spectrum of hazards associated with the transport of metal-bearing sediments during floods. Historical flooding with transport and deposition of mine tailings created the current geomorphological setting in which the Clark Fork River functions. There is anecdotal evidence that the transport of mine wastes and tailings in the late 1800s was affecting the water quality, with associated deleterious impacts on fish habitat (Phillips and Lipton, 1995). Thus, the historical flooding undoubtedly also exacerbated the hazard from contaminants that existed during normal flows. Current conditions along the Clark Fork River demonstrate quite clearly that the hazard to the riparian environment associated with the transport and distribution of metal contaminants during floods extends many years beyond any particular flood. In a system as overwhelmed by contaminated sediments as the Clark Fork River, the hazards from contaminant transport during floods may persist for hundreds of years.

Conclusions

Contaminant hazards during flooding take the form of both inorganic and organic constituents. Inorganic constituents include elevated concentrations of metals that may be introduced to the system from a variety of sources including (1) the uncontrolled release of mine and mill wastes, (2) discharges from urban areas (industrial and sewage), and (3) discharge of metal-bearing water from natural drainages. Organic contaminants are also introduced into the fluvial system but, unlike the inorganic constituents, all organic contaminants are anthropogenic in origin. Agricultural runoff flushes synthetic organic compounds that are applied as herbicides, pesticides, and fungicides. Excess nutrients, such as nitrate, are carried by agricultural runoff as well as by sewage effluent and runoff from livestock waste.

The nature and extent of the contaminant hazard associated with floods are time dependent. On the time scale of the flood, inorganic geochemical processes can affect the distribution of metals among the possible forms (free ion, sediment bound, precipitated, or complexed). Long-term effects to the riparian environment also arise from flood transport and deposition of metal-bearing sediments, as a combination of geochemical reactions and geomorphological evolution of floodplain sediments act to supply deleterious concentrations of metals to the active fluvial ecosystem. In contrast to the inorganic contaminants, synthetic organic compounds typically persist on time scales that are longer than the duration of most floods. Redistribution of sediment-bound organic compounds may represent a long-term hazard in much the same manner as do metal-bearing sediments deposited in the floodplain; the major difference is that synthetic organic compounds normally do not concentrate in sedimentary deposits. There is no question that evolution of human society in the fluvial environment has created potentially significant hazards due to the transport and storage of contaminants during flooding.

References

Appelo, C.A.J. and Postma, D. (1993). *Geochemistry, Groundwater and Pollution*. Rotterdam: Balkema.

Ananyeva, N.D., Naumova, N.N., Rogers, J., and Steen, W.C. (1992). Microbial transformation of selected organic chemicals in natural aquatic systems. In *Fate of Pesticides and Chemicals in the Environment*, ed. J.L. Schnoor, pp. 275–294. New York: John Wiley & Sons Inc.

Andrews, E.D. (1987). Longitudinal dispersion of metals in the Clark Fork River, Montana. In *The Chemical Quality of Water and the Hydrologic Cycle*, eds. R.C. Averett and D.M. McKnight, pp. 179–191. Boca Raton, FL: Lewis Publishers.

Axtmann, E.V. and Louma, S.N. (1991). Large-scale distribution of metal contamination in the fine-grained sediments of the Clark Fork River, Montana, USA. *Applied Geochemistry*, **6**, 75–88.

Bradley, S.B. (1984). Flood effects on the transport of heavy metals. *International Journal of Environmental Studies*, **22**, 225–230.

Bradley, S.B. and Cox, J.J. (1990). The significance of the floodplain to the cycling of metals in the River Derwent catchment, UK. *Science of the Total Environment*, **97/98**, 441–454.

Brook, E.J. and Moore, J.N. (1988). Particle-size and chemical control of As, Cd, Cu, Fe, Mn, Ni, Pb and Zn in bed sediment from the Clark Fork River, Montana (USA). *Science of the Total Environment*, **76**, 247–266.

Butler, D.L., Kruger, R.P., Campbell, B., Thompsion, A.L., and McCall, S.K. (1991). *Reconnaissance Investigation of Water Quality, Bottom Sediment, and Biota Associated With Irrigation Drainage in the Gunnison and Uncompahgre River Basins and at Sweitzer Lake, West-Central Colorado.* U.S. Geological Survey, Water-Resources Investigations Report 91-4103.

Carson, R. (1962). *Silent Spring.* Greenwich, CT: Fawcett.

Cherry, J.A., Morel, F.M.M., Rouse, J.V., Schnoor, J.L., and Wolman, M.G. (1986). Hydrogeochemistry of sulfide and arsenic-rich tailings and alluvium along Whitewood Creek, South Dakota. *Mineral & Energy Resources*, vol. 29, 44 pp.

Ciszewski, D. (1997). Source of pollution as a factor controlling distribution of heavy metals in bottom sediments of Chechlo River (south Poland). *Environmental Geology*, **29**, 50–57.

Davis, J.A. and Kent, D.B. (1990). Surface complexation modeling in aqueous geochemistry. In *Mineral-Water Interface Geochemistry*, eds. M.F. Hochella Jr. and A. F. White, vol. 23, *Reviews in Mineralogy*, pp. 177–260. Mineralogical Society of America, Washington, DC.

Davis, J.A. and Leckie, J.O. (1978). Surface ionization and complexation at the oxide/water interface. II. Surface properties of amorphous iron oxyhydroxide and adsorption of metal ions. *Journal of Colloid Interface Science*, **67**, 90–107.

De Voto, B.A. (1952). *The Course of Empire.* Boston, MA: Houghton Mifflin.

Drever, J.I. (1988). *The Geochemistry of Natural Waters*, 2nd ed. Englewood cliffs, NJ: Prentice-Hall.

Dzombak, D.A. and Morel, F.M.M. (1990). *Surface Complexation Modeling: Hydrous Ferric Oxide.* New York: John Wiley & Sons Inc.

Elzerman, A.W. and Coates, J.T. (1987). Hydrophobic organic compounds on sediments: Equilibria and kinetics of sorption. In *Sources and Rates of Aquatic Pollutants*, eds. R.A. Hites and S.J. Eiesenreich, Advances in Chemistry Series 216, pp. 263–317. Washington, DC: American Chemical Society.

Filipeck, L.H., Nordstrom, D.K., and Ficklin, W.H. (1987). Interaction of acid mine drainage with waters and sediments of West Squaw Creek in the West Shasta mining district, California. *Environmental Science & Technology*, **21**, 388–396.

Garban, B., Ollivon, D., Carru, A.M., and Chesterikoff, A. (1996). Origin, retention and release of trace metals from sediments of the River Seine. *Water Air and Soil Pollution*, **87**, 363–381.

Gonçalves, E.P.R., Soares, H.M.V.M., Boaventura, R.A.R., Machado, A.A.S.C., and da Silva, J.C.G.E. (1994). Seasonal variations of heavy metals in sediments and aquatic mosses from the Cavado river basin (Portugal). *Science of the Total Environment*, **142**, 143–156.

Goolsby, D.A., Battaglin, W.A., and Thurman, E.M. (1993). *Occurrence and Transport of Agricultural Chemicals in the Mississippi River Basin, July Through August 1993.* U.S. Geological Survey Circular 1120-C. 22 pp.

Gore, A. (1993). *Earth in the Balance: Ecology and the Human Spirit.* New York: Plume.

Graf, W.L. (1985). *The Colorado River: Instability and River Basin Management.* Washington, DC: Association of American Geographers.

Graf, W.L. (1994). *Plutonium and the Rio Grande: Environmental Change and Contamination in the Nuclear Age.* New York: Oxford University Press.

Imhoff, K.R., Koppe, P., and Feidrich, D. (1980). Heavy metals in the Ruhr river and their budget in the catchment area. *Water Science and Technology*, **12**, 735–749.

Jafvert, C.T. and Wolfe, N.L. (1987). Degradation of selected halogenated ethanes in anoxic sediment-water systems. *Environmental Toxicology and Chemistry*, **6**, 827–837.

Kemble, N.E., Brumbaugh, W.G., and Woodward, D.F. (1994). Toxicity of metal-contaminated sediments from the Upper Clark Fork River, Montana, to aquatic invertebrates and fish in laboratory exposures. *Environmental Toxicology and Chemistry*, **13**, 1985–1997.

Langmuir, D. (1997). *Aqueous Environmental Geochemistry.* Englewood cliffs, NJ: Prentice-Hall.

Lewin, J. and Macklin, M.G. (1987). Metal mining and floodplain sedimentation, in *International Geomorphology* 1986, Part 1, ed. V. Gardiner, pp. 1009–1027. Chichester: John Wiley & Sons Ltd.

Lewin, J., Davies, B.E., and Wolfenden, P.J. (1977). Interactions between channel change and historic mining sediments. In *River Channel Changes*, ed. K.J. Gregory, pp. 353–367, Chichester: John Wiley & Sons Ltd.

Lewin, J., Bradley, S.B., and Macklin, M.G. (1983). Historical valley alluviation in mid-Wales. *Geological Journal*, **18**, 331–350.

Liao, W., Christman, R.F., Johnson, J.D., Millington, D.S., and Hass, J.R. (1982). Structural characterization of aquatic humic material. *Environmental Science & Technology*, **16**, 403–410.

Macalady, D.L., Trantnyek, P.G., and Grundl, T. (1986). Abiotic reduction reactions of anthropogenic organic chemicals in anaerobic systems: A critical review. *Journal of Contaminant Hydrology*, **1**, 1–28.

Macklin, M.G. (1996). Fluxes and storage of sediment-associated heavy metals in floodplain systems: Assessment and river basin management issues at a time of rapid enviromental change. In *Floodplain Processes*, eds. M.G. Anderson, D.E. Walling, and P.D. Bates, pp. 440–460. London: John Wiley & Sons Ltd.

Macklin, M.G. and Dowsett, R.B. (1989). The chemical and physical speciation of trace metals in fine grained overbank flood sediments in the Tyne basin, north-east England. *Catena*, **16**, 135–151.

Macklin, M.G. and Lewin, J. (1989). Sediment transfer and transformation of an alluvial valley floor: The river South Tyne, Northumbria, UK. *Earth Surface Processes and Landforms*, **14**, 233–246.

Macklin, M.G., Rumsby, B.T., and Newson, M.D. (1992). Historical floods and vertical accretion of fine-grained alluvium in the lower Tyne valley, north-east England. In *Dynamics of Gravel-Bed Rivers*, eds. P. Billi, R.D. Hey, C.R. Thorne, and P. Tacconi, pp. 573–589. Chichester: John Wiley & Sons Ltd.

Macklin, M.G., Ridgway, J., Passmore, D.G., and Rumsby, B.T. (1994). The use of overbank sediment geochemical mapping and contamination assessment: results from selected English and Welsh floodplains. *Applied Geochemistry*, **9**, 689–700.

Marcus, W.A. (1987). Copper dispersion in ephemeral stream sediments. *Earth Surface Processes and Landforms*, **12**, 217–228.

Marron, D.C. (1989). Physical and chemical characteristics of a metal-contaminated overbank deposit, west-central South Dakota, USA. *Earth Surface Processes and Landforms*, **14**, 419–432.

Marron, D.C. (1992). Floodplain storage of mine tailings in the Belle Fourche river system: A sediment budget approach. *Earth Surface Processes and Landforms*, **17**, 675–685.

Moore, J.N. and Louma, S.N. (1990). Hazardous wastes from large-scale metal extraction. *Environmental Science & Technology*, **24**, 1278–1285.

Moore, J.N., Brook, E.J., and Johns, C. (1989). Grain size partitioning of metals in contaminated, coarse-grained river floodplain sediment: Clark Fork River, Montana, USA. *Environmental Geology & Water Sciences*, **14**, 107–115.

Moses, C.O., Nordstrom, D.K., Herman, J.S., and Mills, A.L. (1987). Aqueous pyrite oxidation by dissolved oxygen and ferric iron. *Geochimica et Cosmochimica Acta*, **51**, 1561–1572.

Nimick, D.A. and Moore, J.N. (1991). Prediction of water-soluble metal concentrations in fluvially deposited tailings sediment, Upper Clark Fork Valley, Montana, USA. *Applied Geochemistry*, **6**, 635–646.

Nordstrom, D.K. (1982). Aqueous pyrite oxidation and the consequent formation of secondary iron minerals, In *Acid Sulfate Weathering*, ed. J.A. Kittrick, pp. 37–56. Madison, WI: Soil Science Society of America Special Publication No. 10.

O'Connor, D.J. and Dobbins, W.E. (1958). Mechanisms of reaeration in natural streams. *Transactions of the American Society of Civil Engineers*, **123**, 641–684.

Parkhurst, D.L. (1995). *User's Guide to PHREEQC-A Computer Program for Speciation, Reaction-Path, Advective-Transport, and Inverse Geochemical Calculations.* U.S. Geological Survey Water-Resources Investigations Report 95-4227. Washington, DC: U.S. Geological Survey.

Parks, G.A. (1990). Surface energy and adsorption at mineral-water interfaces: An introduction. In *Mineral-Water Interface Geochemistry*, vol. 23, *Reviews in Mineralogy*, eds. M.F. Hochella Jr. and A.F. White, pp. 133–175. Washington, DC: Mineralogical Society of America.

Parks, G.A. and de Bruyn, P.L. (1962). The zero point of charge of oxides. *Journal of Physical Chemistry*, **66**, 967–973.

Phillips, G. and Lipton, J. (1995). Injury to aquatic resources caused by metals in Montana's Clark Fork River basin: historic perspective and overview. *Canadian Journal of Aquatic Science*, **52**, 1990–1993.

Plumlee, G.S., Smith, K.S., Ficklin, W.H., and Briggs, P.H. (1992). Geological and geochemical controls on the composition of mine drainages and natural drainages in mineralized areas. In *Proceedings of the 7th International Water-Rock-Interaction – WRI-7*, vol. 1, eds. Y.K. Karaka and A.S. Maest, pp. 419–422. Rotterdam: A.A. Balkema.

Rang, M.C. and Schouten, C.J. (1989). Evidence for historical heavy metal pollution in floodplain soils: the Meuse. In *Historical Change of Large Alluvial Rivers: Western Europe*, ed. G.E. Petts, pp. 127–142. New York: John Wiley & Sons Inc.

Rice, P. and Ray, G. (1985). Heavy metals in floodplain deposits along the upper Clark Fork River. In *Proceedings-Clark Fork River Symposium*, eds. C.E. Carlson and L.L. Bahls, pp. 26–45, Butte: Montana Academy of Sciences.

Schnitzer, M. and Khan, S.U. (1972). *Humic Substances in the Environment.* New York: Dekker.

Schnoor, J.L., Sato, C., McKechnie, D., and Sahoo, D. (1987). *Processes, Coefficients, and Models for Simulating Toxic Organics and Heavy Metals in Surface Waters.* Washington, DC: U.S. Environmental Protection Agency Report EPA/600/3-87/015.

Singh, M., Ansari, A.A., Müller, and Singh, I.B. (1997). Heavy metals in freshly deposited sediments of the Gomati River (a tributary of the Ganga River): effects of human activities. *Environmental Geology*, **29**, 246–252.

Stumm, W. and Morgan, J.J. (1981). *Aquatic Chemistry: An Introduction Emphasizing Chemical Equilibria in Natural Waters*, 2nd ed. New York: John Wiley & Sons Inc.

Schwarzenbach, R.P., Gschwend, P.M., and Imboden, D.M. (1993). *Environmental Organic Geochemistry*. New York: John Wiley & Sons Inc.

Taylor, M.P. (1996). The variability of heavy metals in floodplain sediments: a case study from mid Wales. *Catena*, **28**, 71–87.

Thurman, E.M. (1985). *Organic Geochemistry of Natural Waters*. Boston: Martinus Nijhoff.

Wahl, K.L., Vining, K.C., and Wiche, G.J. (1993). *Precipitation in the Upper Mississippi River Basin, January 1 through July 31, 1993*. U.S. Geological Survey Circular 1120-B. Washington, DC.

Wolfe, N.L. (1992). Abiotic transformations of pesticides in natural waters and sediments. In *Fate of Pesticides and Chemicals in the Environment*, ed. J.L. Schnoor, pp. 93–103. New York: John Wiley & Sons Inc.

Wolfenden, P.J. and Lewin, J. (1977). Distribution of metal pollutants in floodplain sediments. *Catena*, **4**, 309–317.

BIOLOGICAL FLOOD PROCESSES AND EFFECTS

CHAPTER 8

Floods, Flood Control, and Bottomland Vegetation

Jonathan M. Friedman
and Gregor T. Auble
Midcontinent Ecological Science Center
United States Geological Survey

Bottomland plant communities are typically dominated by the effects of floods. Floods create the surfaces on which plants become established, transport seeds and nutrients, and remove established plants. Floods provide a moisture subsidy that allows development of bottomland forests in arid regions and produce anoxic soils, which can control bottomland plant distribution in humid regions. Repeated flooding produces a mosaic of patches of different age, sediment texture, and inundation duration; this mosaic fosters high species richness. Because of the many influences of floods on bottomland plant communities, the effects of flood control are pervasive. The response of channel geometry and bottomland vegetation to flood control are conditioned by local fluvial processes, which are determined by the geologic and climatic factors that govern flow variability and sediment load. Construction of upstream reservoirs, bank stabilization, and levee construction generally decrease flow variability and isolate a river from its floodplain. This decreases the physical disturbance on the bottomland and stabilizes water levels, thereby decreasing habitat complexity. Over the long term, pioneer species typically decline and shade-tolerant species increase; however, in the decades after imposition of flood control, adjustments of channel form such as narrowing may cause a transient pulse of establishment of pioneers. Although floods are commonly regarded as a hazard to human communities, they are usually necessary to maintain biotic diversity in bottomlands. By changing the magnitude, frequency, or duration of floods, flood control often constitutes a hazard to riparian vegetation.

Introduction

Bottomland plant communities are diverse, productive, and subject to intense pressure from economic development (Brinson et al., 1981; Gregory et al., 1991). As a result the effects on bottomland vegetation are increasingly included in assessments of proposed flood-control projects. Bottomland vegetation reflects the history of flood timing, frequency, magnitude, and duration. Therefore, altering the flooding regime will affect the vegetation. However, a general prediction of these effects is difficult because of the multiple influences of flooding on bottomland ecosystems.

In this chapter we start by reviewing the principal mechanisms by which floods influence bottomland vegetation. Floods function as transport vectors, distributing plant propagules and importing and exporting sediment and nutrients. Floods affect the moisture regime of bottomland sites in ways that can be either positive or negative for plant growth and survival. In dry landscapes, floods can provide a beneficial moisture subsidy. In other environments, flooding can lead to waterlogged and anoxic soils that produce severe stress for terrestrial plants. The physical disturbance of flooding can damage and remove plants. The character of the vegetation, in turn, is an important component of resistance to flow and thus influences flood hydraulics. The flooding disturbance and channel change that remove some plants also create the bare, moist sites necessary for seedling establishment of many plant species.

Prediction of how flood control will change a particular bottomland community requires quantifying and integrating these multiple influences of flooding on vegetation for a specific location. Nonetheless, general patterns can be identified along several dimensions. We discuss flood-related strategies of bottomland herbs and trees, regional patterns in bottomland forests, and relations between flood disturbance and reproduction of pioneer trees. Finally, we summarize some of the well-documented cases of vegetation response to flood control. Although the scope of the presentation is international and includes the entire plant community, the emphasis reflects our experience working with trees in bottomlands of the western United States.

For the purposes of this chapter, a bottomland is the area receiving surface or subsurface water from a stream for at least part of the time. The rest of the landscape is the upland. All vegetation influenced by surface water or alluvial groundwater from a stream occurs on its bottomland. The term riparian zone, not used in this chapter, is sometimes synonymous with bottomland and sometimes refers more restrictively to that portion of the bottomland immediately adjacent to the stream. The bottomland includes the flood plain, loosely defined here as the alluvial surfaces subject to frequent inundation.

Floods as Transport Vectors

Floods disperse seeds of many bottomland species (Kalliola et al., 1991). For example, along streams in central Amazonia, most species that occupy sites subject to flooding disperse seeds via water (Kubitzki and Ziburski, 1994). Many species, including the willow family (Salicaceae), use both wind and water transport (Johnson, 1994; Kalliola et al., 1991). Some species disperse by means of floating vegetative propagules such as rhizomes or branch fragments (Johansson and Nilsson, 1993; Shafroth et al., 1994). Water dispersal enhances dispersal range, and long distance dispersal may contribute to the high species richness of many bottomland communities (Nilsson et al., 1994). Many bottomland plants time their release of seed to coincide with or to follow flood peaks (White, 1979; Hupp, 1992; Johnson, 1994), which prepare a seedbed and disperse seeds high enough along the bank to avoid future disturbances such as ice scour (Auble and Scott, 1998). Patterns of adult plants in a bottomland are influenced by processes that control deposition of plant propagules (Schneider and Sharitz, 1988; Nilsson et al., 1991; Johansson and Nilsson, 1993; Rood and Kalischuk, 1998).

Flooding promotes exchange of solutes and both organic and inorganic sediments between river and floodplain (Finley, Chapter 7, this volume). Because of nutrients supplied by the river, floodplains typically have high primary productivity (Frangi and Lugo, 1985; Mitsch and Gosselink, 1986). However, the nutrient subsidy depends on both the quantity of nutrients transported by the river and the frequency of flooding. Storage of materials on floodplains is temporary; therefore, during a given flood, the floodplain may act as a source or sink of sediment, nutrients, and contaminants (Mitsch et al., 1979; Sanchez-Perez et al., 1991).

Floodplain plants take up inorganic nutrients and return partially decomposed leaves and stems to the river. Therefore, floodplains can be net importers of inorganic forms of nutrients and net exporters of organic forms (Elder, 1985). Organic matter from floodplains may be important for stream productivity, especially where shade or high-sediment load limits growth of plants and algae in the stream (Vannote et al., 1980). Large woody debris from floodplain trees strongly influences channel geometry and patterns of discharge and provides an important habitat for fish and other aquatic organisms (Harmon et al., 1986). Floodplain bacteria can increase nitrogen in a riverine ecosystem through nitrogen fixation (Walker and Chapin, 1986). On the other hand, periodic wetting and drying of soils releases nitrogen to the atmosphere through denitrification (Mitsch and Gosselink, 1986). Floods may flush away contaminants such as salts that otherwise would accumulate through evaporation on the floodplain surface (Ohmart et al., 1988).

Flooding and Anoxia

Flooding may increase or decrease growth of plants (Stockton and Fritts, 1973; Mitsch et al., 1979). Mortality or reduced growth from prolonged inundation is typically associated with exhaustion of energy reserves or oxygen depletion in the root zone (Stockton and Fritts, 1973). Depletion of oxygen leads to reducing conditions and development of toxic concentrations of metals and organic compounds. Therefore, many species in flood-prone environments have air spaces in roots and stems for transport of oxygen down to the root zone or chemical mechanisms for tolerance or alteration of toxic compounds. The duration of flooding a bottomland plant can survive is influenced by the species and size of the plant; the depth, temperature, and clarity of the water; and the timing of inundation relative to the growing season (Gill, 1970; Whitlow and Harris, 1979; Stevens and Waring, 1985; Friedman and Auble, 1999). Adult trees of North American swamps such as bald cypress (*Taxodium distichum*) or water tupelo (*Nyssa aquatica*) typically survive deep inundation for more than one growing season, whereas shagbark hickory (*Carya ovata*) and white oak (*Quercus alba*) do not survive inundation for more than a few days (Whitlow and Harris, 1979). Within a given species larger plants are usually more tolerant of flooding because of their greater energy reserves and because shallow flooding may not completely submerge the plant (Gill, 1970). Plants are less tolerant of flooding during the growing season and at high temperatures, when relatively rapid metabolism exhausts reserves and depletes soil oxygen (Gill, 1970).

Moisture Subsidy

In a dry environment streams provide a moisture subsidy that allows survival of plants that otherwise could not occur (Hughes, 1990). For example, in much of the Great Plains and Great Basin of North America, forest is mostly restricted to bottomlands. In some situations the moisture subsidy is unrelated to flooding. For example, plants may derive water from sideslope drainage or from an alluvial aquifer controlled largely by base flow. On the other hand, on surfaces high above the channel, survival and growth of plants may depend on a moisture subsidy due to flooding (Reily and Johnson, 1982; Rood et al., 1995). If moisture enters floodplain soils as downward percolation after overbank flooding, then moisture may be provided even by floods of short duration. If moisture enters by lateral infiltration of the bank without overbank flooding, then sustained high flows may be necessary to provide sufficient water. Many bottomland plants in arid climates are sensitive to changes in depth of the alluvial water table (Scott et al., 1999). Floods may cause mortality of bottomland plants by changing the bed level, thereby altering the water-table depth (Cooke and Reeves, 1976).

Physical Disturbance

High flows can physically damage or remove bottomland plants (Ellery et al., 1990; Menges, 1990). Flood debris and floating ice abrade tree bark and provide a point of entry for disease organisms (Sigafoos, 1964; Yanosky, 1982). Debris trapped on the upstream side of a plant magnifies the fluid drag exerted by water against the stem and may cause the plant to be uprooted or the stem to be broken or bent (Osterkamp and Costa, 1987; Oplatka and Sutherland, 1995); many bottomland species resprout from roots or broken or bent stems (Sigafoos, 1964; Hupp, 1988; Krasny et al., 1988; Rood et al., 1994). The plant can be washed away if flood shear stresses are sufficient to mobilize the sediment in which it is rooted (Stevens and Waring, 1985; Osterkamp and Costa, 1987; Friedman and Auble, 1999). A plant may also be injured by deposition of sediment transported by floods (Hupp, 1988). Finally, plants growing on top of a bank can be removed by bank failure after fluvial erosion of sediment at the base of the bank (Thorne, 1990; Hupp, 1992; Mertes, Chapter 5, this volume).

Whereas physical damage resulting from flooding is usually related to the peak discharge, damage from oxygen depletion or exhaustion of energy reserves is more strongly related to flood duration. Because the effects of these two processes differ depending on the species and the position in the bottomland, the influence of controlled floods can be managed by adjusting the flood magnitude and duration (Stevens and Waring, 1985; Friedman and Auble, 1999).

Influence of Vegetation on Flooding

Although the focus of this chapter is the influence of floods on vegetation, living and dead plants in turn can have an important influence on flooding. Vegetation is an important roughness element in channels. The taller, more rigid, and more closely packed the stems, the greater the resistance to flow (Arcement and Schneider, 1989). The reduction in velocity caused by vegetation decreases shear stress on the bank, but it also raises the elevation inundated at a given discharge. Thus vegetation can attenuate flood flows by slowing velocities and causing water to spread higher on the floodplain.

Woody debris in channels can greatly increase channel roughness. However, once this debris has been mobilized by a flood, it can become an agent of erosion (Burkham, 1972; Harmon et al., 1986). Most changes in channel pattern along the Squamish River in British Columbia involve channel abandonment and flow diversion associated with logjams (Hickin, 1984). Removal of woody debris to improve navigability can transform channel geometry even in large rivers (Sedell and Froggatt, 1984; Thorne, 1990).

Reduced velocity around vegetation or woody debris leads to deposition of the fine sediment that is essential for bank cohesion (Nanson and Beach, 1977; Osterkamp and Costa, 1987; McCarthy et al., 1992). Roots bind sediment particles, increasing the critical shear stress necessary to induce erosion. Smith (1976) found that the rate of erosion of a silt bank increased by four orders of magnitude when the mat of willow and graminoid roots was removed and concluded that vegetation was largely responsible for the slow rate of lateral migration in anabranches of an Alberta stream. In New Zealand, willows have been planted to reduce the width of braided channels (Nevins, 1969). In Australia and the western United States, reduction in grass cover by grazing may have helped bring about the development of incised channels by reducing the critical shear stress for sediment transport (Cooke and Reeves, 1976; Prosser et al., 1995).

The influence of plant growth form on channel geometry is scale dependent. Reaches of small streams in grassland are narrower than reaches of the same streams under forest (Zimmerman et al., 1967), possibly because root density of grass sod is higher than that of forest floor and because living and dead wood increase turbulence. On the other hand, woody plants generally have deeper and larger roots than grasses, which give wooded banks along large streams better protection against undercutting (Thorne, 1990). Hey and Thorne (1986) found that rivers with grassy banks were on average 1.8 times as wide as rivers with forested banks. As the forces of the river become large relative to the ability of plants to resist them, the influence of the vegetation on channel pattern should be diminished (Zimmerman et al., 1967; Hickin, 1984). However, vegetation and woody debris can influence the geometry even of large rivers (Sedell and Froggatt, 1984; Harmon et al., 1986).

Flood Disturbance and Bottomland Plant Communities

Damage to some plants provides opportunities for the establishment and growth of others. Thus floods maintain populations of plant species that would decline because of competition in the absence of disturbance (Menges, 1990; Friedman et al., 1996). The character of flood disturbance varies both parallel and perpendicular to the channel. Perpendicular to the channel, from the thalweg to the valley edge, are gradients of decreasing frequency, intensity, and duration of flood disturbance. These disturbance gradients are strongly correlated with vegetation patterns (Bedinger, 1971; Furness and Breen, 1980; Wharton et al., 1982; Bravard et al., 1986; Day et al., 1988; Glavac et al., 1992; Auble et al., 1994, 1997), but their effects are often difficult to distinguish from those of other strongly correlated gradients, including light, nutrients, moisture availability, anoxia, and sediment particle size (Illichevsky, 1933; Robertson et al., 1978; Menges and Waller, 1983; Friedman et al., 1996). Interactions among disturbance, sediment transport, and vegetation growth often lead to development of distinct fluvial surfaces occupied by distinct plant communities (Hupp and Osterkamp, 1985; Harris, 1987). Parallel to the stream, the character of flood disturbance varies in response to changes in sediment supply, lateral constraint, stream gradient, and tributary influence (Hupp, 1988; Nilsson, 1987).

When a flood disturbs part of a bottomland, the resulting new patches of moist, bare ground are colonized by vegetation. As the river moves and sediment is deposited or eroded from these patches, the local flood disturbance regime changes (Wohl, Chapter 6, this volume). Vegetation on such patches develops over time in response to the changing disturbance regime, the changing substrate, and biotic interactions. Although the new area made available by flooding may be small in any given year, the processes of vegetation change begun on such patches continue for decades or centuries, and movement of the channel can result in a complex mosaic of patches with different disturbance histories and vegetation (White, 1979; Pickett and White, 1985; Hupp, 1992; Baker and Walford, 1995; Friedman et al., 1996). This flood-generated mosaic may occupy a large proportion of the land surface, increasing both landscape and species diversity. For example, in a 0.5×10^6 km^2 area of the Peruvian Amazon Basin, 12% of the total land area is occupied by river floodplains and an additional 14.6% is occupied by older fluvial terraces (Salo et al., 1986).

Flood-Related Strategies of Bottomland Herbs

Floods affect bottomland vegetation directly by controlling the frequency, intensity, and duration of physical disturbance and indirectly by influencing sediment particle size, seed availability, and availability of light, moisture, and nutrients. These factors vary greatly in a bottomland in space and time, resulting in a variety of environments, many of which are unique to bottomlands. This variety is responsible for the high species richness typical of bottomlands (Gregory et al., 1991; Pollock et al., 1998).

The influences of flooding lead to a predictable sequence of herbaceous species along a bottomland cross section. Immediately adjacent to the channel, flood frequency, intensity, and duration are too great for long-term plant survival, and the dominant herbs are ruderal species such as smartweed (*Polygonum punctatum*) that are capable of carrying out their life cycle between floods, usually in a single growing season (Grime, 1977; Menges and Waller, 1983). They allocate a large proportion of their resources to producing seeds that are dispersed to other disturbed sites. The rapid growth employed in the ruderal strategy requires abundant nutrients. In some bottomlands of Ontario there is no ruderal community adjacent to the channel, apparently because nutrient scarcity prevents plants from completing their life cycle between floods (Day et al., 1988).

At moderate heights above the channel are plants that can tolerate physical disturbance and anoxia, such as the sedges (*Carex* spp.). These plants survive flood disturbance because of their small size, thin flexible leaves, and placement of rapidly dividing cells close to the ground. To survive the anoxia associated with prolonged inundation, these plants often have hollow stems for transport of oxygen to roots and chemical pathways for detoxification of the products of anaerobic respiration (Gill, 1970).

At greater heights, physical disturbance and anoxia are less important, moisture is abundant, and nutrients have time to accumulate in the soil. Here the dominant herbs are fast-growing and tall and can often spread laterally by root sprouts or runners, characteristics that enable them to monopolize available resources in the absence of physical disturbance (Menges and Waller, 1983; Wilson and Keddy, 1986). Typical species include nettle (*Urtica dioica*) and sunflower (*Helianthus* spp.) (Menges and Waller, 1983). Because trees may be important in this zone, many of the herbs are tolerant of shade.

At still greater heights moisture or nutrient subsidies from the stream may be reduced. In an arid or semiarid climate such surfaces may be prone to drought stress. For example, in the plains of eastern Colorado, the dominant herbs on these surfaces, such as sand dropseed (*Sporobolus cryptandrus*), have adaptations for drought avoidance and water and nutrient conservation including taproots and involute leaves (Friedman et al., 1996).

A. Flood intensity low, flood duration long, climate humid

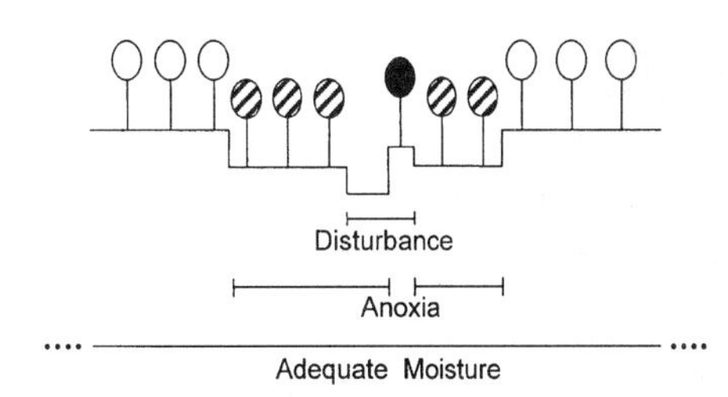

B. Flood intensity high, flood duration short, climate arid

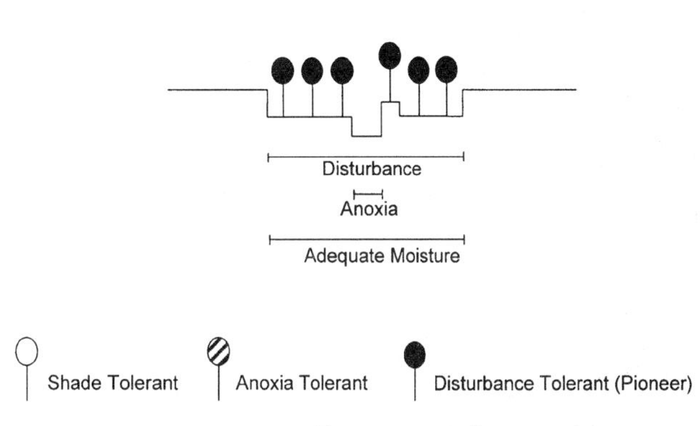

Figure 8.1. Influence of environmental stresses on the distribution of bottomland trees with different strategies. Bars under cross sections show the extent of physical disturbance, anoxia, and adequate moisture for survival of trees. Although the three tree strategies are presented as distinct, many species combine strategies. The term shade tolerant refers to trees that are able to reproduce in the shade of other trees in the absence of physical disturbance. (A) Cross section typical of bottomlands in low-gradient streams in the Coastal Plain of the southeastern United States (Bedinger, 1971). (B) Cross section typical of bottomlands in the western Great Plains (Friedman et al., 1997).

Regional Patterns in Bottomland Tree Communities

The success of any particular plant strategy with respect to floods depends on the frequency, intensity, and duration of flooding, which are controlled by geology, climate, and position in the bottomland. For example, strategies for tolerating anoxia are useful only where flood intensity is low enough to allow plant survival, and flood duration is high enough to produce anoxia.

Regional patterns in flood intensity, flood duration, and drought stress can help to explain regional differences among bottomland forests. In humid regions trees can survive without moisture provided by a stream; therefore, forest occupies almost all the bottomland, an area that is typically much larger than the zone disturbed by the stream in any year (Figure 8.1). Forests in the zone of intense riverine disturbance are dominated by pioneer species: trees and shrubs that release abundant wind- or water-dispersed seeds early every growing season near the time of peak flows. These species grow rapidly, are able to resprout after breakage or burial by floods, and tolerate a moderate amount of anoxia (Sigafoos, 1964; Yanosky, 1982), but they require abundant water and light (White, 1979; Salo et al., 1986). Examples of pioneers in North America include *Acer saccharinum, Betula nigra, Populus deltoides*, and *Salix nigra* (Hupp, 1992). Examples in the western Amazon Basin include *Tessaria integrifolia* and *Gynerium sagittatum* (Kalliola et al., 1991). Away from the zone of heavy disturbance are shade-tolerant trees, often with larger animal-dispersed seeds. In North America these include *Quercus laurifolia* and *Magnolia grandiflora*. These species are slower at colonizing new sites but are able to become established under existing trees in the shade. Where gradients are low and bottomlands wide – for example along rivers in the Coastal Plain of the southeastern United States – there may be a broad area that is not subject to physical disturbance from the stream but that is subject to extended inundation and anoxia. This zone is dominated by trees able to tolerate the stress of anoxia, such as *Taxodium distichum* and *Nyssa aquatica*. In this situation tolerance of anoxia is often considered the most important characteristic affecting bottomland plant distribution (Bedinger, 1971; Wharton et al., 1982). Pioneer tree species occupy a small proportion of the bottomland (Figure 8.1), and many tree species become established in light gaps provided by disturbances other than riverine deposition and erosion, including fire, wind, and extended inundation.

In arid and semiarid regions bottomland forests are narrower because the area capable of supporting trees is limited to a zone adjacent to and receiving moisture from the river (Hughes, 1994). Relatively high flow variability results in a wider zone of riverine disturbance and a narrower zone wet enough throughout the growing season to support trees. In this situation, much or all of the forested zone may be within the area that is frequently disturbed by the river. As a result pioneers such as cottonwood (*Populus* spp.) and willow (*Salix* spp.) can be the dominant and are sometimes the only trees present (Figure 8.1) (Friedman et al., 1997). Plant stress from anoxia occurs in bottomlands of dry regions (Friedman and Auble, 1999) but is often limited in extent (Auble et al., 1994) because flood peaks are of relatively short duration (Baker, 1977; Wolman and Gerson, 1978). Although flow variability in arid watersheds is usually high, many of the large dryland rivers receive much of their flow from relatively humid mountains. The resulting stable flow regime reduces the importance of physical disturbance and increases the importance of shade-tolerant species not requiring disturbance (Hughes, 1990; Friedman et al., 1998).

Figure 8.1 illustrates only two examples of the influence of climate and specific levels of flood intensity and duration on the abundance of tree species with different strategies. Other important combinations occur. For example, many low-energy rivers in humid regions have forest communities like that shown in Figure 8.1A. However, where rivers are higher in energy the zone of disturbance is wider, and pioneer species occupy a larger proportion of the cross section. This combination of high flood intensity, long flood duration, and humid climate is exemplified by the lower Mississippi River before flow regulation and the whitewater rivers of the western Amazon Basin (Kalliola et al., 1991).

Floods and Reproduction of Bottomland Pioneer Trees

Over the past 30 years there have been many studies relating flood disturbance history to the pattern of different-aged trees across a bottomland (Everitt, 1968; Johnson et al., 1976; Bradley and Smith, 1986; Scott et al., 1997). Much of this work has occurred in semiarid interior western North America, where floodplains are dominated by a small number of disturbance-dependent species with distinct annual rings (but see Nanson and Beach, 1977; Gottesfeld and Gottesfeld, 1990; Hupp, 1992; Nanson et al., 1995). This approach is more difficult to apply where floodplain trees reproduce in the absence of flood disturbance or where annual rings are difficult to discern (Hughes, 1990).

Although floods can remove trees, they also promote establishment of pioneer species by providing the necessary bare, moist patches of sediment (White, 1979; Friedman et al., 1997). In nonflood years such patches are formed in locations susceptible to future scouring by ice or water, and long-term survival is therefore low (Figure 8.2) (Auble et al., 1997). Flooding can lead to more long-term survival by forming bare, moist patches in

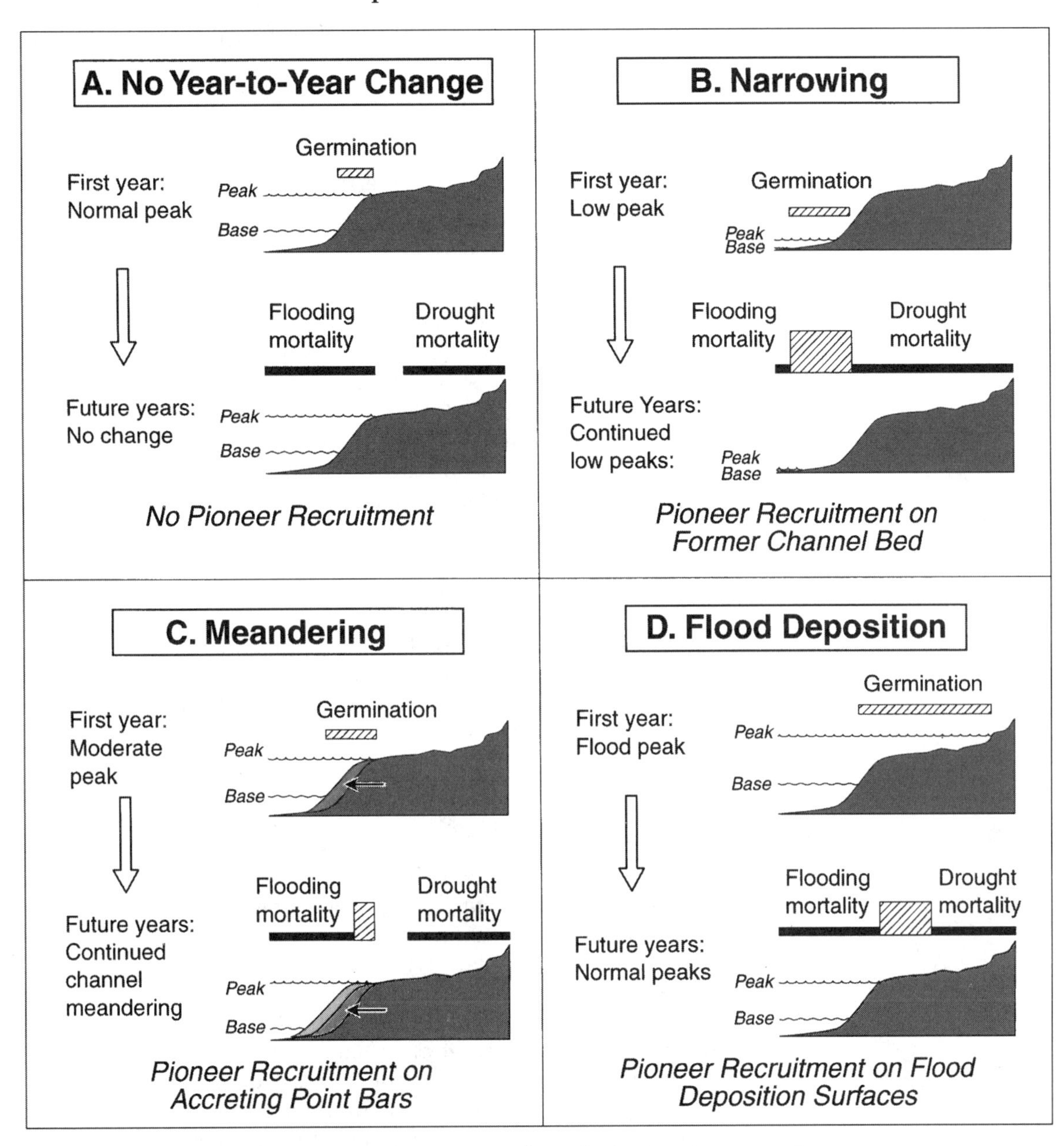

Figure 8.2. Hydrogeomorphic control of recruitment of woody riparian pioneer species. Seed germination, early seedling mortality, and tree recruitment are shown in relation to annual high and low flow lines along a bottomland elevational gradient. Four idealized situations are depicted with a single bottomland cross section. (A) Little or no tree recruitment in the absence of interannual flow variability and channel movement; (B) channel narrowing with recruitment on the former channel bed; (C) recruitment on point bars of a meandering river; and (D) tree recruitment at high elevations associated with infrequent floods and no channel movement. In each of the four situations the cross-hatched area in the upper part indicates the zone of seedling establishment, and the cross-hatched area in the lower part indicates the zone of long-term survival (redrawn from Auble and Scott, 1998).

positions that are relatively safe from future disturbance. Scott et al. (1996) described three examples of flood-related processes that can lead to tree establishment: flood deposition, postflood channel narrowing, and channel meandering (Figure 8.2). These processes are not mutually exclusive and the list is not exhaustive. The purpose of this approach is to emphasize how fluvial processes condition the relation between flow and tree establishment.

Flood Deposition

Because they deposit and erode sediment high above the channel, floods often produce the bare, moist, safe sites suitable for tree establishment (Figure 8.2). This phenomenon occurs on a wide range of streams, but it is especially important where a narrow valley prevents channel movement and the associated creation of relatively low sites safe from future disturbance. In this situation, forests occur as narrow bands, and reproduction occurs only after the largest floods. Along a constrained portion of the Missouri River in central Montana, Scott et al. (1997) excavated and aged plains cottonwood trees, and found that most adult trees were established more than 1.2 m above the lower limit of perennial vegetation and 0–2 years after a flood with a recurrence interval of 9 years or greater. In nonflood years seedlings germinate lower on the bank and are removed by subsequent winter ice and spring floods (Auble and Scott, 1998). The authors argued that maintenance of the frequency and magnitude of extreme floods is essential for reproduction of cottonwoods along this portion of the Missouri River.

Postflood Channel Narrowing

Flood erosion often widens channels. Subsequently, channel narrowing occurs as vegetation, including trees, becomes established on portions of the former channel bed. In this case, vegetation establishment occurs at low positions during periods of relatively low flow (Figure 8.2) (Scott et al., 1996). Postflood narrowing may continue for decades before a narrower equilibrium width is approached (Schumm and Lichty, 1963; Burkham, 1972; Johnson, 1994; Friedman et al., 1996). Thus tree establishment is related to low flows in the short term but is also an indirect result of flood-related widening (Friedman et al., 1996; Stromberg et al., 1997). Postflood channel narrowing is most important along streams subject to large variation in width. Variable width is promoted by a suite of factors that result in low bank strength and high shear stress, especially high flow variability, high channel gradient, high bed sediment load, and a low ratio of valley width to channel width. These characteristics are demonstrated by the braided channels of the High Plains in the southcentral United States (Osterkamp, 1978; Friedman et al., 1998) and by many ephemeral channels in arid regions. Although the focus in this section is on channel narrowing after flood-related widening, narrowing is often caused by other factors including flood control and flow diversion (Petts, 1979; Williams and Wolman, 1984; Friedman et al., 1998), climate change (Schumm, 1969; Nanson et al., 1995), and intrinsic processes of channel change (Nanson, 1986).

Channel Meandering

Where flood shear stresses are low relative to bank strength, channel width is small and stable and postflood channel narrowing is relatively unimportant. This situation is favored by low flow variability, low channel gradient, high suspended sediment load, and a high ratio of valley width to channel width, which decreases flood shear stress by spreading overbank flows over a wide area (Osterkamp, 1978). In these streams channel change occurs less by changes in channel width and more by changes in channel location. A common process in such streams is channel meandering, in which stream bends move progressively outward and downstream while sediment is deposited on the inside of bends on arcuate point bars. Point bars are suitable sites for establishment of bottomland pioneer trees because they are bare and moist, and because progressive channel movement and sediment deposition reduce subsequent local flood disturbance (Figure 8.2) (Bradley and Smith, 1986; Hughes, 1990). Forests consist of parallel arc-shaped bands increasing in age away from the channel (Nanson and Beach, 1977; Bradley and Smith, 1986). Bands of trees are numerous and most are small, reflecting the fact that most channel migration is accomplished by relatively frequent high flows.

Many rivers produce little of the bare, moist habitat suitable for pioneer tree establishment and therefore support only small areas of pioneer forest. For example, some rivers of the coastal plain of the southeastern United States are so low in energy that channel migration is slow, and the rate of formation of bare, moist surfaces is also slow. As a result, pioneer species occupy less area than species that can reproduce in the absence of physical flood disturbance (Bedinger, 1971).

Relative Importance of Floods with Different Recurrence Intervals

Although floods strongly influence the vegetation of most bottomlands, the relative importance of frequent moderate floods and infrequent extreme floods varies with climate and landscape position (Baker, 1977; Wolman and Gerson, 1978). Along large, low-gradient streams in humid climates flow variability tends to be low, and most channel change and pioneer tree establishment occurs during frequent floods. Because any one of these events alters only a small portion of the bottomland (Wolman and Miller, 1960), channel width, forested area, and forest age structure may approach a steady state. When extreme events do occur, channel change is modest because dense vegetation promotes bank strength, and recovery is rapid because flows between extreme events are high enough to readjust the channel (Costa, 1974; Wolman and Gerson, 1978). Meandering channels are typical of streams whose geometry is controlled by frequent, moderate floods (Wolman and Miller, 1960; Hughes, 1994).

Along small or steep streams or in arid climates flow variability is usually higher, and extreme floods tend to have greater and more lasting effects. More channel change and establishment of bottomland forest occurs as a result of infrequent extreme floods that disturb a large portion of the bottomland (Hack and Goodlett, 1960; Burkham, 1972; Wolman and Gerson, 1978; Ish-Shalom-Gordon and Gutterman, 1991). Along these streams, channel width, forested area, and forest age structure are all a function of the magnitude of and time since recent floods (Hughes, 1994; Friedman et al., 1996, 1997). For example, in the late 19th century, the Cimarron River in southwestern Kansas was a 15-m-wide stream with few bottomland trees. A major flood in 1914 widened the channel and destabilized the floodplain, resulting in further widening during subsequent lesser floods. By 1942 the channel was 370 m wide. During the next 12 years, the channel narrowed by half and a large forest of bottomland pioneers was established (Schumm and Lichty, 1963).

Effects of Flood Control on Bottomland Vegetation

Seventy-seven percent of the total discharge of the 139 largest river systems in the northern third of the world is moderately to strongly affected by reservoirs (Dynesius and Nilsson, 1994). Reservoir construction results in immediate and severe impacts on the inundated area upstream and more subtle impacts downstream. In the United States, 5% of the length of major streams has been inundated by large reservoirs (Brinson et al., 1981). Along some extensively regulated rivers, such as the Columbia in Washington, USA, the Missouri in South Dakota and North Dakota, USA (Hitt, 1984), and most large rivers in Sweden (Nilsson, 1984), the river has been transformed into a staircase of reservoirs along which almost all the preexisting bottomland is inundated. Permanent inundation eliminates the bottomland vegetation, replacing it with an aquatic ecosystem (Monosowski, 1986). The new reservoir shoreline includes some bottomland species, but their total area is small because bank slopes are steep. In addition, species richness and cover are typically very low because water-level fluctuations are greater in magnitude than, and different in timing from, those of unregulated rivers (Nilsson, 1984). In sediment-rich systems, long-term reservoir operation produces deltaic deposits that could be managed to promote development of bottomland ecosystems. However, this would require moderation of fluctuations in water levels, which could compromise use of the reservoir for power generation or downstream flood control.

Downstream of reservoirs, the bottomland ecosystem is not immediately eliminated, but it may be altered in many ways. Reservoirs may interfere with plant dispersal by trapping seeds or reducing the high flows that deposit seeds high on the bank in positions suitable for germination and survival (Johansson and Nilsson, 1993). Flood control may alter the timing of peak flows, uncoupling the temporal relation between the period of seed release and the formation of sites suitable for seedling establishment (Fenner et al., 1985; Everitt, 1995). By reducing the exchange between river and floodplain, dam or levee construction may alter temperature, increase salinity, and decrease availability of moisture and nutrients in floodplain soils (Gill, 1973; Ohmart et al., 1988). Reduction of plant growth and decreased habitat quality for animals commonly result (Attwell, 1970; Gill, 1973; Reily and Johnson, 1982; Stromberg and Patten, 1992; Rood et al., 1995).

Dikes and upstream flood-control dams reduce physical disturbance adjacent to a channel, decreasing populations of disturbance-dependent plants (Reich, 1994; Gill, 1973), increasing populations of shade-tolerant species (Johnson et al., 1976; Bravard et al., 1986; Shafroth et al., 1995a) and allowing encroachment of vegetation toward the channel (Turner and Karpiscak, 1980; Nilsson, 1984; Johnson, 1994; Sherrard and Erskine, 1991; Friedman et al., 1997). By reducing flood magnitude and increasing flood duration, flood control may increase the importance of plant stress from anoxia relative to that from physical disturbance (Friedman and Auble, 1999). Decreases in flow variability reduce the spatial and temporal variability of the bottomland environment, resulting in reduced area and diversity of the bottomland plant community (Nilsson, 1984; Auble et al., 1994). Elimination of native species and creation of novel conditions may promote spread of exotic species in bottomlands (Nilsson, 1984).

Channel widening or downcutting may be caused by any flood-control activity that increases the sediment transport capacity of the stream or decreases the sediment load. Such activities include construction of dikes, which increases flow depth; channel straightening, which increases channel gradient; and dam construction, which reduces sediment load. The resulting increase in channel conveyance further isolates the river from its floodplain (Petts, 1979; Williams and Wolman, 1984; Hupp, 1992; Reich, 1994; Marston et al., 1995). In some cases a new floodplain may eventually develop. For example, in the southeastern United States many channels have been straightened, in part to increase flood conveyance. Streams responded by downcutting. Collapse of vertical banks led to channel widening and a new inset floodplain was developed after several decades. The vegetation changes associated with this fluvial process are described by Hupp (1992).

The different effects of flood control may act in combination. For example, along the lower Colorado River and other streams in the dry interior of the western United States, flood control has been associated with the spread of the exotic pioneer shrub saltcedar (*Tamarix* spp.). Like its native counterparts, cottonwood and willow, saltcedar produces abundant wind- and water-dispersed seeds capable of becoming established immediately on bare, moist substrates. Compared with cottonwood and willow, adult saltcedar is more tolerant of high salinity, fire, and drought (Busch and Smith, 1995; Shafroth et al., 1995b) but less tolerant of flood disturbance (Stromberg et al., 1993). Whereas viable seeds of cottonwood and willow are present for only 1–2 months in early summer, saltcedar releases seeds throughout the summer (Warren and Turner, 1975). Flood control and river diversion have fostered increases in saltcedar and decreases in cottonwood and willow by reducing the moisture subsidy and allowing accumulation of salts on the floodplain (Ohmart et al., 1988), decreasing physical disturbance, allowing the accumulation of flammable leaf litter (Ohmart et al., 1988), and delaying peak flows beyond the cottonwood seed viability period (Everitt, 1995). By 1994, saltcedar occupied about 500,000 ha of bottomlands in the United States (Brock, 1994).

Because the influence of floods on bottomlands is mediated by physiographic and climatic factors, the effects of flood control on processes of channel change vary within and among regions. For example, the principal response of wide, shallow braided channels to flood control is channel narrowing (Williams and Wolman, 1984; Johnson, 1994; Church, 1995; Ligon et al., 1995), and the principal response of relatively narrow and deep

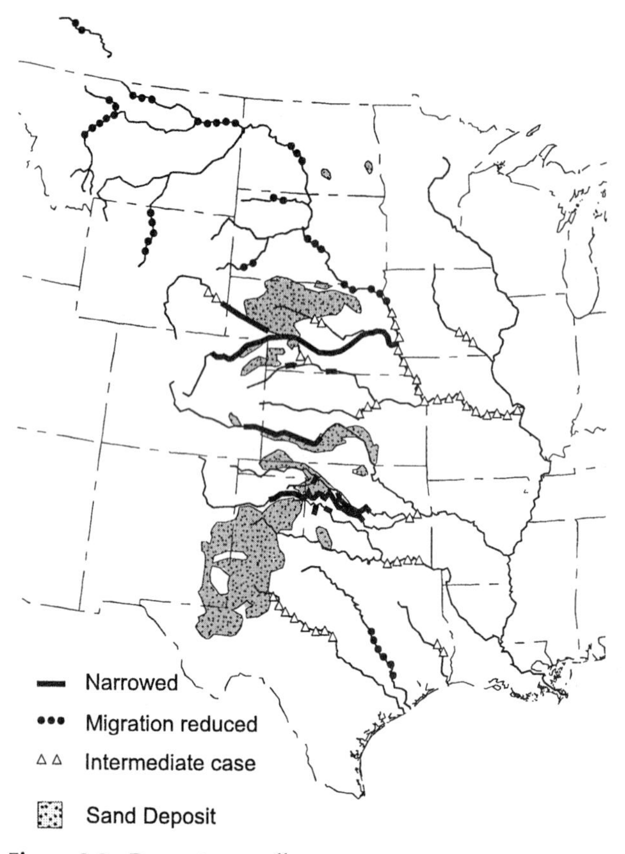

Figure 8.3. Downstream effects of dams on channel width and migration rate in central United States. Location of sand deposits is from Hunt (1979). (Redrawn after Friedman et al., 1998)

meandering channels is a reduction in the channel migration rate (Johnson et al., 1976; Bradley and Smith, 1986; Hughes, 1990). There is a regional pattern in the response of rivers and bottomland forests to flood control in the central United States (Figure 8.3) (Friedman et al., 1998). Before water management, braided channels were most common in the southwestern Plains, where flow variability was high, sand was abundant, and clay and cobbles were relatively scarce; meandering channels were more common to the east and north, where flow variability was low, and clay and cobbles were more abundant. As a result, the principal response to flood control has been channel narrowing in the southwest and reduction in migration rate in the east and north (Figure 8.3) (Friedman et al., 1998).

Channel narrowing is associated with a burst of establishment of native and exotic woody bottomland pioneer species on the former channel bed (Johnson, 1994). Reduction in channel migration rate is associated with a decrease in reproduction of woody bottomland pioneers (Bradley and Smith, 1986). Therefore, bottomland pioneer forests in the southwestern Great Plains have temporarily increased in area after dam construction, while such forests in the northern and eastern plains are declining (Friedman et al., 1998; Johnson, 1998). These regional changes in forest abundance have strongly affected animal communities. For example, almost 90% of the 82 breeding bird species predictably present each spring on the steppe of eastern Colorado were not present in 1900 (Knopf, 1986). The new species use forests that developed during channel narrowing along rivers such as the South Platte as well as trees planted in shelterbelts. At the same time, channel narrowing along the Platte River in Nebraska has degraded the habitat for birds such as the whooping crane (*Grus americana*) and the least tern (*Sterna albifrons*) that use wide braided channels (U.S. Fish and Wildlife Service, 1981).

The current regional contrast in forest development in the central United States is a transient effect of vegetation establishment during the period of channel narrowing along formerly braided streams (Petts, 1979; Church, 1995). The long-term response throughout the Great Plains is a decrease in pioneer species and an increase in shade-tolerant trees, shrubs, and herbs that can reproduce in the absence of physical disturbance (Bravard et al., 1986; Johnson, 1998).

Conclusions

Because of the pervasive influence of floods on bottomland plant communities, the effects of flood control are strong and complex. The response of channel geometry and bottomland vegetation to flood control are conditioned by local fluvial processes, which are determined by the geologic and climatic factors that govern flow variability and sediment load. Construction of upstream reservoirs, bank stabilization, and levee construction decrease flow variability and isolate a river from its floodplain. This decreases the physical disturbance on the bottomland and stabilizes water availability, thereby decreasing habitat complexity on the bottomland. Over the long term, disturbance-dependent forest species typically decline and shade-tolerant species increase; however, in the decades after imposition of flood control, adjustments of channel form such as narrowing may cause a transient pulse of the establishment of species that require bare ground for establishment.

Although the occurrence of floods is commonly regarded as a hazard to human communities, floods are usually necessary for maintenance of biotic diversity on bottomlands. By changing the magnitude, frequency, or duration of floods, flood control often constitutes a hazard to riparian vegetation.

Acknowledgments

M. Scott and P. Shafroth have made major contributions to both conceptual and procedural aspects of the work presented here. V. Lee prepared one of the figures. We thank B. Everitt, C. Hupp, W.C. Johnson, and E. Wohl for constructive reviews. This manuscript was prepared by employees of the U.S. Department of Interior, U.S. Geological Survey, and, therefore, may not be copyrighted.

References

Arcement, G.J. and Schneider, V.R. (1989). *Guide for selecting Manning's roughness coefficients for natural channels and flood plains.* U.S. Geological Survey Water Supply Paper 2339. Washington, DC: U.S. Government Printing Office.

Attwell, R.I. (1970). Some effects of Lake Kariba on the ecology of a flood plain of the mid-Zambezi valley of Rhodesia. *Biological Conservation*, **2**, 189–196.

Auble, G.T., Friedman, J.M., and Scott, M.L. (1994). Relating riparian vegetation to present and future streamflows. *Ecological Applications*, **4**, 544–554.

Auble, G.T. and Scott, M.L. (1998). Fluvial disturbance patches and cottonwood recruitment along the upper Missouri River, Montana. *Wetlands*, **18**, 546–556.

Auble, G.T., Scott, M.L., Friedman, J.M., Back, J., and Lee, V.J. (1997). Constraints on establishment of plains cottonwood in an urban riparian preserve. *Wetlands*, **17**, 138–148.

Baker, V.R. (1977). Stream-channel response to floods with example from central Texas. *Geological Society of America Bulletin*, **88**, 1057–1071.

Baker, W.L. and Walford, G.M. (1995). Multiple stable states and models of riparian vegetation succession on the Animas River, Colorado. *Annals of the Association of American Geographers*, **85**, 320–338.

Bedinger, M.S. (1971). Forest species as indicators of flooding in the lower White River Valley, Arkansas. *U.S. Geological Survey Professional Paper 750-C*: C248-C253.

Bradley, C.E. and Smith, D.G. (1986). Plains cottonwood recruitment and survival on a prairie meandering river floodplain, Milk River, southern Alberta and northern Montana. *Canadian Journal of Botany*, **64**, 1433–1442.

Bravard, J.P., Amoros, C., and Pautou, G. (1986). Impact of civil engineering works on the successions of communities in a fluvial system. *Oikos*, **47**, 92–111.

Brinson, M.M., Swift, B.L., Plantico, R.C., and Barclay, J.S. (1981). *Riparian ecosystems: their ecology and status.* U.S. Fish and Wildlife Service Biological Services Program FWS/OBS-81/17. Washington, DC: U.S. Government Printing Office.

Brock, J.H. (1994). *Tamarix* spp. (Salt Cedar), an invasive exotic woody plant in arid and semi-arid riparian habitats of western USA. In *Ecology and management of invasive riverside plants*, eds. L.C. de Waal, L.E. Child, P.M. Wade, and J.H. Brock, pp. 27–44. New York: John Wiley and Sons, Ltd.

Burkham, D.E. (1972). Channel changes of the Gila River in Safford Valley, Arizona 1846–1970. *U.S. Geological Survey Professional Paper 655-G.*

Busch, D.E. and Smith, S.D. (1995). Mechanisms associated with decline of woody species in riparian ecosystems of the southwestern U.S. *Ecological Monographs*, **65**, 347–370.

Church, M. (1995). Geomorphic response to river flow regulation: case studies and time-scales. *Regulated Rivers: Research and Management*, **11**, 3–22.

Cooke, R.U. and Reeves, R.W. (1976). *Arroyos and environmental change in the American south-west.* Oxford: Clarendon Press.

Costa, J.E. (1974). Response and recovery of a Piedmont watershed from tropical storm Agnes, June 1972. *Water Resources Research*, **10**, 106–112.

Day, R.T., Keddy, P.A., McNeill, J., and Carleton, T. (1988). Fertility and disturbance gradients: a summary model for riverine marsh vegetation. *Ecology*, **69**, 1044–1054.

Dynesius, M. and Nilsson, C. (1994). Fragmentation and flow regulation of river systems in the northern third of the world. *Science*, **266**, 753–762.

Elder, J.F. (1985). Nitrogen and phosphorus speciation and flux in a large Florida river-wetland system. *Water Resources Research*, **21**, 724–732.

Ellery, W.N., Ellery, K., Rogers, K.H., McCarthy, T.S., and Walker, B.H. (1990). Vegetation of channels of the northeastern Okavango Delta, Botswana. *African Journal of Ecology*, **28**, 276–290.

Everitt, B.L. (1968). Use of the cottonwood in an investigation of the recent history of a flood plain. *American Journal of Science*, **266**, 417–439.

(1995). Hydrologic factors in regeneration of Fremont cottonwood along the Fremont River, Utah. In *Natural and anthropogenic influences in fluvial geomorphology*, eds. J.E. Costa, A.J. Miller, K.W. Potter, and P.R. Wilcock, pp. 197–208. Geophysical Monograph 89. Washington, DC: American Geophysical Union.

Fenner, P., Brady, W.W., and Patton, D.R. (1985). Effects of regulated water flows on regeneration of fremont cottonwood. *Journal of Range Management*, **38**, 135–138.

Frangi, J.L. and Lugo, A.E. (1985). Ecosystem dynamics of a subtropical floodplain forest. *Ecological Monographs*, **55**, 351–369.

Friedman, J.M. and Auble, G.T. (1999). Mortality of riparian box elder from sediment mobilization and extended inundation. *Regulated Rivers: Research and Management*, **15**, 463–476.

Friedman, J.M., Osterkamp, W.R., and Lewis, W.M. Jr. (1996). Channel narrowing and vegetation development following a Great-Plains flood. *Ecology*, **77**, 2167–2181.

Friedman, J.M., Osterkamp, W.R., Scott, M.L., and Auble, G.T. (1998). Downstream effects of dams on channel geometry and bottomland vegetation: regional patterns in the Great Plains. *Wetlands*, **18**, 619–633.

Friedman, J.M., Scott, M.L., and Auble, G.T. (1997). Water management and cottonwood forest dynamics along prairie streams. In *Ecology and conservation of Great Plains vertebrates*, eds. F.J. Knopf and F.B. Samson, pp. 49–71. New York: Springer-Verlag.

Furness, H.D. and Breen, C.M. (1980). The vegetation of seasonally flooded areas of the Pongolo River flood plain. *Bothalia*, **13**, 217–231.

Gill, C.J. (1970). The flooding tolerance of woody species – a review. *Forestry Abstracts*, **31**, 671–688.

Gill, D. (1973). Modification of northern alluvial habitats by river development. *Canadian Geographer*, **17**, 138–153.

Glavac, V., Grillenberger, C., Hakes, W., and Ziezold, H. (1992). On the nature of vegetation boundaries, undisturbed flood plain forest communities as an example – a contribution to the continuum/discontinuum controversy. *Vegetatio*, **101**, 123–144.

Gottesfeld, A.S. and Gottesfeld, L.M.J. (1990). Floodplain dynamics of a wandering river, dendrochronology of the Morice River, British Columbia, Canada. *Geomorphology*, **3**, 159–179.

Gregory, S.V., Swanson, F.J., McKee, W.A., and Cummins, K.W. (1991). An ecosystem perspective of riparian zones. *BioScience*, **41**, 540–551.

Grime, J.P. (1977). Evidence for the existence of three primary strategies in plants and its relevance to ecological and evolutionary theory. *The American Naturalist*, **111**, 1169–1194.

Hack, J.T. and Goodlett, J.C. (1960). Geomorphology and forest ecology of a mountain region in the central Appalachians. *U.S. Geological Survey Professional Paper 347.*

Harmon, M.E., Franklin, J.F., Swanson, F.J., Sollins, P., Gregory, S.V., Lattin, J.D., Anderson, N.H., Cline, S.P., Aumen, N.G., Sedell, J.R., Lienkaemper, G.W., Cromack, K., and Cummins, K.W. (1986). Ecology of coarse woody debris in temperate ecosystems. *Advances in Ecological Research*, **15**, 133–302.

Harris, R.R. (1987). Occurrence of vegetation on geomorphic surfaces in the active floodplain of a California alluvial stream. *American Midland Naturalist*, 118, 393–405.

Hey, R.D. and Thorne, C.R. (1986). Stable channels with mobile gravel beds. *Journal of Hydraulic Engineering*, **112**, 671–689.

Hickin, E.J. (1984). Vegetation and river channel dynamics. *Canadian Geographer*, **28**, 111–126.

Hughes, F.M.R. (1990). The influence of flooding regimes on forest distribution and composition in the Tana River floodplain, Kenya. *Journal of Applied Ecology*, **27**, 475–491.

(1994). Environmental change, disturbance and regeneration in semi-arid floodplain forests. In *Environmental change in drylands: biogeographical and geomorphological perspectives*, eds. A.C. Millington and K. Pye, pp. 321–345. New York: John Wiley and Sons, Ltd.

Hunt, C.B. (1979). Surficial geology. Map, 1:7,500,000 scale. *National atlas of the United States of America*. Reston, Virginia: U.S. Geological Survey.

Hupp, C.R. (1988). Plant ecological aspects of flood geomorphology and paleoflood history. In *Flood geomorphology*, eds. V.R. Baker, C. Kochel, and P.C. Patton, pp. 335–356. New York: Wiley Interscience.

(1992). Riparian vegetation recovery patterns following stream channelization: a geomorphic perspective. *Ecology*, **73**, 1209–1226.

Hupp, C.R. and Osterkamp, W.R. (1985). Bottomland vegetation distribution along Passage Creek, Virginia, in relation to fluvial landforms. *Ecology*, **66**, 670–681.

Illichevsky, S. (1933). The river as a factor of plant distribution. *Journal of Ecology*, **21**, 436–441.

Ish-Shalom-Gordon, N. and Gutterman, Y. (1991). Soil disturbance by a violent flood in Wadi Zin in the Negev Desert Highlands of Israel. *Arid Soil Research and Rehabilitation*, **5**, 251–260.

Johansson, M.E. and Nilsson, C. (1993). Hydrochory, population dynamics and distribution of the clonal aquatic plant *Ranunculus lingua*. *Journal of Ecology*, **81**, 81–91.

Johnson, W.C. (1994). Woodland expansion in the Platte River, Nebraska: patterns and causes. *Ecological Monographs*, **64**, 45–84.

(1998). Adjustment of riparian vegetation to river regulation in the Great Plains, USA. *Wetlands*, **18**, 608–618.

Johnson, W.C., Burgess, R.L., and Keammerer, W.R. (1976). Forest overstory vegetation and environment on the Missouri River floodplain in North Dakota. *Ecological Monographs*, **46**, 59–84.

Kalliola, R., Salo, J., Puhakka, M., and Rajasilta, M. (1991). New site formation and colonizing vegetation in primary succession on the western Amazon floodplains. *Journal of Ecology*, **79**, 877–901.

Knopf, F.L. (1986). Changing landscapes and the cosmopolitism of the eastern Colorado avifauna. *Wildlife Society Bulletin*, **14**, 132–142.

Krasny, M.E., Zasada, J.C., and Vogt, K.A. (1988). Adventitious rooting of four *Salicaceae* species in response to a flooding event. *Canadian Journal of Botany*, **66**, 2597–2598.

Kubitzki, K. and Ziburski, A. (1994). Seed dispersal in floodplain forests of Amazonia. *Biotropica*, **26**, 30–43.

Ligon, F.K., Dietrich, W.E., and Trush, W.J. (1995). Downstream ecological effects of dams; a geomorphic perspective. *BioScience*, **45**, 183–192.

Marston, R.A., Girel, J., Pautou, G., Piegay, H., Bravard, J., and Arneson, C. (1995). Channel metamorphosis, floodplain disturbance, and vegetation development: Ain River, France. *Geomorphology*, **13**, 121–131.

McCarthy, T.S., Ellery, W.N., and Stanistreet, I.G. (1992). Avulsion mechanisms on the Okavango Fan, Botswana: the control of a fluvial system by vegetation. *Sedimentology*, **39**, 779–795.

Menges, E.S. (1990). Population viability analysis for an endangered plant. *Conservation Biology*, **4**, 52–62.

Menges, E.S. and Waller, D.M. (1983). Plant strategies in relation to elevation and light in floodplain herbs. *The American Naturalist*, **122**, 454–473.

Mitsch, W.J., Dorge, C.L., and Wiemhoff, J.R. (1979). Ecosystem dynamics and a phosphorus budget of an alluvial cypress swamp in southern Illinois. *Ecology*, **60**, 1116–1124.

Mitsch, W.J. and Gosselink, J.G. (1986). *Wetlands*. New York: Van Nostrand Reinhold.

Monosowski, E. (1986). Brazil's Tucurui Dam: development at environmental cost. In *The social and environmental effects of large dams, volume two: case studies*, eds. E. Goldsmith and N. Hildyard, pp. 191–198. Camelford, Cornwall: Wadebridge Ecological Centre.

Nanson, G.C. (1986). Episodes of vertical accretion and catastrophic stripping: a model of disequilibrium flood-plain development. *Geological Society of America Bulletin*, **97**, 1467–1475.

Nanson, G.C., Barbetti, M. and Taylor, G. (1995). River stabilization due to changing climate and vegetation during the late Quaternary in western Tasmania, Australia. *Geomorphology*, **13**, 145–158.

Nanson, G.C. and Beach, H.F. (1977). Forest succession and sedimentation on a meandering-river floodplain, northeast British Columbia, Canada. *Journal of Biogeography*, **4**, 229–251.

Nevins, T.H.F. (1969). River training–the single-thread channel. *New Zealand Engineering*, December 1969, 367–373.

Nilsson, C. (1984). Effect of stream regulation on riparian vegetation. In *Regulated Rivers*, eds. A. Lillehammer and S.J. Saltveit, pp. 93–106. Oslo: Universitetsforlaget AS.

(1987). Distribution of stream-edge vegetation along a gradient of current velocity. *Journal of Ecology*, **75**, 513–522.

Nilsson, C., Ekblad, A., Dynesius, M., Backe, S., Gardfjell, M., Carlberg, B., Hellqvist, S., and Jansson, R. (1994). A comparison of species richness and traits of riparian plants between a main river and its tributaries. *Journal of Ecology*, **82**, 281–295.

Nilsson, C., Gardfjell, M., and Grelsson, G. (1991). Importance of hydrochory, in structuring plant communities along rivers. *Canadian Journal of Botany*, **69**, 2631–2633.

Ohmart, R.D., Anderson, B.W., and Hunter, W.C. (1988). The ecology of the lower Colorado River from Davis Dam to the Mexico–United States International Boundary: a community profile. *U.S. Fish and Wildlife Service Biological Report* 85 (7.19).

Oplatka, M. and Sutherland, A. (1995). Tests on willow poles used for river bank protection. *Journal of Hydrology (NZ)*, **33**, 35–58.

Osterkamp, W.R. (1978). Gradient, discharge, and particle-size relations of alluvial channels in Kansas, with observations on braiding. *American Journal of Science*, **278**, 1253–1268.

Osterkamp, W.R. and Costa, J.E. (1987). Changes accompanying an extraordinary flood on a sandbed stream. In *Catastrophic flooding*, eds. L. Mayer and D. Nash, pp. 201–224. Boston: Allen and Unwin.

Petts, G.E. (1979). Complex response of river channel morphology subsequent to reservoir construction. *Progress in Physical Geography*, **3**, 329–362.

Pickett, S.T.A. and White, P.S. (1985). *The Ecology of Natural Disturbance and Patch Dynamics*. Orlando, Florida: Academic Press.

Pollock, M.M., Naiman, R.J., and Hanley, T.A. (1998). Plant species richness in riparian wetlands – a test of biodiversity theory. *Ecology*, **79**, 94–105.

Prosser, I.P., Dietrich, W.E., and Stevenson, J. (1995). Flow resistance and sediment transport by concentrated overland flow in a grassland valley. *Geomorphology*, **13**, 71–86.

Reich, M. (1994). Les impacts de l'incision des rivières des alpes bavaroises sur les communautés terrestres de leur lit majeur. *Revue de géographie de Lyon*, **69**, 25–30.

Reily, P.W. and Johnson, W.C. (1982). The effects of altered hydrologic regime on tree growth along the Missouri River in North Dakota. *Canadian Journal of Botany*, **60**, 2410–2423.

Robertson, P.A., Weaver, G.T., and Cavanaugh, J.A. (1978). Vegetation and tree species patterns near the northern terminus of the southern floodplain forest. *Ecological Monographs*, **48**, 249–267.

Rood, S.B., Hillman, C., Sanche, T., and Mahoney, J.M. (1994). Clonal reproduction of riparian cottonwoods in southern Alberta. *Canadian Journal of Botany*, **72**, 1766–1770.

Rood, S.B., Kalischuk, A.R., and Mahoney, J.M. (1998). Initial cottonwood seedling recruitment following the flood of the century of the Oldman River, Alberta, Canada. *Wetlands*, **18**, 557–570.

Rood, S.B., Mahoney, J.M., Reid, D.E., and Zilm, L. (1995). Instream flows and the decline of riparian cottonwoods along the St. Mary River, Alberta. *Canadian Journal of Botany*, **73**, 1250–1260.

Salo, J., Kalliola, R., Häkkinen, I., Mäkinen, Y., Niemelä, P., Puhakka, M., and Coley, P.D. (1986). River dynamics and the diversity of Amazon lowland forest. *Nature*, **322**, 254–258.

Sanchez-Perez, J.M., Tremolieres, M., and Carbiener, R. (1991). Une station d'épuration naturelle des phosphates et nitrates apportés par les eaux de débordement du Rhin: la forêt alluviale à frêne et orme. *Comptes Rendus* Serie III, **312**, 395–402.

Schneider, R.L. and Sharitz, R.R. (1988). Hydrochory and regeneration in a bald cypress-water tupelo swamp forest. *Ecology*, **69**, 1055–1063.

Schumm, S.A. (1969). River metamorphosis. *Journal of the Hydraulics Division, American Society of Civil Engineers*, **95**, 255–273.

Schumm, S.A. and Lichty, R.W. (1963). Channel widening and floodplain construction along Cimarron River in southwestern Kansas. *U.S. Geological Survey Professional Paper 352-D.*

Scott, M.L., Auble, G.T., and Friedman, J.M. (1997). Flood dependency of cottonwood establishment along the Missouri River, Montana, USA. *Ecological Applications*, **7**, 677–690.

Scott, M.L., Friedman, J.M., and Auble, G.T. (1996). Fluvial process and the establishment of bottomland trees. *Geomorphology*, **14**, 327–339.

Scott, M.L., Shafroth, P.B., and Auble, G.T. (1999). Responses of riparian cottonwoods to alluvial watertable declines. *Environmental Management*, **23**, 347–358.

Sedell, J.R. and Froggatt, J.L. (1984). Importance of streamside forests to large rivers: the isolation of the Willamette River, Oregon, U.S.A. from its floodplain by snagging and streamside forest removal. *Verhandlungen der International Verein Limnology*, **22**, 1828–1834.

Shafroth, P.B., Auble, G.T., and Scott, M.L. (1995a). Germination and establishment of the native plains cottonwood (*Populus deltoides* Marshall subsp. *monilifera*) and the exotic Russian-olive (*Elaeagnus angustifolia* L.). *Conservation Biology*, **9**, 1169–1175.

Shafroth, P.B., Friedman, J.M. and Ischinger, L.S. (1995b). Effects of salinity on establishment of *Populus fremontii* (cottonwood) and *Tamarix ramosissima* (saltcedar) in southwestern United States. *Great Basin Naturalist*, **55**, 58–65.

Shafroth, P.B., Scott, M.L., Friedman, J.M., and Laven, R.D. (1994). Establishment, sex structure, and breeding system of an exotic riparian willow, *Salix X rubens*. *American Midland Naturalist*, **132**, 159–172.

Sherrard, J.J. and Erskine, W.D. (1991). Complex response of a sand-bed stream to upstream impoundment. *Regulated Rivers: Research and Management*, **6**, 53–70.

Sigafoos, R.S. (1964). Botanical evidence of floods and flood-plain deposition. *U.S. Geological Survey Professional Paper 485A.*

Smith, D.G. (1976). Effect of vegetation on lateral migration of anastomosed channels of a glacial meltwater river. *Geological Society of America Bulletin*, **87**, 857–860.

Stevens, L.E. and Waring, G.L. (1985). The effects of prolonged flooding on the riparian plant community in Grand Canyon. In *Riparian ecosystems and their management: reconciling conflicting uses, first North American riparian conference*, eds. R.R. Johnson, C.D. Ziebell, D.R. Patton, P.F. Ffolliott, and R.H. Hamre, pp. 81–86. U.S. Department of Agriculture, Forest Service General Technical Report RM-120. Fort Collins, Colorado: Rocky Mountain Forest and Range Experiment Station.

Stockton, C.W. and Fritts, H.C. (1973). Long-term reconstruction of water level changes for Lake Athabasca by analysis of tree rings. *Water Resources Bulletin*, **9**, 1006–1027.

Stromberg, J.C., Fry, J., and Patten, D.T. (1997). Marsh development after large floods in an alluvial arid-land river. *Wetlands*, **17**, 292–300.

Stromberg, J.C., and Patten, D.T. (1992). Mortality and age of black cottonwood stands along diverted and undiverted streams in the eastern Sierra Nevada, California. *Madrono*, **39**, 205–223.

Stromberg, J.C., Richter, B.D., Patten, D.T., and Wolden, L.G. (1993). Response of a Sonoran riparian forest to a 10-year return flood. *Great Basin Naturalist*, **53**, 118–130.

Thorne, C.R. (1990). Effects of vegetation on riverbank erosion and stability. In *Vegetation and erosion*, ed. J.B. Thornes, pp. 125–144. New York: John Wiley and Sons Ltd.

Turner, R.M. and Karpiscak, M.M. (1980). Recent vegetation changes along the Colorado River between Glen Canyon Dam and Lake Mead, Arizona. *U.S. Geological Survey Professional Paper 1132.*

U.S. Fish and Wildlife Service. (1981). *The Platte River ecology study.* Jamestown, North Dakota: Northern Prairie Wildlife Research Center.

Vannote, R.L., Minshall, G.W., Cummins, K.W., Sedell, J.R., and Cushing, C. (1980). The river continuum concept. *Canadian Journal of Fisheries and Aquatic Sciences*, **37**, 130–137.

Walker, L.R. and Chapin, F.S., III. (1986). Physiological controls over seedling growth in primary succession on an Alaskan floodplain. *Ecology*, **67**, 1508–1523.

Warren, D.K. and Turner, R.M. (1975). Salt cedar (*Tamarix chinensis*) seed production, seedling establishment, and response to inundation. *Journal of Arizona Academy of Sciences*, **10**, 135–144.

Wharton, C.H., Kitchens, W.M., Pendleton, E.C. and Sipe, T.W. (1982). *The ecology of bottomland hardwood swamps of the southeast: a community profile.* United States Fish and Wildlife Service FWS/OBS-81/37. Washington, DC: U.S. Government Printing Office.

White, P.S. (1979). Pattern, process, and natural disturbance in vegetation. *Botanical Review*, **45**, 229–299.

Whitlow, T.H. and Harris, R.W. (1979). *Flood tolerance in plants: a state-of-the-art review.* Technical Report E-79-2. Vicksburg, Mississippi: U.S. Army Engineer Waterways Experiment Station.

Williams, G.P. and Wolman, M.G. (1984). Downstream effects of dams on alluvial rivers. *U.S. Geological Survey Professional Paper 1286.*

Wilson, S.D. and Keddy, P.A. (1986). Measuring diffuse competition along an environmental gradient: results from a shoreline plant community. *American Naturalist*, **127**, 862–869.

Wolman, M.G. and Gerson, R. (1978). Relative scales of time and effectiveness of climate in watershed geomorphology. *Earth Surface Processes*, **3**, 189–208.

Wolman, M.G. and Miller, J.P. (1960). Magnitude and frequency of forces in geomorphic processes. *Journal of Geology*, **68**, 54–74.

Yanosky, T.M. (1982). Effects of flooding upon woody vegetation along parts of the Potomac River flood plain. *U.S. Geological Survey Professional Paper 1206.*

Zimmerman, R.C., Goodlett, J.C., and Comer, G.H. (1967). Influence of vegetation on channel form of small streams. In *Symposium on river morphology*, pp. 255–275. Publication No. 75. Gentbrugge, Belgium: International Association of Scientific Hydrology.

CHAPTER 9

Flooding and Aquatic Ecosystems

Richard S. Wydoski
U.S. Fish and Wildlife Service (retired)
Edmund J. Wick
TETRA TECH, Inc.

Periodic natural flooding is important in maintaining the function and integrity of all aquatic ecosystems even though catastrophic events sometimes pose hazards to these systems. The relation of flooding to large river ecosystems is emphasized in this chapter, but the ecological concepts apply to all aquatic systems. Biological, chemical, and physical linkages between terrestrial, riparian, and aquatic ecosystems are described as well as the ecological effects of various human activities on aquatic ecosystems. A case review involves four endemic fishes that are endangered in the upper Colorado River basin of the western United States. During the past century, water development in this basin to serve agriculture, logging, mining, industry, and domestic uses has altered the timing and magnitude of streamflows. This in turn disrupted sediment transport, riparian vegetation, and connectivity of the river and floodplain. River–floodplain connectivity during historically peak spring flows of the upper basin rivers was an important component of the life history strategy of the razorback sucker. A balance must be attained between land-use practices and consumptive uses of water with preservation and maintenance of ecological function, integrity, productivity, biodiversity, and heterogeneity in large rivers and other aquatic systems. Management decisions related to river–floodplain systems must be made with full consideration of the ecological, economical, political, and sociological factors. A systems approach is recommended for applying "adaptive management" that allows actions to be taken based on the best available information while providing for refinement as pertinent new information becomes available.

Introduction

Freshwater constitutes less than 1% of the Earth's water supply and only about 25% of surface freshwater is contained in streams and rivers (Feth, 1973). Because of their paucity, streams and rivers are especially important compared with other aquatic ecosystems.

The natural streamflow regime of virtually all rivers is inherently variable and this variability is critical to maintaining the integrity of river–floodplain ecosystems (Heede and Rinne, 1990; Poff et al., 1997). Fluctuations in streamflows in a river environment provide a wide array of habitats for plants, invertebrates, amphibians, reptiles, fish, birds, and mammals that occupy aquatic and riparian ecosystems (Clark, 1978; Kozlowski, 1984). Instream flows are important to fish for passage, spawning, incubation, rearing, feeding, and resting (Schnick et al., 1982; Stalnaker and Wick, Chapter 16, this volume). Variable streamflows in rivers are linked directly to the physical, chemical, and biological characteristics of flowing water environments (Church, 1992; Davies et al., 1994).

Flooding of natural aquatic ecosystems occurs when a larger-than-average volume of water enters a water body. Flooding can affect all types of aquatic systems, including wetlands, small streams, large rivers, lakes, estuaries, and seacoasts. The primary reservoir for floodwaters is precipitation as rainfall from the atmosphere and the secondary reservoir is accumulated snow and ice on the surface of landscapes (Baker et al., 1988). The magnitude and frequency of flooding are influenced by the size of the drainage area, climate, amount and rate of precipitation, accumulation of snow and ice, air and soil temperature, amount and species of terrestrial or riparian vegetation, and the ability of soil to absorb and retain water (Baker et al., 1988).

Periodic flooding of aquatic ecosystems is a natural phenomenon to which native flora and fauna are adapted, although organisms can be adversely affected by catastrophic events. Ecological functioning of many wetlands depends on periodic flooding or recharge from groundwater. The riparian zone around lakes can become flooded when water inflow is greater than outflow, but this is generally of short duration without serious effects to plants and animals. Tide fluctuations and flows from streams and rivers cause dynamic natural changes in salinity, water temperature, nutrient transfer, and sedimentation in estuaries. Catastrophic flooding of seacoasts occurs from hurricanes and tidal waves (tsunamis) caused by earthquakes or volcanic eruptions. Natural catastrophic flooding in large rivers can occur from formation and failure of lava dams deposited by volcanoes, as in the Grand Canyon of the Colorado River in southwestern United States (Hamblin, 1990), or failure of ice dams that occurred during the Pleistocene in the northwestern United States (Baker, 1981). The power associated with such flooding can dramatically change landscapes (Baker and Costa, 1987). For example, the failure of glacial Lake Missoula in the northwestern United States released as much as 2000 km^3 of water with peak discharges of 15–17 million m^3 s^{-1} that were 30 thousand times greater than that produced by the largest river on Earth, the Amazon River in South America (Baker and Bunker, 1985).

Flooding in small headwater streams is generally minor compared with flooding in larger streams and rivers because drainage areas of the former are smaller. Small streams are often found in forested landscapes where soils can absorb and retain rainfall and meltwater. Flash flooding occurs in small streams of arid regions when a sudden torrent of turbulent water and sediment from summer thunderstorms in mountains causes erosion of canyons, gulches, and arroyos. These streams are generally ephemeral so aquatic ecosystems do not become permanently established and damage is usually physical.

Generally, flooding of the piedmont reaches of large rivers is more common than in other aquatic systems because of the large volume of water collected from an entire drainage basin. This water may cover a large area of floodplain. Floods in large rivers can result in significant effects on landscapes and aquatic systems.

Resiliency of most aquatic ecosystems damaged by flooding is often very rapid. For example, catastrophic flooding of streams may result in the loss of a year-class of trout (Seegrist and Gard, 1972; Hanson and Waters, 1974; Hoopes, 1975). However, Hanson and Waters (1974) documented the rapid recovery in the standing crop and production rate of brook trout (*Salvelinus fontinalis*) in a flood-damaged stream. In California, survival of 0-age brook trout (a fall spawner) was low in years when fall floods occurred and high in years after spring floods (Seegrist and Gard, 1972). Seegrist and Gard reported that this pattern was reversed for spring-spawning rainbow trout, *Oncorhynchus mykiss*, in the same stream.

This chapter discusses (1) the ecological importance of flooding to aquatic ecosystems with an emphasis on large river systems, (2) effects of human perturbations on aquatic ecosystems, (3) a case review of water

Table 9.1. *Aquatic habitats in a river and a floodplain**

Habitats
Main channel habitats
Deep and shallow runs
Deep and shallow pools
Riffles
Shoreline eddies
Shallow glides
Backwaters
Secondary channels
Large woody debris
Large protruding rocks
Rocky, gravel, and sand bottoms
Somewhat narrow riparian areas
Undercut banks
Floodplain Habitats
Backwaters
Embayments
Ephemeral channels
Lagoons
Lakes (Cutoff or Oxbow Lake, Delta Lake, Ephemeral Lake, Gravel-Pit Lake, Intermittent Lake, Levee Lake, Pluvial Lake, Sink Lake, Vernal Lake)
Ponds (Aestival Pond, Beaver Pond, Depression, Gravel-Pit Pond)
Rather broad riparian areas
Wetlands (aquamarsh, backswamp, bog, carr, emergent wetland, forested wetland, lotic wetland, marsh, scrub-shrub wetland, marsh, wet meadow, wooded swamp)

*Habitats are defined in a glossary by Armantrout (1998).

development impacts on some endemic fishes in the upper Colorado River basin of the western United States, and (4) the need to balance preservation and use of river–floodplain environments.

Ecological Importance of Flooding to Aquatic Ecosystems, with an Emphasis on Large River Systems

Flooding is important in maintaining the function and integrity of aquatic ecosystems even though catastrophic events will periodically pose a hazard to those ecosystems. The concepts describing the ecological importance of flooding to large river ecosystems also apply to other aquatic systems.

Major alterations of large river–floodplain systems from human activities occurred before ecological functioning of such systems was studied (Cummins, 1979). However, it was widely recognized that hydrological, physical, chemical, and thermal modifications in regulated rivers have a negative effect on aquatic organisms (Ward and Stanford, 1995). Today, flooding and floodplains are understood to be essential components of large river systems (Junk et al., 1989; Ward and Stanford, 1983, 1995; Bayley, 1991, 1995; Sparks, 1995). Dynamic river–floodplain ecosystems provide a mosaic of aquatic habitats (Table 9.1), as do associated terrestrial upland areas and riparian corridors that support diverse plant and animal communities (Fontaine and Bartell, 1983). Plant and animal communities adapted to seasonal flooding that created productive shallow habitats and promoted nutrient cycling, which is essential to the ecological integrity and functioning of river–floodplain ecosystems (Copp, 1989; Junk et al., 1989; Bayley, 1995; Johnson et al., 1995; Sparks, 1995; Welcomme, 1995).

The ecological integrity of aquatic ecosystems, especially flowing water systems, must be viewed from a landscape perspective (e.g., a watershed) because rivers and streams are ultimately affected by all land-use practices in a watershed (Foreman and Godron, 1986; Ward and Stanford, 1989; Sparks et al., 1990; Schlosser, 1991; Petts et al., 1992; Wesche, 1993; Stanford et al., 1996; Poff et al., 1997; Williams et al., 1997). Watersheds and riparian ecosystems serve as an ecotone or link between terrestrial and aquatic environments (Brinson et al., 1981; Naiman and Decamps, 1990) as well as the ecotone between surface water and groundwater (Gilbert et al., 1990). Annual and periodic overbank flooding form various types of wetlands in floodplains of rivers. Some of these wetlands are maintained by connectivity with rivers for long periods of time, especially during high, prolonged runoff events. Wetlands serve various functions through their linkage to large river ecosystems (Table 9.2) by providing habitats and food resources that are required by communities of aquatic, avian, and terrestrial animals. Wetlands provide natural flood control by absorbing water during peak streamflows and returning it slowly as flows subside. Floodplains also may abate water pollution by absorbing contaminants as water percolates through the floodplain sediments (Clark, 1978; Sparks, 1995; Finley, Chapter 7, this volume).

Nutrient and Energy Transfer within a Large River Ecosystem

The importance of the land–water interface to the productivity of lotic systems has been recognized for over 25 years (Hynes, 1970, 1983; Karr and Schlosser, 1978; Allan,1995). However, interpretation of the complexity of biological responses (e.g., food webs and interactions of invertebrates and vertebrates) (Hildrew, 1992) and the importance of geomorphological or hydrological processes (Bevin and Carling, 1989; Sparks et al., 1990) has occurred only recently. Rivers are characterized by a one-way flow of water that transports nutrients, sediments, pollutants, and organisms downstream, and all upstream activities affect all downstream reaches (National Research Council, 1992). The concepts of the river continuum (Vannote et al., 1980; Sedell et al., 1989) and flood pulse (Junk et al., 1989) apply to large river systems. Longitudinal transfer of nutrients and energy occurs through the river continuum in small, headwater streams and high-gradient, restricted meander canyon reaches of larger streams. Lateral transfer of nutrients and energy occur through flood pulses in low-gradient, unrestricted reaches of floodplains in broad valley reaches. The major zone of productivity in a floodplain is the "moving littoral" (i.e., a shallow zone that extends from the edge of the waterline to several meters in depth) because it covers the maximum area of a floodplain for a given flood as it traverses the floodplain during inundation and draining (Junk et al., 1989; Mertes, Chapter 5, this volume).

Table 9.2. *Wetland functions in a river–floodplain ecosystem*

Hydrology
Source of water supply
Groundwater recharge
Flood storage
Reduction of water velocities
Sediment deposition
Water quality
Erosion control
Nutrient retention and transformation
Contaminant retention and transformation
Fish and wildlife habitat
Habitat diversity
Diverse types of shallow wetlands
Open water wetlands
Riparian zone
Connectivity of lotic to lentic ecosystems
Transition zone from lotic to lentic environments
Biological diversity
Invertebrates
Reptiles
Amphibians
Fish
Birds
Mammals
Ecosystem processes
Provides diverse habitats for flora and fauna
Maintains natural biological diversity
Maintains natural food chain and food web
Allows for habitat protection
Provides for streamflow mediation
Human benefits
Flood control
Water supply
Recreation
Education
Research

Linkage of Nutrient and Energy Transfer from Floodplains to River Ecosystems

Primary production by phytoplankton and periphyton provides the basis for development of a food web (Allee et al., 1949; Odum and Odum, 1959). Phytoplankton and periphyton production and standing crops increase in concert with increases in annual input of nutrients regardless of latitude. Cycling and spiraling of carbon, nitrogen, and phosphorus – key elements for phytoplankton production – are described in detail by Newbold (1992). Nitrogen is the most abundant element in the atmosphere and is generally not limiting. Also, abundant carbon dioxide in the atmosphere provides the necessary carbon. Phosphorus is the most limiting element in north temperate and subarctic waters (Schindler, 1978). Therefore, phytoplankton production and standing crop in north temperate freshwaters are generally proportional to the phosphorus input. Particulate phosphorus, either

Table 9.3. *Zooplankton estimates from standing waters in floodplains of various northern temperate and tropical river systems**

River	Mean number of zooplankton per liter[†]
Amazon	107–738
Danube	2–10,000
Dvina	248–400
Magdelena	25–5,380
Mekong	39–122
Ping and Nan Rivers	136
Missouri	6.7
Nile	0.2–59
Parana	12–24,000
Oka	950–1,281
Sokoto	16–27
Zambesi	110

*Summarized from Table 4.2 and text by Welcomme (1985).
[†]Ranges of mean numbers are provided when available. However, the values in the table do not necessarily reflect true ranges because of variation in the numbers of samples, gear type, and season.

chemically desorbed or actively mobilized by microbiota, is not readily available in rivers with a high sediment load because most of the phosphorus becomes bound to the sediments (Ellis and Stanford, 1988).

The piedmont reaches of large rivers are often turbid from suspended fine sediments that reduce primary and secondary production. Sediments settle in low-velocity floodplain habitats to provide nutrients and allow sunlight to penetrate clearer water, providing favorable conditions for development and maintenance of a food web (Hynes, 1970; Bayley, 1991; Allan, 1995). Water temperatures become warmer in low-velocity floodplain habitats that aid in food production (Kaeding and Osmundson, 1988). The combination of nutrients, sunlight penetration into the water column, and warmer water temperatures in low-velocity floodplain habitats provide better conditions for primary and secondary production than adjacent turbid and cooler main river channels.

Low-velocity habitats in the shallow littoral zone of floodplains produce higher densities of zooplankton (i.e., higher secondary production) compared with the main river channel (Hynes, 1970; Allan, 1995). Various studies reported zooplankton densities that were 30 (Welcomme et al., 1989) to 100 (Hamilton et al., 1990) times greater in floodplain habitats than the adjacent river channels. Welcomme (1985) reported zooplankton densities in floodplains of temperate and tropical rivers that ranged between 0.2 and 24,000 organisms per liter (Table 9.3). Seasonal pulses of total zooplankton numbers occur because of the cyclic nature of nutrients and differences in life cycles of zooplankton species (Welcomme, 1985). Zooplankton produced in floodplains provide an autochthonous food source to main channels in regulated rivers (Eckblad et al., 1984; Kallemeyn and Novotny, 1977).

Various factors affect the microdistribution of benthic insects in streams, but the most important factors are water velocity and substrate size (Blinn et al., 1995; Minshall et al., 1985; Rabeni and Minshall, 1977; Weisberg

Table 9.4. *Comparison of mean fish biomass estimates from the main channel and floodplain habitats of a northern temperate (Danube) and a tropical (Kafue) river**

River	Habitat	Biomass (kg/ha)	
		High water	Low water
Kafue River, Zambia			
	Main channel	337	204
	Open water lagoon	337	426
	Vegetated lagoon	2682	592
Danube River, Austria			
	Main channel	35	
	Backwater†	350	
	Backwater‡	400	

*Summarized from Tables 6.12 and 6.13 by Welcomme (1985).
†Backwater on Czechoslovak bank.
‡Backwater on Hungarian bank.

et al., 1990). Benthic macroinvertebrates in riffles with large cobble substrates produce a higher diversity of species and generally larger numbers of organisms than river reaches that are composed of sand and silt (Hynes, 1970; Minshall et al., 1985; Welcomme, 1985; Resh et al., 1988). Benthic macroinvertebrates provide a critical pathway for energy transport and utilization between productive floodplains and the main channel of stream ecosystems (Gore, 1985). Benthic invertebrates that inhabit riffles colonize downstream reaches by movement through the substrate, displacement by high streamflows, and drift (Hynes, 1970). Upstream movement by benthic invertebrates is generally 5 to 30% of the downstream drift (Bishop and Hynes, 1969). However, upstream movement is hindered by long reaches of sand-silt substrates and streamflows 12 m^3/s or greater (Luedtke and Brusven, 1970). Benthic invertebrates produced in floodplain habitats move or are flushed into the main channel and backwaters, providing a significant link in the food web for aquatic organisms between the river and floodplain (Eckblad et al., 1984).

Relation of Food Resources to Larval Fish Survival

Many riverine fish species exhibit seasonal movements into inundated floodplain habitats for spawning, rearing, and foraging (Lambou, 1963; Ross and Baker, 1983; Finger and Stewart, 1987). Dramatic declines of 0+ fish, correlated to the loss of floodplain habitat, were reported by Coop (1990). Losses of floodplain habitat were considered to be responsible for the reduction of some native fish species in the Danube River, Austria (Schiemer and Spindler, 1989). A higher riverine fish standing stock was associated with high spring floods in the Atchafalaya floodplain in the lower Mississippi River, United States, whereas the standing stock was lower after low spring floods (Bryan and Sabins, 1979). Similarly, a higher mean fish biomass occurred in floodplain habitats of river–floodplain ecosystems in northern temperate zone rivers, such as the Danube River in Austria, as well as in rivers of the tropics, such as the Kafue River in Zambia, Africa (Table 9.4) (Welcomme, 1985). Although flooding has an immediate effect on the composition of fish communities (Bain, 1990) and juvenile fish are especially susceptible to downstream displacement (Harvey, 1987), these changes are transient and fish communities rapidly return to preflood conditions (Bain, 1990). In some fishes, the lateral movement of fish on floodplains decreases exponentially with reductions in streamflow (Kwak, 1988) that may prevent recruitment, particularly if water levels remain low for several years in succession (Starrett, 1951).

Larval fish must initiate feeding during the critical period before they reach a point of no return (Houde, 1987; Li and Mathias, 1982; Miller et al., 1988) or point of irreversible starvation when high mortality occurs from starvation (May, 1974; Li and Mathias, 1982) as larvae convert from endogenous nutrition (i.e., egg sac) to exogenous feeding (i.e., zooplankton and early instars of benthic invertebrates). Adequate densities of zooplankton and benthic invertebrate prey of the right size are important to the growth and survival of larval fish. At low prey densities, predation as well as intra- and interspecific competition results in high mortality (May, 1974; Welker et al., 1994). The growth rate of larval fishes is extremely important to their survival because smaller larvae that are in poor condition (i.e., starved) remain vulnerable to predation for a longer period of time (Leggett, 1986) and are more susceptible to predation because they have less locomotive ability (Rice et al., 1987). However, fish larvae can resume growth quickly after short periods of starvation if they encounter high densities of food organisms of the right size before they reach their point of irreversible starvation (Miglavs and Jobling, 1989).

The year-class strength of fisheries is often determined by environmental conditions such as suitable water temperature as well as the quality and quantity of food organisms during the critical period for fish larvae (Hjort, 1914, 1926). The density, size, timing, and duration of food availability must match the timing of the swimup stage for fish larvae. During years with optimum environmental conditions, high survival of larval and juvenile fish produces strong year classes. The timing, extent, and duration of flooding greatly affect fish species that use floodplains and these factors may exert a moderate-to-strong control in year-class strength of some fishes (Lambou, 1963; Bayley, 1991; Baker and Killgore, 1994).

Aquatic Habitats in Large River Ecosystems

A wide array of habitats occur in a river–floodplain ecosystem (Table 9.1). Clear, concise definitions of aquatic habitats are summarized by Armantrout (1998). The main channels of most rivers have a tendency to meander (Leopold and Wolman, 1957), producing deep habitats on the outside bends and shallow habitats on the inside bends. Streamflow and size of the materials forming the river bed will sort to form riffles from deposition or pools from scouring. Both types of habitats are required by various life stages of fishes and other aquatic organisms. Many adult fish species require riffles with a specific water velocity and substrate composition for spawning and benthic macroinvertebrate production for food. Pools with lower water velocity and depth provide cover and serve as resting or winter habitats. Shallow productive waters such as backwaters and embayments in rivers and floodplain habitats are required for survival and recruitment of larval and juvenile fishes.

Some braided channels become ephemeral as the high flows of a river subside. Deposition of sediments will occur along the shoreline and fill in the head of a secondary channel in a braided reach. Such secondary ephemeral channels become backwaters that are important as low velocity habitats for various life stages of invertebrate and vertebrate animals. Temporary eddies at the mouths of tributaries to the main channel are regularly used by various adult fish species as a refuge from high water velocity in the main channel during high flows. Likewise, embayments are used as feeding and resting areas by juvenile and adult fish.

Water velocities in floodplain habitats are generally low and are particularly important in the life history strategies of fishes adapted for river–floodplain ecosystems. Although the main channel is used by such species during most of the year, aquatic habitats in floodplains provide an essential habitat for adults during floods as refuges, spawning areas, and feeding areas (Baker and Killgore, 1994; Bayley, 1995; Welcomme, 1985; Welcomme et al., 1989). Flooding produces temporary wetlands but other wetlands remain permanent by water percolation through gravel and sand substrates. The diversity of wetlands (e.g., open water, swamps, emergent wetlands, forested wetlands, and lotic wetlands) (Table 9.2) provide a habitat mosaic that is used by many species of plants and animals. Floodplain depressions (ponds and lakes) that connect with the river during high streamflows are important as nursery and rearing areas for juvenile fish. If the connection between the river and floodplain habitat is lost as the river recedes after flooding, larvae and juvenile fishes may remain in depression ponds and lakes until the next flood when connectivity reoccurs between the river and floodplain. In some instances, total mortality of fish may occur in floodplain depressions that dry up. The function and integrity of floodplains are lost through human activities that prevent connectivity of the lotic and lentic systems (Armoros, 1991; Mellquist, 1992; Petts et al., 1992; Ward and Stanford, 1995; Welcomme, 1995; Stanford et al., 1996; Poff et al., 1997).

Human Perturbations of Large River Ecosystems

Humans settled river corridors and floodplains because rivers provided a means of transportation and floodplains were prime areas for agriculture and industry (Stanford et al., 1996; Poff et al., 1997). Human perturbations of aquatic ecosystems through activities in upland areas, the main channel of rivers, floodplains, and riparian zones adversely affected all types of aquatic ecosystems (Nilsson and Jansson, 1995), including wetlands, streams, rivers, lakes, estuaries, and coastal areas (Table 9.5).

Today, nearly all large rivers in the northern third of the Earth have been dammed and regulated to provide water for multiple uses, including irrigated agriculture, flood control, hydroelectric power, industry, and domestic use (Table 9.5) (Gregory and Walling, 1987; Dynesius and Nilsson, 1994). These rivers have been dramatically altered (Wohl, Chapter 4, this volume) through physical changes (Gore, 1994; Gore and Shields, 1995), chemical changes (Sweeting, 1994), and biological changes (Courtenay, 1993, 1995). Some large tropical river systems remain undammed such as

Table 9.5. *Various anthropogenic perturbations in a river–floodplain ecosystem*

Perturbation
Main channel
Fragmentation of riverine habitat by high- and low-head dams
Conversion of reaches from lotic to lentic environments
Regulated streamflow (disruption of daily and seasonal patterns)
Channelization
Changes in hydrograph (e.g., water quantity with time)
Changes in natural water temperature regime
Reduced water quality (pollution through wastewater discharge; increased contaminant levels)
Disruption of natural sediment transport
Disruption of natural nutrient cycling (i.e., continuum concept)
Loss of diverse lotic habitats
Replacement of native fish with nonnative fish species
Predation and competition by nonnative fishes on native species
Potential for introduction of fish diseases and parasites
Alteration of natural food webs
Floodplain
Disruption of natural river–floodplain connectivity
Disruption of natural nutrient cycling (i.e., flood pulse concept)
Loss of diverse wetland habitats
Loss of fish nursery habitats
Floodplain loss places endangered species in jeopardy of extinction
Decreases the integrity and diversity of faunal and floral communities that occupy floodplain habitats
Increases potential for contaminant accumulation
Increases potential for flooding downstream
Riparian zone
Replacement of native riparian vegetation with nonnative species
Reduced riparian corridor along main channel
Reduced riparian area in floodplain
Loss of integrity and diversity of faunal and floral communities that occupy natural floodplain habitats
Upland areas
Agriculture (depletion in natural water regime and increased erosion disrupting natural sedimentation processes)
Logging (increases surface water runoff and decreases groundwater supplies, increased erosion, reduced forest canopy)
Mining (increased erosion, increases leaching of minerals, increase potential for introducing contaminants into aquatic environments)
Recreation (increased erosion and runoff from road construction, potential for increased pollution of aquatic environments through discharge of wastes and contaminants)

the Amazon River in Brazil, South America, but even this mighty river has been adversely affected by indiscrete logging on upland and floodplain forests (Obeng, 1981; Sparks, 1995). Humans will continue to have a dominating influence on watersheds and floodplains of river ecosystems, altering their natural integrity, productivity, biodiversity, and heterogeneity (Frissell et al., 1993; Stanford et al., 1996).

Human perturbations involving stream regulation adversely affected native plants in various countries – e.g., Sweden (Nilsson and Jansson, 1995), France (Decamps et al., 1995), and the United States (Graf, 1978) – while allowing introduced plants to become established, increase in abundance, and expand their range. Changes in the hydrology of a watershed can be especially detrimental to wetland plants in riparian and floodplain habitats (Mitsch and Gosselink, 1986). The relation of flooding to riparian communities is discussed in detail by Kozlowski (1984) and by Friedman and Auble in Chapter 8 of this volume.

Main Channel

Multiple and varied land-use practices in a watershed affect streamflows (i.e., magnitude of discharge, frequency of occurrence, duration of high flows, and rates of change), sediment transport, and water temperature through changes in the natural hydrograph to which native aquatic life adapted (Baxter and Glaude, 1980; Brooker, 1981; Ligon et al., 1995; Welcomme, 1995; Poff et al., 1997). Such practices include, but are not limited to, construction of dams to store water for later release, dewatering of streams for irrigated agriculture, controlled releases of water (i.e., alteration of natural streamflows) for irrigation and generation of hydroelectric power, channelization to control flooding or road construction, dredging to maintain shipping channels, and reduced water quality through discharge of various pollutants and contaminants into flowing waters. Human activities (e.g., agriculture, grazing, logging, mining, and recreation) on upland areas affect streamflows through decreased retention of water in soils, increased runoff, increased sediment loads into streams, water depletion from irrigated agriculture, and decreased water quality through release of wastewater that contains pollutants or contaminants (Wydoski, 1978, 1980; Wohl, Chapter 4, this volume). Consequently, a need for management of large rivers and their floodplains on an ecosystem basis (i.e., watershed or drainage basin) has been recognized by ecologists and conservationists (Welcomme, 1979, 1995; Fontaine and Bartell, 1983; Avies and Walker, 1986; Naiman and Decamps, 1990; Frissell et al., 1993; Wesche, 1993; Petts, 1994; Allan, 1995; Cairns, 1995; Dunne and Leopold, 1995; Sparks, 1995; Stanford et al., 1996).

Dams convert lotic river reaches into lentic environments and the resulting reservoirs do not provide habitats normally used by native riverine fishes. Many stream fishes evolve with life history strategies adapted to the natural hydrograph (Moyle and Herbold, 1987; Welcomme et al., 1989). Nonnative fish species are introduced to utilize the newly created lentic environments and are often considered more detrimental than beneficial to ecosystems (Courtenay, 1995; Magnuson, 1976; Meffe,1985; Moyle et al., 1986). Sometimes small populations of native fishes will persist in short river reaches with suitable habitat (Stacey and Taper, 1992) but less tolerant species will decrease in number and possibly become extinct (Miller et al., 1989). Benthic communities exhibit patch dynamics due to fragmentation of lotic ecosystems (Pringle et al., 1988) in which the structure and integrity of benthic communities are drastically changed (Ward, 1976; Paine, 1980; Bain et al., 1988; Resh et al., 1988; Hildrew, 1992; Shannon et al., 1994) and natural food webs are dramatically disrupted.

The discharge of pollutants and contaminants through domestic, agricultural, and industrial wastewater is a serious problem in many large river–floodplain ecosystems. For example, decreased water quality from wastewater and industrial discharge was identified as the most important problem in the 1320-km Rhine River that drains a catchment basin of 185,000 km^2 in the European countries of Austria, France, Liechtenstein, Luxemburg, and the Netherlands (van Dijk et al., 1995). This river basin is the most densely populated (54 million inhabitants) and industrialized (~10% of the world's chemical industry) in western Europe. Although pollution of the Rhine River remains a major problem, water quality has improved markedly through ecological rehabilitation efforts that have allowed biological communities to partially recover (van Dijk et al., 1995).

Contaminants affect animal communities for a considerable distance downstream from point sources. For example, the number and density of benthic macroinvertebrates immediately downstream of a single uranium-vanadium mill on the Animas River (a tributary of the Colorado River in the southwestern United States) was decimated by the discharge of heavy metals (Tsivoglov et al., 1959). Although some of the more tolerant species were found about 3 km downstream of the discharge, the species composition and abundance did not recover for about 35 km from the point source of the contamination.

Floodplain

The most significant adverse impact of anthropogenic alterations of a river–floodplain ecosystem is undoubtedly the loss of connectivity with the floodplain that alters the productivity of large river systems through disruptions in nutrient cycling and food webs (Junk et al., 1989; Ward and Stanford, 1989; Bayley, 1991; Power et al., 1995a, 1995b; Sparks, 1995; Welcomme, 1995). Fragmentation of riverine habitat by high- and low-head dams in regulated rivers results in the loss of heterotrophic production (Ryder and Pesendorfer, 1989) that is vital to nutrient and energy transfer in the river continuum (Ward and Stanford, 1983, 1995). Construction of levees, dikes, or berms to control overbank streamflows in regulated rivers further reduces the connectivity of rivers with productive floodplains. The frequency and duration of flooding is reduced in regulated streams, affecting the structure and integrity of native fish and invertebrate communities that form a food web (Junk et al., 1989; Bayley, 1991; Sparks, 1995; Welcomme, 1995). Plants and animals with narrow, specific tolerance limits to various biological, chemical, and physical factors (Shelford's Law of Tolerance described by Odum and Odum, 1959) may decrease or disappear.

Although floodplain habitats contribute to the productivity of a river–floodplain ecosystem, they can also become sinks for various contaminants. Low-velocity habitats such as those in floodplains become deposition areas for large amounts of contaminants in sediments (Graf, 1985). In addition to causing direct mortality of aquatic animals, sublethal effects of contaminants can be significant to various life stages of animals (Table 9.6).

Case Review: Impacts of Water Development on Endangered Fishes in the Upper Colorado River Basin

The Colorado River and its tributaries flow through 2317 km of arid land (Carlson and Muth, 1989) and are separated into two distinct basins (an upper and a lower basin) by the 66-m Glen Canyon Dam that forms Lake Powell (see Figure 1.5 of Wohl, Chapter 1, this volume). This river flows

Table 9.6. *Potential effects of contaminants on various life stages of animals (modified from Sheehan et al., 1984)*

Life stage	Vital life process	Critical effect of contaminants
Egg	Meotic division of cells, fertilization, cleavage mitoses of fertilized egg, hatching, respiration	Gene damage, chromosome abnormalities, damage to egg's membrane, direct toxicity to embryo from contaminant, impaired respiration, reduced hatch
Larva	Metamorphosis, morphological development, feeding, growth, avoidance of predators, susceptibility to parasites and disease	Toxicity from bioaccumulated poisions in yolk sac during early feeding, biochemical changes, physiological damage, deformities, behavioral alterations
Juvenile	Feeding, growth, development of immune system and endocrine glands, avoidance of predators susceptibility to parasites and diseases	Direct toxicity, reduced feeding and growth, altered predator–prey relations, impaired chemoreception, reduced resistance to parasites and diseases
Adult	Feeding, growth, sexual maturation	Direct toxicity, adverse alteration of environmental conditions (e.g., dissolved oxygen), physiological and biochemical changes, behavioral alterations

through an arid to semiarid landscape that is supplied by only 2.9 ha-m of precipitation per km^2; this is less than any other major river in the United States.

During the past century, water development in the upper Colorado River basin (Figure 9.1) to serve agricultural, domestic, industrial, and mining activities altered the natural river ecosystem (Carlson and Muth, 1989; Maddux et al., 1993; Miller et al., 1982; Wydoski, 1980; U.S. Fish and Wildlife Service, 1990a, 1990b, 1991). Dam construction and water storage to serve human needs changed the natural hydrograph through dam operations that released water for seasonal irrigation or for generating power during daily peak-use periods. Historic spring peaks in the hydrograph were dampened (Figure 9.2) and streamflows were increased when the rivers would naturally become low after the spring runoff. Changes in the hydrograph have, in turn, altered aquatic habitats, particularly backwater and floodplain habitats that are considered vital to survival in early life stages of native fishes. Cold-water releases from dams reduced water temperatures of the natural and historic warm-water aquatic ecosystem. Nonnative fish species were introduced into the rivers and manmade reservoirs, both intentionally and accidentally, and this changed the species composition of the fish community. Habitat alteration and nonnative fish introductions were considered to be the two most important factors in the extinction of 40 native North American fishes (27 species and 13 subspecies) during the past century (Miller et al., 1989). These two factors also appear to be extremely important in the decline of the four endangered Colorado River fishes.

Altered rivers in the upper Colorado River basin (Upper Basin) had a major negative effect on some native fishes – to the point where the razorback sucker (*Xyrauchen texanus*), bonytail (*Gila elegans*), Colorado squawfish

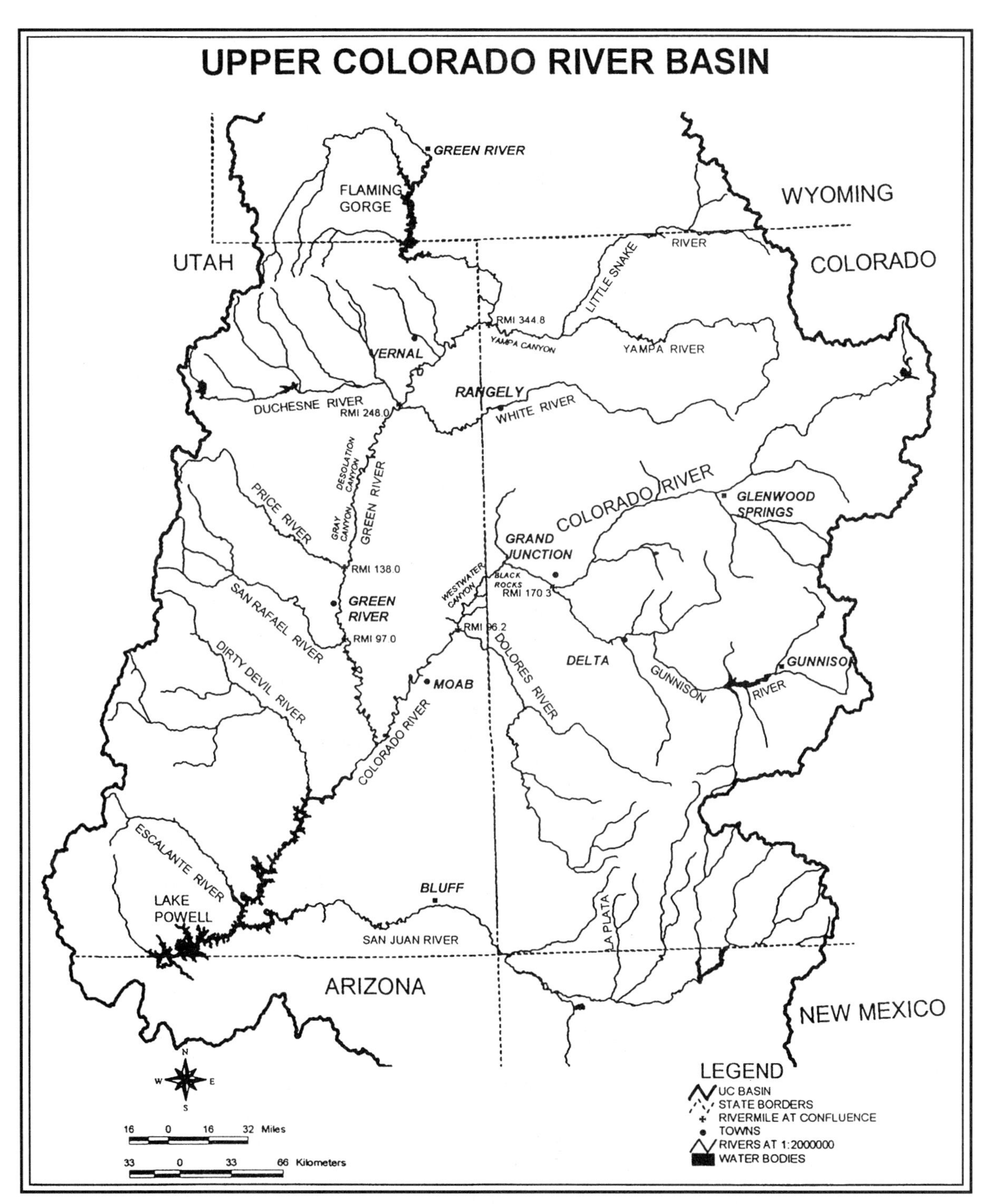

Figure 9.1. Map of the upper Colorado River basin.

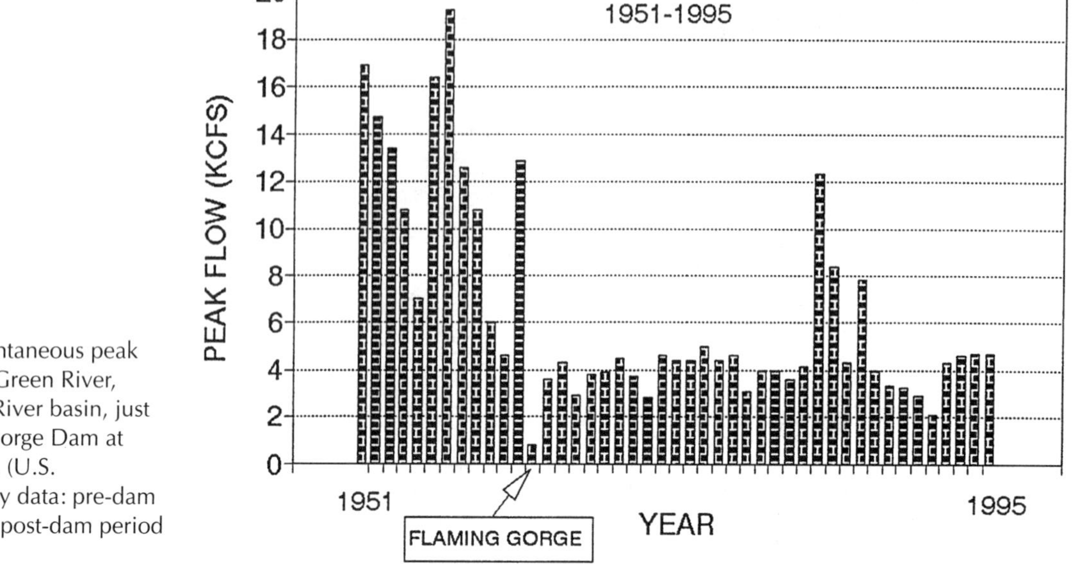

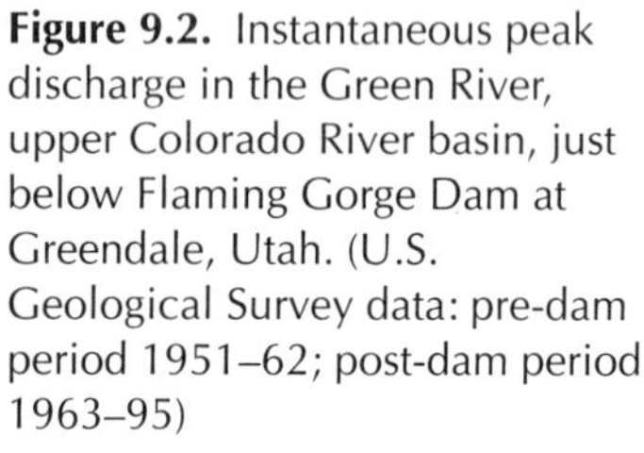
Figure 9.2. Instantaneous peak discharge in the Green River, upper Colorado River basin, just below Flaming Gorge Dam at Greendale, Utah. (U.S. Geological Survey data: pre-dam period 1951–62; post-dam period 1963–95)

(*Ptychocheilus lucius*), and humpback chub (*Gila cypha*) are now listed as endangered under the Endangered Species Act (16 U.S.C. [United States Congress] 1531–1544). Other native fish species (e.g., flannelmouth sucker, *Catostomus latipinnis*; bluehead sucker, *Catostomus discobolus*) are still able to maintain self-sustaining populations, despite the drastically altered condition of the historic river environment. In less-altered river reaches, native fish species still dominate the fish fauna (Anderson, 1997; Burdick, 1995), whereas, in more altered reaches, nonnative fish species are more abundant. This suggests that nonnative species are better suited to the altered ecosystem and can quickly dominate and alter an established native fish community (Li and Moyle, 1993; Meffe, 1985; Stanford et al., 1996).

Importance of Floodplain Habitats to Recovery of Endangered Colorado River Fishes

Although the endangered Colorado River fishes are often considered riverine species, they evolved in a system where lacustrine habitats were part of the natural system. For example, Hamblin (1990) described the formation of natural dams from lava flows in the Grand Canyon. In the Upper Basin, lakes and other lowland areas were regularly flooded. Historically, these off-channel habitats were common in the Upper Basin before dams were constructed and levees built to control the rivers from flooding agricultural lands. Therefore, the endemic Colorado River fishes evolved in a dynamic river system of lotic and lentic aquatic habitats.

The declining numbers of the endangered Colorado River fishes are attributed to the lack of recruitment. For example, the long-lived adult razorback sucker can spawn successfully and produce larvae during some years, but high mortality during the early life stages limits recruitment. In an earlier section, it was emphasized that larval fish must initiate feeding during the critical period after swimup or they will reach a point of irreversible starvation that results in high mortality (Miller et al., 1988). The larvae and juveniles of all four endangered Colorado River fishes feed on zooplankton (Miller et al., 1982) and early instars of benthic invertebrates (Muth et al., 1998). Floodplain habitats are especially important to razorback sucker larvae because this species spawns on the ascending limb of the hydrograph during the spring when small food organisms are scarce in main channels and backwaters of Upper Basin rivers (Wydoski and Wick, 1998). Therefore, emphasis is being placed on razorback sucker in recovery efforts, but all four endangered fishes will benefit from food production in floodplain habitats.

Razorback sucker larvae are 7–9 mm total length (TL) at hatching and 9–11 mm TL at swimup (Muth et al., 1998). The density of zooplankton required for larval razorback sucker survival is between 30 and 60 organisms per fish per day (Figure 9.3) (Papoulias and Minckley, 1990). This number of organisms occurred in floodplain habitats in the Upper Basin but rarely reached that density in backwaters and never reached it in the rivers (Cooper and Severn, 1994a, 1994b, 1994c, and 1994d; Grabowski and Hiebert, 1989; Mabey and Schiozawa, 1993). Thus, floodplain habitats, which produce the highest densities of zooplankton (particularly cladocerans and copepods that are readily eaten), meet the food requirements of razorback suckers during the early life stages in the Upper Basin (Table 9.7).

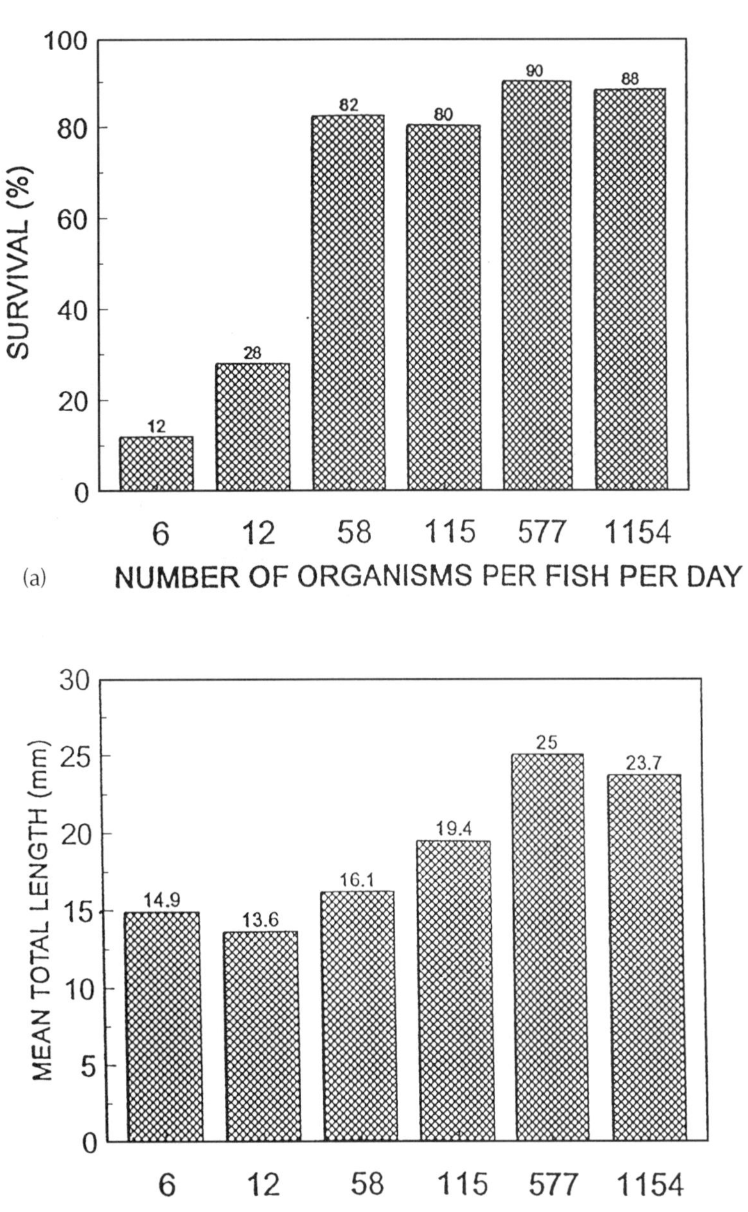

Figure 9.3. Relation of survival (a) and growth (b) of razorback sucker larvae to the density of food organisms after 50 days from swimup in a laboratory. (Adapted from Table 4 of Papoulias and Minckley 1992 [survival] and Table 4 of Papoulias and Minckley 1990 [growth].)

As larval fish become juveniles, larger food organisms become more important. Chironomids constitute a significant part of the diets of juvenile endangered Colorado River fishes (Bestgen, 1990; Grabowski and Hiebert, 1989; U.S. Fish and Wildlife Service, 1990a, 1990b, 1991). The juvenile endangered fishes inhabit low-velocity habitats such as embayments and backwaters (Figure 9.4) along the main channel where the substrate is composed of silt, detritus, and sand – substrates that are generally inhabited by chironomids (Hynes, 1970). The numbers of chironomids in the main channel of the middle Green River ranged between 3500 and 4200 organisms per m^2, between 2300 and 8100 for an ephemeral side channel, and between 9000 and 31,100 for two backwaters (Wolz and Schiozawa, 1995). Floodplain habitats that are inundated for only a short time may contain lower densities

(a)

(b)

Figure 9.4. Various floodplain habitats in the floodplain of the upper Colorado River basin. (a) 1993 flooding of a depression wetland (Wyasket Bottom) on the Ouray National Wildlife Refuge along the middle Green River (potential nursery habitat for larval razorback suckers). (b) Shallow embayment along the Colorado River in Canyonlands (nursery habitat for Colorado squawfish). (c) A 147-ha depression wetland (Old Charley Wash) in the floodplain of the middle Green River (larval razorback suckers survived through the first summer of life in this wetland). (d) Flooded tributary and backwater (Millard Canyon) in the lower Green River (ephemeral habitat for larval razorback suckers).

of chironomids than river backwaters because of a longer life cycle of chironomids compared with the high turnover rate of zooplankton.

Even though survival of larval razorback suckers was high when zooplankton densities were about 60 organisms per fish per day, growth was less until the zooplankton density was 500 organisms per fish per day or more (Figure 9.3). Reduced growth of larval razorback suckers may result in higher mortality through predation by nonnative fishes (Leggett, 1986) because the larvae are in poor condition (i.e., starved) with less locomotive ability (Rice et al., 1987).

Figure 9.4. (*Continued*).

Recruitment of razorback suckers, albeit limited, was documented in floodplain habitats along the middle Green River, downstream of Jensen, Utah. Eight juvenile razorback suckers were captured in a small wetland with abundant aquatic and emergent vegetation that was filled with water from the Green River during a high streamflow in 1993 (Modde and Wick, 1997). The survival of 28 young-of-the-year (YOY) razorback suckers collected in 1995 and 45 YOY collected in 1996 from a 147-ha depression wetland along the Green River (Modde, 1997) demonstrated that larval suckers can survive in floodplain depressions during their first growing

Table 9.7. *Mean number of zooplankton (cladocerans and copepods) from various habitats along the Green River, upper Colorado River basin**

Habitat	Range of mean number of organisms per liter
Main channel†	0–1.3
Backwaters	0–13.1
Floodplain habitats	4.2–81.5

*Mean numbers of zooplankton were obtained from the following studies: Grabowski and Hiebert (1989); Cooper and Severn (1994a, 1994b, 1994c, 1994d) Mabey and Schiozawa (1993). Number of samples taken, sampling gear used, seasons (late spring and summer), and years of sampling differed so ranges were used to illustrate relative productivity of zooplankton.
†Most (Four) investigations were made on the Green River. However, the data include one investigation on the Colorado River and one investigation on the Gunnison River, a major tributary of the Colorado River.

season despite the presence of large numbers of nonnative fishes (631,760 in 1995 and 548,970 in 1996). Over half of these nonnative fish consisted of common carp (*Cyprinus carpio*), fathead minnow (*Pimephales promelas*), red shiner (*Cyprinella lutrensis*), and green sunfish (*Lepomis cyanellus*). Peak zooplankton densities in the wetland produced 43 organisms per liter of water in 1995 and 54 organisms per liter in 1996 (Modde, 1997). This density was within the range required for good survival of razorback sucker larvae (Figure 9.3). The presence of emergent aquatic vegetative cover in this wetland likely increased survival of the YOY razorback suckers by lowering predation by green sunfish, adult red shiners and adult fathead minnows and/or preventing competition by juvenile carp, fathead minnows, and red shiners. Larval razorback suckers could feed on zooplankton with less risk of predation by nonnative fishes in vegetative cover than in open-water habitats. Survival of larval and juvenile razorback suckers in isolated backwaters along Lake Mohave in the lower Colorado River basin was also attributed to dense vegetation (Minckley et al., 1991).

Predation and Competition by Nonnative Fishes on Endangered Fishes

Nonnative fish species compose 76% (42 of 55 species) of the fish community in the Upper Basin (Tyus et al., 1982). In addition, the abundance of nonnative fishes in backwater habitats constitute most of the fish community (96.7 to 99.6%) in the Upper Basin (McAda et al., 1994a, 1994b, 1995).

Predation on larval fish is a significant factor that decreases their survival (Leggett, 1986). Minckley et al. (1991) considered predation by nonnative fishes as the single most important factor in recruitment failure of razorback suckers in Lake Mohave of the lower Colorado River basin. Competition by two species occurs when food is limited, the food is shared, and one of the two species is adversely affected by sharing food (Moyle et al., 1986; Li and Moyle, 1993). Competition is often difficult to document because freshwater fish lack specialization in food habits and there is generally much overlap between species (Larkin, 1956). Low numbers of zooplankton occur in the main channel and backwaters of rivers (i.e., food is limited for early life stages of fish) in the Upper Basin (Table 9.7). This suggests that food for larval razorback suckers is limited in riverine habitats.

Razorback sucker larvae that reach a total length of about 25 mm are too large for minnows such as the red shiner to prey on because the mouths of adult red shiners are gape limited (T. Crowl, Utah State University, 1995, personal communication). Two months or more are required by razorback suckers to become large enough to escape predation by the numerous nonnative minnows in backwaters of the rivers (Figure 9.3). Because of this, it may be necessary to enhance floodplain depressions in the Upper Basin. Periodic inundation (< 2 weeks) under present regulated streamflows may not be enough for sufficient growth by razorback sucker larvae to escape predation by abundant nonnative fishes.

A phenomenon of riverine fishes is the passive downstream drift of larvae after swimup because larvae lack the ability to swim against strong currents (Muth et al., 1998). Drifting larvae would be expected to occupy habitats with low water velocity such as backwaters along the main river channel and off-channel habitats such as floodplain habitats. Predation on razorback sucker larvae by nonnative fishes in backwaters apparently results in extremely high mortality because razorback larvae constitute the largest portion of drifting organisms during the spring.

Physical factors (i.e., habitat) regulate the carrying capacity of a particular aquatic environment (i.e., the ability to support aquatic life). Because the amount of physical habitat available is finite, increases in the number of species present in a particular habitat usually result in smaller populations of most species (Allee et al., 1949). However, native fish species that are adapted to the physical features of natural aquatic habitats are likely to increase if their habitats are enhanced. Many nonnative fishes are extremely adaptable to altered habitats (Li and Moyle, 1993) and have become prevalent in the Upper Basin to the detriment of several native fishes. However, habitat enhancement may provide some recruitment of native fishes even in the presence of well-established nonnative fishes.

Year-Class Strength of Endangered Colorado River Fishes

The four endangered Colorado River fishes are known to be long-lived (Henrickson and Brothers, 1993; Kaeding and Zimmerman, 1983; Miller et al., 1982; McCarthy and Minckley, 1987; Vanicek and Kramer, 1969). Recruitment of all endangered Colorado River fishes was not annual but occurred when environmental conditions favored survival of larval and juvenile fishes. Periodic high survival of fish larvae occurred when the quantity (i.e., density) and quality (i.e., right size) of food organisms were adequate in floodplains during their critical period. This adaptation may have been a life history strategy that evolved in the dynamic natural streamflow regime in rivers of the Upper Basin. Because the endangered Colorado River fishes are long-lived, adequate survival in one of 5 to 10 years should be sufficient to maintain self-sustaining populations after the stocks have become reestablished and adequate spawning, nursery, feeding, and resting habitats are available to them.

Actions Required to Increase Survival of Razorback Sucker Larvae

Various actions involving streamflow management, habitat development and maintenance, management of nonnative fishes, and captive propagation/stocking must be integrated and performed concurrently if self-sustaining populations (i.e., recovery) of razorback suckers are to be realized. Streamflows necessary to provide an adequate frequency of inundation of floodplain habitats must be provided as often as possible to increase survival of razorback sucker larvae during their critical period. Existing floodplain terraces were deposited and maintained by peak spring streamflows that occurred before high dams regulated the flows. Even if existing levees are removed, present managed streamflows in the Upper Basin will not permit inundation of floodplain terraces long enough for razorback sucker larvae to grow sufficiently large to escape predation by nonnative fishes. It may be necessary to excavate existing terraces to allow the rivers to inundate floodplains. If increasing the frequency and duration of flooding through breeching levees, excavating terraces, and increasing streamflows is not possible, then floodplain depressions with abundant aquatic vegetative cover may provide habitat that will prevent extinction of the razorback sucker. Based on the published literature, it is not practical to attempt to control nonnative fishes on a large-scale basis (Wiley and Wydoski, 1994) in Upper Basin rivers. Therefore, nonnative fish management should be emphasized in river reaches that are immediately upstream or downstream of floodplain habitats that are already connected or are reconnected to the rivers. The razorback sucker is nearly extirpated from the upper Colorado River, and captive propagation and stocking will be required to reestablish this species by using genetics conservation guidelines (Williamson and Wydoski, 1994).

Balance of Preservation and Use of River–Floodplain Environments

Water is one of the most precious natural resources on Earth, particularly in arid and semiarid regions. This is reflected in human habitation of river floodplains on all continents. For example, most of the population (65%) in China inhabit river floodplains that constitute about 10% of China's landscape (Dudgeon, 1995). Another example is Bangladesh where 125 million people inhabit 70,000 km^2 of floodplain in the deltas of the Ganges and Brahmaputra Rivers (Wohl, Chapter 1, this volume). In northern temperate regions, river–floodplain ecosystems have been altered so that longitudinal (headwater–river–estuary), lateral (river–floodplain–riparian zone), vertical (surface water–groundwater), and temporal (time scale) dynamic and complex interactive pathways are disrupted (Ward and Stanford, 1989).

Humans will continue to dominate water use of river ecosystems and in the process dramatically reduce the environmental integrity, productivity, biodiversity, and heterogeneity of these ecosystems (Frissell et al., 1993).

Because of strong economic, political, and social pressures from humans for multiple use of large river–floodplain systems for hydroelectric power, agriculture, industry, and municipal uses, it is doubtful that restoration of rivers (i.e., return to a truly original state) (Bradshaw, 1996) is possible. However, it is possible to mitigate (i.e., to moderate), to remediate (i.e., to rectify), or to enhance (i.e., to improve) environmental conditions in a river ecosystem (Stanford et al., 1996). Any habitat enhancement of structure and function of river–floodplain ecosystems will allow improvement of environmental integrity (Stanford et al., 1996; Poff et al., 1997).

Although regulated rivers may be hydraulically optimum for human needs, they are poor ecological systems compared with unaltered rivers. Attempts to enhance or restore ecologically degraded river ecosystems are expensive and often cannot entirely restore all natural ecological conditions (Petts, 1994). Human cultural characteristics, values, perceptions, attitudes, and responses to flooding vary considerably in different parts of the world (White, 1974; Laituri, Chapter 17, this volume) and exclusion of sustainable use of water resources by humans will probably result in failure to maintain or enhance the ecological integrity of aquatic systems (Cairns, 1995). Gore and Shields (1995) provided an example of constructing setback levees along the Danube River in the Czech Republic that provided human safety while recovering some ecological function and value of floodplain habitat. In another example, Heiler et al. (1995) describe how hydrological connectivity and flood pulses will be restored in a free-flowing reach of the Danube River, downstream of Vienna, Austria, as the basis for ecosystem management for a national park.

Power et al. (1995a) pointed out that flooding in the Mississippi River during 1993 "rekindled the national debate [in the United States] over what is to be gained by increasing the height and extent of levees and what has been lost." Aquatic ecosystem management should focus on guiding natural processes of streamflow, erosion, sedimentation, and seasonal flood pulses when making decisions related to river management (Sparks, 1995; Poff et al., 1997). Sparks (1995) stated that "Ecosystem management could actually save money and increase economic efficiency in the long run, because natural functions are restored (e.g., flood storage, conveyance, and moderation; water purification; production of fish and wildlife; and preservation of biodiversity) instead of being maintained by human intervention at great cost and considerable risk of failure." Re-regulation of streamflows and reconnection of floodplains, at least periodically, could reestablish some ecological integrity of river–floodplain ecosystems without significantly compromising human uses (Stanford et al., 1996).

Management decisions related to river–floodplain management must be made with full consideration of the ecological, economical, political, and sociological factors (Table 9.8) to reduce conflicts and avoid polarization of issues regarding human use of water resources and preservation of aquatic systems (Brown, 1992). Effective management of multiple-use rivers through more effective communication among the public, planners, politicians, engineers, and ecologists can benefit people as well as the ecological function and integrity of aquatic systems (Brown, 1992; Hey, 1994). Although uncertainty exists with every management decision because all information required for optimal management decisions usually

Table 9.8. *Primary considerations for maintaining or enhancing the integrity of river–floodplain ecosystems*

- Physical considerations
  - Water supply (quantity of surface water or groundwater)
  - Frequency of inundation
  - River elevation and discharge when floodplain becomes inundated
  - Water depth and duration
  - Control of water exchange and levels in floodplain habitats
  - Water temperature (range and potential for manipulation)
  - Sediment deposition and erosion
- Biological considerations
  - Ecosystem approach
  - Goals and objectives for preservation or enhancement of river–floodplain ecosystem
  - Status and trends of native and nonnative species
  - Reintroductions of captive-reared animals
    - Maintaining genetic integrity and diversity
    - Need for imprinting larval or young fish to source water
    - Batch marks to identify captive-reared animals
    - Size of captive-reared animals at release
    - Timing of release (diel and seasonal)
  - Proper food (e.g., size and abundance) for various life stages; availability for reintroductions
  - Access and exit routes for animals
  - Vegetative cover
  - Control of predators and competitors
    - Nonnative fish, macroinvertebrates, birds, mammals, or other animals
- Chemical considerations
  - Acceptable water quality
    - Chemistry, dissolved oxygen, potential for nitrogen supersaturation from well water
  - Potential contaminants
    - Selenium, heavy metals, pesticides or herbicides
- Economic, political, and sociological considerations
  - Cultural attitudes, perceptions, values
  - Uses of floodplain habitats (human occupation, agriculture, livestock grazing, gravel mining, etc.)
  - Upland activities (agriculture, logging, mining. etc.)

is not available (Ludwig et al., 1993), a systems approach applying adaptive management (Walters and Hillborn, 1978; Walters, 1986) allows actions to be taken based on the best available information while allowing for refinement or adjustments as pertinent new information becomes available.

Intelligent management of ecosystems must involve careful consideration of uses, alternatives, and potential ecological impacts to ensure compatibility among multiple uses of any natural resource (Gore and Petts, 1989; Goldman, 1973; McCloskey, 1973). Natural resource managers can use effective public relations principles to inform the public (Cutler, 1974) and influence politicians in achieving a suitable balance in maintaining ecological function and integrity of aquatic ecosystems while meeting the needs of society (Gilbert, 1971; Dingell, 1972).

References

Allan, J.D. (1995). *Stream ecology: Structure and function of running waters.* London: Chapman and Hall. 388 pp.

Allee, W.C., Emerson, A.E., Park, O., Park, T., and Schmidt, K.P. (1949). *Principles of animal ecology.* Reprinted 1967. Philadelphia: W.B. Saunders Company. 837 pp.

Anderson, R. (1997). *An evaluation of fish community structure and habitat potential for Colorado squawfish and razorback sucker in the unoccupied reach (Palisade to Rifle) of the Colorado River, 1993–1995.* Fort Collins: Colorado Division of Wildlife, Federal Aid Project SE-3. 73 pp.

Armantrout, N.B. (1998). *Glossary of aquatic habitat inventory terminology.* Bethesda, Maryland: American Fisheries Society. 152 pp.

Armoros, C. (1991). Changes in side-arm connectivity and implications for river system management. *Rivers,* **2**, 105–112.

Avies, B.R. and Walker, K.F. (eds.) (1986). *The ecology of river systems.* Dordrecht, Netherlands: W. Junk Publishers. 793 pp.

Bain, M.B. (1990). *Ecology and assessment of warmwater streams: workshop synopsis.* U.S. Department of the Interior, Fish and Wildlife Service, Biological Report **90**(5). 44 pp. Washington, DC.

Bain, M.B., Finn, J.T., and Booke, H.E. (1988). Stream-flow regulation and fish community structure. *Ecology,* **69**, 382–392.

Baker, J.A. and Killgore, K.J. (1994). *Use of a flooded bottomland hardwood wetland in the Cache River system, Arkansas.* U.S. Army Corps of Engineers, Washington, DC: Wetlands Research Program Technical Report WRP-CP-3. 71 pp.

Baker, V.R. (ed.) (1981). *Catastrophic flooding: The origin of the channeled scabland.* Stroudsburg, Pennsylvania: Dowden, Hutchinson, and Ross Publishers. 360 pp.

Baker, V.R. and Bunker, R.L.O. (1985). Cataclysmic late Pleistocene flooding from glacial Lake Missoula: A review. *Quarterly Science Review,* **4**, 1–41.

Baker, V.R. and Costa, J.E. (1987). Flood power. In *Catastrophic flooding,* eds. L. Mayer and D. Nash, pp. 1–24. London: Allen and Unwin Publishers.

Baker, V.R., Kochel, R.C., and Patton, P.C. (eds.) (1988). *Flood geomorphology.* New York: John Wiley and Sons. 503 pp.

Baxter, R.M. and Glaude, P. (1980). *Environmental effects of dams and impoundments in Canada: experience and prospects.* Canadian Fisheries and Aquatic Sciences, Bulletin 205. 34 pp. National Research Council of Canada, Ottawa.

Bayley, P.B. (1991). The flood pulse advantage and the restoration of river-floodplain systems. *Regulated Rivers: Research and Management,* **6**, 75–86.

Bayley, P.B. (1995). Understanding large river-floodplain ecosystems. *BioScience,* **45**, 153–158.

Bestgen, K.R. (1990). *Status review of the razorback sucker, Xyrauchen texanus.* Fort Collins: Larval Fish Laboratory, Colorado State University, Contribution 44. 92 pp.

Bevin, K. and Carling, P. (eds.) (1989). *Floods: Hydrological, sedimentological, and geomorphological implications.* New York: John Wiley and Sons. 290 pp.

Bishop, J.E. and Hynes, H.B.N. (1969). Upstream movement of benthic invertebrates in the Speed River, Ontario. *Journal of the Fisheries Research Board of Canada,* **26**, 279–298.

Blinn, D.W., Shannon, J.P., Stevens, L.E., and Carder, J.P. (1995). Consequences of fluctuating discharge for lotic communities. *Journal of the North American Benthological Society,* **14**, 235–248.

Bradshaw, A.D. (1996). Underlying principles of restoration. *Canadian Journal of Fisheries and Aquatic Sciences,* **53**, Supplement 1, 3–9.

Brinson, M.M., Swift, B.L., Plantico, R.C., and Barclay, J.S. (1981). *Riparian ecosystems: their ecology and status.* U.S. Department of the Interior, Fish and Wildlife Service, FWS/OBS-81-17. 154 pp. Washington, DC.

Brooker, M.P. (1981). The impact of impoundment on the downstream fisheries and general ecology of rivers. In *Advances in applied biology,* ed. T.H. Croaker, **6**, 91–152. London: London.

Brown, C.N. (1992). "Old friends and new kids" in multiobjective management. *Proceedings of the Annual Conference of the Association of State Floodplain Managers,* **16**, 319–320.

Bryan, C.F. and Sabins, D.S. (1979). Management implications in water quality and fish standing stock information in the Atchafalaya River basin, Louisiana. In *Proceedings of the third coastal marsh and estuary management symposium,* eds. J.W. Day et al., pp. 293–316. Baton Rouge: Louisiana State University.

Burdick, B.D. (1995). *Ichthyofaunal studies of the Gunnison River, Colorado, 1992–1994.* Grand Junction, Colorado: Colorado River Fishery Project, U.S. Department of the Interior, Fish and Wildlife Service. 60 pp. + Appendices.

Cairns, J., Jr. (1995). Ecological integrity of aquatic ecosystems. *Regulated Rivers: Research & Management,* **11**, 313–323.

Carlson, C.A. and Muth, R.T. (1989). The Colorado River: Lifeline of the American Southwest. In *Proceedings of the international large river symposium,* ed.

D.P. Dodge, pp. 220–239. Canadian Fisheries and Aquatic Science, Special Publication 106. National Research Council of Canada, Ottawa.

Church, M. (1992). Channel morphology and typology. In *The rivers handbook*, eds. P. Calow and G.E. Petts, **1**, 126–143. Oxford: Blackwell Scientific Publications.

Clark, J. (1978). Freshwater wetlands: habitats for aquatic invertebrates, amphibians, reptiles, and fish. In *Wetland functions and values: the state of our understanding*, eds. P.E. Greeson, J.R. Clark, and J.E. Clark, pp. 330–343. Minneapolis, Minnesota: American Water Resources Association.

Coop, G.H. (1990). Effect of regulation on 0+ fish recruitment in the Great Ouse, a lowland river. *Regulated Rivers: Research and Management*, **5**, 251–263.

Cooper, D.J. and Severn, C. (1994a). *Ecological characteristics of wetlands at the Moab Slough, Moab, Utah.* Denver, Colorado: Recovery Implementation Program, Upper Colorado River Basin. U.S. Department of the Interior, Fish and Wildlife Service. 121 pp.

Cooper, D.J. and Severn, C. (1994b). *Wetlands of the Escalante Ranch area, Utah: Hydrology, water chemistry, vegetation, invertebrate communities, and restoration potential.* Denver, Colorado: Recovery Implementation Program, Upper Colorado River Basin. U.S. Department of the Interior, Fish and Wildlife Service. 97 pp.

Cooper, D.J. and Severn, C. (1994c). *Wetlands of the Escalante State Wildlife Area on the Gunnison River, near Delta, Colorado: Hydrology, water chemistry, vegetation, invertebrate communities, and restoration potential.* Denver, Colorado: Recovery Implementation Program, Upper Colorado River Basin. U.S. Department of the Interior, Fish and Wildlife Service. 94 pp.

Cooper, D.J. and Severn, C. (1994d). *Wetlands of the Ouray National Wildlife Refuge, Utah: Hydrology, water chemistry, vegetation, invertebrate communities, and restoration potential.* Denver, Colorado: Recovery Implementation Program, Upper Colorado River Basin. U.S. Department of the Interior, Fish and Wildlife Service. 79 pp.

Copp, G.H. (1989). The habitat diversity and fish reproductive function of floodplain ecosystems. *Environmental Biology of Fishes*, **26**, 1–27.

Courtenay, W.R., Jr. (1993). Biological pollution through fish introductions. In *Biological pollution: the control and impact of invasive exotic species*, ed. B.N. Knight, pp. 35–61. Indianapolis: Indiana Academy of Sciences.

Courtenay, W.R., Jr. (1995). The case for caution with fish introductions. *American Fisheries Society Symposium*, **15**, 413–424.

Cummins, K.W. (1979). The natural stream ecosystem. In *The ecology of regulated streams*, eds. J.V. Ward and J.A. Stanford, pp. 7–24. New York: Plenum Press.

Cutler, M.R. (1974). New role for government information and education personnel. *Transactions of the North American Wildlife and Natural Resource Conference*, **39**, 397–405.

Davies, B.R., Thoms, M.C., Walker, K.F., O'Keefe, J.H., and Gore, J.A. (1994). Dryland rivers: their ecology, conservation and management. In *The rivers handbook*, eds. P. Calow and G.E. Petts, **2**, 484–511. Oxford: Blackwell Scientific Publications.

Decamps, H., Planty-Tabacchi, A.M., and Tabacchi, E. (1995). Changes in the hydrological regime and invasions by plant species along riparian systems of the Adour River, France. *Regulated Rivers: Research & Management*, **11**, 23–33.

Dingell, J.D. (1972). Political aspects of multiple use. In *River ecology and man*, eds. R.T. Oglesby, C.A. Carlson, and J.A. McCann, pp. 441–453. New York: Academic Press.

Dudgeon, D. (1995). River regulation in southern China: Ecological implications, conservation, and environmental management. *Regulated Rivers: Research & Management*, **11**, 35–54.

Dunne, T. and Leopold, L.B. (1995). *Water in environmental planning.* San Francisco: W.H. Freeman and Company. 818 pp.

Dynesius, M. and Nilsson, C. (1994). Fragmentation and flow regulation of river systems in the northern third of the world. *Science*, **266**, 753–762.

Eckblad, J.W., Volden, C.S., and Weilgart, L.S. (1984). Allochthonous drift from backwaters to the main channel of the Mississippi River. *American Midland Naturalist*, **111**, 16–22.

Ellis, B.K. and Stanford, J.A. (1988). Phosphorus bioavailability of fluvial sediments determined by algal assays. *Hydrobiologia*, **160**, 9–18.

Feth, J.H. (1973). Water facts and figures for planners and managers. *U.S. Department of the Interior, Geological Survey, Circular 601–1.* 30 pp. U.S. Government Printing Office, Washington, DC.

Finger, T.R. and Stewart, E.A. (1987). Response of fishes to flooding regime in lowland hardwood wetlands. In *Community and evolutionary ecology of North American stream fishes*, eds. W.J. Matthews and D.C. Heins, pp. 86–92. Norman: University of Oklahoma Press.

Fontaine, T.D. and Bartell, S.M. (eds.) (1983). *Dynamics of lotic ecosystems.* Ann Arbor, Michigan: Ann Arbor Science. 494 pp.

Foreman, R.T.T. and Godron, M. (1986). *Landscape ecology.* New York: Wiley & Sons. 640 pp.

Frissell, C.A., Liss, W.J., and Bayles, D. (1993). An integrated, biophysical strategy for ecological restoration of large watersheds. In *Proceedings of the symposium on changing roles in water resources management and policy*, eds. N.E. Spangenborg and D.E. Potts, pp. 449–456. Bethesda, Maryland: American Water Resources Association.

Gilbert, D.L. (1971). The sociology of public relations. In *Natural resources and public relations*, ed. D.L. Gilbert, pp. 83–96. Washington, DC: The Wildlife Society.

Gibert., J., Dole-Oliver, M.J., Marmonier, P., and Vervier, P. (1990). Surface water-groundwater ecotones. In *Ecology and management of aquatic-terrestrial ecotones*, eds. R.J. Naiman and H. Decamps, pp. 199–255. London: Partenon Publishers.

Goldman, C.R. (1973). Environmental impact and water development. In *Environmental quality and water development*, eds. C.R. Goldman, J. McEvoy III, and P.J. Richerson, pp. 1–11. San Francisco: W.H. Freeman and Company.

Gore, J.A. (1985). Mechanisms of colonization and habitat enhancement for benthic macroinvertebrates in restored river channels. In *The restoration of rivers and streams: Theories and experience*, ed. J.A. Gore, pp. 81–101. Boston: Butterworth Publishers.

Gore, J.A. (1994). Hydrological change. In *The rivers handbook*, eds. P. Calow and G.E. Petts, v. 2, pp. 33–54. Oxford: Blackwell Scientific Publications.

Gore, J.A. and Petts, G.E. (eds.) (1989). *Alternatives in regulated river management.* Boca Raton, Florida: CRC (Chemical Rubber Corporation) Press. 344 pp.

Gore, J.A. and Shields, F.D., Jr. (1995). Can large rivers be restored? *BioScience*, **45**, 142–152.

Grabowski, S.J. and Hiebert, S.D. (1989). *Some aspects of trophic interactions in selected backwaters and the main channel of the Green River, Utah: 1987–1988.* Denver, Colorado: U.S. Bureau of Reclamation, Research and Laboratory Services Division, Environmental Sciences Section. 130 pp. + an appendix of 155 pp.

Graf, W. (1978). Fluvial adjustments to the spread of tamarisk in the Colorado Plateau Region. *Geological Society of America Bulletin*, **89**, 1491–1501.

Graf, W. (1985). *The Colorado River: instability and river basin management.* Washington, DC: Association of American Geographers. 86 pp.

Gregory, K.J. and Walling, D.E. (eds.) (1987). *Human activity and environmental processes.* New York: John Wiley and Sons. 466 pp.

Hamblin, W.K. (1990). Lake cenozoic lava dams in the western Grand Canyon. In *Grand Canyon geology*, eds. S. Beus and M. Morales, pp. 385–433. New York: Oxford University Press/Museum of Northern Arizona.

Hamilton, S.K., Sippel, S.J., Lewis, W.M., Jr., and Saunders, J.F., III. (1990). Zooplankton abundance and evidence for its reduction by macrophyte matts in two Orinoco floodplain lakes. *Journal of Plankton Research*, **12**, 345–363.

Hanson, D.L. and Waters, T.F. (1974). Recovery of standing crop and production rate of a brook trout population in a flood-damaged stream. *Transactions of the American Fisheries Society*, **103**, 431–439.

Harvey, B.C. (1987). Susceptibility of young-of-the-year fishes to downstream displacement by flooding. *Transactions of the American Fisheries Society*, **116**, 851–855.

Heede, B.H. and Rinne, J.N. (1990). Hydrodynamic and fluvial morphologic processes: implications for fisheries management and research. *North American Journal of Fisheries Management*, **10**, 249–268.

Heiler, G., Hein, T., Schiemer, F., and Bornette, G. (1995). Hydrological connectivity and flood pulses as the central aspects for the integrity of a river-floodplain system. *Regulated Rivers: Research & Management*, **11**, 351–361.

Henrickson, D.A. and Brothers, E.B. (1993). Utility of otoliths of Grand Canyon humpback chub *Gila cypha* for age determinations and as indicators of ecological history of individuals. *Desert Fishes Council 25th Annual Meeting*, November 10–14, 1993, San Nicolas de los Garza, N.L. Mexico. (Abstract)

Hey, R.D. (1994). Environmentally sensitive river engineering. In *The rivers handbook: Hydrological and ecological principles*, eds. P. Calow and G.E. Petts, pp. 337–362. Oxford: Blackwell Scientific Publications.

Hildrew, A.G. (1992). Food webs and species interactions. In *The rivers handbook*, eds. P. Calow and G.E. Petts, **1**, 309–330. London: Blackwell Scientific Publications.

Hjort, J. (1914). Fluctuations in the great fishes of northern Europe viewed in the light of biological research. *Rapports et Proces-Verbaux des Reunions, Conseil Permanent International pour l'Exploration de la Mer*, **20**, 1–228.

Hjort, J. (1926). Fluctuations in the year classes of important food fishes. *Journal du Conseil Permanent International pour l'Exploration de la Mer*, **1**, 5–38.

Hoopes, R.L. (1975). Flooding as a result of Hurricane Agnes and its effect on a native brook trout population in an infertile stream in central Pennsylvania. *Transactions of the American Fisheries Society*, **104**, 96–99.

Houde, E.D. (1987). Fish early life dynamics and recruitment variability. *American Fisheries Society Symposium*, **2**, 17–29.

Hynes, H.B.N. (1970). *The ecology of running waters.* Toronto, Canada: University of Toronto Press. 555 pp.

Hynes, H.B.N. (1983). Groundwater and stream ecology. *Hydrobiologia*, **100**, 93–99.

Johnson, B.L., Richardson, W.B., and Naimo, T.J. (1995). Past, present, and future concepts in large river ecology. *BioScience*, **45**, 134–141.

Junk, W.J., Bayley, P.B., and Sparks, R.E. (1989). The flood pulse concept in river floodplain systems. *Canadian Fisheries and Aquatic Sciences, Special Publication*, **106**, 110–127.

Kaeding, L.R. and Osmundson, D.B. (1988). Interaction of slow growth and increased early-life mortality: A hypothesis on the decline of Colorado squawfish in the upstream regions of its historic range. *Environmental Biology of Fishes*, **22**, 287–298.

Kaeding, L.R. and Zimmerman, M.A. (1983). Life history and ecology of the humpback chub in the Little Colorado and Colorado Rivers of the Grand Canyon. *Transactions of the American Fisheries Society*, **112**, 577–594.

Kallemeyn, L.W. and Novotny, J.F. (1977). *Fish and fish food organisms in various habitats of the Missouri River in South Dakota, Nebraska, and Iowa.* U.S. Department of the Interior, Fish and Wildlife Service, FWS/OBS-77/25. 100 pp. Washington, DC.

Karr, J.R. and Schlosser, I.J. (1978). Water resources and the land-water interface. *Science*, **201**, 229–234.

Kozlowski, T.T. (ed.) (1984). *Flooding and plant growth.* Orlando, Florida: Academic Press. 356 pp.

Kwak, T.J. (1988). Lateral movement and use of floodplain habitat by fishes of the Kankakee River, Illinois. *American Midland Naturalist*, **102**, 241–249.

Lambou, V.W. (1963). The commercial and sport fisheries of the Atchafalaya Basin Floodway. *Proceedings of the Annual Conference of the Southeastern Association of Game and Fish Commissioners*, **17**, 256–281.

Larkin, P.A. (1956). Interspecific competition and population control in freshwater fish. *Journal of the Fisheries Research Board of Canada*, **13**, 327–342.

Leggett, W.C. (1986). The dependence of fish larval survival on food and predator densities. In *The role of freshwater outflow in coastal marine ecosystems*, ed. S. Skiestel, pp. 117–138. New York: Springer-Verlag.

Leopold, L.B. and Wolman, M.G. (1957). Channel patterns: braided, meandering, and straight. *U.S. Geological Survey, Professional Paper 282-B*, 39–85. U.S. Government Printing Office, Washington, DC.

Li, H.W. and Moyle, P.B. (1993). Management of introduced fishes. In *Inland fisheries management in North America*, eds. C.C. Kohler and W.A. Hubert, pp. 287–307. Bethesda, Maryland: American Fisheries Society.

Li, S. and Mathias, J.A. (1982). Causes of high mortality among cultured larval walleyes. *Transactions of the American Fisheries Society*, **111**, 710–721.

Ligon, F.K., Dietrich, W.E., and Trush, W.J. (1995). Downstream ecological effects of dams. *BioScience*, **45**, 183–192.

Ludwig, D., Hilborn, R., and Walters, C. (1993). Uncertainty: Resource exploitation, and conservation: Lessons from history. *Science*, **260**, 17–36.

Luedtke, R.J. and Brusven, M.A. (1970). Effects of sand sedimentation on colonization of stream insects. *Journal of the Fisheries Research Board of Canada*, **33**, 1881–1886.

Mabey, L.W. and Schiozawa, D.K. (1993). *Planktonic and benthic microcrustaceans from floodplain and river habitats of the Ouray Refuge on the Green River, Utah.* Provo, Utah: Department of Zoology, Brigham Young University. 31 pp.

Maddux, H.R., Fitzpatrick, L.A., and Noonan, W.R. (1993). *Colorado River endangered fishes critical habitat: Draft biological support document.* Salt Lake City, Utah: U.S. Department of the Interior, Fish and Wildlife Service, Division of Fish and Wildlife Enhancement. 225 pp.

Magnuson, J.J. (1976). Managing with exotics – a game of chance. *Transactions of the American Fisheries Society*, **105**, 1–9.

May, R.C. (1974). Larval mortality in marine fishes and the critical period concept. In *The early life history of fish*, ed. J.H.S. Blaxter, pp. 3–19. Berlin: Springer-Verlag.

McAda, C.W., Bates, B., Craney, S., Chart, T., Elmblad, B., and Nesler, T. (1994a). *Interagency standardized monitoring program: Summary of results, 1986 through 1992. Recovery Implementation Program, Upper Colorado River Basin.* Denver, Colorado: U.S. Department of the Interior, Fish and Wildlife Service. 73 pp. + Appendices A-D.

McAda, C.W., Bates, B.W., Cranney, J.S., Chart, T.E., Trammell, M.A., and Elmblad, W.R. (1994b). *Interagency standardized monitoring program: Summary of results, 1993.* Denver, Colorado: Recovery Program, Upper Colorado River Basin, U.S. Department of the Interior, Fish and Wildlife Servic. 16 pp. + Appendices A-C.

McAda, C.W., Bates, B.W., Cranney, J.S., Chart, T.E., Trammell, M.A., and Elmblad, W.R. (1995). *Interagency standardized monitoring program: Summary of results, 1994.* Denver, Colorado: Recovery Program, Upper Colorado River Basin, U.S. Department of the Interior, Fish and Wildlife Service. 17 pp. + Appendices A-C.

McCarthy, M.S. and Minckley, W.L. (1987). Age estimation for razorback sucker (Pisces: Catostomidae) from Lake Mohave, Arizona-Nevada. *Journal of the Arizona-Nevada Academy of Sciences*, **21**, 87–97.

McCloskey, M. (1973). Alternatives in water project planning: ecological and environmental considerations. In *Environmental quality and water development*, eds. C.R. Goldman, J. McEvoy III, and P.J. Richerson, pp. 425–437. San Francisco: W.H. Freeman and Company.

Meffe, G.K. (1985). Predation and species replacement in American southwestern fishes. *Southwestern Naturalist*, **30**, 173–187.

Mellquist, P. (1992). River management – objectives and applications. In *River conservation and management*, eds. P.J. Boon, P. Calow and G.E. Petts, pp. 1–10. New York: John Wiley & Sons.

Miglavs, I. and Jobling, M. (1989). Effects of feeding regimes on food consumption, growth rates and tissue nucleic acids in juvenile arctic char, *Salvelinus alpinus*, with particular respect to compensatory growth. *Journal of Fish Biology*, **34**, 947–957.

Miller, R.R., Williams, J.D., and Williams, J.E. (1989). Extinctions in North American fishes during the past century. *Fisheries*, **14**, 22–38.

Miller, T.J., Crowder, L.B., Rice, J.A., and Marshall, E.A. (1988). Larval size and recruitment mechanisms in fishes: toward a conceptual framework. *Canadian Journal of Fisheries and Aquatic Science*, **45**, 1571–1581.

Miller, W.H., Tyus, H.M., and Carlson, C.A. (1982). *Fishes of the Upper Colorado River System: present and future. Proceedings of a symposium.* Bethesda, Maryland: American Fisheries Society. 131 pp.

Minckley, W.L., Marsh, P.C., Brooks, J.E., Johnson, J.E., and Jensen, B.L. (1991). Management toward the recovery of the razorback sucker. In *Battle against extinction: Native fish management in the American West*, eds. W.L. Minckley and J.E. Deacon, pp. 303–357. Tucson: University of Arizona Press.

Minshall, G.W., Cummins, K.W., Petersen, R.C., Cushing, C.E., Bruns, D.A., Sedell, J.R., and Vannote, R.L. (1985). Developments in stream ecosystem theory. *Canadian Journal of Fisheries and Aquatic Sciences*, **42**, 1045–1055.

Mitsch, W.J. and Gosselink, J.G. (1986). *Wetlands*. New York: Van Nostrand and Reinhold Publishing Company. 539 pp.

Modde, T. (1997). *Fish use of Old Charley Wash: Assessment of floodplain wetland importance to razorback sucker management and recovery.* Vernal, Utah: U.S. Department of the Interior, Fish and Wildlife Service, Colorado River Fishery Project. 63 pp.

Modde, T. and Wick, E.J. (1997). *Investigations of razorback sucker distribution, movements and habitat used during spring in the Green River, Utah.* Final report - Recovery Program Project 49. Vernal, Utah: U.S. Department of the Interior, Fish and Wildlife Service, Colorado River Fishery Project. 40 pp.

Moyle, P.B. and Herbold, B. (1987). Life history patterns and community structure in stream fishes of western North America: comparisons with eastern North America and Europe. In *Community and evolutionary ecology of North American stream fishes*, eds. W.J. Matthew and D.C. Heins, pp. 25–32. Norman: University of Oklahoma Press.

Moyle, P.B., Li, H.W., and Barton, B.A. (1986). The Frankenstein effect: Impact of introduced fishes on native fishes of North America. In *Fish culture in fisheries management*, ed. R.H. Stroud, pp. 415–426. Bethesda, Maryland: American Fisheries Society.

Muth, R.T., Haines, G.B., Meismer, S.M., Wick, E.J., Chart, T.E., Snyder, D.E., and Bundy, J.M. (1998). *Reproduction and early life history of razorback sucker in the Green River, Utah and Colorado, 1992–1996.* Final Report - Project 34. Denver, Colorado: Recovery Program, Upper Colorado River Basin, U.S. Department of the Interior, Fish and Wildlife Service. 62 pp.

Naiman, R.J. and Decamps, H. (eds.) (1990). *The ecology and management of aquatic-terrestrial ecotones.* New York: Parthenon Publishing Group. 316 pp.

National Research Council. (1992). *Restoration of aquatic ecosystems: science, technology, and public policy.* Washington, DC: National Academy Press. 552 pp.

Newbold, J.D. (1992). Cycles and spirals of nutrients. In *The rivers handbook*, eds. P. Calow and G.E. Petts, **1**, 379–408. London: Blackwell Scientific Publications.

Nilsson, C. and Jansson, R. (1995). Floristic differences between riparian corridors of regulated and free-flowing boreal rivers. *Regulated Rivers: Research & Management*, **11**, 55–66.

Obeng, L.E. (1981). Man's impact on tropical rivers. In *Perspectives in running water ecology*, eds. M.A. Lock and D.D. Williams, pp. 265–288. New York: Plenum Press.

Odum, E.P. and Odum, H.T. (1959). *Fundamentals of ecology*. Philadelphia: W.B. Saunders Company. 546 pp.

Paine, R.T. (1980). Food webs: Linkage, interaction, strength, and community infrastructure. *Journal of Animal Ecology*, **49**, 667–685.

Papoulias, D. and Minckley, W.L. (1990). Food limited survival of larval razorback sucker, *Xyrauchen texanus*, in the laboratory. *Environmental Biology of Fishes*, **29**, 73–78.

Papoulias, D. and Minckley, W.L. (1992). Effects of food availability on survival and growth of larval razorback suckers in ponds. *Transactions of the American Fisheries Society*, **121**, 340–355.

Petts, G.E. (1994). Rivers: Dynamic components of catchment ecosystems. In *The rivers handbook*, eds. P. Calow and G.E. Petts, **2**, 3–22. Oxford: Blackwell Scientific Publications.

Petts, G.E., Large, A.R.G., Greenwood, M.T., and Bickerton, M.A. (1992). Floodplain assessment for restoration and conservation: Linking hydrogeomorphology and ecology. In *Lowland floodplain rivers: geomorphological perspectives*, eds. P.A. Carling and G.E. Petts, pp. 217–234. Chichester: Wiley and Sons.

Poff, N.L., Allan, J.D., Bain, M.B., Karr, J.R., Prestegaard, K.L., Richter, B.D., Sparks, R.E., and Stromberg, J.C. (1997). The natural flow regime. *BioScience*, **47**, 769–784.

Power, M.E., Parker, G., Dietrich, W.E., and Sun, A. (1995a). How does width affect floodplain river ecology? A preliminary exploration using simulations. *Geomorphology*, **13**, 301–317.

Power, M.E., Sun, A., Parker, G., Dietrich, W.E., and Wootton, J.T. (1995b). Hydraulic food-chain models. *BioScience*, **45**, 159–167.

Pringle, C.M., Naiman, R.J., Bretschko, G., Karr, J.R., Oswood, M.W., Webster, J.R., Welcomme, R.L., and Winterbourn, M.J. (1988). Patch dynamics in lotic systems: The stream as a mosaic. *Journal of the North American Benthological Society*, **7**, 503–524.

Rabeni, C.F., and Minshall, G.W. (1977). Factors affecting microdistribution of stream benthic insects. *Oikos*, **29**, 33–43.

Resh, V.H., Brown, A.V., Covich, A.P., Gurtz, M.E., Li, H.W., Minshall, G.W., Reice, S.R., Shelton, A.L., Wallace, J.B., and Wissmar, R.C. (1988). The role of disturbance in stream ecology. *Journal of the North American Benthological Society*, **7**, 433–455.

Rice, J.A., Crowder, L.B., and Binkowski, F.P. (1987). Evaluating potential sources of mortality for larval bloater: starvation and vulnerability to predation. *Canadian Journal of Fisheries and Aquatic Sciences*, **44**, 467–472.

Ross, S.T. and Baker, J.A. (1983). The response of fishes to periodic spring floods in a southeastern stream. *American Midland Naturalist*, **109**, 1–14.

Ryder, R.A. and Pesendorfer, J. (1989). Large rivers are more than flowing lakes: a comparative review. In *Proceedings of the international large river symposium*, ed. D.P. Dodge, pp. 65–85. Canadian Fisheries and Aquatic Science, Special Publication 106. National Research Council of Canada, Ottawa.

Schiemer, F. and Spindler, T. (1989). Endangered fish species of the Danube River in Austria. *Regulated Rivers: Research and Management*, **4**, 397–407.

Schindler, D.W. (1978). Factors regulating phytoplankton production and standing crop in the world's freshwaters. *Limnology and Oceanography*, **23**, 478–486.

Schlosser, I.J. (1991). Stream fish ecology: a landscape perspective. *BioScience*, **41**, 704–712.

Schnick, R.A., Morton, J.A., Mochalski, J.C., and Beall, J.T. (1982). *Mitigation and enhancement techniques for the upper Mississippi River system and other large river systems.* U.S. Department of the Interior, Fish and Wildlife Service, Resource Publication 149. 714 pp. Washington, DC.

Sedell, J.R., Richey, J.E., and Swanson, F.J. (1989). The river continuum concept: a basis for the expected ecosystem behavior of very large rivers. *Canadian Fisheries and Aquatic Sciences, Special Publication*, **106**, 49–55.

Seegrist, D.W. and Gard, R. (1972). Effects of floods on trout in Sagehen Creek, California. *Transactions of the American Fisheries Society*, **101**, 478–482.

Shannon, J.P., Blinn, D.W., and Stevens, L.E. (1994). Trophic interactions and benthic animal community structure in the Colorado River, Arizona, USA. *Freshwater Biology*, **31**, 213–220.

Sheehan, P.J., Miller, D.R., Butler, G.C., Bouadeau, P., and Rideway, J.M. (eds.) (1984). *Effects of pollutants at the ecosystem level.* New York: John Wiley and Sons. 443 pp.

Sparks, R.E. (1995). Need for ecosystem management of large rivers and their floodplains. *BioScience*, **45(3)**, 168–182.

Sparks, R.E., Bailey, P.B., Kohler, S.L., and Osborne, L.L. (1990). Disturbance and recovery of large floodplain rivers. *Environmental Management*, **14**, 699–709.

Stacey, P.B. and Taper, M. (1992). Environmental variation and the persistence of small populations. *Ecological Applications*, **2**, 18–29.

Stanford, J.A., Ward, J.V., Liss, W.J., Frizzell, C.A., Williams, R.N., Lichatowich, J.A., and Coutant, C.C. (1996). A general protocol for restoration of regulated rivers. *Regulated Rivers: Research and Management*, **12**, 391–413.

Starrett, W.C. (1951). Some factors affecting the abundance of minnows in the Des Moines River, Iowa. *Ecology*, **32**, 13–27.

Sweeting, R.A. (1994). River pollution. In *The rivers handbook*, P. Calow and G.E. Petts, eds., **2**, 23–32. Oxford: Blackwell Scientific Publications.

Tsivoglov, E.C., Schearer, S.D., Shaw, R.M., Jr., Jones, J.D., Anderson, J.B., Spongales, C.E., and Clark, D.A. (1959). *Survey of interstate pollution of the Animas River, Colorado-New Mexico.* Cincinnati, Ohio: U.S. Department of Health, Education, and Welfare, Public Health Service, Robert A. Taft Engineering Center. 116 pp.

Tyus, H.M., Burdick, B.D., Valdez, R.A., Haynes, C.M., Lytle, T.A., and Berry, C.R. (1982). Fishes of the Upper Colorado River Basin. In *Proceedings of a symposium on fishes of the Upper Colorado River System: Present and future*, eds. W.H. Miller, H.M. Tyus and C.A. Carlson, pp. 12–70. Bethesda, Maryland: Western Division, American Fisheries Society.

U.S. Fish and Wildlife Service. (1990a). *Bonytail chub recovery plan*. Denver, Colorado: U.S. Department of the Interior, Fish and Wildlife Service, Region 6. 35 pp.

U.S. Fish and Wildlife Service. (1990b). *Humpback chub recovery plan*. Denver, Colorado: U.S. Department of the Interior, Fish and Wildlife Service, Region 6. 43 pp.

U.S. Fish and Wildlife Service. (1991). *Colorado squawfish recovery plan*. Denver, Colorado: U.S. Department of the Interior, Fish and Wildlife Service, Region 6. 56 pp.

Van Dijk, G.M., Marteijn, E.C.L. and Sculte-Eulwer-Ledig, A. (1995). Ecological rehabilitation of the River Rhine: Plans, progress and perspectives. *Regulated Rivers: Research and Management*, **11**, 377–388.

Vanicek, C.D. and Kramer, R.H. (1969). Life history of the Colorado squawfish, *Ptychocheilus lucius*, and the Colorado chub, *Gila robusta*, in the Green River in Dinosaur National Monument, 1964–1966. *Transactions of the American Fisheries Society*, **98**, 193–208.

Vannote, R.L., Minshall, G.W., Cummins, K.W., Sedell, J.R., and Cushing, C.E. (1980). The river continuum concept. *Canadian Journal of Fisheries and Aquatic Science*, **37**, 130–137.

Walters, C.J. (1986). *Adaptive management of renewable resources*. New York: MacMillan. 374 pp.

Walters, C.J. and Hillborn, R. (1978). Ecological optimization and adaptive management. *Annual Review of Ecological Systems*, **9**, 157–188.

Ward, J.V. (1976). Effects of flow patterns below large dams on stream benthos: a review. In *Proceedings of a symposium and specialty conference on instream flow needs*, eds. J.F. Orsborn and C.H. Allman, **2**, 235–253. Bethesda, Maryland: American Fisheries Society.

Ward, J.V. and Stanford, J.A. (1983). The serial discontinuity concept of lotic ecosystems. In *Dynamics of lotic ecosystems*, eds. T.D.I. Fontaine III and S.M. Bartell, pp. 29–42. Ann Arbor, Michigan: Ann Arbor Science Publishers.

Ward, J.V. and Stanford, J.A. (1989). Riverine ecosystems: The influence of man on catchment dynamics and fish ecology. In *Proceedings of the international large river symposium*, ed. D.P. Dodge, pp. 56–64. Canadian Fisheries and Aquatic Science, Special Publication 106. Ottawa, Canada.

Ward, J.V. and Stanford, J.A. (1995). Ecological connectivity in alluvial river ecosystems and its disruption by flow regulation. *Regulated Rivers: Research & Management*, **11**, 105–119.

Weisberg, S.B., Janicki, A.J., Gerritssen, J., and Wilson, H.T. (1990). Enhancement of benthic macroinvertebrates by minimum flow from a hydroelectric dam. *Regulated Rivers: Research and Management*, **5**, 265–277.

Welcomme, R.L. (1979). *Fisheries ecology of floodplain rivers*. New York: Longman Group, Limited. 317 pp.

Welcomme, R.L. (1985). *River fisheries*. Rome, Italy: Food and Agriculture Organization of the United Nations, Technical Paper 262. 330 pp.

Welcomme, R.L. (1995). Relationships between fisheries and the integrity of river systems. *Regulated Rivers: Research and Management*, **11**, 121–136.

Welcomme, R.L., Ryder, R.A., and Sedell, J.A. (1989). Dynamics of fish assemblages in river systems – a synthesis. *Canadian Fisheries and Aquatic Sciences, Special Publication*, **106**, 569–577.

Welker, M.T., Pierce, C.L., and Wahl, D.H. (1994). Growth and survival of larval fishes: Roles of competition and zooplankton abundance. *Transactions of the American Fisheries Society*, **123**, 703–717.

Wesche, T.A. (1993). Watershed management and land-use practices. In *Inland fisheries management in North America*, eds. C.C. Kohler and W.A. Hubert, pp. 181–203. Bethesda, Maryland: American Fisheries Society.

White, G.F. (1974). Natural hazards research: concepts, methods, and policy implications. In *Natural hazards: Local, national, global*, ed. G.F. White, pp. 3–16. New York: Oxford University Press.

Wiley, R.W. and Wydoski, R.S. (1994). Management of undesirable fishes. In *Inland fisheries management in North America*, eds. C.C. Kohler and W.A. Hubert, pp. 335–354. Bethesda, Maryland: American Fisheries Society.

Williams, J.E., Wood, C.A., and Dombeck, M.P. (eds.) (1997). *Watershed restoration: principles and practices*. Bethesda, Maryland: American Fisheries Society. 549 pp.

Williamson, J.H. and Wydoski, R.S. (1994). *Genetics management guidelines*. Upper Colorado River Basin, Recovery Implementation Program, U.S. Department of the Interior, Fish and Wildlife Service, Denver, Colorado. 40 pp.

Wolz, E.R. and Schiozawa, D.K. (1995). Soft sediment benthic macroinvertebrate communities of the Green River at the Ouray National Wildlife Refuge, Uintah County, Utah. *Great Basin Naturalist*, **55**, 213–224.

Wydoski, R.S. (1978). Responses of trout populations to alterations in aquatic environments: a review. In *Proceedings of the wild trout – catchable trout symposium*, ed. J.R. Moring, pp. 57–92. Corvallis: Oregon Department of Fish and Wildlife.

Wydoski, R.S. (1980). Potential impacts of alterations in streamflow and water quality on fish and macroinvertebrates in the Upper Colorado River Basin. In *Energy development in the Southwest: Problems of water, fish and wildlife in the Upper Colorado River Basin*, eds. W.O. Spofford, Jr., A.L. Parker and A.V. Kneese, pp. 77–147. Research Paper R-18. Washington, DC: Resources for the Future.

Wydoski, R.S. and Wick, E.J. (1998). *Ecological value of floodplain habitats to razorback suckers in the Upper Colorado River Basin*. Denver, Colorado: Upper Colorado River Basin Recovery Program, U.S. Department of the Interior, Fish and Wildlife Service. 55 pp.

EFFECTS OF FLOODS ON HUMAN COMMUNITIES

CHAPTER 10

The Effects of Variable River Flow on Human Communities

Dorothy Merritts
Geosciences Department
Franklin and Marshall College

For millennia, humans have settled along the banks of rivers, depending on their waters for drinking, farming, and navigation. In addition to the benefits of habitation on river floodplains are several costs, the greatest of which is the hazard of flooding when the river stage is higher than usual. Equally costly, however, can be the effects of low river stages, when insufficient water is available for those who have come to depend on minimum flows. In this chapter, the effects of variable flow – from low to high stages – on human communities are examined. Some of the world's largest rivers, such as the Mississippi River discussed in this article, are controlled over hundreds of kilometers in order to protect those living along the floodplains. The problem of the flood protection-development spiral, in which protection itself promotes imprudent development, is examined for the Mississippi River. In contrast, there still are nations in which a large percentage of the populace lives unprotected, susceptible to repeated flooding and its associated disease, trauma, and death. One of these nations, Bangladesh, is described here, along with a review of recent major floods and attempts to mitigate flood hazards in that country. Different aspects of variable river stage are examined, including meander migration, channel bank erosion, overbank deposition, and alluvial fan flooding. The American Southwest is used to illustrate the problem of alluvial fan flooding. Finally, the Colorado River is used as a case study to examine the effects of controlled river flow on human communities.

Introduction

The level, or stage, of water in streams and rivers changes continuously – from day to day, from season to season, and from year to year – in response to changing conditions of weather and climate. In the long history of human habitation of floodplains, people have responded differently to such variability. In many rain-fed agrarian communities, for example, humans have adapted to the rising and falling stages of water, depending on seasonal flooding for replenishment of their farmland. In more arid regions, they have developed technologies to store water in order to release it when needed.

The costs of cohabitation with rivers can be high, though. During times of especially low stage, or drought, crops might fail, resulting in famine. During times of especially high stage, or flooding, people and animals might drown and crops might be destroyed, again resulting in famine. As a result, the human history of interaction with water in rivers often has been one of regulating river flow so that the amount is neither too little nor too much (Cosgrove and Petts, 1990). The earliest hydraulic civilizations, those that relied on great engineering works to store, divert, and appropriate water, date back more than 6000 years. Today, nearly every major river in the world is dammed and its flow is regulated.

In modern industrial societies, changing water level is considered at the least to be a nuisance and in some cases even unacceptable. In the case of the Mississippi River, for example, the dream of American engineers and planners in the 19th century was to make "... the level of the river's surface ... nearly stationary ... to flow forever with a constant, deep, and limpid, stream" (Ellett, 1853). In most 19th and 20th century societies, the consequences of nonvariable flow – many of which are negative – were unanticipated. And yet, attitudes toward natural river flow are changing. Some modern societies cohabiting with large rivers recently have chosen to "live with floods," as in the case of the Bangladeshi living along the Ganges–Brahmaputra system (Franks, 1994). Others are rethinking their relation to variable river flow and are reversing earlier practices that relied on extensive flow regulation.

Several aspects of the effects of variable flow on human communities are considered here and referred to with examples from different parts of the world. Processes of overbank flooding and bank erosion are widely recognized hazards in places such as Bangladesh, but less well understood is the particular suite of flooding processes that occur on alluvial fans. As described here, alluvial fan flooding is now under investigation in the United States, as the Federal Emergency Management Agency (FEMA) works to establish guidelines for development and flood control on alluvial fan landforms. Another hazard to human communities is posed by efforts to control flooding and regulate river flow. A case study for the Mississippi River illustrates the hazards that result from development in areas protected by levees that can be breached during storms exceeding those of levee design. Finally, a case study for the Colorado River is used to illustrate the impact of dams and diminished flow on human communities of the American Southwest. Before presenting these case studies, however, we begin by considering the general nature of the interplay between humans and the variable flow of rivers.

Rivers and Floodplains as Human Habitat

Human Settlement on Floodplains

Human occupancy of valley-bottom waterways, and consequent cohabitation with flood risks, is common throughout the world in arid to humid regions and has been so for all of human history. The reason is simple: humans perceive that the benefits of living along flowing fresh water outweigh the costs. Before the Agricultural Revolution, people relied on streams and rivers for drinking water, travel, fish, and game. Archaeological evidence indicates, for example, that during the Paleolithic period early humans entered what later became Anglo-Saxon England along major rivers, which served as axes from which subsequent settlement spread toward drainage divides.

With development of agricultural practices about 10,000 years ago, river corridors became even more valuable for their fertile floodplain soils and availability of water for irrigating crops. Because of the fine-grained, moist nature of overbank deposits, and the levelness of the floodplain surface, floodplains are excellent for farming. Fertile alluvium revitalized by annual flooding has supported major civilizations along rivers such as the Euphrates, Hwang He, Ganges, Indus, Nile, Tigris, and Yangtse [National Research Council (NRC), 1991a].

One of the best known examples of a civilization that has relied on a major river for thousands of years is that of the Egyptians, who have lived along the Nile in northern Africa since about 4000 B.C. Much of the region drained by the Nile is arid or semiarid; thus, the only part of the watershed suitable for large-scale farming is along the floodplain of the trunk stream itself. Like most other rivers, the stage of the Nile varies throughout the year, beginning to rise in June and overflowing its banks (i.e., flooding) within 3 months. As the river widens and fills the floodplain, its waters are retarded and drop their load of fine silt and organic debris. These fresh deposits provide continuous enrichment to the soil; hence, in ancient Egypt Herodotus called the land "the gift of the river."

A negative aspect of reliance on a river's variability, however, is that too much water can wreak havoc (flooding), and so can too little (drought). Indeed, the changing stages of the Nile were described by Pliny the Elder (A.D. 23–79), the Roman naturalist, in terms of human tragedies and bounties. Pliny's guide to river stage, called a Nilometer, ranged from hunger and suffering at too low a flow to security and abundance in a wet year to disaster in a year so deluged that crops are destroyed and humans are killed. Similarly, the Hwang He (Yellow River) of China, the most lethal river in the world, is called the Mother of China for its annual gift of sediment, but at high flood stages it becomes China's Sorrow. Located in the cradle of Chinese civilization, on a vast alluvial plain with a dense modern population, the swollen river killed at least 900,000 people in 1887. Fewer than 45 years later, in 1931, nearly 4 million people either drowned during flooding, or died soon after from disease, exposure, and starvation.

The long history of human settlement on floodplains has generated a substantial number of flood legends (Laituri, Chapter 17, this volume), the oldest of which is known as The Epic of Gilgamesh and was recorded in cuneiform on clay tablets. Written in Babylon about 1800 B.C., the legend dates to the time of early Sumeria, about 5000 B.C., and tells the tale of a flood sent by a wrathful god determined to destroy humankind (Kovacs, 1985). Some attribute the account of Noah's deluge in the Old Testament (Genesis 7:11–12) to Babylonian roots.

More than 500 legends of great floods have been passed down by some 750 cultures, and in most epics the floods were thought to have covered the entire face of the Earth (Rogers and Feiss, 1998). For example, the Quiche Indians of Mexico believed that gods used a great flood to destroy their first attempts at making humans; the first beings – made of clay – were defective, unable to turn their heads (Clark, 1982). An ancient Norse myth describes a flood caused as blood gushed from an evil god slain by a Norseman and his two brothers (Clark, 1982).

In the modern world, the social and economic benefits of floodplain occupancy are many and varied. The level land along floodplains is used to transport goods by rail and road, and rivers themselves are used to transport cargo and people by barges, boats, and rafts. In the late 20th century, one of the world's most heavily used waterways, the Ohio River (tributary to the Mississippi River), annually transported 100 million tons of coal, 20 million tons of petroleum products, and 60 million tons of chemicals and other goods. Since the Industrial Revolution, waterways have become an important source of cooling water for power-generating stations and of freshwater for manufacturing processes. The potential energy of water falling in streams is harnessed in hydroelectric power plants. Most of the world's cities and factories are located along rivers to take advantage of these assets (Figure 10.1).

Figure 10.1. Many of the world's cities are located along rivers in order to make optimal use of their water for human consumption, commerce, navigation, and industry. The city of Sacramento, state capital of California, is built along the floodplain of the American River. Nearly half a million people and $37 billion worth of property are vulnerable to flooding along this river. (Photograph by Robert Childs, U.S. Army Corps of Engineers)

Human Cohabitation with the Hazard of Changing River Stages

Studies of human behavior indicate that, even with accurate knowledge of flood hazards, people continue to inhabit floodplains and consequently to cohabit with flood risks (Kates, 1962; Parker and Harding, 1979; Laituri, Chapter 17, this volume). Indeed, people often do not cooperate with floodplain managers charged with developing flood reduction programs. In 1963, for example, two planners in Tucson, Arizona, attempted to regulate floodplain development, and an uproarious public hearing resulted, as described in the Tucson Daily Citizen (Cooper, 1963):

> Some 400 irate landowners and residents... swarmed over a floodplain ordinance hearing, hanged two city-county officials in effigy and made one common point clear: They are violently opposed to restrictive zoning.

As described later, substantial bank erosion and collapse during the 1983 Tucson floods served to vindicate the goals of the unfortunate planners.

The Tucson example illustrates that the human desire to cohabit with floods is not completely one of gracious submission. In many cultures, human attitudes toward floodplains and flooding reflect a poor appreciation of the importance of variable water stages and flood corridors to ecosystems and human communities (Cosgrove and Petts, 1990). For example, when Charles Dickens visited the marshy confluence of the Mississippi River and one of its tributaries, the Ohio River, at Cairo, Illinois, in 1842, he wrote of the site with disdain:

> ...on ground so flat and marshy, that at certain seasons of the year it is inundated to the house-tops, lies a breeding place of fever, ague, and death, ...a dismal swamp, ...an ugly sepulchre, a grave uncheered by any gleam of promise... (cited in Stevens et al., 1975).

In 1853, Ellett proposed a new vision for the Mississippi, one in which channel straightening, bank protection, and flood-control reservoirs would transform the "ugly sepulchre" so that

> ...all material fluctuations of the waters will be prevented, and the level of the river's surface will become nearly stationary. Grounds, which are now frequently inundated and valueless, will be tilled and subdued; sandbars will be permanently covered, and, under a uniform regimen of the stream, will probably cease to be produced. The channels will become stationary... The Ohio first, and ultimately the Missouri and Mississippi, will be made to flow forever with a constant, deep, and limpid, stream (Ellett, 1853).

Achieving Ellett's vision took years, much labor and money, and thousands of kilometers of levees. Yet, increasing control of the mighty river only exacerbated the conflict between humans and flooding.

Clear indication of the antagonistic attitude toward flooding is illustrated in a colonel's account of a flood-battle fought along the Mississippi in 1937:

> We have organized our main stem levees as in battle. Flood fighters hold each sector of the river, just as we would assign companies, battalions, brigades and divisions in wartime. Each sector commander must maintain at all costs the

integrity of his levee line (Colonel Eugene Reybold of the U.S. Army Corps of Engineers, quoted in Clark, 1982, p. 82).

Civil engineers have come to recognize that it is too expensive or impractical to protect floodplain communities from all conceivable floods (Foster, 1924; Riggs, 1966). However, cost–benefit analyses of flood protection indicate that in many cases there are economic advantages to providing protection for floods with recurrence intervals of 1 in 50 to 1 in 100 years and sometimes even greater. The problems with such analyses, however, are in determining the risk and uncertainty associated with different meteorological, hydrologic, and hydraulic parameters (U.S. Army Corps of Engineers, 1992; see also Baker, Chapter 13, this volume; Stedinger, Chapter 12, this volume; Hamilton and Joaquin, Chapter 18, this volume). Despite these shortcomings, humans continue to occupy floodplains. More than 20,800 communities in the United States are located in flood-prone areas (Hays, 1981). Between 1975 and 2000, annual flood losses affected an increasing percentage of urban land.

Effects of High River Stage on Human Communities

Floods have detrimental effects on three aspects of the human condition: health, agriculture, and economic systems. In terms of human health, floods can cause death from drowning, impact, or exposure. As population densities on floodplains increase, so too do damage and loss of life during flooding. Loss of lives in the United States has been minimized somewhat by sophisticated flood warning systems that utilize rapid communication and satellite technologies to alert and evacuate residents (Gruntfest, Chapter 15, this volume). Nevertheless, floods are the only storm-related hazard for which the population-adjusted number of deaths in the United States has risen since the 1940s. In addition, annual damage to and loss of property has risen throughout this century, topping more than $12 to $15 billion for the 1993 Mississippi River floods, the greatest disaster in the history of the United States in terms of property damage. It is especially disturbing that property loss and damage from flooding in the United States has risen despite the billions of federal, state, and local dollars spent on flood protection in the 20th century. The increasing loss of property and economic devastation results from the "flood protection-development spiral," as discussed below.

In densely populated, low-lying Third World countries such as Bangladesh, where most of the populace lives unprotected on floodplains and sophisticated flood warning systems are not yet in place, the loss of lives is far more devastating. In 1998, summer storms and flooding inundated more than two-thirds of the country in what is considered to have been the country's most devastating flooding of the 20th century. Nearly 1000 people died, 40 million were left homeless, whole villages were buried in silt and sand, and over 10,000 km of roads were ruined. In 1991, a monsoon and associated wave surges swept over the coastal plain and surrounding islands, killing 131,000 people and causing $2.7 billion in damage, slightly more than 1/10th of the nation's annual gross domestic product (Rogers and Feiss, 1998). Only a few years earlier, in 1987 and 1988, heavy, prolonged periods of rainfall resulted in 1657 and 2379 deaths, respectively. In 1988, 58% of the nation was flooded (Alexander, 1993). In 1988–89, nearly half the nation's development budget was spent to repair flood damage (Alexander, 1993).

Numerous physical injuries are possible as well, and a common effect of flooding – especially in developing nations – is disease and disease transmission (Glass et al., 1977; Beinin, 1979; Bissell, 1983; Gueri et al., 1986; Telleria, 1986; Hederra, 1987). Diseases commonly are associated with disrupted supplies of fresh water; ecological changes in the habitat of rodents, insects, and other organisms that transmit diseases; and respiratory ailments associated with exposure. After the El Niño floods of 1983 in Peru, for example, rates of infant morbidity and mortality increased some 200% to 270% above preflood numbers (Gueri et al., 1986). Finally, when crops and other food sources are destroyed or cannot be shipped because of disrupted transportation, malnutrition and mortality rates rise rapidly. As an example of these multiple and cascading effects of flooding on human health, consider the 1881 and 1931 floods on the Yellow River of China. Of the millions of deaths, most were thought to be the result of starvation, exposure, and disease and not drowning or impact-related trauma.

As with human health, the effects of floods on agriculture and food are multiple and result from direct as well as indirect consequences of the flooding. Crops, seeds, and stored food stocks are destroyed during inundation, causing immediate financial devastation for farmers and subsequent detrimental effects on that part of the population who rely on those farmers for food sources as well as on other sectors of the economy dependent on those products. The Mississippi floods of 1993, for example, destroyed a large percent of the U.S. wheat harvest for that year [National Oceanic and Atmospheric Administration (NOAA), 1994]. Livestock is lost to drowning or dispersed. During the 1987 and 1988 floods of Bangladesh, for example, 236,700 cattle and 616,000 poultry were lost (Brammer, 1990). Finally, paddy and irrigation systems can be washed out by rapidly flowing water, which requires years of repair work to achieve preflood productivity.

The third category of flood damage for human communities is associated with the effects of flooding on housing, buildings, businesses, and property. As water begins to overtop the banks of channels and spread across the floodplain, structures are affected, but the severity of damage depends on how high the stage rises. Ground floors of residences and other buildings are flooded initially, but with increasing stage and higher velocity flows, buildings and loose property (e.g., cars) can be swept away. The floating debris itself can cause further damage as it pounds against structures downstream. During the 1987 and 1988 floods of Bangladesh, 9.7 million homes were destroyed or damaged, 8481 schools were flooded, and 14,000 industrial units were damaged (Brammer, 1990). Shanty towns and squatters' villages on the edges of the world's megacities are especially afflicted by flooding because they typically have high population densities, poorly constructed housing, and precarious building sites (e.g., steep hillslopes, narrow ravines). Cities with substantial shanty towns include Calcutta, India; Dhaka, Bangladesh; Lagos, Nigeria; Monrovia, Liberia; Recife, Brazil; Manila, Philippines; and San Juan, Puerto Rico (Alexander, 1993). In Dhaka, 30 million people were homeless after a 1988 flood, and another 20 million were seriously affected (Alexander, 1993).

Processes that Generate Flood Hazards

The previous section focused on the close relationship between humans, rivers, and floodplains and the detrimental effects of changing water levels on human communities. These effects were related to human health, agriculture, and property. Here, we examine specific flood processes that create hazards, including meander migration with associated channel bank erosion, overbank flooding, and alluvial fan flooding. Two other effects of changing river stage are considered in the next sections. These are the effects of levee breaks, which are considered in a separate case study of the Mississippi River, and the effects of dams and diminished flow, which are considered in a separate case study of the Colorado River.

Meander Migration and Channel Bank Erosion As a Hazard to Human Communities

Floodplains form along both sides of a channel by two important fluvial processes: channel migration, with associated bank erosion and lateral accretion, and overbank deposition. Both processes pose hazards to human communities. During lateral channel migration and shifting, a stream channel erodes its concave banks, where stream energy is high, and drops sediment along its convex banks, where energy is low (see Wohl, Chapter 6, this volume). This process of channel shifting and associated lateral accretion is responsible for most of the continuous building and tearing down of a floodplain, and constructs most of its height (Wolman and Leopold, 1957). Conditions that make rivers prone to bank erosion include high discharge, prolonged flooding, unconsolidated alluvial banks, and diminished vegetation along banks (causing low resistance to shear stress). Falling groundwater tables also can accelerate bank erosion because groundwater is essential for plants growing along channel banks. In addition, a rapidly falling water table can lead to large fluctuations in pore pressure and resultant mass movement along river banks.

Meander migration poses a hazard to human communities on the outside bends of meanders where erosion occurs during channel shifting (Figure 10.2). For example, most of the damage caused by the 1983 flood in Tucson, Arizona, resulted from erosion and collapse of alluvial channel banks (NRC, 1984; Kresan, 1988). In some cases, channel shifting and severe bank erosion occurred before water reached high levels and spilled over its banks, so technically flooding did not even occur (Kresan, 1988). Channel migration was so extreme in places that land outside the boundaries of the designated 100- to 500-year floodplain collapsed into the channel. Although artificial bank protection was effective in some places, it also served to accelerate erosion immediately downstream of protected reaches.

Channel migration is particularly rapid and frequent along rivers within deltaic systems, such as the Brahmaputra–Ganges delta of Bangladesh. The Bangladesh delta system contains some 250 perennial rivers, all of which have changed course numerous times in recent geologic history (Coleman, 1969; Alexander, 1993). Parts of the Brahmaputra River are estimated to be migrating at rates as high as 0.8 km/year (Alexander, 1993). Bedforms within this system are highly mobile, with some kilometer-scale point bars shifting hundreds of meters downstream within a day at times of high water level.

Figure 10.2. Aerial view (downstream toward top) of Pantano Wash, a tributary to the Santa Cruz River, in Tucson, Arizona, taken soon after the October, 1983, flood. Channel bank erosion outside of the meander bend in the foreground has destroyed part of a large mobile home community. (Photograph by Peter L. Kresan)

Overbank Deposition As Benefit and Hazard to Human Communities

In addition to the process of meander migration, floodplains grow by deposition of fine-grained sediment in thin, horizontal sheets on the floodplain surface when streams overflow their banks and spread across the valley bottom (Wolman and Leopold, 1957). The dimensions of stream channels are carefully constructed by streams to convey water for the most frequently experienced low-flow conditions. However, about once every 1 to 2 years, on average, a stream exceeds its bankfull capacity and flows overbank or erodes its banks to create a larger channel (Leopold et al., 1964).

Overbank deposits generally consist of sediments that are light enough to remain suspended in the uppermost parts of the water column. Consequently, they typically are composed of clay and silt and occasionally of fine-grained sand during very large, high-energy floods. Organic matter such as leaves and charcoal are very light and float on water during a flood; thus, they are commonly found at the top of an overbank deposit. Sometimes a stream breaks out of its banks (avulsion) and coarser-grained channel-bed material is deposited on the floodplain (e.g., Ritter, 1988).

Overbank deposits contain nutrients that can increase soil productivity after floodwaters recede. High-frequency, low-magnitude seasonal floods along rivers increase soil fertility by contributing algae, some of which are nitrogen fixing, and decomposed organic debris from floodwaters. In addition, cycles of wetting and drying lead to alternating reducing and oxidizing conditions, respectively, which accelerate rates of chemical weathering of minerals and release nutrients to plants (Alexander, 1993).

Although farmland is replenished with nutrients in areas where annual flooding is part of the farming cycle, in areas of floodplain control land is more likely to become less productive as a result of soil erosion, scour, and sedimentation during flooding (c.f., Ritter, 1988; NOAA, 1994). Most of the damage associated with the 1993 Mississippi River floods was the result of overbank flooding (resulting from levee breaks) and associated erosion and deposition on the floodplain (NOAA, 1994). Some 600 billion tons of soil was lost to erosion, diminishing future farmland productivity. Even in countries where agricultural practices are intimately linked to seasonal cycles of inundation, floods of low frequency and high magnitude can cause substantial devastation (Rasid and Mallik, 1995). During high-magnitude floods nutrient-poor sediments are spread across fields, sometimes in thick deposits, resulting in lower productivity and damaging fields so much that farming can be difficult if not impossible for a number of years.

The need to protect communities from devastating floods, but at the same time to enable the continual overbank replenishment of soils in agricultural cultures adapted to seasonal flooding, has created a dilemma for flood-control planning (c.f., Franks, 1994; Slaa et al., 1994). Recent flood action plans for Bangladesh, for example, recommend a policy of "living with floods" that relies less on major engineering works and more on an integrated approach that considers the needs of multiple social and economic sectors as well as environmental qualities (World Bank, 1990;

Franks, 1994; see also Gruntfest, Chapter 15, this volume). The "green river" concept for the Lower Atrai basin of northwestern Bangladesh (between the confluence of the Ganges and Brahmaputra Rivers) allows for an unimpeded floodway and some overbank flow during peak monsoon season and does not attempt to restrict floodwaters to a prescribed low-flow stage. As described by Franks (1994),

> ...the principle of the green river would require that some part of the local population would live in areas where flooding was expected during the monsoon... [but] this was preferable to the uncertainty of living within an area which enjoyed theoretical protection but which was susceptible to sudden rapid rises of water level when [levee] breaches or cuts took place (p. 372).

Flooding Processes and Hazards on Alluvial Fans

Alluvial fans are similar to deltas in that they are cone-shaped landforms created by deposition at the downstream ends of channels, which debauch onto plains of low gradient. Unlike deltas, they are not restricted to the shores of coastlines or lakes, and they form above rather than below water. However, one feature of alluvial fans is very similar to that of deltas and alluvial plains: the streams that form the fans do not remain fixed in the same place with time. As a consequence, it is very difficult to predict flood hazards for streams that are moving targets and that are as likely to gouge out the sediment in their way and branch into multiple channels as to overflow their banks. Such streams are referred to as having locational instabilities and pose a unique set of problems to human communities (NRC, 1996).

One other aspect of alluvial fans makes them especially hazardous to their occupants. Alluvial fans most commonly are found along mountain fronts, and consequently the sediment excavated from the mountains to build the fans is coarse, as coarse as boulders the size of cars in some cases. The streams that carry such debris usually are steep and prone to debris flows and mudflows, which can be much more destructive than overbank flooding (FEMA, 1989, 1995).

In the United States, alluvial fans pose problems for flood hazard specialists, because most of the standards and guidelines established by FEMA for determining flood-prone areas are based on streams that do not have locational instabilities or debris flows. Stable-channel streams overflow their banks during flooding, but streams on alluvial fans are more likely to enlarge and multiply their channels by bank cutting, aggradation, and incision.

Currently, FEMA is struggling with how to deal with such problems. Rapid growth in sunbelt states such as Arizona, California, and Nevada where alluvial fans are widespread, only worsens the problems, for FEMA must develop new guidelines and regulations while the areas they seek to understand are in the midst of rapid development. In 1994, FEMA sought the help of the National Academy of Sciences through its operating arm, the NRC. FEMA requested that a committee recommend an appropriate strategy for the future. It is likely that this study will lead to new guidelines for FEMA, as did the reports of past task forces that were requested to advise the government on issues of flooding (NRC, 1996).

Effects of the "Flood Protection-Development Spiral" on Human Communities: Case Study of the Mississippi River, USA

The flood protection-development spiral refers to the positive feedback between development and flood protection in flood-prone areas (NRC, 1995). As development on a floodplain proceeds, citizens call for protection after a damaging flood. Construction of levees and storage reservoirs, however, promotes further development in the now protected floodplain (see Watson and Biedenharn, Chapter 14, of this volume). If these structures are inadequate or fail during subsequent floods, even higher levels of protection are demanded. This phenomenon was described in the 1966 Report of the Task Force on Federal Flood Control Policy (U.S. Congress, 1966): "The major purpose of engineering projects is changing from the protection of established property to the underwriting of new development." A prominent example of this scenario has been played out along the Mississippi River in the 19th and 20th centuries, but the most recent floods of 1993 have increased the urgency of getting out of the flood protection-development spiral.

Mississippi River Basin and History of Its Flood-Control Program

The Mississippi River watershed commonly is divided into two parts referred to as the upper and lower basins. The transition between the two at the confluence of the Mississippi trunk stream with its largest tributary, the Ohio River, is dramatic, marked by the sudden change from a narrow channel cascading over bedrock waterfalls to a vast alluvial plain in the lower Mississippi Valley. This alluvial plain is built from more than 300 m of alluvium, forming a nearly featureless lowland 1000 km long and up to 200 km wide. The lower Mississippi Valley formed over millions of years in response to rising and falling sea level and retreating and advancing deltas during the Pleistocene Epoch. When European immigrants arrived in the new world, the Mississippi flowed in sweeping, twisting meander bends across this surface. Today, however, the modern channel of the Mississippi is fettered, constricted, and yoked by thousands of kilometers of levees and hundreds of dams and reservoirs (Watson and Biedenharn, Chapter 14, this volume).

Early efforts to minimize the impact of flooding on human communities relied on levees. The earliest levees – just more than 1 m high – were built in 1727 to protect the first European settlement on the Mississippi River, called Nouvelle (New) Orleans. The Mississippi River Commission was established by Congress in 1879, and until 1927 it relied on a levees only policy to control flooding. With each flood, more levees were constructed, and the height of each levee was raised (Davis, 1982).

In the 20th century, the Mississippi River was substantially straightened, with more than 300 of its original kilometers removed at meander bend cutoffs (Rasmussen, 1994). The call for such shortening resulted from the failure of levees to hold water at more than 120 places during the great Mississippi flood of 1927, which killed several hundred people and left more than half a million homeless. Major General Harley B. Ferguson, the newly appointed president of the Mississippi River Commission, declared that levees alone could not control the great river, because the real problem was that the river could not rid itself quickly enough of its floodwaters while navigating such tortuous hairpin turns. In his own words, "The water wants out. We will give it out" (quoted in Clark, 1982, p. 79).

Within a decade, the lower Mississippi River was straighter and many hundreds of kilometers shorter, and no subsequent flood has caused as many deaths as that of 1927. Still, straightening the river has had unanticipated negative consequences, such as loss of riverine wetlands. Furthermore, after the Great Flood of 1993, the wisdom of large flood control projects that consist of dams, straightened channels, and levees is being reconsidered, as discussed next.

Effects of the 1993 Upper Mississippi River Basin Floods on Human Communities

In mid-June through early August of 1993, widespread and severe flooding occurred throughout the upper Mississippi River basin, destroying millions of hectares of crops, damaging highways and roads, breaking many levees, severely eroding channel banks and hillslopes, and dumping sediment over large areas. More than 30 million ha throughout nine states were catastrophically affected (NOAA, 1994). The number of fatalities was 48, more than 50,000 homes were either damaged or destroyed, and about 54,000 people were evacuated from flooded areas. More than 75 towns were completely inundated, and some of them might never be rebuilt. Total damage was estimated at between $12 and $15 billion, making this single period of flooding the cause of the worst economic disaster in United States history. It also was the greatest flood in modern times in terms of aerial extent of flooding, number of persons displaced, extent of crop and property damage, and duration of flooding (NOAA, 1994).

The Great Flood of 1993 began with a chain of meteorological and hydrologic events. Precipitation was greater than normal for 6 months before the flooding, so soils were saturated throughout the region when a sequence of intense rainstorms began in late June. For this reason, little to no infiltration occurred, and most water became direct stormwater runoff, rapidly raising the water levels of numerous streams that already were at bankfull conditions. Flood-control reservoirs also were at or near capacity. One storm after another pelted the area for nearly 2 months, and one tributary after another swelled and disgorged its load into the main Mississippi River trunk stream. Peak discharges exceeded 100-year flood levels at 46 gauging stations and were greater than any previously recorded values at 42 stations. At St. Louis, Missouri, peak flow was more than 28,000 m^3/s.

Flood damage resulted largely from overbank flooding that occurred where levees broke (Figure 10.3) (NOAA, 1994). Of 229 federal levees, 40 (18%) were overtopped or damaged; of 1347 nonfederal levees, 1043 (77%) were overtopped or damaged (NOAA, 1994). A greater proportion of nonfederal levees failed because they were designed to withstand floods with recurrence intervals of 50 years or less, whereas federal levees were designed to hold back waters of floods with recurrence intervals of 100 to 500 years. As levees were breached, floodwaters spread over urban and suburban areas and inundated millions of hectares of farmland. Even where not inundated, crops on higher ground were damaged as a result of diseases, insects, and weeds that flourished in the moist air and water-logged soils (Figure 10.4).

Figure 10.3. Just north of Valmeyer, Illinois, the Mississippi River breaks through a levee at two spots. Once a levee is breached, it can widen rapidly and enable substantial amounts of river water to spread into protected land behind the levee. This levee failed during the 1993 Mississippi River flooding, after 24 days of attempts to reinforce and save the structure. (Photograph by Scott Dine, St. Louis Post-Dispatch)

Figure 10.4. As hundreds of levees were breached, floodwaters spread onto tens of millions of hectares of farmland during the 1993 Mississippi River flooding, ruining crops and killing livestock throughout nine states in the mid-west. (Photograph by Scott Dine, St. Louis Post-Dispatch)

Lessons from the Flooding of 1993

The U.S. Army Corps of Engineers estimates that flooding would have been far worse if not for the storage of large volumes of water in dozens of flood-control reservoirs, most of which are located within the Missouri River basin. Flood-control reservoirs in the Missouri River basin reduced the discharge at St. Louis by more than 5908 m^3/s in July 1993. Without the reservoirs, many more levees would have been breached and more cities and farms would have been flooded. Yet, adding more flood control reservoirs and raising levees would only promote the flood protection-development spiral.

After the flooding of 1993, the White House directed a major evaluation of its causes and impacts, which was prepared by the Interagency Floodplain Management Review Committee (IFMRC, 1994). The same year, FEMA published the Unified National Program for Floodplain Management (FEMA, 1994). Recommendations in both reports call for a more integrated approach to floodplain management and planning that involves consideration of watersheds as a whole instead of just the main stems of river systems. Examples of specific recommendations are the restoration of wetlands and floodplains and relocation of structures out of floodplains. Interestingly, these recommendations are somewhat similar to the proposed concept of a green river as described above for Bangladesh, in a nation where rivers have not yet been substantially channelized and restricted with levees.

Consequences of Streamflow Altered by Dams: Case Study of the Colorado River

Effect of Dams on Salinity, Discharge, and Sediment Loads in the Colorado River

Modern, large dams have as much effect on rivers and human communities as the floods they are purported to control (c.f., Collier et al., 1996). A classic example is found on the Colorado River in the semiarid southwestern United States, which provides water to residents in many states and northern Mexico. Discharge of the Colorado River has declined continually since the beginning of dam building along the river in the late 1930s, and much of the river's water is now diverted for irrigation and municipal drinking water supplies. Reservoirs of water behind numerous dams provide large surface areas for evaporation losses, as in the Powell Reservoir behind Glen Canyon Dam, where 9% of the river's water is evaporated and the remaining water becomes saltier. After the dam was closed in 1962 and brackish waters were returned to the river from irrigated fields, water salinity at the river's mouth increased from 800 mg/liter to 1500 mg/liter.

Reservoirs also act to trap sediment, eventually filling completely with debris (i.e., the design life of the reservoir). Until that time, however, the sediment load of the dammed river is greatly decreased. The Colorado's sediment load decreased abruptly with completion of the Hoover Dam in the mid-1930s and again in the 1960s as other dams were completed (Dolan et al., 1974; NRC, 1991b).

Diminished in-stream flow, lack of seasonal flooding, reduced sediment loads, and increased salinity have detrimental effects on river channels, associated ecosystems, and human communities (Petts, 1984; NRC, 1991b; see also Friedman and Auble, Chapter 8, and Wydoski and Wick, Chapter 9, this volume). Downstream from a dam, water depleted of its sediment load is released into the channel, and as a consequence the flow has more capacity than needed to transport sediment and is able to scour debris from point bars, the floodplain, and the channel bed (Schmidt and Graf, 1990; Collier et al., 1997). The debris eroded generally is fine grained, however, because large discharges rarely occur after dams and flood control works are built. As a result, rivers such as the Colorado are unable to flush out coarse deposits of gravel and boulders that were moved during previous floods. Each of these types of changes has been documented on numerous rivers throughout the world, including the Colorado, Brazos (Texas), Missouri, and Nile Rivers.

Social and Political Consequences of Diminished Flow on Human Communities Along the Colorado River

The social and economic consequences of diminished flow and increased salinity of the Colorado River have resulted in substantial political fallout in the American Southwest (c.f., Fradkin, 1984; Reisner, 1993). The Colorado River often is merely a salty dribble at its outlet into the Gulf of California in Mexico. In the 1960s, the Mexican government protested that its allocation of water from the Colorado (guaranteed by the United States–Mexican Treaty of 1945) was too brackish to be fit for human consumption or irrigation. As a consequence, the U.S. government agreed to build and maintain a desalination plant in Yuma, Arizona, and passed the high costs to taxpayers.

A number of legislative acts and treaties among southwestern states and Mexico brought the Colorado River to this plight. In 1928, Congress passed the Boulder Canyon Project Act, giving the Bureau of Reclamation and Army Corps of Engineers authorization to construct the giant Hoover Dam, numerous small dams, and several canals. Hoover Dam is the highest in the nation (213 m) and was cited by the American Society of Civil Engineers as one of seven civil engineering wonders in the nation. Electricity generated by dams along the Colorado River is sold throughout the Southwest and used to subsidize the supply of irrigation water to farmers in the surrounding deserts of California, Arizona, Utah, Colorado, and Mexico. The Boulder Canyon Project Act guaranteed a supply of 5.4 billion m^3 of water per year to California, and the Colorado River aqueduct was built to transport water to southern California. Without this allocation, the growth of cities such as Los Angeles would have been extremely curtailed. Today, water from the Colorado supplies more than 12 million people in California.

The state of Arizona protested the building of Hoover Dam, as its allocation of Colorado River water is only 3.5 billion m^3. Furthermore, California was guaranteed to receive its 5.4 billion m^3 even during drought years. Arizona once relied largely on groundwater resources to supplement its water needs, but these resources have been depleted rapidly with booming population growth since World War II. As a result, the states of Arizona and California, as well as Native Americans nations that use water from the Colorado, have been mired in a series of complex court cases since the 1950s as they battle over the rights to each precious cubic meter of water.

Low to High Flows of the Colorado River in the Past 400 Years

Arizona has good reason to worry that drought years might occur, for paleohydrologic analysis of tree rings indicates that the original estimate of how much water is annually available in the Colorado watershed was done during an unusually wet period (Stockton and Jacoby, 1976). Reconstructions of streamflow records from tree rings in the Colorado River region indicate that wet and dry periods have been cyclic over the past four centuries, with average annual flow during wet periods nearly twice that of dry years.

The period of record used to allocate water to users of the Colorado River was the early 1900s, before the Colorado River Compact drafted by states in the watershed in 1922. This was the wettest period in the entire 400-year record. At that time, mean annual flow at Lees Ferry (Grand Canyon) was estimated to be 22 billion m^3 (18 million acre-feet), but this estimate was based on less than a few decades of recorded flow. The total amount of water entitled to users in the United States and Mexico is about 18.5 billion m^3 per year, about 3.7 billion m^3 less than the amount estimated to exist in 1922. Since 1922, the average annual flow has decreased steadily. The long-term mean water yield from tree ring analysis is about 16.5 billion m^3 (13.4 million acre-feet), almost 2.5 billion m^3 less than the amount allocated to modern water users and about equal to the amount consumed in 1978. If the Colorado experiences a period of low flow like those revealed in the tree ring record, the millions of users of Colorado River water could experience a severe water shortage for a number of years.

Conclusions

The effects of variable river flow on human communities are themselves variable, depending on the geomorphic aspects of a given river, the hydroclimatological conditions associated with low or high flows, and the relationship of the local community to the river on which it depends. Along alluvial ephemeral desert streams in southern Arizona, deltaic rivers of Bangladesh, and alluvial fans worldwide, for example, rapid channel bank erosion and migration pose substantial threats to human communities. Ordinary methods of flood control designed to minimize overbank flooding are ineffective in such cases. One response is to protect banks with firm materials (e.g., soil cement) but another is simply to avoid building permanent structures along unstable channel banks.

Even where overbank flooding rather than bank erosion is the major threat, however, as along the restricted, channelized Mississippi River, construction of flood control reservoirs and levees only perpetuates an endless spiral of development with each new increment of protection. Levees cannot be designed to protect against every conceivable flood, so when an unusually high magnitude event does occur – as during the 1993 Mississippi flooding – damage is widespread and extreme.

Other types of grand water regulation efforts also pose hazards to human communities. Along the Colorado River in the American Southwest, for example, regulating the river with dams and diversion channels has resulted in diminished flow (due to evaporation losses) and increasingly saline water. In addition, millions of people who have moved to this arid region and who rely on the Colorado River for water are increasingly vulnerable to the threat of drought and low flow. As learned from paleohydrologic studies, the Colorado – like other rivers – changes stage from year to year, decade to decade, and even century to century, challenging the use of specific 20th century reference years for allocating scarce water resources.

These many examples are reminders of Pliny's Nilometer. As river stage rises from the channel bed, the effects on human communities change from hunger to suffering, happiness, security, abundance, and, finally, disaster. Despite millennia of attempts to deem it otherwise, human communities still are confronted with the hazards of variable flow.

References

Alexander, D. (1993). *Natural Disasters.* New York: Chapman and Hall.

Beinin, L. (1979). Sanitary consequences of inundations. *Disasters*, **3**, 213–216.

Bissell, R.A. (1983). Delayed impact infectious disease after a natural disaster. *Journal of Emergency Medicine*, **1**, 59–66.

Brammer, H. (1990). Floods in Bangladesh, I. Geographical background to the 1987 and 1988 floods. *Geographical Journal*, **156**, 12–22.

Clark, C. (1982). *Planet Earth: Flood.* Alexandria, VA: Time-Life Books.

Coleman, J.M. (1969). Brahmaputra River: Channel processes and sedimentation. *Sedimentary Geology*, **3**, 129–239.

Collier, M.P., Webb, R.H., and Schmidt, J.C. (1996). Dams and rivers: Primer on downstream effects of dams. *U.S. Geological Survey Circular 1126.* U.S. Government Printing Office, Washington, D.C.

Collier M.P., Webb, R.H., and Andrews, E.D. (1997). Experimental flooding in Grand Canyon. *Scientific American*, January, 82–89.

Cooper, J. (1963). Flood plain foes erupt at meeting. *Tucson Citizen*, October 5.

Cosgrove, D. and Petts, G. (1990). *Water, Engineering, and Landscape: Water Control and Landscape Transformation in the Modern Period.* London: Belhaven Press.

Davis, N.D. (1982). *The Father of Waters, a Mississippi River Chronicle.* San Francisco, CA: Sierra Club Books.

Dolan, R., Howard, A.D., and Gallenson, A. (1974). Man's impact on the Colorado River in Grand Canyon. *American Scientist*, **62**, 392–401.

Ellett, C. (1853). *The Mississippi and Ohio Rivers.* Philadelphia, PA: Lippincott, Grambo, and Co.

Federal Emergency Management Agency (FEMA). (1989). *Alluvial Fans: Hazards and Management*, Doc. No. 165. Washington, D.C.: FEMA.

Federal Emergency Management Agency (FEMA). (1994). *A Unified National Program for Floodplain Management.* Washington, D.C.: FEMA (part of a series originating in 1966).

Federal Emergency Management Agency (FEMA). (1995). *Appendix 5: Studies of Alluvial Fan Flooding. Guidelines and Specifications for Study Contractors.* Doc. No. 37. Washington, D.C.: FEMA.

Foster, H.A. (1924). Theoretical frequency curves and their application to engineering problems. *American Society of Civil Engineers Transactions*, **87**, 142–173.

Fradkin, P.L. (1984). *A River No More: The Colorado River and the West.* Tucson, Arizona: University of Arizona Press.

Franks, T.R. (1994). The Green River in Bangladesh: The Lower Atrai Basin. In *Integrated River Basin Development*, eds. C. Kirby and W.R. White, pp. 367–376. Chichester: Wiley and Sons.

Glass, R.I., Urrutia, J.J., Sibony, S., Smith, H., Garcia, B., and Rizzo, L. (1977). Earthquake injuries related to housing in a Guatemalan village. *Science*, **197**, 638–643.

Gueri, M., Gonzales, C., and Morin, V. (1986). The effect of the floods caused by "El Niño" on health. *Disasters*, **10**, 118–124.

Hays, W.W. (1981). Gauging geologic and hydrologic hazards: Earth-science considerations, *U.S. Geological Survey Professional Paper 1240-B.* Washington, D.C.: U.S. Government Printing Office.

Hederra, R. (1987). Environmental sanitation and water supply during floods in Ecuador (1982–1983). *Disasters*, **11**, 297–309.

Interagency Floodplain Management Review Committee (IFMRC). (1994). *Sharing the Challenge: Floodplain Management into the 21st Century.* Washington, D.C.: U.S. Government Printing Office.

Kates, R.W. (1962). *Hazard and Choice of Perception in Flood Plain Management.* Chicago, Illinois: University of Chicago Press.

Kovacs, M.G. (translation) (1985). *The Epic of Gilgamesh.* Stanford, CA: Stanford University Press.

Kresan, P.L. (1988). The Tucson, Arizona, flood of October 1983: Implications for land management along alluvial river channels. In *Flood Geomorphology*, eds. V.R. Baker, R.C. Kochel, and P.C. Patton, pp. 465–489. New York: John Wiley and Sons.

Leopold, L.B., Wolman, M.G., and Miller, J.P. (1964). *Fluvial Processes in Geomorphology.* San Francisco, CA: W.H. Freeman and Company.

National Oceanic and Atmospheric Administration (NOAA). (1994). *Natural Disaster Survey Report: The Great Flood of 1993.* Washington, D.C.: U.S. Department of Commerce.

National Research Council (NRC). (1984). *The Tucson, Arizona, Flood of October, 1983.* Washington, D.C.: National Academy Press.

National Research Council (NRC). (1991a). *Opportunities in the Hydrologic Sciences.* Washington, D.C.: National Academy Press.

National Research Council (NRC). (1991b). *Colorado River Ecology and Dam Management.* Washington, D.C.: National Academy Press.

National Research Council (NRC). (1995). *Flood Risk Management and the American River Basin: An Evaluation.* Washington, D.C.: National Academy Press.

National Research Council (NRC). (1996). *Alluvial Fan Flooding.* Washington, D.C.: National Academy Press.

Parker, D.J. and Harding, D.M. (1979). Natural hazard evaluation, perception and adjustment. *Geography*, **64**, 307–16.

Petts, G.E. (1984). *Impounded rivers: Perspectives for ecological management.* Wiley and Sons: Chichester.

Rasid, H. and Mallik, A. (1995). Flood adaptations in Bangladesh: Is the compartmentalization scheme compatible with indigenous adjustments of rice cropping to flood regimes? *Applied Geography*, **15**, 3–17.

Rasmussen, J.L. (1994). Management of the Upper Mississippi: A Case History. In *The Rivers Handbook (vol. II): Hydrological and Ecological Principles*, eds. P. Calow and G.E. Petts, pp. 441–463. Oxford: Blackwell Scientific Publications.

Reisner, M. (1993). *Cadillac Desert: The American West and Its Disappearing Water.* Revised and updated. Vancouver: Doublas & McIntyre.

Riggs, H.C. (1966). *Chapter 3 – Frequency Curves.* Washington, D.C.: U.S. Geological Survey Surface Water Techniques Series, Book 2.

Ritter, D.F. (1988). Floodplain erosion and deposition during the December 1982 floods in southeast Missouri. In *Flood Geomorphology*, eds. V.R. Baker, R.C. Kochel, and P.C. Patton, pp. 243–259. New York: John Wiley and Sons.

Rogers, J.J.W. and Feiss, P.G. (1998). *People and the Earth: Basic Issues in the Sustainability of Resources and Environment.* Cambridge, U.K.: Cambridge University Press.

Schmidt, J.C. and Graf, J.B. (1990). Aggradation and degradation of alluvial sand deposits, 1965–1986, Colorado River, Grand Canyon National Park, Arizona. *U.S. Geological Survey Professional Paper 1493*, 74 pp. U.S. Government Printing Office, Washington, D.C.

Slaa, B.T., Monar, F.C., and Laboyrie, H. (1994). Erosion control of the Meghna River, Bangladesh. In *Integrated River Basin Development*, eds. C Kirby and W.R. White, pp. 289–298. Chichester: Wiley and Sons.

Stevens, M.A., Simons, D.B., and Schumm, S.A. (1975). Man-induced changes of middle Mississippi River. *Journal of the Waterways Harbors and Coastal Engineering Division, ASCE*, **101**, 119–33.

Stockton, C.W. and Jacoby, G.J. (1976). Long-term surface water supply and streamflow trends in the upper Colorado River basin. *Lake Powell Research Project Bulletin Number 18.* Los Angeles: University of California.

Telleria, A.V. (1986). Health consequences of floods in Bolivia in 1982. *Disasters*, **10**, 88–106.

U.S. Army Corps of Engineers (USACE). (1992). *Guidelines for Risk and Uncertainty Analysis in Water Resources Planning.* Report. 92-R-1. Fort Belvoir, VA: Water Resources Support Center.

U.S. Congress. (1966). *A Unified National Program for Managing Flood Losses.* House doc. no. 465, 89th Congress, 2nd Session. Washington, D.C.: U.S. Government Printing Office.

Wolman, M.G. and Leopold, L.B. (1957). River flood plains: Some observations on their formation. *U.S. Geological Survey Professional Paper 282-C.* U.S. Government Printing Office, Washington, D.C.

World Bank. (1990). *Flood Control in Bangladesh: A Plan for Action.* Technical Paper No. 119. Washington, D.C.: The World Bank, Asia Region Technical Department.

RESPONSES TO FLOODING

CHAPTER 11

Prediction and Modeling of Flood Hydrology and Hydraulics

Jorge A. Ramírez
Civil Engineering Department
Colorado State University

The basic principles underlying the most commonly used physically based models of the rainfall-runoff transformation process are reviewed. A thorough knowledge of these principles is a prerequisite for flood hazard studies; thus, this chapter reviews several physically based methods to determine flood discharges, flow depths, and other flood characteristics. The chapter starts with a thorough review of linear system theory applied to the solution of hydrologic flood routing problems in a spatially aggregated manner: unit hydrograph approaches. The chapter then proceeds to a review of distributed flood routing approaches, in particular the kinematic wave and dynamic wave approaches. The chapter concludes with a brief discussion about distributed watershed models, including single-event models in which flow characteristics are estimated only during the flood and continuous event models in which flow characteristics are determined continuously during wet periods and dry periods.

Introduction

Flood prediction and modeling refer to the processes of transformation of rainfall into a flood hydrograph and to the translation of that hydrograph throughout a watershed or any other hydrologic system. Flood prediction and modeling generally involve approximate descriptions of the rainfall-runoff transformation processes. These descriptions are based on either empirical, or physically based, or combined conceptual–physically based descriptions of the physical processes involved. Although in general the conceptualizations may neglect or simplify some of the underlying hydrologic transport processes, the resulting models are quite useful in practice because they are simple and provide adequate estimates of flood hydrographs.

In modeling single floods, the effects of evapotranspiration as well as the interaction between the aquifer and the streams are ignored. Evapotranspiration may be ignored because its magnitude during the time period in which the flood develops is negligible compared with other fluxes such as infiltration. Likewise, the effect of the stream–aquifer interaction is generally ignored because the response time of the subsurface soil system is much longer than the response time of the surface or direct runoff process. In addition, effects of other hydrologic processes such as interception and depression storage are also neglected. Event-based modeling generally involves the following aspects:

(a) evaluation of the rainfall flux over the watershed $I(\mathbf{x}, t)$ as a function of space and time;
(b) evaluation of the rainfall excess or effective rainfall flux as a function of space and time, $I_e(\mathbf{x}, t)$ – effective rainfall is the rainfall available for runoff after infiltration and other abstractions have been accounted for; and
(c) routing of the rainfall excess to the watershed outlet to determine the corresponding flood hydrograph $Q(t)$.

Hydrologic flood prediction models may be categorized into physical models and mathematical models. Mathematical models describe the system behavior in terms of mathematical equations representing the relationships between system state, input, and output. Mathematical models can, in turn, be categorized as either purely conceptual models or physically based models. Depending on whether the functions relating input, output, and system state are functions of space and time, these models may be further categorized as lumped models or distributed models. Lumped models do not account explicitly for the spatial variability of hydrologic processes, whereas distributed models do. Lumped models use averages to represent spatially distributed function and properties.

Unit Hydrograph Analysis

Sherman (1932) first proposed the unit hydrograph concept. The unit hydrograph (UH) of a watershed is defined as the direct runoff hydrograph resulting from a unit volume of excess rainfall of constant intensity and uniformly distributed over the drainage area. The duration of the unit volume of excess or effective rainfall, sometimes referred to as the *effective duration*, defines and labels the particular UH. The unit volume is usually considered to be associated with 1 cm (1 in.) of effective rainfall distributed uniformly over the basin area.

The fundamental assumptions implicit in the use of UHs for modeling hydrologic systems are as follows:

(a) Watersheds respond as linear systems. On the one hand, this implies that the proportionality principle applies so that effective rainfall intensities (volumes) of different magnitude produce watershed responses that are scaled accordingly. On the other hand, it implies that the superposition principle applies so that responses of several different storms can be superimposed to obtain the composite response of the catchment.
(b) The effective rainfall intensity is uniformly distributed over the entire river basin.
(c) The rainfall excess is of constant intensity throughout the rainfall duration.
(d) The duration of the direct runoff hydrograph – that is, its time base – is independent of the effective rainfall intensity and depends only on the effective rainfall duration.

When the effective rainfall is given as a hyetograph – that is, as a sequence of M rainfall pulses of the same duration Δt – the corresponding direct runoff hydrograph can be expressed as the discrete convolution of the rainfall hyetograph and the UH as,

$$Q_n = \sum_{m=1}^{m^*} P_m U_{n-m+1}, \qquad m^* = \min(n, M) \tag{1a}$$

$$Q_n = Q(n\Delta t), \qquad n = 1, \ldots, N \tag{1b}$$

$$P_m = \int_{(m-1)\Delta t}^{m\Delta t} I_e(\tau)\,d\tau, \qquad m = 1, \ldots, M \tag{1c}$$

where P_m is the volume of the mth effective rainfall pulse, Q_n is the direct runoff, and U_{n-m+1} are the UH ordinates.

Although the above assumptions lead to acceptable results, watersheds are indeed nonlinear systems. For example, UHs derived from different rainfall-runoff events, under the assumption of linearity, are usually different, thereby invalidating the linearity assumption.

The determination of UHs for particular basins can be carried out either by using the theoretical developments of linear system theory or by using empirical techniques. For either case, simultaneous observations of both precipitation and streamflow must be available. These two approaches are presented in more detail in later sections.

Hydrograph Components

Total streamflow during a precipitation event includes the baseflow existing in the basin before the storm and the runoff due to the given storm precipitation. Total streamflow hydrographs are usually conceptualized as being composed of the following:

(a) Direct runoff, which is composed of contributions from surface runoff and quick interflow – UH analysis refers only to direct runoff;
(b) Baseflow, which is composed of contributions from delayed interflow and groundwater runoff.

Surface runoff includes all overland flow as well as all precipitation falling directly onto stream channels. Surface runoff is the main contributor to the peak discharge.

Interflow is the portion of the streamflow contributed by infiltrated water that moves laterally in the subsurface until it reaches a channel. Interflow is a slower process than surface runoff. Components of interflow are quick interflow, which contributes to direct runoff, and delayed interflow, which contributes to baseflow (e.g., Chow, 1964).

Groundwater runoff is the flow component contributed to the channel by groundwater. This process is extremely slow compared with surface runoff.

Schematically in Figure 11.1, the streamflow hydrograph is subdivided into rising limb, rising portion of the hydrograph, composed mostly of surface runoff; crest, zone of the hydrograph around peak discharge; and falling (or recession) limb, portion of the hydrograph after the peak discharge, composed mostly of water released from storage in the basin. The lower part of this recession corresponds to groundwater flow contributions.

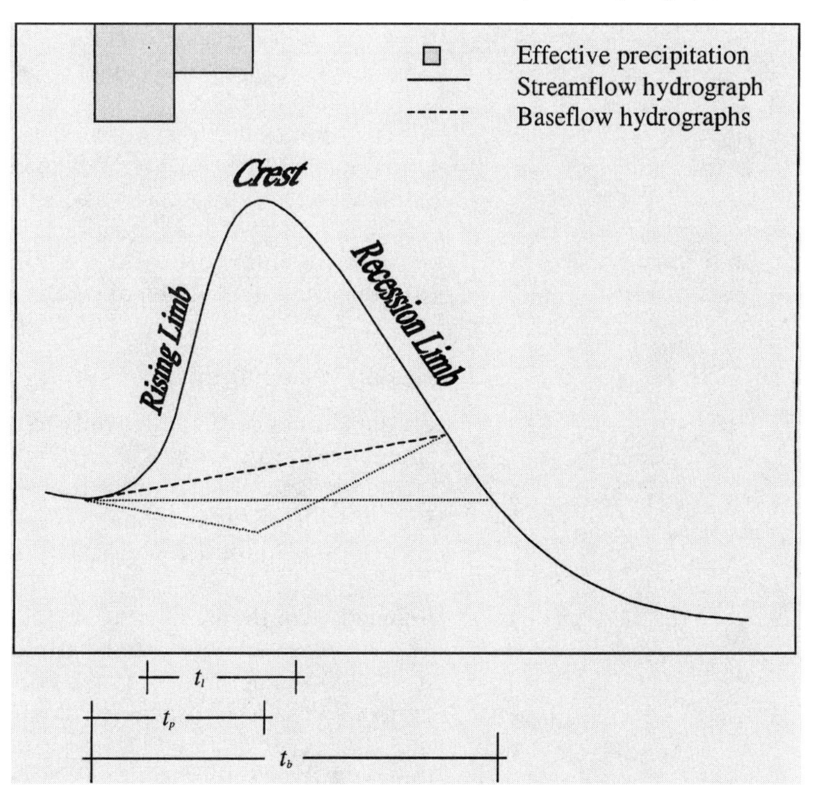

Figure 11.1. Schematic description of hydrograph.

The main factors that affect hydrograph shape are as follows:

(a) Drainage characteristics: basin area, basin shape, basin slope, soil type and land use, drainage density, and drainage network topology. Most changes in land use tend to increase the amount of runoff for a given storm (e.g., Chow et al., 1988; Singh, 1989; Bras, 1990).
(b) Rainfall characteristics: rainfall intensity, duration, and their spatial and temporal distribution, and storm motion, as storms moving in the general downstream direction tend to produce larger peak flows than storms moving upstream (e.g., Chow et al., 1988; Singh, 1989; Bras, 1990).

Hydrographs are also described in terms of the following time characteristics (see Figure 11.1):

(a) Time to peak, t_p: Time from the beginning of the rising limb to the occurrence of the peak discharge. The time to peak is largely determined by drainage characteristics such as drainage density, slope, channel roughness, and soil infiltration characteristics. Rainfall distribution in space also affects the time to peak.
(b) Time of concentration, t_c: Time required for water to travel from the most hydraulically remote point in the basin to the basin outlet. For rainfall events of very long duration, the time of concentration is associated with the time required for the system to achieve the maximum or equilibrium discharge. Kibler (1982) and Chow et al. (1988) summarize several of many empirical and physically based equations for t_c that have been developed. The drainage characteristics of length and slope, together with the hydraulic characteristics of the flow paths, determine the time of concentration.
(c) Lag time, t_l: Time between the center of mass of the effective rainfall hyetograph and the center of mass of the direct runoff hydrograph. The basin lag is an important concept in linear modeling of basin response. The lag time is a parameter that appears often in theoretical and conceptual models of basin behavior. However, it is sometimes difficult to measure in real-world situations. Many empirical equations have been proposed in the literature. The simplest of these equations computes the basin lag as a power function of the basin area.
(d) Time base, t_b: Duration of the direct runoff hydrograph.

Baseflow Separation

As the UH concept applies only to direct runoff, the direct runoff must be separated from the baseflow. Baseflow separation or hydrograph analysis is the process of separating the direct runoff (surface runoff and quick interflow) from the baseflow. This separation is somewhat arbitrary, but it corresponds to theoretical concepts of basin response.

Subjective Methods. Several subjective methods are shown in Figure 11.1. The simplest one consists of arbitrarily selecting the discharge marking the beginning of the rising limb as the value of the baseflow and assuming that this baseflow discharge remains constant throughout the storm duration. A second method consists of arbitrarily selecting the beginning of the groundwater recession on the falling limb of the hydrograph (usually assumed to occur at a theoretical inflection point) and connecting this point by a straight line to the beginning of the rising limb. A third example of subjective methods consists of extending the recession before the storm by a line from the beginning of the rising limb to a point directly beneath the peak discharge and then connecting this point to the beginning of the groundwater recession on the falling limb.

Area Method. The area method of baseflow separation consists of determining the beginning of the baseflow on the falling limb with the following empirical equation,

$$N = bA^{0.2} \tag{2}$$

relating the time in days from the peak discharge N to the basin area A. When A is in square miles b equals 1. When A is in square kilometers b equals 0.8. This equation is unsuitable for smaller watersheds and should be checked for a number of hydrographs before using.

Master Recession Curve Method. This method consists of modeling the response of the groundwater aquifer as a linear reservoir of parameter k. This assumption leads to the following equation for the groundwater recession hydrograph,

$$Q(t) = Q(t_o)e^{-(t-t_o)/k} \tag{3}$$

where $Q(t)$ is the baseflow at time t; $Q(t_o)$ is a reference baseflow discharge at time t_o, and k is the recession constant for baseflow. This method is based on a linear reservoir model of unforced basin response (that is, response from storage) and it can be used to separate the contributions to the recession flow from surface storage, subsurface storage, and groundwater aquifer storage. It involves determination of several recession constants.

Unit Hydrographs: Empirical Derivation

The following are essential steps in deriving a UH from a single storm:

(a) Separate the baseflow and obtain the direct runoff hydrograph.
(b) Compute the total volume of direct runoff and convert this volume into equivalent depth of effective rainfall (in centimeters or in inches) over the entire basin.
(c) Normalize the direct runoff hydrograph by dividing each ordinate by the equivalent volume (inches or centimeters) of direct runoff (or effective rainfall).
(d) Determine the effective duration of excess rainfall. To do this, obtain the effective rainfall hyetograph (e.g., use the ϕ-index; the Horton, Green and Ampt, or Philip equations; or some other method to determine infiltration losses) and its associated duration. This duration is the duration associated with the UH.

UHs are fundamentally linked to the duration of the effective rainfall event producing them. They can be used only to predict direct runoff from storms of the same duration as that associated with the UH or from storms that can be described as a sequence of pulses, each of the same duration as that associated with the UH (see Equation 1).

UHs for Different Effective Duration

A UH for a particular watershed is developed for a specific duration of effective rainfall. When dealing with a rainfall of different duration a new UH must be derived for the new duration. The linearity property implicit in the UH analysis can be used to generate UHs associated with larger or smaller effective rainfall pulse duration. This procedure is sometimes referred to as the S-curve hydrograph method.

S-Curve Hydrograph Method

An S-hydrograph represents the response of the basin to an effective rainfall event of infinite duration. Assume that a UH of duration D is known and that a UH for the same basin but of duration D' is desired. The first step is to determine the S-curve hydrograph by adding a series of (known) UHs of duration D, each lagged by time interval D. The resulting superposition represents the runoff resulting from a continuous rainfall excess of intensity $1/D$.

Lagging the S-curve in time by amount D' and subtracting its ordinates from the original unmodified S-curve yields a hydrograph corresponding to a rainfall event of intensity $1/D$ and of duration D'. Consequently, to convert this hydrograph whose volume is D'/D into a UH of duration D', its ordinates must be normalized by multiplying them by D/D'. The resulting ordinates represent a UH associated with an effective rainfall of duration D'.

Unit Hydrographs: Linear System Theory

A hydrologic system (a basin) is said to be a linear system if the relationship between storage, inflow, and outflow is such that it leads to a linear differential equation. The hydrologic response of such systems can be expressed in terms of an impulse response function (IRF) through a so-called convolution equation. Linear systems possess the properties of additivity and proportionality, which are implicit in the convolution equation. Linear reservoirs, for example, are special cases of a general hydrologic system model in which the storage is linearly related to the output by constant k.

IRF

The IRF of a linear system represents the response of the system to an instantaneous impulse (excitation) of unit volume applied at the origin in time ($t = 0$). The response of continuous linear systems can be expressed in the time domain in terms of the impulse response function via the convolution integral as follows:

$$Q(t) = \int_0^t I_e(\tau)u(t-\tau)d\tau \tag{4}$$

where $u(t)$ is the impulse response function of the system.

In hydrology, it has been customary to assume that watersheds behave as linear systems. When dealing with hydrologic systems, $u(t)$ represents the instantaneous UH (IUH), and $Q(t)$ and $I_e(t)$ represent direct runoff and excess or effective precipitation, respectively. Thus, an IUH represents the response of a watershed (discharge at its outlet as a function of time) to a unit volume of precipitation uniformly distributed over the basin and occurring instantaneously at time $t = 0$.

Unit Step Response Function

The unit step response function (SRF) is the theoretical counterpart to the S-curve hydrograph concept presented earlier in the empirical UH analysis section. It represents the runoff hydrograph from a continuous effective rainfall of unit intensity. As can be seen from its definition, it is the convolution of 1 and $u(t)$ and is obtained as

$$g(t) = \int_0^t u(t)dt \tag{5}$$

Unit Pulse Response Function

The unit pulse response function (PRF) is the theoretical counterpart to the UH concept presented earlier. It represents the runoff hydrograph from a constant effective rainfall of intensity $1/\Delta t$ and of duration Δt.

$$h(t) = \frac{1}{\Delta t}[g(t) - g(t - \Delta t)] = \frac{1}{\Delta t}\int_{t-\Delta t}^{t} u(\tau)d\tau \tag{6}$$

From its definition, the PRF can be seen as the normalized difference between two lagged SRFs (S-curve hydrographs) lagged by amount Δt. This is analogous to the procedure presented earlier in the section on S-hydrograph analysis.

Discrete Convolution Equation

When the effective rainfall is given as a hyetograph – that is, as a sequence of rainfall pulses of the same duration Δt – the corresponding direct runoff hydrograph can be expressed as the discrete convolution of the rainfall hyetograph and a UH,

$$Q_n = \sum_{m=1}^{m^*} P_m U_{n-m+1}, \qquad m^* = \min(n, M) \tag{1a}$$

$$Q_n = Q(n\Delta t), \qquad n = 1, \ldots, N \tag{1b}$$

$$P_m = \int_{(m-1)\Delta t}^{m\Delta t} I_e(\tau)d\tau, \qquad m = 1, \ldots, M \tag{1c}$$

where P_m is the volume of the mth effective rainfall pulse and the UH ordinates are given by

$$U_{n-m+1} = h[(n - m + 1)\Delta t] = \frac{1}{\Delta t}\int_{(n-m)\Delta t}^{(n-m+1)\Delta t} u(\tau)d\tau \tag{7}$$

The UH ordinates correspond to the area under the IUH between two consecutive time intervals.

The discrete convolution can be expressed alternatively as a matrix equation as,

$$[P][U] = [Q] \tag{8a}$$

$$\begin{bmatrix} P_1 & & & \\ P_2 & P_1 & & \\ & P_2 & \cdots & \\ P_M & & \cdots & P_1 \\ & P_M & & P_2 \\ & & \cdots & \\ & & & P_M \end{bmatrix} \begin{bmatrix} U_1 \\ U_2 \\ \vdots \\ U_{N-M+1} \end{bmatrix} = \begin{bmatrix} Q_1 \\ Q_2 \\ \\ Q_n \\ \\ \\ Q_N \end{bmatrix} \tag{8b}$$

where M is the number of effective rainfall pulses, $N - M + 1$ is the number of UH ordinates, and N is the number of direct runoff ordinates. Matrix P is of dimensions $(N \times (N - M + 1))$. Recasting the discrete convolution equation in this manner allows for an objective (and optimal) mathematical derivation of the UH ordinates. Estimation of the UH from simultaneous observations of effective precipitation (P_m) and direct runoff (Q_n) can be seen as a linear static estimation problem for which a method such as least squares, linear programming, or others can be used.

The least-squares approach tries to obtain a set of UH ordinates that minimizes the sum of squares of the errors and leads to the following solution for the UH,

$$[U] = \left[[P]^T[P]\right]^{-1}[P]^T[Q] \tag{9}$$

where $[[P]^T[P]]^{-1}$ is the inverse of matrix $[[P]^T[P]]$, and superscript T stands for transpose. This approach is adequate when the data (precipitation and discharge) are relatively error free or when the errors are expected to be small. When the data are expected to contain large errors a different approach may be more adequate. Such an approach would minimize the sum of the absolute value of the errors. Linear programming would then be the appropriate solution procedure.

Conceptual (Synthetic) Unit Hydrographs

As indicated earlier, sets of concurrent observations of effective rainfall and direct runoff are required for the derivation of UHs. Thus, the resultant UH is specific to the particular watershed defined by the point on the stream where the direct runoff observations were made. When no direct observations are available, or when UHs for other locations on the stream in the same watershed or for nearby watersheds of similar characteristics are required, synthetic UH procedures must be used.

Synthetic UH procedures can be categorized as (e.g., Chow et al., 1988) those based on models of watershed storage (e.g., Nash, 1957, 1958, 1959; Dooge, 1959); those relating hydrograph characteristics (time to peak, peak flow) to watershed characteristics (e.g., Snyder, 1938) [geomorphologic instantaneous UH (GIUH)]; and those based on a dimensionless UH (e.g., Soil Conservation Service, 1972).

Conceptual UHs Based on Models of Watershed Storage

Nash and Dooge Models

Linear Reservoirs. A linear reservoir is characterized by a linear relationship between the storage and the output as

$$S(t) = kQ(t) \tag{10}$$

The impulse response function of such linear reservoir is

$$u(t) = \frac{1}{k} e^{-t/k} \tag{11}$$

Nash (1957, 1958, 1959, 1960) proposed a cascade of n equal linear reservoirs as a model on which to base the derivation of IUHs for natural watersheds. The Nash model is one of the most widely used models in applied hydrology. Using the convolution equation (Equation 4) and the impulse response function for a single linear reservoir (Equation 11), the IUH corresponding to the Nash model can be easily obtained as follows:

$$u_n(t) = \frac{1}{k\Gamma(n)} \left(\frac{t}{k}\right)^{n-1} e^{-t/k} \tag{12}$$

This equation is a two-parameter gamma distribution function, where n is the shape parameter, and k is the scale parameter. When rainfall and runoff data are available, the shape parameter can be estimated as the inverse of the second nondimensional moment of the IUH about the centroid (i.e., the inverse of the square of the coefficient of variation); and the scale parameter can be estimated as the ratio of the first-order moment of the IUH about the origin to n. A general form for the peak discharge $Q_{\max}$ due to a finite pulse of rainfall excess with duration t_r and uniform intensity I_e was suggested by Adom et al. (1989) in the form of a Weibull cumulative distribution function, i.e.

$$Q_{\max} = AI_e \left[1 - e^{-\alpha\left(\frac{t_r}{t_l}\right)^{\beta}}\right] \tag{13}$$

where A is the contributing area, and t_l is the basin lag-time, which is given by $t_l = nk$. The values of a and b, which are parameters depending on n – i.e. the number of reservoirs in the cascade – are given in Table 11.1 (Ramírez et al., 1994).

Linear Channels. A linear channel is characterized by continuity and momentum equations given by

$$\frac{\partial Q}{\partial x} + \frac{\partial A}{\partial t} = 0 \tag{14a}$$

$$A = C(x)Q \tag{14b}$$

Table 11.1. *Parameter values for peak discharge equation of the linear reservoir cascade model*

n	1	2	3	4
a	1	1.017	1.175	1.348
b	1	1.220	1.313	1.370

For conditions of no lateral inflow, the unit impulse response function of such linear channel is

$$u(t) = \delta(t - T) \tag{15}$$

which implies that the downstream outflow is equal to the upstream inflow delayed by lag-time T representing the travel time through the system.

The cascade of linear reservoirs model can be modified to account for translation by using combinations of linear channels and linear reservoirs. Dooge (1959) developed a general UH theory under the assumption that basin response can be represented by a cascade of linear channels and linear reservoirs in series. For this model the drainage area of the watershed is divided into subareas using the isochrones. A linear channel in series with a linear reservoir represents each subarea. If this is done assuming a system of n equal linear reservoirs of parameter k and n equal linear channels of lag-time T, the IUH or impulse response function is

$$u_n(t) = \frac{1}{k\Gamma(n)}\left(\frac{t - nT}{k}\right)^{n-1} e^{-(t-nT)/k} \tag{16}$$

As a general rule, linear channels and linear reservoirs can be combined in series and/or in parallel to model complex hydrologic systems.

Conceptual UHs Based on Relationships between Hydrologic Response and Watershed Characteristics

Flow hydrographs travel through a stream network after runoff is generated on the hillslopes. The drainage network develops so that a certain load of water and sediment (as a result of erosive processes) can be evacuated in an optimal manner [least energy expenditure (e.g., Molnár and Ramírez, 1998a, 1998b)]. Therefore, the characteristics of the outflow hydrograph and the characteristics of the hillslopes and drainage network must be intimately linked as they shape and modify each other.

Geomorphologic Unit Hydrographs. Geomorphologic parameters have been widely used to characterize and model hydrologic response. In particular, geomorphologic parameters have been used in synthesizing UHs for gauged and ungauged watersheds as well as for estimating flood potential indices and flood frequency distributions. Recent efforts have concentrated on establishing a theoretical link between hydrologic response on the one hand and climatic and geomorphologic characteristics on the other. Even though climate and geologic processes are continually changing and affecting hydrologic response at the time scales of interest, climate and geomorphology may be considered constant boundary conditions that uniquely define watershed behavior. The theoretical link was developed in the late 1970s and early 1980s and was based on the ideas of the topologically random model for drainage networks. The fundamental postulates of the random topological model can be summarized as follows (e.g., Shreve, 1967; Gupta and Waymire, 1983; Bras, 1990):

(a) in the absence of natural controls, natural drainage networks are topologically random, and

(b) interior and exterior link lengths, as well as their associated areas, are independent random variables.

GIUH. Rodríguez-Iturbe and Valdés (1979) linked the hydrologic response of watersheds in terms of the IUH to their geomorphologic parameters and to a dynamic parameter. The IUH is interpreted as the probability density function of the travel times to the basin outlet of water droplets randomly and uniformly distributed over the watershed. The resulting IUH or IRF for the basin is a function of the probability density function of the travel times of droplets in streams of a given order and of the transition probabilities of water droplets going from a given state (stream segment) to another of higher order. In this model, streams of a given order define the states of the system. The travel times in the streams are assumed to be exponentially distributed and independent of one another. The initial probabilities of a drop landing anywhere on the basin as well as the transition probabilities, which are required to define the probability function of a particular path that a drop may take to the outlet, can be defined as functions only of geomorphologic and geometric parameters. The initial probability of a drop falling in an area of order i is equal to the percent contributing area for the given order. The initial probabilities are thus related to the average ratio of the average area of sub-basins of a given order to the average area of sub-basins of the next higher order (i.e., the area ratio R_A). Similarly, the transition probabilities from state i (stream order) to state j, where j represents a stream of higher order, are functions of the number of streams of order i draining into streams of order j, divided by the total number of streams of order i. Thus, the transition probabilities are related to the average ratio of the number of streams of a given order to the number of streams of the next order (i.e., bifurcation ratio R_B). These ratios are the parameters of the so-called Horton laws summarized below.

The Horton law of stream numbers states that there exists a geometric relationship between the number of streams of given order N_ω and the corresponding order, ω. The parameter of this geometric relationship is the bifurcation ratio, R_B. This ratio has been found in nature to be between 3 and 5.

$$N_\omega = R_B^{\Omega-\omega} \tag{17}$$

The Horton law of stream lengths states that there exists a geometric relationship between the average length of streams of a given order and the corresponding order ω. The parameter of this relationship is the so-called length ratio R_L.

$$\bar{L}_\omega = \bar{L}_1 R_L^{\omega-1} \tag{18}$$

The Horton law of stream areas states that there exists a geometric relationship between the average area drained by streams of a given order and the corresponding order ω. The parameter of this relationship is the so-called area ratio R_A. This ratio has been observed in nature to vary within a narrow range of 3 to 5.

$$\bar{A}_\omega = \bar{A}_1 R_A^{\omega-1} \tag{19}$$

In the equations above, Ω is the order of the basin, and the overbar indicates the average value of the corresponding variable.

Rodríguez-Iturbe and Valdés (1979) suggested that the probability distribution function of travel time in streams of a given order is an exponential distribution with the parameter λ representing the inverse of the mean travel time. They suggest estimating λ as the ratio of a characteristic velocity, V, and a characteristic length scale given by the mean length of streams of the given order. Valdés et al. (1979) suggest using the peak velocity for the characteristic velocity. Rodríguez-Iturbe and Valdés (1979) obtain expressions for the time to peak t_p and peak discharge q_p of the IUH, which are functions of geomorphology, and of the dynamic parameter represented by the peak velocity. Their results, obtained through numerical integration of the full GIUH and multiple regression analysis are simple, elegant, and easy to use. They are

$$q_p = \frac{1.31}{L_\Omega} R_L^{0.43} V \tag{20}$$

and

$$t_p = \frac{0.44 L_\Omega}{V} \left(\frac{R_B}{R_A} \right)^{0.55} R_L^{-0.38} \tag{21}$$

where L_Ω is the length of the highest-order stream in kilometers, V is in meters per second, and the units of q_p and t_p are the customary units of inverse time in hours, and time in hours, respectively.

Dependence of the GIUH on the dynamic parameter V can be used to address nonlinear watershed response issues. Nonlinear effects in the basin response manifest themselves in the characteristic discharge velocity. The theoretical framework of the GIUH above was verified through numerical experiments with a detailed physically based watershed model on several basins in Venezuela and Puerto Rico with excellent results (Valdés et al., 1979). They indicate that the nonlinear characteristics of the response function of a basin can be modeled with a linear scheme such as the GIUH but with a characteristic velocity that is representative of the discharge velocity for the given event. Rodríguez-Iturbe et al. (1979) show that the dynamic parameter of the GIUH can be taken as the space-time average flow velocity for a given rainfall-runoff event in a basin. This is based on the hypothesis of spatial uniformity of the flow velocity distribution throughout the river network (Pilgrim, 1976 and 1977). However, Agnese et al. (1988) showed that the estimation of the dynamic parameter of the GIUH could also be performed for those basins where flow velocity distribution depends on stream order.

The parameters of the Nash model and the geomorphologic descriptors of drainage networks can be related by relating approximate measures of volume of the IUH for a cascade of linear reservoirs and the volume of the GIUH (Rosso, 1984). Rosso equated the product of the time to peak and peak flow of the GIUH with the product of the time to peak and peak flow of the Nash model IUH. It is simple to show that, for the GIUH, this product equals

$$(t_p q_p)_{\mathrm{GIUH}} = 0.58 \left(\frac{R_A}{R_B} \right)^{0.55} R_L^{0.05} \tag{22}$$

and for the Nash model IUH, this product equals

$$(t_p q_p)_{\mathrm{Nash}} = (\alpha - 1)^{\alpha} e^{1-\alpha} / \Gamma(\alpha) \tag{23}$$

These equations allow estimation of the shape parameter. Using numerical algorithms to solve the resulting equation Rosso obtained

$$\alpha = 3.29\left(\frac{R_B}{R_A}\right)^{0.78} R_L^{0.07} \tag{24}$$

$$k = 0.70[R_A/(R_B R_L)]^{0.48} V^{-1} L_\Omega \tag{25}$$

where α and k are the shape parameter and the scale parameter of the Nash model IUH, respectively. When the parameters of the Nash model are estimated with these equations, and results of their application are compared with other empirical parameter estimation procedures, the adequacy of the geomorphologic connection is clearly demonstrated. These results improve the prediction capability of the Nash model for regions where no hydrologic records exist. The shape parameter of the IUH is shown to be a function only of Horton ratios, whereas the scale parameter is shown to depend on both geomorphology and streamflow velocity.

Geomorphologic IUHs have also been proposed that are based on characterizing drainage networks as a function of network links as opposed to streams (Gupta et al., 1986; Mesa and Mifflin, 1986; Troutman and Karlinger, 1984, 1985, 1986; and Gupta and Mesa, 1988.) In that case, GIUH is expressed in terms of the width function as

$$u(t) = \int_0^\infty g(x, t) N^*(x, t) dx \tag{26}$$

where $g(x, t)$ is the hydrologic response function of a single channel distance x from the basin outlet, and $N^*(x)$ is the normalized width function, which represents the probability density function of the number of links at distance x from the outlet. $N^*(x, t)$ is given by

$$N^*(x, t) = N(x, t) \Big/ \int_0^\infty N(x, t) dx \tag{27}$$

where $N(x, t)$ is the width function, which measures the number of links at distance x from the basin outlet. The link-based GIUH above has been approximated by its conditional expected value, conditional on a topological parameter vector (Karlinger and Troutman, 1985). When the topological parameter is the magnitude M of the network, Karlinger and Troutman have shown that, asymptotically for large M and for $g(x, t)$ representing a simple translation, the expected conditional value of $u(t)$ is

$$E[u(t)/M] = \frac{t}{2M(\bar{\ell}_i/V)^2} \exp\left(\frac{-V^2 t^2}{4M\bar{\ell}_i^2}\right), \qquad t > 0 \tag{28}$$

where M is the magnitude of the basin and equal to the number of first order streams, $\bar{\ell}_i$ is the mean length of the interior links, and V is the velocity of translation. Troutman and Karlinger show that this result is always true regardless of the form of $g(x, t)$.

Finally, Rinaldo et al. (1991) studied hydrologic response by decomposing the process of river runoff into two distinct contributions, one accounting for the travel time within individual reaches (hydrodynamic dispersion) and the other accounting for river network composition (geomorphologic dispersion). Because the analysis showed the latter one to play the major role in determining basin response, models based on the accurate specification of the geometry and the topology of the network and simplified dynamics are theoretically validated regardless of the choice of the travel time probability density function.

Channel Losses in the GIUH. The GIUH has undergone several modifications and enhancements to make it more physically based. One of those modifications consists of incorporating channel infiltration losses. Díaz-Granados et al. (1983a), using previous results of Kirshen and Bras (1983), derived expressions for the probability distribution function of travel times in streams of any given order taking into account channel infiltration losses. The channel response to an instantaneous input anywhere along the channel is interpreted as the conditional probability distribution function of the travel time of a drop traveling a given distance along the channel. This response is obtained as the solution of the linearized equations of motion for unsteady flow in a wide rectangular channel in which the infiltration losses are assumed to be proportional to the instantaneous discharge at any point along the channel. The proportionality coefficient is known as the infiltration parameter. Results indicate that the linear reservoir assumption, implicit in the GIUH (exponential distribution of stream travel times), is in fact adequate.

Studies on the geomorphologic response of river basins have increasingly stressed the importance of providing an accurate description of the quantitative properties of river network systems. Within this context, for example, La Barbera and Rosso (1987, 1989) first indicated the fractal nature of river networks and its relation to quantitative geomorphology as initiated by Horton's studies on river network composition. In addition, general theories of the evolution of drainage systems have been developed. For example, energy dissipation concepts have been used to link the local and global properties of a river network to the hydrologic conditions of its watershed (e.g., Howard, 1990; Rodríguez-Iturbe et al., 1992; Rigon et al., 1993; Molnár and Ramírez, 1998a, 1998b). These general theories allow for a fundamental description of global network properties and thus should be incorporated into improved geomorphologic models of basin response.

The Climatic GIUH (GcIUH). The instantaneous unit hydrograph is derived under the assumption that it is a random function of climatic and geomorphologic characteristics and that it varies with the characteristics of the rainfall excess. Accordingly, Rodríguez-Iturbe et al. (1982a) obtain probability density functions of the peak discharge and the time to peak as functions of the rainfall intensity i and duration t_r as well as geomorphology. Rainfall is characterized by a constant intensity over the storm duration, or a rectangular pulse process. These two characteristics, together with geomorphologic parameters for a first-order basin, define the dynamic parameter of the GIUH.

Using the kinematic wave approximation for flow routing along streams of first order and a derived distribution approach Rodríguez-Iturbe et al. (1982a) obtained analytical expressions for the probability density functions of the time to peak t_p and the peak discharge q_p of the GIUH that depend on the mean rainfall intensity; hence, the name GcIUH. These probability distribution functions assume that the time of concentration of first-order streams is much smaller than the duration of the rainfall. They are

$$f(q_p) = 3.534\Pi q_p^{1.5}\exp\left(-1.412\Pi q_p^{2.5}\right)$$
$$f(t_p) = \frac{0.656\Pi}{t_p^{3.5}}\exp\left(\frac{-0.262\Pi}{t_p^{2.5}}\right) \tag{29}$$

where

$$\Pi = \frac{L_\Omega^{2.5}}{\bar{i}A_\Omega R_L \alpha_\Omega^{1.5}} \tag{30}$$

The first moments of peak discharge and time to peak of the GcIUH can be obtained as

$$E(q_p) = \frac{0.774}{\Pi^{0.4}} \qquad E(t_p) = 0.858\Pi^{0.4}$$
$$\sigma q_p = \frac{0.327}{\Pi^{0.4}} \qquad \sigma t_p = 0.915\Pi^{0.4} \tag{31}$$

where L_Ω is the length of the highest order stream in kilometers, A_Ω is the basin area in square kilometers, $\bar{i}$ is the mean rainfall intensity in centimeters per hour, and α_Ω is in $s^{-1}\,m^{-1/3}$. The units of q_p and t_p are the customary units of inverse time in hours and time in hours, respectively.

For individual storm events, the above equations can be manipulated to obtain expressions for the peak discharge and time to peak as functions of the particular storm intensity and duration. These expressions lead to

$$q_p = \frac{0.871}{\Pi^{0.4}} \tag{32a}$$

$$t_p = 0.585\Pi^{0.4} \tag{32b}$$

$$\Pi_i = \frac{L_\Omega^{2.5}}{iA_\Omega R_L \alpha_\Omega^{1.5}} \tag{32c}$$

The GcIUH establishes a link between climate, watershed geomorphology, and the hydrologic response of the basin. Furthermore, the GcIUH allows estimation of the IUH for a given particular rainfall input, so that problems associated with nonlinear basin behavior are avoided. For example, runoff from a given rainfall event can be computed by using an IUH derived from the given storm event characteristics, thereby avoiding the well-known errors incurred when using an IUH based on a different event.

Rodríguez-Iturbe et al. (1982b) compare results by the GcIUH theory with results obtained from IUHs derived from simulated rainfall–runoff events for one basin and three different climates. The theoretical distributions of the time to peak and peak discharge compare very well with the experiments. Variations in q_p and t_p resulting from nonlinear basin behavior induce large uncertainty in the predicted peak discharge and time to peak for individual events when traditional UH approaches are used (e.g., Caroni et al., 1986; Caroni and Rosso, 1986; Rosso and Caroni, 1986). These problems are avoided if the GcIUH is used, because of its functional dependence on the rainfall characteristics.

Snyder's Synthetic UH

The synthetic UH of Snyder (1938) is based on relationships found between three characteristics of a standard UH and descriptors of basin morphology. These relationships are based on a study of 20 watersheds located in the Appalachian Highlands and varying in size from 10 to 10,000 square miles. The hydrograph characteristics are the effective rainfall duration t_r, the peak direct runoff rate q_p, and the basin lag time t_l. From these relationships, five characteristics of a required UH for a given effective rainfall duration may be calculated (e.g., Chow et al., 1988; Bras, 1990): the peak discharge per unit of watershed area q_{pR}, the basin lag t_{lR}, the base time t_b, and the widths W (in time units) of the UH at 50% and 75% of the peak discharge.

Standard UH. A standard UH is associated with a specific effective rainfall duration, t_r, defined by the following relationship with basin lag t_l

$$t_l = 5.5t_r \tag{33}$$

For a standard UH the basin lag t_l and the peak discharge q_p are given by

$$t_l = C_1 C_t (LL_c)^{0.3} \tag{34}$$

$$q_p = \frac{C_2 C_p A}{t_l} \tag{35}$$

The basin lag time of the standard UH (Equation 34) is in hours, L is the length of the main stream in kilometers (miles) from the outlet to the upstream divide, L_c is the distance in kilometers (miles) from the outlet to a point on the stream nearest the centroid of the watershed area, and $C_1 = 0.75$ (1.0 for English units). The product LL_c is a measure of watershed shape. C_t is a coefficient derived from gauged watersheds in the same region and represents variations in watershed slopes and storage characteristics. The peak discharge of the standard unit hydrograph (Equation 35) is in m^3/s (cfs), A is the basin area in km^2 (mi^2), and $C_2 = 2.75$ (640 for English units). As C_t, C_p is a coefficient derived from gauged watersheds in the area and represents the effects of retention and storage.

To compute C_t and C_p for a gauged watershed, the values of L and L_c are measured. From a derived UH of the watershed, values of its associated effective duration t_R in hours, its basin lag t_{lR} in hours, and its peak discharge q_{pR} in m^3/s are obtained. If $t_{lR} = 5.5t_R$, then the derived UH is a standard UH and $t_r = t_R$, $t_l = t_{lR}$, and $q_p = q_{pR}$, and C_t and C_p are computed by the equations for t_l and q_p given above corresponding to the standard UH.

If t_{lR} is quite different from $5.5t_R$, the standard basin lag is computed using

$$t_l = t_{lR} + \frac{t_r - t_R}{4} \tag{36}$$

This equation must be solved simultaneously with the equation for the standard unit hydrograph lag time, $t_l = 5.5t_r$, in order to obtain t_r and t_l. The value of C_t is then obtained using the equation for t_l corresponding to the standard unit hydrograph. The value of C_p is obtained using the expression for q_p corresponding to the standard UH but using $q_p = q_{pR}$ and $t_l = t_{lR}$.

When an ungauged watershed appears to be similar to a gauged watershed, the coefficients C_t and C_p for the gauged watershed can be used in the above equations to derive the required synthetic UH for the ungauged watershed.

Required UH. The peak discharges of the standard and required UHs are related as follows,

$$q_{pR} = \frac{q_p t_l}{t_{lR}} \tag{37}$$

Assuming a triangular shape for the UH and given that the UH represents a direct runoff volume of 1 cm (1 in.) the base time of the required UH may be estimated by

$$t_b = \frac{C_3 A}{q_{pR}} \tag{38}$$

where C_3 is 5.56 (1290 for the English system).

As an aid in drawing an adequate UH, the U.S. Army Corps of Engineers developed relationships for the widths of the UH at values of 50% (W_{50}) and 75% (W_{75}) of q_{pR}. The width in hours of the UH at a discharge equal to a certain percent of the peak discharge q_{pR} is given by Chow et al. (1988) as

$$W_{\%} = C_w \left(\frac{q_{lR}}{A} \right)^{-1.08} \tag{39}$$

where the constant C_w is 1.22 (440 for English units) for the 75% width and equal to 2.14 (770 for English units) for the 50% width. Usually, one-third of this width is distributed before the peak time and two-thirds after the peak time, as recommended by the U.S. Army Corps of Engineers. However, several others have recommended different distribution ratios. For example, Hudlow and Clark (1969) recommend a partition of 4/10 and 6/10, respectively.

Figure 11.2 illustrates the form of Snyder's synthetic UH. Note that the time lag is not the same as the time to peak. Also, note that the widths of the hydrograph at 50% and 75% of the peak flow are distributed so that the longer time is to the right of the time to peak.

Conceptual UHs Based on Dimensionless Hydrographs

Soil Conservation Service (SCS) Dimensionless Hydrograph. The dimensionless UH developed by the SCS (Figure 11.3) has been obtained from the UHs for a great number of watersheds of different sizes and for many different locations. The SCS dimensionless hydrograph is a synthetic UH in which the discharge is expressed as a ratio of discharge, q, to peak discharge, q_p, and the time by the ratio of time t to time to peak of the UH t_p. Given the peak discharge and the lag time for the duration of the excess rainfall, the UH can be estimated from the synthetic dimensionless hydrograph for the given basin.

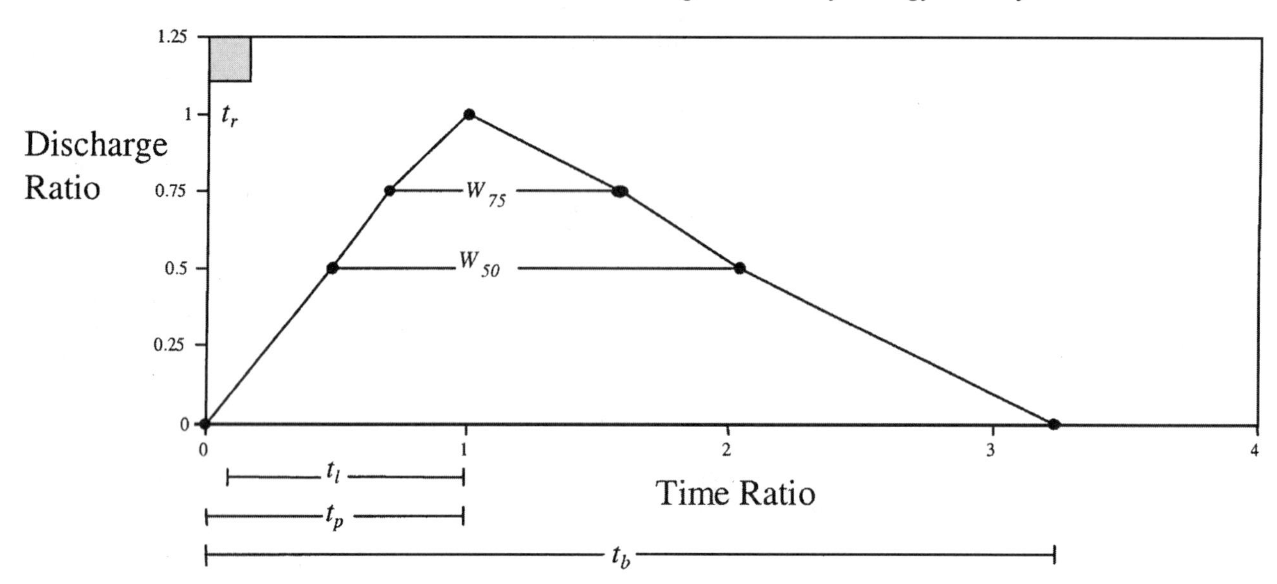

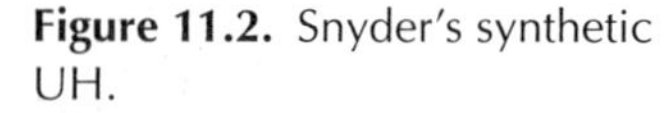
Figure 11.2. Snyder's synthetic UH.

The SCS suggests that the dimensionless UH can be described in terms of an equivalent triangular hydrograph. The values of q_p and t_p can then be estimated with this simplified triangular UH whose height is equal to q_p and whose time base t_b is equal to 2.67 t_p. The time is usually expressed in hours (SCS) and the discharge is expressed in $m^3/s/cm$ (or cfs/in.). After analysis of a great number of UHs, the SCS recommends a recession duration of 1.67 t_p. Because the volume of direct runoff must equal 1 cm, it can be shown that $q_p = CA/t_p$ where $C = 2.08$ (483.4 in the English system) and A is the drainage area in square kilometers (square miles). From a study of many large and small rural watersheds the basin lag is $t_l = 0.6t_c$, where t_c is the time of concentration of the watershed. The time to peak t_p is then equal to $t_r/(2 + t_l)$ (e.g., U.S. Soil Conservation Service, 1985).

Figure 11.3. SCS dimensionless UH.

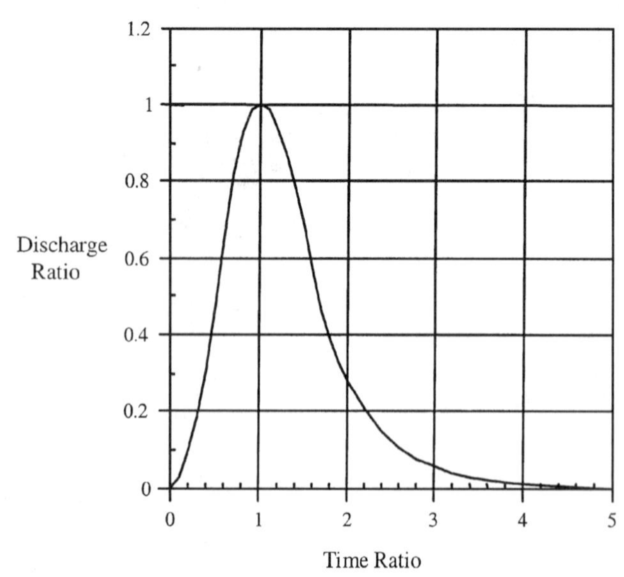

Flood Frequency Distributions and Climatic GIUH. Eagleson (1972) presents a general framework for the development of flood frequency distributions, $F(Q_p)$ (see also Stedinger, Chapter 12, this volume). Based on Eagleson's framework (see Figure 11.4), several authors have developed flood frequency distributions that include various assumptions about the characteristics of the rainfall process in terms of its probability density function $f_I(i)$ and the watershed processes transforming rainfall into runoff encoded in the function $g(i, t; \theta)$. Using either numerical or analytical procedures these authors have arrived at different flood frequency distributions. One of the most recent and most interesting developments has been the use of the GIUH concept to quantify and analyze the effect of geomorphologic characteristics in determination of regional flood frequency.

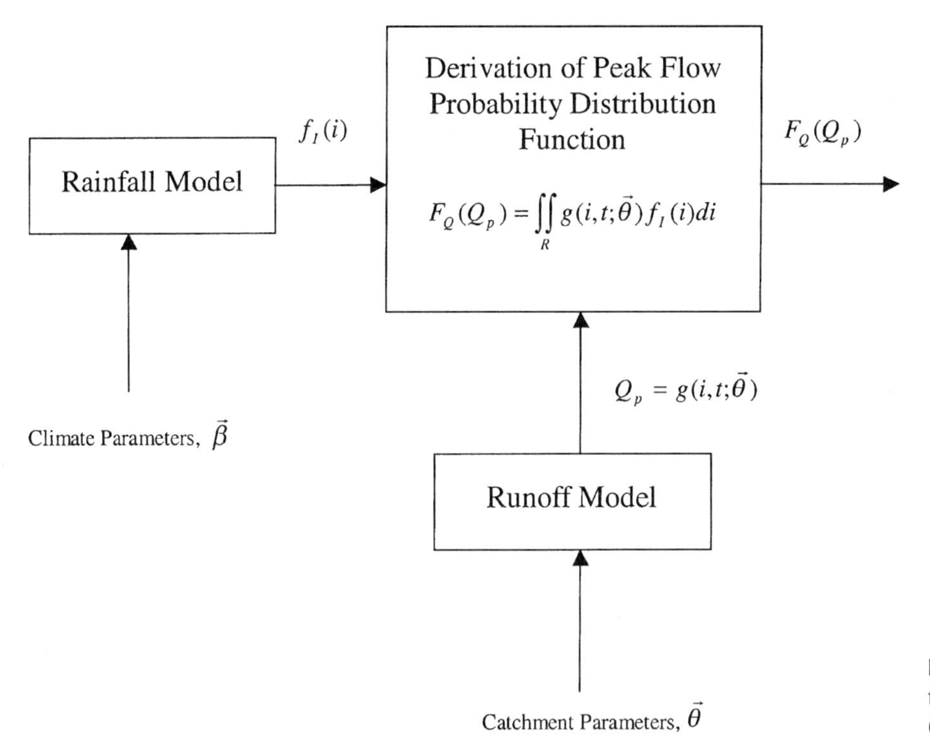

Figure 11.4. General framework for flood frequency analysis (adapted from Eagleson, 1972).

Hebson and Wood (1982) derived a flood frequency distribution function from assumed climatic distributions for the rainfall parameters and used the GIUH as the transformation function of rainfall into runoff. They assumed a model of rectangular pulses for storms whose intensity and duration are independent and exponentially distributed random variables. In terms of the equivalent recurrence interval T_E Hebson and Wood obtain

$$\ln T_E = -\ln \left[\lambda \int_0^{t_k} \exp(-\lambda t_{re} - \beta^*(Q_p - Q_b)/A_c g(t_{re})) dt_{re} + \exp(-\lambda t_k - \beta^*(Q_p - Q_b)/A_c) \right] - \ln n \quad (40)$$

where n is the average number of annual direct runoff events, Q_p is the flood peak under consideration, Q_b is the base flow, A_c is the contributing area, $g(t)$ is the area under the GIUH as a function of time, t_{re} is the duration of excess rainfall, t_k is the duration of the GIUH or kernel length, λ is the inverse mean duration of rainfall events, and β^* is a modified areal inverse mean intensity of rainfall events. They applied this concept to two third-order basins, the Bald Eagle Creek in Pennsylvania and the Davidson River in North Carolina. Comparisons of their flood frequency distributions with the one derived by Eagleson (1972), which is based on kinematic wave concepts for overland flow, indicate that the best-fit GIUH produces better representation of the observed flood frequency distributions.

Díaz-Granados et al. (1983, 1984) derived a flood frequency distribution based on the GcIUH. Effective rainfall intensity i_e and effective rainfall duration t_e were obtained by assuming precipitation to be a process of rectangular pulses and runoff to be generated only by the Horton infiltration excess mechanism. Runoff is then generated only when the precipitation intensity exceeds the soil infiltration capacity. Infiltration capacity was defined by using Philip's solution to the one-dimensional, concentration- dependent equation of the diffusion process in unsaturated porous media. Storm intensity and duration were assumed to be independent and exponentially distributed random variables. They used a derived distribution approach to link the joint probability distribution function of storm duration and storm intensity and a physically based representation of the infiltration process to obtain the joint probability density function of effective storm duration and effective storm intensity. This derived distribution was used together with a triangular representation of the GcIUH to obtain a flood frequency distribution. The flood frequency distributions obtained by this method depend on climatic and geomorphologic parameters of the watershed as well as on soil properties governing the infiltration process. In terms of the equivalent recurrence interval, their derived flood frequency distribution is

$$T_E^{-1} = n[1 - F_Q(Q_p)] \tag{41}$$

$$F_Q(Q_Q) = 1 - \delta \exp(\beta a - 2\sigma)\Gamma(\sigma + 1)\sigma^{-\sigma} \left[I + \sum_{i=1}^{4} J_i \right] \tag{42}$$

where I and J_i are integral equations that cannot be solved analytically and that depend on the mean intensity and the mean duration of precipitation events as well as on the infiltration sorptivity of the soil, the discharge under consideration, and geomorphologic parameters of the watershed (see Díaz-Granados et al., 1984).

Given that the above distribution depends on the elusive initial soil moisture, Díaz-Granados et al. (1984) resort to using arguments of ecological optimality for water-limited natural systems to define a long-term average soil moisture, s_o. Using this approach, they verified their flood frequency distribution against observed distributions from two very different climatic regimes. Their results indicate good agreement for both wet and arid climates, although the agreement is better for the arid climate of Santa Paula Creek in California than for the wet climate of the Nashua River basin in New Hampshire.

However, obtaining the analytical form of the flood frequency distribution from climate and geomorphology often results in a cumbersome derivation, sometimes yielding implicit equations to be solved numerically. To overcome this problem, Adom et al. (1989) introduced the method of approximate moments by Taylor series to estimate second-order statistics of both flood peak and volume. For this purpose, they used the SCS's curve number method to describe basin rainfall excess and the GIUH to model surface runoff response. After a given type of frequency distribution has been identified at the regional scale by analyzing annual flood series, this method can be used to estimate the flood frequency distribution for ungauged catchments.

Flood Routing

The term flood routing refers to procedures to determine the outflow hydrograph at a point downstream in a river (or reservoir) as a function of the inflow hydrograph at a point upstream.

As flood waves travel downstream they are attenuated and delayed; that is, the peak flow of the hydrograph decreases and the time base of the hydrograph increases. The shape of the outflow hydrograph depends on the channel geometry and roughness, bed slope, length of channel reach, and initial and boundary flow conditions.

The propagation of flood waves in a channel is a gradually varied unsteady flow process, which is governed by conservation of mass and momentum equations. The solution of these equations in a distributed manner is referred to as distributed routing of flood waves. When no spatial variability is taken into account and when the channel reach or reservoir is considered as a black box, the corresponding routing procedure is referred to as lumped routing.

Lumped Flood Routing

Lumped or hydrologic routing is governed by the continuity equation and a storage-discharge functional relationship. The continuity equation is

$$\frac{dS(t)}{dt} = I(t) - O(t) \tag{43}$$

In this equation, $S(t)$ is the storage within the system (channel reach or reservoir), $I(t)$ is the inflow hydrograph at the upstream end of the reach, and $O(t)$ is the outflow hydrograph at the downstream end of the reach. The continuity equation can be integrated over a given Δt to obtain

$$S(t_{i+1}) - S(t_i) = \int_{S(t_i)}^{S(t_{i+1})} dS(t) = \int_{t_i}^{t_{i+1}} I(t)dt - \int_{t_i}^{t_{i+1}} O(t)dt \tag{44}$$

where the subscripts i and $i+1$ refer to the beginning of two consecutive time intervals. Assuming a linear variation of input and output fluxes during the Δt leads to

$$S(t_{i+1}) - S(t_i) = \frac{\Delta t}{2}[I(t_{i+1}) + I(t_i)] - \frac{\Delta t}{2}[O(t_{i+1}) + O(t_i)] \tag{45}$$

In Equation 46 there are two unknown quantities: $S(t_{i+1})$ and $O(t_{i+1})$. Thus, a second equation relating $S(t)$ and $O(t)$ is required. The nature of this relationship may be linear, or nonlinear, or one to one, or hysteretic. The complexity of the routing procedure depends on the nature of this relationship.

Reservoir or level pool routing refers to flood routing for systems whose storage and outflow are related by a function of the type $S(t) = f[O(t)]$ which is one-to-one (i.e., unique, nonhysteretic). This characteristic of their S vs. O relationships implies that for a given set of conditions (e.g., stage or storage) the outflow is unique, independent of how that storage is achieved. Such systems have a pool that is wide and deep compared with its length in the direction of flow, low flow velocities, and horizontal

water surfaces or negligible backwater effects. Reservoirs or systems with horizontal water surfaces have relationships of the one-to-one type.

The solution procedure involves rearranging the continuity equation so that all unknown quantities are on the left-hand side of the equation,

$$\frac{2S(t_{i+1})}{\Delta t} + O(t_{i+1}) = [I(t_{i+1}) + I(t_i)] + \left[\frac{2S(t_i)}{\Delta t} - O(t_i)\right] \qquad (46)$$

For level pool systems, the storage and the outflow are one-to-one functions of elevation (see Figures 11.5 and 11.6).

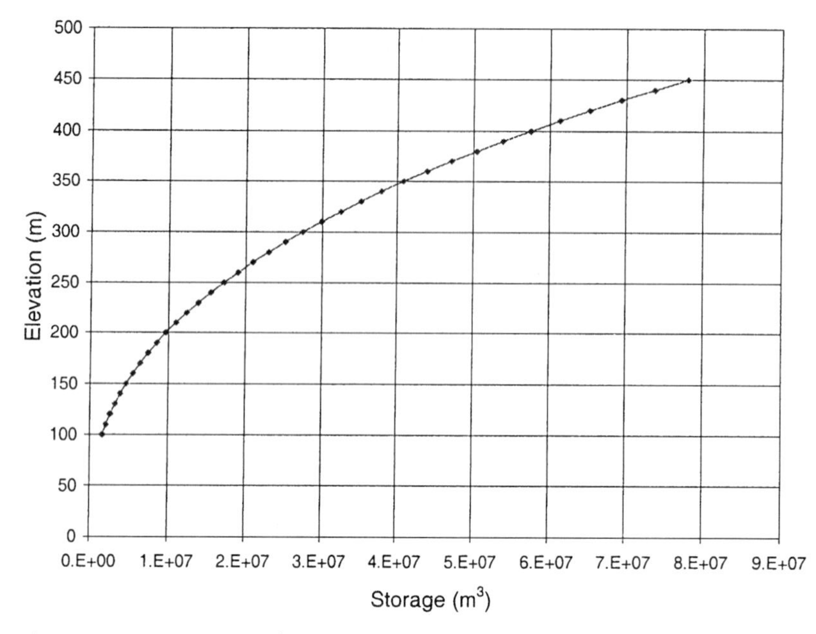

Figure 11.5. Storage vs. elevation relationship.

Thus, the left-hand side of the equation above is a unique function of water surface elevation in the system only. Usually, the storage–elevation relationship is available from topographic surveys, and the outflow–elevation relationship is available from hydraulic considerations with respect to the outlet structures (e.g., spillways, etc.).

The solution involves development of the function $2S/\Delta t + O = f[O(h)]$ (see Figure 11.7) and then using the function sequentially to solve for $O(t_{i+1})$ for every time step. In this equation, h is water surface elevation in the reservoir.

As observed in Figure 11.8, the effect of storage is to redistribute the flow hydrograph by shifting its centroid from the position of that of the inflow hydrograph to that of the outflow hydrograph. For these systems, the peak outflow occurs when the outflow hydrograph intersects the inflow hydrograph, at which time the maximum storage in the system is also achieved (see Figure 11.8).

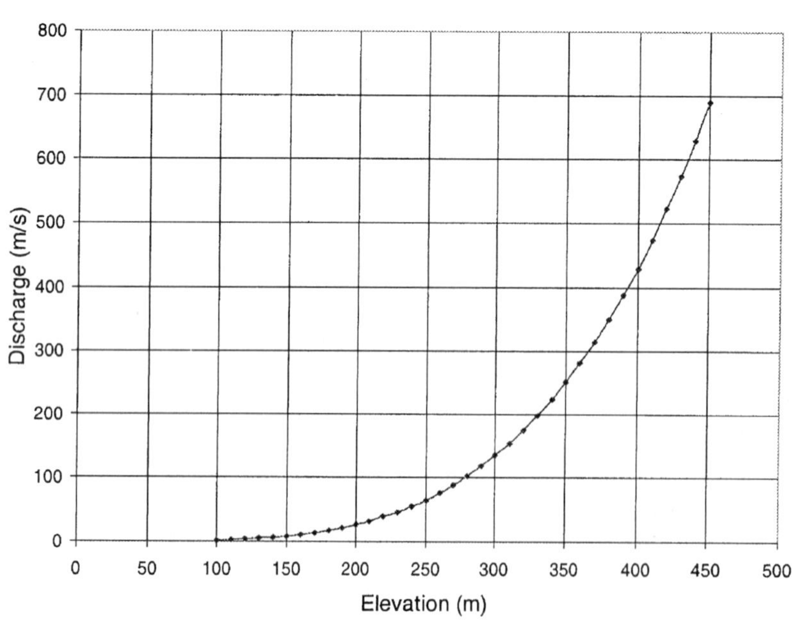

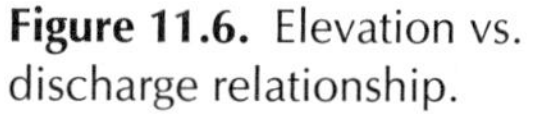

Figure 11.6. Elevation vs. discharge relationship.

Channel Routing. The storage–discharge relationship in the case of channel routing is not one-to-one, but hysteretic. In other words, the outflow from a given reach is not a single valued function of the storage in the reach. In its most simple interpretation, the magnitude of the outflow discharge depends on whether the outflow is occurring during the rising limb or the recession limb of the inflow hydrograph.

This behavior is a result of backwater effects, which are important in long and narrow flow routing conditions such as occur in channels.

The Muskingum method is one of the most widely used methods of lumped, channel flow routing. In addition to the lumped form of the continuity equation

$$\frac{dS(t)}{dt} = I(t) - O(t) \quad (43)$$

this method assumes a linear storage discharge relationship as given below

$$S(t) = k[xI(t) + (1 - x)O(t)] \quad (47)$$

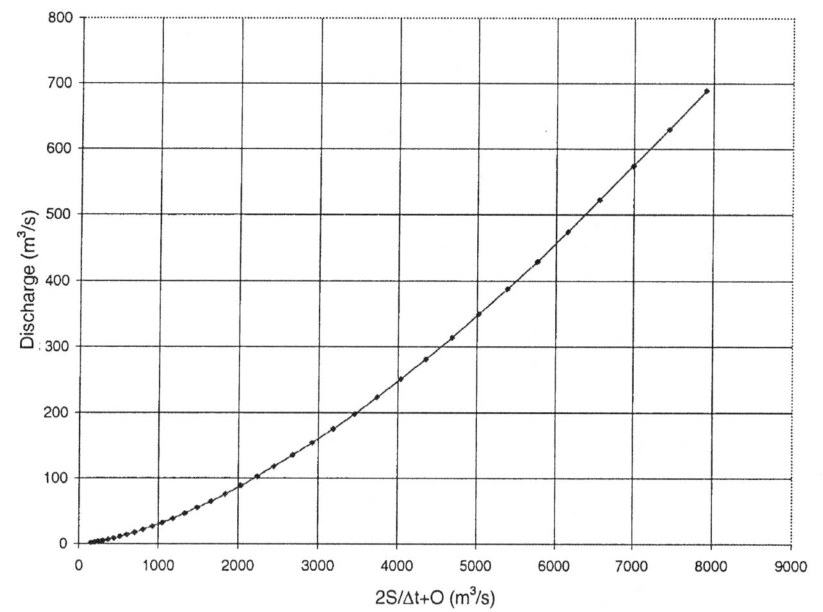

Figure 11.7. Storage-indication function.

This relationship specifies that the storage is proportional to a convex linear combination of inflow and outflow rates. The Muskingum method has two parameters, k and x. Parameter k has units of time and represents the average travel time of a kinematic wave through the reach, and it is known as the storage time constant. Parameter x is dimensionless and represents the relative importance of the effects of inflow rates on defining the storage in the reach. Theoretically this parameter varies between 0 and 1.

Combining the continuity and storage equations and integrating over a short time interval, the outflow at the end of a given time interval, $O(t_{i+1})$, can be expressed as a linear combination of the inflow rates at the beginning and end of the given interval $I(t_i)$ and $I(t_{i+1})$ as well as the outflow rate at the beginning of the interval $O(t_i)$ as

$$Q(t_{i+1}) = C_o I(t_{i+1}) + C_1 I(t_i) + C_2 O(t_i) \quad (48)$$

The coefficients or weights of this linear combination can be computed as a function of k, x, and Δt as indicated below

$$C_o = (-kx + 0.5\Delta t)/C_3 \quad (49a)$$

$$C_1 = (kx + 0.5\Delta t)/C_3 \quad (49b)$$

$$C_2 = (k - kx - 0.5\Delta t)/C_3 \quad (49c)$$

$$C_3 = k - kx + 0.5\Delta t \quad (49d)$$

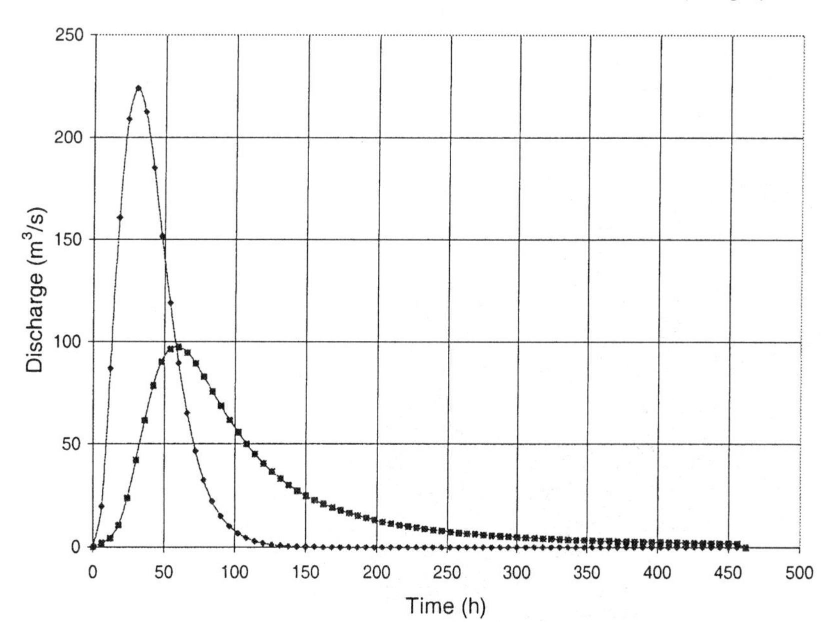

Figure 11.8. Level pool routing: inflow and outflow hydrographs.

It is simple to show that the Muskingum routing solution can also be expressed as a discrete convolution

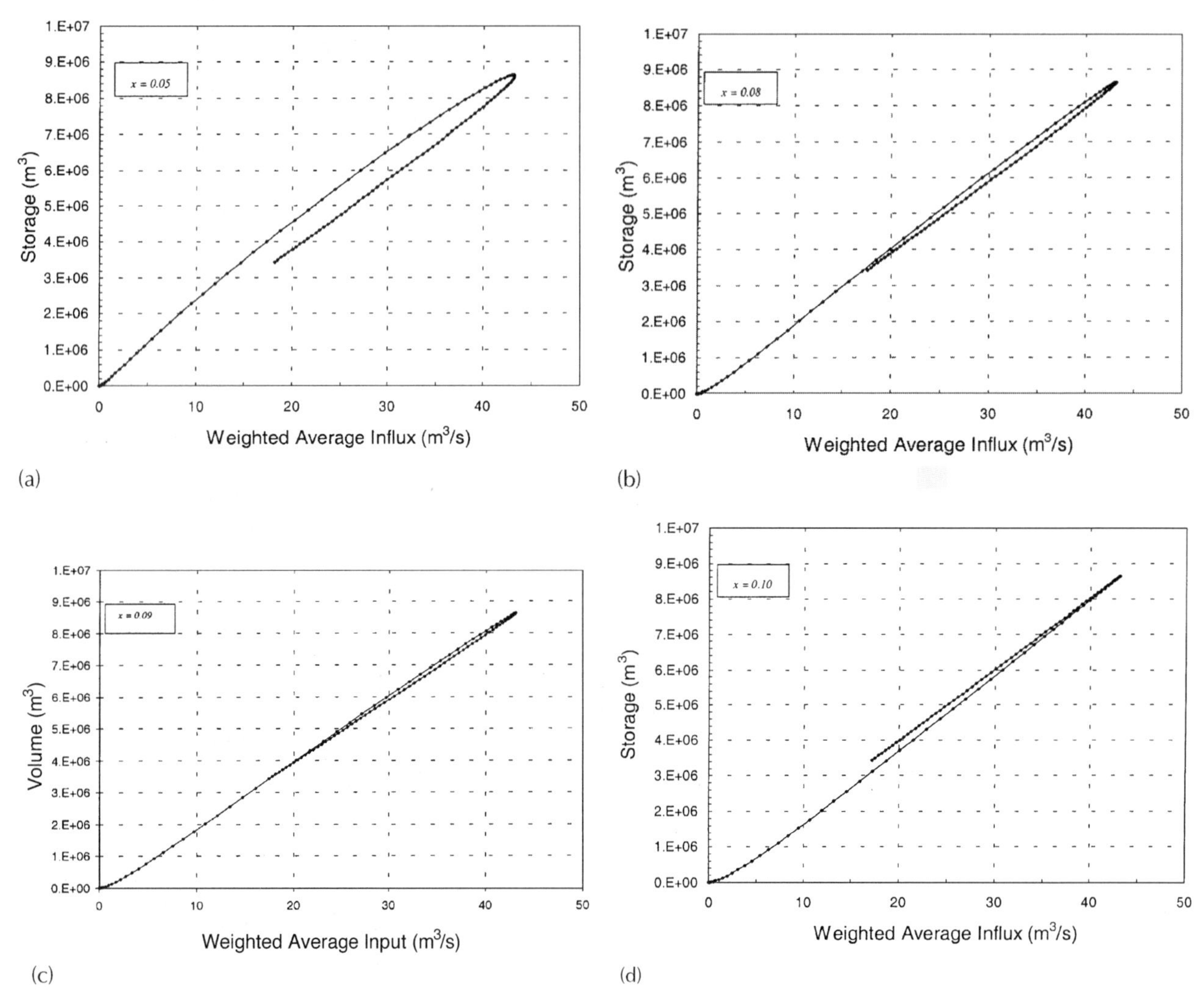

Figure 11.9. (a) Weighted average influx vs. storage function ($x = 0.05$), (b) weighted average influx vs. storage function ($x = 0.08$), (c) weighted average influx vs. storage function ($x = 0.09$), (d) weighted average influx vs. storage function ($x = 0.10$).

equation as

$$Q_n = \sum_{m=1}^{n} I_{n-m+1} U_m \tag{50}$$

where U_m is the UH of the Muskingum method whose ordinates are given by the following equations

$$\begin{aligned} U_1 &= C_o \\ U_2 &= C_o C_2 + C_1 \\ U_m &= U_{m-1} C_2 \qquad m > 2 \end{aligned} \tag{51}$$

The Muskingum parameters k and x can be easily estimated graphically or by the least-squares method. The graphical method consists of choosing x so that the loop resulting from the plot of S vs. $[xI + (1 - x)O]$ becomes as close to a straight line as possible (see Figure 11.9a–d). The slope of the resulting straight line is an estimate of the parameter k (see Equation 47).

To ensure positivity of the UH ordinates, Δt must be selected so that $x < 0.5\Delta t/k < 1 - x$.

This method results in a trial and error procedure, which, when properly carried out, constitutes a graphical implementation of the least-squares method. In other words, the least-squares method is a numerical expression of the graphical method. Observe that routing in systems with hysteretic S vs. O relationships leads to additional modifications of the inflow hydrograph represented in a delay of the response of the system and in asynchronous occurrences of the maximum storage and maximum outflow, with the latter occurring after the former (see Figure 11.10).

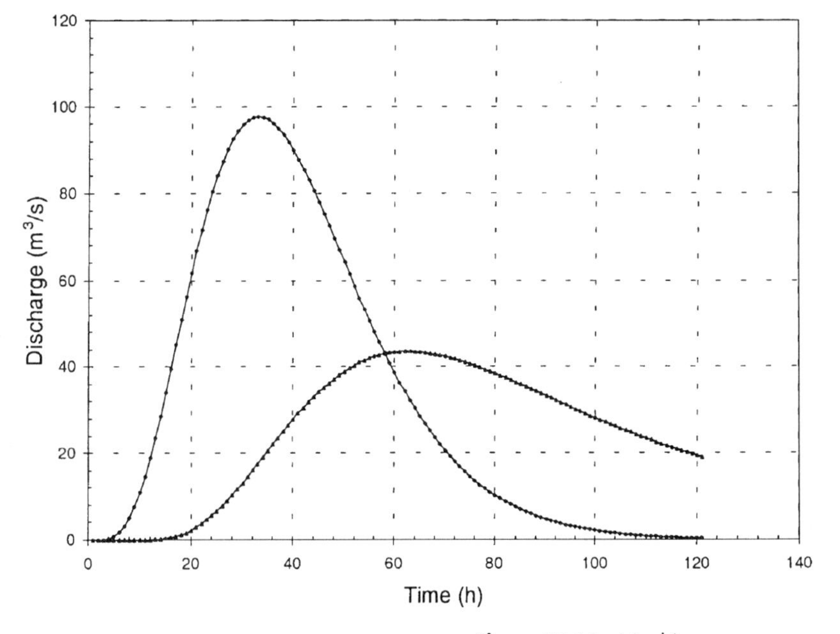

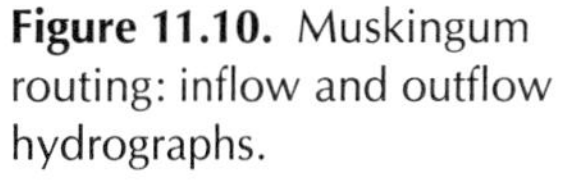
Figure 11.10. Muskingum routing: inflow and outflow hydrographs.

Using a hydraulic analogy and applying the Muskingum equations over a short Δx, the Muskingum routing equation constitutes an approximate solution to the kinematic wave equations (see Distributed Flood Routing section below) (Cunge, 1969). Expressing the solution in a space-time domain, the Muskingum routing equations can be rewritten as

$$Q_{i+1}^{j+1} = C_o Q_i^{j+1} + C_1 Q_i^j + C_2 Q_{i+1}^J \tag{52}$$

where the flow discharge Q_i^j refers to position i in space and j in time. In this solution, the parameters k and x required to obtain C_o, C_1, C_2, and C_3 are estimated as

$$K = \frac{\Delta x}{c_k} \tag{53a}$$

$$x = \frac{1}{2}\left(1 - \frac{Q}{c_k B S_o \Delta x}\right) \tag{53b}$$

In addition, Cunge also demonstrated that this solution constitutes an approximate solution of a modified diffusion equation if the parameters k and x are estimated as expressed above. In those equations, c_k is the celerity of the kinematic wave corresponding to Q and B, where B is the width of the water surface, Q is the discharge, A is the cross-sectional area of flow, S_o is the bottom slope, and Δx is the increment in space. Based on this analysis, it can be shown that x should be between 0 and 0.5.

Distributed Flood Routing

The characteristics of hydrologic systems are extremely variable in space and time. The excitation of the system (i.e., precipitation), the initial conditions, and the boundary conditions are, in general, functions of space and time. Therefore, the response of the system (that is, the flow of water through the soil and in the channels) is a distributed process in which the characteristics of the flow change in both time and space. Flow characteristics such as flow discharge, flow depth, and flow velocity can be obtained as a function of space and time by using a dynamic, distributed flow routing procedure.

The governing equations for one-dimensional, unsteady flow in an open channel are

$$\frac{\partial Q}{\partial x} + \frac{\partial A}{\partial t} = q \tag{54}$$

$$\frac{\partial Q}{\partial t} + \frac{\partial(\beta Q^2/A)}{\partial x} + gA\left(\frac{\partial y}{\partial x} - S_o + S_f\right) - \beta q v_x = 0 \tag{55}$$

where Q is discharge, A is the cross-sectional flow area, y is flow depth, S_o is the channel bottom slope, S_f is the friction slope, β is the momentum correction coefficient for nonuniform flow velocity distribution, q is lateral inflow per unit length of flow, and v_x is the magnitude of the lateral inflow velocity component in the direction of flow.

These two equations are known as the de Saint-Venant equations. The main assumptions encoded in these equations are (1) the flow is one-dimensional; (2) the vertical accelerations are negligible so that the vertical pressure distribution is hydrostatic (gradually varied flow); (3) the slope of the channel bottom is small and the channel bottom is fixed; (4) resistance coefficients for steady uniform turbulent flow are applicable; and (5) the fluid is incompressible and of constant density.

The momentum equation includes terms for changes in momentum due to (1) changes in velocity over time (local acceleration); (2) changes in velocity in space along the channel (convective acceleration); (3) water level differences along the channel and consequent imbalance of pressure forces; (4) gravitational acceleration; (5) frictional forces; and (6) lateral inflow. Local and convective accelerations represent the effect of inertial forces in the flow. Depending on the flow conditions, some of these terms may be negligible compared with others in the equation giving rise to either kinematic, diffusive, or full dynamic wave situations. This sometimes leads to simplifications of the flood routing problem (e.g., Chow et al., 1988; Singh, 1989; Bras, 1990).

Kinematic Waves. Kinematic waves govern the flow when inertial and pressure forces are not important. In natural flood wave conditions both kinematic and dynamic waves are present. However, for situations where the Froude number is less than 2, dynamic waves tend to attenuate more rapidly than kinematic waves (e.g., Eagleson, 1970). Although there is no single set of criteria to define when a kinematic wave solution is acceptable, the magnitude of the so-called kinematic parameter is useful. The kinematic parameter is defined as

$$K = \frac{S_o L}{y_o F^2} \tag{56}$$

where S_o is the bottom slope, L is the length of (overland or channel) flow, y_o is the normal flow depth at the downstream end of the flow length, and F is Froude number for normal flow at the downstream end of the flow length. Woolhiser and Liggett (1967) have shown that, for K values greater than 10, the flow dynamics tend to be dominated by kinematic waves. In overland flow situations the value of K usually exceeds this threshold, $[K \sim O(10^2 - 10^3)]$, warranting a kinematic wave approach for such flow routing problems.

For a kinematic wave, the momentum equation reduces to

$$S_o = S_f \tag{57}$$

implying that the energy grade line is parallel to the channel bottom and that the flow is steady and uniform. The above momentum equation can be shown to be equivalent to the following stage-discharge relationship

$$Q = \alpha A^{\beta} \tag{58}$$

which, for example, can be satisfied by Manning equation

$$Q = \frac{\sqrt{S_f}}{n} R^{2/3} A \tag{59}$$

Together with the continuity equation

$$\frac{\partial Q}{\partial x} + \frac{\partial A}{\partial t} = q \tag{54}$$

these equations represent the routing equations for the kinematic wave flow routing approach.

If an observer moves with the kinematic wave at a speed equal to the kinematic wave celerity, the observer would see the flow rate increase at a rate equal to the lateral inflow rate q as shown below

$$\frac{dQ}{dx} = \frac{\partial Q}{\partial x} + \frac{\partial Q}{\partial t}\frac{dt}{dx} = \frac{\partial Q}{\partial x} + \frac{1}{c_k}\frac{\partial Q}{\partial t} = q \tag{60}$$

It can be seen that for conditions of no lateral inflow, that is, for $q = 0$, $dQ/dt = 0$. Thus, kinematic waves do not attenuate; they simply translate downstream without dissipation. Given that at any cross section Q and A are functionally related as $Q = \alpha A^{\beta}$, the continuity equation can be rewritten as

$$\frac{\partial Q}{\partial x} + \frac{dA}{dQ}\frac{\partial Q}{\partial t} = q \tag{61}$$

and

$$\frac{\partial A}{\partial t} = \frac{dA}{dQ}\frac{\partial Q}{\partial t} = \frac{1}{\alpha\beta A^{\beta-1}}\left(\frac{\partial Q}{\partial t}\right) \tag{62}$$

$$\frac{\partial Q}{\partial x} + \frac{1}{\alpha\beta A^{\beta-1}}\left(\frac{\partial Q}{\partial t}\right) = \frac{\partial Q}{\partial x} + \frac{Q^{1/\beta-1}}{\alpha^{1/\beta}\beta}\frac{\partial Q}{\partial t} = q \tag{63}$$

From Equations 60 and 63, it is easy to obtain the following expression for the kinematic wave celerity

$$c_k = \frac{dQ}{dA} = \alpha\beta A^{\beta-1} \tag{64}$$

Assuming that overland flow situations can be described as flow over planes, the kinematic wave approximation has been widely applied to construct models of the rainfall–runoff process. A basin is described as a collection of such overland flow planes representing the hillslope processes feeding into a network of collector channels, which feed into a main channel. In this application, the lateral inflow is taken to be equal to the rainfall excess, and the channel inflow is taken to be flow per unit width of plane. The kinematic wave model offers the advantage over the UH method that it is a one-dimensional, spatially distributed solution of the physical equations governing the surface flow. Furthermore, for overland flow waves in the case of homogeneous planes, the kinematic wave equations have explicit analytical solutions. However, more complex representations of the flow process must sometimes be used – two-dimensional and fully dynamic.

In general, the de Saint-Venant equations do not have explicit analytical solutions, except in a few particular cases (e.g., kinematic waves for homogeneous planes). Thus, numerical solution procedures must be used, like finite-differences methods, or the method of characteristics. Finite-difference methods solve the partial differential equations on a grid placed over the x–t plane. For the kinematic wave problem presented above, these solutions can be linear or nonlinear.

A simple implementation of the linear and nonlinear solutions suggested by Chow et al. (1988) is adapted below. For the linear solution, the equation is linearized by substituting an average of known solutions for the coefficient of the nonlinear term. This leads to the following solution

$$\frac{Q_{i+1}^{j+1} - Q_i^{j+1}}{\Delta x} + \frac{1}{\alpha^{1/\beta}\beta}\left(\frac{Q_{i+1}^{j} + Q_i^{j+1}}{2}\right)^{1/\beta-1}\left(\frac{Q_{i+1}^{j+1} - Q_{i+1}^{j}}{\Delta t}\right) = \left(\frac{q_{i+1}^{j+1} + q_{i+1}^{j}}{2}\right) \tag{65}$$

$$Q_{i+1}^{j+1} = \frac{\left[\frac{\Delta t}{\Delta x} Q_i^{j+1} + \frac{1}{\alpha^{1/\beta}\beta} Q_{i+1}^{j}\left(\frac{Q_{i+1}^{j}+Q_i^{j+1}}{2}\right)^{1/\beta-1} + \Delta t\left(\frac{q_{i+1}^{j+1}+q_{i+1}^{j}}{2}\right)\right]}{\left[\frac{\Delta t}{\Delta x} + \frac{1}{\alpha^{1/\beta}\beta}\left(\frac{Q_{i+1}^{j}+Q_i^{j+1}}{2}\right)^{1/\beta-1}\right]} \tag{66}$$

In this linear solution, the subscript refers to the space coordinate and the superscript refers to the time coordinate. The solution advances on a time line from upstream to downstream.

For the nonlinear kinematic wave solution, the partial differential equations are recast in finite difference form as shown below. The same subscript and superscript convention applies as for the linearized solution.

$$\frac{Q_{i+1}^{j+1} - Q_i^{j+1}}{\Delta x} + \frac{A_{i+1}^{j+1} - A_{i+1}^{j}}{\Delta t} = \frac{q_{i+1}^{j+1} + q_{i+1}^{j}}{2} \tag{67}$$

$$\frac{\Delta t}{\Delta x} Q_{i+1}^{j+1} + \alpha^{-1/\beta}\left(Q_{i+1}^{j+1}\right)^{1/\beta} = \frac{\Delta t}{\Delta x} Q_i^{j+1} + \alpha^{-1/\beta}\left(Q_{i+1}^{j}\right)^{1/\beta} + \Delta t\left(\frac{q_{i+1}^{j+1} + q_{i+1}^{j}}{2}\right) \tag{68}$$

$$C = \frac{\Delta t}{\Delta x} Q_i^{j+1} + \alpha^{-1/\beta}\left(Q_{i+1}^{j}\right)^{1/\beta} + \Delta t\left(\frac{q_{i+1}^{j+1} + q_{i+1}^{j}}{2}\right) \tag{69}$$

An iterative solution based on Newton's method is shown below

$$f\left(Q_{i+1}^{j+1}\right) = \frac{\Delta t}{\Delta x} Q_{i+1}^{j+1} + \alpha^{-1/\beta}\left(Q_{i+1}^{j+1}\right)^{1/\beta} - C \tag{70}$$

$$f'\left(Q_{i+1}^{j+1}\right) = \frac{\Delta t}{\Delta x} + \frac{1}{\alpha^{1/\beta}\beta}\left(Q_{i+1}^{j+1}\right)^{1/\beta-1} \tag{71}$$

$$\left(Q_{i+1}^{j+1}\right)_{k+1} = \left(Q_{i+1}^{j+1}\right)_{k} - \frac{\frac{\Delta t}{\Delta x}\left(Q_{i+1}^{j+1}\right)_k + \alpha^{-1/\beta}\left[\left(Q_{i+1}^{j+1}\right)_k\right]^{1/\beta} - C}{\frac{\Delta t}{\Delta x} + \frac{1}{\alpha^{1/\beta}\beta}\left[\left(Q_{i+1}^{j+1}\right)_k\right]^{1/\beta-1}} \tag{72}$$

Newton's method is an iterative numerical solution procedure, which progressively determines improvements to a trial solution as a function of the gradient of the function at the trial solution. The method converges relatively quickly. A convergence criterion can be defined in advance, so that a solution is reached when this criterion is met (e.g., the magnitude of the improvement is less than a threshold).

Dynamic Waves. Dynamic waves govern the flow whenever inertial and pressure forces are significant compared with the gravitational and frictional forces. This condition occurs for unsteady, nonuniform flows for which backwater or tidal effects are important as, for example, during the movement of a flood wave in a system of channels of shallow slopes and may be induced by downstream reservoirs or other downstream controls, by tributaries, by tides, by storm surges (e.g., caused by hurricanes), or by large discontinuities in the flow characteristics as a result of dam breaks or large reservoir releases. As indicated above for the case of hydraulic river routing, backwater effects induce rating curves that are hysteretic, that is, rating curves for which the discharge is not a single valued function of stage (i.e., storage per unit length of channel) in the section.

Full dynamic wave flood routing models can be both steady-state and unsteady state models. The Hydrologic Engineering Center's River Analysis System (HEC-RAS) model is an example of a steady-state flow model used to compute water surface profiles associated with one-dimensional, gradually varied, steady flows. HEC-RAS is designed to perform one-dimensional hydraulic simulations for a network of channels and computes surface water profiles associated with subcritical, supercritical, and mixed-flow regimes. Surface water profiles are solved sequentially by solving the energy equation with an iterative algorithm called the standard step method. The energy equation is

$$Y_2 + Z_2 + \frac{\alpha_2 v_2^2}{2g} = Y_1 + Z_1 + \frac{\alpha_1 v_1^1}{2g} + h_e \tag{73}$$

where h_e is the energy head loss, Y_i is the surface water depth at section i, Z_i is the elevation of the channel bottom at section i, v_i is the average velocity at cross section i, and g is the acceleration of gravity. The energy head loss between two cross sections accounts for friction losses and for losses induced by contractions or expansions and is evaluated as

$$h_e = LS_f + C\left|\frac{\alpha_1 v_1^2}{2g} - \frac{\alpha_2 v_2^2}{2g}\right| \tag{74}$$

where L is the reach length weighted by discharge, S_f is a representative friction slope between the two sections, and C is a coefficient for expansion–contraction losses. For situations where flow regimes change abruptly from supercritical to subcritical or vice versa, the HEC-RAS model uses the steady-state momentum equation instead, as the energy equation is not applicable under those conditions. The momentum equation is written in this case as

$$P_2 - P_1 + W_x - F_f = \rho Q \Delta v_x \tag{75}$$

where P_i is the hydrostatic pressure force at section i, W_x is the component of the weight of the water in the section in the direction of flow, F_f is the force due to friction in the reach between sections, Q is the discharge, ρ is the density of water, and Δv_x is the change in average velocity within the reach. In addition, the model allows for specification of variable cross-sectional characteristics both at a section and along the channel. Common applications of the HEC-RAS model include floodplain delineation and analysis for flood protection and insurance purposes, modeling of single and/or multiple culverts and bridges, modeling of gated spillways and weirs, modeling of floodway encroachment, and modeling of ice-covered rivers.

In the context of fully dynamic, unsteady flows, the models of the National Weather Service developed by Fread and collaborators are well documented and commonly used (e.g., Fread, 1974, 1976, 1978, 1988; Fread and Lewis, 1988). Among these models are the DWOPER model (Dynamic Wave Operational Model) (Fread, 1978), the DAMBRK model for dam failure studies (Fread, 1988), and the FLDWAV model (Fread and Lewis, 1988), which synthesizes DWOPER and DAMBRK. These and similar models solve the fully unsteady dynamic wave equations by using finite differences. However, the treatment and presentation of the details of the solution is beyond the scope of this chapter. For those details the reader is referred to the above references and those given therein.

Distributed Watershed Models

Distributed and semidistributed watershed models are generally categorized into single-event simulation models and continuous simulation models. The former models are concerned with simulating a single flood as a result of the occurrence of an individual precipitation or flood-inducing event. The latter models are concerned with the simulation of watershed processes during a period of time that encompasses more than one precipitation or flood-causing event. Thus, continuous models are also concerned with evaporation and subsurface soil moisture redistribution during the interstorm periods. Most of these models include elements or components that are described by the concepts of previous sections in this chapter. However, it is implied that watershed models deal with transformation of precipitation into runoff, whereas some of the approaches presented earlier in this chapter are concerned simply with the translation of a flood wave, and not necessarily with the complete precipitation–runoff transformation. These models are then referred to as flood hydraulics models (e.g., HEC-RAS).

Single-Event Simulation Models

There are a great variety of single event watershed models. Singh (1989) provides a summary table listing 15 such models and reviews some of them. The following sections are based on the review presented by Ramírez et al. (1994). Reference is made to two models developed by the U.S. Geological Survey and one developed by the Hydrologic Engineering Center of the U.S. Corps of Engineers. However, many alternative models are available, such as the TR-20 (U.S. Soil Conservation Service, 1973), SWMM (Metcalf and Eddy Inc., 1971), IHM (Morris, 1980), and FLEA (Ranzi and Rosso, 1991a).

The USGS rainfall–runoff model (Dawdy and O'Donnell, 1965) was one of the first single-event models developed to predict flood volume and peak rates of runoff for small drainage areas. The model has been used by the U.S. Geological Survey in over 20 states to develop peak flows and hydrographs for state highway agencies. The process is triggered by the precipitation input P, which augments surface storage R, which in turn is depleted by infiltration F and evaporation E_R. The channel outflow Q begins when the threshold value R^* is exceeded. The infiltration is evaluated by a Horton-type equation. The infiltration F and capillary rise C augment the soil storage M, which is depleted by transpiration E_M. Deep percolation D occurs when the soil moisture threshold M^* is exceeded, augmenting groundwater storage G, which in turn is depleted by capillary rise C and base flow B. If G exceeds the threshold value G^*, M is absorbed into G; then C and D disappear and E_M and F operate on G. The channel storage S and groundwater storage G are assumed to be linear reservoirs with storage constants K_S and K_G, respectively. The channel flow Q is routed through a linear reservoir S yielding the flow Q_S, which added to the base flow B gives the total flow Q.

An alternative USGS model is the distributed routing rainfall–runoff model developed by Dawdy et al. (1972). This model uses a modified version of the Philip's infiltration equation and the infiltration capacity concept of Crawford and Linsley (1966) to determine the excess rainfall. A kinematic wave model component is used for routing. The model was specifically designed for routing urban flood discharges through a branched system of pipes or natural channels with rainfall used as input. The drainage basin is presented as a set of segments (overland flow, channel reservoir, and nodal) that jointly describe all sub-basins of the entire basin.

A single event watershed model that has been widely used for estimation of surface runoff and river/reservoir flow in a basin during floods is the HEC-1 flood hydrograph package. It was originally developed in 1967 by the staff of the Hydrologic Engineering Center at Sacramento, California (now at Davis, California). The current version of the HEC-1 package includes computational capabilities of dam breaks, project optimization, and kinematic wave programs (HEC, 1981).

The hydrologic simulation capabilities of HEC-1 include several techniques to input and distribute the precipitation, treat the precipitation as rainfall or snowfall, compute rainfall and snowmelt losses and excess, and determine sub-basin outflow hydrographs by various hydrologic routing techniques. The model may be used to simulate a simple single-basin watershed or a very complex basin with practically unlimited number of sub-basins and river reaches. The HEC-1 model can account for temporal and spatial variability of the precipitation runoff process in a semidistributed sense. That is, within a sub-basin, HEC-1 uses spatially and temporally lumped parameters to simulate the precipitation–runoff process. The precipitation hyetograph (rain and/or snow) is input over the sub-basin, and the losses are computed, leaving an excess precipitation hyetograph, which in turn is transformed into surface runoff hydrograph through a specified UH. The subsurface runoff hydrograph is computed separately and added to the surface runoff hydrograph to yield the total subbasin runoff hydrograph. In addition to being capable of hydrologic simulation, the HEC-1 package also has a provision for evaluating reservoir and channel development plans for flood control purposes by performing the economic analyses of flood damages for existing and post-development conditions. An additional application of the calibrated model is for impact assessment studies of basin modifications and channel improvements. This can be accomplished by modifying the physical parameters of the sub-basin or routing reach in which the change is expected or proposed and by reanalyzing the response of the watershed.

Continuous Simulation Models

Three continuous watershed models developed in the United States are summarized in this section. The review is based on that of Ramírez et al. (1994). The models are the Stanford watershed model (SWM), the National Weather Service model (NWS), and the precipitation runoff modeling system (PRMS) of the USGS. However, many alternative models exist, such as the SSARR (U.S. Army Corps of Engineers, 1972) for daily streamflow simulations and forecasting for large watersheds, the SHE model developed in Europe (Abbott, 1986a, 1986b), and the LBRM developed by the U.S. Great Lakes Environmental Research Laboratory (Croley, 1985) for simulation of daily runoff in the Great Lakes basins.

The SWM was originally developed by Crawford and Linsley (1966) for continuous simulation of the hydrologic processes in a watershed at hourly time scales. Since then, the original model has undergone a number of revisions and expansions to satisfy specific needs and requirements (e.g., Claborn and Moore, 1970; Liou, 1970; Anderson, 1971; Ricca, 1972). Several models, such as the agricultural runoff management model, ARM, (Donigian and Crawford, 1979) and the hydrologic simulation program, HSPF, of EPA (Johanson et al., 1980), use the basic structure of the original SWM for runoff simulation but also simulate water quality and sediment erosion.

The SWM model partitions the watershed into land segments whose boundaries are established according to hydrologic characteristics and user's needs. The segmentation of land is usually based on meteorologic, soil, topographic, and drainage considerations. Thus, each land segment is assumed to be homogeneous with respect to hydrologic characteristics, and is assumed to produce a homogeneous hydrologic response. Streamflow may be computed at several locations in the stream network.

Two storage zones control overland flow, infiltration, interflow, and inflow to the groundwater storage. The upper zone storage may be physically interpreted as primarily consisting of depression storage, whereas the lower zone represents the soil moisture storage above the capillary zone. The role of the upper zone is to simulate the initial response to rainfall, and therefore it is important for small storms and for the first few hours of major storms. The lower zone plays a major role during large storms in controlling temporal variation of infiltration rates. Infiltrated water moves to the lower zone and groundwater storage. Additional infiltration may occur from depression storage or from runoff, giving rise to delayed infiltration. A fraction of total infiltrated water reaches the groundwater storage, which then contributes to baseflow. Moisture may be lost from the system as evapotranspiration from interception, upper zone, lower zone, groundwater storage, and stream surfaces. Direct inflows from the impervious areas, overland flow, interflow, and groundwater flow are combined to become total inflow into the channel network, which contributes to the streamflow hydrograph.

The U.S. National Weather Service (NWS) developed a computerized system of operational river forecast procedures, including data acquisition and processing, computation of mean basin precipitation, snow accumulation and ablation, explicit soil moisture accounting, channel routing, parameter optimization and verification, and operational forecasting. This set of models, known as the National Weather Service River Forecast System, (NWSRFS), has been extensively described (e.g., Monro and Anderson, 1974; Peck, 1976). The soil moisture accounting model of the NWSRFS has been developed by the NWS Sacramento, California River Forecast Center (e.g., Burnash et al., 1973). The NWS model can be partitioned into two major components, the Sacramento soil moisture accounting model and the flow routing model. The Sacramento soil moisture accounting model is a continuous, conceptual, lumped input and lumped parameter model that transforms the rainfall input in the land phase of the hydrologic cycle into channel inflows. The flow routing model is responsible for routing the channel inflows to the basin or sub-basin outlet by using unit hydrograph and Muskingum routing methods.

The Sacramento soil moisture accounting (SMA) model assumes that a catchment consists of two reservoirs, the upper zone and the lower zone. This conceptualization is intended to provide a simple but effective representation of the hydrologic process of the catchment starting with precipitation, subsequent vertical and horizontal movement of water through and over the soil, and finally production of runoff. The upper zone represents the upper soil layer and interception storage, and the lower zone represents the bulk of the soil moisture and groundwater storage. Each zone stores water in two forms, tension water and free water. Tension water is depleted only by evapotranspiration. Depletion of free water occurs vertically as percolation and horizontally as evapotranspiration and interflow. In the lower zone, there are two types of free water storage: primary storage, which is slow draining and provides baseflow over long periods of time, and supplementary storage, which is fast draining and supplements baseflow after a period of relatively recent rainfall. Movement of water from the upper zone to the lower zone is achieved through percolation, which is a nonlinear function of upper zone free available water and lower zone moisture deficiency.

Two levels of watershed partitioning are required in the model. The first level partitions the watershed into soil moisture accounting areas where each area is a homogeneous unit in terms of the SMA model parameters. Rainfall and evapotranspiration input data are assumed uniform over each SMA area. The second level of partitioning divides an SMA area into smaller homogeneous units representing individual flow planes (overland and channel). Each flow plane is assumed to have homogeneous unit graph-Muskingum model parameters. Although there are two levels of watershed partitioning, one can assume a flow plane to coincide with the SMA area. Streamflow is computed at the outlet or flow points of each flow plane.

The model is set up to simulate basin hydrology on hourly or longer time intervals. Rainfall input data can be given only on an hourly basis, whereas evapotranspiration and streamflow input data can be given at longer time intervals as an integer multiple of the time interval of rainfall. In such cases, the model uniformly distributes the evapotranspiration data and computes streamflow over the time basis of rainfall data and over the time interval specified for streamflows.

Finally, the model generates five components of flow: (1) direct runoff from permanent and temporary impervious areas; (2) surface runoff, which occurs when the upper zone free water storage is full and precipitation intensity exceeds the rate of percolation and interflow; (3) interflow resulting from lateral drainage of the upper zone free water storage; (4) supplemental baseflow; and (5) primary baseflow. The first three runoff components represent the total channel inflow, and the latter two are the total baseflow. In the NWS model, the total channel inflow constitutes the surface runoff contribution to the streamflow hydrograph routed via unit hydrograph/Muskingum routing methods, and a portion of the total baseflow is the subsurface runoff contribution to streamflow. This subsurface flow contribution is added to the routed streamflow at the basin or sub-basin outlet by using a linear decay weighting function, which is similar to the hydrograph routing method.

The USGS developed a precipitation–runoff modeling system (PRMS) to simulate surface water runoff, sediment yields, and general basin hydrology under various combinations of precipitation, climate, and land use (Leavesley et al., 1983). The system is physically based so that each component of the hydrologic cycle is expressed as functional relationships of known physical laws or empirical knowledge of the hydrologic process. The PRMS model can be a lumped or a distributed parameter model capable of simulating mean daily flows and storm flow hydrographs.

The watershed is partitioned into hydrologic response units (HRUs), which may or may not be associated with individual subwatersheds. HRUs are defined based on characteristics such as slope, orientation of hillslope or aspect, soil type, vegetation type, elevation, and precipitation distribution (Leavesley et al., 1983), which imply similar hydrologic response behavior. Water and energy balances are determined daily for each HRU. The inputs are daily precipitation, maximum and minimum daily air temperature, daily solar radiation, and daily streamflow (for calibration). Precipitation is reduced by interception. The energy inputs of air temperature and solar radiation drive the processes of evaporation, transpiration, sublimation, and snowmelt.

The model is represented by a series of storage (reservoirs). Each HRU has a soil-zone reservoir. This represents the soil mantle that can lose water by evaporation and transpiration. Average root depth of the predominant vegetation on the HRU defines the depth of the soil-zone reservoir. Water storage in this zone is increased by infiltration and is decreased by evapotranspiration. The soil zone is represented by two layers. The upper layer (recharge layer) loses water by evaporation and transpiration, and the lower layer loses water only by transpiration.

When the soil zone reservoir reaches field capacity, excess water percolates to the subsurface and groundwater reservoirs. The subsurface reservoir responds faster than the groundwater reservoir. The subsurface reservoir can be linear or nonlinear, whereas the groundwater reservoir responds linearly. Subsurface and groundwater reservoirs can be defined for each HRU or for larger areas composing several HRUs. The snow component of PRMS simulates the accumulation and depletion of a snowpack on each HRU.

The USGS model can simulate basin hydrology on a daily or shorter time interval for flood simulation. The daily mode simulates hydrologic components on daily averages or total values. Shorter time simulation, with time steps on the order of minutes, computes water discharge and sediment yield from selected rainfall events. Normally the model operates on a daily mode until it reaches dates of storm events. In these cases, the model shifts to the shorter time interval mode until the storm period terminates. The model then returns to the daily mode. Leavesley and Stannard (1990) used ARC/INFO with PRMS to simulate hydrographs in the East Fork Carson River, and Luellwitz (1991) used ARC/INFO with PRMS to simulate hydrographs in a subalpine watershed in Colorado. This model has been incorporated into the modular modeling system of the USGS, which is briefly described in the next section.

Current Developments

Flood hydrology developed very rapidly after the work of Sherman and Horton in the 1930s and 1940s. This has been the result of improvements in our understanding of the fundamental laws governing hydrologic process behavior and of improvements in computational power and observational capabilities. Parallel to the development of increasingly more powerful computer hardware has been development of computer programming languages, database systems, and graphics and image processing tools, which have made it easier to use large amounts of input data and have enhanced analysis and interpretation of output.

Development of digital terrain models (DTM) and database systems for spatial data, so-called Geographic Information Systems (GIS), has had a tremendous effect on our ability to describe and understand the effects of highly heterogeneous boundary conditions on hydrologic response. GIS makes it possible to integrate efficiently not only the topography but also the geomorphology, soil type, vegetation, and land-use characteristics of the basin with physically based hydrologic models of watersheds. These geomorphoclimatic and hydraulic characteristics, whose extreme variability in space and time has profound nonlinear effects on hydrologic response, have become increasingly available (e.g., NEXRAD radar precipitation fields) with the advent and rapid development of remote sensing. GRASS of the U.S. Corps of Engineers and ARC/INFO of Environmental Systems Research Institute of California, USA, are two examples of GIS packages widely used in conjunction with DTMs, remotely sensed data (i.e., precipitation), and hydrologic modeling.

The Modular Modeling System (MMS) (Leavesley et al., 1996) is an integrated system of computer software that has been developed with financial support from the USGS to provide the research and operational framework needed to support development, testing, and evaluation of physical-process algorithms and to facilitate integration of user-selected sets of algorithms into operational physical-process models. As such, the MMS integrates all the improvements referred to above in computer software and hardware within a modeling framework aimed at facilitating and improving hydrologic modeling. MMS uses a module library that contains modules for simulating a variety of water, energy, and biogeochemical processes (e.g., it includes the PRMS model). A model is created by selectively coupling the most appropriate modules from the library to create a suitable model for the desired application. Where existing modules do not provide appropriate process algorithms, new modules can be developed. The MMS modeling system also provides installation capabilities for new module development, and model development and application using the MMS graphical user interface. The MMS enhances the ability of the modeler to address the specific conditions under consideration, as it allows building of different model configurations either from a bank of hydrologic simulation modules or by an explicit coding of new modules. For example, Epstein and Ramírez (1993, 1994) used the GRASS GIS package with PRMS in an earlier version of the MMS to simulate streamflow hydrographs in the Rio Grande basin. Their work addresses issues of climatic variability and its effect on hydrologic response.

During the past decade, recognition of the significant effect of spatial variability on hydrologic response has prompted renewed interest in development of spatially distributed hydrologic models for flood routing and for flood risk analyses. In addition to describing the complex interactions of local hillslope processes, variable soil and vegetation characteristics, and channel network, spatially distributed modeling has the advantage of providing predictions at any location throughout the river network and not only at the outlet. This is an important advantage in planning and design problems where predictions are required at a specific river cross section at which perhaps no historical observations are available.

Mancini and Rosso (1989) used a GIS to investigate the spatial variability of SCS curve number within the basin, thus assessing the effects of such variability on basin scale estimates. Brath et al. (1989) analyzed scale effects in distributed rainfall–runoff modeling, Pilotti and Rosso (1990) presented SHELL (i.e., a general framework for modeling the distributed response of drainage basins to storm rainfall), and Ranzi and Rosso (1991b) reported a physically based approach to modeling distributed snowmelt in alpine catchments.

Burlando et al. (1994) developed a methodology for evaluating flood risk in a distributed modeling framework. Their flood risk analyzer (FLORA) combines a geomorphologic model of flood frequency with a distributed description of the basin given in terms of soil types, land use, and topography. Runoff potential and infiltration characteristics are described as a function of the SCS curve number model. The geomorphologic model of flood frequency is based on the ideas of Eagleson (1972) introduced earlier in this chapter. In their application, Burlando et al. use the approach of Bacchi and Rosso (1988) to derive the second-order properties of the frequency distribution of flood peaks for any point of the basin as a function of the stochastic properties of precipitation and of the soil-type, land-use, and geomorphologic characteristics of the section in question. Assuming an independent Poisson marks model for precipitation (Eagleson, 1972), and the SCS curve number model for infiltration losses, they show that the frequency distribution of the annual maxima is an extreme value type

II distribution and obtain expressions for its second-order moments as a function of climate and geomorphology. They applied this methodology successfully to the Sansobbia River basin in northern Italy.

Physically based distributed modeling of watershed processes has become increasingly feasible in recent years. In addition to development of improved computational capabilities, DTMs and GIS tools, this has been brought about by advances in our understanding of the fundamental physical processes underlying the hydrologic cycle and of the solution of the mathematical equations representing those processes. Generally, the various hydrologic processes governing the transformation of rainfall into runoff are extremely complex, nonlinear, and spatially and temporally variable, giving rise to complex sets of strongly coupled nonlinear differential equations. Analytical solutions for these equations have been found for the simplest cases only, and one dimensional and two-dimensional numerical solution schemes have been developed and applied. For example, Saghafian (1992), Ogden (1992), Ogden and Julien (1993), Ogden et al. (1995), and Saghafian et al. (1995) developed two-dimensional numerical schemes for routing the excess rainfall through the overland phase of a watershed. The CASC2D model (e.g., Saghafian et al., 1995) is a distributed-parameter, two-dimensional, unsteady diffusion wave model that routes overland flow by using a diffusion wave approximation for the flow dynamics. The equations of motion are solved with a finite difference scheme. The model includes continuous soil-moisture accounting, rainfall interception, infiltration, surface and channel runoff routing, soil erosion, and sediment transport. Over each pixel of the domain, flow is allowed to take place only one-dimensionally and in the direction of the steepest slope of either of two orthogonal directions in an orthogonal cascade of planes. Input to each pixel is in the form of precipitation and upstream inflow from all directions. Infiltration losses are diagnosed at each pixel by using the Green-Ampt infiltration equations. CASC2D is one of the surface-water hydrologic models supported by the watershed modeling system (WMS) under development at Brigham Young University. More recently, Fiedler and Ramírez (1998) developed a fully unsteady, fully dynamic, two-dimensional hydrodynamic wave model for simulation of the rainfall–runoff process that allows for fully interactive infiltration. In addition to the explicit consideration of fully interactive infiltration, Fiedler and Ramírez developed a numerical scheme capable of describing the flow dynamics as affected by microtopographic conditions and other rapidly varying soil and vegetation characteristics. As such, this capability allows for consideration of issues of scale and takes into account the effect of microchannel formation (e.g., rills, etc.) on runoff production and flood dynamics.

References

Abbott, M.B. (1986a). An introduction to the European Hydrological System – Système Hydrologique Européen, SHE – 1: history and philosophy of a physically-based, distributed modelling system. *Journal of Hydrology*, **87**, 45–59.

Abbott, M.B. (1986b). An introduction to the European Hydrological System – Système Hydrologique Européen, SHE – 2: structure of a physically-based, distributed modelling system. *Journal of Hydrology*, **87**, 61–77.

Adom, D.N., Bacchi, B., Brath, A., and Rosso, R. (1989). On the geomorphoclimatic derivation of flood frequency (peak and volume) at the basin and regional scale. In *New Directions for Surface Water*

Modeling, ed. M.L. Kavvas, pp. 165–176, IAHS Publ. no. 181, Wallingford, UK.

Agnese, C., D'Asaro, F., and Giordano, C. (1988). Estimation of the time scale of the geomorphologic instantaneous unit hydrograph from effective streamflow velocity. *Water Resources Research*, **24**(7), 969–978.

Anderson, E.A. (1971). *FORTRAN-IV Program for Stanford Watershed Model IV*. NOAA, NWS, Office of Hydrology, Silver Spring, MD, February.

Bacchi, B. and Rosso, R. (1988). Analisi Geomorfoclimatica dei Modeli di Regionalizzazione della Frequenza delle Piene. In *Proc. XXI Ital. Conf. On Hydraulics and Water Engineering*, L'Aquila, September 5–8, **1**, 15–27.

Bras, R.L. (1990). *Hydrology, an Introduction to Hydrologic Science*. Addison Wesley, Reading, Massachusetts.

Brath, A., La Barbera, P., Mancini, M., and Rosso, R. (1989). The use of distributed rainfall-runoff models based on GIS at different scales of information. *Proceedings 3rd National Conference on Hydraulic Engineering*, pp. 448–453, American Society of Civil Engineers, New Orleans, August 14–18.

Burnash, R.J.C., Ferral, R.L., and McGuire, R.A. (1973). *A Generalized Streamflow Simulation System: Conceptual Modeling for Digital Computers*. U.S. Department of Commerce, National Weather Service and State of California, Department of Water Resources, Sacramento, CA.

Burlando, P., Mancini, M., and Rosso, R. (1994). FLORA: a distributed flood risk analyzer. In *Computer Support for Environmental Impact Assessment* eds. G. Guariso and B. Page, Elsevier Science B.V. (North Holland), pp. 91–102, Amsterdam.

Caroni, E. and Rosso, R. (1986). A comparison between direct and indirect estimation of the Nash model of catchment response. *Proceedings International Conference on Hydrological Processes in the Catchment*, Cracow, May 8–11, **2**, 27–34.

Caroni, E., Rosso, R., and Siccardi, F. (1986). Nonlinearity and time variance of the hydrologic response of a small mountain creek. In *Scale Problems in Hydrology*, eds. V.K. Gupta, I. Rodriguez-Iturbe, and E.F. Wood, pp. 19–38, D. Reidel Publishing Co., Boston, MA.

Chow V.T. (ed.) (1964). *Handbook of Applied Hydrology*. McGraw Hill, New York.

Chow, V.T., Maidment, D., and Mays, L.W. (1988). *Applied Hydrology*. McGraw Hill, New York.

Claborn, B.J. and Moore, W. (1970). *Numerical Simulation in Watershed Hydrology*. Hydraulic Engineering Laboratory, University of Texas, Rep. No. HYD 14-7001, Austin, Texas.

Clark, C.O. (1945). Storage and The Unit Hydrograph. *Proceedings of the American Society of Civil Engineers*, **9**, 1333–1360.

Crawford, N.H. and Linsley, R.K. Jr. (1966). *Digital Simulation in Hydrology*. Stanford Watershed Model IV. Dept. of Civil Engineering, Stanford University, Stanford, CA., Tech. Rep. No. 39, July.

Croley, T.E. II. (1985). Forecasting Great Salt Lake levels. In *Proceedings, Problems and Prospects for Predicting Great Salt Lake Levels*, eds. P.A. Kay and H.F. Diaz, pp. 179–188. University of Utah, Center for Public Affairs and Administration. Salt Lake City, UT, March 26–28.

Cunge, J.A. (1969). On the subject of a flood propagation method (Muskingum method). *Journal of Hydraulic Research, International Association of Hydraulic Research*, **7**(2), 205–230.

Dawdy, D.R. and O'Donnell, T. (1965). Mathematical models of catch water behavior. *Journal of the Hydraulics Division*, **91**(HY4), 123–137.

Dawdy, D.R., Lichty, R.W., and Bergman, J.M. (1972). A rainfall-runoff simulation model for estimation of flood peaks for small drainage basins. *U.S. Geological Survey Professional Paper 506-B*, 28 pp, U.S. Government Printing Office, Washington, DC.

Díaz-Granados, M.A., Bras, R.L., and Valdés, J.B. (1983a). Incorporation of channel losses in the geomorphologic IUH. *R.M. Parsons Laboratory* T. R. No: 293, MIT, Cambridge, MA.

Díaz-Granados, M.A., Valdés, J.B., and Bras, R.L. (1983). A derived flood frequency distribution based on the geomorphoclimatic IUH and the density function of rainfall excess. *R.M. Parsons Laboratory* T.R. No. 292, MIT, Cambridge, MA.

Díaz-Granados, M.A., Valdés, J.B., and Bras, R.L. (1984). A physically-based flood frequency distribution. *Water Resources Research*, **20**(7), 995–1002.

Donigian, A.S., Jr. and Crawford, N.H. (1979). *User's Manual for the Nonpoint Source (NPS) Model*. Environmental Resources Information Center, U.S. Environmental Protection Agency, Cincinnati, OH.

Dooge, J.C.I. (1959). A general theory of the unit hydrograph. *Journal of Geophysical Research*, **64**(2), 241–256.

Eagleson, P.S. (1970). *Dynamic Hydrology*. McGraw Hill, New York, NY.

Eagleson, P.S. (1972). Dynamics of flood frequency. *Water Resources Research*, **8**(4), 878–898.

Epstein, D. and Ramírez, J.A. (1993). A daily spatial disaggregation approach and its application in hydrologic sensitivity analysis of the upper Rio Grande to climate variability. *American Geophysical Union Front Range Meeting*, Boulder, February 8–10.

Epstein, D. and Ramírez, J.A. (1994). Spatial disaggregation for studies of climatic hydrologic sensitivity.

ASCE Journal of the Hydraulics Division, **120**(12), 1449–1467.

Fread, D.L. (1974). *Numerical Properties of Implicit Four-Point Finite Difference Equations of Unsteady Flow.* HRL-45, NOAA Tech. Memo NWS HYDRO-18, Hydrologic Research Laboratory, National Weather Service, Silver Spring, MD.

Fread, D.L. (1976). *Theoretical Development of Implicit Dynamic Routing Model.* HRL-113, Hydrologic Research Laboratory, National Weather Service, Silver Spring, MD.

Fread, D.L. (1978). NWS operational dynamic wave model. In *Verification of Mathematical and Physical Models.* Proceedings of 26th Annual Hydraulics Division Specialty Conference, ASCE, College Park, MD, pp. 455–464.

Fread, D.L. (1988). The *NWS DAMBRK Model: Theoretical Background/User Documentation.* HRL-256, National Weather Service, Silver Spring, MD.

Fread, D.L. and Lewis, J.M. (1988). FLDWAV: A generalized flood routing model. *Proc. National Conference on Hydraulic Engineering*, ASCE, Colorado Springs, Colorado, pp. 668–673.

Fiedler, F.R. and Ramírez, J.A. 2000. A numerical method for hydrodynamic modeling of overland flow. *International Journal for Numerical Methods in Fluids*, **32**(2), 219–239.

Gupta, V.K. and Mesa, O.J. (1988). Runoff generation and hydrologic response via channel network geomorphology: recent progress and open problems. *Journal of Hydrology*, **102**(1–4), 3–28.

Gupta, V.K. and Waymire, E. (1983). On the formulation of an analytical approach to hydrologic response and similarity at the basin scale. *Journal of Hydrology*, **65**, 95–123.

Gupta, V.K., Waymire, E., and Rodríguez-Iturbe, I. (1986). On scales, gravity, and network structure in basin runoff. In *Scale Problems in Hydrology*, eds. V.K. Gupta, I. Rodríguez-Iturbe, and E.F. Wood, Dordrecht, Holland, D. Reidel.

Hebson, C. and Wood, E.F. (1982). A derived flood frequency distribution using Horton order ratios. *Water Resources Research*, **18**(5), 1509–1518.

Howard, A.D. (1990). Theoretical model of optimal drainage networks. *Water Resources Research*, **26**(9), 2107–2117.

Hudlow, M.D. and Clark, R.A. (1969). Hydrologic synthesis by digital computer. *Journal of the Hydraulics Division, Proceedings of the American Society of Civil Engineers*, **95**(HY3), 839–860.

HEC – Hydrologic Engineering Center (1981). *The New HEC-1 Flood Hydrograph Package.* Tech. Paper No. 82, U.S. Army Corps of Engineers, Davis, CA.

Johanson, R.C., Imhoff, J.C., and Davis, H.H. (1980). *User's Manual for Hydrologic Simulation Program-FORTRAN (HSPF).* EPA-600/9-80-015, U.S. Environmental Protection Agency, Athens, GA.

Karlinger, M.R. and Troutman, B.M. (1985). Assessment of the instantaneous unit hydrograph derived from the theory of topologically random networks. *Water Resources Research*, **21**(11), 1693–1702.

Kibler, D.F. (1982). Desk-top methods for urban stormwater calculation. Chapter 4. In *Urban Stormwater Hydrology*, ed. D.F. Kibler. Water Resources Monograph 7, American Geophysical Union, Washington, DC.

Kirshen, D.M. and Bras, R.L. (1983). The linear channel and its effect on the geomorphologic IUH. *Journal of Hydrology*, **65**, 175–208.

La Barbera, P. and Rosso, R. (1987). Fractal geometry of river networks (abstract). *EOS Trans. AGU*, **68**(44), 1276.

La Barbera, P. and Rosso, R. (1989). On the fractal dimension of stream networks. *Water Resources Research*, **25**(4), 735–741.

Leavesley, G.H., Lichty, R.W., Troutman, B.M., and Saindon, L.G. (1983). *Precipitation-Runoff Modeling System: User's Manual*, U.S. Geological Survey Water-Resources Investigations Report 83-4328, 207 pp, U.S. Government Printing Office, Washington, DC.

Leavesley, G.H., Restrepo, P.J., Markstrom, S.L., Dixon, M., and Stannard, L.G. (1996). *The Modular Modeling System (MMS): User's Manual, (Updated March 1998).* U.S. Geological Survey-Open File Report 96–151, U.S. Government Printing Office, Washington, DC.

Leavesley, G.H. and Stannard, L.G. 1990. *Application of Remotely Sensed Data in a Distributed Parameter Watershed Model.* U.S. Geological Survey Water Resources Investigations.

Liou, E.Y. (1970). *Opset: Program for Computerized Selection of Watershed Parameter Values of the Stanford Watershed Model.* Research Dept. No. 34, Water Resources Institute, University of Kentucky, Lexington.

Luellwitz, T. (1991). *The Application of a Deterministic Hydrologic Model and GIS to Simulate Water Yield Increase due to Timber Harvest in a Subalpine Watershed.* Master of Science Report, Department of Earth Resources, Colorado State University, Fort Collins.

Mancini, M. and Rosso, R. (1989). Using GIS to assess spatial variability of SCS curve number at the basin scale. In *New Directions for Surface Water Modeling*, ed. M.L. Kavvas, IAHS Publ. no.181, pp. 435–444, IAHS Pubs, Wallingford, UK.

Mesa, O.J. and Mifflin, E.R. (1986). On the relative role of hillslope and network geometry in hydrologic response. In *Scale Problems in Hydrology*, eds.

V.K. Gupta, I. Rodríguez-Iturbe, and E.F. Wood, D. Reidel., Dordrecht, Holland.

Metcalf and Eddy. (1971). *Storm Water Management Model. Vol. 1.* Environmental Protection Agency, Water Resources Engineers, University of Florida, Gainesville, FL.

Molnár, P. and Ramírez, J.A. (1998a). Energy dissipation theories and optimal channel characteristics of river networks. *Water Resources Research.* **37**(7), 1809–1818.

Molnár, P. and Ramírez, J.A. (1998b). An analysis of energy expenditure and downstream hydraulic geometry at Goodwin Creek. *Water Resources Research*, **37**(7), 1819–1829.

Monro, J.C. and Anderson, E.A. (1974). National Weather Service River Forecasting System. *Journal of the Hydraulics Division*, **100**, 621–630.

Morris, E.M. (1980). Forecasting flood flows in grassy and forested basins using a deterministic distributed mathematical model. *International Association of Scientific Hydrology Publication*, **129**, 247–255, IAHS Pubs, Wallingford, UK.

Nash, J.E. (1957). The form of the instantaneous unit hydrograph. *International Association of Scientific Hydrology Publication*, **45**(3), 114–121, IAHS Pubs, Wallingford, UK.

Nash, J.E. (1958). Determining runoff from rainfall. *Proceedings of the Institution of Civil Engineers*, **10**, 163–184.

Nash, J.E. (1959). Systematic determination of unit hydrograph parameters. *Journal of Geophysical Research*, **64**(1), 111–115.

Nash, J.E. (1960). A unit hydrograph study, with particular reference to British catchments. *Proceedings of the Institution of Civil Engineers*, **17**, 249–282.

Ogden, F.L. (1992). *Two-Dimensional Runoff Modeling with Weather Radar Data.* Ph.D. Dissertation, Dept. of Civil Engineering, Colorado State University, Fort Collins, CO.

Ogden, F.L. and Julien, P.Y. (1993). Runoff sensitivity to temporal and spatial rainfall variability at runoff plane and small basin scales. *Water Resources Research*, **29**(8), 2589–2597.

Ogden, F.L., Richardson, J.R., and Julien, P.Y. (1995). Similarity in catchment response 2. Moving rainstorms. *Water Resources Research*, **31**(6), 1543–1547.

Peck, E.L. (1976). *Catchment Modeling and Initial Parameter Estimation for the National Weather Service River Forecasting System.* NOAA Technical Memorandum NWS Hydro-31, Washington, DC.

Pilgrim, D.H. (1976). Travel times and non linearity of flood runoff from tracer measurements on a small watershed. *Water Resources Research*, **12**(4), 487–496.

Pilgrim, D.H. (1977). Isochrons of travel time and distribution of flood storage from a tracer study on a small watershed. *Water Resources Research*, **13**(3), 587–595.

Pilotti, M. and Rosso, R. (1990). Shell: A general framework for modeling the distributed response of a drainage basin. In *Computational Methods in Surface Hydrology*, eds. G. Gambolati, A. Rinaldo, C. Brebbia, W.G. Gray, and G.F. Pinder. Springer Verlag, Berlin, pp. 517–522.

Ramírez, J.A., Salas, J.S., and Rosso, R. (1994). In *Determination of Flood Characteristics by Physically-Based Methods*, NATO advanced study institute on coping with floods. eds. G. Rossi, N. Hamancioglu, and V. Yevjevich, Kluwer Academic Publishers, pp. 77–110, Dordrecht.

Ranzi, R. and Rosso, R. (1991a). Flea: flood event analyzer. In *Guida al Software Ambientale*, ed. G. Guariso, Patron, Bologna, pp. 121–126, Wallingford, UK.

Ranzi, R. and Rosso, R. (1991b). A physically based approach to modeling distributed snowmelt in a small alpine catchment. In *Snow Hydrology and Forests in High Alpine Areas*, eds. H. Bergmann, H. Lang, W. Frey, D. Issler, and B. Salm. IAHS Publ. no. 205, pp. 141–150, Wallingford, UK.

Ricca V.T. (1972). *The Ohio State University Version of the Stanford Streamflow Simulations Model, Part I – Technical Aspects.* Ohio State University, Columbus May.

Rigon, R., Rinaldo A., Rodríguez-Iturbe, I., Bras, R.L., and Ijjász-Vásquez, E.J. (1993). Optimal channel networks: a framework for the study of river basin morphology. *Water Resources Research*, **29**(6), 1635–1646.

Rinaldo, A., Marani, A., and Rigon, R. (1991). Geomorphologic dispersion. *Water Resources Research*, **27**(44), 513–525.

Rodríguez-Iturbe, I., Rinaldo, A., Rigon, R., Bras, R.L., Marani, A., and Ijjász-Vásquez, E.J. (1992). Energy dissipation, runoff production, and the three-dimensional Structure of river basins. *Water Resources Research*, **28**(4), 1095–1103.

Rodríguez-Iturbe, I., González Sanabria, M., and Bras, R.L. (1982a). A geomorphoclimatic theory of the instantaneous unit hydrograph. *Water Resources Research*, **18**(4), 877–886.

Rodríguez-Iturbe, I., González Sanabria, M., and Caamaño, G. (1982b). On the climatic dependence of the IUH: a rainfall-runoff analysis of the Nash model and the geomorphoclimatic theory. *Water Resources Research*, **18**(4), 887–903.

Rodríguez-Iturbe, I., Devoto, G., and Valdés, J.B. (1979). Discharge response analysis and hydrologic similarity: the interrelation between the geomorphologic IUH and the storm characteristics. *Water Resources Research*, **15**(6), 1435–1444.

Rodríguez-Iturbe, I. and Valdés, J.B. (1979). The geomorphologic structure of hydrologic response. *Water Resources Research*, **15**(6), 1409–1420.

Rosso, R. (1984). Nash model relation to Horton order ratios. *Water Resources Research*, **20**(7), 914–920.

Rosso, R. and Caroni, E. (1986). Analysis, estimation and prediction of the hydrological response from catchment geomorphology. *Excerpta*, **1**, 93–108.

Saghafian, B. (1992). *The Effect of Spatially Varied Characteristics on Watershed Response: A Two-Dimensional Distributed Approach*. Ph.D. dissertation, Civil Engineering Department, Colorado State University, Fort Collins, CO.

Saghafian, B., Julien, P.Y., and Ogden, F.L. (1995). Similarity in catchment response 1. Stationary rainstorms. *Water Resources Research*, **31**(6), 1533–1541.

Sherman, L.K. (1932). Streamflow from rainfall by the unit graph method. *Engineering News Record*, **108**, 501–505.

Shreve, R.L. (1967). Infinite topologically random channel networks. *Journal of Geology*, **75**, 178–186.

Singh, V.P., (1989). *Hydrologic Systems. Watershed Modeling. Volume II.* Prentice Hall, Englewood Cliffs, NJ.

Snyder, F.F. (1938). Synthetic unit graphs. *Transactions of the American Geophysical Union*, 19, 447–454.

Soil Conservation Service, Hydrology. (1972). *National Engineering Handbook, Section 4*. Soil Conservation Service, U.S. Department of Agriculture, Washington, DC.

Troutman, B.M. and Karlinger, M.R. (1984). On the expected width function for topologically random channel networks. *Journal of Applied Probability*, **21**, 836–884.

Troutman, B.M. and Karlinger, M.R. (1985). Unit hydrograph approximation assuming linear flow through topologically random channel networks. *Water Resources Research*, **21**(5), 743–754.

Troutman, B.M. and Karlinger, M.R. (1986). Averaging properties of channel networks using methods in stochastic branching theory. In *Scale Problems in Hydrology*, eds. V.K. Gupta, I. Rodríguez-Iturbe, and E.F. Wood. D. Reidel, Dordrecht, Holland.

Valdés, J.B., Fiallo, Y., and Rodríguez-Iturbe, I. (1979). A rainfall-runoff analysis of the geomorphologic IUH. *Water Resources Research*, **15**(6), 1421–1434.

U.S. Army Corps of Engineers. (1972). *Program Description and User's Manual for SSARR Model Streamflow Synthesis and Reservoir Regulation.* Program 724-K5-G0010, U.S. Army Corps of Engineers, Davis, CA.

U.S. Soil Conservation Service. (1973). *Computer Program for Project Formulation Hydrology.* Technical Release No. 20, U.S. Department of Agriculture, Washington, DC.

U.S. Soil Conservation Service. (1985). *National Engineering Handbook. Section 4, Hydrology.* U.S. Department of Agriculture, Washington, DC.

Woolhiser, D.A. and Liggett, J.A. (1967). Unsteady one-dimensional flow over a plane: The rising hydrograph. *Water Resources Research*, **3**(3), 753–771.

CHAPTER 12

Flood Frequency Analysis and Statistical Estimation of Flood Risk

Jery R. Stedinger
School of Civil and Environmental Engineering, Cornell University

Planners and the public require reliable estimates of the risk of large floods to support flood risk management. In river basins in which human activities and natural processes have not resulted in significant changes in the distribution of floods, statistical procedures are commonly used to estimate flood risk probabilities. This chapter reviews the statistical methods most often used to describe flood risk as well as the statistical and hydrologic issues that arise. Recent research in flood frequency analysis documents the advantages of a modeling/data paradigm that combines at-site systematic flood records and regional and historical flood information when estimating flood risk at a site. An example is the index-flood method, which has been the subject of much research activity. Historical and paleoflood information provide additional sources of information about the distribution of floods. At ungauged sites, regional regression models are often developed to estimate flood quantiles from physiographic basin characteristics. Development of hydrographs for studying river basin and reservoir operations pose special problems because it is not clear what the critical duration will be or how critical inflows would be distributed spatially.

Introduction

Floods from natural processes around the world continue to take thousands of lives and to cause billions of dollars in damages. Floodplain management plans and designs for bridges, flood control works, reservoirs, and nonstructural flood-damage reduction measures need to reflect the likelihood or probability of large flood flows. Structural measures including dams, levees, and channel modifications often can reduce flood flows and their consequences (Watson and Biedenharn, Chapter 14, this volume). However, they cannot eliminate the risk of flooding, and society needs to address this residual risk. Other flood risk management tools that address residual risk include nonstructural actions such as flood-proofing dwellings to mitigate damages, restricting development, and implementing flood warning and response measures (Gruntfest, Chapter 15, this volume). Whatever strategy is adopted, sound planning and system design benefit from accurate estimates of flood risk. Such estimates allow a quantitative balancing of flood control efforts and the resultant benefits and also enhance the credibility of floodplain development restrictions (Hamilton and Joaquin, Chapter 18, this volume).

At locations with statistically stationary flood hydrologies, flood-frequency analysis basically poses a problem of insufficient information. Even with a 100-year record, there is tremendous uncertainty in estimates of flood levels exceeded with a probability of 1 or 2%. Unfortunately, available records are generally less than 30 years in length and may not even exist at sites of interest. As a consequence, hydrologists need to use practical knowledge of catchment processes with efficient and robust statistical techniques to develop the best estimates of flood risk that they can. If change over time due to urbanization or natural processes alters the flood-frequency relationship, then other methods need to be used based on routing "natural flows" or computing possible flood flows from rainfall series (Ramírez, Chapter 11, this volume).

This chapter provides an overview of traditional flood-frequency analysis ideas and methods. It begins with a very brief introduction to the notation used to describe flood flows (instantaneous annual maxima or annual maximum volumes over n days for $n = 1, 2, 3, \ldots$) as random variables. The second section discusses the value of using mathematical distributions to describe the distribution of flood flows and presents the most commonly used distributions. That is followed by a discussion of the issues that should be considered when selecting a distribution and a corresponding fitting procedure. Also of concern are opportunities for using regional hydrologic information and available historical and paleoflood information to improve estimated flood-flow frequency relationships. Later sections discuss problems with measurement errors, mixed flood populations, and rogue and low outliers in annual flood series. Partial duration series are an alternative and attractive framework for evaluation of flood risk. Regional regression methods are also discussed for deriving estimates of flood quantiles at ungauged sites. The chapter concludes with a discussion of concerns when computing flood hydrographs for operation studies. More detailed discussions of flood-flow frequency methods are available (NERC, 1975; Kite, 1988; Chow et al., 1988; Stedinger et al., 1993; Kottegoda and Rosso, 1997; McCuen, 1998).

Statistics for Flood Frequency Analysis

The value of the flood flow X that occurs in any year is thought of as a random variable drawn from some probability distribution determined by natural phenomena. Generally, a lowercase letter x is used to denote a possible value of X. For a random variable X, its cumulative distribution function (cdf) $F_X(x)$ is defined as:

$$F_X(x) = Pr[X \leq x] \tag{1}$$

The probability density function (pdf) $f_X(x)$ describes the relative likelihood that a continuous random variable X takes on different values. The pdf is the derivative of the cdf

$$f_X(x) = dF_X(x)/dx \tag{2}$$

In flood-frequency analysis, particular percentiles or quantiles of a distribution are often used as design flows. The pth quantile of the X distribution is denoted x_p. The pth quantile of the X distribution is the flow value exceeded in any year with probability $1 - p$. The *return period* of a flood is often specified instead of the exceedance probability. For example, the annual maximum flood flow $x_{0.99}$ exceeded with a 1% probability in any year, or a chance of 1 in 100, is called the 100-year flood. The relationship between the return period and the exceedance probability $q = (1 - p)$ is

$$T = 1/q \tag{3}$$

Several arguments lead to this definition. For example, if the magnitudes of floods that occur in different years are independent, then the probability that the first exceedance of level x_p occurs in year k is

$$Pr[\text{exactly } k \text{ years until } X \geq x_p] = p^{k-1}q \tag{4}$$

This is a geometric distribution with mean $1/q$. Thus, the average waiting time until the level x_p is exceeded equals T years. Similarly, on average there will be one 100-year flood in a 100-year period.

Return periods have been incorrectly understood to mean that one and only one T-year event should occur every T years. This is not true. Actually, the probability of the T-year flood being exceeded is $1/T$ in every year. Hydrologists can attempt to avoid the awkwardness of small probabilities and the incorrect implication of return periods by reporting odds ratios: for example, the 100-year flood can be described as a value with a 1 in 100 chance of being exceeded each year. Alternatively, it is often called the 1%-chance exceedance flood. (See also Hamilton and Joaquin, Chapter 18, this volume.)

Traditional Moments

The mean, variance, and coefficient of skewness are used to describe the character of a probability distribution. The mean is defined as the expected value of the random variable

$$\mu_x = E[X] \tag{5}$$

and the variance, denoted Var(X) or σ_x^2, is defined as

$$\sigma_x^2 = \text{Var}(X) = E[(X - \mu_x)^2] \tag{6}$$

The standard deviation σ_x is the square root of the variance.

A dimensionless description of the variability of flood flows is provided by the coefficient of variation (CV). Similarly, the coefficient of skewness γ_x and the coefficient of kurtosis κ_x provide dimensionless descriptions of the shape of a flood flow distribution X. Table 12.1 contains the definitions of these variables. The CV is a measure of the relative variability of flood flows; values in the range 0.20 to 1.5 bracket most annual maximum flood series (Landwehr et al., 1978). The skewness coefficient describes the relative asymmetry of a distribution. Common values range from +0.5 to

Table 12.1. *Definitions of dimensionless product-moment and L-moment ratios*

Name	Denoted	Definition
	Product-moment ratios	
CV	CV_x	σ_x/μ_x
Coefficient of skewness	γ_x	$E[(X-\mu_x)^3]/\sigma_x^3$
Coefficient of kurtosis	κ_x	$E[(X-\mu_x)^4]/\sigma_x^4$
	L-moment ratios	
L-CV*	L-CV, τ_2	λ_2/λ_1
L-coefficient of skewness	L-skewness, τ_3	λ_3/λ_2
L-coefficient of kurtosis	L-kurtosis, τ_4	λ_4/λ_2

*Hosking and Wallis (1997) use τ instead of τ_2 to represent the L-CV ratio.

+6 for annual flood series. Finally, the kurtosis, which is used less often, describes the thickness of a distribution's tails.

L-Moments

L-moments are another way to summarize the statistical properties of hydrologic data based on linear combinations of the original data (Hosking, 1990). Recently, hydrologists have found that regionalization methods that use L-moments are superior to methods that use traditional moments (Hosking and Wallis, 1997). L-moments have also proved useful for constructing goodness-of-fit tests (Hosking et al., 1985, Chowdhury et al., 1991; Fill and Stedinger, 1995), measures of regional homogeneity, and distribution selection methods (Vogel and Fennessey, 1993; Hosking and Wallis, 1997; Wang, 1997). The first L-moment is the arithmetic mean

$$\lambda_1 = E[X] \tag{7a}$$

Let $X_{(i|n)}$ be the ith largest observation in a sample of size n ($i = 1$ corresponds to the largest). Then, for any distribution, the second L-moment is a description of scale based on the expected difference between two randomly selected observations

$$\lambda_2 = (1/2)E\left[X_{(1|2)} - X_{(2|2)}\right] \tag{7b}$$

Similarly, L-moment measures of skewness and kurtosis use

$$\lambda_3 = (1/3)E\left[X_{(1|3)} - 2X_{(2|3)} + X_{(3|3)}\right] \tag{7c}$$

$$\lambda_4 = (1/4)E\left[X_{(1|4)} - 3X_{(2|4)} + 3X_{(3|4)} - X_{(4|4)}\right] \tag{7d}$$

Sample estimates are often computed with an intermediate statistic called probability weighted moments (PWMs) defined as

$$\beta_r = E\{X[F(X)]^r\} \tag{8}$$

where $F(X)$ is the cdf of the random variable X. Unbiased PWM estimators (Landwehr et al., 1979) computed as

$$\begin{aligned} b_0 &= \bar{X} \\ b_1 &= \sum_{j=1}^{n-1}(n-j)X_{(j)}/[n(n-1)] \\ b_2 &= \sum_{j=1}^{n-2}(n-j)(n-j-1)X_{(j)}/[n(n-1)(n-2)] \end{aligned} \tag{9}$$

are recommended (Hosking and Wallis, 1995). These are examples of the general formula

$$\begin{aligned} \hat{\beta}_r = b_r &= \frac{1}{n}\sum_{j=1}^{n-r}\binom{n-j}{r}X_{(j)}\Big/\binom{n-1}{r} \\ &= \frac{1}{(r+1)}\sum_{j=1}^{n-r}\binom{n-j}{r}X_{(j)}\Big/\binom{n}{r+1} \end{aligned} \tag{10}$$

for $r = 1, \ldots, n-1$. L-moments are easily calculated in terms of PWMs using

$$\begin{aligned} \lambda_1 &= \beta_0 \\ \lambda_2 &= 2\beta_1 - \beta_0 \\ \lambda_3 &= 6\beta_2 - 6\beta_1 + \beta_0 \\ \lambda_4 &= 20\beta_3 - 30\beta_2 + 12\beta_1 - \beta_0 \end{aligned} \tag{11}$$

Formulas for directly calculating L-moment estimators are provided by Wang (1996). Measures of CV, skewness, and kurtosis of a distribution can be computed with L-moments as they can with traditional product moments. Table 12.1 provides definitions.

Use of Mathematical Probability Distributions

In practice, the true mathematical form for the distribution that describes flood flows is not known. Moreover, even if it were, its functional form may have too many parameters to be of much practical use. The real issue is how to select a physically reasonable and simple distribution to describe the frequency of large flood flows to estimate that distribution's parameters and ultimately to obtain risk estimates of satisfactory accuracy for the problem at hand.

Fitting a continuous mathematical distribution to data sets yields a compact and smoothed representation of the flood frequency distribution revealed by the available data and a systematic procedure for extrapolation to frequencies beyond the range of the data set. Because floodplain management regulations and restrictions can be unpopular, it is important for flood risk management agencies to have legally defensible procedures (Thomas, 1985). The sections below review some of the methods adopted for fitting distributions as well as distributions used to describe annual flood-flow distributions.

Parameter Estimation Methods

An important task is estimation of the parameters of a proposed flood-flow distribution so that required quantiles and expectations can be calculated with the fitted model. The field of statistics provides several general approaches for estimating the parameters of a probability distribution. A simple approach that has often been used in hydrology is the method of moments. It uses the available sample to compute the parameters of a distribution so that the theoretical moments of the distribution of X exactly equal the corresponding sample estimators of the mean, variance, and perhaps the coefficient of skewness. Alternatively, parameters can be estimated by using the sample L-moments discussed above, corresponding to the method of L-moments. Still another method that has strong statistical motivation is the method of maximum likelihood (Kite, 1988; Kottegoda and Rosso, 1997).

Finally, a simple method of fitting flood frequency curves is to plot the ordered flood values on special probability paper and then draw a line through the data (Gumbel, 1958). Even today, that simple method remains attractive when some of the smallest floods are zero or have unusually small values (Kroll and Stedinger, 1996). In any case, plotting the ranked annual maximum series against a probability scale is an excellent and recommended practice for seeing what the data look like and for ensuring that a fitted curve is consistent with the data (Stedinger et al., 1993).

Statisticians and hydrologists have invested a great deal of time investigating these and other methods of fitting different distributions. In general, when fitting a distribution that describes the data well with large samples, the maximum-likelihood estimators are best. However, with flood records of 50 years or less, and when fitting distributions that do not exactly match the data, the choice is generally less clear. We return to this topic below.

A related issue is how to evaluate which of several estimation methods provides the best estimator of a distribution's parameters. "Best" can be measured in terms of the accuracy with which the parameters themselves are estimated, the accuracy of estimated quantiles of the distribution, or in terms of the accuracy of estimated damages and flood risk (Stedinger, 1997). One also needs to determine how accuracy should be measured. Some studies have used average squared loss, some have used average absolute loss with different weights on under- and overestimation, and some have used the squared loss of the log-quantile estimator (Slack et al., 1975; Kroll and Stedinger, 1996). In almost all cases one is also interested in the bias of an estimator, which is the average value of the estimator minus the true value of the parameter or quantile being estimated. Special estimators have been developed to compute design events, which on average are exceeded with the specified probability and have the anticipated risk of being exceeded (Beard, 1960, 1997; Rasmussen and Rosbjerg, 1989, 1991a; Stedinger, 1997; Rosbjerg and Madsen, 1998).

Lognormal Distribution

The normal (N) or Gaussian distribution is the most commonly used distribution in statistics. It is also the basis of the lognormal (LN) and three-parameter lognormal (LN3) distributions, which have seen many applications in flood frequency analysis in the United States and Japan. Reasonably efficient estimators of the parameter of both distributions are available (Stedinger, 1980; Hoshi et al., 1984), and the corresponding models are easy to use. If floods X have a lognormal distribution, then

$$Y = \ln(X) \tag{12}$$

is normally distributed. A lognormal distribution has probability density function

$$f_X(x) = \frac{1}{x\sqrt{2\pi\sigma_Y^2}} \exp\left\{-\frac{1}{2}\left[\frac{\ln(x)-\mu_Y}{\sigma_Y}\right]^2\right\}, \quad x > 0 \tag{13a}$$

with mean and variance

$$\mu_X = \exp\left[\mu_Y + \sigma_Y^2/2\right] \quad \text{and} \quad \sigma_X^2 = \mu_X^2\left\{\exp\left[\sigma_Y^2\right] - 1\right\} \tag{13b}$$

and skewness coefficient $\gamma_X = 3\text{CV}_X + \text{CV}_X^3$ where the coefficient of variation CV_X is σ_X/μ_X. The three-parameter lognormal distribution is obtained by modeling $Y = \ln(X - \xi)$ with a normal distribution where ξ is a lower bound parameter for X. As a result

$$\mu_X = \xi + \exp\left[\mu_Y + \sigma_Y^2/2\right] \quad \text{and} \quad \sigma_X^2 = (\mu_X - \xi)^2\left\{\exp\left[\sigma_Y^2\right] - 1\right\} \tag{14}$$

where $\gamma_X = 3\phi + \phi^3$ with $\phi^2 = \{\exp[\sigma_Y^2] - 1\}$. (See Stedinger et al., 1993).

Gumbel and Generalized Extreme Value Distributions

The annual flood is the largest flood flow during a year, and thus one might expect its distribution to belong to the set of extreme value distributions (Gumbel, 1958; Kottegoda and Rosso, 1997). These are the distributions obtained in the limit, as n becomes large, by taking the largest of n independent random variables. The extreme value (EV) type I distribution or Gumbel distribution has often been used to describe flood flows. It has the cumulative distribution function

$$F(x) = \exp\{-\exp[-(x - \xi)/\alpha]\} \tag{15a}$$

with mean and variance

$$\mu_X = \xi + 0.5772\alpha \quad \text{and} \quad \sigma_X^2 = \pi^2\alpha^2/6 \approx 1.645\alpha^2 \tag{15b}$$

Its skewness coefficient is a constant, $\gamma_X = 1.1396$.

The generalized extreme value (GEV) distribution is a general mathematical expression that incorporates Gumbel's type I, II, and III EV distributions for maxima (Gumbel, 1958; Hosking et al., 1985). In recent years it has seen use as a general model of extreme events including flood flows (NERC, 1975; Hosking and Wallis, 1997). The GEV distribution has cdf

$$F(x) = \exp\left\{-[1 - \kappa(x - \xi)/\alpha]^{1/\kappa}\right\}, \quad \text{for } \kappa \neq 0 \tag{16}$$

where for $\kappa > 0$, floods must be less than the upper bound $x < (\xi + \alpha/\kappa)$, whereas for $\kappa < 0$, floods are greater than the lower bound $x > (\xi + \alpha/\kappa)$. The mean, variance, and skewness coefficient are ($\kappa > -1.3$):

$$\begin{aligned}
\mu_X &= \xi + (\alpha/\kappa)[1 - \Gamma(1 + \kappa)] \\
\sigma_X^2 &= (\alpha/\kappa)^2\{\Gamma(1 + 2\kappa) - [\Gamma(1 + \kappa)]^2\} \quad \text{and} \\
\gamma_X &= \text{Sign}(\kappa)[-\Gamma(1 + 3\kappa) + 3\Gamma(1 + \kappa)\Gamma(1 + 2\kappa) \\
&\quad - 2\Gamma^3(1 + \kappa)]/[\Gamma(1 + 2\kappa) - \Gamma^2(1 + \kappa)]^{3/2}
\end{aligned} \tag{17}$$

where $\Gamma(1 + \kappa)$ is the classical gamma function. The Gumbel distribution is obtained when $\kappa = 0$. For $|\kappa| < 0.3$, the general shape of the GEV distribution is similar to the Gumbel distribution, although the right-hand tail is thicker for $\kappa < 0$ and thinner for $\kappa > 0$. The parameters of the GEV distribution are often computed by using L-moments and the relationships

$$\begin{aligned}
\kappa &= 7.8590c + 2.9554c^2 \\
\alpha &= \kappa\lambda_2/[\Gamma(1 + \kappa)(1 - 2^{-\kappa})] \\
\xi &= \lambda_1 + (\alpha/\kappa)[\Gamma(1 + \kappa) - 1]
\end{aligned} \tag{18}$$

where

$$c = 2\lambda_2/(\lambda_3 + 3\lambda_2) - \ln(2)/\ln(3)$$

developed by Hosking et al. (1985). Moment estimators can provide more precise quantile estimators (Madsen et al., 1997a).

Log-Pearson Type 3

Another family of distributions used in hydrology is based on the Pearson type 3 (P3) distribution (Bobée and Ashkar, 1991). The log-Pearson type 3 distribution (LP3) describes a random variable whose logarithms have a Pearson type 3 distribution. It has a probability density function given by

$$f_X(x) = |\beta|\{\beta[\ln(x) - \xi]\}^{\alpha-1} \exp\{-\beta[\ln(x) - \xi]\}/\{x\Gamma(\alpha)\} \tag{19}$$

where for $\beta < 0$, floods are restricted to the range $0 < x < \exp(\xi)$. For $\beta > 0$, floods have a lower bound so that $\exp(\xi) < x$. The LP3 distribution has mean and variance

$$\mu_X = e^{\xi}\left(\frac{\beta}{\beta - 1}\right)^{\alpha} \quad \text{and} \quad \sigma_X^2 = e^{2\xi}\left\{\left(\frac{\beta}{\beta - 2}\right)^{\alpha} - \left(\frac{\beta}{\beta - 1}\right)^{2\alpha}\right\} \tag{20}$$

for $\beta > 2$ or $\beta < 0$. These expressions are seldom used. Instead, analysis generally focuses on the distribution of the logarithms $Y = \log(X)$ of the flows, which would have a Pearson type 3 distribution with moments μ_Y, σ_Y, and γ_Y (IACWD, 1982; Bobée and Ashkar, 1991). As a result, flood quantiles are calculated as

$$x_p = \exp\{\mu_Y + \sigma_Y K_p[\gamma_Y]\} \tag{21}$$

where $K_p[\gamma_Y]$ is a frequency factor corresponding to cumulative probability p for skewness coefficient γ_Y. ($K_p[\gamma_Y]$ corresponds to the quantiles of a gamma distribution with zero mean, unit variance, and skewness coefficient γ_Y.) For $\gamma_Y = 0$, the log-Pearson type 3 distribution reduces to the two-parameter lognormal distribution discussed above.

Since 1967 the recommended procedure for flood-frequency analysis by federal agencies in the United States uses this distribution; current guidelines in Bulletin 17B (IACWD, 1982) suggest that the skew γ_Y be estimated by a weighted average of the at-site sample skewness coefficient and a regional estimate of the skewness coefficient. Bulletin 17B also includes tables of frequency factors, a map of regional skewness estimators, checks for low outliers, confidence interval formula, a discussion of expected probability, and a weighted-moments estimator for historical data.

A Modeling/Data Paradigm

Frequency analysis is a problem in hydrology because sufficient information is seldom available at a site to determine adequately the frequency of rare events, even if one knows what family of distributions to use. At some sites no information is available. When one has, for example, 30 years of data to estimate the event exceeded with a chance of 1 in 100 (the 1% chance exceedance event), extrapolation is required. Given that sufficient data are seldom available at the site of interest, it makes sense to use climatic and hydrologic data from nearby and similar locations. In this context the National Research Council (NRC, 1988, p. 6) proposed three principles for hydrometeorological modeling: "(1) 'substitute space for time'; (2) introduction of more 'structure' into models; and (3) focus on extremes or tails as opposed to, or even to the exclusion of, central characteristics."

One substitutes space for time by using hydrologic information at different locations to compensate for short records at a single site. This is easier to do for rainfall, which, in regions without appreciable topographical relief, should have fairly uniform characteristics over large areas. It is more difficult for floods because of the effects of catchment topography and geology. A successful example of regionalization is the index-flood method discussed in the next section. Regional regression is also discussed below and provides a means for adjusting regional experience to reflect the likely flood risk at a particular site of interest. Regional regression is particularly useful for developing models for use at ungauged sites for which a flood record is not available. Cunnane (1988) discusses these and other regionalization procedures. Also discussed below are sources of historical and paleoflood data, which provide yet another source of information on the frequency of floods in a basin.

Much of the published flood-frequency analysis literature addresses how to estimate the parameters of a distribution given a systematic flood record of independent observations from that distribution. Those studies have contributed greatly to our understanding of the sampling variability of flood-quantile estimates, particularly when estimating three parameters (Potter, 1987; Bobée and Rasmussen, 1995). Still they fail in two respects. First, they seldom discuss or address how the NRC committee's ideas apply to the problem and thus how systematic records of annual floods should be combined with regional hydrologic information and possible historical information for a basin. There is only so much information in a short 30-year record of annual maximum floods, and substantial improvements in the precision of quantile estimates will require bringing additional information to bear on the problem. Regional hydrologic experience with an understanding of how flood distributions change with catchment size and other characteristics of a basin is critical to this effort. Similarly, if information can be obtained on the magnitude and frequency of large floods that occurred before a gauge was established, or after one was discounted, then if used appropriately that information can be of great value in flood-frequency analyses (Cohn and Stedinger, 1987).

These ideas are represented in Figure 12.1, which indicates that development of a flood frequency distribution should rely on the available systematic or gauged streamflow record for a site, historical and paleoflood information that can be identified for periods when a gauge was not operational or did not exist, and information obtained by analysis of flood records in the region.

Many studies assume that one is primarily interested in how well estimation methods work when fitting the correct distribution. In hydrology, one hopes that our simple

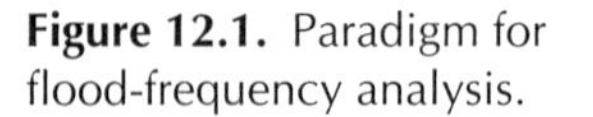
Figure 12.1. Paradigm for flood-frequency analysis.

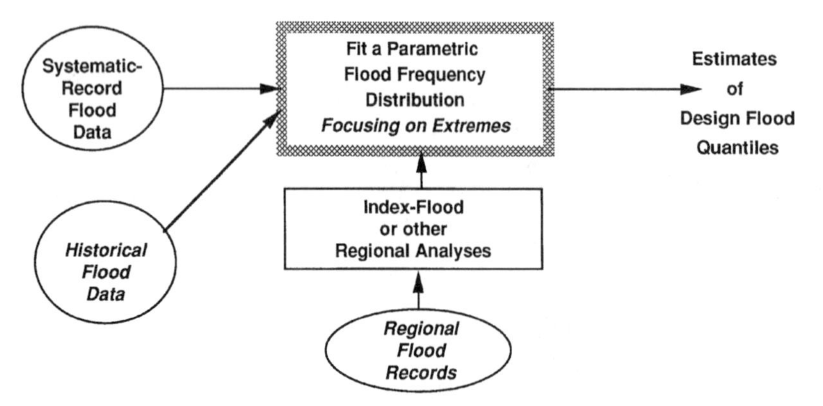

parametric probability distributions are reasonable approximations of those actually used in nature. However, it is unlikely that they are correct. As a result (Stedinger, 1989):

The issue is not how best to fit the right distribution, but how when fitting the wrong distribution to avoid doing it poorly.

For example, Slack et al. (1975) consider the consequences of fitting any one of four distributions (normal, Gumbel, three-parameter lognormal, and Weibull) to data drawn from those four families. More accurate quantile estimates were often obtained by intentionally fitting the wrong two-parameter distribution instead of the correct three-parameter distribution. Lu and Stedinger (1992) and Rosbjerg et al. (1992) have shown that fitting a two-parameter Gumbel distribution to many reasonable flood series yields more precise flood quantiles than fitting the more general three-parameter GEV distribution, when data are drawn from the more general GEV distribution with $\kappa \neq 0$. As a general rule, in a 30-year record there is insufficient information to resolve the shape parameter of the GEV distribution or the skewness coefficient of the log-Pearson type 3 distribution.

We also seek parameter estimation methods that will generally do well, even when we are fitting a distribution that deviates from that which generated the flood record. This is not an easy task. Stedinger (1980, fig. 2) shows that, when fitting a two-parameter lognormal distribution to log-Pearson data, maximum-likelihood estimators are generally more precise, unless the data have a negative log-space skew of -0.5 or less, in which case the method of moments estimators can do much better.

Selecting a Distribution

Several fundamental issues arise when selecting a distribution for use in flood-frequency analysis. Kelman (1987) notes that one should distinguish among the following questions:

1) What is the true distribution from which the observations are drawn?
2) Is a proposed distribution consistent with the available data for a site?
3) What distribution should be used to obtain reasonably accurate and robust estimates of design quantiles and hydrologic risk?

Question 1 is often asked. Unfortunately, the true distribution is probably too complex to be of practical use. Thus, strictly speaking, this is not really a useful question from a practical point of view. Question 3 more directly addresses the real management and engineering question.

Question 2 can be addressed by analytical goodness-of-fit criteria that are useful for gaining an appreciation for whether observed lack of fit is likely due to sample-to-sample variability or whether a particular departure of the data from a model is statistically significant (Vogel and McMartin, 1991; Chowdhury et al., 1991). Almost always several distributions provide statistically acceptable fits to the available data so that goodness-of-fit tests are unable to identify the "true" or "best" distribution to use. Stedinger et al. (1993) discuss standard goodness-of-fit statistics, including the probability plot correlation test and several others based on the L-skewness coefficient.

Question 3 addresses what distribution provides a reasonable description of the flood flows in a basin to serve as the basis of an accurate and robust flood risk estimation procedure. This is important in hydrologic applications and has been the subject of many studies (Cunnane, 1989; Kuczera, 1982; Slack et al., 1975; IACWD, 1982, appendix by Beard; Potter and Lettenmaier, 1990; Stedinger and Lu, 1995). At one time, the distribution that best fitted each data set was used for frequency analysis for that basin, but this approach has now been largely abandoned. Such a procedure is too sensitive to sampling variations in the data. Recently, L-moment diagrams, which display sample estimators of the L-kurtosis versus the L-skewness, have been used successfully to select a reasonable family of distributions (lognormal, Pearson, log-Pearson, GEV, generalized Pareto) to describe flood distributions and other phenomena (Hosking and Wallis, 1997). L-moments have the advantage that they are much less biased than traditional product moments estimators (Vogel and Fennessey, 1993), and thus if records from many stations are used in an analysis, the L-moment diagram can point to an appropriate family for modeling the distribution of the observations. LH-diagrams proposed by Wang (1997) place greater weight on the larger observations.

Operational procedures adopted by different national flood-frequency studies for use in their respective countries have been based on a combination of regionalization of some parameters and split-sample/Monte Carlo evaluations of different estimation procedures to find distribution-estimation procedure combinations that yield reliable flood quantile and risk estimates (Hosking and Wallis, 1997). In the United States, the log-Pearson type 3 distribution was adopted in 1967 with Bulletin 15, "A Uniform Technique for Determining Flood Flow Frequencies"; in 1976 use of a weighted skew coefficient was prescribed with the publication of Bulletin 17 "Guidelines for Determining Flood Flow Frequencies" (IACWD, 1982; Thomas, 1985). An index-flood GEV procedure was selected for the British Isles (NERC, 1975).

Regionalization and Index-Flood Method

The index-flood procedure is a simple regionalization technique with a long history in hydrology and flood-frequency analysis (Dalrymple, 1960; NERC, 1975). It uses data sets from several sites in an effort to construct more reliable flood-quantile estimators. A similar regionalization approach in precipitation frequency analysis is the station-year method, which combines rainfall data from several sites without adjustment to obtain a large composite record to support frequency analyses.

Research has demonstrated the potential advantages of index-flood procedures (Lettenmaier et al., 1987; Stedinger and Lu, 1995; Hosking and Wallis, 1997; Madsen, and Rosbjerg, 1997a). The idea behind the index-flood approach is to use the data from many hydrologically similar basins to estimate a dimensionless flood distribution (Wallis, 1980). Thus this method "substitutes space for time" by using regional information to compensate for having relatively short records at each site. The concept underlying the index-flood method is that the distributions of floods at different sites in a region are the same except for a scale or index-flood parameter, which reflects the size, rainfall, and runoff characteristics of each watershed. Research is revealing when this assumption may be reasonable. Often a more sophisticated multiscaling model is appropriate (Gupta and Dawdy, 1995a; Robinson and Sivapalan, 1997).

Generally the mean is used as the index flood. The problem of estimating the *p*th quantile x_p is then reduced to estimation of the mean for site μ_x and the ratio x_p/μ_x of the *p*th quantile to the mean. The mean can often be estimated adequately with the record available at a site even if that record is short. The indicated ratio is estimated with regional information. The British *Flood Studies Report* (NERC, 1975) calls these normalized flood distributions growth curves.

An index-flood methodology could be used with many different distributions. Outlined below is an L-moment/GEV version of the algorithm that has been used in many studies.

Let there be K sites in a region with records $\{x_t(k)\}$, $t = 1, \ldots n_k$; and $k = 1, \ldots, K$. The L-moment/GEV index-flood procedure is

1) Compute unbiased estimators of the L-moment coefficient of variation $\hat{\tau}_2(k)$ and L-moment coefficient of skewness $\hat{\tau}_3(k)$ for each site k.
2) To obtain a normalized frequency distribution for the region, compute the regional average of the normalized L-moments ratios, $\bar{\tau}_2$ and $\bar{\tau}_3$, where

$$\bar{\tau}_j = \frac{1}{K}\sum_{k=1}^{K} \hat{\tau}_j(k) \quad \text{for } j = 2 \text{ and } 3 \tag{22}$$

   If the length of records vary substantially, it may be advantageous to weight the different at-site values to reflect their relative precision (Stedinger et al., 1993; Hosking and Wallis, 1997). Such a weighted estimator can also be used to obtain an unbiased estimator of the site-to-site variability of such parameters after accounting for the sampling variability in the individual $\hat{\tau}_j(k)$ (Tasker and Stedinger, 1986; Fill and Stedinger, 1998).
3) Using the average L-moment ratios $\bar{\tau}_2$ and $\bar{\tau}_3$ with a mean λ_1 of 1, determine the parameters and quantiles x_p^R of the normalized regional GEV distribution by using Equation 18.
4) The estimator of the 100p-percentile of the flood distribution at any site k is

$$\hat{x}_p(k) = \bar{x}(k)x_p^R \tag{23}$$

   where $\bar{x}(k)$ is the at-site sample mean for site k. Stedinger et al. (1993) discuss estimation of the precision of this quantile estimator.

Key to the success of the index-flood approach is identification of sets of basins that have similar coefficients of variation and skew. Basins can be grouped geographically as well as by physiographic characteristics, including drainage area and elevation. Regions need not be geographically contiguous. Each site can potentially be assigned its own unique region consisting of sites with which it is particularly similar (Zrinji and Burn, 1994), or regional regression equations can be derived to compute normalized regional quantiles as a function of a site's physiographic characteristics and other statistics (Fill and Stedinger, 1998).

Clearly the next step for regionalization procedures, such as the index-flood method, is to move away from estimates of regional parameters that do not depend on basin size and other physiographic parameters. Gupta et al. (1994) argue that the basic premise of the index-flood method, that the coefficient of variation of floods is relatively constant, is inconsistent with the known relationships between CV and drainage area (see also Robinson and Sivapalan, 1997). Recently, Fill and Stedinger (1998) built such a relationship into an index-flood procedure by using a regression model to explain variations in the normalized quantiles. Tasker and Stedinger (1986) illustrated how one might relate log-space skew to physiographic basin characteristics (see also Gupta and Dawdy, 1995b); Madsen and Rosbjerg (1997b) did the same for a regional model of κ for the GEV distribution. In both studies, only a binary variable representing region was found useful in explaining variations in these two shape parameters.

Once a regional model of alternative shape parameters is derived, there may be some advantage to combining such regional estimators with at-site estimators by using an empirical Bayesian framework or some other weighting scheme. For example, Bulletin 17B recommends weighting at-site and regional skewness estimators but almost certainly places too much weight on the at-site values (Tasker and Stedinger, 1986). Examples of empirical Bayesian procedures are provided by Kuczera (1982), Madsen and Rosbjerg (1997b), and Fill and Stedinger (1998). Madsen and Rosbjerg's (1997b) computation of a κ model with a New Zealand data set demonstrates how important it can be to do the regional analysis carefully, taking into account the cross-correlation among concurrent flood records.

Other Regionalization Schemes

When one has relatively few data at a site, the index-flood method is an effective strategy for deriving flood-frequency estimates. However, as the length of the available record increases, there soon comes a point when it becomes advantageous to also use the at-site data to estimate the CV as well. Stedinger and Lu (1995) found that the L-moment/GEV index-flood method did quite well for humid regions ($C_v \approx 0.5$) when $n < 25$ and for semiarid regions ($C_v \approx 1.0$) for $n < 60$, if reasonable care is taken in selecting the stations to be included in a regional analysis. However, with longer records it became advantageous to use the at-site mean and L-CV with a regional estimator of the shape parameter for a GEV distribution. In many cases, this would be roughly equivalent to fitting a Gumbel distribution corresponding to shape parameter $\kappa = 0$. Gabriele and Arnell (1991) develop the idea of having regions of different size for different parameters. For realistic hydrologic regions, these and other studies illustrate the value of regionalizing estimators of the shape and often the CV of a distribution.

Use of Historical and Paleoflood Information

Over the past several decades, water management agencies around the world have collected streamflow records at thousands of sites. This information provides a vast resource for flood-frequency studies and other investigations. Still, when frequency analyses are required at particular sites, there often is no streamflow gauge located near that site and as a result regional estimators are required. In fact, even with a gauged record, it is almost always the case that available streamflow records are shorter than would be advantageous to estimate the 50-year or 100-year flood commonly used in floodplain planning and design of bridges, culverts, and flood control structures.

As discussed by Baker in Chapter 13 of this volume, other sources of information on the frequency of floods at a site can include historical, botanical, and physical paleoflood records documenting the occurrence of floods or their failure to exceed thresholds of various magnitudes (Benson, 1950; Stedinger and Baker, 1987). Here historical information refers either to human records and human memories of the occurrence of large floods and related hydrologic information or, more generally, to information gathered on any floods outside the period for which there is a gauged record. Paleoflood records generally refer to records of floods that occurred before continuous station records were available at a site and are not the result of human observations of the events (Costa, 1986; Jarrett, 1991). Such evidence may be botanical, corresponding to damaged or deformed trees (Hupp, 1988), or it may be physical paleoflood evidence, corresponding to scour lines, high water marks, or slack water deposits, which reveal the stages of floods (Baker et al., 1988; O'Connor et al., 1994; Ostenaa et al., 1996; Baker, Chapter 13, this volume). Although there is a long history of use of historical information describing observed large floods, botanical and physical paleoflood records have seen less use.

One can often determine that a flood is the largest to have occurred over some period. This can be because it left the high-water mark, or the case can be documented from written records that extend beyond the date when that flood occurred (Hirsch and Stedinger, 1987). In other cases, several floods may be recorded (or none at all), because they exceed some perception level defined by the location of dwellings and economic activities, and thus sufficiently disrupted people's lives for their occurrence to be noted, or for the resultant botanical or physical damage to document the event. In statistical terms, historical information represents a censored sample because only the largest floods are recorded, either because they exceeded a threshold of perception for the occupants of the basin or because they were sufficiently large to leave physical evidence that was preserved. To interpret correctly such data, hydrologists should understand the mechanisms or reasons that historical, botanical, or geophysical records document that floods of different magnitudes either did or did not occur. The historical record should represent a "complete catalog" of all events that exceeded various thresholds so that it can serve as the basis for frequency analyses.

The use and value of such information in flood-frequency analyses has been explored in several studies (Leese, 1973; Condie and Lee, 1982; Hosking and Wallis, 1986a, 1986b; Hirsch and Stedinger, 1987; Salas et al., 1994; Cohn et al., 1997; England, 1998). Research has confirmed the value of historical and paleoflood information when properly used (Jin and Stedinger, 1989). In particular, Stedinger and Cohn (1986) and Cohn and Stedinger (1987) have considered a wide range of cases using the effective record length and average gain to describe the value of historical information.

They began by computing the mean square error (*MSE*) of alternative estimators of the 99th percentile of a flood distribution as

$$MSE[\hat{X}_{0.99} \mid s, h] = E\{(\hat{X}_{0.99} - X_{0.99})^2\} \tag{24}$$

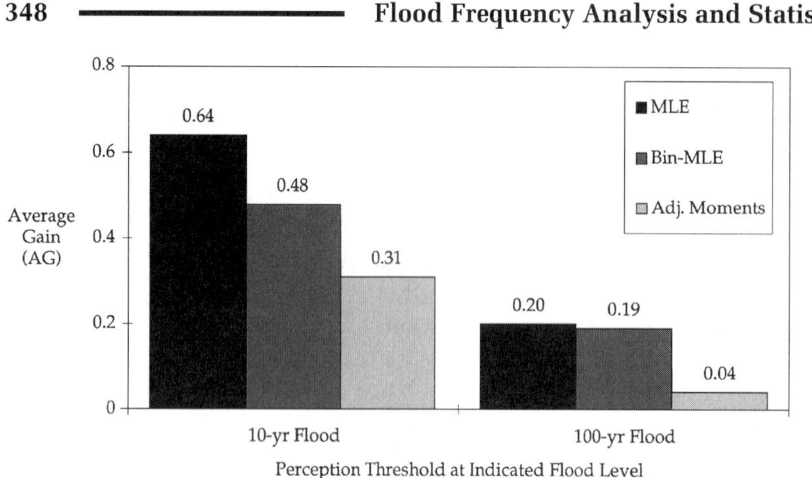

Figure 12.2. Average Gains when estimating 100-year flood quantile of a two-parameter lognormal distribution ($s = 20$ years; $h = 100$ years).

for a sampling experiment employing an s-year systematic (gauged record) and an additional h-year historical period. The historical-record period corresponded to a period when all floods above some perception threshold would have their magnitude recorded by human observers or by physical features that could later be interpreted.

The effective record length (*ERL*) is proportional to the reciprocal of the *MSE* statistic:

$$ERL[s, h] = s\, MSE[X_{0.99} \mid s, h = 0]/MSE[X_{0.99} \mid s, h] \tag{25}$$

The average gain (*AG*) is defined as the average increase (gain) in the effective record length per year in the historical period:

$$AG = (ERL[s, h] - s)/h \tag{26}$$

The average gain measures the improvement in each estimator's precision from the addition of historical information.

Figure 12.2 reports the *AG* from using historical information with the two-parameter lognormal distribution for perception thresholds at the 10- and 100-year flood. The first estimator, denoted MLE, is the maximum likelihood estimator of the distribution's parameters using the magnitudes of the over-threshold historical peaks. The second estimator, denoted Bin-MLE, is the MLE with binomial censored data so that only the number of exceedances of the threshold is used in the likelihood function to estimate the parameters. Finally, the adjustment-moments estimator corresponds to the procedure used in the United States by federal agencies based on the Water Resource Council's Bulletin 17B (IACWD, 1982). The experiment included an $s = 20$-year systematic record and an $h = 100$-year historical record.

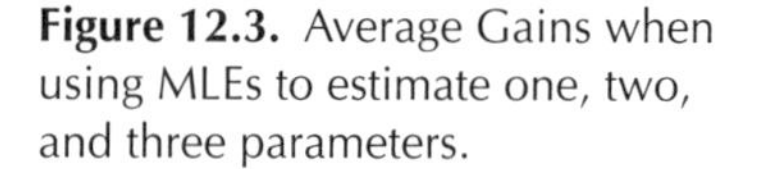

Figure 12.3. Average Gains when using MLEs to estimate one, two, and three parameters.

For high thresholds there is relatively little difference between the performance of the MLE and Bin-MLE procedures. The adjusted-moment estimator corresponding to the Bulletin 17B weighted-moments estimator performs only modestly well in the best of cases (see also Stedinger and Cohn, 1986; Lane, 1987; England, 1998).

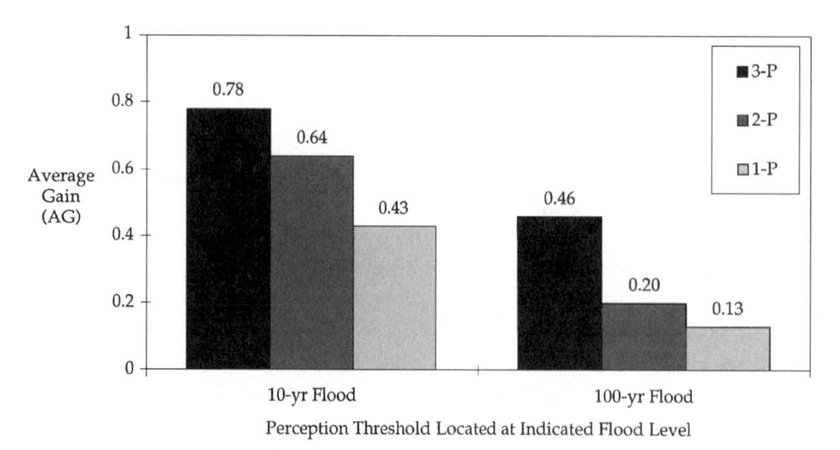

Figure 12.3 illustrates the effect of the number of parameters estimated on the average gain from using historical information with MLEs (see also Cohn and Stedinger, 1987). For a perception threshold at the 100-year flood magnitude, the average gains are 13%, 20%, and 46%, respectively, when estimating one, two, and three parameters with

the MLE procedure; for a threshold at the 10-year flood magnitude, the average gains increased to 43%, 64%, and 78%. The gains achieved when estimating the 100-year flood for three-parameter distributions are remarkable.

A hydrologist can never be sure and has no basis to believe that a fitted distribution is the same as the one nature uses to generate flood flows. Stedinger and Cohn (1986) tested the performance of two-parameter MLE estimators for the lognormal distribution when flood peaks were drawn from log-Pearson type III distributions. Historical information substantially reduced the bias and MSE of flood-quantile estimates. Jin and Stedinger (1989) observed the same results with index-flood procedures that used historical information at those sites where it was available.

Recently Cohn et al. (1997) developed an expected moment estimation method that appears to be as efficient as the maximum-likelihood approach and that works well with the log-Pearson type 3 distribution (see also England, 1998). In practice, historical, botanical, and physical-paleohydrologic flood data can be described by the estimated magnitudes of flood peaks, by ranges, or just by threshold exceedances (corresponding to binomial censoring). The basic maximum-likelihood and expected moment estimation methodologies are flexible and should be adapted to reflect the hydrologic and physiographic characteristics of a region, the available systematic and historical databases, and the analytical capabilities and resources for a study.

Measurement Errors

Statistical flood-frequency analyses depend on measured values of the maximum instantaneous flood flow or the annual maximum volume over n days ($n = 1, 2, 3, \ldots$). Particularly for the largest flows, such measurements are subject to error because they rely on extension of rating curves, and sometimes slope-area calculations (with estimates of model coefficients), if direct estimates of the discharge were not made. Measurements are also subject to the instability of stage-discharge relationships in some channels during large events. In river basins with reservoir and other flow controls, significant analysis may be required to compute the uncontrolled or natural flood flows using differences in reservoir storage levels and routing calculations.

Several studies have considered whether measurement errors in flood flows are likely to have a significant effect on the precision and bias of estimated flood quantiles (Potter and Walker, 1985; Kuczera, 1992, 1996). The answer depends on the relative variability of the flood flows and on the magnitude, correlation, and character of the errors. The issue of measurement error is often raised in the context of historical and paleoflood information because errors are likely to be larger in those cases (Hosking and Wallis, 1986ab; Kuczera, 1992, 1996). For that reason, many studies employing historical and paleoflood flows use bounds and ranges to describe likely flow values (Stedinger et al., 1988; Ostenaa et al., 1996).

Mixtures

A common problem in hydrology is that annual maximum flood series include events from distinctly different processes (Hirschboeck et al.,

Chapter 2, this volume), such as summer precipitation and spring snowmelt. As a result the annual maximum flood series X corresponds to the maximum of the maximum summer flood S and the maximum winter flood W:

$$M = \max\{S, W\} \tag{27}$$

Sometimes a more reliable estimate of the flood-flow frequency relationship for the annual maximum can be obtained by modeling the two-component flood series separately and then mathematically combining those distributions to derive the annual risk (Waylen and Woo, 1982; Stedinger et al., 1993). If two or more flood series are modeled, then more parameters must be estimated, but more data are available than if the annual maximum series for each type of event is used (Rasmussen and Rosbjerg, 1991b). This is often advantageous when thunderstorms cause the largest floods, but in most years the annual flood is a snowmelt event. As shown by the example in Waylen and Woo (1982), in such a situation the annual maximum series is dominated by the less important and often well-behaved snowmelt events, the distribution of which does not describe the significant risk of large thunderstorm-generated floods.

Rogue Observations and Low Outliers

Most of the commonly used probability distributions employed to describe annual flood series have two or three adjustable parameters, as do the log-normal, GEV, and log-Pearson distributions introduced above. In many cases, such simple descriptions of the flood-flow frequency relationships cannot describe irregularities in the distribution of the smallest recorded annual floods. For example, in the arid southwestern part of the United States and many other parts of the world, the maximum annual flood in some years is zero. Remembering the charge by the NRC (1988) that flood-frequency analysis should focus on the large values of concern, it is important to ensure that unusually small observations do not distort the description of the risk of large floods.

Unusually small annual floods are sometimes called low outliers. Bulletin 17B defines outliers as "Data points which depart significantly from the trend of the remaining data." Because Bulletin 17 uses the logarithms of the observed flood peaks to fit a Pearson type 3 distribution with a generalized skew coefficient, one or more unusual low-flow values can distort the entire fitted frequency curve. To avoid such problems Bulletin 17B contains a statistical test of whether a low outlier is statistically significant (IACWD, 1982; Stedinger et al., 1993). Stedinger et al. (1988) advocated censoring any small floods that appear to have an unusual effect on the fitted flood curve. The argument they make is that, because a selected three-parameter distribution seldom describes the true distribution of flood flows, it is unwise to let the smallest observed annual floods have a significant influence on the derived distribution of extremely large floods. The two populations almost assuredly arise from different meteorological processes (Hirschboeck et al., Chapter 2, this volume).

The challenge is how to fit a flood-flow distribution after one or more of the smallest observations are "censored" or equivalently treated as a "zero." Bulletin 17B uses a conditional probability adjustment that fits a log-Pearson type 3 distribution $G(x)$ to the uncensored "nonzero" floods and then estimates the unconditional distribution for the annual flood as

$$F(x) = p_0 + (1 - p_0)G(x) \qquad x > x_0 \tag{28}$$

where floods less than x_0 are treated as zeros and are assumed to occur with probability p. Kroll and Stedinger (1996) compare maximum-likelihood, regression using a probability plot and partial PWM methods for directly fitting a lognormal distribution to a censored series using only the fact that the censored values were less than the threshold value x_0. Probability plots have often been used to derive flood-frequency curves and that method did very well in the Kroll-Stedinger experiments. Plotting positions for censored data sets are discussed by Hirsch and Stedinger (1987). Finally, a more sophisticated approach would be to use the maximum-likelihood or expected moments method discussed in the section on historical information. However, those methods have relatively little advantage over the simpler methods when there are relatively few zero floods.

Partial Duration Series

Two general approaches are available for modeling flood series (Langbein, 1949). An annual maximum series considers only the largest event each year. A partial duration series (PDS) or peaks-over-threshold approach includes all "independent" peaks above a truncation or threshold level. An objection to using annual maximum series is that it employs only the largest event each year, regardless of whether the second largest event in a year exceeds the largest events of other years. Moreover, the largest annual flood flow in a dry year in some arid or semiarid regions may be zero or so small that calling it a flood is misleading.

Use of a PDS framework avoids such problems by considering all independent peaks that exceed a specified threshold, as shown in Figure 12.4. And one can estimate annual exceedance probabilities from the analysis of PDS. Arguments in favor of PDS are that relatively long and reliable PDS records are often available, and if the arrival rate for peaks over the threshold is large enough (1.65 events per year for the Poisson-arrival with exponential-exceedance model) PDS analyses should yield more accurate estimates of extreme quantiles than the corresponding annual-maximum frequency analyses (NERC, 1975; Rosbjerg, 1985). A drawback of PDS analyses is that one must have criteria to identify only independent peaks (and not multiple peaks corresponding to the same event); thus, PDS analysis can be more complicated than analyses that use annual maxima. Partial duration models, perhaps with parameters that vary by season, are often used to estimate expected damages from hydrologic events when more than one damage-causing event can occur in a season or within a year (North, 1980).

A model of a PDS series has at least two components: first, one must model the arrival rate of events larger than the threshold level; second, one must model the magnitudes of those events. For example, a Poisson distribution has often been used to

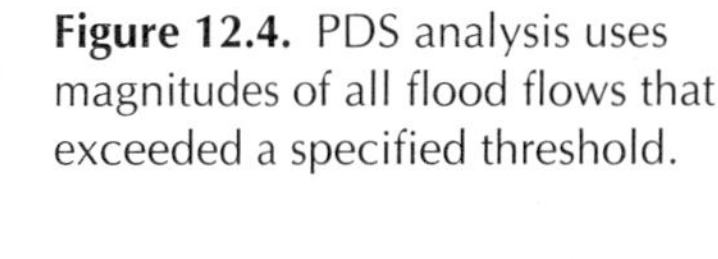

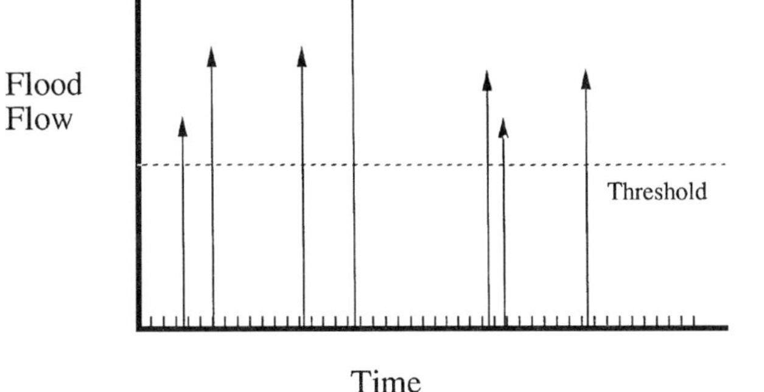

Figure 12.4. PDS analysis uses magnitudes of all flood flows that exceeded a specified threshold.

model the arrival of events, and an exponential distribution has been used to describe the magnitudes of peaks that exceed the threshold.

There are several general relationships between the probability distribution for annual maximum and the frequency of events in a partial duration series. For a PDS model, let λ be the average arrival rate of flood peaks greater than the threshold x_0 and let $G(x)$ be the probability that flood peaks when they occur are less than x, and thus those peaks fall in the range $[x_0, x]$. The annual exceedance probability for a flood, denoted $1/T_a$, corresponding to an annual return period, T_a, is related to the corresponding exceedance probability $q_e = [1-G(x)]$ for level x in the partial duration series by

$$1/T_a = 1 - \exp\{-\lambda q_e\} = 1 - \exp\{-1/T_p\} \tag{29}$$

where $T_p = 1/(\lambda q_e)$ is the average return period for level x in the PDS.

Many different choices for $G(x)$ may be reasonable. In particular, the generalized Pareto distribution (GPD) is a simple distribution that is useful for describing floods that exceed a specified lower bound. The cumulative distribution function for the GPD is

$$F_X(x) = 1 - [1 - \kappa(x - \xi)/\alpha]^{1/\kappa} \tag{30}$$

with mean and variance

$$\mu_x = \xi + \alpha/(1+\kappa)\kappa \quad \text{and} \quad \sigma_x^2 = \alpha^2/[(1+\kappa)^2(1+2\kappa)] \tag{31}$$

where for $\kappa < 0$, $\xi \leq x$, whereas for $\kappa > 0$, $\xi \leq x \leq \xi + \alpha/\kappa$ (Hosking and Wallis, 1987). A special case of the GPD is the two-parameter exponential distribution (for $\kappa = 0$).

Use of a GPD for $G(x)$ with a Poisson arrival model yields a GEV distribution for the annual maximum series greater than x_0 (Smith, 1984; Stedinger et al., 1993; Madsen et al., 1997a). The Poisson-Pareto and Poisson-GPD models are a very reasonable and powerful description of flood risk (Rosbjerg et al., 1992). They have the advantages that they focus on the distribution of the larger floods and that regional estimates of the GEV distribution's shape parameter κ from annual maximum and PDS analyses can be used interchangeably.

Madsen and Rosbjerg (1997a) use a Poisson-GPD model as the basis of a PDS index-flood procedure; Madsen et al. (1997b) show that the estimators are fairly efficient. Information from many sites was pooled to estimate the single shape parameter κ and the arrival rate, where the threshold was a specified percentile of the daily flow duration curve at each site; then at-site information was used to estimate the mean above-threshold flood. Alternatively, one could use the at-site data to estimate the arrival rate as well.

This PDS model is consistent with the principles advocated by the NRC (1988) in that (1) it uses regional information to compute κ, which is the hardest parameter to resolve with at-site data; (2) it uses a mathematical structure that yields a reasonable distribution for extreme floods; and (3) it focuses on extremes by ignoring small annual floods that are less than a threshold defined by a specified percentile of the daily flow duration curve. Madsen and Rosbjerg (1997b, table 3) introduce additional structure into the regional analysis by building a model that describes how the three parameters are likely to vary with catchments characteristics and average annual rainfall; with their two regions in New Zealand, the regional

κ-model was a constant that appeared to be much more precise than at-site estimators.

Regional Regression

The methods described above work well on the major branches of a river for which there are gauged records of floods and perhaps also historical information. Even the index-flood method, as it was described, requires some data to estimate the mean flow. In many situations, estimates of flood quantiles are needed on small tributaries and other catchments for which almost no hydrologic records are available. If rainfall records are available and data are available to calibrate a rainfall-runoff model, then a synthetic flood record can be computed. However, such rainfall-runoff models often require a significant effort to calibrate and may not accurately reproduce the most extreme flood flows. Regional regression models are commonly used in such situations to predict flood quantiles at ungauged sites. In a test in the United States this method did as well as or better than more complex rainfall-runoff modeling procedures (Newton and Herrin, 1982).

Consider the traditional log-linear model for estimating y_i, the logarithm of a flood quantile for site i, using watershed characteristics such as drainage area and slope:

$$y_i = \alpha + \beta_1 \log(\text{area}_i) + \beta_2 \log(\text{slope}_i) + \cdots + \epsilon_i \tag{32}$$

A challenge in analyzing this model and estimating its parameters with available records is that one obtains only sample estimates $\hat{y}_i$ of the hydrologic statistics y_i. Thus the observed error ε is a combination of (1) the time-sampling error in sample estimators of y_i (these errors at different sites can be cross-correlated if the records are concurrent), and (2) underlying model error (lack of fit) due to failure of the model to accurately predict the true value of y_i at every site. Often these problems have been ignored and standard ordinary least-squares regression has been used (Thomas and Benson, 1970).

To address these problems, Stedinger and Tasker (1985, 1986a,b, 1992) developed a generalized least-squares (GLS) regional regression methodology. Tasker and Stedinger (1989) discussed operational extensions including use of historical information, leverage and influence statistics, and network design applications. One can use the available record lengths n_i, regional estimates of the standard deviations s_i, and cross correlations to estimate the sampling errors of the $\hat{y}_i$ and their cross correlations. These, along with the true model-error variance γ, determine the covariance matrix $\Lambda(\gamma)$ of the residual errors ε. Let $\mathbf{Y}$ be the vector with components $\hat{y}_i$, β the vector coefficients in Equation 32, and X the matrix of independent variables in Equation 32. Then the GLS regression estimates of the model parameters β are determined by solving

$$\{X^T \Lambda(\gamma)^{-1} X\}\beta = X^T \Lambda(\gamma)^{-1} \mathbf{Y} \tag{33}$$

with model-error γ determined by the method-of-moments so that

$$(\mathbf{Y} - X\beta)^T \Lambda(\gamma)^{-1} (\mathbf{Y} - X\beta) = N - k \tag{34}$$

where N is the dimension of $\mathbf{Y}$ and k is the dimension of β.

The GLS procedure assigns different weights to the y_i estimators from different sites based on the record lengths at each site and the regional model-error variance γ so that it can make more efficient use of information from sites with different length records. As a result the GLS has more efficient estimates of the model parameters (Kroll and Stedinger, 1998) as well as providing an almost unbiased model-error variance estimator (Stedinger and Tasker, 1986b). Madsen and Rosbjerg (1997b) show that GLS estimators are the natural generalization of the record-length-weighted average used in some index-flood procedures.

Hydrographs and River Basin Operation

In general, this chapter addressed the issue of flood-frequency analysis as if planners were concerned only with instantaneous peaks at one site. In practice, other complications arise. For example, when considering communities and damage sites along a single river, if record lengths are different for different sites, the resulting flood-quantile estimators may not be consistent; that might be the case if the 1% chance flood did not increase as one moved downstream. Thus, it may be appropriate to do some smoothing of computed flood quantiles to obtain a consistent set of flood-frequency relationships.

For a single large reservoir, critical inflow hydrographs are often constructed by determining the critical flow (with risks of 1 in 10, 1 in 100, and more extreme values) for different durations; a synthetic hydrograph can then be constructed that, over a range of durations, has a particular exceedance probability. However, it often is necessary to make some adjustment to the flow-frequency curves for different durations so that they remain consistent (Balocki and Burges, 1994); for example, the 3-day flood volume for each exceedance probability should always be less than the 5-day volume.

In the operation of large multireservoir systems, such as those on the Colorado River, the joint operation of several reservoirs is of concern. The development of critical inflow hydrographs for multiple reservoir systems can be difficult. If reservoirs on two major tributaries were both to receive floods with exceedance probabilities of 1 in 100, with realistic timing for the two events, the resulting hydrograph for a main-stem reservoir may represent a more extreme event. On the other hand, the operation of different reservoirs may be sensitive to floods of different durations. It can be a relatively difficult, if not impossible, task to construct a single hydrologic event that represents at every site a risk of 1 in 100 or another value. Thus, development of design events for such systems can require considerable engineering judgment and trial-and-error to determine what represents the most critical events for the system in addition to ensuring that flow-frequency relationships at different sites and for different durations are consistent.

These problems are resolved if the design flow is generated by a particular design rainfall for the region; however, construction of such a rainfall event would pose a significant challenge and must address the issue of the appropriate spatial and temporal distribution of rainfall. One can use many storms in combination with other data and regulation policies for flood-control facilities to compute unregulated and regulated frequency curves (Bradley and Potter, 1992; Naghettini et al., 1996).

Conclusions

Rational flood risk management requires reliable estimates of the risk of large floods in which planners and the public have reasonable confidence. In catchments with reasonably stationary hydrologic processes, there are a number of ways to construct estimates of flood risk depending on the character and quality of the data available. In particular, recent research in flood-frequency analysis has created new possibilities for bringing at-site systematic flood records, as well as regional and historical flood information, to bear on the problem of estimating flood risk at a site. The index-flood method using the average of L-moment estimators of the L-CV and L-skewness has been found to work well for reasonably homogeneous basins when record lengths are short. As records get longer, it can be advantageous to use an at-site estimator of the CV of floods with a regional estimator of the shape of the flood distribution. Other research has developed methods for using cultural records of large floods and other historical and paleoflood information in flood-frequency analysis. Historical and paleoflood information can be of tremendous value in flood-frequency analysis. The more parameters that are estimated, the greater the value of such information. Even with index-flood procedures, historical information is valuable at the sites where it is available. GLS regional regression methods can be used to develop sophisticated regional models of flood quantiles and the normalized flood quantiles (growth curve) employed by index-flood procedures, which depend on a basin's physiographic characteristics. PDS methods have potential advantages. Future research may document benefits from modeling separately storms from different meteorological processes. Overall, methods available for flood-frequency analysis continue to improve to make better use of available sources of information on the risk of flooding.

References

Baker, V.R., Kochel, R.C., and Patton, P.C. (eds.) (1988). *Flood Geomorphology.* New York: John Wiley and Sons.

Balocki, J.B. and Burges, S.J. (1994). Relationship between n-day flood volumes for infrequency large floods. *Journal of Water Resources Planning and Management*, **120**(6), 794–818.

Beard, L.R. (1960). Probability estimates based on small normal-distribution samples. *Journal of Geophysical Research*, **65**(7), 2143–2148.

Beard, L.R. (1997). Estimating flood frequency and average annual damages. *Journal of Water Resources Planning and Management*, **123**(2), 84–88.

Benson, M.A. (1950). Use of historical data in flood frequency analysis. *EOS, Trans. AGU*, **31**, 419–424.

Bobée, B. and Ashkar, F. (1991). *The Gamma Distribution and Derived Distributions Applied in Hydrology.* Littleton, Colorado: Water Resources Press.

Bobée, B. and Rasmussen, P. (1995). Recent advances in flood frequency analysis, U.S. National report to IUGG, 1991–1994. *Reviews in Geophysics*, **33**, suppl.

Bradley, A.A. and Potter, K.W. (1992). Flood frequency analysis of simulated flows. *Water Resources Research*, **28**(9). 2375–2385.

Chow, V.T., Maidment, D.R., and Mays, L.W. (1988). *Applied Hydrology.* New York: McGraw-Hill Corp.

Chowdhury, J.U., Stedinger, J.R., and Lu, L.-H. (1991). Goodness-of-fit tests for regional GEV flood distributions. *Water Resources Research*, **27**(7), 1765–1776.

Cohn, T.A. and Stedinger, J.R. (1987). Use of historical information in a maximum-likelihood framework. *Journal of Hydrology*, **96**(1–4), 215–23.

Cohn, T.A., Lane, W.L., and Baier, W.G. (1997). An algorithm for computing moments-based flood quantile estimates when historical flood information is available. *Water Resources Research*, **33**(9), 2089–96.

Condie, R. and Lee, K. (1982). Flood Frequency Analysis with Historical Information. *Journal of Hydrology*, **58**(1/2), 47–61.

Costa, J.E. (1986). A history of paleoflood hydrology in the United States, 1800–1970. *EOS*, **67**, 425–430.

Cunnane, C. (1988). Methods and merits of regional flood frequency analysis. *Journal of Hydrology*, **100**, 269–290.

Cunnane, C. (1989). *Statistical Distributions for Flood Frequency Analysis*, Operational Hydrology Report No. 33. Geneva, Switzerland: World Meteorological Organization WMO-No. 718.

Dalrymple, T. (1960). Flood frequency analysis. U.S. Geological Survey, *Water Supply Paper* 1543-A. US Government Printing office, Washington, DC.

England, J.F. (1998). *Assessment of Historical and Paleohydrologic information in flood frequency analysis.* MS Thesis, Colorado State University, Colorado.

Fill, H. and Stedinger, J. (1995). L-Moment and PPCC Goodness-of-Fit tests for the Gumbel Distribution and Effect of Autocorrelation. *Water Resources Research*, **31**(1), 225–229.

Fill, H. and Stedinger, J. (1998). Using regional regression within index flood procedures and an empirical Bayesian estimator. *Journal of Hydrology*, **210** (1–4), 128–145.

Gabriele, S. and Arnell, N. (1991). A hierachical approach to regional flood frequency analysis. *Water Resources Research*, **27**(6), 1281–1289.

Gumbel, E.J. (1958). *Statistics of Extremes.* New York: Columbia University Press, 375 pp.

Gupta, V.K., Mesa, O.J., and Dawdy, D.R. (1994). Multiscaling theory of flood peaks: Regional quantile analysis. *Water Resources Research*, **30**(12), 3405–3412.

Gupta, V.K. and Dawdy, D.R. (1995a). Physical interpretation of regional variations in the scaling exponents of flood quantiles. *Hydrology Processes*, **9**(3/4), 347–361.

Gupta, V.K. and Dawdy, D.R. (1995b). Multiscaling and skew separation in regional floods. *Water Resources Research*, **31**(11), 2761–2776.

Hirsch, R.M. and Stedinger, J.R. (1987). Plotting positions for historical floods and their precision. *Water Resources Research*, **23**(4), 715–727.

Hoshi, K., Stedinger, J.R., and Burges, S. (1984). Estimation of log normal quantiles: Monte Carlo results and first-order approximations. *Journal of Hydrology*, **71**(1/2), 1–30.

Hosking, J.R.M., Wallis, J.R., and Wood, E.F. (1985). Estimation of the generalized extreme-value distribution by the method of probability weighted moments. *Technometrics*, **27**(3), 251–261.

Hosking, J.R.M. and Wallis, J.R., (1986a). Paleoflood hydrology and flood frequency analysis. *Water Resources Research*, **22**(4), 543–550.

Hosking, J.R.M. and Wallis, J.R. (1986b). The value of historical data in flood frequency analysis. *Water Resources Research*, **22**(11), 1606–1612.

Hosking, J.R.M. and Wallis, J.R. (1987). Parameter and quantile estimation for the generalized Pareto distribution. *Technometrics*, **29**(3), 339–349.

Hosking, J.R.M. (1990). L-Moments: Analysis and estimation of distributions using linear combinations of order statistics. *Journal of the Royal Statistical Society, B*, **52**(2), 105–124.

Hosking, J.R.M. and Wallis, J.R. (1995). A comparison of unbiased and plotting-position estimators of L moments. *Water Resources Research*, **31**(8), 2019–25.

Hosking, J.R.M. and Wallis, J.R. (1997). *Regional Frequency Analysis: An Approach Based on L-moment.* Cambridge: Cambridge University Press.

Hupp, C.R. (1988). Plant ecological aspects of flood geomorphology and paleoflood history. In *Flood Geomorphology*, eds. V.R. Baker, R.C. Kochel, and P.C. Patton. New York: John Wiley and Sons.

IACWD (Interagency Advisory Committee on Water Data). (1982). *Guidelines for Determining Flood Flow Frequency.* Bulletin #17B. Reston, Virginia: U.S. Department of the Interior, U.S. Geological Survey, Office of Water Data Coordination.

Jarrett, R.D. (1991). Paleohydrology and its value in analyzing floods and droughts – Floods and Droughts. In *National Water Summary 1988–89*, U.S. Geol. Survey Water Supply Paper 2375, p. 105–116. US Government Printing office, Washington, DC.

Jin, M. and Stedinger, J.R. (1989). Flood frequency analysis with regional and historical information. *Water Resources Research*, **25**(5), 925–936.

Kelman, J. (1987). *Cheias E Aproveitamentos Hidrelétricos.* Rio de Janeiro, Brazil: Revista Brasileria de Engenharia.

Kite, G.W. (1988). *Frequency and Risk Analysis in Hydrology.* Littleton, Colorado: Water Resources Publication.

Kottegoda, M. and Rosso, R. (1997). *Statistics, Probability, and Reliability for Civil and Environmental Engineers.* New York: McGraw-Hill.

Kroll, K. and Stedinger, J.R. (1996). Estimation of moments and quantiles with censored data. *Water Resources Research*, **32**(4), 1005–1012.

Kroll, C.N. and Stedinger, J.R. (1998). Regional hydrologic analysis: Ordinary and generalized least squares revisited. *Water Resources Research*, **34**(1), 121–128.

Kuczera, G. (1982). Combining site-specific and regional information, an empirical Bayes approach. *Water Resources Research*, **18**(2), 306–314.

Kuczera, G. (1992). Uncorrelated measurement error in

flood frequency inference. *Water Resources Research*, **28**(1), 183–188.

Kuczera, G. (1996). Correlated rating curve error in flood frequency inference. *Water Resources Research*, **32**(7), 2119–2127.

Landwehr, J.M., Matalas, N.C., and Wallis, J.R. (1978). Some comparisons of flood statistics in real and log space. *Water Resources Research*, **14**(5), 902–920.

Landwehr, J.M., Matalas, N.C., and Wallis, J.R. (1979). Probability weighted moments compared with some traditional techniques in estimating Gumbel parameters and quantiles. *Water Resources Research*, **15**(5), 1055–1064.

Lane, W.L. (1987). Paleohydrologic data and flood frequency estimation. In *Regional Flood Frequency Analysis*, ed. V.P. Singh, pp. 287–298. Dordrecht: D. Reidel Publishing Co.

Langbein, W.B. (1949). Annual floods and the partial duration flood series. *EOS, Trans. AGU*, **30**(6), 879–881.

Leese, M.N. (1973). Use of censored data in the estimation of Gumbel distribution parameters for annual maximum flood series. *Water Resources Research*, **9**(6), 1534–1542.

Lettenmaier, D.P., Wallis, J.R., and Wood, E.F. (1987). Effect of regional hereogeneity on flood frequency estimation. *Water Resources Research*, **23**(2), 313–324.

Lu, L. and Stedinger, J.R. (1992). Variance of 2- and 3-parameter GEV/PWM Quantile Estimators: Formulas, Confidence Intervals and a Comparison. *Journal of Hydrology*, **138**(1/2), 247–267.

Madsen, H. and Rosbjerg, D. (1997a). The partial duration series method in regional index flood modeling. *Water Resources Research*, **33**(4), 737–746.

Madsen, H., Pearson, C.P., Rasmussen, P.F., and Rosbjerg, D. (1997a). Comparison of annual maximum series and partial duration series methods for modeling extreme hydrologic events 1. At-site modeling. *Water Resources Research*, **33**(4), 747–758.

Madsen, H., Pearson, C.P., and Rosbjerg, D. (1997b). Comparison of annual maximum series and partial duration series methods for modeling extreme hydrologic events 2. Regional modelling. *Water Resources Research*, **33**(4), 759–770.

Madsen, H. and Rosbjerg, D. (1997b). Generalized least squares and empirical Bayes estimation in regional partial duration series index-flood modeling. *Water Resources Research*, **33**(4), 771–782.

McCuen, R. (1998). *Hydrologic Analysis and Design*, 2nd ed. Prentice Hall. Upper Saddle River, NJ.

Naghettini, M., Potter, K.W., and Illangasekare, T. (1996). Estimating the upper tail of flood-peak frequeny distributions using hydrometeorological information. *Water Resources Research*, **32**(6), 1729–1740.

NRC (National Research Council). Committee on Techniques for Estimating Probabilities of Extreme Floods. (1988). *Estimating Probabilities of Extreme Floods, Methods and Recommended Research.* Washington, DC: National Academy Press.

NERC (Natural Environmental Research Council). (1975). *Flood Studies Report. London, England: vol. I, Hydrological Studies.* 550 pp. Washington, DC: National Academy Press.

Newton D.W. and Herrin, J.C. (1982). Assessment of commonly used flood frequency methods. *Transportation Research Record* **TRR896**, Washington, DC: Transportation Research Board.

North, M. (1980). Time-dependent stochastic model of floods. *Journal of the Hydraulic Division, ASCE*, **106**(HY5), 649–665.

O'Connor, J.E., Ely, L.L., Wohl, E.E., Stevens, L.E., Melis, T.S., Kale, V.S., and Baker, V.R. (1994). A 4500-year record of large floods on the Colorado River in the Grand Canyon, Arizona. *Journal of Geology*, **102**, 1–9.

Ostenaa, D.A., Levish, D.R., and O'Connell, D.R.H. (1996). *Paleoflood Study for Bradbury Dam.* Denver, Colorado: Seismotectonic Report 96-3, Seismotectonic and Geophysics Section, U.S. Bureau of Reclamation.

Potter, K.W. and Lettenmaier, D.P. (1990). A comparison of regional flood frequency estimation methods using a resampling method. *Water Resources Research*, **26**(3), 415–424.

Potter, K.W. (1987). Research on flood frequency analysis: 1983–86. *Reviews of Geophysics*, **25**(2), 113–118.

Potter, K.W. and Walker, J.F. (1985). An empirical study of flood measurement error. *Water Resources Research*, **21**(3), 403–406.

Rasmussen, P.F. and Rosbjerg, D. (1989). Risk estimation in partial duration series. *Water Resources Research*, **25**(11), 2319–2330.

Rasmussen, P.F. and Rosbjerg, D. (1991a). Evaluation of risk concepts in partial duration series. *Stochastic Hydrology and Hydraulics*, **5**(1), 1–16.

Rasmussen, P.F. and Rosbjerg, D. (1991b). Prediction uncertainty in seasonal partial duration series. *Water Resources Research*, **27**(11), 2875–2883.

Robinson, J.S. and Sivapalan, M. (1997). Temporal scales and hydrologic regimes: Implications for flood frequency scaling. *Water Resources Research*, **33**(12), 2981–2999.

Rosbjerg, D. (1985). Estimation in partial duration series with independent and dependent peak values. *Journal of Hydrology*, **76**, 183–195.

Rosbjerg, D., Madsen, H., and Rasmussen, P.F. (1992). Prediction in partial duration series with generalized Pareto-distributed exceedances. *Water Resources Research*, **28**(11), 3001–3010.

Rosbjerg, D. and Madsen, H. (1998). Design with uncertain design values. In *Hydrology in a Changing Environment, Vol. III*, eds. H. Wheater and C. Kirby, pp. 155–163. New York: John Wiley & Sons.

Salas, J.D., Wohl, E.E., and Jarrett, R.D. (1994). Determination of flood characteristics using systematic, historical and paleoflood data. In *Coping with Floods*, eds. G. Rossi et al., pp. 111–134. The Netherlands: Kluwer Academic Publishers. Dordrecht.

Slack, J.R., Wallis, J.R., and Matalas, N.C. (1975). On the value of information in flood frequency analysis. *Water Resources Research*, **11**(5), 629–648.

Smith, R.L. (1984). Threshold methods for sample extremes. In *Statistical Extremes and Applications*, ed. J. Tiago de Oliveira, pp. 621–638. Dordrecht, Holland: D. Reidel Publishing Company.

Stedinger, J.R. (1980). Fitting log normal distributions to hydrologic data. *Water Resources Research*, **16**(3), 481–490.

Stedinger, J.R. (1989). Using historical and regional information. In *Flood Frequency Analyses*. Hokkaido, Japan: Pacific International Seminar on Water Resource Systems.

Stedinger, J.R. (1997). Expected probability and annual damage estimators. *Journal of Water Resources Planning and Management*, **123**(2), 125–135. [With discussion, Leo R. Beard. (1998). *Journal of Water Resources Planning and Management*, **124**(6), 365–366.]

Stedinger, J.R. and Tasker, G.D. (1985). Regional hydrologic regression, 1. Ordinary, weighted and generalized least squares compared. *Water Resources Research*, **21**(9), 1421–1432.

Stedinger, J.R. and Tasker, G.D. (1986a). Correction to 'regional hydrologic analysis, 1. Ordinary, weighted and generalized least squares compared.' *Water Resources Research*, **22**(5), 844.

Stedinger, J.R. and Tasker, G.D. (1986b). Regional hydrologic analysis, 2. Model error estimates, estimation of sigma, and log-Pearson Type 3 distributions. *Water Resources Research*, **22**(10), 1487–1499.

Stedinger, J.R. and Cohn, T.A. (1986). Flood frequency analysis with historical and paleoflood information. *Water Resources Research*, **22**(5), 785–793.

Stedinger, J.R. and Baker, V.R. (1987). Surface water hydrology: historical and paleoflood information. *Reviews of Geophysics*, **25**(2), 119–124.

Stedinger, J.R., Surani, R., and Therivel, R. (1988). MAX Users Guide: *A Program for Flood Frequency Analysis Using Systematic-Record, Historical, Botanical, Physical Paleohydrologic and Regional Hydrologic Information Using Maximum Likelihood Techniques*. Department of Environmental Engineering, Cornell University. Ithaca, NY.

Stedinger, J.R. and Tasker, G.D. (1992). Generalized least squares analyses for hydrologic regionalization. Invited paper in Symposium on Regionalization in Hydrology, *Hydraulic Engineering, Saving a Threatened Resource in Search of Solutions*, ASCE Water Forum '92, Baltimore, MD.

Stedinger, J.R., Vogel, R.M., and Foufoula-Georgiou, E. (1993). Frequency analysis of extreme events. In *Handbook of Hydrology*, ed. D. Maidment. New York: McGraw-Hill.

Stedinger, J.R. and Lu, L. (1995). Appraisal of regional and index flood quantile estimators. *Stochastic Hydrology and Hydraulics*, **9**(1), 49–75.

Tasker, G.D. and Stedinger, J.R. (1986). Estimating generalized skew with weighted least squares regression. *Journal of Water Resources Planning and Management*, **112**(2), 225–237.

Tasker, G.D. and Stedinger, J.R. (1989). An operational GLS model for hydrologic regression. *Journal of Hydrology*, **111**, 361–375.

Thomas, D.M. and Benson, M.A. (1970). Generalization of streamflow characteristics from drainage-basin characteristics. U.S. Geological Survey, *Water-Supply Paper 1975*. US Government Printing office, Washington, DC.

Thomas, W.O. (1985). A uniform technique for flood frequency analysis. *Journal of Water Resources Planning and Management*, **111**(3), 321–337.

Vogel, R.M. and McMartin, D.E. (1991). Probability plot goodness-of-fit and skewness estimation procedures for the Pearson type III distribution. *Water Resources Research*, **27**(12), 3149–3158.

Vogel, R.M. and Fennessey, N.M. (1993). L-moment diagrams should replace product moment diagrams. *Water Resources Research*, **29**(6), 1745–1752.

Wallis, J.R. (1980). Risk and uncertainties in the evaluation of flood events for the design of hydraulic structures. In *Piene e Siccita*, eds. E. Guggino, G. Rossi, and E. Todini, pp. 3–36. Catania, Italy: Fondazione Politecnica del Mediterraneo.

Wang, Q.J. (1996). Direct sample estimators of L-moments. *Water Resources Research*, **32**(12), 3617–3619.

Wang, Q.J. (1997). LH Moments for statistical analysis of extreme events. *Water Resources Research*, **33**(12), 2841–2848.

Waylen, P. and Woo, M.-K. (1982). Prediction of annual floods generated by mixed processes. *Water Resources Research*, **18**(4), 1283–1286.

Zrinji, Z. and Burn, D.H. (1994). Flood frequency analysis for ungauged sites using a region of influence approach. *Journal of Hydrology*, **153**, 1–21.

CHAPTER 13

Paleoflood Hydrology and the Estimation of Extreme Floods

Victor R. Baker
Department of Hydrology and Water Resources, The University of Arizona

Paleoflood hydrology is the study of past or ancient flow events that occurred before direct measurement by modern hydrological procedures. Effects of ancient floods on natural recording systems may include flood deposits, damage to vegetation (botanical paleoflood data), and erosion of channel-margin materials. Flood stages recorded naturally (paleoflood data) are transformed by hydraulic theory at the sites where the flood effects are indicated to generate water-surface profiles for the corresponding paleoflood discharges. As a practical matter, in the public debate over responses to potential hazards, the documented occurrence of an ancient (but real) cataclysmic process is likely to have more impact than is discussion of various hypothetical frequency distributions. Thus, the confirmation of a flood design number with paleoflood data is not merely proper science; it also serves to increase public confidence in any proposed solution that ultimately will lead to great economic or social expense for hazard mitigation.

Introduction

Paleoflood hydrology is the study of past or ancient flow events that occurred before direct measurement by modern hydrological procedures. The definition of a paleoflood varies with the site of investigation. At a gauge site there will be a systematic record of measured streamflows with a duration equal to the time period over which measurements were taken. There also may be a record of historical flood events that precede the gauge record but for which stages were observed and recorded (Benson, 1950; Thomson et al., 1964). Historical floods are those interpreted from human records other than the modern hydrological instrumentation used at gauge sites. Human recording, either by modern instrumentation or by other means of direct observation, is not required in paleoflood hydrology. Rather, it is the lasting effects of paleofloods on the landscape and on sediments that are interpreted by investigators long after the causative action of the responsible floods. A great variety of signs, or indices, of past floods are the subject of paleoflood hydrological interpretation.

Conventional hydrological streamflow measurements are possible because of the effects of water stage on a mechanical recording device. These effects are subsequently transformed by hydraulic theory to values of discharge at controlled study sites (gauging stations). For historical flood data, human observation is required, but the modern hydrological procedures employed with conventional data do not apply. Paleoflood data derive from the effects of ancient floods on natural recording systems. These effects may include flood deposits, damage to vegetation (botanical paleoflood data), and erosion of channel-margin materials. Flood stages observed nonconventionally (historical data) or recorded naturally (paleoflood data) are transformed by hydraulic theory at the sites where the flood effects are indicated. Note that paleofloods can be of any age; they are distinguished only by the lack of human observation or conventional gauging.

The basic principles of paleoflood hydrology can be applied to any geological time period. However, the time scale of interest for most flood hazard studies is limited to the last 10,000 years. Before this period, major changes in global climate produced a flood-generating regime very different from that at present. As a matter of experience, most paleoflood records used in applied studies have record lengths that vary from several hundred to a few thousand years. Some of the longer records of extreme flood peaks include those of the Pecos River (Patton and Dibble, 1982) and the Colorado River (O'Connor et al., 1994).

Paleoflood Hydrological Methodology

Philosophical Issues

Historical, botanical, and physical paleoflood data are all interpreted through an inverse form of scientific inference. This mode of inference provides a complement, not an alternative, to conventional forms of hydrological investigation. In conventional hydrological study, there are two approaches. In one of these, theoretical relationships are derived by extrapolation from hydrological measurements. This extrapolation depends on assumptions made about the causative hydrological conditions. Alternatively, theoretical hydrological relationships may be used to predict, via modeling, those specific variables that cannot be directly measured. Causative conditions are likewise assumed without testing. These two conventional approaches involve inductive and deductive reasoning, respectively.

In paleoflood hydrology, the specific effects of past floods are directly observed. These effects are then used to infer the causative conditions – that is, the responsible flood magnitudes. Basic hydraulic relationships are assumed to be valid in making the connection between observed paleostage indicators and calculated causative paleodischarges. This approach, in contrast to those noted above, involves retroduction, which is a distinctive reasoning process of the earth sciences (Baker, 1996). Retroduction infers cause from effect. In this case, the effects constitute a kind of measurement device consisting of the preserved effects of past flood flows on the landscape, on vegetation, or on past human observation (historical floods). As in the case of flow effects on a stream gauge, hydraulic principles must be used to transform the observed effect into its causative parameter, discharge. Although there may be considerable variation in the preservation of effects, and on the accuracy of their transformation, paleoflood and conventional gauge information share the status of being data. As such, they constitute the reality against which explanatory-predictive models must be judged and from which explanations of hydrological reality must derive.

Types of Paleoflood Investigation

Many techniques are available for retroduction of past flood-flow parameters by principles of geomorphology and related aspects of Quaternary stratigraphy, geochronology, and sedimentology. The most commonly used procedures can be divided into the following categories: (1) regime-based paleoflood estimates (RBPE), (2) paleocompetence studies, and (3) paleo-stage estimates. Failure to distinguish among the characteristics of data produced by these and other varieties of paleohydrological reconstruction has led to misleading generalizations erroneously applied to all paleoflood information (e.g., Hosking and Wallis, 1986; Yevjevich and Harmancioglu, 1987).

RBPE. These procedures utilize empirically derived relationships to relate relatively high-probability flows, such as the mean annual flood or bank-full discharge, to paleochannel dimensions, sediment types, paleochannel gradients, and other field evidence. The relationships apply to alluvial channels, which adjust their width, depth, sediment transport, and slope to relatively high-probability flood discharges. These variables are related to discharge by regression equations derived from observations of relationships in modern alluvial channels. RBPE studies have been summarized by Dury (1976), Ethridge and Schumm (1978), and Williams (1984). Because nearly all RBPE relationships apply only to relatively frequent floods, they are of little use in evaluating extreme floods.

Paleocompetence Studies. In these procedures, empirical regression or theoretical expressions are used to relate very large sedimentary particles (Figure 13.1) to the hydraulic conditions responsible for their transport – generally the paleoflood shear stress, velocity, or stream power (SS-V-SP). Paleohydrological applications usually apply to maximum particle sizes. Where local controls on sediment transport are not known, SS-V-SP studies can be of very low accuracy (Church, 1978; Maizels, 1983). However, limiting curves for probable causative conditions (Williams, 1983) and combinations of calculation procedures (Costa, 1983) can be effective in constraining possible error.

Figure 13.1. Very large boulders transported by high-energy floods in Katherine Gorge, Northern Territory, Australia (Baker and Pickup, 1987). Note people at bottom center of photograph for scale.

Figure 13.2. Silt lines (light streaks on stone at top and bottom) emplaced by ancient floods of Boulder Creek, south-central Utah. Floods emplaced these silts over petroglyphs, carved by ancient people between 700 and 1000 years ago.

Paleostage Estimates. Paleostages can be estimated by documenting flood-induced erosion or deposition up to the water level reached by a flow. The most accurate and best-preserved records of paleostage flow data have been found to occur along stable-boundary fluvial reaches characterized by slackwater deposits and paleostage indicators (SWD-PSI). Slackwater deposits consist of sedimentary particles with high settling velocities, so that they accumulate relatively rapidly from suspension during major floods. Paleostage indicators are other paleo-high-water marks, which include silt emplaced on bedrock channel walls (Figure 13.2), scour of regolith materials marginal to channels, the tops of cobble-boulder bars, and tree scars.

Preservation of slackwater deposits during successive flows occurs because of effective flow boundaries for extreme floods in river reaches optimum for SWD-PSI accumulation. Such river reaches generally have relatively narrow, deep cross sections, so that small changes in flood discharge yield large changes in stage. Floods in these reaches will also produce very high rates of energy dissipation, or power (ω), per unit area of streambed, which can be expressed as:

$$\omega = \gamma QS/W = \tau \overline{V} \tag{1}$$

where W is bed width, Q is discharge, S is energy slope, γ is the specific weight of the transporting fluid, $\overline{V}$ is the mean flow velocity, and τ is the bed shear stress. The flood features of very deep, high-gradient flows are generated by power values many orders of magnitude larger than those produced in alluvial rivers (Baker and Costa, 1987). Bedrock gorges (Figure 13.3) generate high values of power per unit area and bed stress during rare, high-magnitude floods. The erosional bedforms, boulder transport, and slackwater deposition produced by the intense hydraulics in these situations all afford opportunities to understand the causative processes without the difficulty or impossibility of direct measurement.

At very high values of power per unit area, both water and sediment move at high stage along the zone of maximum velocity in the flowing

Figure 13.3. Aravaipa Canyon in southern Arizona, showing the narrow, deep cross-sectional shape of bedrock gorges ideal for SWD-PSI paleoflood hydrological investigations.

water. However, separated from this zone of flood flow are areas of slackwater at the mouths of tributaries to the channel or at various channel boundary irregularities (Figure 13.4). An effective flow boundary separates the high-velocity, energetic, sediment-charged floodwater from local slackwater zones. Any sediment moving across the boundary is rapidly deposited up to the level of the maximum flood stage. These relationships have been confirmed both in the field and in the laboratory (Baker and Kochel, 1988).

Although large floods may be very erosive within their effective flow boundaries, removing the evidence of preceding flows, they only add more deposits to the slackwater sites lying outside the effective flow boundaries. Thus, the slackwater site (Figure 13.5) will preserve a sedimentary sequence of deposits representing the largest floods that occurred along the adjacent river reach. Much of SWD-PSI analysis consists of interpreting the sequence and timing of flows from the preserved geomorphological record.

SWD-PSI Paleoflood Investigations

Study Phases

A SWD-PSI paleoflood study can have up to six major phases: (1) discovery in the field of key streams and individual sites that preserve appropriate SWD-PSI evidence (Figure 13.6), (2) stratigraphic and geochronological analysis of individual flood-sedimentation units or paleostage indicators in order to associate the paleofloods with a specific sequence and record length, (3) hydraulic analysis of the chosen study reach, (4) relation of the SWD-PSI evidence to the channel hydraulics for paleodischarge determination, (5) performance of a flood magnitude-frequency analysis of

Figure 13.4. Extreme flood inundation of Katherine Gorge, Northern Territory, Australia. The thread of high velocity was moving down the center of the channel, from left to right, center of photograph. A slackwater area is at the top, center of the photograph. (February 1979 monsoonal flooding; photographed by V.R. Baker)

data established in phases (2) through (4), and (6) analysis of temporal and spatial patterns in paleoflood data at multiple sites.

A reconnaissance phase is essential before the investment of time and resources necessary for SWD-PSI analysis. A considerable body of experience is now available for identifying appropriate reaches (Baker, 1987; Baker and Kochel, 1988). Regional factors of importance include (1) confined canyons or gorges in resistant geological materials, (2) adequate concentrations of relatively coarse sediment in transport, (3) appropriate sites for slackwater accumulation outside effective flow boundaries, (4) no significant channel bed instabilities over the time scale of interest, and (5) stability of SWD-PSI sites over the time scale of interest. In general, arid, semiarid, and savanna climates have proven best relative to factors (2) and (5), but certain optimum sites provide useful records in humid regions as well.

Stratigraphy and Geochronology

Accuracy in determining paleoflood ages is largely a function of the geochronologic techniques applied. Each dating technique has an associated uncertainty. Radiocarbon dating of organic material entrained in or associated with the flood deposits (Figure 13.7) is the standard approach (Kochel and Baker, 1982, 1988). However, other tools have been used in relative dating and correlation. Deposits can

Figure 13.5. Slackwater site on the Salt River, central Arizona. Note layered sandy deposits at lower center and right.

be placed in a hierarchy by age based on the stratigraphic position of the deposit (Costa, 1978), on the presence of archeological materials (Patton and Dibble, 1982; Ely and Baker, 1985), and on the development of material properties that change as a function of time, such as soil color and degree of induration (Baker et al., 1983a; Kochel, 1980).

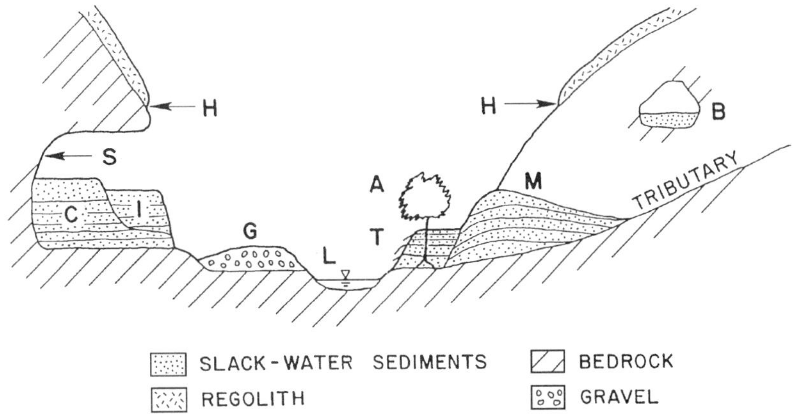

Floods in the time period between 10,000 and 350 years ago are most amenable to radiocarbon dating. Uncertainties inherent in the counting techniques range from about 40 to 90 years. Because of secular changes in $^{12}C/^{14}C$ reservoir conditions, calibration curves must be used to relate radiocarbon years to calendar years (Damon et al., 1974). Appropriate consideration must be given to the effects of counting errors, sample sizes, contamination, and calibrations on paleoflood information.

A special technique is available to date floods younger than A.D. 1950 to the precise year. Paleoflood studies that employ this technique may be useful in remote areas where conventional hydrologic data are lacking. Baker et al. (1985) discuss applications in the Northern Territory, Australia. There is a widely applicable potential use of this tool to document 40+ year records of modern floods at appropriate ungauged sites.

Floods during the period A.D. 1650 to 1950 are difficult to date accurately by radiocarbon analysis because of loading of the atmosphere with CO_2 during the industrial revolution (Stuiver, 1982). However, numerous other dating procedures may be used for such relatively young floods, including historical documents, distinctive archeological materials, and dendrochronology.

H
H
B
S
A
M
TRIBUTARY
C
I
G
T
L
SLACK-WATER SEDIMENTS
BEDROCK
REGOLITH
GRAVEL

Figure 13.6. Idealized cross section of a bedrock river channel showing typical slackwater deposits and palaeostage indicators (SWD-PSI). M, Mounded slackwater deposit at tributary mouth; T, slackwater terrace; A, flood-damaged tree; G, gravel bar; C, slackwater filling of a rock-shelter cave; I, inset slackwater deposit; S, silt line; H, scour of soil or regolith on channel margin; B, cave deposit; L, low-water flow in channel.

Figure 13.7. Slackwater deposits (sand) interbedded with tributary gravel, Aravaipa Canyon, southern Arizona (Figure 13.3). The researcher is collecting leaf material for radiocarbon dating.

Hydraulic Analysis

The hydraulics of SWD-PSI reaches pose the same problem as indirect measurement of any large recent flood discharge. In conventional practice, water-surface profiles for various candidate discharges are deductively predicted from a flow model fit to the channel geometry of a reach (O'Connor and Webb, 1988). Actual paleoflow magnitude determination is an inverse solution in which the SWD-PSI field evidence of each preserved paleoflood is related to the appropriate water-surface profile responsible for that evidence.

The computer flow models used in SWD-PSI studies determine water-surface profiles by step-backwater analysis (Feldman, 1981). These models compute energy-balanced water-surface profiles for steady, gradually varied flow on the basis of the one-dimensional energy equation (a form of the Bernoulli equation) in conjunction with the step-backwater method of profile computation. For known flow conditions (discharge, stage, and channel geometry) specified at one cross section, stage is calculated at an adjacent cross section of known geometry and reach length by conserving mechanical energy and accounting for estimated flow-energy losses associated with boundary roughness and channel expansions and contractions. Calculated conditions at the second cross section are then taken as "known" for the next cross section. In this manner, the stage can be established for each cross section, and the water-surface profile can be determined for an indefinite length of channel. For application to paleoflood channel geometries, various water-surface profiles are computed with different combinations of discharge and energy-loss coefficients until a profile is achieved that most closely matches the water-surface profile suggested by paleostage indicators (O'Connor and Baker, 1992).

O'Connor and Baker (1992) and O'Connor (1993) provide detailed discussions of the error sources and other limitations on paleohydraulic calculations for ancient floods. Important assumptions intrinsic to step-backwater analysis are that the flow is steady, gradually varied, and one dimensional. Another assumption is that the existing measured cross sections closely approximate the channel geometry at peak flood stages; i.e., there has not been significant post-peak channel downcutting, filling, or widening. This assumption must be evaluated at each site. For a given study reach, cross sections are required and are positioned and spaced to (1) characterize downstream changes in channel geometry, and (2) minimize changes in flow conditions between sections.

Figure 13.8 shows a SWD-PSI study reach on the Verde River in central Arizona. Figure 13.9 shows the computed water-surface profiles fit to the channel geometry of this reach. Note that elevations of flood deposits can be related to the water stages and discharges necessary for their emplacement.

Data Structuring

At selected SWD-PSI study sites, it is possible to obtain information to various tolerance levels specifying the magnitudes (paleodischarges) and ages of multiple ancient floods. These data may be structured (classified) in various ways, depending on the intended use of the information. The structuring scheme required by the intended use of the data will then influence the field program. Experience has shown that it is relatively easy to reconstruct the largest paleoflood in a given time period by SWD-PSI techniques. However, reconstruction and completeness of record for smaller paleoflows requires greater time and resources. A balance may need to be achieved between the value of added information and the cost of obtaining such additional data.

Baker (1989) discusses the use of multiple-site studies and correlation techniques in relation to the concept of event censoring levels. A censoring level is a threshold discharge (or stage) above which the paleoflood record can be assumed to be complete over a specified time interval. Completeness means that no actual discharge (or stage) occurred in the time interval that exceeded the censoring level without leaving its evidence in the paleoflood record. In practice, the largest paleofloods leave the most durable records, so setting the censoring level depends on the information extracted about the smaller floods. The geometry of natural SWD-PSI sites ensures that complete samples of floods immediately smaller than the maximum will be preserved (Baker, 1989). However, each field situation differs in how low the censoring level can be set with confidence.

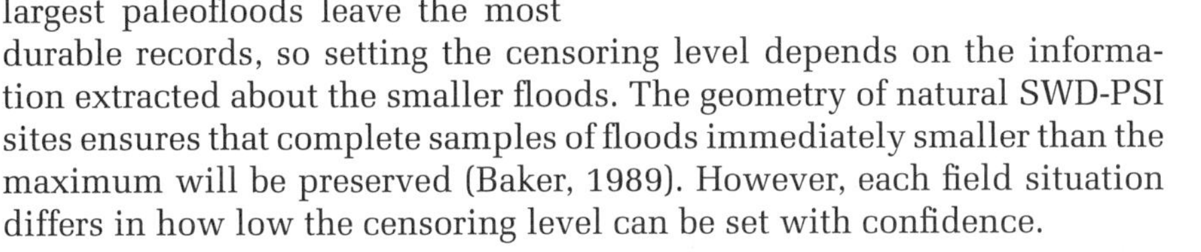

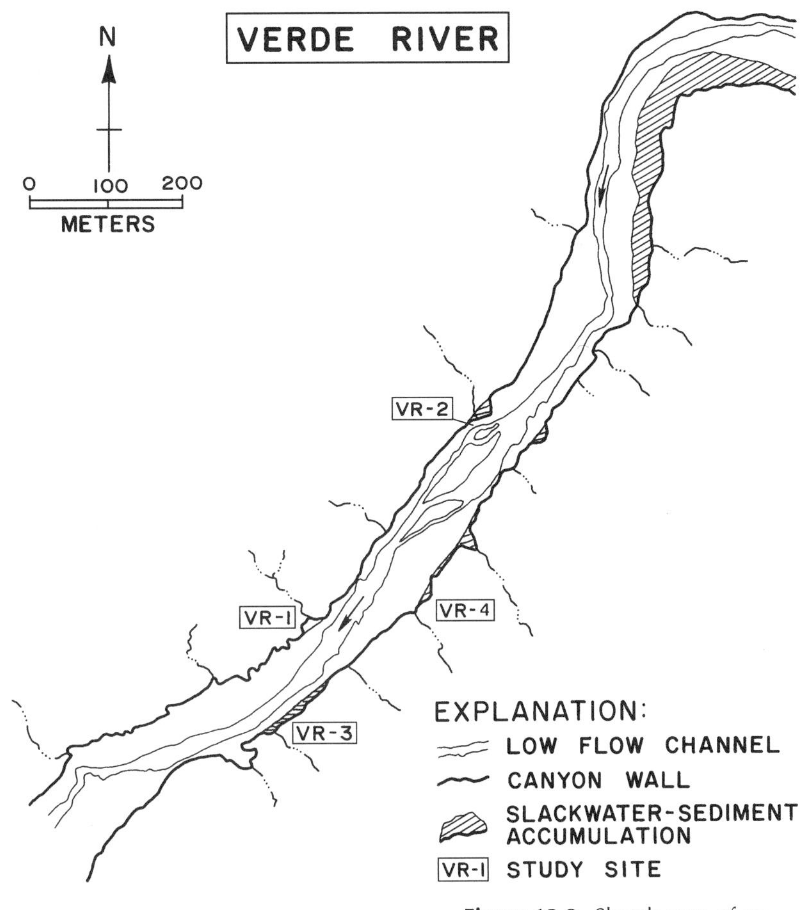

Figure 13.8. Sketch map of an SWD-PSI study reach on the Verde River in central Arizona. This reach is located 13 km downstream of the U.S. Geological Survey stream gauge at Tangle Creek. The site is described in more detail by O'Connor et al. (1986).

Flood-Frequency Analysis

The problem of frequency analysis for paleofloods is closely related to that for historical floods and for conventional gauge records (see Stedinger, Chapter 12, this volume). Appropriately structured paleoflood data can be combined with annual peak gauged flow records in the maximum-likelihood method of Stedinger and Cohn (1986) and Stedinger et al. (1988). A related Bayesian methodology has been applied to the problem by O'Connell (see Ostenaa et al., 1996, 1997). These methods explicitly incorporate the information provided by the paleoflood data (Stedinger and Baker, 1987), avoiding various misconceptions about paleoflood data that have appeared in the literature (e.g. Hosking and Wallis, 1986).

Temporal and Spatial Patterns

Ideally, multiple, long-term paleoflood records should be developed for an entire hydrological region (Figure 13.10). These records can then be

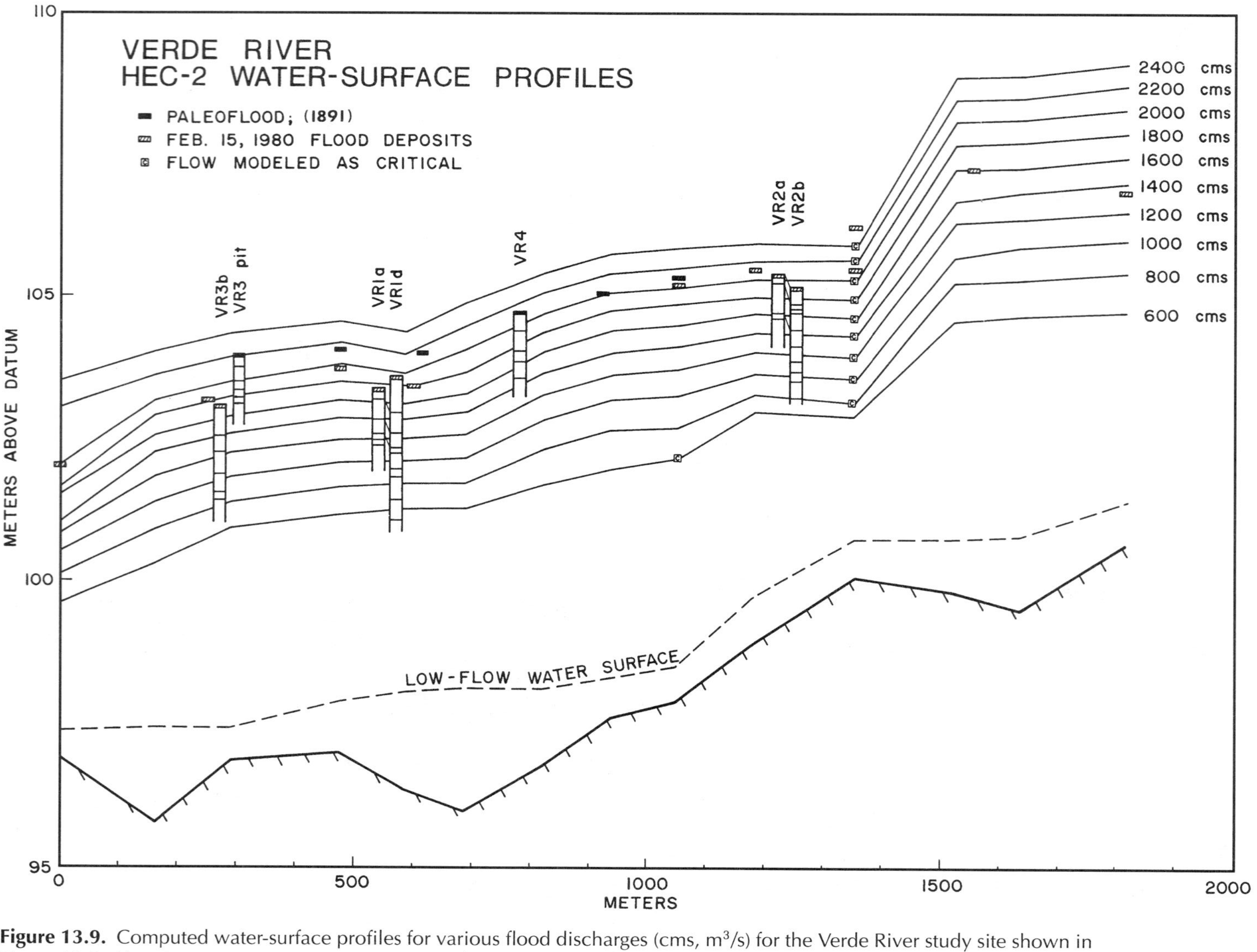

Figure 13.9. Computed water-surface profiles for various flood discharges (cms, m^3/s) for the Verde River study site shown in Figure 13.8. Also shown are representative stratigraphic sections of flood deposits and top layers of other flood deposits in their elevational relationship to the palaeoflood discharges. Note also elevations reached by historical floods in 1891 and 1980. (From O'Connor et al., 1986)

compared to discover regional patterns of peak floods in time and space. Such patterns have been described for the lower Colorado River basin (Ely, 1997; Ely et al., 1993; Enzel et al., 1993, 1996). In this region, the paleoflood data document upper boundary to flood magnitudes (Enzel et al., 1993). They also show that the largest floods are not completely random occurrences on long time scales. Instead, these largest floods cluster in certain time periods of several centuries (Ely et al., 1993).

Studies of paleofloods in multiple small drainage basins may be possible in certain regions (Martinez et al., 1994; House, 1996). These studies reveal the long-term influences of extreme storms in generating maximum flood peaks. Comparison of paleoflood records to problematic conventional flow estimates can be used to improve determinations of rare peak flood measurements (House and Pearthree, 1995).

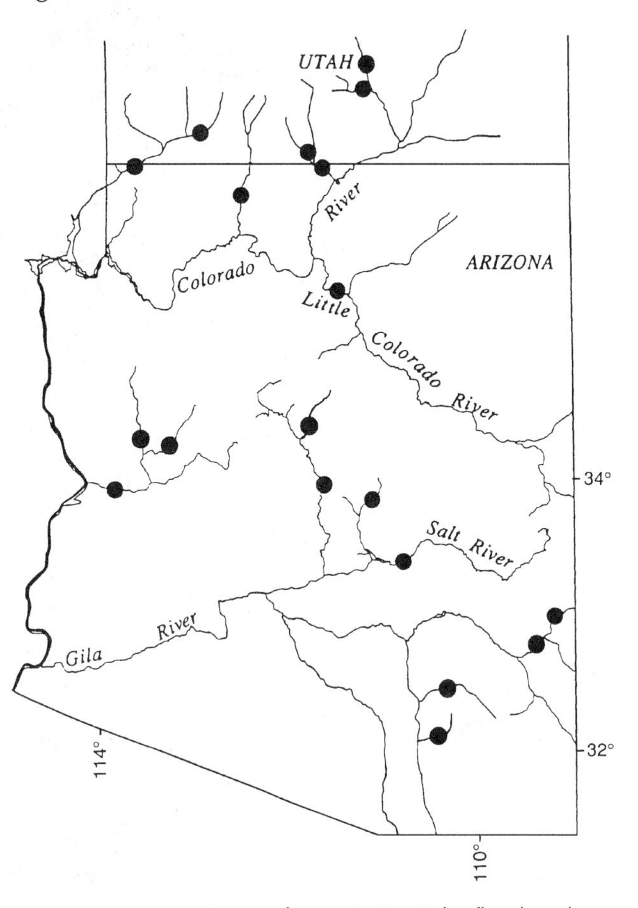

Figure 13.10. Paleoflood study sites in Arizona and southern Utah. See Enzel et al. (1996) for discussion of these sites.

Predicting Extreme Floods

Philosophy of Flood Prediction

Risk of extreme floods can be considered the product of adverse flood consequences (usually related to flood magnitude) multiplied by the probability of flood occurrence. Flood-risk assessment requires statements to be made about specific extreme floods that are expected to occur in the future. For matters of public safety, such statements (predictions) should compel belief (1) by those who certify the safety of a potentially hazardous structure, and (2) by those at risk from the hazard. Type 2 credibility can be ignored only in totalitarian societies.

Maximum credibility for prediction of extreme floods can be achieved by truthful statements. Unfortunately, the truth of a statement about the future can be verified only by the fulfillment of a prediction. This affords credibility in hindsight but not for the prospective problem intrinsic to risk assessment. Alternative strategies for achieving credibility in risk assessment are to make predictions (1) reasonable, (2) realistic, or (3) preferably both reasonable and realistic. Strategy 1 is used when experts say the most reasonable things possible about future extreme floods, given the available theoretical knowledge about flood behavior. Strategy 2 is used when experts interpret (observe) what nature says to us about the real occurrences of extreme floods. Strategy 3 is commonly achieved for relatively small, common floods, because such floods occur in conventional stream gauge records (strategy 2). Because of their rarity, however, extreme floods are not adequately represented in these conventional means of allowing nature to speak to us about its flood behavior.

The only available source for enhancing credibility for extreme flood estimates via strategy 2 is to seek evidence of past occurrences of extreme floods (paleoflood hydrology). There is no other source of reality for such floods.

Strategy 1 is completely under the control of an expert. It utilizes symbolic logic to achieve efficiency and reproducibility of results. However, its claims to credibility reside solely in the reasonableness of the expert opinion. The opinion of nature (reality) is not consulted in strategy 1.

Claims are sometimes made that strategy 2 cannot be used in the case of extreme floods. For example, it might be asserted that nature is highly variable in what it has to say to us, or that our capability for interpreting what nature has to say is both difficult and expensive, or that nature can tell us only of its past and that the future may be very different from past behavior. The one thing that is certain in these claims is that we will know nothing about them unless we actually see what nature has to say to us about extreme floods.

The common approach to strategy 1 is flood-frequency analysis, described by Stedinger (Chapter 12, this volume). For example, the concept of the 100-year flood was envisioned to provide a uniform standard for flood-hazard zonation. By using replicable, accepted procedures to available stream gauge data, one can extrapolate information temporally (at a site) or regionally to yield estimates of long-term frequencies of occurrence for various discharges. These procedures explicitly incorporate uncertainties for the lack of information that commonly exists on the sizes and occurrences of past floods. The procedures also incorporate assumptions about the behavior of floods over time and space. Unfortunately, these assumptions have not been as clearly stated and elaborated upon as have the procedures and their mathematical rationalizations.

The issue of assumptions underlying flood-frequency analysis is an involved one. One of the simplest assumptions is that instantaneous probabilities are equated to historic frequencies of occurrence (Klemes, 1989). Thus, the 100-year flood can occur at any time; specifically, it has a 1% probability of occurrence in any year. Studies of long records of paleofloods (e.g., Knox, 1993; Ely et al., 1993) show that this assumption can be wrong for long time scales. The largest floods preferentially cluster in certain time intervals, perhaps influenced by long-term trends in atmospheric circulation or oceanic sea surface temperatures (Hirschboeck, 1987). For short time scales, the assumption is also wrong. Consider seasonal climate regimes, such as the monsoonal tropics affected by tropical storms. Big floods cannot occur at all times but only in storm seasons. However, any resident of such regions knows that storm seasons vary from year to year. It is common sense to observe that the damaging 100-year flood does not have the same chance of occurrence each year. Its chances of occurrence are higher in years with more pronounced storm seasons.

Nonstationarity

Nearly all mathematical treatments of hydrologic processes assume that the underlying probability distribution for a random variable, such as flood recurrence, remains constant through time. Stationarity is the required assumption of most techniques for estimating flood exceedence probabilities. The U.S. Water Resources Council (1982, p. 6), for example, requires the following: "*. . . the array of flood information is a reliable and representative time sample of random homogeneous events.*" However, changes in climate, vegetation, and watershed conditions may dramatically alter flood responses, as documented by periods of differing flood-frequency characteristics in some very long paleoflood records (Patton and Dibble, 1982; Knox, 1985; Webb, 1985; Webb and Baker, 1987; Webb et al., 1988).

For paleoflood records extending back beyond the past century, the key stationarity questions center on climate variability. Deviations from purely random event generation occur because catastrophic floods are generated in many cases by nonrandom occurrences of atmospheric phenomena (Hirschboeck, 1987, 1988). The scientifically interesting goal is to document such occurrences by developing long-term paleoflood data sets at multiple localities. Without such information, it is impossible to evaluate the validity of any statistical flood-frequency analysis, especially those based only on temporally restricted, conventional data sets. Ultimately, the nonstationarity "problem" in flood-frequency analysis will have to be viewed in the context of hydroclimatological models for long-period climatic variations. Such models hold the promise of meshing paleoflood studies and other aspects of flood geomorphology with the relatively short time scales of gauged flow records (Hirschboeck, 1987).

Public Perception

As a practical matter, in the public debate over responses to potential hazards, the documented occurrence of an ancient (but real) cataclysmic process is likely to have more impact than is discussion of various hypothetical frequency distributions. Thus, the confirmation of a flood design number with paleoflood data is not merely proper science; it also serves to increase public confidence in any proposed solution that ultimately will lead to great economic or social expense for hazard mitigation.

The designation, via strategy 1 of a given flood prediction, such as the 100-year flood, is clearly replicable and follows acceptable engineering practice. Nevertheless, this approach has many elements that defy common sense, which derive in part from the unreality of underlying assumptions. The importance of this comes in the perception of the public, which must bear the economic consequences for societal actions that use the 100-year flood designation as their scientific basis. Such actions include mandated flood insurance rates, zoning restrictions, and required adjustment to potential hazards. Although these actions are certainly desirable in the face of real threats to public safety, the conventionally argued scientific basis for identifying such threats is not to appeal to reality but rather to invoke the 100-year flood or similar conceptualizations that are widely misunderstood by the public.

The public perception issue of the 100-year flood is simply stated. The concept has nothing to do with real years, and it has very little to do with real floods (Baker, 1994). The 100-year flood is an idealization. Human perception is grounded, not in idealizations, but in experience of real years and real floods (Baker, 1998a). Of course, this is why societal interest in adjustment to floods is so much more intense after a flood disaster. How is it best to maintain at least some of this commitment over the long time spans during which the memory of calamity fades? Present policy does this through the 100-year flood conceptualization, but there are alternatives.

At any flood-prone site, there are evidences of past floods of great magnitude. These are real floods that may have occurred in the recorded human history for the site, or their evidence may be preserved in geological records. The latter are paleofloods, and zones of inundation by these real events provide clear evidence of danger. Common sense holds that what has happened can indeed happen again. Although the probabilities of real phenomena may not be as easy to specify as those for idealized conceptualizations of those phenomena, the consequences in human perception of hazard are immensely greater.

Applied Paleoflood Hydrology

In a sense, paleoflood data can be considered to have the same body of applications as conventional data obtained for floods at gauging stations or by indirect measurement shortly after the event. On the other hand, paleoflood data will likely include information about past flows much larger in magnitude than those documented by the conventional record. A long time period of recording is provided as well as information at ungauged sites. The net effect is to expand the completeness of data coverage in time (to longer flow records), in space (to ungauged sites), and in scale (to larger events recorded in the increased data space). Note that this constitutes a time-space extension of real data coverage and not an idealized regionalization of limited conventional data.

With due consideration to its variable accuracy and completeness, discussed below, paleoflood data may be combined with conventional data, both for flood-based statistical analysis (Stedinger and Baker, 1987) and as a check on assumptions implicit in model-based estimates of flood parameters. The best paleoflood sites produce accurate estimates of the largest flows experienced over centuries to millennia on a given river reach. This is information from the upper tail of the parent distribution of floods and not from the many small flows near the mean of the distribution.

Safety of Dams

A major area of potential application is in relation to high-risk project design, as in the case of dams, nuclear facilities, hazardous waste sites, and critical infrastructure. An example of evolving application is Colorado State House Bill No. 1186 (April 4, 1986), which requires, when appropriate, the use of geologic and vegetative studies to establish probable future hazardous surface-water flows in relation to reservoir design and construction. This act followed from the use of paleocompetence-based paleoflood data in evaluating flood hazards and dam safety in the Colorado Front Range (Costa, 1978; Jarrett and Costa, 1982; Jarrett, 1987).

The National Research Council report *Safety of Dams* advocates analysis of physical evidence of large paleofloods to provide objective evidence of the likelihood and frequency of larger floods than can be documented by gauged flow records (NRC, 1985, p. 235). The report notes that stratigraphic and geomorphic evidence of extraordinary paleofloods has the potential of illustrating what size floods can occur. Such paleoflood data should be used to demonstrate that calculated probable maximum flood (PMF) values, used in dam safety design, are credible and are neither unreasonably large nor small (NRC, 1985, p. 235).

The United States has generally adopted hydrometeorological procedures for dam spillway design. U.S. dams must be designed to withstand the discharge of a PMF. A PMF is determined by using a rainfall-runoff model for a particular drainage basin receiving the probable maximum precipitation (PMP). There is considerable criticism, by many hydrological researchers, that reliance solely on PMF-based methods leads to false security and/or misallocation of resources for dam safety. Recent work by the U.S. Bureau of Reclamation (Ostenaa et al., 1996) incorporates paleoflood data in dam safety evaluations.

Spillway design and nuclear safety applications center around specific design numbers for floods with very small annual exceedence probabilities (10^{-2} to 10^{-7}). There are also important science issues that relate to these searches for design numbers. One is the open question of whether natural conditions result in an upper bound to flood magnitudes in various physiographic/climatological regions. Assumptions made with regard to this question cannot be tested with conventional data in many regions, but paleoflood data provide the potential for such a test (Baker et al., 1990). Other assumptions relate to potential occurrences of postulated events. If a 10^{-3} probability event is postulated by a statistical flood model, then it is likely that such an event will have occurred in the last 10^4 years. By regionalizing paleoflood data, it will be possible to test for the occurrence of even rarer floods.

Discussion and Conclusions

Paleoflood hydrological studies of the type reviewed in this chapter have been extensively used in several parts of the world. These areas include (1) the southwestern United States (Baker, 1975; Baker et al., 1979; Kochel et al., 1982; Enzel et al., 1996), (2) northern and central Australia (Baker et al., 1983b, 1987; Patton et al., 1993; Pickup, 1989; Wohl et al., 1994b), (3) India (Kale et al., 1993, 1996, 1997; Ely et al., 1996; Baker, 1998b), (4) Israel (Wohl et al., 1994a; Greenbaum et al., 1998), and (5) South Africa (Zawada, 1997; Smith and Zawada, 1990). Some paleoflood work has also been done in Spain (Benito et al., 1998), China (Shih Fucheng et al., 1985), Greece (Lewin et al., 1991), and Japan (Jones et al., 1998). Clearly, the applications have been in semiarid and seasonal tropical or subtropical regions. Applications are also possible in humid temperate areas (Patton, 1988; Knox, 1993), and a great potential exists to extend paleoflood hydrology to many parts of the world as part of a grand global experiment (Baker, 1995).

Paleoflood hydrology has achieved accuracies and rigor of methodology comparable to those of more conventional approaches to the study of extremely rare, high-magnitude floods. Given that paleoflood studies afford the only source of information of real (actually occurring) floods over very long time scales (several decades to millennia), such studies are having increased practical application. However, when paleoflood and conventional hydrological approaches are compared many misconceptions remain. The foregoing has been a brief introduction to paleoflood hydrology, more extensive descriptions of which can be found in the references cited.

Acknowledgments

I thank Dan Cenderelli, Bob Jarrett, and Ellen Wohl for their reviews. My research in paleoflood hydrology has been supported over the years by the U.S. National Science Foundation. This chapter is contribution No. 51 of the Arizona Laboratory for Paleohydrological and Hydroclimatological Analysis, Department of Hydrology and Water Resources, The University of Arizona.

References

Baker, V.R. (1975). *Flood Hazards along the Balcones Escarpment in Central Texas.* Austin: University of Texas Bureau of Economic Geology Circular No. 75-5, 22 pp.

Baker, V.R. (1987). Paleoflood hydrology and extraordinary flood events. *Journal of Hydrology*, **96**, 79–99.

Baker, V.R. (1989). Magnitude and frequency of palaeofloods. In *Floods, Their Hydrological, Sedimentological and Geomorphological Implications*, eds. K. Beven and P. Carling, pp. 171–183. Chichester: Wiley.

Baker, V.R. (1994). Geomorphological understanding of floods. *Geomorphology*, **10**, 139–156.

Baker, V.R. (1995). Global paleohydrological change. *Quaestiones Geographicae* Special Issue, **4**, 27–35.

Baker, V.R. (1996). Discovering the future in the past: Paleohydrology and global environmental change. In *Global Continental Changes*, eds. J. Branson, A.G. Brown, and K.J. Gregory, pp. 73–83. London: The Geological Society Special Publication 115.

Baker, V.R. (1998a). Hydrological understanding and societal action. *Journal of the American Water Resources Association*, **34**, 819–825.

Baker, V.R. (1998b). Future prospects for past floods in India. In *Flood Studies in India*, ed. V.S. Kale, pp. 219–228. Bangalore: Geological Society of India Memoir 41.

Baker, V.R. and Costa, J.E. (1987). Flood power. In *Catastrophic Flooding*, eds. L. Mayer and D. Nash, pp. 1–21. Boston: Allen and Unwin.

Baker, V.R. and Kochel, R.C. (1988). Flood sedimentation in bedrock fluvial systems. In *Flood Geomorphology*, eds. V.R. Baker, R.C. Kochel, and P.C. Patton, pp. 123–137. New York: Wiley.

Baker, V.R. and Pickup, G. (1987). Flood geomorphology of the Katherine Gorge, Northern Territory, Australia. *Geological Society of America Bulletin*, **98**, 635–646.

Baker, V.R., Ely, L.L., and O'Connor, J.E. (1990). Paleoflood hydrology and design decisions for high-risk projects. In *Proceedings of the 1990 National Hydraulic Engineering Conference*, pp. 433–438. New York: American Society of Civil Engineers.

Baker, V.R., Kochel, R.C., and Patton, P.C. (1979). Long-term flood-frequency analysis using geological data. *International Association of Hydrological Sciences Publication*, **128**, 3–9.

Baker, V.R., Kochel, R.C., Patton, P.C., and Pickup, G. (1983a). Paleohydrologic analysis of Holocene flood slack-water sediments. In *Modern and Ancient Fluvial Systems*, eds. J.D. Collinson and J. Lewin, pp. 229–239. International Association of Sedimentologists. Blackwell Scientific Publications, Oxford, England.

Baker, V.R., Pickup, G., and Polach, H.A. (1983b). Desert paleofloods in central Australia. *Nature*, **301**, 502–504.

Baker, V.R., Pickup, G., and Polach, H.A. (1985). Radiocarbon dating of flood events, Katherine Gorge, Northern Territory, Australia. *Geology*, **13**, 344–347.

Baker, V.R., Pickup, G., and Webb, R.H. (1987). Paleoflood hydrologic analysis at ungauged sites, central and northern Australia. In *Regional Flood Frequency Analysis*, ed. V.P. Singh, pp. 325–338. Boston: D. Reidel.

Benito, G., Machado, M.J., Perez-Gonzalez, A., and Sopeña, A. (1998). Palaeoflood hydrology of the Tagus River, central Spain. In *Paleohydrology and Environmental Change*, eds. G. Benito, V.R. Baker, and K.J. Gregory, pp. 317–333. Wiley and Sons, Chichester: U.K.

Benson, M.A. (1950). Use of historical data in flood frequency analysis. *Transactions of the American Geophysical Union*, **31**, 419–424.

Church, M.A. (1978). Paleohydrological reconstructions from a Holocene valley fill. In *Fluvial Sedimentology*, ed. A.D. Maill, pp. 743–772. Canadian Society of Petroleum Geologists Memoir 5. Calgary, Alberta, Canada.

Costa, J.E. (1978). Holocene stratigraphy in flood frequency analysis. *Water Resources Research*, **14**, 626–632.

Costa, J.E. (1983). Paleoconstruction of flash flood peaks from boulder deposits in the Colorado Front Range. *Geological Society of America Bulletin*, **94**, 986–1004.

Damon, P.E., Ferguson, C.W., Long, A., and Wallick, E.I. (1974). Dendrochronologic calibration of the radiocarbon time scale. *American Antiquity*, **39**, 350–365.

Dury, G.H. (1976). Discharge prediction present and former, from channel dimensions. *Journal of Hydrology*, **30**, 219–245.

Ely, L.L. (1997). Response of extreme floods in the Southwestern United States to climatic variations in the late Holocene. *Geomorphology*, **19**, 175–201.

Ely, L.L. and Baker, V.R. (1985). Reconstructing paleoflood hydrology with slackwater deposits: Verde River, Arizona. *Physical Geography*, **6**, 103–126.

Ely, L.L., Enzel, Y., Baker, V.R., and Cayan, D.R. (1993). A 5000-year record of extreme floods and climate change in the southwestern United States. *Science*, **262**, 410–412.

Ely, L.L., Enzel, Y., Baker, V.R., Kale, V.S., and Mishra, S. (1996). Changes in the magnitude and frequency of the Late Holocene monsoon floods on the Narmada River, central India. *Geological Society of America Bulletin*, **108**, 1134–1148.

Enzel, Y., Ely, L.L., House, P.K., Baker, V.R., and Webb, R.H. (1993). Paleoflood evidence for a natural upper bound to flood magnitudes in the Colorado River Basin. *Water Resources Research*, **29**, 2287–2297.

Enzel, Y., Ely, L.L., House, P.K., and Baker, V.R. (1996). Magnitude and frequency of Holocene paleofloods in the Southwestern United States: A review and discussion of implications. In *Global Continental Changes: The Context of Palaeohydrology*, eds. J. Branson, A.G. Brown, and K.J. Gregory, pp. 121–137. London: The Geological Society Special Publication 115.

Ethridge, F.G. and Schumm, S.A. (1978). Reconstructing paleochannel morphologic and flow characteristics: Limitations and assessment. In *Fluvial Sedimentology*, ed. A.D. Miall, pp. 703–721. Calgary: Canadian Society of Petroleum Geologists Memoir 5.

Feldman, A.D. (1981). HEC models for water resources system simulation: Theory and practice. *Advances in Hydroscience*, **12**, 297–423.

Greenbaum, N., Margalit, A., Schwartz, U., Schick, A.P., and Baker, V.R. (1998). A high-magnitude storm and flood in a hyperarid catchment, Nahal Zin, Negev Desert, Israel. *Hydrological Processes*, **12**, 1–23.

Hirschboeck, K.K. (1987). Catastrophic flooding and atmospheric circulation anomalies. In *Catastrophic Flooding*, eds. L. Mayer and D. Nash, pp. 25–56. London: Allen and Unwin.

Hirschboeck, K.K. (1988). Flood hydroclimatology. In *Flood Geomorphology*, eds. V.R. Baker, R.C. Kochel, and P.C. Patton, pp. 27–49. New York: Wiley.

Hosking, J.R.M. and Wallis, J.R. (1986). Paleoflood hydrology and flood frequency analysis. *Water Resources Research*, **22**, 543–550.

House, P.K. (1996). *Reports on Applied Paleoflood Hydrological Investigations in Western and Central Arizona.* Tucson: University of Arizona Ph.D. Dissertation.

House, P.K. and Pearthree, P.A. (1995). A geomorphic and hydrologic evaluation of an extraordinary flood discharge estimate: Bronco Creek, Arizona. *Water Resources Research*, **31**, 3059–3073.

Jarrett, R.D. (1987). *Flood Hydrology of Foothill and Mountain Streams in Colorado.* Fort Collins: Colorado State University Ph.D. Dissertation.

Jarrett, R.D. and Costa, J.E. (1982). Multidisciplinary approach to the flood hydrology of foothill streams in Colorado. In *International Symposium on Hydrometeorology*, pp. 560–565. Denver: American Water Resources Association.

Jones, A.P., Shimazu, H., and Oguchi, T. (1998). Holocene slack-water deposits on the Nakagawa River, Tochigi Prefecture, Japan. In *Third International Meeting on Global Continental Paleohydrology: Abstracts of Conference Papers*, eds. M. Grossman, T. Oguchi, and H. Kadomura, pp. 59–60. Kumagaya: Rissho University.

Kale, V.S., Mishra, S., Baker, V.R., Rajaguru, S.V., Enzel, Y., and Ely, L.L. (1993). Prehistoric flood deposits of the Choral River, central Narmada, India. *Current Science*, **65**, 877–878.

Kale, V.S., Ely, L.L., Enzel, Y., and Baker, V.R. (1996). Palaeo and historical flood hydrology, Indian Peninsula. In *Global Continental Changes: The Context of Palaeohydrology*, eds. J. Branson, A.G. Brown, and K.J. Gregory, pp. 155–163. London: The Geological Society Special Publication 115.

Kale, V.S., Mishra, S., and Baker, V.R. (1997). A 2000-year palaeoflood record from Sakarghat on Narmada, central India. *Journal of the Geological Society of India*, **50**, 283–288.

Klemes, V. (1989). The improbable probabilities of extreme floods and droughts. In *Hydrology of Disasters*, ed. O. Melder, pp. 43–51. London: James and James.

Knox, J.C. (1985). Responses of floods to Holocene climatic change in the upper Mississippi Valley. *Quaternary Research*, **23**, 287–300.

Knox, J.C. (1993). Large increases in flood magnitude in response to modest changes in climate. *Nature*, **361**, 423–425.

Kochel, R.C. (1980). *Interpretation of Flood Paleohydrology Using Slackwater Deposits, Lower Pecos and Devils Rivers, Southwestern Texas*. Austin: University of Texas Ph.D. Dissertation.

Kochel, R.C. and Baker, V.R. (1982). Paleoflood hydrology. *Science*, **215**, 353–361.

Kochel, R.C. and Baker, V.R. (1988). Paleoflood analysis using slackwater deposits. In *Flood Geomorphology*, eds. V.R. Baker, R.C. Kochel, and P.C. Patton, pp. 357–376. New York: Wiley.

Kochel, R.C., Baker, V.R., and Patton, P.C. (1982). Paleohydrology of southwestern Texas. *Water Resources Research*, **18**, 1165–1183.

Lewin, J., Macklin, M.G., and Woodward, J.C. (1991). Late Quaternary fluvial sedimentaiton in the Voidomatis basin, Epirus, N.W. Greece. *Quaternary Research*, **35**, 103–115.

Maizels, J.K. (1983). Palaeovelocity and palaeodischarge determination for coarse gravel deposits. In *Background to Palaeohydrology*, ed. K.J. Gregory, pp. 101–139. Chichester: Wiley.

Martinez-Goytre, J., House, P.K., and Baker, V.R. (1994). Spatial variability of paleoflood magnitudes in small basins of the Santa Catalina Mountains, southeastern Arizona. *Water Resources Research*, **30**, 1491–1501.

National Research Council. (1985). *Safety of Dams: flood and Earthquake Criteria*. Washington, DC: National Academy Press.

O'Connor, J.E. (1993). *Hydrology, hydraulics, and sediment transport of Pleistocene Lake Bonneville flooding on the Snake River, Idaho*. Geological Society of America Special Paper 274. Boulder, CO.

O'Connor, J.E. and Baker, V.R. (1992). Magnitudes and implications of peak discharges from Glacial Lake Missoula. *Geological Society of America Bulletin*, **104**, 267–279.

O'Connor, J.E., Fuller, J.E., and Baker, V.R. (1986). *Late Holocene Flooding Within the Salt River Basin, Central Arizona*. Phoenix: Report to Salt River Project.

O'Connor, J.E. and Webb, R.H. (1988). Hydraulic modeling for paleoflood analysis. In *Flood Geomorphology*, eds. V.R. Baker, R.C. Kochel, and P.C. Patton, pp. 383–402. New York: Wiley.

O'Connor, J.E., Ely, L.L., Wohl, E.E., Stevens, L.E., Melis, T.S., Kale, V.S., and Baker, V.R. (1994). A 4500-year record of large floods on the Colorado River in the Grand Canyon, Arizona. *Journal of Geology*, **102**, 1–9.

Ostenaa, D.A., Levish, D.R., and O'Connell, D.R.H. (1996). *Paleoflood Study for Bradbury Dam, Cachuma Project, California*. Denver: U.S. Bureau of Reclamation Seismotectonic Report 96–3.

Ostenaa, D.A., Levish, D.R., O'Connell, D.R.H., and Cohen, E.A. (1997). *Paleoflood Study for Causey and Pineview Dams, Weber and Ogden River Projects, Utah*. Denver: U.S. Bureau of Reclamation Seismotectonic Report 96–6.

Patton, P.C. (1988). Geomorphic response of streams to floods in the glaciated terrain of southern New England. In *Flood Geomorphology*, eds. V.R. Baker, R.C. Kochel, and P.C. Patton, pp. 261–277. New York: Wiley.

Patton, P.C. and Dibble, D.S. (1982). Archaeologic and geomorphic evidence for the paleohydrologic record of the Pecos River in West Texas. *American Journal of Science*, **282**, 97–121.

Patton, P.C., Pickup, G., and Price, D.M. (1993). Holocene paleofloods of the Ross River, central Australia. *Quaternary Research*, **40**, 201–212.

Pickup, G. (1989). Palaeoflood hydrology and estimation of the magnitude, frequency and areal extent of extreme floods – An Australian perspective. *Civil Engineering Transactions*, **CE31**, 19–29.

Shih Fucheng, Yi Yuanjun, and Han Manhua (1985). Investigation and verification of extraordinarily large floods of the Yellow River. Paper presented at the *U.S.-China Bilateral Symposium on the Analysis of Extraordinary Flood Events*, Nanjing, China.

Smith, A.M. and Zawada, P.K. (1990). Palaeoflood hydrology: A tool for South Africa? – An example from the Crocodile River near Brits, Transvaal, South Africa. *Water South Africa*, **16**, 195–200.

Stedinger, J.R. and Cohn, T.A. (1986). Flood frequency analysis with historical and paleoflood information. *Water Resources Research*, **22**, 785–793.

Stedinger, J.R. and Baker, V.R. (1987). Surface water hydrology: Historical and paleoflood information. *Reviews of Geophysics*, **25**, 119–124.

Stedinger, J.R., Surani, R., and Therival, R. (1988). *The MAX Users Guide*. Ithaca: Cornell University Department of Environmental Engineering.

Stuiver, M. (1982). A high-precision calibration of the A.D. radiocarbon time scale. *Radiocarbon*, **24**, 1–26.

Thomson, M.T., Gannon, W.B., Thomas, M.P., and Hayes, G.S. (1964). Historical floods in New England. *U.S. Geological Survey Professional Paper*, **1779-M**, 1–105.

U.S. Water Resources Council. (1982). *Guidelines for Determining Flood Flow Frequency*. Washington, DC: United States Water Resources Council, Bulletin 17B.

Webb, R.H. (1985). *Late Holocene Flooding on the Escalante River, South-Central Utah*. Tucson: University of Arizona Ph.D. Dissertation.

Webb, R.H. and Baker, V.R. (1987). Changes in hydrologic conditions related to large floods on the Escalante River, south-central Utah. In *Regional Flood Frequency Analysis*, ed. V.P. Singh, pp. 309–323. Boston: D. Reidel.

Webb, R.H., O'Connor, J.E., and Baker, V.R. (1988). Paleohydrologic reconstruction of flood frequency on the Escalante River. In *Flood Geomorphology*, eds. V.R. Baker, R.C. Kochel, and P.C. Patton, pp. 403–418. New York: Wiley.

Williams, G.P. (1983). Paleohydrological methods and some Swedish fluvial environments. I-cobble and boulder deposits. *Geografiska Annaler*, **65A**, 227–243.

Williams, G.P. (1984). Paleohydrologic equations for rivers. In *Developments and Applications of Geomorphology*, eds. J.E. Costa and P.J. Fleisher, pp. 353–367. Berlin: Springer-Verlag.

Wohl, E.E., Greenbaum, N., Schick, A.P., and Baker, V.R. (1994a). Controls on bedrock channel morphology along Nahal Paran, Israel. *Earth Surface Processes and Landforms*, **19**, 1–13.

Wohl, E.E., Webb, R.H., Baker, V.R., and Pickup, G. (1994b). *Sedimentary Flood Records in the Bedrock Canyons of Rivers in the Monsoonal Region of Australia*. Fort Collins: Colorado State University Water Resources Paper No. 107, 102 pp.

Yevjevich, V. and Harmancioglu, N.B. (1987). Research needs on flood characteristics. In *Application of Frequency and Risk in Water Resources*, ed. V.P. Singh, pp. 1–21. Boston: D. Reidel.

Zawada, P.K. (1997). Paleoflood hydrology: Method and application in flood-prone Southern Africa. *South African Journal of Science*, **93**, 111–132.

FLOOD HAZARD MITIGATION STRATEGIES

CHAPTER 14

Comparison of Flood Management Strategies

Chester C. Watson
Department of Civil Engineering
Colorado State University

David S. Biedenharn
Engineering Research and Development Center, U.S. Army Corps of Engineers

Case studies from the Mississippi River basin in the United States and the Tone River basin of Japan are used to illustrate structural approaches to flood hazard mitigation. Both drainage basins have extensive levees, channelization, and reservoirs. Management strategies on the Tone River began in 1590. Management strategies on the lower Mississippi River date to the 1700s but have changed significantly since the federal government became involved in 1824. On both rivers, management strategies evolved in response to the changing demands of the public. Study of the evolution of these management strategies offers an excellent opportunity to review flood-management strategies.

Introduction

The primary purpose of this paper is to document the evolutionary process of the flood management strategy on the lower Mississippi River and to compare these efforts to similar efforts on the Tone River in Japan. Most of the major changes in each plan did not occur at random but were directly related to external factors such as major floods. The Mississippi River project has evolved from a single-purpose navigation project to a multipurpose project, which incorporates all aspects of water and related resources within an environmentally sustainable framework. The Tone River project is primarily a navigation and flood-control project. Structural flood-control measures along the Mississippi and Tone Rivers are representative of similar measures undertaken along other rivers, such as the extensive levees on the major rivers of China or the channelization and levee systems of many rivers in Europe.

Categories of Flood-Control Measures

There are four principal categories for structural flood-control measures: levees, floodways, channel improvement and stabilization, and reservoirs. Artificial levees constructed for flood control are linear earthen mounds constructed in the floodplain to confine flood discharges. Figure 14.1 provides a historical comparison of the cross-sectional size and shape of Mississippi River levees. Winkley (1971) states that the essential conditions governing construction include an adequate height to prevent overtopping, a wide base to protect against foundation seepage, and a massive cross section to prevent seepage through the levee.

A floodway is a controlled distributary channel constructed to divert flood discharge, which allows the remaining main channel flood to be conveyed at a lower elevation. For example, on the Mississippi River over half of the design flood is diverted through floodways in the lower 500 km of the river to protect infrastructure.

Channel improvement and stabilization include numerous types of channel modifications such as artificial cutoffs, bank armoring, changes in channel alignment, and removal of debris. The purpose of channel

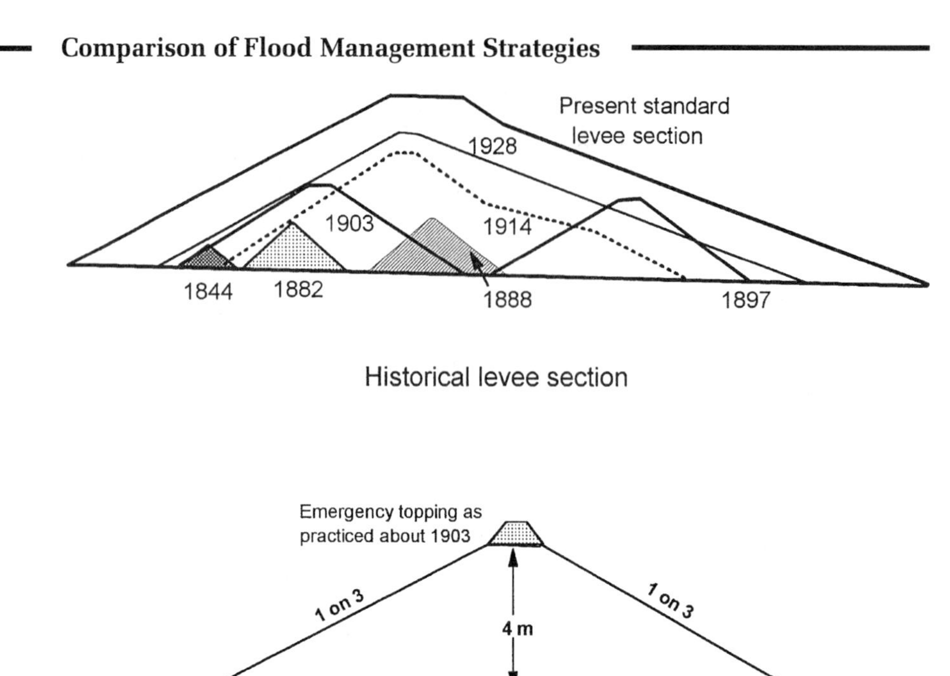

Figure 14.1. Schematic cross-sectional comparison of levee sections along the lower Mississippi River (after Winkley, 1971, Figure 19.5).

improvement and stabilization is to provide a channel that will convey the flood discharge more efficiently. During the period 1929 to 1942, the Mississippi River was shortened by approximately 245 km by artificial cutoffs. This series of cutoffs resulted in a dramatic reduction of flood levels. Biedenharn and Watson (1997) discuss the long-term effects of these cutoffs on the Mississippi River.

Reservoirs are usually constructed for multiple purposes, including recreation, water supply, and flood control. The effect of reservoirs on flood control in large rivers is to regulate tributary discharge into the river and to limit sediment that would be transported into the river naturally.

Description of Case Study Areas

The lower Mississippi River extends from Cairo, Illinois, to the Gulf of Mexico. Figure 14.2 shows the primary tributaries and outlets of the Mississippi River. The total drainage area of the Mississippi River and tributaries is about 2,975,000 km^2, which includes 41% of the continental United States and a small portion of Canada. The alluvial valley width of the lower Mississippi River varies from about 48 to 200 km. A major flood in 1927 inundated most of the valley. Most of the river floodplain is now protected by levees that confine the flood flows within an average width of about 5 km. The mean annual flow at Natchez, Mississippi, is 17,000 m^3/s, and the highest recorded discharge is 61,200 m^3/s. The design flood flow at Natchez, Mississippi, is 77,000 m^3/s.

The Tone River basin is the largest in Japan, draining 16,840 km^2. More than 90% of the floods have been caused by typhoons. The most disastrous flood, in September 1947, occurred as a result of 300 mm of precipitation throughout the basin. Uzuka and Tomita (1993) indicated that the design flood flow near the mouth is 16,000 m^3/s.

Evolution of Management Strategies on the Mississippi River

In this section, we divide the evolution of management strategies for the lower Mississippi River project into four time periods: (1) The early period, before 1824; (2) the period from about 1824 to 1927, when the federal government began to exercise control over the river; (3) the period from 1927 to the early 1960s, when the project was modified and expanded considerably with a primary emphasis on flood control and navigation; and (4) the period from the early 1960s to the present, when the project adopted a much broader perspective that included many environmental issues.

Early History, before 1824

On May 2, 1803, a treaty and convention were signed between the United States and France that transferred the Louisiana Territory to the United States. At this point, the United States obtained total control over the entire lower Mississippi River.

Navigation was not the only concern of the early settlers of the lower Mississippi River. As these early settlers began to develop the lower Mississippi Valley in the early 1700s, the need for flood protection was immediately recognized, particularly in the low-lying areas around New Orleans, Louisiana. Local landowners constructed levees in an effort to protect their property. The construction of these local, often discontinuous levees naturally led to many conflicts of interest among neighboring landowners. As conflicts continued to escalate with increasing development of the area, it became apparent that a coordinated effort to provide flood control was needed. Unfortunately, it

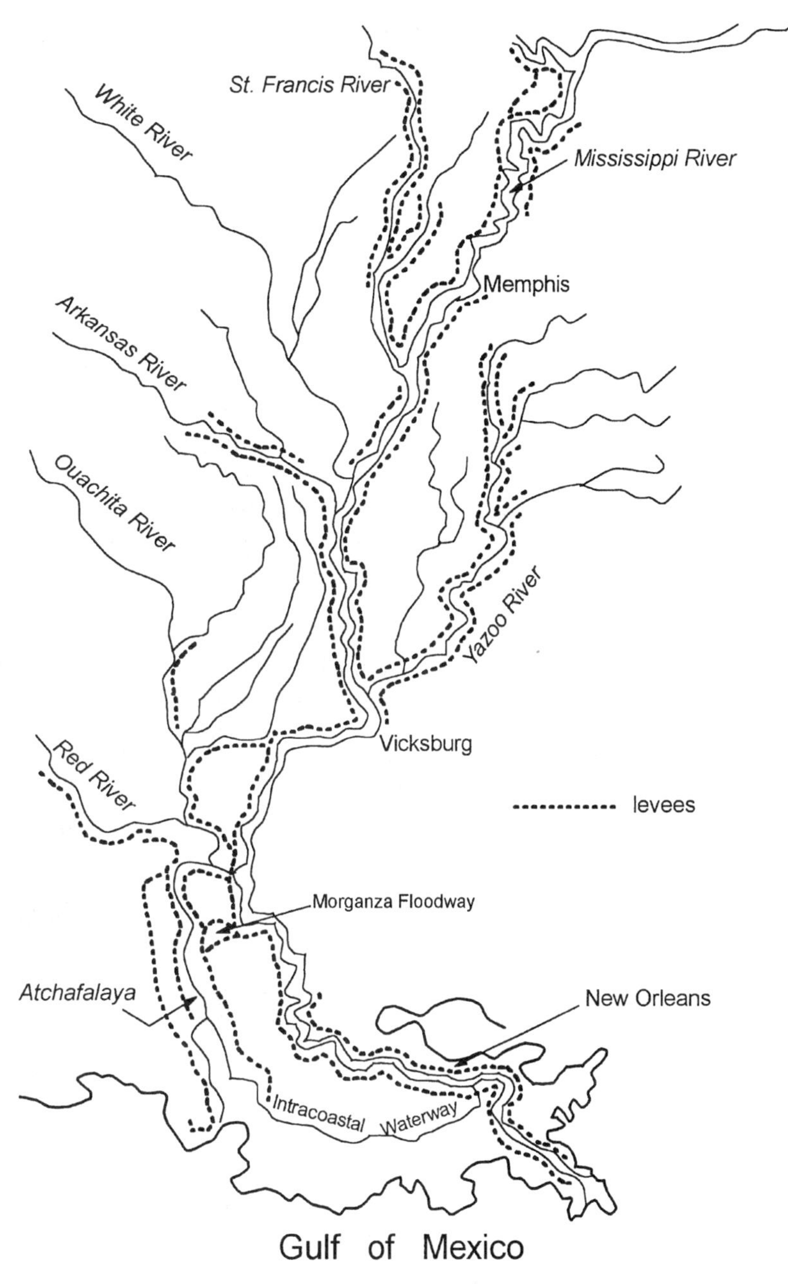

Figure 14.2. Lower Mississippi River flood-control plan (after Winkley, 1971, Figure 19.2).

took more than a hundred years before federal involvement occurred, and then only for navigation.

1824 to 1927

With the appearance of steamboats in the early 1800s, the need for navigation improvements on the nation's waterways began to receive increasing attention. One of the major lessons learned from the War of 1812 was that the defense of the United States depended on development of improved infrastructure systems. In 1819, Secretary of War John C. Calhoun recommended that the Corps of Engineers be authorized to improve waterways navigation and other transportation systems. The intent of his proposal was directed at facilitating the movement of the Army and of materials while contributing to national economic development. Congress accepted Calhoun's recommendation in 1824 by passing a General Survey Act that authorized the President to use Army engineers to survey road and canal routes of national importance in a commercial or military view (U.S. Army Corps of Engineers, 1991). The first federal operations in the lower Mississippi Valley also occurred in 1824 when Congress appropriated funds for improving the river for navigation under the first Rivers and Harbors Bill. The 1824 Rivers and Harbors Bill authorized the removal of sandbars and snagging of the Mississippi and Ohio Rivers and is usually considered the beginning of the planning responsibilities of the U.S. Army Corps of Engineers for the lower Mississippi River (Holmes, 1979). As the nation expanded, Congress directed more and more attention to river improvements as a federal responsibility but still almost exclusively for navigation.

Floods in 1849 and 1850 caused widespread distress and focused national attention on the flooding problems of the lower Mississippi River. As a result of this public concern, Congress passed the Swamp Acts of 1849 and 1850, which granted to the states all unsold swamp lands and provided that funds from the sale of those lands be used for flood protection. By 1858, levee development had reached a point that would not be exceeded for many years. Unfortunately, these levees often were not continuous and were of deficient height and cross section. Consequently, a total of 32 crevasses formed in the levees during the flood of 1859. The first attempt by Congress to secure effective flood protection through the Swamp Acts failed for two reasons (U.S. Army Corps of Engineers, 1972): (1) the enormity and engineering difficulty of the task were not fully appreciated, and (2) there was a lack of coordination among the different states and the levee districts. Thus, even at this early stage of river development it was clear that an urgent need existed for improved river engineering competence and for better coordination, cooperation, and communication among the various landowners and the local, state, and federal agencies to overcome the conflicts.

In 1850, after a further expression of national interest, Congress appropriated $50,000 to initiate a survey of the river by Captain A. A. Humphreys and Lieutenant H. L. Abbot, officers of the Corps of Engineers. In 1861 Humphreys and Abbot submitted their report to Congress entitled "Report Upon the Physics and Hydraulics of the Mississippi River" (Humphreys and Abbot, 1861). This hallmark report represents the first comprehensive attempt to understand the physical characteristics of the river system. The report rejected the use of reservoirs, cutoffs, and outlets and recommended levees as the sole protection measure. This policy later became known as the levees only policy, which was followed for many years. It is important to recognize that the focus of the nation at this time was on providing navigation and flood control with little or no concern for the environment. For this reason, the Humphreys and Abbot report was restricted to engineering with little or no mention of environmental features.

By 1879, the nation recognized that improvements of the river for navigation and flood control could not be accomplished without substantial federal participation. The necessity for coordination of engineering operations through a centralized organization was also apparent. In response to these needs, Congress established the Mississippi River Commission (MRC) on June 28, 1879, which became the central agency responsible for management of the entire lower Mississippi River valley. Section 4 of the act prescribed the duties of the Commission as follows:

> ... It shall be the duty of said Commission to take into consideration and mature such plan or plans and estimates as will correct, permanently locate, and deepen the channel and protect the banks of the Mississippi River; improve and give safety and ease to the navigation thereof; prevent destructive floods; promote and facilitate commerce, trade, and the postal service; and when so prepared and matured, to submit to the Secretary of War a full report of their proceedings and actions, and of such plans with estimates of the cost thereof, for purposes aforesaid, to be by him transmitted to Congress: Provided, that the Commission shall report in full upon the practicability, feasibility, and probable cost of the various plans known as the jetty system, the levee system, and the outlet system, as well as upon such others as they deem necessary ....

Although the need for flood control was one of the principal factors in the creation of the MRC, development of navigation remained the primary emphasis of the MRC. The MRC began levee work in 1882, thus initiating the first coordinated levee system on the lower Mississippi River. This work was primarily aimed at improving the channel for navigation, with flood control benefits listed as incidental to navigation.

The Rivers and Harbors Act of 1899 was by far the single most important piece of legislation concerning navigation in the history of the United States. This act gave the U.S. Army Corps of Engineers the authority to require permits for construction across any navigable waterway (Section 9), depositing material in a navigable water (Section 10), depositing refuse (Section 13), anchoring or causing obstruction to navigation (Section 15), and sunken vessels (Sections 19 and 20) (Hydrologic Engineering Center, 1986). This legislation strengthened the ability of the Corps of Engineers to provide an adequate navigation project on the lower Mississippi River.

By the early 1900s the MRC was building levees, constructing bank protection works, and dredging channels. Major floods in 1912, 1913, and 1916 prompted Congress to pass the first Flood Control Act on March 1, 1917, which authorized construction of levees for control of floods. Flood control was now a primary objective of the MRC, with levees serving as the sole line of defense. This levees only policy was soon to be proven totally inadequate.

1927 to Early 1960s

By 1927, with the MRC acting as the central agency responsible for flood control and navigation along the lower Mississippi River, things appeared to be well under control. However, the events of 1927 changed the history of the MRC and the lower Mississippi River forever. The flood of 1927 was the most disastrous flood in U.S. history. This catastrophic flood inundated an area of about 67,000 km^2, flooding cities, towns, and farms, with property damages in excess of $1,000,000,000 in 1976 dollars (U.S. Army Corps of Engineers, 1972). Even more devastating was the loss of 214 lives and the displacement of 637,000 people. This flood received so much national attention that a new national awareness of the critical need for flood control in the lower valley once again surfaced. As a consequence of this national concern the Flood Control Act of 1928 was passed, which committed the federal government to a definitive program for flood control.

The flood of 1927 forced the MRC to recognize that the levees only philosophy was no longer adequate to meet the flood-control demands of the lower valley. Consequently, a new plan was adopted on May 15, 1928, which contained four basic components (U.S. Army Corps of Engineers, 1972): (1) levees, (2) floodways, (3) channel improvement works, and (4) work in the tributary basins. The overall project is often referred to as the Mississippi River and Tributaries (MR&T) Project.

The 1928 Flood Control Act also established the responsibilities of local interests. Under this act, locals were not required to contribute to the construction cost of the project but had to maintain the minor flood control works and provide rights-of-way without cost to the federal government.

Another unique aspect of the 1928 Flood Control Act was that it required the MRC to conduct regular inspection trips of the project and to hold public hearings on these trips (U.S. Army Corps of Engineers, 1972). These public hearings are a critical component in the management of the MR&T Project because they provide a forum for local citizens and interest groups to receive information, express their concerns, and influence specific aspects of the project.

With the passage of the 1928 Flood Control Act, the scope of the project was broadened considerably. The MRC had the authority to address flood control and navigation in a comprehensive manner, including work in the tributary basins. There was also a definite policy for local cooperation, and a mechanism for coordinating and communicating with the public was established through the public hearings. However, at this point there still was no public outcry for assessing the environmental impacts of the project. Basically the public was content to simply get the water off the land.

However, on August 12, 1958, Congress passed Public Law 85-624, known as the Fish and Wildlife Coordination Act, which provided that wildlife conservation should be a consideration and should be coordinated with other features of water-resource development programs (U.S. Army Corps of Engineers, 1959). The 1958 Act authorized federal agencies to modify or add projects to accomodate the conservation of wildlife resources as long as the modifications were compatible with the project purposes and as long as the cost of the modifications or land acquisitions was an integral part of the project. This act also provided that agencies should coordinate with the Fish and Wildlife Service to minimize damages to wildlife resources. In accordance with this provision the Corps of Engineers and the Fish and Wildlife Service began to hold meetings aimed at improving the project plans to restore or preserve fish and wildlife resources. Although this act had very little real effect on execution of the project, it is important because it opened the doors of communication for the first time between the Corps of Engineers and the Fish and Wildlife Service.

In 1958 the Corps recognized the need to review the MR&T Project after 30 years of experience. The result was a report entitled, *Comprehensive Review of the Mississippi River and Tributaries Project* (U.S. Army Corps of Engineers, 1959). This report not only addressed the traditional engineering aspects of the project but also considered possible modification to avoid conflict or to serve additional purposes. The Fish and Wildlife Service participated in this study and made several recommendations to be included in the main stem project plan. The MRC agreed to continue to cooperate in the limited development and maintenance of water areas in borrow pits to serve as a habitat for production of fish and wildlife. Although this was the only area of wildlife conservation in which the Corps agreed to participate, it is significant because it represents the first definitive attempt to address the environmental concerns of the project.

By the early 1960s, the MR&T Project still focused almost exclusively on the engineering aspect of flood control and navigation. However, the nation was slowly starting to recognize the importance of protecting the environment and, as a result, various interest groups and agencies such as the Fish and Wildlife Service were just beginning to make their presences known as important players in management of the system.

Figure 14.3 diagrams the project design flood on the Mississippi River. Tributaries combine to result in 77,000 m^3/s at Natchez, Mississippi. Old River diversion and the Morganza Floodway divert a total of 35,000 m^3/s to the Atchafalaya River, leaving 42,000 m^3/s in the Mississippi River. Upstream of New

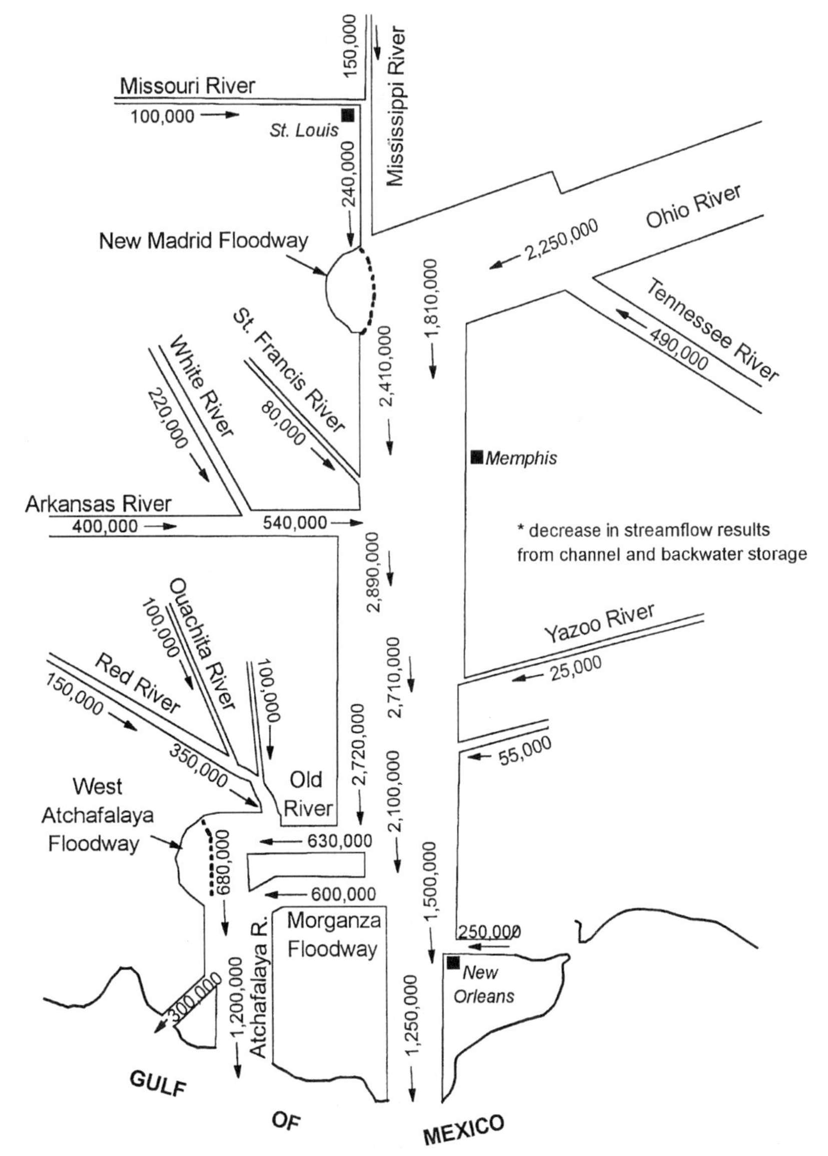

Figure 14.3. Schematic illustration of the structure of the project design flood for the lower Mississippi River basin. Discharges in cubic feet per second (after Winkley, 1971, Figure 19.3).

Orleans, Louisiana, the Bonnet Carre Spillway diverts 7,000 m^3/s to the Gulf of Mexico through Lake Pontchartrain, leaving 35,000 m^3/s in the Mississippi River at the mouth.

1960s to Present

The 1960s ushered in a new wave of environmental awareness that had a significant effect on the MR&T Project as well as on all water resource projects in the United States. One of the first and most significant pieces of legislation was passage of the Water Resources Planning Act in July 1965. The primary purpose of the act was to provide for optimum development of the nation's water resources through coordinated planning of water and related land resources. This act created the Water Resources Council (WRC), which had the responsibility for instituting a program of comprehensive planning for all major river basins in the United States.

The passage of Public Law 91-190, the National Environmental Policy Act, in 1969 had perhaps as much impact on the project as any legislation since the 1928 Flood Control Act (U.S. Army Corps of Engineers, 1986). This act declared that it is the national policy to:

> ... encourage productive and enjoyable harmony between man and his environment and to promote efforts which will prevent or eliminate damage to the environment and biosphere and stimulate the health and welfare of Man ....

Another act that had profound impacts on the MR&T Project was Public Law 93-205, the Endangered Species Act of 1973 (U.S. Army Corps of Engineers, 1986). This act provided a means for conserving endangered and threatened species and a program for conservation of ecosystems upon which those species depend. As a result of these and numerous other environmental laws and regulations passed during this period, along with growing demands of the public for more environmentally sound projects, the Corps' philosophy for the MR&T Project began to change.

In February 1968, the WRC designated the MRC as the lead agency for conducting a comprehensive study of the lower Mississippi River. This study had been authorized by the Flood Control Act of 1966. The primary purpose of the Lower Mississippi Region Comprehensive Study was to develop a program that could be used as a guide to the best use or combination of uses of water and related land resources in the region (U.S. Army Corps of Engineers, 1972). This was a comprehensive study that recognized all aspects of the project including environmental impact, water quality, impacts of land use on water resources, water supply, flood control and navigation, sediment and erosion control, municipal water quality, hydropower, coastal and estuarine processes, archeological and historical aspects, and public health. Although the recommendations of this study did not significantly alter the flood control and navigation elements of the main stem MR&T Project, it was the first time these features were considered in the broader context of a comprehensive systems analysis.

The public demands for increased inclusion of environmental features in the main stem MR&T Project continued to grow throughout the 1970s and 1980s. These changes did not occur overnight or without considerable conflict between the Corps and other agencies and interest groups and even within the Corps itself. Change is often difficult, particularly when it involves agencies and groups who have different agendas and little or no appreciation of the other's function. In 1978 President Carter issued executive order 12075 to ensure interagency cooperation (Office of Federal Register, 1978). These interagency agreements required the agencies to sign Memorandums of Understanding that outlined the jurisdiction and authorities of the agencies with regard to the project. This executive order had a tremendous effect on management of water resources projects, and the MR&T Project was no exception. This basically forced the different agencies to begin to communicate with each other and to try to negotiate their differences. As communication, coordination, and cooperation among the various agencies and groups improved, some of the inherent mistrust and antagonism was reduced. Consequently, there is a much more harmonious and cooperative atmosphere among the various groups today than there was 20 years ago. This does not suggest that this is a perfect marriage of the various interests or that there are no more serious disagreements or conflicts, but rather that the lines of communication are at least open, thereby providing a mechanism for resolving conflicts.

One very promising environmental feature that has been used on the lower Mississippi River is the concept of creating a notch or gap in new or existing stone dikes. These notches are designed to create more open water habitat by allowing more flow to pass through the dike field. Because of the limited information on the environmental effects of dike notching, the Lower Mississippi Valley Division (LMVD) has initiated studies to determine precisely how these features affect the aquatic habitat along the river. The LMVD is also trying to develop criteria for the design of environmental notches. The Corps is working closely with the Fish and Wildlife Service in studying these structures.

Articulated Concrete Mattress (ACM) is the primary form of bank stabilization on the lower Mississippi River. ACM basically consists of individual concrete slabs tied together to form a flexible mattress that is placed along the eroding bank. The LMVD has recently initiated studies to determine whether roughening or grooving the surface of the ACM will significantly improve the quality of benthic habitat. As part of this study, the Corps recently installed a revetment consisting of grooved mats at a study site near Baton Rouge, Louisiana. Studies by the LMVD, coordinated with the Fish and Wildlife Service, will be aimed at determining whether grooving the ACM will significantly improve the benthic habitat without sacrificing the structural integrity of the ACM.

Endangered species are a significant concern in the lower Mississippi River. Perhaps the most notable is the Interior Least Tern, a bird whose prime nesting habitat is the sand bars and islands in the Mississippi River. A major concern in recent years has been the loss of this habitat, much of which has been attributed to construction of the dike fields in the river. Thus a conflict has arisen between the need for dikes to provide flood control and navigation and the resulting loss of an endangered species habitat. This type of conflict can be handled only through close coordination between the Fish and Wildlife Service and the Corps. As a consequence, many dike plans have recently been significantly modified (sometimes resulting in higher construction cost) to minimize the loss of the Least Tern habitat.

Although the above examples are very brief and represent only a small sample of the many efforts being undertaken in the MR&T Project, they are presented to illustrate two important aspects of the new strategies for environmental management. The first is that to improve the environmental quality of a river system one must understand the physical system. Unfortunately, there are severe gaps in our knowledge in both the engineering and the environmental arenas. The above examples illustrate how the Corps, working with other groups such as the Fish and Wildlife Service, are attempting to advance the state of knowledge of these systems. The second point illustrated by these examples is the importance of coordination among the various interest groups. Without this type of coordination, agencies and diverse groups would be gridlocked.

Evolution of Management Strategies on the Tone River

Uzuka and Tomita (1993) documented the evolution of navigation and flood control on the Tone River. Flood control work on the Tone River began after establishment of the Tokugawa Shogunate in 1590. At that time new channels were constructed, diverting a distributary to Tokyo Bay, and the main flow was diverted to Choshi, on the Pacific side of the Boso Peninsula. The distributary to Tokyo Bay was named the Edo River. The goals of the original project included:

a. Prevention of flooding in the basin and irrigation for rice cultivation,
b. Flood control in the city of Edo,
c. Development of navigable channels,
d. Military defenses of Edo, and
e. Improvement of land routes.

Initially the new diversion channel was excavated to 5.4 m in width, was widened in 1654 to 12.6 m, in 1688 and 1703 to 50 m, and in 1809 to about 70 m. Navigation of the river system was a primary means of commerce until about 1900. Major river improvement work of that period stressed the building of ports at estuaries and channels for navigation. During the 1800s, civil engineering technology and expertise were introduced from the Netherlands and were used to implement integrated flood control systems. These techniques included river gauging and a common survey datum, fascine bank stabilization along the Edo River in 1875, systemwide planning for navigation in 1885, and dredging of the Tone River canal in 1890. However, some of the low-water channel work of 1890 was seriously damaged by floods in 1890, 1895, and 1896. The response to flood damage was the River Law enacted in 1896, which stipulated that the nation was responsible for river management. River work then turned to focus on flood control instead of providing navigation passage at low flows.

In 1900, the Tone River improvement plan was developed and construction was initiated. This plan has been revised four times as major floods affected the region. The overall plans include channel stabilization, reservoirs, floodway diversions, and levees. Table 14.1 provides design discharges for each of the plans. Figure 14.4 shows the distribution of the design flood discharge. Uzuka and Tomita (1993) explain the distribution of the design discharge. The design flood discharge at Yattajima is set at 16 km^3/s and the flow of small tributaries is 1 km^3/s. The

Table 14.1. *Tone River improvement plans*

Plan	Date	Design discharge (m^3/s)	Diversion discharge (m^3/s)
Meiji Improvement Plan	1901	3,750	970
Tone River Improvement Plan	1912	5,570	2,230
Tone River Expansion Plan	1939	10,000	5,800
Revised River Improvement Plan	1949	14,000	8,000
New Tone River Basic Plan	1980	17,000	9,000

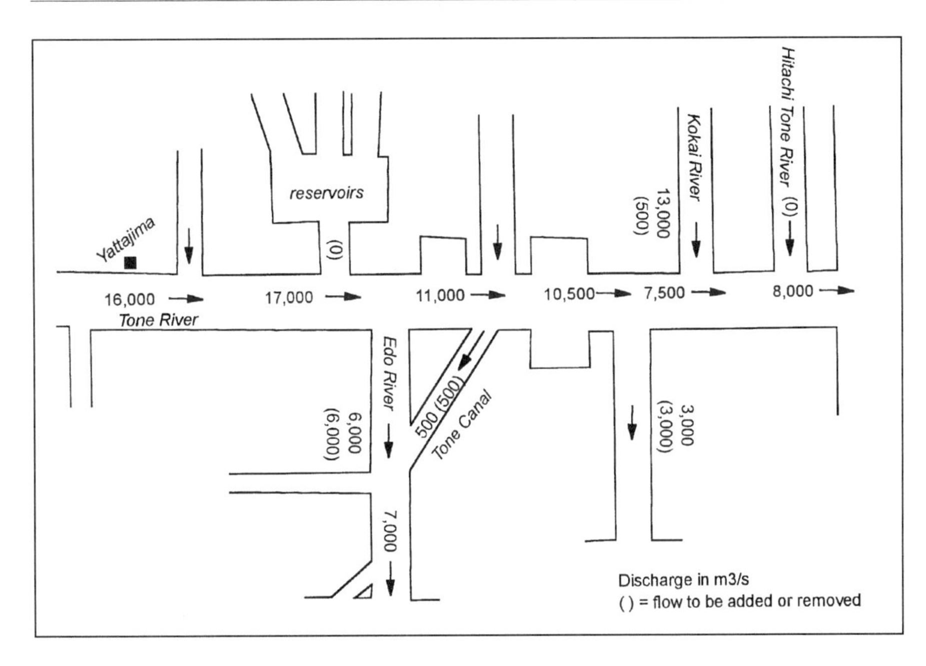

Figure 14.4. Schematic illustration of the structure of the Tone River design flood discharge (after Uzuka and Tomita, 1993, Figure 2.1.13).

Watarase Reservoir controls any additional discharge to the main stream of the Tone River, setting the maximum flood discharge at 17,000 m^3/s. At Sekiyado, 6,000 m^3/s are diverted by the Edo River, decreasing the Tone River to 11,000 m^3/s. Three reservoirs, the Tanaka, Sugou, and Inadoi, prevent the Kinu River from affecting the design discharge on the Tone. The Tone canal reduces the discharge to 10,500 m^3/s by diversion, and a balance of 2,500 m^3/s is diverted at other sites to result in a final discharge of 8,000 m^3/s.

The 1949 flood control plan for the Tone River was developed as a result of the September 1947 flood. Industrialization and rapid urbanization of the upper watershed of the river basin have resulted in an increased runoff, and this increase resulted in the 1980 Plan. In their 1993 paper, Uzuka and Tomita stated that additional improvements are recommended. They also pointed out that increasing the elevations of the existing levees is difficult because of heavy urbanization. Expected increases in discharge are being managed by maintaining levees and by dredging river beds to maintain conveyance.

Summary

The management strategies on the lower Mississippi River and the Tone River evolved over a period of years in response to public need. Both projects initially emphasized navigation, defense, and commerce. As population along the rivers increased, drawn by commerce and farming opportunities, the emphasis of the projects shifted to flood control. The plans for both rivers include the same features:

a. Levees to confine the flood waters,
b. Floodways to divert the flow from the main channel,
c. Reservoirs to control the flood magnitude, and
d. Channel stabilization to protect flood-control features along the river.

However, even with considerable infrastructure and resources, flood control on both river systems requires constant vigilance. The possibility of moving populations and commerce out of the floodplain (Gruntfest, Chapter 15, this volume) for either the Mississippi River or the Tone River is an option of limited viability. Therefore, society in general, and scientists and engineers in particular, are given the tasks of working together to identify and solve problems within a range of practical solutions.

It has been only since about 1970 that environmental considerations have begun to be incoporated into the planning process on the Mississippi River, and Uzuku and Tomita (1993) reported no environmental emphasis on the Tone River. Consequently, river engineering is in its infancy in terms of trying to establish effective ways to incorporate environmental concerns into water resource management decisions.

With a large percentage of the channel improvement features already completed on both rivers, this does not signal an end to the need for effective water resource management. On the contrary, increasing demands on the system from conflicting interest groups, growing regional conflicts, population growth and development, coastal wetland issues, identification of more endangered species, limited funding, and the possibility of future catastrophic floods or earthquakes are just a few of the complex issues that may demand the attention of water resource managers in the future. The long-term effects of comprehensive flood-control works on a natural river are uncertain (e.g., Friedman and Auble, Chapter 8, this volume; Wydoski and Wick, Chapter 9, this volume), and these effects require consideration. Many developed nations are shifting the emphasis in flood-hazard mitigation from solely structural measures to a combination of structural and nonstructural measures (Gruntfest, Chapter 15, this volume; Hamilton and Joaquin, Chapter 18, this volume). Engineers of the future will have to be proactive, flexible, and able to deal with short-term crises and to anticipate long-term demands.

References

Biedenharn, D.S. and Watson, C.C. (1997). Stage adjustment in the lower Mississippi River, USA. *Regulated Rivers: Research and Management*, **13**, 517–536.

Holmes, B.H. (1979). *A History of Federal Water Resources Program and Policies, 1961–1972*. Washington, DC: U.S. Government Printing Office, U.S.D.A. Miscellaneous Publication No. 1379.

Humphreys, A.A. and Abbot, H.L. (1861). *Report upon the Physics and Hydraulics of the Mississippi River.* Washington, DC: U.S. Army Corps of Engineers Professional Paper No. 13.

Hydrologic Engineering Center. (1986). *Planning for Hydrologic Engineers with Emphasis on Flood Control.* Davis, California: U.S. Army Corps of Engineers, Hydrologic Engineering Center.

Office of Federal Register. (1978). *Code of Federal Regulations.* Office of Federal Register, Washington, DC: National Archives and Records Service.

U.S. Army Corps of Engineers, Mississippi River Commission. (1959). *Comprehensive Review of the Mississippi River and Tributaries Project.* Vicksburg, Mississippi.

U.S. Army Corps of Engineers, Mississippi River Commission. (1972). *Improvement of the Lower Mississippi River and Tributaries* – 1932 to 1972. Vicksburg, *Mississippi.*

U.S. Army Corps of Engineers, Mississippi River Commission. (1986). *Planner Orientation for Water Resources.* Huntsville, Alabama: U.S. Army Corps of Engineers, Huntsville Division.

U.S. Army Corps of Engineers. (1991). *The History of the U.S. Army Corps of Engineers. U.S.* Vicksburg, Mississippi: Army Corps of Engineers, Pamphlet EP-360-1-22.

Uzuka, K. and Tomita, K. (1993). Flood control planning – case study of the Tone River. In *Research and Practice of Hydraulic Engineering in Japan*, ed. T. Nakano, no. 4. Special issue, *Journal of Hydroscience and Hydraulic Engineering*, Japan Society of Civil Engineering, pp. 5–22.

Winkley, B.R. (1971). Practical aspects of river regulation and control. In *River Mechanics*, edited & published by H. W. Shen, Ft Collins, Colorado, vol. 1, pp. 19-1–19-79.

CHAPTER 15

Nonstructural Mitigation of Flood Hazards

Eve Gruntfest
Department of Geography and Environmental Studies
University of Colorado

Introduction

Extent of Flood Vulnerability – Advantages of Floodplain Occupation

Rivers play a critical role in world history. They allow cities to develop. They serve as major transportation routes linking inland regions with the seas. River floodplains provide some of the most productive farmland. They serve as significant ecological systems and offer recreation opportunities.

Riverine floods are extremely expensive in terms of lost lives and lost property. However, floods also have benefits. They carry deposits of fertile alluvium. Floodplains have supported settlements along the Euphrates, Ganges, Indus, Nile, Tigris, and Yangtse Rivers for thousands of years (Alexander, 1993). Floodplain communities support river transport and support activities dependent on the river including fishing, barge traffic, and power-generating stations. They also have scenic value.

Goals for floodplain management can be conflicting. The flood loss reduction goal may contradict goals of real estate and industrial development or increased agricultural production. Floodplains are not always well managed. In the United States, although only about 7% of the country's total land area lies in floodplains, over 7 million structures and billions of dollars worth of community facilities and private property are vulnerable to floods (Owen, 1981).

In 1955, U.S. floodplains had 10 million occupants. Thirty years later the number doubled to 20 million, and by the mid-1990s about 12% of the national population lived in areas of periodic inundation. In fact, one-sixth of the nation's floodplains are urbanized and they contain more than 20,000 communities that are susceptible to flooding. Half of these communities have been developed since the early 1970s (Burby, 1985; Montz and Gruntfest, 1986; Alexander, 1993).

Flood Losses

Many of the people at-risk do not understand the potential consequences of the hazards they face. In the United States flood damages exceed $2 billion annually. Only 20–30% of eligible structures are insured against flooding. Federal and state disaster assistance accounts for most of the difference. In the United States almost two-thirds of the residential flood losses result from events that occur once every 1–10 years, even though the 100-year floodplain regulation is standard (Alexander, 1993).

In 1972, 215 people died in the Rapid City, South Dakota, flood. In 1976, 139 people were killed by a flash flood in Big Thompson Canyon, Colorado. In 1997, 12 tourists died in a flash flood in Antelope Canyon in Arizona. El Niño-related floods killed more than 900 people between 1997 and 1998 in Latin America (*Economist*, 1998).

Losses from the 1993 midwestern floods on the Mississippi and Missouri Rivers included 52 deaths. More than 56,000 homes and 2900 businesses were severely damaged and 8.5 million farm acres were affected. Transportation systems suffered more than $1.9 billion in losses. The overall event total was an estimated $18 billion (Chagnon, 1996). Structural damages exceeded expectations partially because of record fast flood flows and prolonged submersion (Chagnon, 1996).

Federal flood-control efforts in the Mississippi basin are credited with preventing nearly $20 billion in potential damages during the 1993 floods (Galloway, 1994). In spite of vast federal flood damage reduction efforts, people and property remain at risk to flooding (Galloway, 1994). Although loss of life from flooding in the United States is decreasing, property damages continue to rise.

In the United States floods tend to be repetitive phenomena. From 1972 to 1979, 1900 communities were declared disaster areas by the federal government more than once, 351 were inundated at least three times, 46 at least four times, and 4 at least five times. As of 1993, the United States was said to spend $9 billion on flood control a year and $300 million on flood forecasting (Alexander, 1993; Conrad, 1998).

Definitions of Structural and Nonstructural Measures

Adjustments to floods can be broadly classified into structural and nonstructural. Nonstructural approaches involve adjustment to human activity to accommodate the flood hazard (James, 1975), whereas structural methods are based on flood abatement or protection of human settlement and activities against the ravages of inundation (Watson and Biedenharn, Chapter 14, this volume). In Europe almost all measures that are taken have elements of combined structural and nonstructural measures. There has also been a move to be antistructural. For example, some dikes are being removed in favor of nonstructural or more environmentally sensitive techniques. Figure 15.1 shows the range of possible adjustments to flooding divided into structural and nonstructural categories (Smith and Ward, 1998).

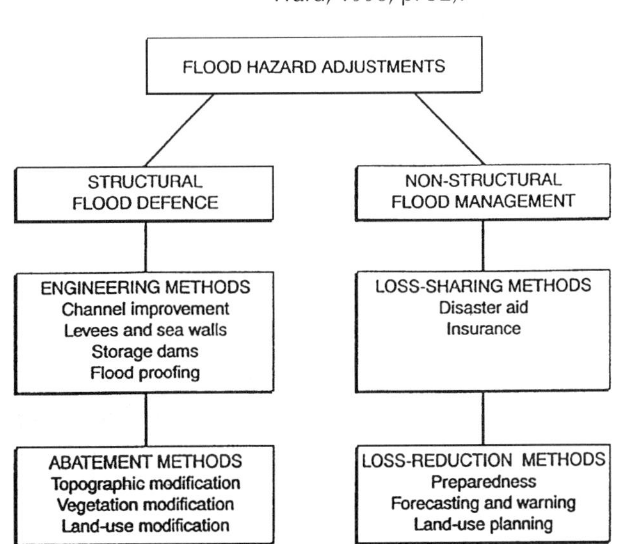

Figure 15.1. Range of flood hazard adjustments (Smith and Ward, 1998, p. 32).

Structural change involves modification to the built environment to minimize or eliminate flood damage directly or flood channel construction changes. Structural measures are expensive. They may give the illusion of security but the record shows otherwise (Alexander, 1993). The security can be temporary. A flood can occur that is bigger than the design of the channel or levee, and changing priorities in flood control projects that require higher reservoir levels for recreation or water supply can diminish the efficacy of structural measures (Williams, 1998). In the United States, the U.S. Army Corps of Engineers is primarily responsible for structural mitigation of flood hazards. Taken as a whole, Corps projects have returned more dollars than they cost.

After the 1927 Mississippi River floods, when the extensive levees built along the river partially collapsed, "the Corps did not abandon its dream of controlling all floods. Instead, it proposed building large dams upstream to reduce flood peaks to the capacity of the floodway between the levees" (Williams, 1998, p. 51). Two hundred people died in the 1927 floods, 700,000 people were displaced, and more than 135,000 buildings were damaged (Moore and Moore, 1989).

The failure of structural flood-control works poses a significant threat to the lives of the people who live downstream of a massive structural project such as a dam. More than 2000 people died in 1969 in Italy when the Vaiont Dam collapsed (Blaikie et al., 1994). Because of stringent engineering standards and a system of inspections, the United States has seen few major failures.

Nonstructural measures include floodproofing, land-use planning, soil-bioengineering, warning systems, preflood mitigation efforts, and insurance. The simplest nonstructural measure is to accept the loss. Another nonstructural measure is to provide postflood relief. Protection of floodplain residents and users and the supply of relief when they suffer damage, are forms of hidden subsidy (Alexander, 1993, p. 136). This category includes aid provided by the Red Cross, voluntary organizations, and governmental agencies.

Until the 1970s, most flood loss reduction efforts involved structural solutions (Laituri, Chapter 17, this volume). Although nonstructural measures were discussed as alternatives, they were rarely implemented. The shift from mostly structural to mixed structural/nonstructural measures began in the 1970s and it will continue into the next century. The mix of adjustments varies for each situation.

The rest of this chapter is divided into three parts. First, it discusses the major nonstructural measures and how they are being applied, principally in the United States. These measures include flood insurance and land-use management, acquisition and relocation, soil bioengineering, acquisition and relocation, floodproofing, preflood mitigation preparedness, and outdoor warning systems. Second, the chapter provides case studies of community experience with a range of nonstructural mitigation measures and flood case studies highlighting the danger of relying heavily on structural measures in California's central valley and the value of nonstructural measures in Fort Collins, Colorado. Finally, the chapter discusses the trends toward more adoption of nonstructural measures with recommendations for combined structural/nonstructural methods for most effective flood hazard reduction.

Discussion of Nonstructural Measures

Floodplain Mapping, Land-Use Ordinances, and Flood Insurance

In 1968 the U.S. National Flood Insurance Program (NFIP) was launched. It made affordable insurance available to residents in flood-prone areas. As of 1999 more than 18,000 communities belong to the program. Participating local governments require developers to meet minimum standards designed to avoid damages that might be inflicted by a catastrophic 100-year flood. The program also requires property owners to purchase flood insurance to receive a federally insured mortgage (Myers, 1996). Flood insurance is a means for placing some of the burden of losses onto the people who take (or make) the risk – namely, the floodplain users and residents (Alexander, 1993). Communities can participate in a community rating system, established by the Federal Emergency Management Agency (FEMA), that allows them to show innovative strategies to reduce flood losses in return for lower insurance premiums for floodplain residents.

Before a community can participate in the flood insurance program, the flood hazard must be recognized, assessed, and mapped. These assessments include flood history, cost and types of past flood damages, maps of the limits of the 100-year flood (or other design flood) on a topographic map, compilations of profiles and cross sections of the river to show the levels of past floods, and compilations of flood-frequency curves and locally representative hydrographs. A modeling study is normally essential in this process with models prepared for the situations at recorded times of earlier floods.

In the United States, FEMA works with state and community governments to identify their flood hazard areas and publishes a Flood Hazard Boundary Map of those areas. When a community joins the NFIP, it must require permits for all construction or other development in these areas and ensure that the construction materials and methods used will minimize flood damage. However, there is not careful monitoring to ensure that reducing flood hazard in a particular area does not increase flood potential elsewhere. Often, the problems are just shifted to different locales. In return the federal government makes subsidized flood insurance available to those whose structures were in the flood hazard area before the flood maps were issued. All others are eligible for flood insurance at actuarial rates. FEMA issues a flood insurance rate map after the flood insurance study of risk zones and elevations has been prepared (http://floodplain.org/Jan32.htm).

Acquisition and Relocation

One of the most promising strategies for reducing flood losses is public acquisition of land susceptible to flooding. The authorization for U.S. federal cost sharing for relocation is more than 30 years old. However, only recently have communities tired by chronic flooding taken advantage of funding packages and relocated. David Conrad's 1998 report for the National Wildlife Federation is the most comprehensive overview of the use of voluntary buyouts and relocations as a floodplain management option. Since the 1993 Mississippi River floods, nearly 20,000 properties in 36 states and one territory have been bought out (http://www.nwf.org/nwf/pubs/higherground/intro.html). Examples of acquisition projects include the following: (www.fema.gov/mit/ homsups.htm)

- In Hopkinsville, Kentucky, FEMA bought 23 houses flooded in 1997.
- In Bismarck, North Dakota, 391 substantially damaged properties are being acquired with 75% FEMA funding. Four local governments are engaged in acquiring flood-damaged homes.
- By 1998, 5 years after the catastrophic floods in the midwestern United States, more than 12,000 homes in the upper Mississippi Basin were scheduled for voluntary relocation and thousands of hectares of marginal farmland have been transferred by willing sellers or lessors to the state and federal government. The entire town of Valmeyer, Illinois, was relocated. The town had a long history of floods. In 1943, 1944, and 1947 unusually high levels of the Mississippi River caused flooding in the nearby bottomlands affecting Valmeyer. After 1947 floods the U.S. Army Corps of Engineers raised the levees protecting the reach of the floodplain to 14 m. On August l, 1993, the flood overtopped the levees. (Nationwide between 1993 and 1998 over 25,000 families moved from floodplains.)
- Other communities in Illinois also have had recent buyouts. In Montgomery 48 flood-prone homes will be purchased and the land will be left as open space. The village and homeowners contributed 25% of the project cost and FEMA contributed $4 million.
- The Corps of Engineers is also engaged in buyout plans. In Birmingham, Alabama, a $30 million flood damage reduction project is under way. Businesses and residences are being moved and segments of the floodplain are being acquired. Approximately 642 structures in six neighborhoods are being acquired and demolished. Eleven churches and three businesses were removed. Study for this project was completed in 1981 and it is finally being implemented nearly 20 years later. Each of these small efforts contributes to reducing the housing stock located in floodplains. However, urban developments elsewhere intensify overall vulnerability to floods.

Floodproofing

Floodproofing is a range of adjustments aimed at reducing flood damages to a structure or to the contents of buildings. There are three categories: 1) raising or moving the structure, 2) constructing barriers to stop floodwater from entering a building, and 3) wet flood proofing (U.S. Army Corps of Engineers, 1997). Figure 15.2 shows various methods of floodproofing. Figure 15.3 shows examples of floodproofing from Japanese home designs.

In March 1998 FEMA approved $1,602,600 for building elevation and has funded the cost share of $796,950 to Placer County, California. The 28 homes were flooded during 1983, 1986, 1995, and 1997. The homes will be elevated 0.6 m above the 100-year flood level (the elevation that has a 1% or better chance of being reached in any given year) (www.fema/govreg-ii/1998/batavia.htm).

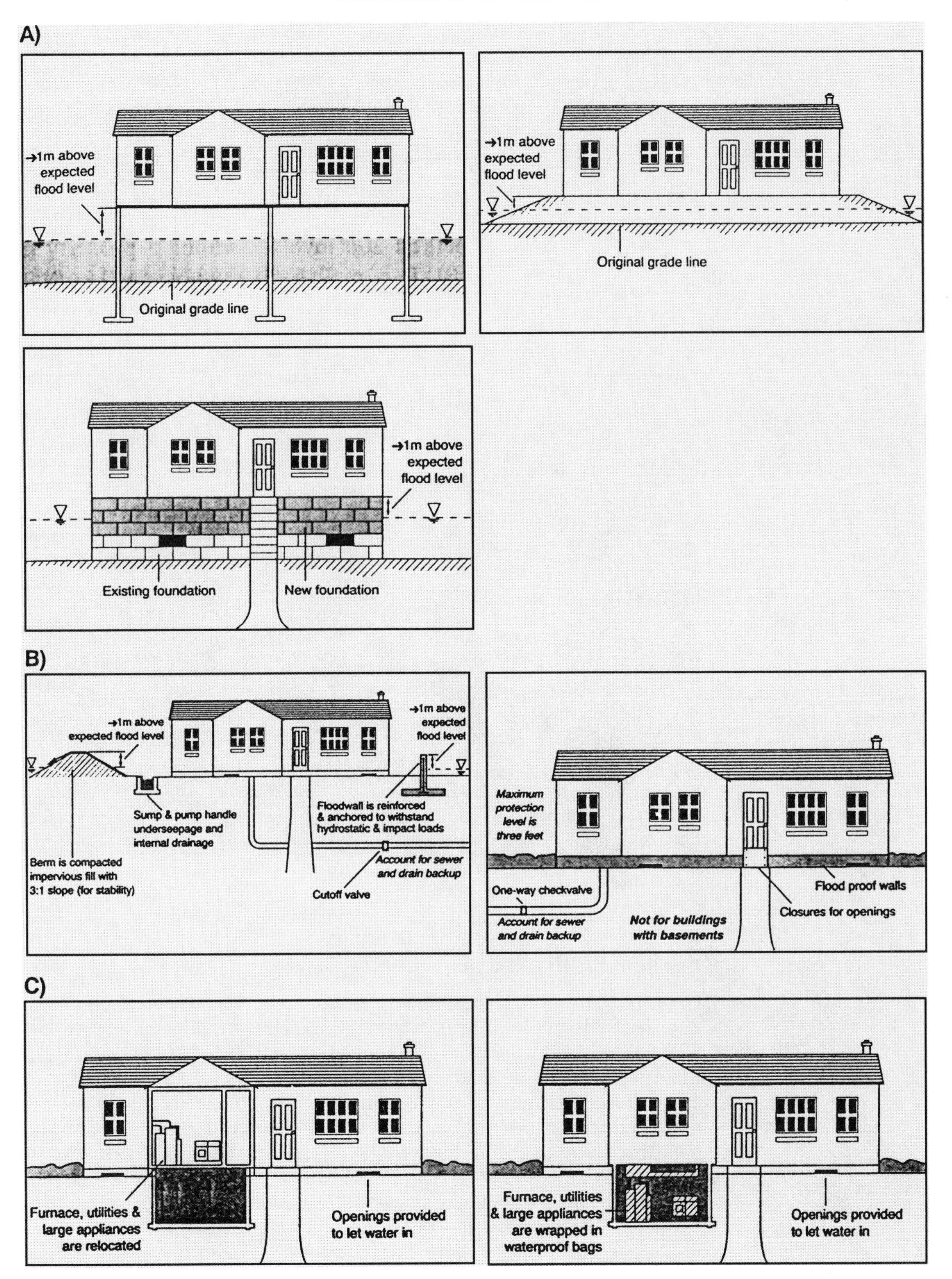

Figure 15.2. Examples of floodproofing techniques (Smith and Ward, 1998, p. 235).

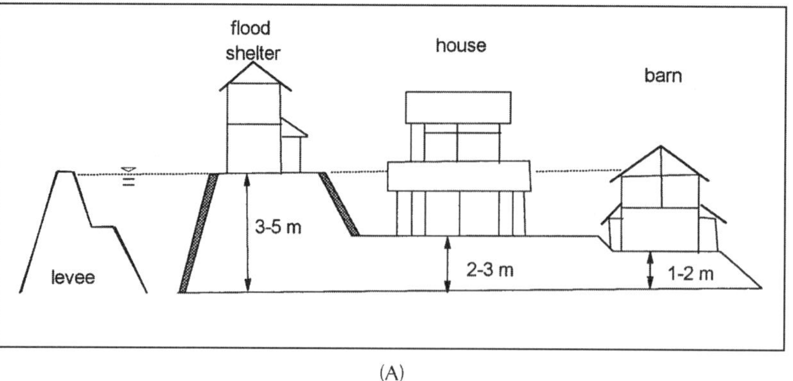

(A)

(B)

Figure 15.3. (A) Schematic drawing of higher elevation flood shelter traditionally used by residents along the Tone River, Japan. (B) Photograph of a complex of houses and barns next to a rice paddy near the river; the flood shelter rises above the other structures at left. (Photograph courtesy of Ellen Wohl)

Preflood Mitigation and Preparedness before the 1998 El Niño Storms in California

Oakland, California, committed $3 million to preventive measures. The city cleared hillside spillways of brush and debris to allow the stormwater channels to work. Sandbags were used so that Lake Merritt would not flood adjacent stores. Sandbags were distributed to residents – more than 110,000 were picked up from local fire stations. Residents agreed to keep drains clear of debris during heavy rains. The city issued rain slickers, hats, and rakes to the volunteers; critical areas subject to flooding were identified, and the city agencies worked hard to keep fallen trees from blocking drains (http://www.fema.gov/nwz98/eln0304.htm).

Many other examples of similar projects were carried out to reduce the El Niño storm damages throughout California. In addition many people bought flood insurance. The number of policies statewide increased from 264,914 to 365,000 by the end of 1997.

Flood loss reports reinforce the emphasis on structural measures by not fully taking into account the savings that nonstructural measures bring. In the 1998 storms in California, flood damage during this El Niño year was no worse than in 1997, a non-El Niño year. However, the figures do not recognize massive efforts to reduce vulnerability by clearing channels of debris and brush. Nonstructural measures adopted after the predictions of serious flooding from El Niño but before the storms began did reduce vulnerability. Accurate records on the extent to which these efforts reduced flood losses are not available. The public perception continues that these El Niño-related storms were actually less serious than those in previous years (California Department of Water Resources, 1997).

(C)

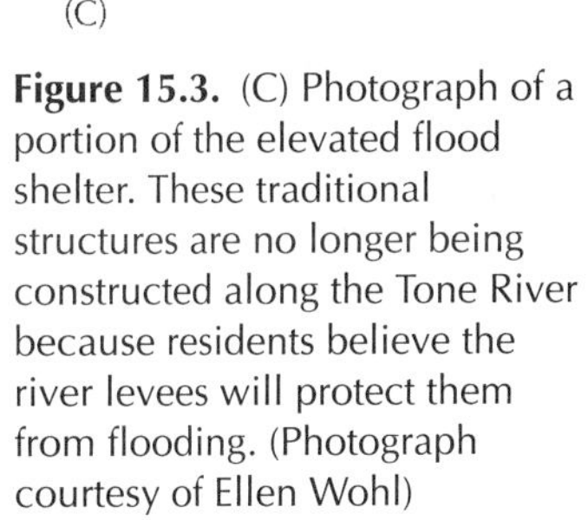

Figure 15.3. (C) Photograph of a portion of the elevated flood shelter. These traditional structures are no longer being constructed along the Tone River because residents believe the river levees will protect them from flooding. (Photograph courtesy of Ellen Wohl)

Flood-Warning Systems

New technological advances in stream and rain gauge networks and the increased regional floodplain management efforts have led to adoption of thousands of local flood-warning systems. Many are simple detection systems and do not provide any mechanism for alerting the population at-risk. Until the 1990s warning or detection systems were planned and administered primarily at the local level.

Since then, the federal government, including the Bureau of Reclamation, the U.S. Army Corps of Engineers, the National Oceanic and Atmospheric Administration, and FEMA have actively participated in installation and maintenance of detection and warning systems. Many systems are still managed by regional or local entities, but the percentage of federal dollars has increased substantially. Standards have also been established to help make the systems more compatible across regions (U.S. Department of Commerce, 1997).

An automated integrated network of stream and rain gauges is being used in more than 1000 communities in the United States to help provide lead time for floods. Most of the systems are developed through collaborative efforts of many agencies. These ALERT systems (automated local evaluation in real time) have performed many functions other than flood warning, including helping in water supply decision making, fire weather forecasting, pollution monitoring, and providing data for river recreationists (Gruntfest and Huber, 1991). The availability of real-time data on the Internet also has increased interest in these monitoring systems (Gruntfest and Weber, 1998). Many communities have raw data available so residents can monitor the rivers and rainfall upstream from where they live. Raw data are rarely of much use. A graphical user interface is usually essential to communicating the situation: this is easily done on the Internet, but it requires a backup center.

South central Kentucky received FEMA funding to help install a severe weather warning system consisting of sirens (http://www.fema.gov/reg-iv/r4mit47/htm). The state of Arizona is developing a network for flood warning throughout the state. More than 30 agencies and communities are working together on the comprehensive ALERT system (http://nimbo.wrh.noaa.gov/Alert/).

Warning systems may be nothing more than "cheap payoffs of the raingods." Too often communities install raingauge/streamgauge monitoring systems without a plan for disseminating the warning message. A warning system is necessary only after poor land use decisions have been made, allowing people to settle in harm's way. Many of the systems being built are not being maintained adequately enough to be reliable (Gruntfest and Huber, 1991; Parker and Fordham, 1996). Public education encouraging people to heed environmental cues is also being used. It is particularly difficult to provide adequate lead times for flash floods. Signs (Figure 15.4) aim to give people in Colorado's Front Range canyons a suggestion for what to do if they perceive flash flooding. Some communities have drills to test the reliability and completeness of their systems to be sure the systems will operate when conditions warrant.

In 1999 a combination of factors increased the likelihood that automated detection systems may become more popular and more valuable. More powerful, less expensive computers, and World Wide Web access provided opportunities for inexpensive real-time weather data. Although real-time stream and rain gauge networks may be originally installed for flash flood forecasting, many agencies and users find the data useful for other purposes. For example, river recreationalists are active consumers of U.S.

Figure 15.4. Warning sign at the entrance to Big Thompson Canyon in the Front Range of Colorado (USA). (Elevated pipe in the background is part of a water diversion system.)

Geological Survey stream gauge data (Gruntfest and Waterincks, 1998). This trend may increase awareness and dissemination of real-time data.

Soil Bioengineering

Anchored plantings along streambanks serve as the basis for this technique. Soil bioengineering and biotechnical engineering are cost-effective and environmentally compatible ways to protect slopes against surficial erosion and shallow mass movement. These approaches provide alternatives to structural channel improvements. In fact, they raise questions about the notion of why engineers ever considered that concrete-lined channels should be considered improved (Gray and Sotir, 1996). Generally, bioengineering solutions also must include a strategy to convey floodwaters downstream.

The bioengineering technique is gaining support throughout the United States and Europe. It is less expensive to install and less expensive to maintain as well. The broader adoption of soil bioengineering may radically alter floodplain management. The approach is still considered new and it is not well established in civil engineering. However, greater knowledge of the effectiveness of the technique will further diffuse its application. Soil bioengineering combined with other structural and nonstructural measures holds great promise for meeting the sustainability goals and improving the beauty of floodplains.

Case Studies

The next sections highlight some innovative examples of the use of structural and nonstructural measures and how recent floods have shown the value of the integrated approach.

Urban Drainage and Flood-Control District, Metropolitan Denver, Colorado

Since 1969 when the Urban Drainage and Flood-Control District was established by the Colorado legislature, the District has assisted local government in the metropolitan area with multijurisdictional drainage and flood-control problems. The District covers an area of 4100 km^2 and includes Denver, parts of the five surrounding counties, and all or parts of 33 incorporated cities and towns. The major drainageways are defined as draining at least 405 hectares. As of 1998 about 2 million people live in the District (http://www.udfcd.org/activity.html). The District is funded by property tax mill levies. The District employs 18 full-time employees. The District has five programs: master planning, design and construction, maintenance, floodplain management, and managing the South Platte River.

The District has been involved in both structural and nonstructural measures to reduce flood losses in the 29 communities that make up Denver's metropolitan area. The nonstructural measures the District performs include an integrated warning system and multiobjective greenway projects that offer a range of recreation opportunities.

The District's warning system includes extensive ALERT systems of gauges. It also incorporates the skills of a private meteorologist and several other private companies that provide software to help identify flood threats. The software providers give the District the opportunity to quickly interpret meteorological and hydrologic data. The Denver District has an international reputation for innovative flood mitigation strategies. The waterways of Denver have changed from eyesores in the 1970s to being prime recreational corridors.

Pima County, Arizona, Flood Control District

The District protects and tries to expand the protected riparian habitats. This helps to minimize flood threats to floodplain land upstream and downstream of development and to protect the natural resource values of floodplain areas. The riparian ecosystems also recharge valuable groundwater resources. The District enforces floodplain regulations that are included in the county's Floodplain and Erosion Hazard Management Ordinance. The county's webpage provides information for residents interested in learning about local flood hazards and what can be done to reduce losses and provides a useful model for other communities to follow (http://www.dot.pima.gov/flood/fpm/index.htm).

Floodproofing Efforts in Bangladesh

In Bangladesh, there are many benefits as well as costs associated with floods. Farmland replenishment with water and nutrients is essential for agriculture. Therefore major structural flood protection measures that eliminate floods are not necessarily the most valuable approach to flood hazard mitigation.

During serious flooding in 1988 many families evacuated their homes. Some residents floodproofed by raising the floor level of their homes. Floor raising is viable if houses do not have to be relocated more than once every 4 years (Thompson and Tod, 1997). Community flood shelters also appear to be economically viable if buildings are used and funded as schools or other community facilities and if the site will be secure from erosion for at least four years. Along the Brahmaputra–Jamuna efforts are also under way to improve flood warnings, improve access to boats in emergencies, and improve access to postflood loans to flood-affected citizens. In the charlands of the three main rivers, the Jamuna–Brahmaputra, Ganges–Padma, and Meghna, the estimated population is over 4 million people (Thompson and Tod, 1997; Paul, 1997).

Floodplain Management in Argentina

Argentina suffers from increasingly frequent, severe floods. The country is trying to shift to long-term sustainable responses (Penning-Rowsell, 1996). As has been mentioned in other locales, officials will resist zoning that delimits areas in which development is prohibited or scaled back. Flood losses in Argentina will increase because the population is concentrated in flood-prone urban areas, a consequence of rural depopulation and unregulated or weakly regulated land use (Penning-Rowsell, 1996).

Argentina is developing flood hazard mitigation plans that incorporate structural and nonstructural measures. Argentina has been working with help from the World Bank to reduce vulnerability. The Bank provides loans and emphasizes the need for an integrated, sustainable approach to floodplain management. The goal is to "shift away from a reliance on engineered solutions toward land-use controls on floodplain development" (Penning-Rowsell, 1996:72).

In the Parana–Paraguay River corridor there has been substantial geomorphologic and hydrologic change and urbanization has intensified flooding. In light of the major changes in the dimension and location of the flood hazards, structural flood solutions, such as the ring embankments that function as an encircling levee surrounding main urban areas, may not work (Penning-Rowsell, 1996). The floods may rise above these dikes and overtopping would cause catastrophic flooding (Penning-Rowsell, 1996).

Structural Measures Intensify Impacts of the 1997 Northern California Floods

Before the 1997 northern California floods, communities in California's Central Valley felt safe building and investing in the floodplain. For 60 years the federal government invested billions of dollars in constructing huge multipurpose dams on the rivers flowing in the Central Valley. This region has immense evaporation losses. "To control floods, the reservoirs were designed to be drawn down in the winter to capture the peak of the flood and to keep flows downstream within their floodway capacities" (Williams, 1998:51). However, over the years priorities changed and flood control was only one of several priorities for dam function.

As an example, after the New Don Pedro Dam was authorized in 1944, politicians argued over the uses of the dam. The struggles over power generation, irrigation, and flood control left flood control as the lowest priority. Because the reservoir was so full (any water released was considered wasted because of the other demands for water), the 60-year flood spilled over the reservoir and the flood wave broke many levees. More than 600 homes were destroyed or damaged in the 1997 floods. Williams (1998) points out the difficult position of dam operators who must try to balance all the interests in the multipurpose dams.

Preflood Nonstructural Measures Prevent Losses in Fort Collins, Colorado

On July 27, 1997 more than 35 cm of rain fell in southwestern Fort Collins. The flash flood killed five people and destroyed a great deal of property (Water Center, 1998). Nonstructural measures implemented before the 1997 flood reduced vulnerability in Fort Collins, Colorado. The progressive stormwater management agency had carried out recommendations made in the 1989 Spring Creek master plan and had removed structures from the floodplain. The city estimates that it saved up to 98 lives by removing 86 structures from the floodplain. These included a retirement home, nine houses, a mobile home park, and one business from the Spring Creek floodplain. More than $5 million had been spent since 1988 to reduce vulnerability. In addition to removing the structures, drainage and bridge improvements were made, and a railroad embankment was reinforced (Grimm, 1998). Based on the estimated project cost of $5 million, the benefit cost ratio for predisaster mitigation is between 1.67 and 2.91 not including lives saved (Grimm, 1998).

Other nonstructural measures are in place whose benefits are more difficult to quantify including a flood awareness week, a mass mailing on local flood hazards to all residents living in or near the floodplain, and a comprehensive open space planning preservation plan. The city's master plans and floodplain regulations exceed minimum NFIP criteria. Fort Collins has been named by FEMA as a Project Impact community. The Project Impact designation indicates that FEMA considers Fort Collins a city worthy of financial support for its exemplary preflood mitigation efforts. Further innovations are expected as the city develops a warning system and tries to integrate flood detection information with its Geographic Information System. The city's web page offers a wide range of hazard mitigation information. (http://www.ci.fort-collins.co.us/csafety/oem/index.htm). This page can serve as a model for other communities that are considering web page development.

Combined Structural and Nonstructural Measures to Reduce Flood Losses

From the first attempts to reduce flood losses in the United States, structural measures were preferred for three main reasons: 1) their benefits appeared to be relatively easy to measure, 2) they did not require extensive and politically controversial land-use planning, and 3) the federal cost-sharing agreements encouraged communities to select the most expensive engineering projects. These reasons were supported by a faith in the technology of structural measures to protect people and property from floods. However, the record now shows that, in spite of massive expenditures, flood losses have continued to increase. Further discussion of the philosophical contexts can be found in Laituri, Chapter 17, this volume.

Since the 1960s, especially in the United States, there has been a call for a shift from primarily structural measures to control floods to nonstructural measures. Throughout the 1970s there was optimism that the U.S. Water Resources Council and its publications would significantly shift flood control from nearly entirely structural measure reliance to a consideration of structural and nonstructural measure combinations. The Council offered the possibility of comprehensive floodplain management. The Council's publications by Owen (1981) and Owen and Wall (1981) still stand as excellent overviews of the range of flood loss reduction possibilities. However, during President Reagan's assault on the federal budget, particularly the environmental agencies, the Council lost funding and was effectively put out of business in the 1980s.

Land-use control is one of the most effective ways of reducing flood hazards. Statutes, ordinances, regulations, and compulsory purchases can be used and relocation can be subsidized. A floodway left undeveloped through the city can become beautiful public open space.

Not only is there a conflict in goals among agencies, but there is conflict within the lead emergency agency, FEMA. NFIP does take in more money than it pays out annually. But this conclusion ignores the added money that the federal government pays out in disaster relief.

Engineering solutions are still prevalent but the move toward increased emphasis on nonstructural measures continues. Williams (1998), a civil engineer, believes that flood management rather than exclusive reliance on flood control will reduce flood hazards significantly. If flood management strategies are made up of structural measures and also of land-use controls, insurance, floodproofing, and emergency preparedness, we are more likely to see a reduction in damages (Williams, 1998). One weakness in the United States is that there is no particular agency with the mission to implement a national flood management policy. Williams recommends an independent audit of the effectiveness of our existing flood-control systems.

In 1989 the Chief of the U.S. Army Corps of Engineers, H. J. Hatch, said that the environment "is the most significant engineering issue of the next decade" and that "engineers must look at (their) work in a broad social and environmental context." He embraced the idea of "sustainable" development as an ethic for guiding the Corps work. He even suggested renaming the Corps. However, their traditional sense of "wise use" still determines most efforts nearly 10 years later (Riley, 1998).

The United States is coming to appreciate the full significance of the fragile ecosystems that border rivers. The widespread loss of riverine habitat indicates that the nation faces dire ecological consequences (Wohl, Chapter 4, this volume). When development takes place in floodplains, when river channels are straightened, and when locks and dams are built, wetlands and aquatic habitats are eliminated and species are lost (Galloway, 1994).

Does increased federal interest in warning systems result from recognition of the efficacy of warning systems? No, it results primarily from the low costs involved in detection systems relative to the structural alternatives. Also, cost-effectiveness of warning systems can be readily demonstrated. Methodologies are being tested to evaluate how much reliability a warning system might have, but the results are not yet conclusive enough to abandon structural measures in favor of warning systems (Krzysztofowicz et al., 1994).

Even after the 1993 Mississippi and Missouri River floods, the Galloway report (1994) recommended that the division of decision and cost-sharing responsibilities among federal, state, and local governments be more clearly defined and that there be a national strategy of avoiding inappropriate use of the floodplain, minimizing vulnerability to damage through both nonstructural and structural means, and mitigating damages as they occur (Galloway, 1994). The report argued for full consideration of the economic, social, and environmental costs and benefits for all future floodplain management activity. Many of the nonstructural recommendations in this report were considered extremely controversial. The recommendations have not been implemented in any cohesive ways.

In 1998 the U.S. Army Corps of Engineers developed its Challenge 21 to focus on more sustainable approaches. "Through its focus on nonstructural alternatives to flood protection, it will move families and businesses out of harm's way and strive to return the floodplains of rivers and creeks to a condition where they can naturally moderate floods as well as provide other benefits to communities and the environment." Federal and local governments will share the costs. The federal share is 50% for the cost of the studies and 65% for project implementation. The annual federal allocation for selected communities will be limited to $10,000,000 (U.S. Army Corps of Engineers, press release, 1998). Maybe the new program will make a difference.

Other countries besides the United States are also exploring means of combining structural and nonstructural flood hazard mitigation measures. For example, the Danish Hydraulic Insitute uses real-time data and Geographic Information System applications to develop "hydroinfomatics" for more effective information dissemination, modeling, and floodplain management (Danish Hydraulic Institute, 1994, 1998). In addition to dams and structural measures, one television channel in Norway is devoted to daily flood forecasting at a regional scale, with the forecasts based on rainfall runoff, river routing, and reservoir operation and routing models (Killingtveit, 1997). Along the Po River basin of Italy, levee systems are supplemented by the use of multisensor data for accurate regional forecasting of meteorological patterns and geomorphic mapping of the spatial distribution of hazards from the resulting floods, leading to different levels of flood warning throughout the basin (Marchi et al., 1997).

Conclusion

At the millennium, efforts by the World Bank, the U.S. Army Corps of Engineers under Challenge 21, and FEMA as part of Project Impact are aimed at more sustainable approaches to reducing flood losses. These initiatives require bold cooperation among agencies and thoughtful creative management of structural and nonstructural measures. Although the rhetoric has changed toward placing more value on nonstructural alternatives, legislative changes will have to take place before a serious shift in policy occurs (Riley, 1998). After the disastrous 1997 floods in Grand Forks, North Dakota, the Corps of Engineers is going ahead with an enormous $250 million structural flood-control project (Findlay, 1998).

Finally, the federal government also must integrate its disaster policy and flood damage reduction policy. Larry Larson, executive director of the Association of State Floodplain Managers, states "We can expect little change in the way that the federal government responds to flood disasters until taxpayers demand that states and communities take steps to provide additional protection to ward off these likely disasters. Until it costs state and local governments more to do nothing than to prevent flood disasters, floodplain managers will continue to struggle to maintain effective programs" (1996, p. 104). A similar dilemma holds outside the United States.

References

Alexander, D. (1993). *Natural disasters.* New York: Chapman and Hall, 632 pp.

Blaikie, P., Cannon, T., Davis, I., and Wisner, B. (1994). *At risk natural hazards, people's vulnerability and disasters.* New York: Routledge, 282 pp.

Burby, R.J. (1985). *Flood plain land use management: a national assessment.* Boulder, CO: Westview Press.

California Department of Water Resources. (1997). *Final report of the flood emergency action team.* Sacramento, CA.

Chagnon, S.A. (1996). *The great flood of 1993 causes, impacts and responses.* Boulder, CO: Westview Press, 321 pp.

Conrad, D. (1998). Higher ground: voluntary property

buyouts. In *The nation's floodplains, a common ground solution serving people at risk, taxpayers and the environment.* National Wildlife Federation, Washington, DC.

Danish Hydraulic Institute. (1998). *MIKE 11 GIS – A flood management system for rivers and floodplains.* version 2.1. May.

Danish Hydraulic Institute. (1994). *Mike 11 FF short description. Real-time flood forecasting modeling.*

Economist. (1998). The season of El Niño: 347, no. 8067, May 9, pp. 35–38.

Findlay, L.R. (1998). Soaked, burned and salvaged, Grand Forks, North Dakota rebuilds after its epic flood and fire. *Architecture*, **87**, 49.

Galloway, G. (1994). *A blueprint for change, sharing the challenge: floodplain management into the 21st century.* Report of the Interagency Floodplain Management Review Committee to the Administration Floodplain Management Task Force, Washington, DC.

Gray, D.H. and Sotir, R.B. (1996). *Biotechnical and soil bioengineering slope stabilization a practical guide for erosion control.* New York: Wiley, 378 pp.

Grimm, M. (1998). Floodplain management. *Civil Engineering*, March, pp. 62–64.

Gruntfest, E. and Huber, C.J. (1991). Toward a comprehensive national assessment of flash flooding in the United States. *Episodes*, **14**, 26–34.

Gruntfest, E. and Waterincks, P. (1998). *Beyond flood detection – alternative applications of real-time data.* U.S. Bureau of Reclamation http://web.uccs.edu/geogenvs/work/Eve/Beyond%20Flood%20Detection%20Final.html.

Gruntfest, E. and Weber, M. (1998). Internet and emergency management prospects for the future. *International Journal of Mass Emergencies and Disasters*, **16**, 1.

James, L.D. (1975). Formulation of nonstructural flood control programs. *Water Resources Bulletin*, **11**, 668–705.

Killingtveit, A. (1997). Flood regimes and flood prevention in Norway: lessons learnt from the 1995 flood. In *International seminar on recent trends of floods and their preventive measures*, Hokkaido River Disaster Prevention Research Center, Hokkaido, Japan, pp. 1–19.

Krzysztofowicz, R., Kelley, K.S., and Long, D. (1994). Reliability of flood warning systems. *Water Resources Planning and Management*, **120**, 906–926.

Larson, L. (1996). Lessons drawn from the 1993 flood. *Forum for Applied Research and Public Policy*, fall, no. 3, 102–104.

Marchi, E., Roth, G., and Siccardi, F. (1997). Flood control and forecasting in the Po River basin. In *International seminar on recent trends of floods and their preventive measures*, Hokkaido River Disaster Prevention Research Center, Hokkaido, Japan, pp. 73–86.

Montz, B.E. and Gruntfest, E. (1986). Changes in American urban floodplain occupancy since 1958: the experiences of nine cities. *Applied Geography*, **6**, 325–338.

Moore, J.W. and Moore, D.P. (1989). *The Army Corps of Engineers and the evolution of federal floodplain management policy.* Boulder, CO: Institute of Behavioral Science, special publication no. 30.

Myers, M.F. (1996). Midwest floods channel reforms. *Forum for Applied Research and Public Policy*, **11**, 88–97.

Owen, H.J. (1981). *Cooperative flood loss reduction a technical manual for communities and industry.* Washington, DC: U.S. Water Resources Council, and others, 105 pp.

Owen, H.J. and Wall, G.R. (1981). *Floodplain management handbook.* Washington, DC: U.S. Water Resources Council, 69 pp.

Parker, D. and Fordham, M. (1996). Evaluation of flood forecasting, warning and response systems in the European Union. *Water Resources Management*, **10**, 279–302.

Paul, B.K. (1997). Flood research in Bangladesh in retrospect and prospect: a review. *Geoforum*, **28**, 121–131.

Penning-Rowsell, E.C. (1996). Flood hazard response in Argentina. *The Geographical Review*, **86**, 72–90.

Riley, A.L. (1998). *Restoring streams in cities a guide for planners, policymakers, and citizens.* Washington, DC: Island Press, 423 pp.

Smith, K. and Ward, R.C. (1998). *Floods: physical processes and human impacts.* New York: John Wiley.

Thompson, P. and Tod, I. (1997). Mitigating losses in Bangladesh's active floodplains. In *Proceedings of a Conference*, eds. G.M. Housner and R. Chung, pp. 213–224. American Society of Civil Engineers, New York, NY.

U.S. Army Corps of Engineers. (1997). *Flood proofing techniques, programs and references.* Available from Corps at CECW-PF, 20 Massachusetts Avenue, NW, Washington, DC, 26 pp.

U.S. Department of Commerce. (1997). *Automated local flood warning systems handbook.* Weather Service Hydrology Handbook no. 2, February, Silver Spring, MD: National Oceanic and Atmospheric Administration, National Weather Service, Office of Hydrology, unpaged.

Water Center, Colorado State University. (1998). *Fort Collins Flood 1997, assessing the July 28, 1997 extreme event that hit Fort Collins and Colorado State University.* Conference Proceedings, November 6, 257 pp.

Williams, P. (1998). Inviting trouble downstream. *Civil Engineering*, February, 50–53.

Web Pages Related to Nonstructural Measures

Flood warnings – integrated network of stream and rain gauges.
http://nimbo.wrh.noaa.gov/Alert/ or http://www.io.com/~rooke/alert/
http://floodplain.org/p=basics.htm
http://FEMA.gov – U.S. Federal Emergency Management Agency
http://ceres.ca.ca.gov/ – State of California Water Resources Agency
http://www.usace.army.mil/inet/functions/cw/cwfpms/fpms.htm – U.S. Army Corps of Engineers with emphasis on floodplain management activities
http://www.nwf.org/nwf/pubs/higherground/intro.html – National Wildlife Federation site for manuscript Higher Ground
http://www.udfcd.org/activity.html – Urban Drainage and Flood Control District, Denver, CO
http://member.aol.com/damsafety/homepage.htm – Association of State Dam Safety officials
(http://www.fema/govreg-ii/1998/batavia.htm) – Source on innovative mitigation
http://www.ci.fort-collins.co.us/csafety/oem/index.htm – Comprehensive emergency preparedness homepage from Ft. Collins, CO: an excellent reference.

CHAPTER 16

Planning for Flow Requirements to Sustain Stream Biota

Clair B. Stalnaker
Midcontinent Ecological Science Center
U.S. Geological Survey

Edmund J. Wick
TETRA TECH, Inc.

When planning flow requirements necessary to sustain native stream biota, historic patterns of annual flow magnitude and interannual variability, including floods, should be examined closely to determine their potential role in life history and habitat requisites. Each stream's biota is uniquely adapted to habitat created by fluvial geomorphic processes associated with the geologic setting and climate of a region. When watershed habitat and flows are altered by human development, complex changes often occur in river physical processes that disrupt life history strategies. Because of the complexity associated with multiplicity of change in sediment transport, disruption of energy and nutrient inputs, introduction of nonnative biota, and blocked access to critical habitat, a high degree of uncertainty surrounds flow management decisions that will adequately restore needed physical function of the stream and maintain self-sustaining populations of native biota. This chapter discusses case studies on the Trinity River, California, and Green River, Utah, that are examples of rivers to which floods were important to river process and native biota. Flow planners for these rivers have struggled to meet flood flow needs and flow timing that could improve recruitment of native biota. We offer a logical framework for the appropriate use of simulation modeling for exploration and experimentation in adaptive flow management of streams. Adaptive environmental assessment and management (AEAM) is presented as a formal systematic and rigorous approach that emphasizes learning from past experience, data analysis, and experimentation, along with the operational flexibility to respond to monitoring data, research findings, and varying resource and environmental conditions. Developing and effectively applying appropriate models within a functional framework such as AEAM is suggested as a key to moving management agencies toward managing rivers to sustain native riverine biota.

Rivers Must Be Managed as Integrated Systems of Dynamic Physical, Chemical, and Biological Processes

Impacts on stream biota due to human perturbation are often noticed only after severe declines in species abundances have already occurred. Unfortunately, much of the dam construction and river modification that have affected river systems throughout the world were done without adequate understanding of their potential impacts on native biota. Efforts should be made to study relatively unaltered systems in advance of future development and to put in place adequate physical and biological monitoring programs. Once river systems have been ecologically altered by a combination of changed flow magnitude or pattern, modified sediment transport, blocked migratory routes, and disruption of energy and nutrient inputs, the complexity of determining cause-and-effect relationships rises to a point that uncertainty prevails about designs for remedial action. In other words, it is easier to preserve a healthy river system than to rehabilitate a degraded one.

When protection or rehabilitation flows for river systems are being planned, the localized adaptations of stream biota are often overlooked. In the process of evolution, genetic material of local stream biota is molded over time to best adapt to the available river habitat niches provided by the prevailing geologic setting and climate of the region (Friedman and Auble, Chapter 8, this volume; Wydoski and Wick, Chapter 9, this volume). For example, through the process of evolution, spawning concentrations of stream fishes develop where the physical geologic setting best provides for egg incubation and nursery habitat leading to highest survival through the juvenile stage. In addition, imprinted information obtained by stream biota during development and used during migratory reproductive behavior, home range determination, and feeding behavior plays a critical role in maintaining self-sustaining aquatic populations. Imprinted cues that initiate migratory spawning behavior can originate from odors imparted by local springs or tributaries and dissolved minerals and native vegetation.

The river continuum concept has been developed to assist in understanding trophic production and nutrient cycling in the longitudinal direction. This concept is often used in describing healthy river systems (Vannote et al., 1980). What are often overlooked, however, are the lateral contributions to river productivity provided by seasonal connectivity to the river floodplain.

The flood pulse spawning and rearing strategy by which fishes seasonally invade the floodplain have been reported for fishes of both tropical and temperate regions (Crance, 1988; Junk et al., 1989). The flood pulse and floodplain connectivity are essential components of healthy alluvial river ecosystems (Bayley, 1991; Petts and Maddock, 1994). Energy dynamics and food production are strongly dependent on floodplain habitats in which shallow littoral zones rich in organic matter yield much higher densities of zooplankton than habitats in river channels (Hynes, 1970; Welcomme, 1985, 1989). If sustained populations of locally adapted stream biota are an objective of flow management in a river system, those physical processes that are critical in creating the habitat conditions necessary for successful life functions of the native biota must be identified and managed. Reproductive bottlenecks created by human alteration of the system must be identified, and plans for remedial action must be developed, tested, and implemented. Resources must be carefully allocated and balanced to achieve integrated physical and biological studies of riverine systems (Ligon et al., 1995).

Often changes in stream biota are complicated because of introductions of nonnative fishes and vegetation, placing further uncertainty in defining remedial actions. Better understanding of geomorphological changes has been suggested as the key to understanding long-term ecological consequences of dams and other stream disturbances (Ligon et al., 1995). Therefore, an interdisciplinary and integrated science approach that maintains close linkage between biological and physical studies is mandatory.

Although concern for single aquatic species often drives river management plans, care should be taken to consider maintaining ecological integrity through the entire river system. River channel narrowing or widening may have profound effects on critical habitat areas for some species. Alteration of rivers likely results in varied responses longitudinally, and areas that can be rehabilitated are often limited based on flow and sediment dynamics.

Although restoration of regulated river systems to pristine conditions is not practical, we agree with Gore and Shields (1995) that "there is considerable potential for rehabilitation, that is, the partial restoration of riverine habitats . . . renewal of physical and biological interactions between the main channel, backwaters, and floodplains is central to the rehabilitation of large rivers."

Simulation modeling linking physical and biological concepts is proposed as the appropriate integrative science mechanism for evaluating river reaches for the purpose of exploring the interactive processes and likely habitat states achieved under alternative schemes for rehabilitating impacted reaches through flow and sediment management. Areas that are critical for reproduction and growth should be identified and evaluated in terms of the physical processes that drive the sediment dynamics that create the physical habitat characteristics needed throughout spawning, incubation, and emergence. The spatial distribution of nursery areas, including the floodplain, should be identified and evaluated for connectivity with changing flow stage so that use for key biota is sustained during critical life history periods. Management of healthy functioning river ecosystems requires maintenance of the inherent inter- and intraannual variability in patterns of flow and sediment movement under which the fauna evolved. Implications of altered sediment transport on instream flow modeling are often ignored (Carling, 1995). Because of the complex interactions among discharge, sediment transport, and water quality in altered systems, simply mimicking the predevelopment hydrograph alone does not renew the habitat building processes. River flows must be managed to provide enough variability and energy to build and maintain the necessary channel morphologic features that provide the habitat critical to the life cycles of the native fauna.

Holling (1998) suggests that ecological science is in transition as reflected by the tension between two very different streams of science. One stream of biological science is "essentially experimental, reductionist, and disciplinary in character.... The other stream is the integration of parts. It uses the results of the first, but identifies gaps, develops competing causative hypotheses, and constructs and uses simulation models as devices for exploration and experimentation over scale ranges that are impossible to achieve by experiments in nature."

The approach presented here to understanding, sustaining, or redirecting the physical changes in a river system is through assembling a suite of conceptual and predictive physical (flow routing and sediment transport) and biological models that can be used (through simulation) to effectively assess dynamic equilibrium conditions and biotic habitat suitability at key locations throughout a river system.

Shortcomings of Many Instream Flow Studies

In water allocation decisions, it is desirable to know the relations between benefits and streamflow for the various uses. In general, the benefits of most out-of-stream uses can be quantified in terms of some type of production function that relates the quantity of water used to the benefits produced. Methods for identifying instream flow requirements have been in a continual state of evolution since the mid-1950s (Stalnaker, 1982, 1994). Unfortunately, little work has been done to quantify the benefits produced by managed instream flows. Before the mid-1970s, instream flow assessments typically resulted in a single streamflow value – a "minimum flow" above which all flows were considered available for out-of-stream use. These flow recommendations were determined from analyses of hydrologic records and/or fish distribution and relative abundance studies.

Instream flow methods were sought that could assist the state engineer, state water administration office, or state legislature to identify some socially acceptable compromise for water use permitting planning and development of out-of-stream uses to continue. Within this political and institutional framework, inexpensive methods were developed to identify region-wide minimum flows that could be "reserved" to satisfy legal obligations to maintain some existence level of the instream fishery resources (Bayha, 1978). These methods later became known as standard-setting methods (Trihey and Stalnaker, 1985). In practice, the degree of protection actually afforded to aquatic species by application of these early instream flow methods varied considerably, both within and among states (Lamb and Meshorer, 1983). Minimum streamflow standards have been viewed as regulatory constraints imposed on a river system. Most river system analysts have used water supply simulations to determine the degree of impact to traditional consumptive water uses in a river basin brought about by imposing minimum flows. In some cases, simulation is used to evaluate the maximum level to which the "minimum streamflow" can be set and sustained under drought conditions. Thus, simulation is used to determine the effect the minimum streamflow constraints may have on long-term water supply potential (Yun-Sheng and Xiao-Dong, 1991).

The problem with basing the well-being of aquatic biota or fish populations on minimum flow standards is that maintaining these low streamflows throughout much of the year over the long term never has the same effects as the infrequent, natural occurrence of seasonally occurring low streamflows. As large water projects were built and operated in compliance with minimum flow standards, it became apparent that the downstream aquatic habitats and fishery resources were being decimated by adherence to a minimum streamflow that became manifest as an abnormally low and constant streamflow throughout much of the year (Collier et al., 1996).

During the late 1970s and 1980s, methods were developed that attempt to evaluate instream habitat conditions on the basis of the hydraulic, structural, and thermal aspects of the stream. These methods are incremental in nature because of the need to examine components of in-channel habitat at different streamflows (Trihey and Stalnaker, 1985). The typical product when an incremental method is applied is a quantitative relation between the suitability of some attribute of fish habitat and streamflow conditions. Despite the continual evolution of instream flow methods, the ability to reliably forecast the number of fish produced in response to a given streamflow regime has been attempted only for salmonid fishes and with limited success (Bovee, 1988; Jowett, 1993; Nehring and Anderson, 1993; Bartholow, 1996).

Of considerable importance to instream flow assessments are the goals and perceptions of society, which differ regionally. What is seldom recognized by many practitioners is that today's instream flow management decisions are no longer only about sustaining low levels of flowing water capable of supporting some fish. They reflect the river corridor goals of society, often resulting in reallocation of water, and necessitate evaluation of the effects of reallocation on existing users. The current review of the operation of large federal water projects on the Colorado, Columbia,

Klamath-Trinity, Missouri, Mississippi, and Sacramento Rivers could lead to potentially large changes in long-established water management practices in order to rehabilitate these severely degraded systems. The purpose of instream flow studies has evolved from the preservation of a wetted channel and maintenance of marginal levels of sport fish populations through the enhancement objectives in previously dewatered stream reaches associated with hydropower relicensing to the reallocation of water to instream uses and habitat rehabilitation concepts of today. To responsibly address these present-day interests, instream flow practitioners must be capable of applying a variety of models in an integrated manner to define much more than a relation between streamflow and temperature or the hydraulic suitability of the fish habitat.

The integration of various quantitative instream flow and institutional decision-making techniques into an overarching instream flow methodology is the foundation of the continuously evolving instream flow incremental methodology (IFIM) [Stalnaker et al. (1995); Bovee et al. (1998)]. A pioneering application of the IFIM was the definition of a monthly flow regime necessary to maintain an existing salmon population in the Terror River, Alaska (Wilson et al., 1981). The effectiveness of this application of IFIM at maintaining salmon production while licensing a 20-MW hydroelectric project in a previously undisturbed watershed was documented through a 6-year monitoring study (Railsback et al., 1993). This is one of the few documented efforts to evaluate the biological consequences of a permitted change in flow regime.

The theme developed in the remainder of this chapter is that fluvial habitat studies must be designed to test hypotheses about flow and sediment management appropriate to the needs of threatened or otherwise desirable aquatic fauna. This should result in management proposals that confront uncertainty by proceeding with experimental flow regimes rigorously evaluated through adaptive environmental assessment.

There are four major components of riverine habitat that determine aquatic biological productivity:

1. Hydrologic regime (stage, duration, and timing),
2. Channel form (fluvial processes, hydraulics, sediment, substrate distribution, and riparian and aquatic vegetation),
3. Water quality (including temperature, dissolved oxygen, and pH), and
4. Watershed contributions (inorganic sediments, nutrients, and particulate organic matter).

Complex interactions of these components determine the primary production, secondary production, and fish production of the stream reach (see also Karr, 1991; Poff et al., 1997).

Most attention has been directed toward components 1 and 3 above. Both of these, in conjunction with empirical measurement of component 2, have led to two somewhat independent classes of routinely applied river habitat simulation models: habitat hydraulics models [see IAHR (1994) and Leclerc et al. (1996); proceedings of two recent international symposia on habitat hydraulics] and water-quality models. To date, virtually no attention has been given to component 4 (energy inputs and distribution) in instream flow studies and river management decisions. As we look to watershed management in conjunction with streamflow regulation, this neglected component may become critical, especially for floodplain rivers.

Component 2 (channel form) has primarily been input through empirical measurement of the channel bathymetry along a series of transects. The dynamic nature of the channel geometry in alluvial streams requires numerous sets of measurement over time or simulating the changing channel form. The science of fluvial processes is rapidly maturing, so that simulation models are available and could be routinely applied to river management.

For alluvial streams, the state-of-the-art allows for the dynamic character of stream channels to be described and predictions to be made of their future state after flow and sediment alterations. Currently, many empirical data must be obtained on site for use with the available analytical models. Many of the stream habitat studies and water management decisions of the past three decades were in headwater and bedrock controlled streams where the channel form could be measured and safely assumed to be fixed or unchanging. However, when this logic is applied to alluvial streams, any predictions of habitat quality under altered flow regimes are flawed. Over the past 50 years, major changes have been documented to alluvial channels below impoundments (Collier et al., 1996). Managed flood flows must now become a part of operational procedures of water control structures on alluvial rivers in order to maintain or restore the channel dynamics and sustain physical habitat and biological integrity.

Most hydraulic habitat models simulate a relationship between streamflow and physical habitat for various life stages of a species of fish or for a recreational activity. The basic objective of hydraulic habitat simulation is to obtain a dynamic representation of the stream habitat so that the streamflow regime may be linked, through biological considerations, to the social, political, and economic decision process. One of the most widely used hydraulic habitat models is PHABSIM, which was developed to analyze the spatial distribution of some hydraulic attributes within a specified channel and to display the relation between streamflow and physical habitat quality or between streamflow and suitable recreational river space (Milhous et al., 1989). Habitat (or recreational) suitability is given as a continuous function of streamflow. This relation is then linked with streamflow gauge records, or simulated hydrologic time series, to yield a habitat time series.

The two basic components of hydraulic habitat models are the hydraulic simulation and the habitat quality rating of given stream cells (spatially distributed in a stream reach) utilizing defined hydraulic parameters and species or guild habitat suitability criteria. Thus, hydraulic simulation is used to describe the area of a stream having suitable combinations of depth and velocity as a function of flow.

The most comprehensive simulation of fluvial habitat is accomplished by using the physical structure of the stream, hydraulic and water-quality models, and time series of streamflow. The modification of stream habitat quality and quantity due to changes in temperature, dissolved oxygen, water chemistry, and channel form is analyzed separately from hydraulic habitat simulations. Temperature in a stream varies with the seasons, local meteorological conditions, stream network configuration, and the flow regime. Thus, the temperature influences on habitat must be analyzed at a subbasin scale. Water quality under natural conditions is strongly influenced by climate and catchment geology, with the result that there is considerable natural variation. When we impose human effects, the range of water-quality possibilities becomes rather large.

Time-related changes in channel form vary from being insignificant in high-gradient bedrock controlled streams (typical headwaters) to dramatic in low-gradient alluvial rivers. Energy inputs are directly related to the hydrologic connectivity between the channel and its floodplain or riparian zone.

Addressing Integrated Science Predictions through Simulation Modeling and Testing

River system management *must* involve scientists from the disciplines of hydrology, geomorphology, water chemistry, aquatic and riparian ecology, and engineering (e.g., hydraulics, sediment transport, ice formation). Social concerns are such that leaving river management decisions to one or two scientific disciplines is no longer acceptable. Mistakes of omission can no longer be tolerated. The lack of communication among scientific disciplines, managers, and stakeholders has become the primary impediment to multiple objective management incorporating various ecosystem processes. Simulation modeling tempered with rigorous empirical monitoring is the vehicle to bring integrated science to the management of fluvial systems. A common vocabulary and definition of terms for the modeling process are given followed by a suite of models for evaluating river channel equilibrium and response to directed management actions. Suggested approaches are offered based on the ability to translate model output to habitat needs of aquatic (fishes and invertebrates) and riparian (woody and herbaceous plants) biota.

Modeling as a Process for Describing the State of Fluvial Habitat

The American Society for Testing and Materials (ASTM, 1994) proposed a standard protocol for evaluating environmental chemical-fate models, including proposed standard definitions for selected modeling terms. Although not all modelers, or model users, will agree completely with these specific definitions, the standardization they provide may help to facilitate communication among modelers and scientists from different disciplines as well as other professionals.

After a model has been selected (for application), a series of steps or procedures are followed that compose the model application process. These steps, listed in Table 16.1, can be grouped into three phases: Phase I includes data collection, model input preparation, and parameter evaluation; Phase II is model testing; and Phase III is analysis of alternatives. Model testing is a critical and often neglected part of the model application process.

Phase I provides the groundwork for the modeling effort by developing, collecting, discovering, and preparing the data needed for model

Table 16.1. *The modeling/testing/validation process*

Phase I – Model building	Data collection Input preparation Parameter evaluation
Phase II – Model testing	Calibration Validation, verification (Postaudit), confirmation
Phase III – Model comparisons	Analysis of alternatives

application. For rivers, this includes the observed bathymetry, meteorologic, hydrologic, and water quality data in addition to site characterization information and biologic habitat needs and goals. Data collection may also include sample collection and analysis if adequate historical data are not available. Input preparation is the process of preparing the collected data in a format acceptable to the chosen model. Parameter evaluation is the process of estimating the specific parameters required by the model based on site characteristics, application guidelines, objectives, model, and prior experience. At the conclusion of Phase I, model execution runs can be initiated.

Model testing is Phase II of the process. The confusion surrounding the model testing phase occurs largely because different meanings have been attached to the terms calibration, verification, validation, and postaudit in the technical literature. The process of model testing ideally includes all three steps shown in Phase II. "Ideally" because in many applications existing data will not support performance of all steps. In fluvial habitat modeling, measured data for validation are often lacking, and postaudit analyses have been rare for most modeling exercises on regulated rivers. Integrating the diverse sciences involved in managing fluvial systems now requires effort in model building, testing, and comparison of alternatives before the necessary shared *trust* among scientists of all disciplines, managers, and stakeholders can be reached.

Calibration is a necessary procedure for all simulation models and is probably the most misunderstood of all the model application components. Calibration is the process of adjusting selected model parameters within an expected range until the differences between model predictions and field observations are within acceptable criteria for performance (i.e., it is the manipulation of the independent variables to obtain a match between the observed and simulated distribution of dependent variables). Calibration attempts to account for spatial variations not represented by the model formulation; functional dependencies of parameters that are nonquantifiable, unknown, and/or not included in the model algorithms; and extrapolation of laboratory measurements of parameters to field conditions. Calibration increases the user effort and data required to appropriately apply a model. Unfortunately, many models can be operated without calibration depending on the extent to which critical model parameters (usually refined through calibration) can be estimated from past experience and other data or simply run with built-in default values. Nevertheless, all model applications should incorporate some form of testing of the model output.

Validation – a Management Responsibility

Validation is the complement to calibration; model predictions are compared with field observations that were not used in model development or calibration. For hydrologic and water quality models this is often the second half of split-sample testing procedures, in which the universe of data has been divided (either in space or time), with half the data used for calibration and the other half used for validation. Validation is an independent test of how well the model (with its calibrated parameters) is representing the important processes that occur in the fluvial system. Although field and environmental conditions are often different during the validation step, parameters determined during calibration are not adjusted during validation. Validation means that a model meets specified performance requirements and is acceptable for its intended use (Rykiel, 1996).

Many investigators have used validation and verification interchangeably. Validation means determining the model's ability, after calibration, to represent a specific site and/or a specific model application, whereas verification is restricted to verifying the operation of the numerical procedures in the code (ASTM, 1994). Such verification is typically a model development activity and should not be of concern in management applications unless research and new model development are necessary as part of a specific river system management program. According to Oreskes et al. (1994), the practice of comparing numerical and analytical solutions in model development is also referred to as bench-marking.

Model Confirmation – No Longer Exclusively a Research Activity But a Management Necessity

Postaudit analyses are the ultimate tests of a model's predictive capabilities for management decisions. Model predictions for a proposed operating procedure are compared with empirical observations after implementation of the procedure. The degree to which agreement is obtained, based on the acceptance criteria, reflects on both the model capabilities *and the assumptions made by the model user* to represent the proposed alternative. In the past, postaudit analyses have been performed in a few regulated river situations. This should become routine in river management situations. It is the essential assessment ingredient to the adaptive management paradigm discussed below.

The final and perhaps most critical phase of the modeling process is the use of a model as a decision aid for assessing management or regulatory alternatives. In analyzing various alternatives, the confirmed and therefore "trusted" model is used as a tool to predict the changes in system response resulting from proposed alternatives; these alternatives may be modeled by adjustments (changes) to model input, parameters, and/or system representation. During the postaudit phase, the model results are compared with observed data for selected time periods, whereas in analysis of alternatives the model results are used for sensitivity analyses – for exploring "what if" questions related to each proposed alternative and appropriate baseline or reference conditions. In this way, the relative changes in system response associated with a proposed alternative can be identified, analyzed, and evaluated for accuracy of predictions.

Oreskes et al. (1994) suggest that "In areas where public policy and public safety are at stake, the burden is on the modeler [those applying the models to management alternatives] to demonstrate the degree of correspondence between the model and the material world it seeks to represent, and to delineate the limits of the correspondence."

Rykiel (1996) refers to this postaudit testing as predictive validation in which the model is used to forecast the system behavior and comparisons are made to determine whether the system's behavior and the model's predictions are the same. The strongest case occurs when the model output is generated *before* the data are collected.

Mitigation planning attempts to quantify the habitat difference (losses) between the baseline conditions (status quo) and the evaluated alternatives (especially the selected preferred alternative). Mitigation for this habitat difference is often accomplished through purchase or management of off-site habitats compensated for (through trades for water storage, out-of-kind habitats, land, research, or money) or simply foregone.

Integrated Science: Renewing the Call for Adaptive Environmental Assessment and Management

Basic Premises

Recent actions aimed at sustaining the biological integrity of river corridor habitats in large part have failed, and a shift in perspective is currently occurring within the natural resource community. Emerging strategies primarily center around an adaptive approach to ecosystem management. Single species management, the complexity of species interactions and interrelationships, and limited scientific knowledge about the interactions of abiotic and biotic factors are all evidence that traditional approaches to management of rivers are inadequate to preserve biotic community diversity. Although the concept of ecosystem management is not new, its implementation in regulated rivers is. Given this, it is important to stress not just flow recommendations and nonflow channel alterations but also implementation of a new paradigm of river management built on the two-decades-old concept of adaptive environmental assessment and management (see also Hilborn and Walters, 1992).

Adaptive environmental assessment and management (AEAM) is a formal, systematic, and rigorous program of learning from the outcomes of management actions accommodating change and improving management (Holling, 1978). Such a program combines assessment and management. Most agency and task force structures do not allow both to go on simultaneously (International Institute for Applied Systems Analysis, 1979). The basis of adaptive environmental assessment and management is the need to learn from past experience, data analysis, and experimentation. AEAM combines experience with operational flexibility to respond to future monitoring and research findings and varying resource and environmental conditions. AEAM uses conceptual and numerical models and the scientific method to develop and test management choices. Decision makers use the results of the AEAM process to manage environments characterized by complexity, shifting conditions, and uncertainty about key system component relationships (Haley, 1990; McLain and Lee, 1996).

The AEAM approach to management relies on teams of scientists, managers, and policymakers to jointly identify and bound management problems in quantifiable terms (Holling, 1978; Walters, 1986). In addition, the adaptive approach "to management recognizes that the information on which we base our decisions is almost always incomplete" (Lestelle et al., 1996). This recognition encourages managers to treat management actions as experiments whose results can be used to better guide future decisions. AEAM need not only monitor changes in the ecosystem but also should develop and test hypotheses of the causes of those changes in order to promote desired changes. The result is informed decisions and increasing certainty within the management process.

Modern management strategies must have explicit and measurable outcomes. There are not many clear-cut answers to complex hydraulic, channel structure, and water-quality changes, but the AEAM process allows managers to adjust management practices (such as reservoir operations) and integrate information relating to the riverine habitats and the system response as new information becomes available.

Alluvial river systems are complex and dynamic. Our understanding of these systems and predictive capabilities are limited. Together with changing social values, these knowledge gaps lead to uncertainty about how to best implement habitat maintenance or restoration efforts on regulated rivers. Resource managers must make decisions and implement plans despite these uncertainties. AEAM promotes responsible progress in the face of uncertainty. AEAM provides a sound alternative to either "charging ahead blindly" or being "paralyzed by indecision."

Holling (1978) states that "AEAM avoids the pitfall of requiring the costly amassing of more descriptive data before proceeding with policy initiatives. Instead, strategies are adopted as learning experiments in a fluid feedback structure that mandates vigorous self-critiquing and peer review at every stage, such that evaluation and corrective information is disclosed quickly and strategies modified or discontinued accordingly."

A well-designed AEAM program (1) defines goals and objectives in measurable terms; (2) develops hypotheses, builds models, compares alternatives, and designs system manipulations and monitoring programs for promising alternatives; (3) proposes modifications to operations that protect, conserve, and enhance the resources; and (4) implements monitoring and research programs to examine how selected management actions meet resource management objectives. The intention of the AEAM program is to provide a process for cooperative integration of water control operations, resource protection, monitoring, management, and research.

AEAM assesses the results and effects of reservoir operations and water resource allocations on biotic resources. The results of the assessments sustain or modify future operations. Outlined in Figure 16.1 is a 10-step AEAM process.

Resource agencies and stakeholders form the ecosystem restoration goals through a watershed planning process. A key to successful watershed planning and ecosystem restoration is a combination of democratic stakeholder processes, technical input, and leadership. It is an error to assume that people will protect a stream if they are "educated." Management

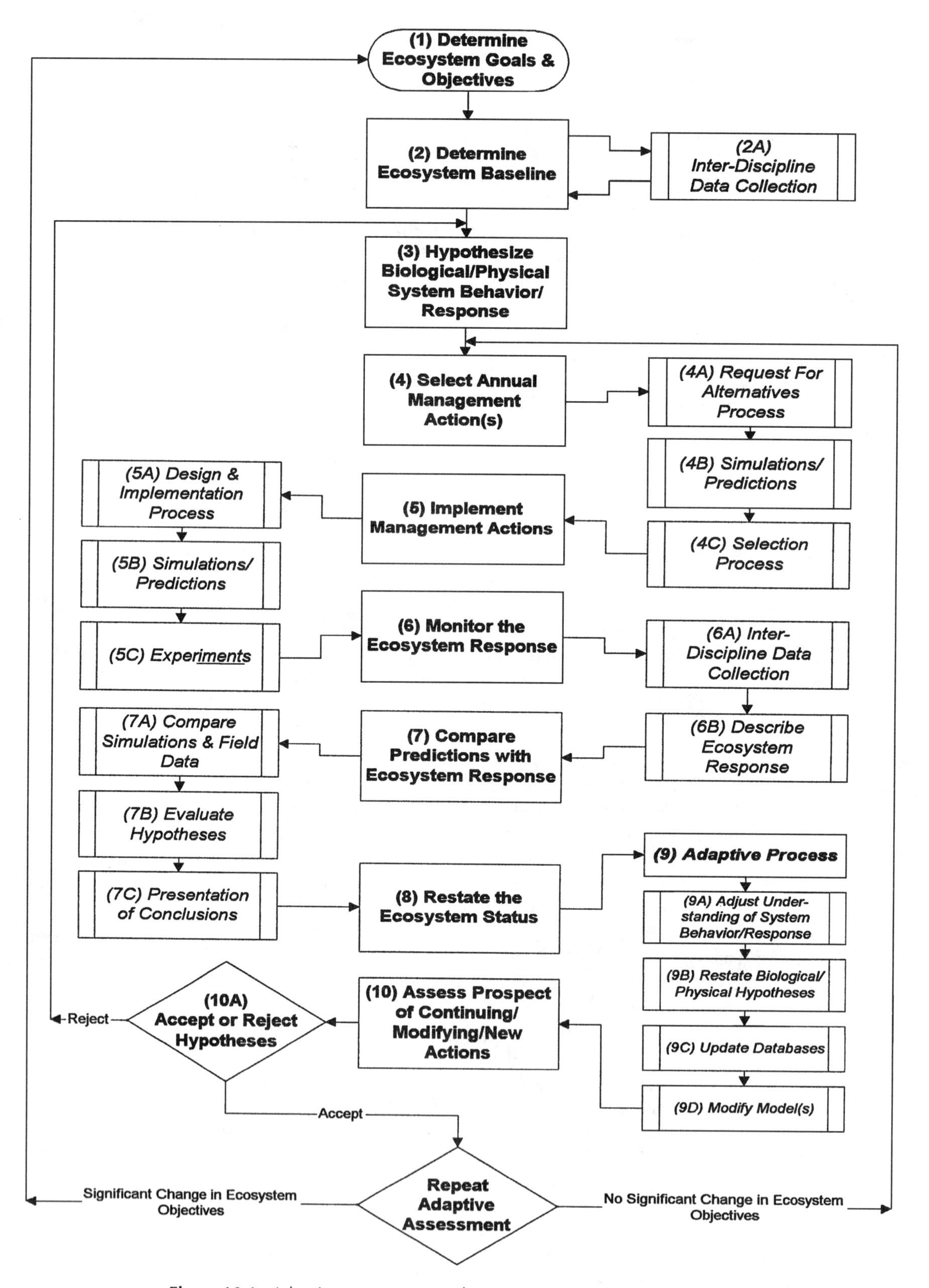

Figure 16.1. Adaptive assessment and management process as proposed for the annual selection and evaluation of river system operation criteria and procedures.

should work toward creating common ground where there are win/win outcomes; consider competitiveness, environmental soundness, and social/political issues; clarify areas of conflict and view conflict as an opportunity to learn; maintain a policy evaluation framework that assumes, and is adaptable to, changing objectives; and address clearly stated conflicting alternatives instead of a single, presumed true social goal (Holling, 1977). Once goals for restoring or sustaining the ecosystem are firmly in view, the technical process may begin by implementing Steps 2 through 7 (Figure 16.1).

An ongoing element of the process is to constantly challenge the stated system hypotheses and improve the ability to predict the behavior and response of the ecosystem so that progress toward the management objective is positive and rapid. If certain hypotheses of system response are not supported then new hypotheses must be proposed, modeled, and in turn tested. Learning to guess better is an ongoing management endeavor, using all available tools and monitoring, based on accepting or rejecting hypotheses of system response to management actions.

The scientists offer an annual statement of the system hypotheses presenting evidence in support or rejection of tested hypotheses.

Annual management actions are designed as operating criteria and procedures. If system hypotheses are supported (not rejected), then the process is repeated by going back to Step 4 and selecting annual operating criteria and procedures for the forecasted water supply (water year class).

If system hypotheses are rejected, recycle through the process by going back to Step 3 and stating alternative hypotheses to achieve the same management goals.

On occasions such as natural disasters, toxic spills, or major legislative actions the ecosystem management (social) goals may change. In such events recycle through the adaptive process by going back to Step 1. Restate the system goals, perhaps requiring a different or modified baseline and certainly the generation of new hypotheses of system response translated to new measurable system objectives.

Case Example: Trinity River, California

The Trinity River, a wild and scenic river in northern California, is the largest tributary to the Klamath River (Figure 16.2). Trinity and Lewiston Dams, operated by the U.S. Bureau of Reclamation, regulate 1863 km^2 (719 mi^2) of the basin. Lewiston Dam is the present upstream limit of salmon and steelhead migration, keeping anadromous salmonids from accessing over 175 km (109 mi) of historic spawning, rearing, and holding habitat [U.S. Fish and Wildlife Service (USFWS), 1994].

The mean annual water yield from the basin over the 1912–1995 period of record was 1.5×10^9 m^3 (1,249,000 acre-feet). After construction of the Trinity River Project, annual releases downstream to the Trinity River ranged from 8% of the annual yield in 1965 to 63% of the annual yield in 1994. From 1961 to 1995, the average downstream release was 28% of the mean annual yield, whereas 72% was exported to the Sacramento River basin (USFWS and Hoopa Valley Tribe, 1999).

Before construction of the Trinity River Division of the California Central Valley Project, the Trinity River was a meandering, unregulated, mixed alluvial river within a broad floodplain, characterized by repeated

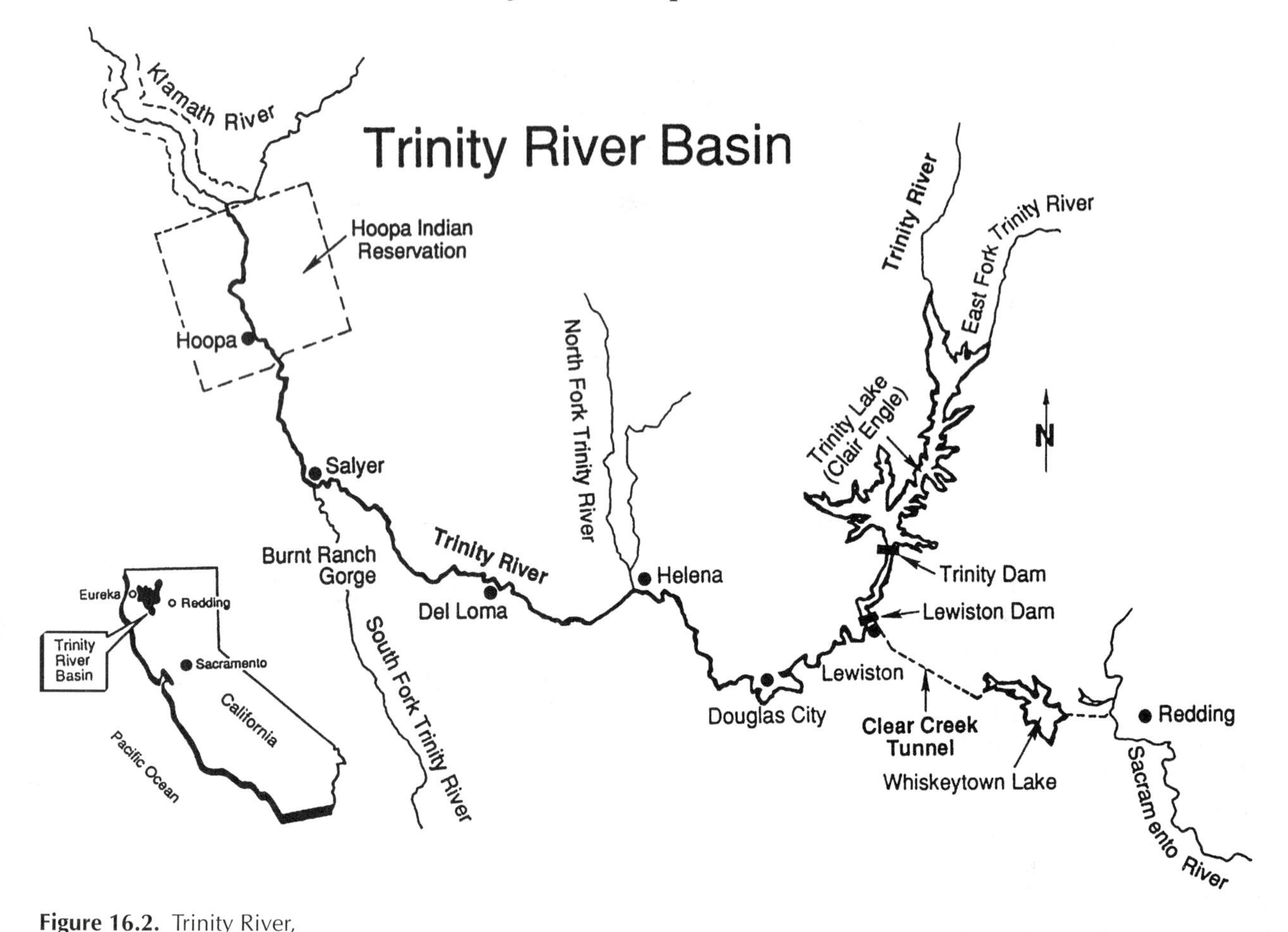

Figure 16.2. Trinity River, California.

alternate bar sequences. Flow regulation removed nearly all high flows that were responsible for forming and maintaining a dynamic alternate bar morphology. No longer scoured by winter floods downstream from the dams, riparian vegetation encroached onto the river channel and formed riparian berms along the channel margins. Reduced flows, loss of coarse sediment, and riparian encroachment caused the mainstem river downstream from the Lewiston Dam to change from a series of alternating riffles and deep pools that provided high-quality salmonid habitat to a rectangular channel confined between riparian berms. The lost alluvial features and diverse riverine habitats compromised salmonid habitats and the populations that relied on them. Operations of the dam also changed the thermal regime of the Trinity River, providing warmer water temperatures during the winter and colder water temperatures during the summer than were present before development.

Although predevelopment spawning escapement data are sparse, the data that are available indicate that naturally spawning populations of anadromous salmonids have declined dramatically since construction of the Trinity Project. Average postdevelopment escapement data for naturally produced spring- and fall-run chinook salmon represent a 68% reduction compared with the predevelopment average. Large chinook salmon spawning escapements since 1978 have been typically dominated by hatchery-produced fish that spawned in the Trinity River and are not indicative of healthy spawning and rearing conditions. The in-river coho salmon spawning population is predominantly of hatchery origin, with only 3% of the spawning coho attributable to natural production. Although naturally produced fall-run steelhead make up a large proportion of the natural spawners (70%), this still represents a 53% reduction from predevelopment estimates. Coho salmon that return to Klamath and Trinity Rivers, the southern Oregon/northern California coast evolutionary unit, have been listed as threatened pursuant to the Federal Endangered Species Act. The final rule recognized that various habitat declines affected coho salmon populations, including channel morphology changes, substrate changes, loss of off-channel rearing habitats, declines in water quality (e.g., elevated water temperatures), and altered streamflows. The steelhead and chinook salmon populations of the Trinity River are also candidate species and may warrant listing in the future (USFWS and Hoopa Valley Tribe, 1999).

Sediment Transport

Completion of Trinity and Lewiston Dams in 1961 eliminated coarse sediment supply from upstream sources. Downstream tributaries continue to supply sediment, and in some cases (e.g., Grass Valley Creek) the sediment delivery has increased because of intensive land use and deforestation. The absence of high mainstem flows has permitted tributary sediments to accumulate, forming tributary fans at tributary confluences. The excessive sand supply and decreased transport capacity have reduced pool volumes and infiltrated spawning gravels. In many downstream reaches, a layer of fine particles is evident on top of the larger predam bed surface (McBain and Trush, 1997).

Riparian Encroachment

Predam riparian plant communities were maintained in early seral stages by floods. High winter floods undercut banks, toppled mature trees, and scoured out younger trees. Cottonwood forest occupied floodplains and lower terraces. After dam construction, the actively scoured channel width decreased, and the edges of alternate bars were rapidly colonized by riparian vegetation, which trapped fine sediment and eventually "fossilized" the banks, creating a trapezoidal, fixed channel. The main factors influencing the river segment below Lewiston Dam are the reduction in peak flows and subsequent decrease in sediment transport capacity and stabilization by vegetation. Currently, sand berms (up to 2 m above the low water surface) are nearly continuous from Lewiston Dam downstream to the confluence of the North Fork Trinity River. This has effectively eliminated most of the rearing habitat for salmon in this portion of the Trinity River. Congressional actions identify restoration of the salmon stocks as social goals (USFWS and Hoopa Valley Tribe, 1999).

Applying AEAM to the Trinity River

The Trinity River Flow Evaluation Report (TRFE) (USFWS and Hoopa Valley Tribe, 1999) concludes that the river channel has degraded to such an extent that simply managing flow releases from the existing reservoirs cannot achieve the salmonid restoration goals mandated by Congress. The

Table 16.2. *Recommended annual water volumes available for instream release to the Trinity River in thousands of acre-feet (TAF)*

Water-year class	Instream allocation (TAF)
Extremely wet	815.2
Wet	701.0
Normal	646.9
Dry	452.6
Critically dry	368.6
Average (weighted by water year probability)	594.5

primary hypothesis is that a combination of managed high-flow releases, mechanical disturbance, and vegetation removal will redirect geomorphic processes so that a more complex channel form will evolve, creating the shallow, low-velocity, edge-associated rearing habitats necessary to enhance freshwater salmonid production.

Reservoir releases and channel rehabilitation projects have been designed to produce a substantial increase in carrying capacity (usable salmonid rearing habitat area) within the rehabilitated channel. What is not known is the rate of change or length of time needed to achieve this new channel equilibrium. AEAM coupled with environmental assessment (annual monitoring) is proposed for achieving the salmonid restoration goals. The management actions prescribed include channel rehabilitation in combination with annual reservoir releases based on forecast water supply and the recommended allocation for the water year class. Five water-year volumes (see Table 16.2), each with unique hydrograph components, provide the interannual variability necessary to drive the fluvial processes toward a new channel configuration, while maintaining the hydraulic and temperature conditions at levels that are equal to or greater in quality than that existing since the closure of the dams. The remainder of this section discusses a specific application of the AEAM process to the Trinity River.

Step 1. Goals and Objectives for the Trinity River. One of the stated goals for the Trinity River is "... the development of recommendations regarding permanent instream fishery flow requirements and Trinity River Division operating criteria and procedures for restoration and maintenance of the Trinity River fishery" (Central Valley Project Improvement Act, Title XXXIV of P.L. 102-575). The report recommendations include a series of five flow allocations, one for each class of water year. Primary objectives of the recommendations are as follows:

- Manage the reservoir releases to provide a much improved (near optimum) temperature regime. An optimum temperature regime increases fish residence time and growth rates, resulting in larger smolts exiting the system. Larger smolts have better survival, which leads to an increase in numbers of returning adults.
- Manage the river corridor to achieve a 2- to 4-fold (based on modeling results) increase in the shallow margin backwater habitat also necessary for many anadromous young-of-year salmonids.

- Manage reservoir releases to control vegetation establishment on alluvial features. Schedule reservoir releases to scour seedlings on bars after the seed fall during the spring-summer period. Investigate superimposing reservoir releases on tributary flows when the opportunity is present.
- Manage reservoir releases within the evolving channel to optimize hydraulic conditions for spawning, incubation, and young-of-year production for a given water year and channel form. As the channel changes from the present straight trapezoidal form toward the desired alternating point bar configuration, the habitat hydraulics should be adjusted annually to maximize suitable conditions for a given year.

Step 2. Hypotheses. The premise of the TRFE is that a combination of mechanical disturbance and vegetation removal in addition to managed high-flow releases in the spring will promote stream channel migration leading to a new channel form that is expected to provide significantly increased spawning and rearing habitat for anadromous salmonids.

One of the central hypotheses to be tested is that habitat diversity in the upper river, on both the meso- and microhabitat scales, will increase after the managed releases within the TRFE recommended water-year allocations. Although the changes in habitat diversity are expected to be obvious, there will remain a question of degree. A methodology must be embraced to quantify the existing habitat diversity and the annual change created as the management recommendations are implemented. This will enable comparative evaluations to be made and elucidate the effectiveness or noneffectiveness of specific restoration measures.

A second hypothesis central to the TRFE recommendations is that the juvenile salmonid rearing habitat, believed to be limiting smolt production in the Trinity River, will increase in quantity and quality after creation of a more complex and dynamic channel form. Rearing habitat area, which currently is highly variable depending on streamflow, will increase (at least a 4-fold increase) and become more stable over a wide range of flows.

The third central hypothesis is that salmonid smolt survival will increase because of better temperature conditions that increase growth and promote extended smoltification and reduced travel time associated with emigration.

Before proceeding with AEAM this set of hypotheses and series of events is transformed into a set of measurable responses. By way of examples, we offer three initial quantification steps.

First, describe the existing channel geometry in two dimensions by subsampling along surveyed transects or grids. Subsampling should be sufficient to describe the bathymetry of the crossing bar pool sequences at upper, lower, and middle portions of the river from Lewiston Dam to the North Fork. Transects should be georeferenced so monitoring measurements can be repeated. These measurements are needed to quantify the degree of bar formation, lateral movement, and removal of established woody vegetation attained on an annual basis. The straight trapezoidal channel should evolve toward a more sinuous alternate bar type form having increased shallow-water area and low-velocity backwaters, which are critical for rearing young salmonids.

Second, the amount of habitat area available to provide suitable spawning and rearing conditions associated with channel margins should be measured annually. Annual measurements along habitat transects or grids quantify the amount, and simulation modeling quantifies the duration, of usable spawning and rearing habitat area over the biological year (September to July) and for the forecast water year. Geomorphology, vegetation conditions, and salmonid habitat must be quantified by the same sampling strategy. The same strategy allows extrapolation describing 64 km between Lewiston Dam and the North Fork.

Third, the length and weight of Chinook salmon young-of-the-year can be sampled every few weeks from hatching through emigration from the stream study segment. Substantial trapping effort at the downstream end of the study segment is needed to estimate the total number of Chinook salmon presmolts leaving the segment. These two sets of measurements can estimate the annual growth increments through the season and young-of-year production within the river.

Step 3. Document Channel Form, Riparian Vegetation, and Salmonid Population Trends. Through comparison of annual measurements and use of simulation modeling, progress toward the habitat and production objectives can be quantitatively expressed. Progress toward the program objectives and any trends identified should be reported annually to the stakeholders. This report may address the following questions:

- Are salmonid population numbers improving?
- Quantify as population estimates, not just abundance indices.
- Is anadromous salmonid habitat improving?
- Are native riparian communities establishing on different geomorphic surfaces? Are reservoir releases removing germinated vegetation?
- Are the riparian berms continuing to build, are they remaining stable, or are they beginning to break down from Lewiston Dam to the North Fork Trinity River confluence?
- Are channel reaches migrating laterally and becoming more dynamic?
- Are floodplains forming?
- Are alternate bars forming?
- How does Trinity River water affect water quality of the Klamath River?
- There is evidence that water-quality conditions in the Klamath River may be substantially worse than in the Trinity River. Will the difference in water quality occur during spring outmigration, especially in dry years? If so, how is this affecting smolt survival? What about other life stages?

Step 4. Management Actions. The recommendations for management actions incorporate different schedules for flow releases under five defined water-year allocations (determined by snow pack, inflows, and reservoir levels measured each spring from mid-February through April). All year classes include a recommendation of high-flow reservoir releases in late April to mid-July and a program of gravel placement in the mainstem. These releases follow proposed berm removal projects. The intent of 43 potential berm removal project sites is to knock down the densely vegetated channel berms at selected sites along the river from Lewiston Dam downstream to the North Fork confluence.

Different April to July flow release schedules are proposed for normal, wet, and extremely wet years so that in 6 of 10 years the channel is predicted to change in cross section and plan form. The goal is a meandering point-bar–pool–point-bar configuration within the old floodplain. These water-year classes, each with unique hydrograph components, provide the interannual variability necessary to drive the fluvial processes. A rehabilitated channel, although at a smaller scale than the predam channel, could sustain salmonid rearing habitat two to perhaps four times the current amount. Analyses with the Chinook salmon production model (SALMOD) indicate that young-of-year production can be significantly increased (perhaps doubling) if the rehabilitated channel attains a 4-fold increase in the total available rearing habitat throughout the 64-km reach below Lewiston Dam, all other things being equal (same average ocean survival and number of returning spawners and no further degradation of water quality, etc.).

The existing baseline conditions can be quantitatively expressed as historical time series, starting with streamflow and reservoir release records. The resulting hydrologic time series is input to the Stream Network Temperature Model (SNTEMP) (Theurer et al., 1984), the Physical Habitat Simulation Model (PHABSIM) (Milhous et al., 1989) and the Time Series Library (TSLIB) (Milhous et al., 1990) to produce a weekly estimate of the total usable habitat available throughout the study segment. The habitat time series is input to the Salmon Production Model (SALMOD) (Bartholow et al., 1997) to produce a weekly time series of salmonid production estimates. This includes estimates of growth, downstream distribution, and number exiting the study segment.

Although the habitat response hypotheses could be tested with one-dimensional hydraulic and habitat models within PHABSIM, an alternative now exists. This alternative utilizes two-dimensional hydraulic models that provide major advancements in the state-of-the-art of riverine habitat assessments. Many in the instream flow modeling community believe that two-dimensional *hydraulic* models are superior to their one-dimensional counterparts for simulating velocity distribution throughout river channel reaches (Ghanem et al., 1994; Leclerc et al., 1996). These advantages are particularly evident in complex river channels of the type, it is hypothesized, the Trinity River will become as a result of the proposed management. These models are spatially explicit, allowing for calculation of different measures of habitat environmental heterogeneity, and offer the potential to describe both spatial and temporal heterogeneity in a single habitat metric. This new technology is recommended for evaluating the habitat response to the proposed Trinity River AEAM actions.

Step 5. Implement Actions. The AEAM team should convene each year in mid-February after initial water supply forecasts of a normal or wetter water year. The team objective would be to estimate the magnitude and duration of reservoir releases that year. The goals of the release schedule include mobilizing alluvial features established the previous year, scouring emergent riparian vegetation, and sediment transport. Physical process modeling can help optimize the reservoir release necessary to mobilize alluvial features and optimize lateral bank cutting. After the water year has been declared by the Bureau of Reclamation, these physical process models can be used to simulate conditions for the remainder of the year based on the allocated water volume and the reservoir operation criteria.

The degree of channel change can then be projected by using the HEC-6 or other physical process models that predict aggradation or degradation of the channel. Kondolf and Micheli (1995) presented a protocol for documenting changes in channel form. Reservoir release temperatures, downstream water temperature, usable habitat, and young-of-year Chinook salmon production are all then simulated by using the assumed reservoir release schedule and the physical model predicted channel responses. The annual estimates of returning adult Chinook spawners and the habitat state during the previous fall are important inputs to these simulations. Therefore, each annual production run is based on the latest empirical data (September to May) and simulated conditions for the remainder of the biological year (May to July).

Step 6. Monitoring. Physical process numerical models are useful in two ways. First they require a systematic collection of data inputs. A well-designed monitoring program will yield the correct type, quality, quantity, and frequency of data. Second, they indicate where significant physical changes may occur, serving to focus monitoring activity in new and perhaps unexpected locations.

The run mesohabitat type currently dominates the upper river. These runs are generally long and straight and are confined by berms on either side. At the targeted rehabilitation sites, removal of the berm on one side of the river and implementation of the prescribed flow regimes should produce transverse riffles between point bars with adjacent pools. Besides these major mesohabitat features, it is expected that additional mesohabitat types will also result, such as backwaters and riffle-pool transitional zones. The number of different mesohabitat types and the proportion each represents should change significantly over present conditions, as should the range of hydraulic conditions present.

The annual evaluation of habitat changes at the mesohabitat level is straightforward. The types of preproject mesohabitats present, the stream area each encompasses, and the proportion each represents in the reach can be compared with the previous year.

A more detailed evaluation of habitat diversity is needed at the microhabitat scale. The premise is that all habitat types are potentially important to the health of the anadromous salmonid community. Therefore, the monitoring objective is to quantitatively describe the mix of heterogeneous microhabitat types without regard to which species or life stage may or may not use a particular type. This is done by defining discrete, nonoverlapping combinations of microhabitat characteristics and treating them in the same manner as individual species in developing community metrics.

Leonard and Orth (1988) introduced habitat guilds and Bain and Boltz (1989) used the concept of developing habitat suitability criteria to define habitat use guilds. The same concept can be applied to defining microhabitat types. For example, depth can be classified as shallow, moderate, and deep; likewise, velocities can be partitioned into slow, medium, and fast classifications; cover can be designated by function (e.g., velocity shelter) or simply by presence or absence. Each combination would describe a unique microhabitat type (e.g., shallow, slow, no cover).

Because each combination of habitat attributes is unique, it can be treated much the same as a species in traditional ecological community analyses. Thus, for a given streamflow, one could derive values for habitat richness (the number of unique microhabitat types present), habitat diversity (an index of the heterogeneity among microhabitat types present), and habitat evenness (the ratio between calculated microhabitat diversity and the maximum microhabitat diversity possible).

The habitat diversity-discharge relationships, displayed graphically, allow comparative evaluations to determine whether microhabitat diversity is increasing in the rehabilitation reaches. These relationships also provide insight into the stability of microhabitat diversity. That would be an indicator of the constancy in abundance of diverse microhabitat conditions as stream discharge changes. A time series analysis will show the temporal variability of habitat diversity. Using the habitat diversity-discharge function and a hydrologic time series, an annual chronology of habitat diversity and area could be evaluated.

On an annual basis assess the abundance and health (size, growth, diseases, adenosine triphosphatase activity) of salmon smolts utilizing cooler water temperature conditions. Fish samples for measurement with rotary screw taps or other capture techniques at key locations (upper Trinity River, lower Trinity River and near the estuary) could be taken. On a longer time scale the annual number of returning adults serves as a measure of success.

Under controlled and natural settings, monitoring is used to examine how water temperature affects smoltification of Trinity River parr and smolts. There also may be a need to examine the effects of low concentrations of dissolved oxygen on parr and smolts, particularly during dry and critically dry years.

Step 7. Compare Predictions and Observations. During early winter, model simulations are run again using the actual preceding 12 months of flow releases and downstream tributary inflows. Seldom do meteorological and precipitation patterns follow seasonal patterns exactly as in the past. Therefore, the physical process and biological models are more fairly tested by comparing outputs (predictions) based on actual (as near as it can be determined) streamflow distribution through the river segment. Habitat and salmonid production output are compared with measured channel form, smolt growth, and production.

Step 8. Restate System Status. The system state and the degree of progress toward the stated management objectives are determined by comparison with observations from previous years.

Step 9. Adaptive Process. Scientific evidence is presented to the managers and stakeholders in support of, or refuting, the original hypotheses. Scientists revisit the hypotheses [or develop new hypotheses if the original(s) is rejected] and recalibrate models awaiting the next round of forecasts, decisions, and simulations. If certain hypotheses are rejected or alternatives are proposed, experimental flow releases or other management actions are designed (within the bounds of the water-year allocation) and submitted to management before the winter-spring forecast period. Table 16.3 lists the models and the monitoring data needs as proposed for the Trinity River.

Table 16.3. *Suite of models for predicting habitat and chinook salmon production responses to regulated flood flows on the Trinity River, California*

	Hydrology	Geomorphology	Sedimentation	Riparian vegetation	Temperature	Habitat	Chinook-Salmon production
Data needs	Reservoir inflow Winter snow depth Reservoir elevation and volume	Bar structure Bar mobilization Bar migration Berm/riparian destruction	X section and aerial Gradations Transport by size fraction Fate: scour/fill	Density Age Type Germination	Reservoir boundary Condition Width/depth Shade Temperature recording	Bathymetry Depth, velocity, substrate and cover distribution x section	No. spawners Presmolt outmigration Spawning locations Size of outmigrants Rearing habitat
Techniques	Empirical forecast Annual declaration of water year type (extra wet, wet, normal, dry, extra dry) Mass balance	Annual videography Ground survey Aerial topography X section at reference sites Longitudinal WS and bed profiles Particle size fractions Gradations	Bed load/suspended load Discharge recording Tributary measurements Particle size fractions Gradations	Seedling counts Ground survey	Estimate of Tributary inputs	Hydraulic simulations Estimates of usable habitat area	Estimates of escapement and smolt production Weekly estimates of survival and growth
Models	PROSIM TRNMOD Riverware	HEC-2 (HEC-RAS) HEC-6	HEC-6 CH3D-SED	Vegetation Establishment model, Mahoney (96)	SNTEMP BETTER	PHABSIM TSLIB	SALMOD
Simulations/ predictions	Updated biweekly Longitudinal distribution of discharge	Areas of bars/ pools by reach Sinuosity Water surface profiles	Scour & deposition at x sections	Area of bars with seedlings Durability by reach	Longitudinal profile Historical and forecast time series Area of stream with suitable temperature	Time series of spawning and rearing habitat	Weekly estimates of number and size leaving specific areas

Step 10. Assess Prospect of Continuing/Modifying/New Actions. Design annual management actions (operating criteria and procedures). If system hypotheses are supported (not rejected), then recycle through the process by going back to Step 4 and selecting annual operating criteria and procedures for the forecasted water supply (water-year class).

If system hypotheses are rejected, recycle through the process by going back to Step 3, stating alternative hypotheses to achieve the same management goals.

Organization

Implementation of an AEAM program is considered critical to the success of the Trinity River fishery restoration and maintenance effort. The following implementation scenario has been recommended to the Secretary of the Interior (USFWS and Hoopa Valley Tribe, 1999). Underlying principles are that "the best science" underpin yearly and within-year operating decisions and that all Trinity River AEAM program activities comply with applicable laws and permitting requirements. Additionally, independent peer review of all technical studies, analyses, and evaluations generated by the program would be conducted.

The proposed program would be directed by the Secretary of the Interior through a designee, who would serve as the principal contact for issues and decisions associated with the program. His or her responsibility would include ensuring that the Department of the Interior fulfills its obligation to restore and maintain the Trinity River fishery.

Components of the Trinity AEAM program would include a Trinity Management Council (TMC) supported by a Technical Modeling and Analysis Team (TMAT) and a rotating Scientific Advisory Board (SAB). The program would include consultation with other agencies and interest groups through periodic interaction through a stakeholders group. Scientific credibility would be ensured through external peer review of operating plans, models, sampling designs, and biological monitoring. The general roles and responsibilities of these groups are summarized below.

The TMC would be composed of fishery agency representatives. The Secretary's designee would serve as executive director. This TMC would approve fishery restoration plans and proposed changes to annual operating schedules submitted by a TMAT. The TMC would be the focal point for issues and decisions associated with the program. The executive director's responsibilities would include ensuring that the Department of the Interior fulfills its obligations for streamflow releases and rehabilitation of the river corridor habitats. The executive director in consultation with the council members would review, modify, accept, or remand the recommendations from the TMAT in making decisions about changes in reservoir releases, dam operations, and other management actions.

The TMAT would consist of a permanent group of scientists selected to represent the interdisciplinary nature of the decision process. Collectively, they should possess the skills and knowledge of several disciplines: water resources, engineering, geomorphology, water quality, fish population biology, riparian ecology, computer modeling, and data management. The TMAT responsibilities include design for data collection, methodology, analyses, modeling, predictions, and evaluating hypotheses and model improvements. This team would have delegated from the executive director a budget and the responsibility for preparing requests for proposals to conduct specialized data collections for model input and validation. Spatial coverage and sampling designs for long-term monitoring for status and trends would be developed in consultation with the management agencies and specific recommendations made to the TMC for funding. Funding for long-term monitoring would remain with the TMC.

The SAB would be appointed by the executive director. This group would be composed of prominent scientists appointed and appropriately compensated for 2- to 3-year rotating terms. The SAB would be responsible for semiannual review of the analyses, models, and projections of the TMAT as well as providing a science review of the overall management plans and implementation of the annual operating criteria and procedures as directed by the TMC. The SAB would also select outside peer reviewers and conduct the review and selection process for contracted data collection, research, or model development.

Case Example: Green River, Utah

The Green River is the major tributary of the Colorado River (Figure 16.3), which drains approximately 8% of the continental land mass of the United States. Because the Colorado River was isolated from other river basins for millions of years, 74% of its fish fauna are endemic species (Behnke and Benson, 1983). Extreme variability in flow, turbidity, and temperature created a harsh aquatic ecosystem. In the upper portions of the Colorado River basin, above Lake Powell, only 13 fish species coevolved. Four of the 13 native fishes inhabiting the middle Green River are now federally listed as endangered. The endangered Colorado squawfish (*Ptychocheilus lucius*) and razorback sucker (*Xyrauchen texanus*) evolved unique adaptations to the historic river flows and geomorphic structure but are now affected by water development activities that lead to altered flows, temperature regimes, sediment discharges, blocked access to floodplain habitat, and introduction of nonnative species.

Although basinwide effects have been considerable, there is great potential to manage river flow and redirect physical processes on the Green River, improving conditions for natural recruitment of endangered fishes. Flaming Gorge Dam, which now controls main-stem flows on the Green River system, is due for a revised Biological Opinion in 1999 to reassess winter and spring flow regimes for the benefit of endangered fishes. Biological and physical studies of the native fishes of the Colorado River basin have been conducted for over 20 years by various state and federal agencies, and hypotheses about key life-history needs have been developed.

Although changes to base flow levels have resulted in some positive biological response, peak flow management has been limited in scope. Adaptive management is proposed to provide flood flows to establish floodplain connectivity and redirect fluvial processes to enhance spawning bar conditions. The Green River with its highly variable and diverse geologic conditions and uniquely adapted fish fauna provides a classic example of how biological, physical, and chemical processes must be managed conjunctively to rehabilitate the ecosystem. We suggest that sufficient understanding of the life-history habitat requirements of the endangered Colorado squawfish and razorback sucker exists to design and implement an adaptive

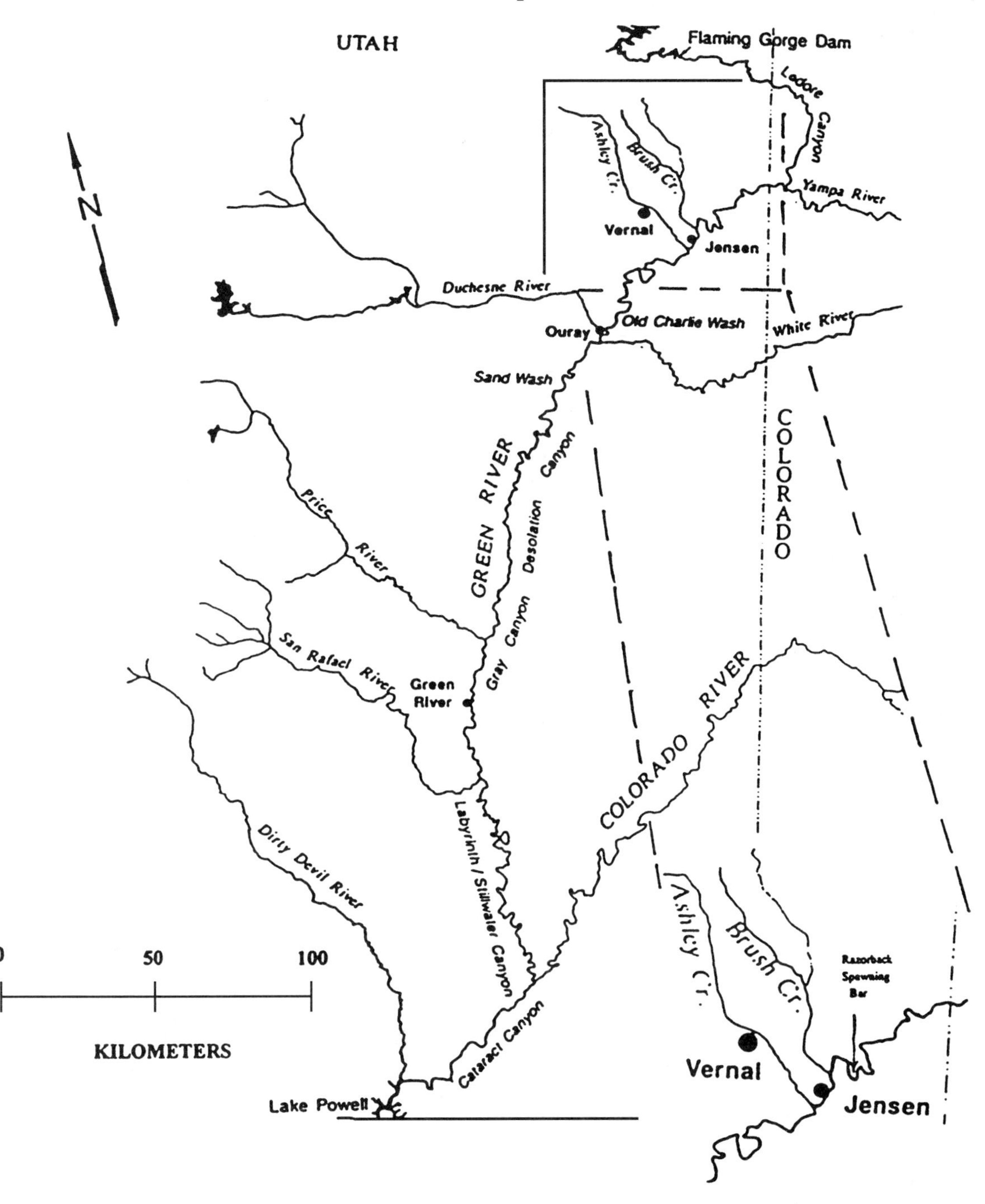

Figure 16.3. Green River, Utah and Colorado.

environmental assessment and management program for the Green River much the same as that formerly proposed for the Trinity River.

System Instability and Alteration of Fish Habitat

The Green River channel was historically wider, with a more active channel and floodplain than today (Graf, 1978). Gellis et al. (1991) noted that suspended-sediment loads decreased after the early 1940s in the Colorado Plateau portion of the Colorado River basin, although discharge of the major rivers did not change significantly. They attributed the reduced sediment loads to decreased sediment production resulting from improved land use conditions and to sediment storage in channels of tributary basins. Salt cedar (tamarisk) progressively colonized banks in an upstream direction at a rate of about 20 km yr^{-1} (Graf, 1978).

During moderate flow periods and in the absence of high scouring flows, sediment deposits were quickly stabilized by vegetation (including introduced tamarisk), and natural levees developed (Graf, 1978, 1985). Flooding frequency temporarily increased as even moderate flows filled narrowed, vegetation-encroached river margins. Sediment was deposited in low-velocity vegetated surfaces, on top of riverbank margins and midchannel islands, raising surface elevations from 1 to 3.5 m above low water level in the canyonlands reach of the Green River. Widths in this reach were reduced on average 27%, with a range of 13–55% (Graf, 1978).

Water depletions coupled with increased bank height have reduced overbank flooding frequency, severing the hydrological connection between the river channel and the floodplain. Dam construction in the 1960s further reduced seasonal flow variability and led to further reduction in floodplain connectivity. On the middle Green River after the closure of Flaming Gorge Dam, hydrologic connection to the highly productive and low velocity rearing habitat for native fishes has been reduced to a one in 10-year event instead of the pre-development regular recurrence of about once every 2 to 3 years. Instantaneous peak flows on the upper Green River at Greendale were reduced on average about 63% after closure of Flaming Gorge in 1962. Interestingly, the only peak flows exceeding the power plant capacity of 130 $m^3 s^{-1}$ (4700 $ft^3 s^{-1}$) occurred in 1983, 1984, 1986, and 1997, when higher than expected reservoir inflows forced spills and use of the jet tube bypass. These high flows were not planned releases to benefit endangered fish but were done to draw the reservoir level down.

Sediment Discharge Pattern Altered

Sediment supply from the upper Green River as measured at Greendale decreased by 3.6 million tons per year (reduced to 0 at the Greendale gauge) (Andrews, 1986). Effects on channel morphology were rapid. Fine sediments were scoured from the river bed near the dam, and channel widths decreased. However, sediment input from Vermillion Creek and Red Creek in and above Browns Park, located 70 km upstream of the confluence with the Yampa River, remained a significant source of sediment to the Lodore Canyon reach immediately downstream. Studies by Merritt (1997) have shown that geomorphic changes at Browns Park since Flaming Gorge Dam regulation have proceeded as a series of complex responses resulting in channel widening instead of narrowing. Merritt (1997) predicted that under current Flaming Gorge operations, island bar formation and eventual channel shelf formation on river margins may lead to development of a meandering system nested within the higher banks of the former floodplain.

Studies of the geomorphology of Lodore Canyon indicate that channel narrowing and channel bed aggradation are the present trends in this reach (Grams and Schmidt, 1997). Factors influencing the channel form in this reach are reduction in peak flows, subsequent reduction in sediment transport capacity, and stabilization of sediment by vegetation.

Mean annual sediment discharge decreased by 54%, from 6.9 to 3.21 million tons at Jensen (Andrews, 1986). Andrews (1986) calculated the mean annual supply of sediment to this reach to be 3.31 million tons. Therefore, the reach upstream from Jensen maintains a near balance of sediment supply and transport. However, a century or more may be required for the river to adjust to the effects of Flaming Gorge Dam. The computed effective discharge of the Green River at Jensen gauge decreased from 580 $m^3 s^{-1}$ (20,500 $ft^3 s^{-1}$) to 325 $m^3 s^{-1}$ (11,500 $ft^3 s^{-1}$) during the postdam period (1964–1981).

Colorado Squawfish's Unique Life History Adaptations to Flow Regime

Two river reach spawning locations have been confirmed for Colorado squawfish on the Yampa–Green River system. Colorado squawfish spawn during the lower third of the recessional limb of the annual snowmelt hydrograph on the Yampa River at midchannel bars located 26.6 and 29.8 km upstream of the confluence with the Green River within Dinosaur National Monument, respectively, and in Desolation Canyon on the Green River, 248.4 km from the Colorado River confluence. The movement of squawfish to specific spawning locations in Yampa and Desolation Canyons suggests that suitable spawning habitat in the basin is limited to locations with unique geomorphic conditions and that special imprinting (chemoreceptive) mechanisms may enable Colorado squawfish to return to specific natal spawning sites (Wick et al., 1983; Tyus, 1985; Harvey et al., 1993).

Recessional-flow bar dissection has been hypothesized as a primary process for creation of suitable spawning substrate (Harvey et al., 1993). The cleansing and reworking of spawning cobbles during the recessional limb of the hydrograph, combined with ponding above the primary bars, increases the local hydraulic slope, resulting in recessional-flow bar dissection. Shear stress increases over the bar deposits at lower discharges and enables motion of the bed and redistribution of cobbles into secondary and tertiary bars (Harvey et al., 1993).

The combination of geomorphic process and flow timing results in the unique quality of spawning sites on the Green–Yampa system. Colorado squawfish congregate here in large numbers, making populations vulnerable to severe decline if physical process or access to the sites is disrupted.

Nursery habitats for rearing squawfish are formed as backwaters and side channels in the active channel, when recessional flows decline to base flow. Sand deposits form embayments and slackwaters in eddy complexes. Side channels often become cut off at their upper ends, forming side-channel slackwaters. If flow regulation maintains discharge significantly above normal base flow levels during the rearing season, these normally low-velocity habitats remain through-flowing and young fishes are then transported downstream greater distances. Understanding spatial and temporal trends in channel aggradation and degradation and the role of Flaming Gorge Dam operations in providing stable low-flow rearing conditions are critical to sustaining the species.

This Green River nursery habitat extends from the Yampa River confluence downstream to the Colorado River and on to the inflow to Lake Powell, a distance of over 600 km. Modeling of key nursery sites after annual runoff forecasts may prove useful in prescribing channel maintenance flows for areas that are threatened by continued vegetation encroachment, channel narrowing, and habitat degradation.

Stable flow management during the base-flow period has been considered an important tool for enhancing Colorado squawfish young-of-year recruitment. However, flood flow management strategies are needed that renew Colorado squawfish spawning bars and rearing areas and also provide spawning and rearing habitat for other native species such as the razorback sucker.

Flood Pulse Spawning Adaptation of Endangered Razorback Sucker

Known spawning aggregations of razorback sucker in the upper Colorado River basin are located upstream of unconfined river reaches (Delta on the Gunnison River; Rifle, Grand Valley, and Moab on the Colorado River; and Rainbow Park, Jensen, and Ouray on the Green River). These unconfined river reaches on the Colorado River system had historic (predam) hydrologic connections with floodplain wetlands, providing potentially rich food resources for both young and adult fish. Wick (1997) hypothesized that the razorback sucker spawning strategy is correlated strongly, not only in location but also in timing, with the annual flood pulse. A geologic setting that provides a combination of cobble spawning sites upstream of productive floodplain nursery areas with suitable flow hydraulics and floodplain connectivity appears to be required for successful young-of-year recruitment.

On the middle Green River, razorback sucker congregate at the spawning bars just before peak spring flow at river mile 500, about 16 km upstream of Jensen, Utah (Figure 16.3), and on the lower Yampa River 0.8 km above its confluence with the Green River (Tyus and Karp, 1990). Water temperature averages 14°C (range 9 to 17°C) through the spawning period (Tyus and Karp, 1990). The Green River spawning site appears to be the most heavily used, which accounts for the fact that over 71% of all captures of razorback sucker in spawning condition from the Green River are from this location (USFWS database, Vernal Field Station, B. Haines, personal communication, 1997).

The Green River spawning site is located along the right side of a split channel around a large alluvial island. This is the first gravel bed reach extending across the entire channel upstream of the relatively unconfined sand bed river reach upstream of Ouray. The river channel at the spawning site varies in width from 115 m at the downstream constriction to 345 m at its broadest expansion around an island in the split flow reach.

Recruitment Success Related to Fluvial Geomorphic Processes and Floodplain Connection

Because sand deposition has been noted over the main razorback sucker spawning site on the middle Green River (Tyus and Karp, 1990), concern was raised about possible effects of sedimentation on spawning success. This concern prompted initiation of studies of the primary spawning site to determine the dynamics of sediment movement (Wick, 1997). Most spawning observations have been associated with gravel and cobble bars. Razorback suckers appear to work the substrates to clean them of sediments, as evidenced by observations of worn caudal and anal fins during this period. Sedimentation over spawning cobble-bar sites during or immediately after this period is hypothesized as a key factor contributing to razorback sucker recruitment failure some years.

Successive cross-section measurements in 1993 and 1994 and HEC-2 model analyses of the primary spawning site indicated that it is subject to backwater effects and sedimentation during early spring, when Flaming Gorge releases are high (Wick, 1997). Surface sand deposits located just upstream of the spawning channel are inundated and mobilized when flows exceed the calculated effective discharge of 325 $m^3 s^{-1}$ (11,500 $ft^3 s^{-1}$). At that time sand is scoured off these deposits and transported directly onto the spawning channel.

After closure of Flaming Gorge Dam in 1962 until the high water years 1983–1986, there was little detectable evidence of new recruitment into the Green River adult razorback population. Minckley (1989) aged 10 razorback suckers captured from the middle Green River in the early 1980s and found them to average 29 years of age.

Recently, recruitment of a few young adults into the spawning population has been associated with Flaming Gorge Dam bypass flows during 1983, 1984, and 1986 when high reservoir levels combined with high peak runoff flows resulted in inundation of bottomlands adjacent to the river (Wick, 1997). The capture of these young recruits in 1990–1993 provided encouraging evidence that recruitment could be assisted by peak flow management. Based on their size and growth increments between recaptures, their years of origin were back-calculated to the 1983–1986 high flow years (Modde et al., 1996).

Sampling since 1993 has revealed that larva capture is highly variable year to year (Muth, 1995; Muth et al., 1998). Loss of connectivity between the river and suitable floodplain nursery habitat is suspected of contributing to recruitment failure. This hypothesis was supported recently in the finding of 28 naturally recruited juvenile razorback in 1995 (Modde, 1996) and another 52 young razorback in 1996 as a result of reconnecting Old Charlie Wash to the river during spring runoff (USFWS Vernal Field Office, T. Modde, 1997, personal communication). Old Charlie Wash is a specially modified floodplain habitat with fish harvest facilities.

Successful recruitment of razorback suckers through their first year of life has two necessary components: (1) successful spawning, egg deposition, and hatching of larvae depend on relatively clean gravel and cobble on the primary spawning bars, with little deposition of fine sediments before emergence; (2) after emergence, larvae must have access to suitable floodplain nursery sites and conditions that provide adequate food and cover. Sufficient duration and proper timing of hydrologic connectivity to achieve access to floodplains and provide primary production for larvae growth (size) sufficient to avoid predators when forced to return to the active channel are critically important.

Both habitat features must be provided sequentially (in terms of timing and duration of flows) in a given water year to produce successful spawning and rearing, leading to successful recruitment. This successful sequence has not occurred very frequently, based on stream gauge records and recent estimates of the razorback population size and age structure on the middle Green River (Modde et al., 1996).

Recruitment success was determined by establishing a standardized larva sampling program with light traps. Field deployment occurred in various nursery habitat types, including flooded tributary streams, cutoff secondary channel backwaters, and wetland complexes (Muth, 1995; Muth et al., 1998).

The biological information in the form of larval razorback sucker light-trap capture data support the sedimentation hypothesis at the primary spawning site. Very low larval catch rates in 1995 provided reason to suspect razorback sucker production declines during high flow years when peak flow timing is altered on the Green River by Flaming Gorge releases. A relatively low larval catch in 1996, observations of sedimentation on the primary spawning bar in mid-May 1996, and HEC-6 (Hydrologic Engineering Center, 1992) simulation modeling that agreed with observed sedimentation patterns caused by early Flaming Gorge releases confirmed the problem. Wick (1997) concluded that razorback sucker production at the primary site is particularly vulnerable during wet water years when flow release patterns at Flaming Gorge are higher than natural inflow to the reservoir during the early spring runoff period. Flows then exceed the present effective discharge, inundating river margin sand deposits, and moving the previously deposited sediments onto the spawning channel.

Applying AEAM to the Green River

Adaptive management has been initiated by the U.S. Bureau of Reclamation as part of Flaming Gorge Dam operations. The flow management actions taken to date have been to reduce postrunoff releases from Flaming Gorge Dam and to limit daily fluctuation to increase the duration, quantity, and quality of suitable backwater nursery habitat within the active channel to benefit Colorado squawfish rearing during late summer and fall base flows. The recent findings of Wick (1997) illustrate the importance of the timing of flood flow releases and the sequencing of events within and across years in order to manage Green River flows to benefit Colorado squawfish and razorback sucker. Implementation of the AEAM approach to managing Flaming Gorge Dam is suggested to improve understanding of this timing issue.

The required AEAM steps are outlined below.

Steps 1, 2, and 3 – Goal Setting. Determine baseline conditions and develop hypotheses. These already exist to some degree as described above and by Wick (1997). The basic premise is that recruitment of larvae and juvenile razorback sucker can be enhanced by flow management, which reduces sedimentation over spawning sites (gravel bars), and restores access to and productivity of floodplain habitat. The Green River example is further developed through Steps 4–10 of the process (Figure 16.1).

Steps 4 and 5 – Select and Implement Management Actions. When entering a potential wet water year during which daily flows approaching 430 $m^3 s^{-1}$ (15,000 $ft^3 s^{-1}$) on the Yampa River are projected, the middle Green River peak should be enhanced by building on the Yampa River peak with flow support (jet tube or bypass flows $\geq 244\ m^3\ s^{-1}$) from Flaming Gorge Dam. These release flows should coincide with the natural flood timing of the upper Green River flows and build on the later portion of the Yampa River peak flow. For the predam period of record of 1947 to 1962, the mean annual peak at Jensen was 680 $m^3\ s^{-1}$ (24,000 $ft^3\ s^{-1}$) [FLO Engineering, Inc., 1997]. This peak flow level is suggested as a minimum target flow. Spawning habitat conditions for razorback sucker may be improved in wet water years by managing Flaming Gorge releases to achieve this flow as measured at the Jensen gauge.

In subsequent flow years the rising limb of the hydrograph should parallel natural inflow patterns, but Flaming Gorge releases should be managed to keep flows below the level of inundation of sediment deposits established the previous year. This flow management sequence should distribute sediment higher along the channel margins on the Green River during wet water years and subsequently reduce sediment movement over the spawning channel during the early spring period in subsequent water years. With this scenario, the conditions in the spawning channel should improve and production of razorback sucker larvae should increase.

With this approach, a minimum of two successive years of flood flow management will be needed: the first year to move sediment from channel beds to higher elevations on the river margins and the second year to provide clean cobble bars during the spawning period. During the second (high recruitment) year, flows should be managed during spawning to be below the level at which previously formed sediment deposits are inundated (present level is 325 $m^3\ s^{-1}$ or 11,500 $ft^3\ s^{-1}$) until razorback larvae begin emerging from spawning sites. Flows should then be brought up to levels that allow larvae access to the floodplain nursery habitat. These levels are estimated to range between 450 $m^3\ s^{-1}$ (15,800 $ft^3\ s^{-1}$) and 643 $m^3\ s^{-1}$ (22,700 $ft^3\ s^{-1}$).

Step 6 – Monitoring. Considerable studies have been conducted over the past 20 years on the Green River and ongoing biological and physical monitoring programs are in place. However, to date tested procedures or protocols for assessing downstream effects of Flaming Gorge Dam to reoperation releases or experimental flow patterns have not been established. Baseline physical data are needed to complement existing biological information and to facilitate development of a suite of predictive models to help design and evaluate flow regimes that will provide suitable river habitat conditions when necessary for a fish's life-stage success. Flow routing models are needed to estimate the hydrograph at key spawning and rearing sites. Then models that evaluate detailed flow characteristics, channel equilibrium, aggradation, and degradation (i.e., HEC-2, HEC-6) can be used to estimate river and channel response to flow management.

Many of these activities are ongoing or completed for the known spawning sites and nearby rearing areas for Colorado squawfish and razorback sucker. However, complete characterization of the Green River down to the confluence with the Colorado River, and of the Colorado River down to Lake Powell, is needed.

Discharge. Gauges for water discharge measurement are in place at key locations (Greendale, Jensen, Green River, and just above and below the Little Snake confluence on the Yampa River and at Lily on the Little Snake). Other temporary water-stage loggers have been added at Mitten Park, Ouray, and Canyonlands on the Green River to calibrate a flow routing model such as FLO-2D (FLO Engineering, Inc., 1998). Additional discharge measurements may be needed to facilitate accurate flow routing downstream past the Colorado River and into Lake Powell.

Sediment. Sediment measurements were resumed at Jensen in 1996 after suspension in 1979. A sediment review panel met in February 1997 to make recommendations on locations, frequency, and types of sediment analysis needed. A sediment measurement program was recommended to assess sediment transport and bed material conditions. Although resumption of historical suspended sediment measurements and initiation of bedload measurements were recommended for 1998, budget limitations enabled only periodic measurements at Jensen and at two sites near Deerlodge Park on the Yampa River. These data are needed to evaluate sedimentation at the razorback spawning sites using HEC-6. Other models, such as RMA2 with SED-2D, should be evaluated for possible use. Sediment transport evaluation could be integrated into the FLO-2D routing model. Routing model applications eventually should be expanded to include sediment contribution by all major tributaries.

Step 7 – Compare Predictions and Observations. The approach taken for the Green River has been to develop a flow routing model (FLO-2D) that can account for overbank water storage in floodplain wetlands and water stored in the channel from Flaming Gorge Dam to the confluence with the Colorado River. FLO-2D is a finite difference model using a diffusive wave approximation to the full dynamic momentum equation to route a flow hydrograph. This model can be used to evaluate flow releases from Flaming Gorge Dam to optimize water delivery to key habitat areas and to ensure proper stage, timing, and duration of flows to achieve conditions that favor larval production and access to hydrologically connected floodplain areas.

Local areas of concern can be modeled in greater detail by using the one-dimensional HEC-2 and HEC-6 models. If the output of these models agrees with empirical support data, the models can be considered credible and useful to generate annual predictions. Established geomorphic and habitat transects should be resurveyed after experimental releases and compared with model predictions. Future utilization of the HEC-6 model could be made by using estimated suspended load and bedload data for experimental flow years.

Step 8 – Restate System Status. The status and the degree of progress toward meeting the stated management objectives can be determined by comparison with the previous year's observations.

Steps 9 and 10 – Adapt and Modify Actions as Needed. Scientific evidence is presented to the managers and stakeholders in support of or refuting the original hypotheses. The hypotheses are restated and the models recalibrated awaiting the next round of forecasts, decisions, and simulations. If certain hypotheses are rejected or alternatives are proposed, experimental flow releases or other management actions should be designed and submitted for management consideration before the winter-spring forecast period. Table 16.4 lists the models and major monitoring data needs as described for the Green River.

This case illustrates the importance of managed flood flows and interannual flow variability for creating and sustaining suitable habitat conditions for successful young-of-year Colorado squawfish and razorback sucker production (each with unique reproductive strategies). In practice several other aspects of the Green River ecosystem (i.e., maintenance of native riparian vegetation) should be evaluated in a similarly rigorous manner and factored into the annual program of establishing Flaming Gorge Dam operating criteria and procedures for a given year.

A New Paradigm for Management of Rivers for Human Use and Ecological Integrity

We conclude that (1) managed floods are important for sustaining stream biota in alluvial streams; (2) existing flow-planning frameworks do a poor job of establishing flood needs; (3) AEAM is suggested as a framework for incorporating flood needs in flow planning; and (4) the Trinity and Green Rivers provide examples to show that an integrated science approach to flow and flood management is much needed and feasible.

A Record of Decision on operations of the Trinity River reservoirs is due in 1999, and a new biological opinion on Flaming Gorge Dam is also due in 1999. These decisions provide the opportunity to use physical and biological models to facilitate the adaptive environmental assessment and management approach to restoring the unique fish fauna by rehabilitating the river channels to provide habitats much improved over existing conditions. The concept of restoring the natural hydrograph pattern discussed by Poff et al. (1997) is becoming generally accepted, especially the role of hydrologic variability in sustaining the ecological integrity of river ecosystems. This is also discussed by Stanford et al. (1996). An adaptive management approach of experimental test flows should be accompanied by physical process modeling and an evaluation program to monitor the physical and biological responses. Such a program, similar to the recommendations of Ligon et al. (1995), has been initiated for the Green River but needs to be supported by a rigorous prediction, monitoring, and model validation program. The opportunity exists to establish a more formal AEAM program on the Trinity River in 1999. In both cases, creation of an interdisciplinary team of scientists – who run simulations, design and carry out monitoring programs, and offer recommendations to management and the stakeholders – is critical to successful implementation of the AEAM philosophy. The most common problem with the use of models in decision making is failure to state the validation criteria for acceptability. Rykiel (1996) proposed that model outputs fall within the 95% confidence interval 75% of the time for the most important variables in dynamic models.

Most biological estimates, such as fish population size, have confidence intervals that may exceed ±50%, and many are simply indices to indicate trends. Likewise, information on sediment transport is very limited for most large river systems. Certainly a more focused monitoring effort is needed for the two cases presented here. To adequately manage river systems for multiple use and conserve the biotic resources, ongoing

Table 16.4. *Suite of models for predicting habitat responses to regulated flood flows on the Green River, Utah*

	Hydrology	Geomorphology	Sedimentation	Habitat	Razorback sucker recruitment
Data needs	Stage recorders Reservoir elevation and storage volume Precipitation records Digitized surface topography for overland flow	Survey of x sections in all alluvial reaches and potential floodplain connection areas Bed material size	Transport by particle size fraction Fate: scour and fill x sections Bed material at restoration site	Aerial photography Airborne videography x sections	Field sampling of larval abundance at reference site below spawning areas
Techniques	Empirical Forecast Annual Declaration of water year type (wet, dry, normal) Mass balance Finite difference flood routing	Videography Aerial photography comparisons X section comparison at reference sites	Bed load/suspended load Discharge recording Tributary measurements Particle size fractions Gradations	Computed areas of low velocity associated with side channels and backwaters Computed area and duration of floodplain inundation	Standardized young-of-year sampling protocol Light trapping for larvae near known spawning
Models	CRDSS FLO-2D Riverware	SAM HEC-6 (RMA2 with SED-2D)	HEC-6 (RMA2 with SED-2D)	HEC-2 PHABSIM TSLIB	Individual-based models (Crowder et al., 1992)
Simulations/ predictions	Water surface elevations Longitudinal distribution of discharge Overbank flooding	River segment estimates of channel stability down to confluence	Scour & deposition estimates at x sections	Empirical observation of extent of floodplain inundation Digitized area of inundation from aerial surveys	Annual index to young-of-year success

monitoring of flow, sediment, and biological status is essential. With such data and the use of simulation models, river systems can be adaptively managed. Such informed decision making, utilizing water supply forecasting, and predictions of system response, is within the state-of-the art. What is needed is a focused interdisciplinary effort that involves physical and biological scientists. Peer review of all analyses, experimental design, and monitoring are essential to establish scientific credibility.

References

Andrews, E.D. (1986). Downstream effects of Flaming Gorge Reservoir on the Green River, Colorado and Utah. *Geological Society of American Bulletin*, **97**, 1012–1023.

ASTM. (1994). *Compilation of ASTM standard definitions*, 8th ed. Philadelphia, PA: American Society for Testing and Materials. 600 pp.

Bain, M.B. and Boltz, J.M. (1989). *Regulated streamflow and warmwater fish: A general hypothesis and research agenda.* U.S. Fish and Wildlife Service Biological Report, **89**(18), 28 pp.

Bartholow, J.M. (1996). Sensitivity of a salmon population model to alternative formulations and initial conditions. *Ecological Modeling*, **88**, 1, 215–226.

Bartholow, J.J., Sandelin, J., Coughlan, B.A.K., Laake, J., and Moos, A. (1997). *SALMOD, a population model for salmonids: User's Manual*, Beta Test Version 2.0, U.S. Geological Survey, Fort Collins, CO. Unpublished computer documentation. September 1997. 86 pp.

Bayha, K. (1978). *Instream flow methodologies for regional and national assessments.* Instream Flow Information Paper No. 7, FWS/OBS 78/61, U.S. Fish and Wildlife Service, Washington, DC. 98 pp.

Bayley, P.B. (1991). The flood pulse advantage and the restoration of river-floodplain systems. *Regulated Rivers: Research and Management*, **6**, 75–86.

Behnke, R.J. and Benson, D.E. (1983). *Endangered and threatened fishes of the Upper Colorado River Basin.* Colorado State University Cooperative Extension Service Bulletin 503A, Ft. Collins, CO. 38 pp.

Bovee, K.D. (1988). *Use of the instream flow incremental methodology to evaluate influences of microhabitat variability on trout populations in four Colorado streams.* Proceedings of the Western Division of the American Fisheries Society, Albuquerque, NM. 31 pp.

Bovee, K.D., Lamb, B.L., Bartholow, J.M., Stalnaker, C.B., Taylor, J., and Henriksen, J. (1998). *Stream habitat analysis using the Instream Flow Incremental Methodology.* U.S. Geological Survey, Biological Resources Division Information and Technology Report USGS/BRD-1998–0004. Washington, DC: U.S. Govt. Printing Office, viii + 131 pp.

Carling, P. (1995). Implications of sediment transport in instream flow modelling of aquatic habitat, Chapter 2. In *The ecological basis for river management*, eds. D.M. Harper and A.J.D. Ferguson, pp. 17–31. John Wiley and Sons, Ltd., London.

Collier, M., Webb, R.H., and Schmidt, J.C. (1996). Dams and rivers: Primer on the downstream effects of dams. *U.S. Geological Survey Circular*, 1126. 94 pp.

Crance, J.H. (1988). *Relationships between palustrine wetlands of forested riparian floodplains and fishery resources: A review.* U.S. Fish and Wildlife Service Biological Report, **28**(32). Washington, DC. 27 pp.

Crowder, L.B., Rice, J.A., Miller, T.J., and Marschall, E.A. (1992). Empirical and theoretical approaches to size-based interactions and recruitment variability in fishes. In *Individual-based models and approaches in ecology: populations, communities and ecosystems*, eds. D.L. DeAngelis and L.G. Gross, pp. 237–255. Chapman and Hall, New York.

FLO Engineering, Inc. (1997). *Green River floodplain habitat restoration investigation – Bureau of Land Management sites and Ouray National Wildlife sites near Vernal, Utah.* Report prepared for Recovery Program for the Endangered Fishes of the Upper Colorado. FLO Engineering, Inc., Breckenridge, CO. 62 pp.

FLO Engineering, Inc. (1998). *Green River FLO-2D discharge routing model Flaming Gorge Dam to Colorado River confluence.* Report prepared for National Park Service and Fish and Wildlife Service. FLO Engineering, Inc., Breckenridge, CO. 15 pp with appendices.

Gellis, A., Hereford, R., Schumm, S.A., and Hayes, B.R. (1991). Channel evolution and hydrologic variations in the Colorado River basin: Factors influencing sediment and salt loads. *Journal of Hydrology*, **124**, 317–344.

Ghanem, A., Steffler, P., Hicks, F., and Katopodis, C. (1994). Two-dimensional finite element flow modeling of a physical habitat. In *Proceedings of the 1st International Symposium on Habitat Hydraulics*, pp. 84–88. The Norwegian Institute of Technology, Trondheim, Norway.

Gore, J.A. and Shields, F.D. Jr. (1995). Can large rivers be restored? *BioScience*, **45**, 142–152.

Graf, W.L. (1978). Fluvial adjustments to the spread of tamarisk on the Colorado Plateau region. *Bulletin of the Geographical Society of America*, **89**, 1491–1501.

Graf, W.L. (1985). *The Colorado River: instability and river basin management.* Association of American Geographers, Washington, DC, 86 pp.

Grams, P.E. and Schmidt, J.C. (1997). *Geomorphology of the Green River in Dinosaur National Monument.* Draft final report. Submitted to Dinosaur National Monument, Colorado. National Park Service. Project Number DINO-R93-0039.

Haley, K.B. (1990). Operational research and management in fishing. In *Operations research and management in fishing*, ed. A.G. Rodrigues, pp. 3–7. NATO ASI Series F: Applied Sciences, vol. 189, Dordrecht, The Netherlands.

Harvey, M.D., Mussetter, R.A., and Wick, E.J. (1993). A physical process-biological response model for spawning habitat formation for endangered Colorado squawfish. *Rivers*, **4**, 114–131.

Hilborn, R. and Walters, C.J. (1992). Quantitative fisheries stock assessment choice. In *Dynamics and uncertainty.* Chapman and Hall, New York. 570 pp.

Holling, C.S. (1977). *Adaptive environmental assessment and management.* University of British Columbia, Vancouver, Canada, 597 pp. with appendices.

Holling, C.S. (1978). *Adaptive environmental assessment and management.* John Wiley and Sons, New York.

Holling, C.S. (1998). Two cultures of ecology. *Conservation Ecology* [online]2, 2, 4. Available from the Internet. URL: http://www.consecol.org/Journal/vol2/iss2/art4

Hydrologic Engineering Center. (1992). *Guidelines for the calibration and application of computer program HEC-6*, Training Document No. 13. U.S. Army Corps of Engineers, Davis, CA.

Hynes, H.B.N. (1970). *The ecology of running waters.* University of Toronto Press, Toronto, Canada, 555 pp.

IAHR. (1994). *Proceedings of the First International Symposium of Habitat Hydraulics.* The Norwegian Institute of Technology, Trondheim, Norway, 637 pp.

International Institute for Applied Systems Analysis. (1979). *Expect the unexpected: An adaptive approach to environmental management.* Executive Report 1, International Institute for Applied Systems Analysis, Luxemburg, Austria, 16 pp.

Jowett, I.G. (1993). Models of the abundance of large brown trout in New Zealand rivers. *North American Journal of Fisheries Management*, **12**, 417–432.

Junk, W.J., Bayley, P.B., and Sparks, R.E. (1989). The flood pulse concept in river floodplain systems. *Canadian Journal of Fisheries and Aquatic Sciences*, Special Publication, **106**, 110–127.

Karr, J.R. (1991). Biological integrity: A long-neglected aspect of water resource management. *Ecological Applications*, **1**, 66–84.

Kondolf, G.M. and Micheli, E.R. (1995). Evaluating stream restoration projects. *Environmental Management*, **19**(1), 1–15.

Lamb, B.L. and Meshorer, H. (1983). Comparing instream flow programs: A report on current status. In *Proceedings Conf. Advances in Irrigation and Drainage: Surviving External Pressures*, pp. 435–443. American Society of Engineers, Irrigation and Drainage Division, Jackson, WY.

Leclerc, M., Capra, H., Valentin, S., Boudreault, A., and Cote, Y., eds. (1996). *Ecohydraulics 2000.* Proceedings Second International Symposium on Habitat Hydraulics, Quebec, Canada. INRAS-Eau, Vol. A, 893 pp; Vol B, 995 pp.

Leonard, P.M. and Orth, D.J. (1988). Use of habitat guilds of fishes to determine instream flow requirements. *North American Journal of Fisheries Management*, **8**, 399–409.

Lestelle, L.C., Mobrand, L.E., Lichatowich, J.A., and Vogel, T.S. (1996). *Ecosystem diagnosis and treatment (EDT). Applied ecosystem analysis – a primer.* Prepared for U.S. Department of Energy, Bonneville Power Administration, Environmental Fish and Wildlife, Portland, OR, 95 pp.

Ligon, F.K., Dietrich, W.E., and Trush, W.J. (1995). Downstream ecological effects of dams: A geomorphic perspective. *Bioscience*, **45**, 183–192.

McBain, S. and Trush, W. (1997). *Trinity River channel maintenance flow study final report.* Prepared for the Hoopa Valley Tribe, Trinity River Task Force.

McLain, R.J. and Lee, R.G. (1996). Adaptive management: Promises and pitfalls. *Environmental Management*, **20**, 437–448.

Merritt, D.M. (1997). *Riparian vegetation and geomorphic features on regulated and unregulated rivers: Green and Yampa, northwestern Colorado.* Unpublished MS thesis, Ft. Collins, CO, 65 pp.

Milhous, R.T., Updike, M.A., and Schneider, D.M. (1989). *Physical habitat simulation system reference manual – Version II.* Instream Flow Information Paper 26. FWS/OBS-89(16).

Milhous, R.T., Bartholow, J.M., Updike, M.A., and Moos, A.R. (1990). Reference manual for the generation and analysis of habitat time series – Version II. Instream Flow Information Paper No. 27. *U.S. Fish and Wildlife Service Biological Report 90*(16). 249 pp.

Minckley, W.L. (1989). *Aging of Colorado squawfish/razorback sucker by otolith examination.* Final report for U.S. Fish and Wildlife Service. Arizona State University, Tempe, 9 pp.

Modde, T. (1996). Juvenile razorback sucker (*Xyrauchen texanus*) in a managed wetland adjacent to the Green River. *Great Basin Naturalist*, **56**, 375–376.

Modde, T., Burnham, K.P., and Wick, E.J. (1996). Population status of the razorback sucker in the middle Green River. *Conservation Biology*, **10**, 110–119.

Muth, R.T. (1995). *Development of a standardized monitoring program for basin wide evaluation of restoration activities for razorback sucker in the Green and Upper Colorado River systems.* Final Report to the Recovery Implementation Program. Larval Fish Laboratory, Colorado State University, Fort Collins, CO, 48 pp.

Muth, R.T., Haines, G.B., Meismer, S.M., Chart, T.E., Synder, D.E., and Bundy, J.M. (1998). *Reproduction and early life history of razorback sucker in the Green River, Utah and Colorado, 1992–1996.* Draft final report for Project 34. Submitted to the Recovery Implementation Program for the Endangered Fish Species in the Upper Colorado River Basin. Contribution 101. Colorado Sate University Larval Fish Laboratory, Fort Collins, CO.

Nehring, R.B. and Anderson, R.M. (1993). Determination of population-limiting critical salmonid habits in Colorado streams using the Physical Habitat Simulation System. *Rivers*, **4**, 1–19.

Oreskes, N., Shrader-Frechette, K., and Belitz, K. (1994). Verification, validation, and confirmation of numerical models in the earth sciences. *Science*, **263**, 641–646.

Petts, G.E. and Maddock, I. (1994). Flow allocation for in-river needs. In *The rivers handbook*, eds. P. Calow and G.E. Petts, pp. 289–307. Blackwell Scientific Publications, Oxford.

Poff, L.N., Allan, J.D., Bain, M.B., Karr, J.R., Prestegaard, K.L., Richter, B.D., Sparks, R.E., and Stromberg, J. C. (1997). The natural flow regime: A paradigm for river conservation and restoration. *BioScience*, **47**, 769–784.

Railsback, S. F., Blackett, R. F., and Pottinger, N. D. (1993). Evaluation of the fisheries impact assessment and monitoring program for the Terror Lake hydroelectric project. *Rivers*, **4**, 312–327.

Rykiel, E.J. Jr. (1996). Testing ecological models: The meaning of validation. *Ecological Modeling*, **90**, 229–244.

Stalnaker, C.B. (1982). Instream flow assessments come of age in the decade of the '70's. In *Research and scientific accomplishments for enhancement of fish and wildlife resources in the decade of the '70's*, eds. W.T. Mason and S. Iker, pp. 119–141. Environmental Protection Agency Habitat Protection Monograph. EPA 600/8-82-022. Washington, DC.

Stalnaker, C.B. (1994). Evolution of instream flow habitat modeling. In *The rivers handbook*, eds. P. Calow and G.E. Petts, pp. 276–286. Blackwell Scientific Publications, Oxford.

Stalnaker, C.B., Lamb, B.L., Henriksen, J., Bovee, K., and Bartholow, J. (1995). *The instream flow incremental methodology: a primer for IFIM.* National Biological Survey, Biological Report 29. Washington, DC. 45 pp.

Stanford, J.A., Ward, J.V., Liss, W.J., Frissell, C.A., Williams, R.N., Lichatowich, J.A., and Coutant, C.C. (1996). A general protocol for restoration of regulated rivers. *Regulated Rivers: Research and Management*, **12**, 391–413.

Theurer, F.D., Voos, K.A., and Miller, W.J. (1984). Instream water temperature model. *Instream Flow Information Paper No. 16.* U.S. Fish and Wildlife Service (FWS/OBS-84/15). Washington, DC.

Trihey, E.W. and Stalnaker, C.B. (1985). Evolution and application of instream flow methodologies to small hydropower development. In *Symposium on small hydropower and fisheries*, eds. F.W. Olson, R.H. White, and R.H. Hamre, pp. 176–183. American Fisheries Society, Bethesda, MD.

Tyus, H.M. (1985). Homing behavior noted for Colorado squawfish. *Copeia*, **1985**, 213–215.

Tyus, H.M. and Karp, C.A. (1990). Spawning and movements of razorback sucker, (*Xyrauchen texanus*) in the Green River Basin of Colorado and Utah. *Southwestern Naturalist*, **29**, 289–299.

U.S. Fish and Wildlife Service. (1994). *Restoration of the mainstem Trinity River: Background report.* Trinity River Fishery Resource Office, Weaverville, CA, 14 pp.

U.S. Fish and Wildlife Service and Hoopa Valley Tribe. (1999). *Trinity River flow evaluation. Final report.* Prepared by FWS, Coastal California Fish and Wildlife Office, Arcata, CA.

Vannote, R.L., Minshall, G.W., Cummins, K.W., Sedell, J.R., and Cushing, C.E. (1980). The river continuum concept. *Canadian Journal of Fisheries and Aquatic Sciences*, **37**, 130–137.

Walters, C. (1986). *Adaptive management of renewable resources.* Macmillian, New York.

Welcomme, R.L. (1985). *River fisheries.* Food and Agriculture Organization of the United Nations, Fisheries Technical Paper 262. Rome, Italy, 330 pp.

Welcomme, R.L. (1989). Floodplain fisheries management. In *Alternatives in regulated river management*, eds. J.A. Gore and G.E. Petts, pp. 209–233. Chemical Rubber Corporation (CRC), Press, Inc., Boca Raton, FL.

Wick, E.J., Stoneburner, D.L., and Hawkins, J.A. (1983). *Observations on the ecology of the Colorado squawfish (Ptychocheilus lucius) in the Yampa River*. Report Number 83-7 Colorado Water Resources Field Support Laboratory. National Park Service, Fort Collins, CO, 55 pp.

Wick, E.J. (1997). *Physical process and habitat critical to the endangered razorback sucker on the Green River, Utah*. Unpublished PhD Dissertation. Department of Earth Resources, Colorado State University, Ft. Collins, CO, 145 pp.

Wilson, W.J., Trihey, E.W., Baldrige, J.E., Evans, C.D., Thiele, J.G., and Trudgen, D.E. (1981). *An assessment of environmental effects of construction of the proposed Terror Lake Hydroelectric Facility, Kodiak, Alaska instream flow studies*. Final report to the Kodiak Electric Association. Anchorage, Alaska.

Yun-Sheng, Y. and Xiao-Dong, J. (1991). Evaluating sustainable instream flow and water supply potential of a river by simulation. In *Hydrology for the water management of large river basins*. Proceedings of the Vienna Symposium, August 1991. IAHS Publ. No. 201.

SOCIETAL CONTROLS ON HUMAN RESPONSES TO FLOOD HAZARDS

CHAPTER 17

Cultural Perspectives of Flooding

Melinda J. Laituri
Department of Earth Resources
Colorado State University

This chapter examines the complex relationships that exist between culture, technology, poverty, development, and science. Each offers an explicit frame of reference from which to examine the effect of floods on humans and their environment. Throughout the past humans have responded to floods either by avoiding hazardous areas or by structurally modifying the landscape. Both types of solutions have had a significant impact on societal structures. Technology and policy have profoundly influenced the effects of floods that have resulted in unintended consequences for people living with floods. Alternative solutions may provide new methods for solving hazardous situations associated with flooding, but globalization of the world economy provides another frame of reference that will influence the human relationship with the environment in general and with floods in particular.

Introduction

Hazardous situations, such as floods, pose risks that demand that social systems respond and attempt to cope with disastrous outcomes. Comparative analyses of these social systems reveal the complexities of social interrelationships and local adaptations to extreme events. Placed in a cultural context, questions such as what is a flood, what are flood hazards, and what risks are associated with floods allow us to study risk perception and response strategies due to societal, political, and economic differences and similarities.

Culture is a complex, dynamic concept. Marston and Knox define culture as "a shared set of meanings that are lived through the material and symbolic practices of everyday life" (1998, p. 190). An essential part of one's culture is a worldview: the frame of reference for organizing life's activities. This frame of reference is an essential cultural ingredient of every society and reflects the need to establish an order to explain the hows and whys of daily existence. Culture implies that we learn from our elders and neighbors a way of living in a place that is more refined or better adapted than our genes alone can offer (Nabhan, 1997, p. 4). Therefore, culture necessarily includes tools, technology, and local knowledge. Collectively, culture is subject to reevaluation and redefinition by the group as other influences such as social, political, economic, and historical factors change.

Culture provides socially constructed myths about Nature. Myth refers to a system of beliefs that are plausible rather than demonstrably true. These myths or systems of belief become part of a worldview and influence human interpretations of natural phenomena. Worldviews provide powerful cultural lenses, magnifying one danger, obscuring another threat, and selecting others for minimal attention or even disregard. To an extent, culture mediates the human relationship with Nature. It influences behavior and is manifested in the artifacts and organization of society as well as society's molding of the environment. This is a recursive relationship in that, as we change Nature to provide for our material culture, we change ourselves.

This recursive relationship is exhibited in the human response to hazardous events. Humans react to their environment as they perceive and interpret it through previous experience and knowledge. Understanding hazardous situations is a cognitive process made up of several components: defining the hazard, assessing the risk, and determining the response. Hazard is defined as the source of danger, the threat or adverse consequences posed to humans and nature by phenomena transmitted by the environment (Bjonness, 1986; Hiessl and Waterstone, 1986). Identifying a hazard is based on local knowledge or a collective "common sense" informed by a temporal dimension (i.e., relative frequency and magnitude of a flood) situated in a particular locale (Kirby, 1990). The notion of risk involves uncertainty and the possibility of some kind of loss, damage, or adverse effect (Hiessl and Waterstone, 1986); it is assessed according to the type and degree of threat posed to humans, animals, and property. Mitigative action, preventive strategies, and imaginative belief systems are developed by societies based on hazard identification and perception.

Hazardous situations are further mediated by technology, globalization, and levels of development. More-developed and less-developed countries experience different dynamics of risk (Smith, 1990). For example, technological projects (e.g., large-scale dams for hydroelectric power and flood control) in less-developed countries are designed to assist in the development process yet may compound hazardous situations through forced relocation. In these situations, migrating populations are exposed to risk in areas undergoing rapid economic, social, and environmental change. Introduction of new technology influences local knowledge and how human culture negotiates the human relationship with Nature. New technologies and development programs are fueled by the process of globalization – the increasing interconnectedness of different parts of the world primarily through economic, trade, and market strategies. Worldviews, in particular indigenous belief systems, are threatened by the homogenizing influence of globalization (Shiva, 1997). Local strategies for hazard mitigation may be improved, displaced, or removed because of these global influences.

This chapter examines the cultural context in which flood hazards are framed and debated and how risk is socially constructed. The role of belief systems or myth and social cognition inform the perception of risk. This perception contributes to and assists in maintaining a particular way of life. However, risk is often politically negotiated because of competing, distinctive, and often mutually exclusive sets of beliefs and values. Mixed cultural orientations, complex social alliances, and evolving worldviews are fundamental to the debate about the social and political meanings of solutions to hazards and their implications for alternative ways of life. How does one myth of Nature – one set of beliefs about what the world is like, what its risks are, and who is to blame for untoward events – come to seem more sensible than another? How are perceptions about hazards affected by perspectives such as science, development, and poverty; each provides a distinctive cultural lens or frame of reference with which to assess and understand flooding. Increasingly, the cultural milieu is no longer linked solely to place but is conditioned by the effects and outcomes of technology and globalization. In turn, globalization and the changing nature of the economic underpinnings of society influence ways in which risk is socially constructed. This chapter examines how the frame of reference has shifted with regard to floods and how this shift has influenced cultural contexts and responses.

Myth and Culture

Societies have contended with floods throughout history. This is evidenced not only by the geologic record but also through technological adaptations societies have developed. Roman aqueducts, Egyptian agricultural practices, and Persian urban water systems are all efforts to control water for irrigation, provide a more reliable water supply, and protect societies from extreme events (Worster, 1985). Extreme events were interpreted as Acts of God, beyond control or predictability, and explained through stories. The deluge myth is one example of the widespread fear of flooding and its repercussions. However, embedded in such myths is recognition of the beneficial and regenerative power of floods.

The deluge myth is a story of too much water and it is one of the oldest problems humans have faced, "wiping out life and property with such completeness that it has seemed like divine retribution for some monumental evil people have committed" (Worster, 1985, p. 19). Cultures throughout the world are replete with tales of deluges of epic proportions. In fact, individual myths are strikingly similar among cultures separated by geography (Bierlein, 1994). Vishnu, the god of protection, used one of his 10 lives to save Mother Earth from a flood (India). Yu the Great is a hero who protects people from the floods of the Yellow River (China). Origin myths of many Native Americans, such as the Mojave, Apache, Cree and Algonquin, begin with a flooded Earth. Hawaii had its own indigenous flood myth called "Kai-a-ka-hina-li'I" (sea that made the chiefs fall down) that predated the arrival of the missionaries (Bierlein, 1994, p. 127).

A common theme of these flood myths is the destruction of humankind because of wickedness and evil-doing – a moral message of divine punishment. Humans are deserving of the retribution of the gods. Embedded in the deluge myth is not only the notion of Nature as uncontrollable but of humans as puny by comparison. This societal frame of reference concedes the inevitability of destructive acts of God. Upon Ra's instructions, Hathor slew thousands of humans so that "so much blood poured into the Nile that it overflowed its banks, and the mixture of blood and water inundated the land, destroying everything in its path" (Bierlein, 1994, p. 135).

Another common thread is that the flood brings new life. Survivors of the flood repopulate the earth and begin anew. The flood has washed away the evil, leaving behind the new world. The sun god Ra was honored during the annual flooding of the Nile for renewing the soil (Egypt). The Gilgamesh Epic (Babylonia) and the Book of Genesis (Hebrew) both tell of great floods with an individual singled out (Utnapishtim and Noah, respectively) to build an ark and survive the deluge. The Incan flood myth, predating the arrival of Europeans, is the story of two shepherd families that repopulated the earth after an intense rainstorm and ensuing flood, resulting in "human beings living everywhere; llamas, however, remember the flood and prefer to live only in the highlands" (Bierlein, 1994, p. 134).

Such myths of retribution and renewal served to explain the natural vagaries and unpredictability of floods as well as the extraordinary disastrous event. These collective stories help humans develop responses to hazardous situations through the recitation of past experiences that explain recurring events and identify both risk-avoiding strategies and ritual response. The Sherpas of the Khumbu Himal in eastern Nepal interpret hazards as supernatural when events actually harm the village. The response to such threats may be in the form of ritual or religious activity. Alternatively, hazards may be due to natural events, the consequences of the earth's physical processes. These types of events warn the village and the response is in the form of some type of mitigation activity (Bjonness, 1986). However, the forces behind real threats of flood and torrent are considered to be invisible forces, which the Sherpas classify as demons, deities, or gods whom one must be careful not to offend (Bjonness, 1986). Bjonness (1986) states: "It is important to note that although the Sherpas seek explanations by consulting Lamas, they also perform practical maintenance when any mountain hazard occurs, as well as carry out risk-avoiding strategies" (p. 288).

This arrangement of practical and religious responses to floods is exhibited in numerous cultures and serves to reveal not only belief systems but also the attendant ways of life that are manifested in particular locations. By approximately 5000 years before present (BP) technologies were developed in Mesopotamia to regulate flow through the use of irrigation canals and retention basins in Mesopotamia (Gibson, 1974). Flood water irrigation – the diversion of occasional flash floods during the rainy season – was a significant component of early irrigation systems throughout Mesopotamia (Adams, 1974), the Anasazi southwest (Vivian, 1974), coastal Peru (Moseley, 1974), and Mexico (Hunt and Hunt, 1974). Control of the hydrologic regime was necessary in arid regions to entrap water in basins from rain and floods and to divert water for irrigation via canals and ditches at the appropriate time. The hazardous situation was twofold: the prospect of uncontrolled flooding that could affect lives and property and the washing away of previously water-deprived plants – the harvest – which could affect the food supply.

Natural disasters may accelerate and promote major social changes or even initiate evolutionary transformations (Torry, 1979). Specific adaptations to flood regimes have been a critical element of human development; technology and innovation have played an important role in the emergence of early civilizations. Indigenous views of the environment and resources were evidenced through religious or ritually symbolic aspects affecting irrigation practices. These practices were linked to seasonal changes throughout the calendar year and were celebrated by sacred feasts and holidays dedicated to water, planting, and harvesting (Hunt and Hunt, 1974). Wittfogel (1957) argued that the rise of hydraulic societies, a social order based on the intensive, large-scale manipulation of water, assisted in creating religions that worshipped river gods and political classes that organized and maintained water conveyances. Local knowledge coupled with technological advances created particular social responses to floods in places such as ancient Egypt and Sumer.

Innovation and Societal Change

Many earth processes that are hazardous, such as fire and flood, have been used by humans to improve their environment. Both Sumer (located in the southern region of Mesopotamia) and Egypt were adept at controlling the rivers that flowed through their countries by using sophisticated irrigation and flood control. However, the particular geography of each region created very different conditions for the long-term success of their practices.

Sumer: Solution to Flooded Soils

Over the course of several thousand years, the Tigris/Euphrates River valley was transformed into a desert with increased siltation, water logging, and salinization. Farmers had to contend with flooding of a different sort; a higher water table that led to saturation or flooding of the subsurface and allowed for increased concentration of salts on the soil surface because of the high evaporation rates and temperatures. The archaeological record reveals that the inhabitants adapted to increased salinization and a rising water table by replacing wheat with the more salt-tolerant barley plant. Also, they expanded agricultural practices into marginal lands by constructing an intensive and complex canal system. Ultimately, environmental degradation forced the people of Sumer to migrate to other regions (Ponting, 1991; Gibson, 1974).

The unreliability of floods, the dependence on flow irrigation, the need for canal construction, and the problem with salinization are all factors that forced an extensive rather than intensive form of agriculture in Sumer. Areal expansion of farming was adopted instead of intensification of irrigation to increase agricultural production (Neely, 1974). An important aspect of this type of agriculture was the use of fallow to allow land to rest and retard salinization. Fernea (1970) argued that, in Iraq, "the traditional tribal system of land tenure and use was also well suited to traditional methods of extensive cultivation; indeed the two aspects of agriculture must have evolved together in this region" (p. 54). Fernea (1970) and Gibson (1974) described how new social structures were imposed and interwoven with old ones in prerevolutionary Iraq and in ancient Mesopotamia. They described the increasing role of the state to control labor, the creation of a central administration and attendant bureaucrats, landlordism, and increasingly large land holdings to create "the evolution and elaboration of social stratification" (Gibson, 1974, p. 17). In turn, these changing social conditions contributed to deterioration of tribal methods of cultivation and damage to the Mesopotamian plain, leading to collapse and reconstruction of the relationships, organizations, and civilizations of the region.

Egypt: Changes to the Hydraulic Regime

Egypt exploited annual floods for 7000 years. An implicit recognition of the ambivalent nature of floods was embedded in cultural practice. The benefits derived from annual flooding were acknowledged and specific strategies were adopted to take best advantage of the flood regime (Worster, 1985; Reisner, 1986; Ponting, 1991). The particular geologic structure of the Nile River valley allowed development of agriculture variously termed flood-retreat agriculture, recession agriculture, or basin irrigation, which existed on many rivers throughout Africa and southern Asia (Worster, 1985; Ponting, 1991; Blaikie et al., 1994). With only minimal human interference and a relatively low level of technology, the Egyptians capitalized on the annual flood regime of the Nile by evening out the variable natural flood level, retaining water in basins for later use, and using the silt to fertilize the land. Religion was critical for societal control and establishing traditions and practices compatible with irrigation and flood control (Wortser, 1985). The major problem with the system was the fluctuations of the actual flood from year to year. Not until the arrival of the British were large-scale dams introduced on the Nile to regulate floods with the construction of the initial Aswan Dam in 1902.

In 1964, the High Aswan Dam was completed. A consequence of the High Aswan Dam was control of the natural annual Nile flood and entrapment of the nutrient-laden silt that naturally fertilized the floodplain. The High Aswan Dam not only affected the annual Nile River flood regime but also created a permanent flood in the form of Lake Nasser, which provides hydroelectricity production, fishing, and irrigation. Year-round irrigation permitted three harvests a year instead of a single harvest. Significant environmental problems ensued. Egypt experienced problems similar to those of Sumer: salinization, water logging, and siltation (Rogers and Lydon, 1994). As a result of salinization, two-thirds of the land reclaimed over the past 40 years is now out of production. In addition, fisheries and water quality have declined (Jones, 1998).

The High Aswan Dam was built within the changing cultural context of a newly emerging state that profoundly affected the Egyptian worldview. In 1952, the postcolonial monarchy was deposed and 72 years of British occupation came to an end. President Nassar nationalized the Suez Canal in 1956 and the High Aswan Dam became a nationalistic symbol for an emerging modern Egypt (Reisner, 1986; Jones, 1998). Construction of the High Aswan Dam evoked the technological imperative associated with the era of the building of large dams. This technological imperative was further reinforced by the removal of Ramses II's temples of Abu Simbel. Carved into a sandstone cliff approximately 1200 BP, this important historical site was threatened with inundation by Lake Nasser. In a remarkable engineering feat sponsored by an international effort, the temples (with four statues 20 m tall and numerous smaller ones) were cut apart and reassembled 64 m above the river. A further consequence of Lake Nasser was the forced removal of 120,000 people who did not receive the same level of international attention as Ramses' temples. Where did this flood of people go?

In the examples of Sumer and Egypt the flood hazards – waterlogging of the soil and changes in the hydrologic regime – were largely induced by human activity, introduction of new technologies, and changing social structures. Unintended outcomes resulted in degraded environments in the form of soil erosion, decreased water quality, and loss of agricultural land and, in some instances, increasing susceptibility to extreme events and hazardous situations. Technology and social change represent intertwined aspects of culture. In the case of Sumer and Iraq, change occurred largely from within and among adjacent communities. However, the meeting of different cultures was critical to changes wrought in Egypt. Structural flood control introduced from Britain allowed for the establishment of agricultural production, primarily for cotton year-round, replacing an ancient system of flood adaptation. The newly regulated flow increasingly did not allow for traditional flood-retreat agriculture or for production of traditional crops for local consumption, further disrupting cultural patterns and societal relationships (Horowitz, 1989). Similar changes occurred in India after the introduction of new technologies and cultural perspectives.

Cultural Conflict and Technology

A significant solution to controlling water where and when humans would like to have it has been through the construction of dams. One of the first dams was built approximately 4000 BP in Egypt. Earth dams were part of the irrigation systems of ancient civilizations of Mexico, Mesopotamia, Peru, and the southwestern United States. The construction of large-scale, resilient dams did not occur until after the development of portland cement (1845), explosives (nytroglycerine, which was discovered in 1846), and the mechanization of earth-moving equipment (the steam engine was created in 1769). The era of megadam building was initiated after Hoover Dam (1936) was completed on the Colorado River in the United States.

The introduction of large-scale dams represents an important cultural shift throughout the world. Cultures' adaptive strategies for living with floods, celebrating the floods through ritual ceremonies or attributing them to supernatural forces, were displaced by changes in technology predicated on advances in science. Traditional belief systems were replaced or came into conflict with scientific perspectives and new technologies to control Nature. This control of Nature was driven by forces nonindigenous to the regions, introducing a different frame of reference – that of the superiority and ethnocentrism of Euro-American solutions and perspectives.

It is important to note that dams built for irrigation, hydroelectricity, or even water supply do not always include flood control. However, there is an embedded assumption about dams that by their very presence they will protect people from floods, regardless of their primary purpose or whether the floodway has been properly managed downstream (i. e., encroachment of development into the floodway) (Williams, 1998; Powledge, 1982).[1] Multipurpose dams are problematic to manage because of conflicting demands and public perception. Williams (1998) states: "while irrigation and power interests promoted most large multi-purpose dams, public support and voter approval stemmed from their promise of flood control" (p. 52). In developing countries, large-scale dams are built not only for hydroelectricity but also for flood control and irrigation; they have also been a prestige symbol representing industrialization and development (Ekins, 1992).

Colonial Influence

During the colonial period, European powers and the emerging United States bestowed upon the subjugated regions of the world a shared set of

[1] In 1975 the Bureau of Reclamation mission statement said: "Probably the least heralded feature of many multi-purpose dams and reservoirs is their ability to control flood waters. Virtually all regulating facilities on Bureau of Reclamation projects provide some flood protection, even though they may not have been initially authorized nor designed for that purpose" (Powledge, 1982, p. 156).

beliefs and frame of reference with regard to technology: there is absolute benevolence of progress brought about by technology and with this progress comes civilization. This period reflected the incipient state of a universal way in which Nature was to be managed. For example, during the 1860s, an expansive engineering feat for controlling and managing the Mahanadi River basin in India was put into effect. Copying patterns used in Europe and the United States based on rivers in temperate and dry climatic zones, the Public Works engineers relied on Euro-American technology to control the rivers of India. Hill states: "The administrative engineers of the industrialized world were convinced that the generic nature of deltaic rivers make it unnecessary to attend to the unique characteristics of the particular environment within which the rivers were situated" (1995, p. 52).

A doctrine of environmental management was devised and exported worldwide based on the historical and social developments and cultural biases peculiar to Europe and the United States. These hydraulic engineers were steeped in engineering ideology and the superiority of technological solutions. They were trained in countries that had undergone the Industrial Revolution, where progress was predicated on controlling nature and where nature was no match for "civilized," technically proficient people. Worster writes: "They [hydraulic engineers] understood that they were engaged in a mission of conquest that was going on in all parts of the world – in India, Egypt, the Sahara, Australia" (1985, p. 143). Additionally, the control of nature and the subjugation of the indigenous populations of the colonies were linked to the religious validity of Christianity as a superior religion. Sir Arthur Cotton, the Chief Engineer of Bengal during the Maharandi River Project stated: "In [any] district where our Western knowledge and energy have been brought to bear, the people freely acknowledge that it is to Europeans and Christians that they are indebted for benefits which they never received from their own Government and their own gods" (Hill, 1995, p. 64). This doctrine and these managers all originated in areas where the amount of rainfall was moderate and the river systems were considered predictable and manageable.

Whereas local knowledge was critical to development of the particular adaptations to the environment in places such as Sumer, Egypt, and India, in the colonial period, local knowledge was displaced and traditional solutions were replaced with engineering solutions and technology from the West. In India, this "arrogance of power" encouraged the replacement of flood control measures, which had relied on embankments and channels, with dams and reservoirs (Hill, 1995). The results of the Mahanadi River Basin Project were exacerbated flooding, increased environmental damage, and loss of lives (Hill, 1995).

Coupling colonial power with technological advances created a particular frame of reference with regard to the natural environment that has ignored the unique characteristics of place. These solutions were the precursors to modern efforts to development strategies. Dams by their very nature create floods of their own and, in turn, create unique hazards associated with permanent inundation (i.e., losses of habitat, of land, of livelihoods). Dams built for hydroelectric power, irrigation, and flood control result in permanent flooding of inhabited areas, forcibly moving millions of people, as in the case of large dams built on the Narmada River in India. National economic priorities, such as increased hydroelectricity for industry and agricultural production, driven by the culture of global competition, have come into direct conflict with traditional ways of life. Completion of the Tawa dam, a smaller dam that is part of the larger Narmada Valley Project, resulted in "a veritable nation of beggars and moving migrants who have not found a piece of land to settle on ever since" (Ekins, 1992, p. 92). Incomplete implementation of development plans sponsored by the World Bank has played a critical role in creating disastrous outcomes for displaced people (Ekins, 1992; Blaikie et al., 1994).

Dams for development can and have shifted the floodwater problem elsewhere. For example, India's Farakka Barage diverts Ganges River peak flows through Bangladesh (Monan, 1989). This permanent flooding of the reservoir creates a flood of people to urban regions in search of employment, homes, and a new way of life. Resettlement sites experience widespread problems with food shortages, fragmentation of families, conflicts with host populations, and deficient compensation (Ekins, 1992). The imposition of nonindigenous strategies for flood control and prevention has affected the local culture, transforming the environment and placing demands on people to devise new strategies for living.

Development, Poverty, and Floods

Development and poverty offer yet another frame of reference from which to assess and understand human responses to floods. Floods and other types of natural disasters may be immediate causes of disasters and they may or may not be predicted. "However, the severity and form of damage depend primarily upon the pre-existing state of society and its environmental relations" (Hewitt, 1997, p. 22). Natural disasters can be interpreted as the coincidence between natural events and conditions of vulnerability (Maskrey, 1989, p. 1). Vulnerability has been related to concepts such as resilience, marginality, susceptibility, adaptability, fragility, and risk in connection with climate impact and global change studies (Liverman, 1990; Hewitt, 1984; Kates et al., 1985; Wilhite and Easterling, 1987). Vulnerability analysis has been conducted from the perspective of biophysical conditions that define human vulnerability (Mitchell, 1989): political economy – understanding the social, political, and economic context of society (Susman et al., 1984) and development and poverty issues relative to the environment or political ecology (Blaikie et al., 1994). From this last perspective, vulnerability has been defined as "the degree to which different classes of society are differentially at risk" and considers the role of development in increasing social inequality, levels of poverty, and vulnerability of the disadvantaged (Hewitt, 1984; Hamilton and Joaquin, Chapter 18, this volume). Critical to many vulnerability studies is the notion that populations are differentially vulnerable within societies, in particular women and children (Hanchett et al., 1998).

The Bangladesh Flood Action Plan (FAP)

Bangladesh has severe problems with flood disasters. It is also largely rural and one of the poorest and most densely populated countries. In addition,

Bangladesh has been an independent nation since 1971, a relatively short period of time, and has been plagued by both political unrest and natural disasters. Indigenous methods exist for protection from flooding, such as building houses on artificial mounds that raise them above normal flood levels (Blaikie et al., 1994). But the term "normal flood levels" becomes problematic in a location undergoing rapid population growth, changes in land tenure and land uses, migration, and increases in refugee populations. Part of the problem in Bangladesh is historical. An unstable economy was inherited from the rule by Britain and later West Pakistan. There exists a legacy of lack of cooperation between India and Bangladesh in the management of the Ganges through Bangladesh (Monan, 1989). Additionally, the unequal pattern of ownership and income and an elite dependent on foreign aid tip the balance of power toward landowners against the poor.

Bangladesh's FAP began as a response to severe floods that covered nearly three-quarters of the country in 1984, 1987, and 1988. These floods attracted significant media attention, disrupted people's lives, and aroused international concern for the people of Bangladesh. The FAP was a program composed of a series of studies of flood problems and issues in Bangladesh funded by a consortium of donors and coordinated (at times) by the World Bank. The FAP framework, as defined by the government and many of the donors, emphasized the mechanical aspects of flooding and flood control (Hanchett et al., 1998). The FAP reflects a particular frame of reference of many development strategies that are increasingly being questioned: a "tech-fix" approach or "project culture." Water will be contained and moved downstream between large, long embankments through the use of flood control structures in an effort to "end all floods" (Blaikie et al., 1994, p. 138).

The tech-fix approach maintains a culture of development that favors the development community and the power elite within Bangladesh and echoes the perspectives introduced during the colonial period. Large-scale engineering contracts, foreign lenders, consultancy fees, and purchase of equipment all favor foreign constituents of aid and development policies. Local elites and politicians benefit through protection of their land and property as well as gaining local brokerage fees. Social programs, such as flood proofing – the development of ways to help local populations survive floods with minimal disruption of work and social life – were included in the FAP but were never given high priority (Hanchett et al., 1998).

In the FAP, the definition of flooding was widely contested: was it riverine flooding, flash flooding, or flooding associated with cyclonic storms? The Bengali language distinguishes between normal seasonal heavy rain and flooding (*barsha*) and unusually deep and prolonged severe flooding (*banna*) (Hanchett et al., 1998; Blaikie et al., 1994). In addition, floods and reactions to floods vary throughout the country and from year to year; the very nature of the flood problem proved difficult to define for the FAP.

Critics argue that cheaper alternatives are being ignored, specifically in connection with benefits derived from seasonal flooding (*barsha*). Ponds that remain behind after flooding provide areas for fish and shrimp to spawn and create a local source of income and nutrition. A "living with flood" strategy would depend on some of the solutions presented in the

FAP, specifically flood warning systems (Rogers et al., 1989). This alternative strategy would address the existence of massive squatter settlements and refugee camps – "invisible communities" (Kaplan, 1994) – in order to "assess what is actually needed to reduce vulnerability rather than some grand design which supposedly prevents all flooding without considering who benefits and who loses, who pays for it, and whether it is needed anyway" (Blaikie et al., 1994, p. 143).

Solutions for flooding in places such as Bangladesh need to address invisible communities. Parker and Thompson write: "The poor often live in over-crowded conditions in substandard dwellings unable to withstand the effects of storms, winds and flood waters. Slum settlements are often located in marginal areas – areas avoided by the better off – including floodplains or steep slopes, both of which present hazards in storm events" (1991, p. 42). Such communities tend to bear the brunt of many environmental disasters because response mechanisms and mitigation strategies fail to take into account their existence. Nor is relief assistance to squatter settlements supplemented by disaster planning by local and central governments (Mulwanda, 1989). Invisible communities are poorly understood and are often outside the realm of disaster response plans (Gruntfest, Chapter 15, this volume), mitigation projects, and prevention strategies as well as many development plans.

Poverty creates its own culture, in which past mistreatments fuel a distrust of institutions despite dependence on foreign and governmental aid. Unresolved cultural conflicts between the disadvantaged and exploiters as well as language barriers strengthen cultural biases and strain social relations. This particular frame of reference fuels specific reactions in times of disaster: a greater reluctance to accept "official" warnings and aid, greater resistance to cooperative relationships among disaster-related agencies, and a greater dependence on the kin group as a source of advice and help (Clifford, 1956; Dynes, 1975; Wisner, 1993).

The role of gender is also particularly important in understanding vulnerability and responses to disaster in general and to floods in particular (Fordham, 1998). A gender study that was part of Bangladesh's FAP illuminates issues surrounding poor women in less-developed countries (Hanchett et al., 1998):

- A weaker economic position in society can exacerbate the effects of floods because of women's employment, assets, and uses of credit.
- Men's and women's responsibilities change during and after a flood. Women tend to shoulder a greater burden of household flood-coping activities than do men.
- Female-headed households have additional responsibilities and unique problems. Poor female heads of households have the potential to become destitute after a flood. Cultural mores may prevent officials from conducting business directly with women.

The FAP gender study makes an important contribution to understanding women's experiences and roles during disastrous events. Studies need to be conducted to understand the role of gender and women's contribution during and after floods.

Conventional thinking about disasters has taken the benignity and rationality of state response for granted – the state mobilizes its resources to protect its citizens. However, relief aid and the process of development assistance tend to reinforce the political, socioeconomic hierarchy. Elite interests in society capture a large part of the benefits of both financial aid for recovery and the continuation of development projects (Wisner, 1993). In addition, severe floods may intensify economic effects that are the result of economic and social inequities. For example, the poor and landless in Bangladesh experience hunger during the monsoon season every year because of limited day-labor opportunities; severe floods exacerbate this situation and intensify hunger problems over a longer period of time (Hanchett et al., 1998). Socioeconomic class plays a pivotal role in people's ability to cope with severe floods and their aftermath and is further compounded by gender.

Generally, the state reaction to floods has focused on disaster and development planning. The underlying causes of floods remain unexamined. In the case of the FAP, gender and social studies suggested that social change was necessary for long-term solutions to floods where water and human society are closely interconnected. An official of Bangladesh responded, "What do you want us to do, change our whole society?" (Hanchett et al., 1998, p. 228). The underlying cause of the hazard remains only superficially understood. The flood may have occurred because of extreme weather events but the outcomes and impacts often depend on historical circumstances or social relations that may have exacerbated the effect on vulnerable populations and marginal environments (Merritts, Chapter 10, this volume).

The role of the world economy in creating vulnerable locations needs to be better understood and considered as part of the planning process for disaster prevention. The core industrial regions of the world economy draw on the ecological capital of locations that supply both raw materials and cheap labor. These "shadow ecologies" reflect the ecological consequences of global economic connectedness (MacNeill et al., 1995). Development plans predicated on an economic culture that privileges free trade strategies and large-scale projects hearken back to the colonial policies of the last century. Globalization of the world economy tends to homogenize solutions to production practices and utilization of the environment and reflects an imposition of top-down solutions from other places. Alternative strategies for flood control demand a holistic approach, locally based, that recognizes the interconnections not only between economies and environments but also the landscapes that people have created. Wisner (1993) states: "While physical hazards arise locally, the constraints on humans that create their vulnerability to these hazards can originate in the influence of structures located on another continent" (p. 129).

Solutions imposed from distant locations are not necessarily problematic, as evidenced by the process of diffusion where new ideas and technologies have affected cultures throughout the past. Large-scale dams have created solutions to flood control, water supply, and irrigation. However, all the primary locations for large-scale dams have been used (Gleick, 1993). The recognition that dams can have destructive repercussions for local populations has led to innovative perspectives – a cultural shift – regarding dams and a reassessment of how floods in fact may be useful for both society and the environment.

Redefining Floods: Scientific Perspectives

The notion of living with floods is an important concept. Humans have rationalized floods through myths of retribution and renewal; tolerated floods; and attempted to limit, prevent, and control floods. Yet, humans have also lived with floods, as in Egypt and Bangladesh. A local saying from the Brahmaputra floodplain region states: "People do not die if there is a flood, but people die if there is no flood" (Hofer, 1998, p. 137). The terms controlled flooding and restoration flood refer to efforts to capitalize on the regenerative powers of floods (Anderson et al., 1996). Controlled flooding of the Colorado River by releases from the Glen Canyon Dam was undertaken as a way to manage sediment and other resources in the Grand Canyon. A Bureau of Land Reclamation study states: "Scientific research conducted during the U.S. Bureau of Reclamation's Glen Canyon Environmental Studies program has led to a consensus that floods are necessary to maintain the Colorado River's geomorphic structure and related ecosystems downstream from Glen Canyon Dam" (Bureau of Reclamation, 1996). Between 26 March and 3 April 1996 an experimental controlled flow release with a high steady discharge of 1274 m^3/s was released from the Colorado River's Glen Canyon Dam. "This marked the first time that dam managers have used a large flood to renew the health of a river system" (Adler, 1996, 188).

A controlled flood on the Colorado River represents a fundamental change in how rivers should be managed. It represents the recognition of the importance of floods for maintaining ecosystems, rejuvenating beaches, and preserving habitats for native plants and animals. Scientists learned from the experimental Grand Canyon flood that the potential exists to mimic scour and fill processes to maintain large sand bars along the river banks, to keep sand bars clear of vegetation, and to clear debris fans from the mouths of tributaries to prevent constriction of the river (Anderson et al., 1996). Additionally, the experimental flow acknowledged the serious environmental consequences that had occurred downstream after construction of the Glen Canyon Dam.

This flood was designed for environmental benefits and the type of studies conducted reflect this: flow of water, transport of sediment, erosion of debris fans, and water chemistry. This collection of a series of very specific parameters for measurement is dictated by models of how floods occur based on knowledge about the physical world and to "provide information for science-based decision-making" (Stocking, 1987). However, if the words "environmentally," "ecologically," "economic," and "engineering" were substituted for "science," the purpose of the studies would have an entirely different meaning based on the disciplinary perspective. Each discipline creates its own language, frame of reference, tools, methods, vocabulary, conceptual frameworks, and biases. Each discipline creates its own culture. However, "our culture generally accepts the hegemony of Western empirical science, if for no other reason than because of its practical achievements in applied technology over the last few centuries" (Wisner, 1993, p. 128).

This flood demanded a high level of cooperation and planning among a variety of agencies and researchers: the U.S. Geological Survey, the Bureau of Reclamation, and the National Park Service. In addition, many factors were assessed and incorporated into the flood scenario: effects on both native and nonnative fish, the extent of beaches before and after, location of archaeological sites and recreational uses, and future plans for creating a long-term flood regime (Collier et al., 1997).

The experimental flood illuminates the need to reevaluate and redefine floods. There has been much written and implemented into policy since the 1950s that focuses on managing floodplains instead of controlling floods (White, 1986). It is important to note that there is a difference between managing a floodplain and creating a flood. Controlled floods might be an outgrowth of changing attitudes toward floodplain management, which also reflects recognition of environmental change when floodplains are not flooded regularly. David Wegner, the manager of the Glen Canyon Environmental Studies Project, comments: "There are dams all over the world that are nearing the end of their useful lives. We need to start looking at what happens next" (DiLeo, 1997). Government officials in Japan and Turkey are interested in organizing such floods in their own countries (Adler, 1996).

Discussions have begun among environmentalists, scientists, and public policy makers to address the possible dismantling of dams. This is a significant shift in perception with regard to river basin management. However, it is important to recognize the importance of such issues on society. How to live with floods to support traditional lifestyles, local economies, and ecosystems is one set of questions. What types of flood-control solutions and disaster management plans are necessary is another set of issues (Gruntfest, Chapter 15, this volume). But these issues are not mutually exclusive. The intersection of social activity and physical processes requires complex models for understanding floods, science, disaster, and vulnerability. Potentially, the 1996 experimental flood in the Grand Canyon may provide vital information for the importance of seasonal flooding; its scientific basis may lend increased legitimacy from World Bank supporters to the Bangladeshi plan of living with floods.

Conclusion: Globalization of Culture

A final arena to consider in discussing cultural aspects in response to flooding is the role that social institutions play in disaster planning and in framing the risk calculus for a given society. Generally, social institutions fall into two categories: traditional institutions, which have existed for a long time organized around activities such as family, religion, and political power; and recently evolved institutions organized around activities such as science, technology, medicine, sports, and mass media (Quarantelli and Wenger, 1991). Both the traditional and newer social institutions differ in how much they are culturally bound. Medical procedures, scientific experiments, and international disaster relief generally will be undertaken in the same way regardless of where they occur. Mass communication systems represent one of the least culturally bound of all social institutions (Jacobson and Deutschmann, 1962). These researchers comment: "Necessarily all of these phenomena occur in conjunction with a standardization of the content of communication so that there is neither the variety nor the unique flavor in subject matter that would be expected from the immense diversity of peoples . . . what is news in London seems also to be news in Buenos Aires" (p. 152).

In a study comparing disaster reporting in the United States and Japan, Quarantelli and Wenger (1991) found far more similarities than differences between the two countries. Disasters were viewed as major news stories, there was a reliance on official sources for information about the disaster and civil disturbance, there was initially a lack of accurate information about the disaster, and there was a diminution of the gatekeeping process in which information did not undergo the normal editing process (Sood et al., 1987). The results of this study suggest that institutionalized social and cultural patterns of mass media reporting exert a powerful influence. The process of disaster reporting reflects the subculture of the work environment of the mass media and journalism. This subculture is derived from the mass communication systems of Western developed countries, which in turn will provide the model for societies elsewhere (Quarantelli and Wenger, 1991).

The new social institutions of science, technology, and mass communication are powerful mediators of culture. Increasingly, they define societal context and response. Arguably, economic globalization and development policies are hybrids of traditional and new institutional arrangements. Their historical roots are in the colonial past and industrial development. The worldwide trend of linking development strategies to economic "shock therapies" and free market ideology has resulted in not only economic globalization but also cultural homogenization. Research will need to be conducted on how these global strategies affect disaster incidence, planning, and response as related to cross-boundary issues, multinational water planning, and human resettlement and migration. Increased understanding of the relationship of disastrous outcomes with natural hazards and their historical roots is an important component in addressing the differential risk calculus in different places.

The role of culture in understanding the impact of floods is not only a function of geography but also of the larger narratives of science, development, technology, and globalization. Local solutions and responses to floods continue to occur and need to be better understood, but the influence of global narratives on the societal response also needs to be examined along with the perceptions and beliefs they engender (Table 17.1). Global narratives create their own set of myths or beliefs about floods that influence responses, mitigation, and solutions. The extent to which global perspectives affect local geographies needs to be examined to determine the relationship between global influence and local outcomes.

Several cross-cultural studies in the disaster-risk area indicate the need for more explicit research on culture and disaster in general and culture

Table 17.1. *Global narratives with regard to floods*

Narrative	Frame of reference	Approach	Belief
Science	Physical phenomenon	Differing disciplinary perspectives	Assists in understanding uncertainty
Technology	Structural solutions	Hard vs. soft technologies	Controlling flood hazards
Development	Poverty and aid	Indigenous vs. non-indigenous strategies	State as benign
Globalization	Free market ideology and global trade	Homogenize response and perspective	Economy as benign

and flooding specifically: 1) What is the relationship between how individuals perceive risks and how culture shapes the societal or group response to disaster (Dake, 1992), 2) how do different levels of economic development among countries affect the societal response to disasters (Dynes, 1975), 3) what role do social relationships and political centralization play in framing disaster and for disaster response (Clifford, 1956; Dynes, 1975), and 4) what role does globalization in the form of new economic relationships and new social institutions have for understanding and solving disaster (Hoberg, 1986; Wisner, 1993; Quarantelli and Wenger, 1991)?

The frame of reference remains critical in how floods are perceived and how cultural contexts and responses are determined. There exist several basic responses to floods: 1) controlling floods through structural solutions (Watson and Biedenharn, Chapter 14, this volume); 2) reducing the anthropogenic impact of flooding, such as changes in land use (Wohl, Chapter 1, this volume); and 3) removal of people from hazardous areas (Gruntfest, Chapter 15, this volume) or strategies of living with floods. Although the latter approach to floods is still practiced in many parts of the world, the former has a long history (e.g., Nile and Indus River valleys). Structural solutions have accelerated during the past two centuries, driven initially by U.S. and European colonialism, followed by development and aid projects sponsored by the World Bank, and most recently dictated by globalization of the world economy endorsed by the International Monetary Fund. The Three Gorges project in China epitomizes trends associated with megadams – larger impoundments, resettlement of large numbers of people, permanent flooding of farmland, and loss of archaeological sites. The project will provide hydropower, improved water supply, and flood control (Gleick, 1993). However, structural schemes to control and engineer rivers have been questioned. The 1993 Mississippi floods revealed how the coupling of natural events in the form of heavy rain falling on a saturated basin and structural solutions in the form of a flood containment policy and channel modification program precipitated "the most costly and devastating flood to ravage the United States in modern history" (Jones, 1998, p. 236). Compounding the situation was the multiple management aspect of the reservoirs for flood control: to guarantee water supply and to provide recreational opportunities (Williams, 1998; Jones, 1998). There is recognition of widespread environmental change associated with such projects and this has led to increased interest in searching for alternative solutions.

New approaches to flood management may not be entirely innovative. Rather such solutions may represent increased recognition of the local environment and indigenous or traditional strategies to contend with floods. The 1994 report, "Sharing the Challenge: Floodplain Management into the 21st Century" recommended that nonstructural solutions were critical for flood damage protection and that people and structures should be moved from frequently flooded areas (Western Water Policy Review Advisory Commission, 1998). The report of the Western Water Policy Review Advisory Commission (1998) recommended that a watershed approach embracing local participation would enhance sustainable management of water resources linked to basin-level objectives (p. xvi). Such changes in policies provide fertile ground to study how they will be implemented, what their effect will be on preventing floods, and how risk is framed within different cultural milieus.

References

Adams, R. (1974). Historic patterns of Mesopotamian agriculture. In *Irrigation's Impact on Society*, eds. T. Downing and M. Gibson, pp. 1–6. Anthropological Papers of the University of Arizona, No. 25. Tucson, AZ: University of Arizona Press.

Adler, T. (1996). Healing water: flooding rivers to repent for the damage done by dams. *Science News*, **150**, 188–89.

Anderson, M., Graf, J., and Marzolf, G. (1996). U.S. Department of Interior, U.S. Geological Survey, Fact Sheet FS-089-96. Washington, DC: U.S. Government Printing Office.

Bierlein, J.F. (1994). *Parallel Myth*. New York: Ballentine Books.

Bjonness, I. (1986). Mountain hazard perception and risk-avoiding strategies among the Sherpas of Khumbu Himal, Nepal. *Mountain Research and Development*, **6**(4), 277–292.

Blaikie, P., Cannon, T., Davis, I., and Wisner, B. (1994). *At Risk: Natural Hazards, People's Vulnerability, and Disasters*. London: Routledge.

Bureau of Reclamation. (1996). *Glen Canyon Dam Beach/Habitat-building Test Flow: Environmental Assessment and Finding of No Significant Impact*. Salt Lake City, UT: Bureau of Reclamation, 64 pp.

Clifford, R. (1956). *The Rio Grande Flood: A Comparative Study of Border Communities in Disaster*. Washington, DC: National Academy of Sciences.

Collier, M., Webb, R., and Andrews, E. (1997). Experimental flooding in Grand Canyon. *Scientific American*. January, 82–9.

Dake, K. (1992). Myths of nature: culture and the social construction of risk. *Journal of Social Issues*, **48**(4), 21–37.

DiLeo, M. (1997). The undoing of a dam. *American Way*, 15 November, 62–185.

Dynes, R. (1975). The comparative study of disaster: a social organizational approach. *Mass Emergencies*, **1**, 21–31.

Ekins, P. (1992). *A New World Order: Grassroots Movements for Global Change*. London: Routledge.

Fernea, R. (1970). *Shaykh and Effendi: Changing Patterns of Authority Among the El Shabana of Southern Iraq*. Cambridge, MA: Harvard University Press.

Fordham, M. (1998). Making women visible in disasters, problematising the private domain. *Disasters*, **22**(2), 126–143.

Gibson, M. (1974). Violation of fallow and engineered disaster in Mesopotamian civilization. In *Irrigation's Impact on Society*, eds. T. Downing and M. Gibson, pp. 7–20, Anthropological Papers of the University of Arizona, No. 25. Tucson, AZ: University of Arizona Press.

Gleick, P. (ed.) (1993). *Water in Crisis: A Guide to the World's Freshwater Resources*. Oxford: Oxford University Press.

Hanchett, S., Akhter, J., and Akhter, K. (1998). Gender and society in Bangladesh's flood action plan. In *Water, Culture and Power: Local Struggles in a Global Context*, eds. J. Donahue and B. Johnston, pp. 209–34, Washington DC: Island Press.

Hewitt, K. (ed.) (1984). *Interpretations of Calamity*. Boston: Allen and Unwin.

Hewitt, K. (1997). *Regions of Risk: A Geographical Introduction to Disasters*. Singapore: Longman.

Hiessl, H. and Waterstone, M. (1986). *Issues with Risk*. Water Resources Research Center: University of Arizona.

Hill, C. (1995). Ideology and public works: "managing" the Mahanadi River in colonial north India. *Capitalism Nature Socialism*, **6**(4), 51–64.

Hoberg, G. (1986). Technology, political structure and social regulation: a cross-national analysis. *Comparative Politics* April, 357–376.

Hofer, T. (1998). Do land use changes in the Himalayas affect downstream flooding? – traditional understanding and new evidences. In *Flood Studies in India*, ed. V. Kale, pp. 119–41, Bangalore: Geological Society of India.

Horowitz, M. (1989). Victims of development. *Development Anthropology Network*, **7**(2), 1–8.

Hunt, E. and Hunt, R. (1974). Irrigation, conflict and politics: a Mexican case. In *Irrigation's Impact on Society*, eds. T. Downing and M. Gibson, pp. 129–158, Anthropological Papers of the University of Arizona, No. 25. Tucson, AZ: University of Arizona Press.

Jacobson, E. and Deutschmann, P. (1962). Introduction. *International Social Science Journal*, **14**, 151–161.

Jones, J. (1998). *Global Hydrology: Processes, Resources and Environmental Management*. Essex, England: Addison Wesley Longman Limited.

Kaplan, R. (1994). The coming anarchy. *The Atlantic Monthly*. February, 44–76.

Kates, R., Ausubel, H., and Berberian, M. (eds.) (1985). *Climate Impact Assessment*. SCOPE 27. New York: Wiley.

Kirby, A. (ed.) (1990). *Nothing to Fear*. Tucson, AZ: Arizona University Press.

Liverman, D. (1990). Drought in Mexico: climate, agriculture, technology and land tenure in Sonora and Puebla. *Annals of the Association of American Geographers*, **80**(1), 49–72.

MacNeill, J., Winsemius, P., and Yakushiji, T. (1995). The shadow ecologies of western economics. In *Green Planet Blues: Environmental Politics from*

Stockholm to Rio, eds. K. Conca, M. Alberty, and G. Dabelko, Boulder, CO: Westview Press, pp. 91–93.
Marston, S. and Knox, P. (1998). *Human Geography*. Upper Saddle River, NJ: Prentice Hall.
Maskrey, A. (1989). *Disaster Mitigation: A Community Based Approach*. Development Guidleines No. 3. Oxford: Oxfam.
Mitchell, J. (1989). *Risk Assessment of Environmental Change*. Working Paper 13, Environment and Policy Institute, East West Center, Honolulu: East West Center.
Monan, J. (1989). *Bangladesh: The Strength to Succeed*. A Report from Oxfam. Oxford: Oxfam.
Moseley, M. (1974). Organizational preadaptation to irrigation: the evolution of early water-management systems in coastal Peru. In *Irrigation's Impact on Society*, eds. T. Downing and M. Gibson, pp. 77–82, Anthropological Papers of the University of Arizona, No. 25. Tucson, AZ: University of Arizona Press.
Mulwanda, M. (1989). Squatter's nightmare: the political economy of disasters and disaster response in Zambia. *Disasters*, **13**(4), 345–350.
Nabhan, G.P. (1997). *Cultures of Habitat: On Nature, Culture, and Story*. Washington, DC: Counterpoint.
Neely, J. (1974). Sassanian and early Islamic water-control and irrigation systems on the Deh Luran plain, Iran. In *Irrigation's Impact on Society*, eds. T. Downing and M. Gibson, pp. 21–42, Anthropological Papers of the University of Arizona, No. 25. Tucson, AZ: University of Arizona Press.
Parker, D. and Thompson, P. (1991). Floods and tropical storms. In *WHO/UNITAR: The Challenge of African Disaster*, pp. 38–59, New York: WHO and UNITAR.
Ponting, C. (1991). *A Green History of the World*. London: Sinclair-Stevenson Limited.
Powledge, F. (1982). *Water: The Nature, Uses and Future of Our Most Precious and Abused Resource*. New York: Farrar Straus Giroux.
Quarantelli, E. and Wenger, D. (1991). A cross-societal comparison of disaster news: reporting in Japan and the United States. In *Risky Business: Communicating Issues of Science, Risk and Public Policy*, eds. L. Wilkins and P. Patterson, pp. 97–112, New York: Greenwood Press.
Reisner, M. (1986). *Cadillac Desert: The American West and its Disappearing Water*. New York: Viking.
Rogers, P. and Lydon, P. (eds.) (1994). *Water in the Arab World: Perspectives and Prognosis*. Cambridge, Mass.: Harvard University Press.
Rogers, P., Lydon, P., and Seckler, D. (1989). *Eastern Waters Study: Strategies to Management Flood and Drought in the Ganges-Brahmaputra Basin*. Report prepared by Irrigation Support Project for Asia and the Near East. Washington, DC: USAID.
Shiva, V. (1997). *Biopiracy: The Plunder of Nature and Knowledge*. Boston, MA: South End Press.
Smith, K. (1990). The risk transition. *International Environmental Affairs*, **2**(3), 227–251.
Sood, R., Stockdale, G., and Roger, E. (1987). How the news media operate in natural disaster. *Journal of Communication*, **37**(3), 27–41.
Stocking, M. (1987). Measuring land degradation. In *Land Degradation and Society*, eds. P. Blaikie and H. Brookfield, pp. 49-63. London: Methuen.
Susman, P., O'Keefe, P., and Wisner, B. (1984). Global disasters: A radical interpretation. In *Interpretations of Calamity*, ed. K. Hewitt, pp. 264–283, Boston: Allen and Unwin.
Torry, W. (1979). Hazards, hazes and holes: a critique of the environment as hazard and general reflection on disaster research. *Canadian Geographer*, **XXIII**, 368–383.
Vivian, R. (1974). Conservation and diversion: water-control systems in the Anasazi southwest. In *Irrigation's Impact on Society*, eds. T. Downing and M. Gibson, pp. 95–112, Anthropological Papers of the University of Arizona, No. 25. Tucson, AZ: University of Arizona Press.
Western Water Policy Review Advisory Commission. (1998). *Water in the West: Challenge for the Next Century*. Springfield, VA: National Technical Information Service.
White, G. (1986). Natural hazards research. In *Geography, Resources and Environment: Volume 1 – Selected Writings of Gilbert F. White*, eds. R. Kates and I. Burton, pp. 324–347, Chicago, IL: University of Chicago Press.
Wilhite, D. and Easterling, W. (eds.) (1987). *Planning for Drought*. Boulder: Westview Press.
Williams, P. (1998). Inviting trouble downstream. *Civil Engineering*, February, 51–53.
Wittfogel, K. (1957). *Oriental Despotism: A Comparative Study of Total Power*. New Haven: Yale University Press.
Wisner, B. (1993). Disaster Vulnerability: scale, power and daily life. *Geo-Journal*, **30**(2), 127–140.
Worster, D. (1985). *Rivers of Empire: Water, Aridity and the Growth of the American West*. New York: Pantheon Books.

CHAPTER 18

Urban Planning for Flood Hazards, Risk, and Vulnerability

Douglas Hamilton and Alejandro Joaquin
Hydrologic Consultants
Irvine, California

Urban planning is the process of implementing principles that enable society to achieve certain goals. Evaluation is an integral part of the planning process because it provides a way to assess how a society responds to being harmed by floods. Identifying a current flood hazard problem is the first step in the planning procedure; the second step involves comparing the recurring outcome with a beneficial and viable solution; and the third step in the planning process is to diagnose the causes and effects of the current flood condition. The life cycle of a typical emergency management strategy includes response, recovery, mitigation, risk reduction, prevention, and preparedness. The United States' Federal Emergency Management Agency has an organizational structure that mirrors these processes. Historical examples of urban flood-planning measures come from Egypt, China, Europe, Japan, the United States, and Iran. Unwanted outcomes of planning are illustrated in a case study from Bangladesh.

Introduction

This chapter examines urban planning as a societal response to being harmed by floods. From this perspective, the history of the way we have responded to floods is the history of planning. Understanding this record provides insight into why planning is implemented the way it is today. The historical perspective also helps to identify the milestones in which planning has been ineffective (such as in Bangladesh flood shelters turned into storage sheds) and in which it has achieved success (as in China's Forbidden City, which has not been flooded since it was constructed).

The objective of this chapter is to identify ways that urban planning has successfully accomplished its purpose with respect to the reduction of flood hazard, risk, and vulnerability. Planning is not a purpose in itself. It is rather the process of implementing principles that enable society to achieve certain goals. This process has evolved significantly during the past few decades. For example, the role of structural flood-control measures in attracting people to live nearer zones of flood risk has been identified an important number of times this century and was reinforced most recently in the United States with the flooding of the Mississippi River in 1993. One of the reasons for having a chapter like this in a volume on inland flood hazards is to discuss how the goals of planning and even society's recognition for the need to plan have changed in recent years. When the Mayor of Shanghai was looking for someone to head the city planning commission he invited a skilled colleague – ironically the present Mayor of Shanghai Xu Kuangedi. When Mr. Xu responded, "I don't believe in planning." The mayor replied, "That's just what we need, someone with new ideas." The authors hope that this chapter results in new ideas.

The Planning Process

When goals are measurable and clearly expressed, the effectiveness of the process used to reach an audience can be evaluated. Evaluation is an

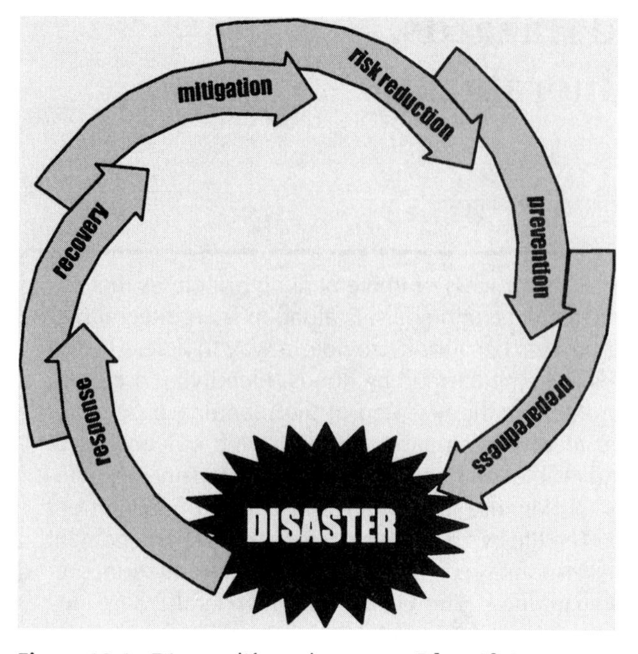

Figure 18.1. Disaster life cycle (adapted from FEMA, 1996).

integral part of the planning process because it is a way to assess how a society responds to being harmed by floods. Generally speaking, flood-planning programs retain a generic mold that cover essential elements such as a predisaster information plan. The life cycle of a typical emergency management strategy includes response, recovery, mitigation, risk reduction, prevention, and preparedness (Figure 18.1). The United States Federal Emergency Management Agency (FEMA) has an organizational structure that mirrors the functions that take place in this life cycle of emergency management (FEMA, 1997.) FEMA's disaster strategy includes rescue education and ongoing outreach efforts to provide the public at risk with a hazard readiness program. Effectiveness of the programs boils down to the continuing development of vulnerability and risk education about recurrent flood hazards [Interagency Floodplain Management Review Committee (IFMRC), 1994].

Identifying a current flood hazard problem is the first step in the planning procedure. Once specific data are established, the second step involves comparing the recurring outcome with a beneficial and viable solution. This step is a difficult process because it often demands dovetailing diverse objectives, such as the maintenance of floodways versus preservation of natural stream courses. In the United States, some government agencies that engage in such planning create working groups or task forces that consist of multiple parties representing different and opposing interests. The goal of the group is to determine the preferred outcome of the planning process.

The third step in the planning process is to diagnose the causes and effects of the current flood condition. This means examining the symptoms of the problem in order to understand why past efforts have failed. Interestingly, most of the literature available on urban flood planning falls into this category. These references commonly focus on cultural problems with implementing a mitigation project, structural measures that served to attract people nearer to flood hazards, political barriers to classification of hazardous areas, and transfer of technology into unbefitting situations. Although such studies are useful in terms of diagnoses, few are comprehensive enough to consummate a change in the outcome of the overall planning process.

Fisher et al. (1991) suggest that the first three steps in the planning process (identification, comparison, and diagnosis) serve only to establish what is wrong. They state that two additional steps are necessary to bring about change. The fourth step generates future options including proven strategies used elsewhere, academic or theoretical approaches, and development of broad ideas. Fisher distinguishes between options, or that which can be accomplished only with the cooperation of other parties, and independent alternatives, or that which can be done alone in the event that mutual agreement is improbable. In general, the outcome of the planning process can be more productive through cooperation than by acting independently, although it is frequently more difficult, time-consuming, and complex. The fifth step is to translate the better options into action by defining specific steps to be taken by each party.

FEMA's Mitigation Action Plan serves as a platform for action and discussions among mitigation partners. This type of strategy depends on individuals, government agencies, and the private sector, all three acknowledging their vulnerability and accepting their responsibilities for reducing exposure to risk from natural hazards (FEMA, 1997). FEMA acts as the catalyst for mitigation of flood control in the United States, enabling federal agencies to ultimately spearhead a national public awareness about the reduction of risk due to floods. This, in turn, allows the private sector to develop business interruption plans and implement public incentives from insurance to job security. Once the bridge to the public is secured, individual citizens, aware of flood hazards, are more inclined to support enforcement of measures designed to reduce their personal vulnerability, thus protecting communities from the impacts of natural disasters.

"As a nation, we must protect our people and our built and natural environments from the risks posed by natural hazard events. We must support new and ongoing efforts that are effective in reducing damage and injury from these events and that allow our communities to recover from their impacts as quickly as possible. The National Mitigation Goal describes our destination; the National Mitigation Strategy points the way" (FEMA, 1996).

Risk Vocabulary

Although risk communication is not the main topic of this chapter, dialogue plays an important role in the effectiveness of flood planning. At some point within the planning process, a dialogue of purpose occurs with those living at risk. Serious communication problems have been diagnosed in this aspect of flood planning. "Floodplain information should be available to the general public in formats that the average person can understand and use" (IFMRC, 1994).

Much of the work of flood risk is devoted to gathering technical information and improving the way it is analyzed. Those involved in this process have created a vocabulary applying distinct disciplines in order to deal systematically with a given problem. For example, to a hydrologist preparing a floodplain map, the term 100-year flood means the maximum instantaneous flow of water that, during any given year, has a 0.01 probability of being equaled or exceeded at a concentration point in the watershed. To some members of the risk-bearing public, this term has lost its technical definition. The 100-year flood becomes an extraordinary event that may never materialize because of the probability that the risk-bearing public at large will not exist on this planet more than another 50 years. In other words, it's not their problem. According to John McPhee, the 100-year flood is "a mixture of hydrological and human events" (McPhee, 1989).

Because of mistaken interpretations about the intended concept the term has been renamed. To cure the public of their misunderstanding, terms such as the 1% chance event, or the design flood, or the level of protection have been created. Although these new terms avoid the original confusion, they also move the dialogue to a pseudo-mathematical level that is beyond the reach of many. Redefining risk vocabulary removes erroneous concepts, but it also reveals that professionals treat flood risk primarily as a technical problem. This example identifies the fact that our lexicon can cause individuals to detach themselves from the reality of being at risk. The following paragraphs discuss various authors' interpretations of terms used in urban planning.

Urban planning is the process of decision making to enact laws, ordinances, and regulations to achieve social, economic, and environmental objectives (Johnson, 1979). With respect to natural hazards, Tobin and Montz (1997) write that urban planning is a response whereby goals are set for modifying the effects of the hazard on society, and methods of attaining these goals are selected, implemented, and enforced. The general purpose of urban planning for floods is to reduce exposure and vulnerability. This purpose can be achieved by using structural measures, such as building dams and levees, as well as nonstructural measures along with zoning, subsidizing insurance, and relocation (Gruntfest, Chapter 15, and Stalnaker and Wick, Chapter 16, this volume).

Risk is exposure to an undesirable event, or "the downside of a gamble" (Merriman and Browitt, 1993). Often linked with the discussion of risk is probability, which is the quantification of risk. Urban planning for flood hazards involves risk assessment and risk management. Risk assessment "combines hazard information with information on human activity, structures, and natural resources to determine the likely impacts of a hazardous event" (National Research Council, 1991). Risk management is "the process of minimizing, distributing, and sharing potentially adverse consequences" (Tobin and Montz, 1997). Part of the urban-planning process is to identify and reduce the gap between perceived risk, or the subjective values to which people react and respond, and statistical or predicted risk, which are calculated quantities based on observations and theory (Tobin and Montz, 1997).

A hazard can be viewed as a natural or manmade process that is "a potential threat to humans and their welfare" (Smith, 1992). A nuclear reactor meltdown, for example, is a manmade hazard, whereas a hurricane is a hazard caused by forces outside our control. Gilbert White wisely observed "people transform the environment into resources and hazards" (Burton et al., 1978). Disasters are the intersection of hazards and human activity. They occur when "a significant number of people experience a hazard, and suffer severe damage and/or the disruption of their livelihood system in a way that recovery is unlikely without external aid" (Blaikie et al., 1994).

Vulnerability is a "predisposition to adverse consequences" (Merriman and Browitt, 1993) or at its base, "difficult to defend." Vulnerability, in terms of flood planning, is a measure of a population's exposure to risk and of its ability to anticipate, cope with, resist, and recover from harm caused by a flood hazard (Blaikie et al., 1994). Various components of vulnerability are shown in Figure 18.2.

With respect to the occurrence of a natural disaster, two populations at the same level of risk may have different levels of vulnerability. For example, although the risk of flood hazard may be similar for an affluent community in Laguna Beach, California, and for a rural farming village in the Ghatail province of Bangladesh, the outcomes are probably different.

The Laguna Beach population is much more likely to have properties that are engineered to resist failure, access to telecommunications, transport facilities, early warning detection systems, and local medical treatment. Moreover, because their livelihood does not depend on their land and location, this group has more options for relocating, so they are better able to recuperate from incurred losses. Additionally, most of this populace would be able to seek legal recourse for damaged property. In view of these lifestyle differences, the underdeveloped Bangladesh village has an increased vulnerability over the Laguna Beach community. This example illustrates how socioeconomic status within a given community at flood risk is closely correlated with vulnerability (Blaikie et al., 1994).

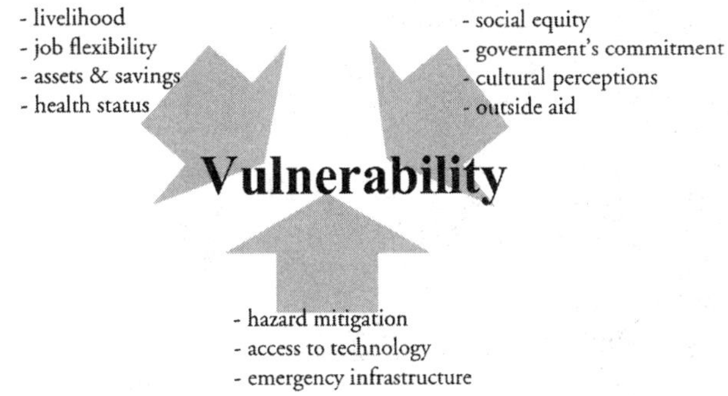

- livelihood
- job flexibility
- assets & savings
- health status

- social equity
- government's commitment
- cultural perceptions
- outside aid

Vulnerability

- hazard mitigation
- access to technology
- emergency infrastructure

Figure 18.2. Factors to be considered when evaluating a population's flood hazard vulnerability.

Hazard Mitigation

Hazard mitigation is a term that includes a range of activities. FEMA (1996) defines hazard mitigation as "sustained action taken to reduce or eliminate the long-term risk to people and property from hazards and their effects." The elements of mitigation are hazard identification and risk assessment (FEMA, 1997). Essentially, mitigation is a proactive approach that reduces vulnerability through the application of risk-management principles. This is described in FEMA's National Mitigation Strategy (1996). The document provides a mitigation philosophy for the United States. Its ultimate goals are to increase public awareness of natural hazard risk and to significantly reduce losses from natural hazards by the year 2010. Within the document, a timetable is outlined for the completion of mitigation-related federal projects, and a framework is established for setting, evaluating, and revising national goals.

FEMA's strategy focuses on awareness, incentives, and coordination of hazard management. Public awareness is achieved through dissemination of information in a way that "impels action." This national strategy is based on the premise that every mitigation dollar spent now will prevent the loss of a significantly greater amount in the future. It is a response to the "unacceptable" losses from recent disasters in the United States (Figure 18.3). Floods, which have been a factor in over 80% of presidentially declared disasters in the United States (FEMA, 1996), are especially amenable to mitigation.

In an article entitled *Why the United States Is Becoming More Vulnerable to Natural Disasters* (Van der Vink et al., 1998), an improved understanding of hazard vulnerability is discussed. To develop a more resilient society, it is necessary to attain "public awareness that most natural disasters are not random acts, but rather the direct and predictable consequence of inappropriate land use" (Van der Vink et al., 1998). In so doing the idea that natural disasters are just that, natural, will be adopted into our vocabulary, thus redefining a harmful event into a natural process that serves to build our landscape and shape our environment (Van der Vink et al., 1998).

In the United States urban-planning tools such as zoning, building codes, sanitary ordinances, capital improvement programs, and acquisition or relocation of flood-prone buildings are becoming more and

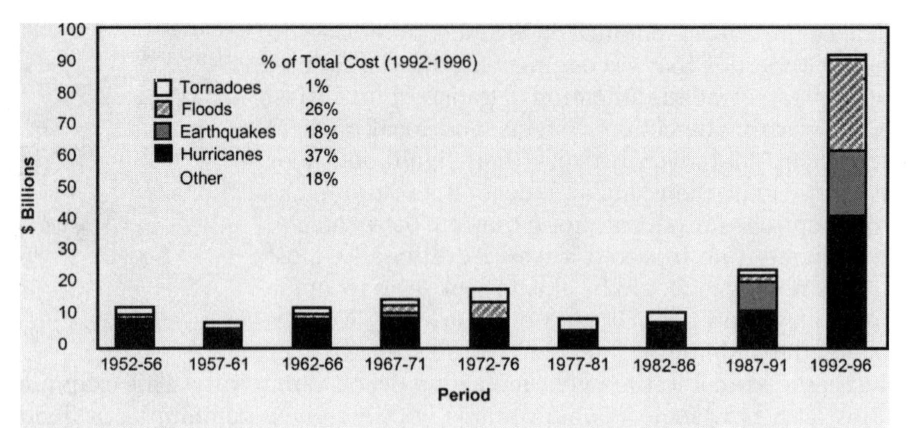

Figure 18.3. Five-year costs of natural disasters. There are many ways to calculate cost; these figures reflect the direct costs for repairing damage caused by natural disasters. In the United States, these costs average $20 billion per year, of which $15 billion is due to hurricanes, floods, earthquakes, and tornadoes. The "other" category shown above includes extreme weather, drought, and wildfires (adapted from *American Geophysical Union*, 1998).

more prevalent. There have been occasions to point out the success of urban-planning measures. Houses that complied with building codes in Florida, for example, survived floods caused by Hurricane Opal (FEMA, 1997). Other disasters have demonstrated planning deficiencies, such as the massive losses due to levee failures in the Midwest floods of 1993. These events have played a crucial role in shaping FEMA's mitigation strategy, which emphasizes long-term economic success for the community as a whole instead of short-term compromises that lead to repetitive future losses. These principles recognize that some "hazard control" (structural) measures can worsen the long-term hazard in certain situations. More permanent solutions that do more than simply delay the inevitable effects of hazards must be sought and implemented. For example, a component of the National Mitigation Strategy is to increase public awareness of natural hazard risk. The aim is to stir up communities into taking steps toward building safer living and working spaces (FEMA, 1996). This push for public awareness is consistent with the basic principle of the National Flood Insurance Program, which is that "those who choose to assume greater risk must accept responsibility for that choice" (FEMA, 1996).

Over the years, U.S. strategy has evolved away from a reliance on structural flood-control projects and toward local land-use regulation. The same has been true in Europe, but there has not been a comparable period of reliance on flood-control structures abroad. In Bayern, Germany, a formalized planning model that incorporates engineering with nature, resulted in a system called *Biotop* (smallest self-sustaining, integrated unit of biological space). Biotop contains principal elements developed by Bayern's river reconstruction project (see Table 18.1).

The Japanese have adopted Biotop engineering and spliced the ideas of natural harmony mixed with cutting-edge technology within the confines of a natural environment. The German perspective adopted by Japan views river reconstruction as "an artful agent of human peace of mind, as well as a way of moderating climate and creating a more pleasing and functional landscape" (Japan River Bureau, 1998). The Japanese are using the idea that nature, even during chaos, must be allowed to shape our environment, and communities at risk must learn to exist within hazard limitations.

Conservation technology goes hand in hand with Japan's ongoing harmony and concern with the environment. The idea that Japan "must enter

Table 18.1. *Principles of Biotop river reconstruction planning*

A river basin, including the farmland intended as flood control must be regarded as an ecological complex.
Preexisting species of flora and fauna should be preserved to the greatest possible extent, and species diversity in the ecosystem should be enhanced.
The characteristics of each species habitat should be developed in harmony with the natural dynamic alterations that take place in a river, such as erosion, sedimentation, and floods.
The unique character of each river should be determined by a balance of natural and human influences according to the condition of surroundings.
Profound consideration must be given to consistent maintenance of the natural habitat and its unique characteristics.
Given the length of time required to create a new Biotop, it is essential to first preserve what remains of the original biostructure.
Repair and maintenance should be done in harmony with ecological rhythms and cycles.
New planting technology must be incorporated into traditional civil engineering techniques.
Any plan to reconstruct a natural river environment must afford space sufficient to the dictates of ecological system requirements.

a new age where (they) live carefully and humbly in the earth's limited natural system" (Japan River Bureau, 1998) has birthed a revolution in Japanese river construction. The Japanese government has been attracted by a technology for river maintenance that has been developing in Germany since the 1960s. This technology, termed "sustainable development" contains harmonious elements understood, if not embraced, by the Japanese culture. "There is a need in Japan for the exploration of such development/conservation technology in terms of both planning and implementation" (Japan River Bureau, 1998). In a spirit of international cooperation, Japan signed the Convention of Biological Diversity at the Earth Summit in 1992. This treaty ensures the development of neonatural river construction techniques blended with traditional Japanese river-management methods.

With total damage caused by river overflow and water collecting behind levees on the increase, concentration of urban flood planning has caused flood damage costs to soar. In Japan, the Ministry of Construction (MOC) has improved flood mitigation through projects that include construction of dams, building river channels and underground rivers, as well as regulation of basins. The MOC's aggressive "Comprehensive Flood Control Measures" consolidates the "use of facilities to maintain the water-retaining and retarding functions of river basins, the creation of incentives to use land safely and to build flood-resistant buildings, and the establishment of flood warning and evacuation systems" (Japan River Bureau, 1998). In addition, the MOC has formulated basin development plans that include improvement of the environment.

An important disaster mitigation measure implemented in Japan is the river information system, which has been developed to quicken decisions for actions during floods as well as actions for proper day-to-day river management. The system also supplies data on an as-needed basis. The MOC uses information from the river information system to "predict floods and issue forecasts and warnings, which the agencies involved can use to organize flood watches, flood-fighting, and evacuation operations in order to prevent or mitigate flood damage" (Japan River Bureau, 1998).

Historical Examples of Urban Flood-Planning Measures

When it comes to dealing with the deluge, different river systems have spawned different ideas for flood planning. Gilbert White noted that "each disaster generates measures to prevent its repetition" (Burton et al., 1978). For floods, perhaps because they are more frequent than other natural hazards, such as earthquakes or tornadoes, age-old common sense has dictated that important structures be erected above flood levels, where they are much less prone to devastation. Since ancient history, there are documented urban-planning measures for cultures around the globe.

Egypt and The Nile

Protection of mummified bodies from floodwaters was a design criterion that helped locate the pyramids on Egyptian highlands. For thousands of years, Egyptians referred to the Nile River's annual flooding as the "Gift of the Nile." Egypt's "River of the Gods," the longest river in the world, was noted for its fertile deposits along the banks where civilizations depended on the annual floods to rejuvenate the soil.

The Aswan High Dam was completed on the upper Nile in 1970. Although the dam successfully traps the Nile waters in a reservoir, the downside is that the dam also traps 98% of the Nile's rich sediments, resulting in erosion and dangerously high levels of soil salinity (Fahim, 1981). Although flood and drought risk have been drastically reduced, a new vulnerability to health hazards from exposure to water-borne diseases has been created in the region.

China: Rivers and Tributaries

Urban development has prevailed in flood-prone China for 5000 years. Through various historical periods, the Chinese have discovered numerous effective measures to weather disastrous floods. A book written in China during the period of the Warring States (475–221 BC) advised that an ideal site for a capital was at the foot of a mountain or near a large river, high enough to avoid flooding, and low enough to use the water. The same book noted that protective embankments should be erected for low-built cities (Qingzhou, 1990).

Flood protection systems historically incorporated into Chinese cities include moats, canals and rivers within cities, drainage ditches, and sluices. Communication systems were also developed early on for emergency evacuation as well as flood control. Many cities with flood-control devices had designated city officials to be in charge of their maintenance. For example, as early as 500 BC, flood-control officials at the Dujiang Yan irrigation system in southwestern China placed metal bars across the bed of the Min River at the beginning of each flood season. After the floods passed, they would excavate the sediment deposits until the metal bars were revealed, keeping the river at a constant flood capacity.

In the annals of China's city planning efforts, site selection has been key. Four useful aspects of site selection can be identified. First, location

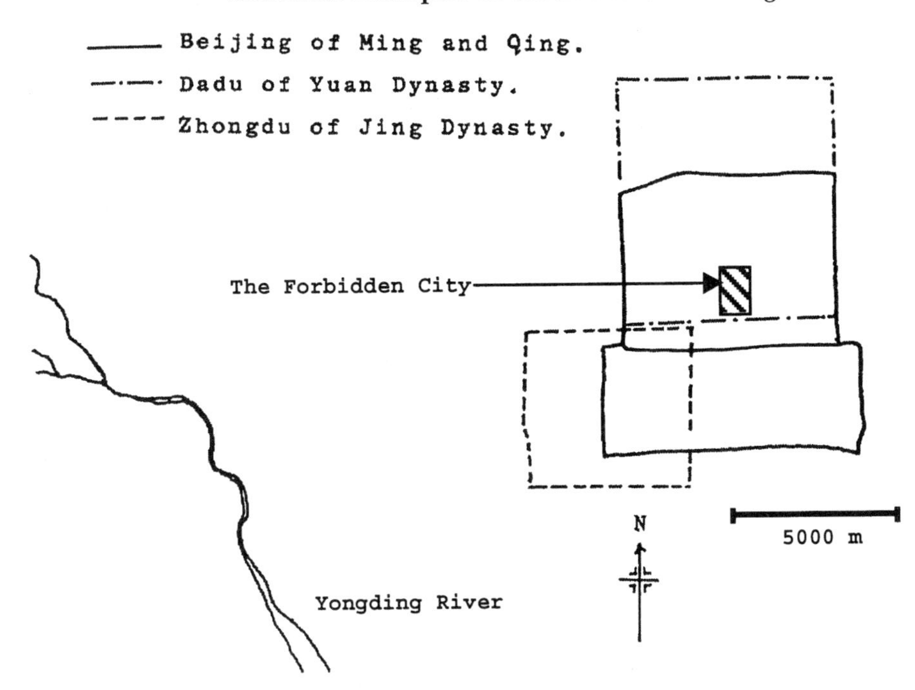

Figure 18.4. Map showing the changing sites of Beijing, reflecting the planners' compulsion to maintain a safe distance from the shifting course of the Yongding River (adapted from Qingzhou, 1990).

on high ground decreases vulnerability to floods. Second, the high ground should be distant from unstable riverbanks (Figure 18.4). Third, situating cities on convex banks minimizes erosion from floods. Finally, settlements situated by rivers with rocks acting as spurs projecting into their upper reaches are naturally protected from flood erosion.

A remarkable success in China's flood-planning history is the design of the Forbidden City (Figure 18.5), built in 1420. The city was so well developed and reinforced that there have been no flood problems to date. The Forbidden City is surrounded by a moat 50 m wide and a wall 10.4 m high. The surrounding palace walls are 10 m high with a total length of 3400 m, protected further by a 52-m-wide moat (ChinaVista, 1999).

During the Song Dynasty (960–1911), an early flood-warning strategy was established by imperial edict. When the water level in the river rose above 7.5 *Chi* (approximately 2.3 m), 3000 palace guards would be sent to protect the riverbanks (Qingzhou, 1990). The water wheel was invented chiefly for irrigation by Ma Jun during the Three Kingdoms (220–265), but during the Song Dynasty it was adapted for urban drainage and flood control as well. It was during this period that China developed a more scientific understanding of the drainage function with respect to canals. The flood-control purpose of canals was explicitly recognized. The Song Dynasty Book notes that Poet Su Shi organized flood-control planning and implementation (Quingzhou, 1990).

Historically, when an undesirable outcome from a flood occurs, urban planning has focused on physical measures to prevent the same outcome. After 5000 people died in Ankang, China, in 1583, a special evacuation dike was built so the city's inhabitants could reach higher ground quickly (Figure 18.6). When the city's walls collapsed from floods, stonefacing was added and water-resistant materials were incorporated into the mortar. Throughout China's history the construction of city

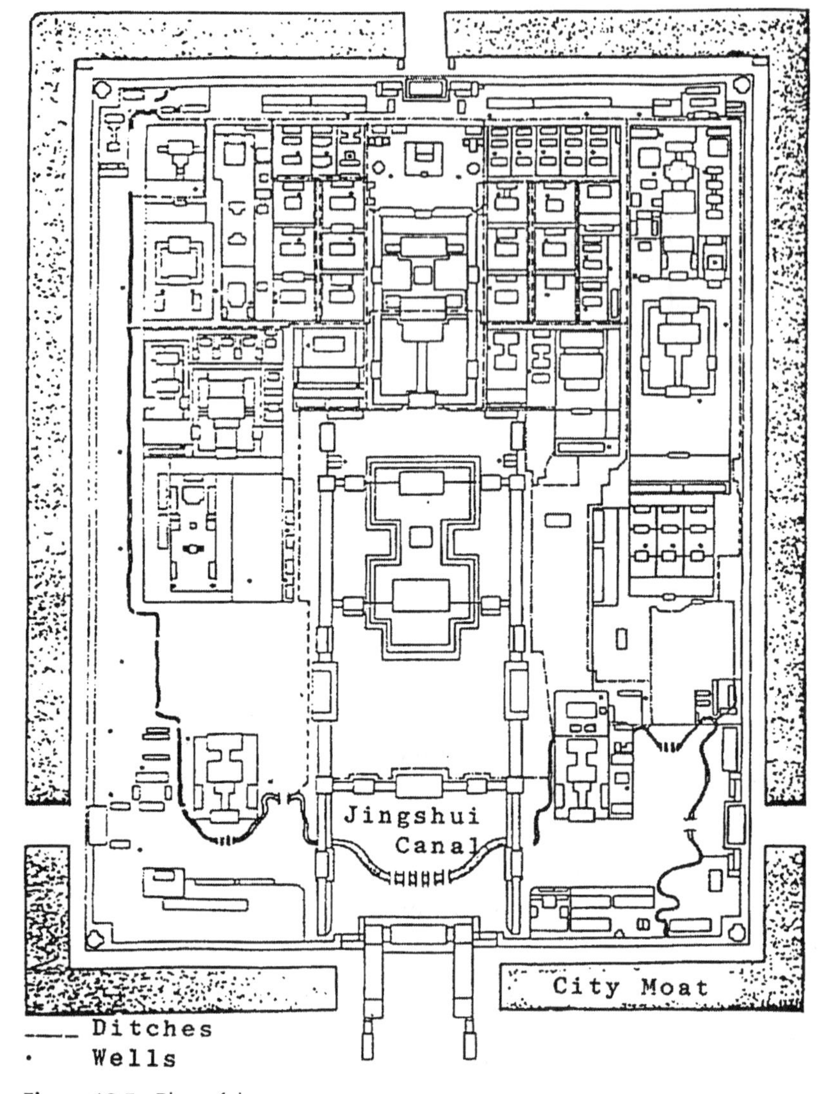

Figure 18.5. Plan of the Forbidden City. The city is surrounded by a moat system 52 m wide, 6 m deep, and 3.8 km long, for a capacity of 1,185,600 m^3. There has been no flooding problem in the Forbidden City since it was constructed in 1420 (Qingzhou, 1990).

walls has served the dual purpose of military defense and flood control. Additives such as glutinous rice-lime were mixed in with the building material to improve durability (Qingzhou, 1990). This practice, although substantially refined, continues today.

Along the Yangtze River, structural measures such as spurs, moats, and levees have been used with little success. Records indicate that the Yangtze has burst through its banks over 1500 times and changed its course 26 times between 50 BC and AD 1949 (Qingzhou, 1990). The Three Gorges Dam, a 185-m-high dam, China's most ambitious project since the Great Wall, "will displace nearly two million people as it swallows up cities, farms, and the canyons of the Yangtze River" (Zich, 1997). Dambuilders are diverting the river along 600 km of affected waterway. Thirteen replacement cities are under way along the Yangtze with a completion date of AD 2009.

The Hwang He, or "Yellow River" (coined because of its tons of yellow channel-choking mud) has been responsible for China's most catastrophic floods. This river chain has been dubbed "China's Sorrow," because throughout history the Yellow River has killed more people than any other river in the world (NOVA, 1998). In 1887, nearly two million died, in 1931 almost four million were killed, and in 1938, about one million people perished because of the overflowing banks of the Yellow River. As early as the Third Century BC, flood control was attempted. In fact, when damming failed to control the floods, an engineer named Xia Yu (a.k.a. Rong Yu) developed the idea of dredging to encourage the Yellow River to flow in its proper channel. Engineer Yu was promoted to Emperor of China for his contribution (NOVA, 1998) (Asiapac, 1999).

In an ongoing effort to control the Yellow River, the immediate flood strategy, under construction by the Chinese government, heralds an innovative undertaking commissioned as the Xiaolangdi Multipurpose Dam Project. "Boasting ten intake towers, nine flood and sediment tunnels, six power tunnels and an underground powerhouse, the structure may finally mitigate *China's Sorrow*," a recent television documentary reports (NOVA, 1998).

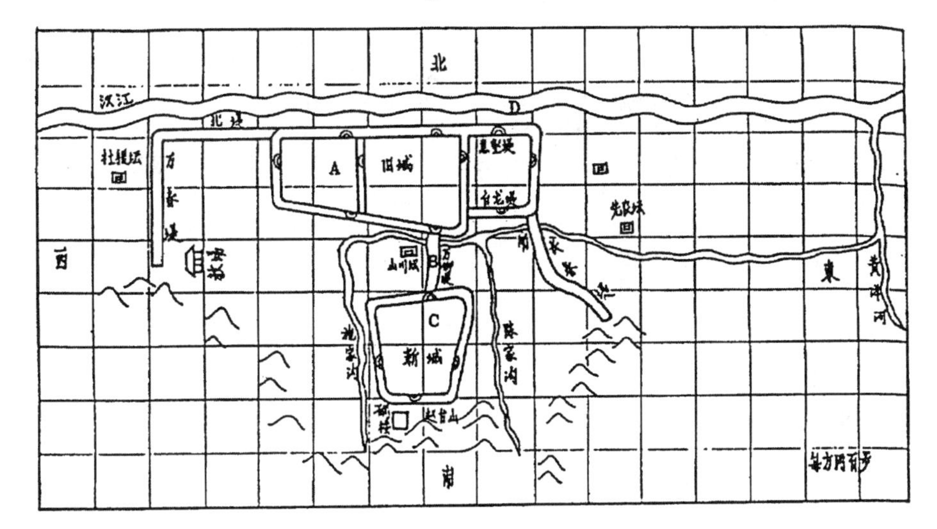

Figure 18.6. Plan of Ankang. The city has been completely submerged many times by the Hanshui River, which characteristically floods sharply and suddenly. In 1583, after 5000 people died in one such flood, the mayor of the city built an evacuation dike leading to higher ground. A = the northern city or old town; B = evacuation dike; C = southern city or new town; D = Hanshui River (Qingzhou, 1990).

Europe

An early example of European flood control planning is the dredging of a Po River tributary. In the 1400s, the tributary's junction with the Po became clogged with sediment and subsequently overflowed into the fields, diverting the river. At that time, it was up to the Pope to decide whether to keep its new course or put the river back on its original course (Levi, 1995). More comprehensive action was taken when the Danube River flooded Hungary in 1838. Formation of the Hungarian National Water Authority resulted in a formally organized "flood fighting service," which eventually constructed the largest span of levees and dikes in Europe during that period (Burton et al., 1978).

More recently, advances in floodplain mapping and record keeping have allowed formal delineation of areas prone to flood hazard and this, in turn, has spurred development of legislation controlling land use. The 1947 Town and Country Planning Acts of England and Wales, for example, give the government power to "refuse planning permission" for land in zones subject to flooding (Blaikie et al., 1994).

Japan

Throughout its long history of river flooding, Japan's key objectives have been to control flooding and to ensure the availability of river water for daily and industrial use (Japan River Bureau, 1998). At the close of the NARA Era (710–794), people who moved near the big rivers built levees to control floods. A system of *shoen*, or manors governed by aristocrat landowners, evolved. During the Shoen Era (Ninth to Fifteenth Centuries), most of Japan's water was drawn from small ponds or catchment reservoirs.

During the Sengoku Era (Sixteenth Century) to the Edo Era (Seventeenth to Nineteenth Centuries) feudal lords from ancient times became interested in expansion and it was during these growing periods that migration to alluvial fan areas took place. In what is Yamanashi Prefecture today, one feudal lord began river improvements and levee construction to relocate channels and control flooding.

In the early Seventeenth Century, the first Shogun, Ieyasue Tokugawa, commissioned a plan to divert the Tone-gawa River system. The Tone River (greater Tokyo area) originally flowed into Tokyo Bay via the Ara and Edo Rivers. At this stage in its history, the Tone was diverted to the Watarase River. Eventually, the troubled Tone River was diverted to the Kinu River to finally empty into the Pacific Ocean. Channel excavation works added to flood prevention by creating inland navigation routes. Dutch engineers assisted during the 1800s.

In 1896, heavy floods motivated formulation of the River Law that has become the mainstay in Japanese society for administration and improvement of rivers to alleviate flood disasters. Big floods occurred in 1902, 1907, and 1910. "The flood of 1910 precipitated a new epoch in the promotion of anti-flood measures, and an extraordinary flood control investigation council was implemented in order to overcome flood disasters" (Japan River Bureau, 1998). This culminated in Japan's First Flood Control Plan in 1911. The Second Flood Control Plan was formulated in 1921, and the Third Plan was instituted in 1933. World War II removed flood control work from priority projects because of war efforts.

The first Five-Year Plan for Flood Control was developed after the Ise-wan Typhoon of 1960 claimed 5000 lives. In 1964, the River Law was amended, securing a national budget allocation for flood control. Today, river improvement funding is grouped into classifications. The Tone River is considered a Class A river system and because it is deemed important for the national economy in conjunction with people's livelihood and vulnerability to a flood prone river system (Japan River Bureau, 1998). Other Class A rivers and their characteristics are shown in Table 18.2.

United States

During the Industrial Age, flood control in the United States consisted mostly of levee districts and quasi-public groups building and maintaining structures in order to capture the Mississippi River. At times, individual property owners were responsible for their own flood control. In 1724, a law was enacted that ordered homeowners to raise their own levees in order to "flood-proof" their river bank homes.

An early example of nonstructural mitigation can be found in the record books of the Laona Township of Illinois. Although no formal study was conducted to determine flood levels, a city ordinance forbade structures and fences within 45 m of the centerline of every stream. They also declared that no building may be built with a floor elevation less than 4.6 m above regular stream levels.

In 1936, a Flood Control Act was passed in response to recurrent loss of life and property damage caused by periodic flooding. It marked the decision of the U.S. Congress to take responsibility for protecting communities from catastrophe. The primary means of achieving this goal were through the configuration of well-engineered structures. Nonstructural methods for reduction of flood hazard, such as acquisition and evacuation, were also mentioned.

In the late 1940s, the Truman Commission recommended that zoning and warning be included as part of federal flood-management programs. This initiated the process of identifying flood hazards and informing the

Table 18.2. *Partial list of Class A Rivers in Japan. There are a total of 13,798 Class A Rivers in Japan with a combined catchment area of 239,947 km^2. In 1995, the Japanese River Bureau budget exceeded US$ 20 billion for flood control projects and US$ 600 million for disaster rehabilitation (Japanese River Bureau, 1995)*

Major Rivers in Japan

River	Catchment area (km^2)	Main stream length (km)	Discharge: Mean annual m^3/s	Discharge: Maximum m^3/s	Observation period
Tone	16,840	322	190	1207	1939–1992
Isikari	14,330	268	520	4482	1954–1992
Sinano	11,900	376	451	2094	1951–1992
Kitakami	10,150	249	252	1788	1952–1992
Kiso	9,100	227	211	1984	1951–1992
Tokati	9,010	156	206	2938	1954–1992
Yodo	8,210	75	210	2308	1951–1992
Agano	7,710	210	328	2200	1951–1992
Mogami	7,040	229	296	2446	1959–1992
Tesio	5,590	256	223	2302	1971–1992
Abukuma	5,400	239	112	2389	1956–1992
Tenryu	5,090	213	190	1054	1939–1992
Omono	4,710	133	194	1765	1939–1992
Yonesiro	4,100	136	183	1609	1956–1992
Fuzi	3,990	128	47	619	1960–1992

population at risk so that they could make prudent decisions about their well-being. In 1966, the United States House Document 465 was passed with the goal of preventing "uneconomic use and development of floodplains" (Federal Interagency Floodplain Management Task Force, 1992). In 1969, the National Flood Insurance Act sought to enable vulnerable communities to mitigate their flood hazards as well as to improve the national nonstructural response to floods (Alexander, 1993). Conditional, subsidized insurance was implemented as a chief means of meeting this goal. These actions marked a shift in philosophy about urban planning for floods in the United States. There was a change in focus from structural to nonstructural means of flood hazard control and mitigation, and this became undeniably the worldwide trend for the last third of the Twentieth Century.

At the end of the Twentieth Century, FEMA's goals included instituting a rapid response capability, computerizing the disaster assistance application process, establishing a national finance center, and applying technological advances to streamline the delivery of assistance. The overall goal of FEMA is "establishing public trust and keeping President Clinton's commitment that the (American) government would be there when the public was most in need, in the aftermath of disasters" (FEMA, 1997).

Iran

Tabriz, Iran, receives about 30 cm of rain annually. The Mihran-rud, normally a dry streambed, runs through the heart of the town (Figure 18.7).

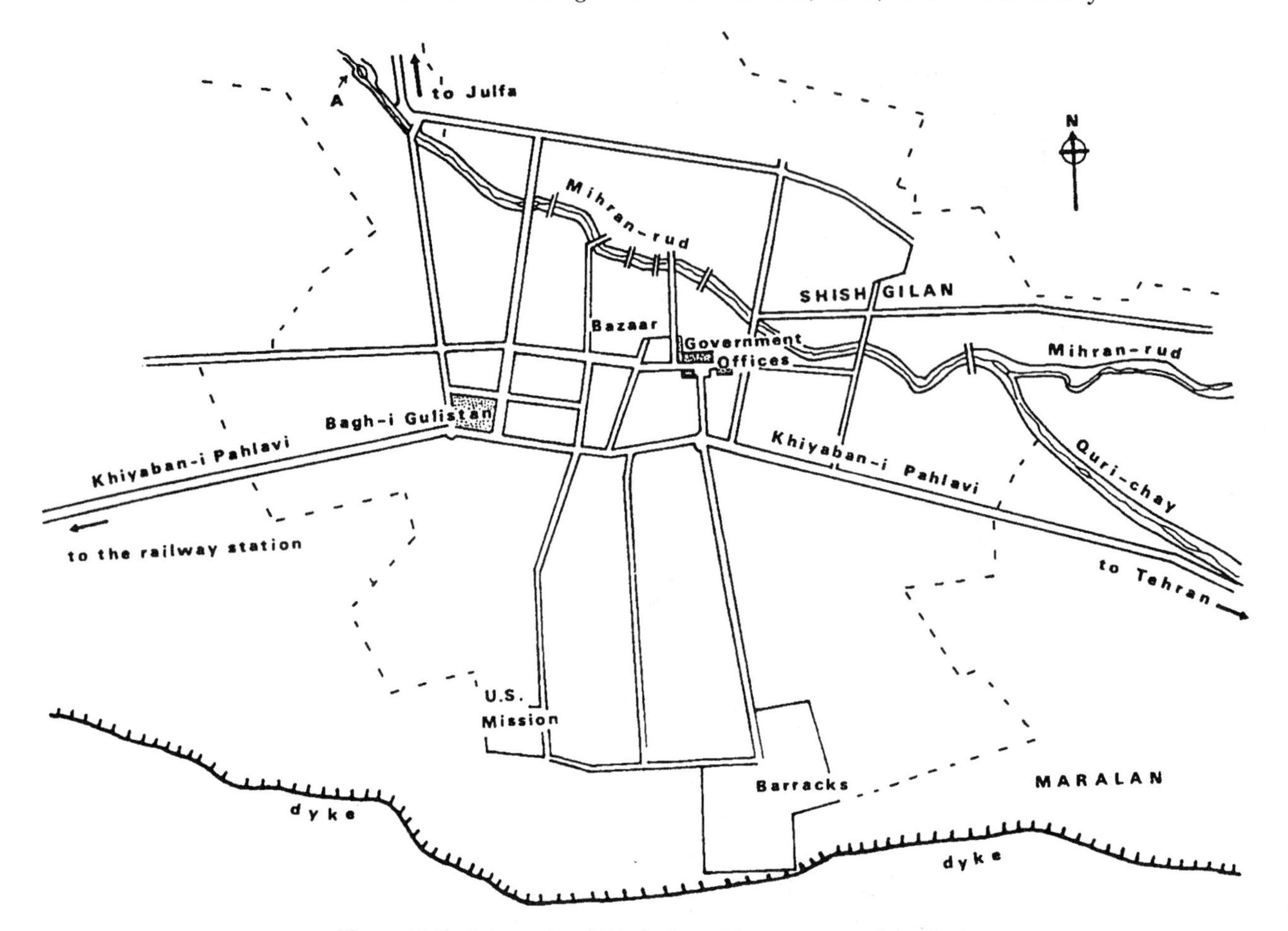

Figure 18.7. Schematic of Tabriz, Iran. Rivers running through the town are dry for most of the year and are used as thoroughfares. Although highly embanked, they have a history of overtopping and flooding the town. In 1934, the Quri-shay burst its banks and was diverted by the Khiyaban-i Pahlavi into the center of the commercial district, bending light posts and leaving 1-m boulders in its wake (Melville, 1993).

Flood season is during the early summer when maximum precipitation combines with the final snowmelt to produce violent flash floods.

Flood hazard in Tabriz, (northwestern) Iran was first incorporated into a legal petition to a Mongol ruler during the first half of the fourteenth century. The petition zeroed in on villagers upstream who had constructed a dam on the Mihran-rud to draw off water for crops. The dam was thought to cause siltation in the river by diminishing the flow, resulting in a raised channel bed downstream that was more likely to overflow in a flood. The petition sought the ruler's permission to carry out the fatva (legal opinion) that the offending dam should be destroyed. At the same time, it was suggested that "an embankment and deeper canals" be constructed for the benefit and prosperity of the community (Melville, 1983).

Early reference to flood mitigation in Tabriz dates to the Islamic year 1044 (AD 1633). "The Shah ordered the construction of a deep canal (nahr) and a strong embankment (sadd) of stone and plaster, to carry and divert flood waters" (Melville, 1983). The job was quickly carried out by the army. Soon after completion of the canal and embankment, there was a destructive flood caused by unprecedented rain, and it was concluded that had the canal and barrier not been engineered, the city would have suffered great damage. Nevertheless, observations by later visitors suggest that the floodworks were not maintained, and catastrophic floods continued to occur in Tabriz, notably one in 1872 that is commemorated in a poem.

A policy of methodical urbanization was instituted by the Shah in the 1920s, and the existing embankments of the Mirhan-rud were revamped at this time. Capital improvement projects, including a municipal garden, were created and a broad new avenue, the Khiyaban-I Pahlavi, was routed through the heart of the city. Throughout this period, monies were not properly allocated for dike repair and maintenance. In 1934, an incessant downpour burst through a dike east of Tabriz and was funneled into the center of town by the new Khayaban-I Pahlavi, depositing 1-m boulders and animal corpses in front of government offices. In this case, the town's vulnerability to flooding was inadvertently increased by poor prioritizing and planning. In exasperation and protest, the residents cut down the trees in the municipal garden.

Planning for relief and reconstruction is documented after the 1934 floods in Tabriz. All officials whose salary exceeded a certain amount had 10% automatically deducted. A fund-raising performance with compulsory attendance was held, and taxes were put on provisions, food, and fodder entering the city. Funds from public and private sources were used to construct stronger dikes, deeper beds to carry the river, and a masonry dam (Melville, 1983). In the end, the flood disaster resulted in public pressure for an investment in physical measures to prevent repetition of the event.

Unwanted Outcomes and Some Solutions from the Planning Process

Urban planning for flood hazards, risk, and vulnerability has been successful and active in societies where leadership is scrutinized and well defined. Goals of urban planning have been attainable where the structure of society is uncomplicated. However, as society grows more complex, interests that conflict with reducing vulnerability materialize.

Case Study: Bangladesh

Bangladesh occupies a network of 250 rivers known as the Bengal delta. Within this system of plentiful waters are three main flooding sources – the Ganges, the Brahmaputra-Jumuna, and the Meghna. Geographical and historical surveys confirm that the Brahmaputra-Jumuna "swept to the eastward and washed away several villages on the bank of the river, but afterwards retired towards the west forming a new channel, and leaving a number of sandbanks and alluvial accretions on the east of its bed... but, in the consequence of numerous alterations in the current, they change their appearance almost yearly" (Hunter, 1877, p. 385).

The Brahmaputra-Jumuna provides "an interesting historical insight into the development of social stratification and eventual land consolidation by the powerful zamindars (land proprietors) of colonial Bengal" (Zaman, 1996). Most early settlers were agricultural tribes and castes whose livelihood depended on harvesting forests and jungles along the riverine corridor. Zamindars who needed their flood-prone lands attended took advantage of the labor and developed a slave system along the river. Eventually, the British instigated the Permanent Settlement of 1793 that established a hierarchy of landlords over sharecroppers, which persists in some form to this day.

Throughout its disaster-prone history, riparian ownership has deemed inundation a country-wide necessity. Floods bring displacement, and displaced households survive by moving to new lands. Because of this type of collective fatalism, "Bengalis themselves are coping astonishingly well with the enemy they've always known" (Haque and Zaman, 1993).

Bengalis have always shown awareness and adaptability to floods. In fact, floods are so much a part of their existence that the Bengali language has two words for flood: *barsha* (normal beneficial floods) and *bona* (infrequent, destructive large floods). As in India, elevated mounds have been built for flood protection, and the centers of towns in flood-prone areas are placed on higher ground. Large-scale institutional efforts at flood control in Bangladesh took root in the 1950s and have burgeoned into an international endeavor as floods persist.

In a country as flood prone as Bangladesh, where one-fifth of the land area is subject to flooding during a normal monsoon season (Alexander, 1993), strict measures are needed to avert catastrophes. During the exceptional back-to-back floods of 1987 and 1988, it is estimated that two-thirds of the country of Bangladesh was inundated (Thompson and Sultan, 1996). Floods of this nature, which exceed the magnitude of the design event, are seldom addressed in planning because of the general acceptance that some level of risk will invariably prevail.

On the Ganges-Brahmaputra floodplain of Bangladesh, there is a long history of building embankments for agricultural protection. These range from the ancient levees built by *char* (accretion) land squatters, to the 3.6-m-high embankments observed by European travelers along the rivers in the 1760s (Merriman and Browitt, 1993), to the massive projects that have been proposed by the World Bank (see Table 18.3).

In planning, the needs of the present sometimes outweigh consequences in the future. This is true for the rural farmers and fishermen of Bangladesh. In many ways, fishing and farming activities have been disrupted by the increasing number of embankment construction programs (Figure 18.8). River silt no longer fertilizes cropland in many areas, and numerous seasonal ponds that provided spawning grounds for fish have been eliminated. Water that was once easily accessible has been cut off, regulated by those controlling the sluice gates.

Because of deprivation, farmers and fishermen have been known to sabotage embankments through berm cuts. Although intentional public cuts made in the earthen berms allow river water to irrigate fields and to form habitat for fish, these breaches diminish usefulness for flood protection overall. In the flood embankment areas, those berms that are not purposely cut are often breached by floodwaters because of poor construction and a lack of maintenance. Therefore, the hazard mitigation that is meant to provide flood protection becomes jeopardized (Thompson and Sultan, 1996). The situation is aggravated by the fact that it is also common for squatters to build homes on the embankments, thus weakening and degrading them.

Several studies have noted the need to provide a suitable berm or platform to minimize damage caused by the above settlements. It is clear in

Table 18.3. *Projects that make up the massive World Bank-sponsored Bangladesh Flood Action Plan (FAP), published in 1990. Expected cost, US$ 10.2 billion over 20 years*

FAP	Activities	Donors
1	Brahmaputra RB strengthening	IDA
2	Brahmaputra right bank study	UK-Japan
3	Brahmaputra left bank study	EEC-France
4	Ganges right bank study/embankment	ADB-UNDP
5	Meghna left bank study	Canada
6	Northeast regional study and rehabilitation	EEC-IDA
7	Cyclone protection project	Japan/ADB
8A	Greater Dhaka protection project	ADB/IDA
8B	Dhaka town protection/environmental improvement	ADB/IDA
9A	Secondary town protection schemes	UNDP, Japan
9B	Meghna bank protection study	UNDP, Japan
10	Flood forecasting and early warning project	UNDP
11	Flood preparedness	UNDP
12	Agricultural review	UK-Japan
13	Operation and maintenance study	UK-Japan
14	Flood response study – active floodplain	USA
15	Land acquisition and resettlement project	Sweden
16	Environmental study	USA
17	Fisheries study and pilot project	UK
18	Topographic mapping	Finland, Switzerland, etc.
19	Geographic information system	USA
20	Compartmentalization pilot project	NL, Germany
21	Bank protection pilot project	Germany-France
22	River training/AFPM pilot	Germany-France
23	Flood modeling-management project	USA
24	River survey programme	EEC
25	Flood modeling-management project	Denmark, etc.
26	Institutional development programme	UNDP, France, etc.

these cases that local government cannot always successfully manage embankments. A 1989 study by the United Nations Development Program cites the "total disinterest often demonstrated by the population to maintain the structures and works that are to protect their life and property" as a critical problem of embankment projects (Sklar, 1993). Without mechanisms to police and to maintain embankment structures, this type of flood-control project will not fully accomplish the objectives of urban planning.

Without realizing it, planners in Bangladesh may have taken for granted the existence of a certain type of government infrastructure capable of managing flood-control projects while ensuring equity and viability. Some rural areas in Bangladesh do not have effective governance structures for implementing projects. Failure to address this element of vulnerability is a deficiency in planning that reduces the effectiveness of flood-control programs. For instance, areas with informal local government structures have increased vulnerability because hazard mitigation programs cannot be protected from selfish individual acts, which jeopardize the general population and invalidate the program.

An example of this type of vulnerability is shown in a study on the impact of local elites on the effectiveness of flood shelters (Khan, 1993). Close to half a billion U.S. dollars were allocated for building cyclone/flood shelters between 1973 and 1978. In this case, shelters were typically two-story

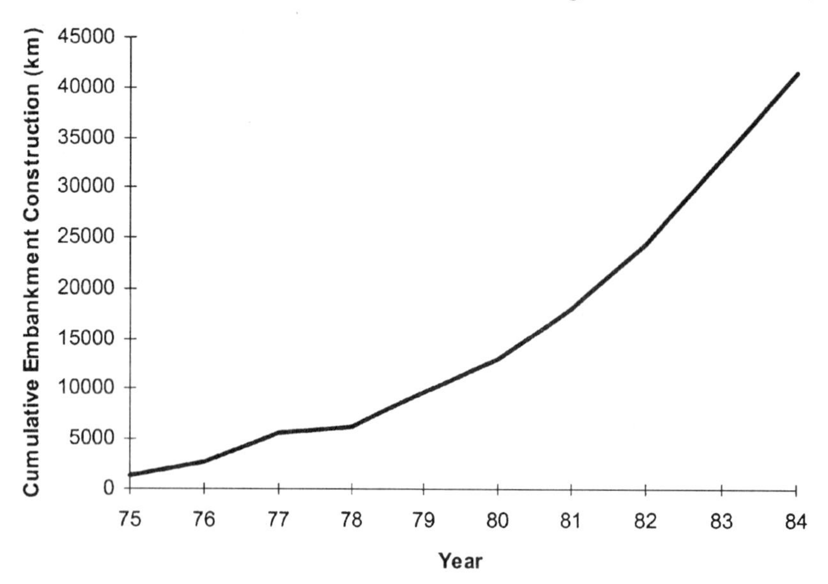

Figure 18.8. Cumulative length of embankment construction in Bangladesh from 1975 to 1984. The increasing rate in later years continues today (adapted from Haque and Zaman, 1993).

structures that were 36 m long, 8 m across, and 5.5 m high. These shelters were designed to house 300 people on the second level, with the ground floor reserved for animals. Instances abounded of local elites acquisitioning the shelters for their personal use, dictating where they were built, and controlling who was granted access (Figure 18.9).

These actions reduced the effectiveness of the shelter as a hazard mitigation measure. In surveys taken in the study area, many flood-affected people did not use the shelters because they were too far from their homes or because of the bad treatment that they received at the shelter. In this case, planning failed to take into consideration the effects of the local power structure. Thus, the provision of shelters was of value to some but made others more vulnerable because limited government resources were incorrectly allocated to a single means of mitigation.

Postdisaster surveys show that the people of Bangladesh were not equally affected. The most vulnerable areas involved landless poor, marginalized farmers, and agricultural producers. In addition, it is noted that floods more severely affect woman-headed families and children. Nongovernmental organizations (NGOs) play a crucial role in providing emergency food relief, homesteading, roads and embankment repair, agricultural rehabilitation, and health services to populations where local governments are not geared for the task.

Figure 18.9. Chart showing how some elites in Bangladesh exaggerated to bias location of flood shelters to their villages (adapted from Khan, 1993).

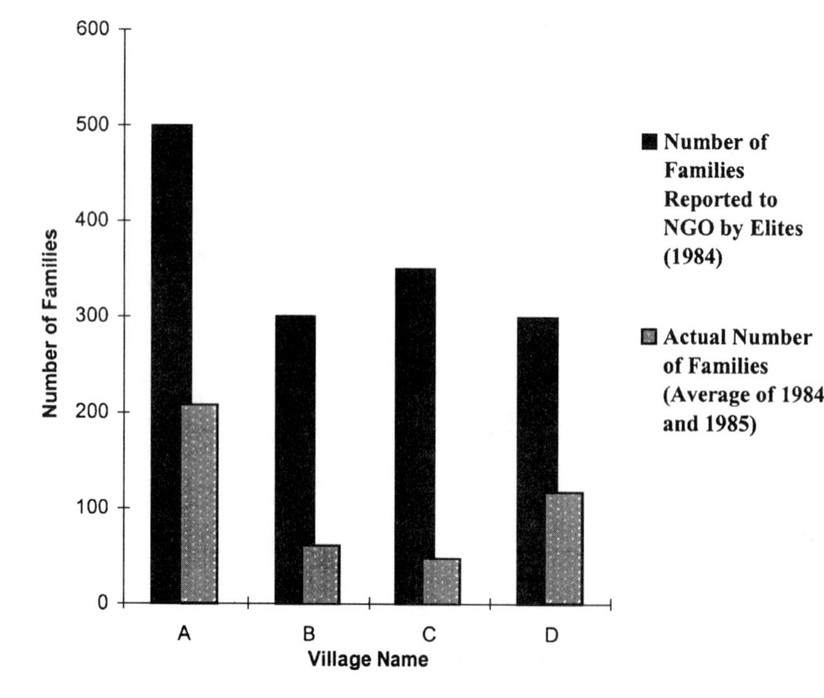

"Proshika" is an example of a NGO whose mission is to help Bengalis cope with disaster. This organization assisted close to 414,000 affected families in the floods of 1984 and the twin back-to-back floods of 1987 and 1988 and of 1995 and 1996. In addition, they are currently providing relief and rehabilitation from the century's worst flood, which took place in 1998. Proshika's disaster management includes education and distributing information through radio and television broadcast and support of smaller NGOs.

Conclusion

In conclusion, the steps of the planning process are applied to the flood shelter case in order to achieve a better outcome. Fisher et al. (1991) propose a model as an essential tool for the planning process. The process (Figure 18.10) maps out agreements of various strengths and how those strengths multiply options through the process of shuttling between specific and general ideas (Fisher et al., 1991).

Multiple options examine a problem from the perspective of various professions and diverse disciplines. "Consider in turn how each expert would diagnose the situation, what kinds of approaches each might suggest, and what practical suggestions would follow from those approaches" (Fisher et al., 1991).

In addition, this approach is applicable when differences include values placed on time, in forecasts, and in aversion to risks. This directly relates to flood mitigation as a different value is placed on time and predictions involving vulnerability. Emphasis on present and future prevention of flood risk varies because of the different circumstances each area represents.

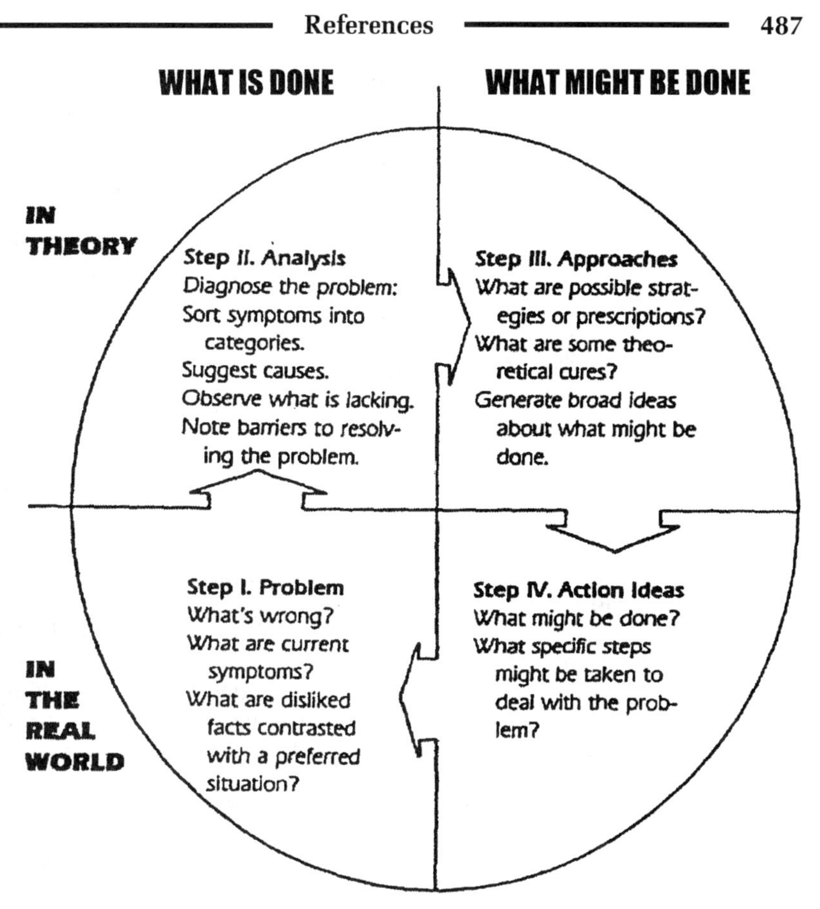

Figure 18.10. Four basic steps in inventing options (Fisher et al., 1991).

For the Bangladesh flood shelter problem the process consists of the following steps: First, the facts associated with this problem are that shelters were built but were not fully used for their intended purpose. Second, from the NGO perspective, the desired outcome was that the shelters would be fully and equitably used for protection from floods. They were not. From the elite's perspective, the actual outcome is acceptable. Third, a possible cause of the problem might be that there is a utilitarian need for the shelters during nonflood periods. Fourth, options to dealing with this problem include: Enforcement by the NGOs acting alone, coordinate with local governing bodies to have multiple uses of the shelters during the dry season, setting up a council to manage the use of the shelters, etc. Fifth, depending on how the options are generated, specific groups or individuals could be assigned the task of implementing one or more of the options. After implementation has occurred, the problem as originally identified in the first step can be monitored.

References

Alexander, D. (1993). *Natural Disasters.* London: University College London Press.

Asiapac. (1999). 100 Chinese Emperors. http://www.span.com.au/100emperors/5.html.

Blaikie, P., Cannon, T., Davis, I., and Wisner, B. (1994). Bangladesh – A 'Tech-fix' or People's Needs Based Approach to Flooding? In *At Risk: Natural Hazards, People's Vulnerability, and Disasters.* New York: Routledge.

Burton, I., Kates, R.W., and White, G.F. (1978). *The Environment as Hazard.* New York: Oxford University Press.

China Vista/Beijing Tour. (1999). University of Washington. http://www.chinavista.com/tour.html.

Fahim, H. (1981). *Dams, People and Development: The Aswan High Dam Case.* New York: Pergamon Press.

Federal Interagency Floodplain Management Task Force (FEMA). (1992). *Floodplain Management in the United States: An Assessment Report.* Washington, DC: U.S. Government Printing Office.

FEMA. (1996). National Mitigation Strategy. *About FEMA.* "What We Do." Online. (August 28, 1996): http://www.fema.gov/mit/ntmstrat.html.

FEMA. (1997). *Report on Costs and Benefits of Natural Hazard Mitigation.* Washington, DC: U.S. Government Printing Office.

Fisher, R., Ury, W., and Patton, B. (1991). *Getting to Yes.* New York: Houghton Mifflin Company.

Haque, C.E., and Zaman, M.Q. (1993). Human Responses to Riverine Hazards in Bangladesh: A Proposal for Sustainable Floodplain Development. *World Development*, **21**, 93–107.

Hunter. (1877). 385 pp. (See Zaman, M.Q.)

Interagency Floodplain Management Review Committee. (1994). *Sharing the Challenge: Floodplain Management into the 21st Century.* Washington DC.

Japan River Bureau. Online. (1998). http://www.moc.go.jp/river/english/bureau/p02.html; http://www.moc.go.jp/river/english/bureau/p16.html.

Johnson, W.K. (1979). *Introduction to Civil Engineering Planning.* Lecture Notes.

Khan, M. I. (1993). The Impact of Local Elites on Disaster Preparedness Planning: The Location of Flood Shelters in Northern Bangladesh. *Disasters*, **15**, 341–353.

Levi, E. (1995). *The Science of Water: The Foundation of Modern Hydraulics.* New York: ASCE Press.

McPhee, J. (1989). *The Control of Nature.* New York: Farrar Straus Giroux.

Melville, C.P. (1983). The 1934 floods in Tabriz, N.W. Iran. *Disasters*, July 2, 1983, pp. 107–116.

Merriman, P.A. and Browitt, C.W.A. (eds). (1993). *Natural Disasters: Protecting Vulnerable Communities, Proceedings of the London Conference.* London: Thomas Telford.

National Research Council. (1991). *A Safer Future: Reducing the Impacts of Natural Disasters.* Washington DC: National Academy Press.

NOVA. (1998). *Flood!* Dealing with the Deluge. Online. http://www.pbs.org/wgbh/nova/flood/deluge.html.

Qingzhou, W. (1990). The Protection of China's Ancient Cities from Flood Damage. *Disasters*, **13**, 193–225.

Sklar, L. (1993). Drowning in Aid: The World Bank's Flood Action Plan. *Multinational Monitor*, April 1993, pp. 8–13.

Smith, K. (1992). *Environmental Hazards: Assessing Risk and Reducing Disaster.* New York: Routledge.

Thompson, P. and Sultan, P. (1996). Distributional and Social Impacts of Flood Control in Bangladesh. *Geographical Journal*, **162**, 1–13.

Tobin, G.A. and Montz, B.E. (1997). *Natural Hazards: Explanation and Integration.* New York: The Guilford Press.

Van der Vink, G., et al. (1998). *American Geophysical Union.* Department of Geosciences. Princeton University Princeton, NJ: Newsletter, November 3, 1998, pp. 533–537.

Zaman, M.Q. (1996). Social Structure and Process in Char Land Settlement in the Brahmaputra-Jamuna Floodplain. *University of Lethbridge*, **26**, 673–690.

Zich, A. (1997). China's Three Gorges Before the Flood. *National Geographic*, **192**, 4–33.

CONCLUSIONS

CHAPTER 19

Floods in the 21st Century

Ellen E. Wohl
Department of Earth Resources
Colorado State University

By the end of the 20th century, at least two trends had become apparent in flood hazards and flood-hazard mitigation. The first of these was an increase in flood damages in most countries. In the United States, flood hazards at the end of the 20th century were emphasized by the 1993 flood on the upper Mississippi River, which caused 38 deaths and 12 to 16 billion dollars in damage (Galloway, 1994). This flood was the most prominent in a series of floods; more than 80% of presidentially declared disasters in the closing decades of the century resulted from floods (National Mitigation Strategy, 1995). Van der Vink et al. (1998) attributed the increase in flood damages to changes in population and national wealth density, instead of to an increase in flood magnitude and/or frequency. Although improvements in flood forecasting, warning systems, and building codes greatly reduced the number of fatalities resulting from floods in the United States during the 1990s, the percentage of national wealth spent on homeowners' and property/casualty insurance remained approximately constant with respect to gross national product (Council of Economic Advisors, 1994; U.S. Census Bureau, 1997; Van der Vink et al., 1998). This increase in flood damages has also occurred in other countries, although in developing nations both deaths and damage to structures during floods have increased.

Four of the primary challenges in reducing escalating flood damages are (1) estimating flood magnitude for a given recurrence interval, (2) accurately forecasting the occurrence of a flood resulting from rapidly changing weather conditions, (3) effectively operating flood-warning and evacuation procedures, and (4) establishing and enforcing land-zoning regulations that reduce the potential for people and structures to be exposed to flood hazards. The first two challenges are strictly scientific in that they require accurate models of the physical processes that cause floods. Accurate estimation of the 100-year flood, for example, requires systematic, historical, and geologic records of past floods (Baker, Chapter 13, this volume) as well as statistical distributions that adequately characterize the likelihood of flood occurrence over a specified time interval (Stedinger, Chapter 12, this volume). Accurate forecasts of flood occurrence during an evolving meteorological situation require records of past meteorological patterns and resultant flooding and hydrological models that adequately characterize rainfall-runoff and flood-routing processes (Hirschboeck et al., Chapter 2, this volume; Ramirez, Chapter 11, this volume). As summarized in the preceding chapters of this volume, there remain many obstacles to accurate flood modeling and prediction, in part because the processes operating to produce floods are highly variable, stochastic, and interrelated, and in part because human impacts on climate and land surface are altering the patterns of flood occurrence (Wohl, Chapter 4, this volume).

The second set of challenges – flood warning and evacuation, and flood zoning – are potentially more complex in that they involve subjective human attitudes and governmental policies (Gruntfest, Chapter 15, this volume; Laituri, Chapter 17, this volume; Hamilton and Joaquin, Chapter 18, this volume). Implementation of flood-warning and zoning procedures requires translating the uncertainties currently inherent in flood estimation and forecasting into language and procedures that society as a whole can understand and accept. Assuming that many individuals do not evaluate probability and risk in a logical, objective manner and are reluctant to assume personal responsibility for their actions, this process of translating technical understanding into societal response is extremely difficult.

The second basic trend in flood hazards and flood-hazard mitigation during the last decades of the 20th century was an increasing recognition (1) that the structure and functioning of aquatic and riparian ecosystems is highly dependent on flow regime, particularly on the timing, magnitude, frequency, and duration of floods; and (2) that human manipulation of flow regimes has substantially altered many of the world's aquatic and riparian ecosystems. In developed nations, essentially every major drainage network has been altered by dams, diversions, channelization, levees, and other structures that affect the downstream flow of water and the connectivity of channels and floodplains. At the start of the 21st century, major projects such as China's Three Gorges Dam are rapidly creating similar conditions in developing nations.

The primary challenges in reducing the impacts of flow regulation on aquatic and riparian ecosystems are (1) understanding how the structure and functioning of these ecosystems respond to changes in flow regime across various spatial and temporal scales within a drainage network, and (2) developing and implementing flow regimes along regulated rivers in response to this understanding. Studies of the response of riparian vegetation and aquatic communities to flow regime are relatively few and recent, having begun mostly in the late 1960s (Friedman and Auble, Chapter 8, this volume; Wydoski and Wick, Chapter 9, this volume). Only during the 1980s and 1990s were sufficient case studies available to permit the development of broader conceptual models such as the river continuum concept (Vannote et al., 1980), the flood-pulse concept (Junk et al., 1989), and Scott et al.'s (1996) models of contrasting patterns of flood-related vegetation establishment as a function of channel planform. Along many rivers, investigators have little evidence of the character of the riparian and aquatic ecosystems before flow regulation or the construction of structures and must attempt to determine process-response relationships in a highly altered system that still may be responding to human effects. Limited understanding of the relations between river ecosystems and flow regime makes it very difficult to recommend and implement changes to existing flow regulation patterns, because such changes usually have economic repercussions and consequently may be strongly opposed by some members of a society (Stalnaker and Wick, Chapter 16, this volume).

The three drainage basins highlighted in this volume exemplify these issues of flood hazards and flood-hazard mitigation. As one of the most densely populated and highly developed drainage basins on Earth, the Tone River of Japan had increasing loss of human life and property damage from floods during the last decades of the 20th century (Wohl, Chapter 1, this volume). The drainage basin has nearly 100 hydroelectric power plants and extensive intake and diversion structures for irrigation. The extent and diversity of riparian vegetation have been substantially reduced, and formerly abundant and widespread fish species are now endangered.

Although the Colorado River basin of the United States has a much lower human population density than the Tone basin, the Colorado supplies water to several major cities and extensive agricultural lands in the arid southwestern United States (Wohl, Chapter 1, this volume). Twelve major dams in the basin have substantially altered the river's flow regime, affecting endemic riparian and aquatic communities in 13 national parks and monuments. Four of the six dominant native fish species in the Colorado River will be threatened or endangered at the start of the 21st century.

Like the Tone River basin, Bangladesh has an extremely high density of human population (Wohl, Chapter 1, this volume). Wet season floods, tropical cyclones, and storm surges spread floodwaters across vast expanses of Bangladesh, killing hundreds of thousands of people and destroying buildings and farmlands. Flow regulation and channelization is fairly minimal at the start of the 21st century, and natural flow regimes are fairly intact. Human response to flooding is primarily at the level of individuals or small communities, rather than regional or national government, and is based on traditional responses to frequent, severe flooding. In the absence of extensive flood warning or zoning procedures, flood damages will likely continue to increase as population density increases.

In the case of the Tone River and Bangladesh, perceived flood hazards result primarily from the presence of floods and the effect on human communities. In the case of the Colorado River, perceived flood hazards result primarily from the absence of floods along the regulated river and the effect on riparian and aquatic communities. Riparian and aquatic communities have also been significantly affected by flow regulation in the Tone River basin, but at the start of the 21st century environmental degradation is not given as much attention in Japan as in the United States. The condition of riparian and aquatic communities in Bangladesh was largely unstudied. For each of these river basins, mitigation of flood hazards will require a major effort on the part of governmental agencies, communities, and individuals. Where individuals or communities must be removed from hazardous areas, or where existing flow regulation patterns must be altered to restore a more natural flow regime, some segments of a society will be unwilling to assume what they perceive to be an unfair burden or cost. Where tremendous time and energy have already been invested in infrastructure that will make it difficult to restore natural flood regimes, as in Japan, society as a whole may decide that the destruction of riparian and aquatic communities is an acceptable loss if human risks from floods are reduced. However, the increasing flood damages in Japan and other countries suggest that intensively engineered, densely populated, biologically impoverished river corridors are not an acceptable response to flood hazards. Most of the world's cultures have a tradition of aggressively interfering in natural systems and altering natural processes to the extent possible in order to produce immediate human benefits. To deliberately turn aside from these conditions by recognizing river corridors as natural systems that must be left minimally populated by humans, with unregulated or minimally regulated flows, will be a major shift in human response to flood hazards. If such a shift is to be accomplished, we must continue to study flood hazards from the broadest, most interdisciplinary perspective possible. The concept of interdisciplinary approaches to understanding natural systems, as exemplified by Earth System Science, became increasingly popular at the end of the 20th century. Implementation of this ideal during the 21st century with respect to flood hazards requires that we acknowledge that both the presence and absence of floods may create hazardous conditions and that past approaches to flood-hazard mitigation have not always been successful for various reasons, indicating the need to reevaluate past assumptions. To effectively mitigate flood hazards during the coming decades, we need both better understanding and more effective implementation of that understanding.

References

Council of Economic Advisors. (1994). *1994 economic report of the President.* Washington, DC: Government Printing Office.

Galloway, G.E. (1994). *Sharing the challenge: floodplain management into the 21st century.* Washington, DC: A report of the Interagency Floodplain Management Review Committee.

Junk, W.J., Bayley, P.B., and Sparks, R.E. (1989). The flood-pulse concept in river-floodplain systems. In *Proceedings of the International Large River Symposium,* ed. D.P. Dodge, pp. 110–127. Ottawa, Canada: Canadian Special Publications of Fisheries and Aquatic Sciences 106.

National Mitigation Strategy. (1995). *Partnerships for building safer communities.* Washington, DC: Federal Emergency Management Agency Publications Office.

Scott, M.L., Friedman, J.M., and Auble, G.T. (1996). Fluvial process and the establishment of bottomland trees. *Geomorphology,* **14**, 327–339.

U.S. Census Bureau. (1997). *Statistical abstracts of the United States.* Washington, DC: Department of Commerce.

Van der Vink, G., Allen, R.M., Chapin, J., Crooks, M., Fraley, W., Krantz, J., Lavigne, A.M., LeCuyer, A., MacColl, E.K., Morgan, W.J., Ries, B., Robinson, E., Rodriguez, K., Smith, M., and Sponberg, K. (1998). Why the United States is becoming more vulnerable to natural disasters. *EOS, Transactions, American Geophysical Union,* **79**, 533–537.

Vannote, R.L., Minshall, G.W., Cummins, K.W., Sedell, J.R., and Cushing, C.E. (1980). The river continuum concept. *Canadian Journal of Fisheries and Aquatic Sciences,* **37**, 130–137.

Index